Introduction to Statistics and Probability

Edward J. Dudewicz
The Ohio State University

HOLT, RINEHART AND WINSTON
New York · Chicago · San Francisco · Atlanta
Dallas · Montreal · Toronto · London · Sydney

To my wife
Patricia Anne Scott Dudewicz

Library of Congress Cataloging in Publication Data
Dudewicz, Edward J
Introduction to statistics and probability.
(International series in decision processes)
1. Mathematical statistics. 2. Probabilities.
I. Title.
QA276.D813 519.5 75-26827
ISBN 0-03-08-6688-X
Printed in the United States of America
678 038 123456789

PREFACE

This book is intended as a basic text in mathematical statistics with probability. The projected audience is junior and senior undergraduate students in mathematical sciences, as well as beginning graduate students in these fields and also the better prepared (mathematically) and motivated graduate students in related fields. Judicious use by the instructor of the copious problems and separate honors problems should allow the text to lend itself well to the situations wherein diverse groups of students must take their basic course together. The honors problems should give the teacher of even a homogeneous body of students an unusual degree of flexibility, making it easy to assign special projects, and so on.

The prerequisites needed will vary according to such local characteristics as depth of coverage desired, length of course, and ability and motivation of the students; as a usual minimum I recommend a two semester course in integral and differential calculus (including multiple integration, partial differentiation, and infinite series), plus the student's ability and desire to fill in any gaps in his preparation which may appear as the course progresses. (While some more advanced topics from mathematics are at least mentioned in the text, for example, arbitrary unions, indexing sets, matrix algebra, topology of Euclidean spaces, and infima and suprema and their limits, most of these are either in optional sections or are given a brief explanation sufficient for our use of them at the point where they occur. The latter yield mathematical precision, yet a student or class will lose little if they—as we note in the text—for example, read "max" for "sup.")

It is intended that this text should fill the gap between mathematically precise (but applications devoid) and data-oriented (but theory inadequate) texts. Here the reader will find that, while Chapters 1–6 in themselves can constitute the basis for a quarter or semester course in probability, they have been written so as to furnish a relatively complete probabilistic basis for the study of mathematical statistics (Chapters 7–14) where they find their fulfillment. (The attention to applications permeates the text not through copious examples of data, but through formulation of the theory with a view to its application and with respect for applications, crisp discussions of application-relevant factors, and a few key homework problems and examples.) The problems are an integral part of the text, and are designed to appropriately reward effort with understanding and insight, rather than as "busy work," and also probe and extend the theory.

This book is far more ambitious than first intended for several reasons. First, to allow course content to vary from year to year both for variety and to suit the needs of diverse student groups. Second, to allow material to be either skimmed or delved into more deeply as seems appropriate. Third, to allow the more able and interested students to delve into the subject deeply, go to the literature, and do introductory research. The detail of explanation is intended to be enough so that those teaching the course outside the statistics department will find the motivation and application enlightening, yet not so extensive as to jade the senses. A detailed *Teacher's-Solutions Manual*[1] is intended to help instructors in bringing modern probability and statistics to institutions where the staff has little previous experience or training in these areas, but does have mathematical ability.

This book started out as a set of notes which I developed during one academic year for Introduction to Probability and Statistics and Statistical Inference, which I was teaching at The University of Rochester. Those courses were found to be very suitable preparation for further work in both applied and theoretical statistics and probability. (During the early stages several of the usual texts were used as well as these notes.) The student population varied in preparation and level over this period, from freshmen to graduate students,[2] and this helped greatly in preparing a text which is both mathematically precise and develops a "feel" for statistics as a method of real-world decision making. At The Ohio State University a final draft of this text was used recently in an advanced statistics course. This course is offered for students with some applied background in probability and statistics who, upon entering graduate school, find they need a stronger background in probability and statistics (for example, new statistics and industrial engineering graduate students).

A year (two semesters of 3 credit hours) course based on this text could probably expect to cover Chapters 1–6 in the first semester and Chapters 7–12 in the second. The dependence in text is usually linear through Chapters 1–6 and can (if you are willing to ignore and tolerate brief comments designed to tie together the various chapters) branch thereafter as illustrated in the

[1] Available, upon adoption, from the publisher.

[2] Although this text can therefore be used for students of very limited backgrounds (for example, some of my students lacked multiple integration and partial differentiation) and at the Freshman-Sophomore level, such use is for talented students accompanied by unusual instructor effort.

following figure:

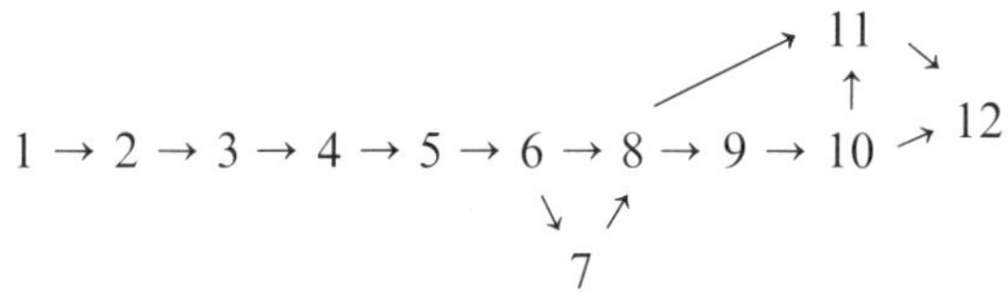

Thus, to cover Chapters 1–6 at various levels you can vary the depth (omitting optional material and skimming the multivariate cases of Sections 3.4, 3.5, 3.6, 4.4, 4.5, 4.6, and 5.3) rather than the topics. You can also approach Chapters 7–12 in this manner, or can omit chapters or sections, in order to achieve various levels. Although it will usually be found possible to cover Chapters 1–6 in a semester at any level, only perhaps half of Chapters 7–12 can be covered thoroughly in that time at the lower levels, or you can cover the material briefly after building a solid base in Chapter 8.

In the context of a quarter system, the above paragraph applies to two quarters of 5 credit hours each, or to three quarters of 3 credit hours each. (In the latter case the middle quarter would bridge probability and statistics.)

A three-semester (3 credit hours per semester) sequence should be able to cover the entire text fairly thoroughly, as should a three-quarter (5 credit hours per quarter) sequence. Chapters 13 (Nonparametric Statistical Inference) and 14 (Regression and Linear Statistical Inference) have been provided in order to allow coverage of these topics in situations where this is important in light of the aims of the course and the students' goals. In Section 12.6 we note where these two important subfields of statistics fit into the fabric of the subject along with other subfields.

Some unusual features have already received special note in this preface, for example, the problems and honors problems. Other general features of merit include the numbering system that will enable quick location of theorems, examples, and so on, which are referred to in the text, as well as a good selection of tables, an unusually detailed subject index, a complete author index, an extensive glossary of symbols, and detailed answers to selected problems. The basic topics are covered with untraditional completeness and accuracy. Interval estimation is presented in uncommon detail (Chapter 10), not relegated to a few pages in another chapter. The decision-theoretic approach to statistics is referred to throughout as an integrating factor, is introduced in Chapter 9 (where it gives unity to hypothesis-testing considerations) and is detailed in Chapter 12. Ranking and selection procedures receive treatment in a separate chapter (Chapter 11). Other features include points which should delight the student, for example Simpson's Paradox, and/or the teacher, for example the proof of the independence of $\overline{X}$ and s^2, and the treatment of the Behrens-Fisher problem. Now that statistics has matured to a point where no introduction can introduce students to all of its branches without being either shallow, superficial, or incomprehensible, it is especially important to cover the basic principles and the decision-theoretic integration in a first course. Then we can easily add our own "favorite topic" as the course concludes (and, in fact, there are solid pedagogical reasons why you should do so).

NUMBERING SYSTEM

The numbering system is designed to enable quick location of theorems, examples, and so on, which are referred to in the text. Theorems, Lemmas, Definitions, and so on, are numbered sequentially along with equations and expressions within each Section (for example: Definition 1.2.1 is in Chapter 1, Section 2; it is followed by expression (1.2.2); and so on). [This is simpler to use than the common system wherein Theorems, Lemmas, Definitions, and so on are numbered sequentially in a *separate* system from the equations and expressions; such a system can have a Definition 1.2.1 as well as an expression (1.2.1), often separated by several pages.]

Figures in the text have a separate numbering system (necessitated by the fact that their placement in the text must often differ from their placement in the manuscript which preceeded it); for example Figure 1.2–2 is the second figure in Section 2 of Chapter 1. Tables, except for those appearing at the back of the book, are numbered similarly in a separate system. Footnotes are numbered sequentially (1, 2, . . .) within each chapter.

Thanks are due to my family, teachers, colleagues, publishers, editors (Rosalind Sachoff and Robert de Villeneuve), reviewers, students (Michael C. Mitchelmore in particular), and assistants (Vijay Agnihotri, Thomas A. Bishop, Hubert J. Chen, Ahtsu Huang, and Richard Tomasino in particular) for their contributions from which this book has benefited. Comments received from Professors Hubert J. Chen (Memphis State University), Robert Goosey (Eastern Michigan University), Alan M. Kirch (Macales College), Ingram Olkin (Stanford University), John S. Ramberg (University of Iowa), and Yung L. Tong (University of Nebraska) were especially helpful. Final thanks are due to my excellent typists, Nancy Hufsmith, Linda Levkulich, Phyllis Giles, Mrs. Edna G. Hughes, Sandra Cantrell, Dorothy Monnin, Denise Balduff, and Kath Narcross for their superb work.

The Ohio State University
Columbus, Ohio
October, 1975

Edward J. Dudewicz

CONTENTS

Preface **iii**

Mathematical Notations and Greek Alphabet xi

1 Naive Set Theory **1**

1.1 Basic Definitions 2
1.2 Additional Definitions 4
1.3 Simple Laws and Some Extensions 5
1.4 Sets in Probability and Statistics 7
Problems 9
Honors Problems 11

2 Probability **12**

2.1 Axioms of Probability; Elementary Properties 12
2.2 Finite Discrete Sample Space and Combinatorics 16
2.3 Conditional Probability; Bayes' Rule; Additional Properties 23
2.4 Independent Events 30
Problems 34
Honors Problems 40

3 Random Variables: Discrete Case **45**

3.1 Discrete Probability Functions and Distribution Functions 45
3.2 Special Discrete Distributions 50
3.3 Relations among Certain Discrete Distributions 56
3.4 Multivariate Distributions; Independence 59
3.5 Special Multivariate Discrete Distributions 64
3.6 Distributions of Functions of Discrete Random Variables 66
Problems 67
Honors Problems 72

4 Random Variables: Continuous and Mixed Cases **76**

4.1 Classification of Distribution Functions 77
4.2 Special Continuous Distributions 80

4.3 Relations among Certain Continuous Distributions 90
4.4 Multivariate Continuous Distributions; Independence 90
4.5 Special Multivariate Continuous Distributions 91
4.6 Distributions of Functions of Continuous Random Variables 93
Problems 103
Honors Problems 108

5 Moments 110

5.1 Expectation and Other Moments; Median 110
5.2 Mean and Variance of Special Random Variables 116
5.3 Multivariate Moments 119
5.4 Moment-Generating Functions and Characteristic Functions 127
Problems 133
Honors Problems 139

6 Sums of Random Variables and Limit Laws 141

6.1 Characteristic Functions and the Distributions of Sums 141
6.2 Chebyshev and Markov Inequalities 143
6.3 Modes of Convergence; Laws of Large Numbers 145
6.4 Characteristic Function Methods for Limits; Central Limit Theorem 147
Problems 152
Honors Problems 155

7 An Example of Statistical Analysis 157

7.1 (Nonstochastic) Approximation 157
7.2 Point Estimation 159
7.3 Stochastic Approximation 160
Problems 168
Honors Problems 171

8 Point Estimation and Statistical Inference Principles 172

8.1 Bias and Consistency with Examples 172
8.2 Maximum-Likelihood Estimation 184
8.3 Consistency of MLE's 193
8.4 Asymptotic Normality; Asymptotic Relative Efficiency 205
8.5 Quantiles 214
8.6 Cramér-Rao Inequality 222
8.7 Efficiency 229
8.8 Sufficiency 233
8.9 Completeness 246
8.10 Other Estimation Methods 248
Problems 251
Honors Problems 260

9 Tests of Hypotheses 268

9.1 Simple and Composite Null and Alternative Hypotheses 268
9.2 Tests; Errors of Type I and Type II; Power 270
9.3 The Neyman-Pearson Lemma; Its Generalizations and Uses 277
9.4 A Decision-Theoretic View of the Hypothesis Testing Problem 290
9.5 Testing Normal Means (One Population) 297
9.6 Testing Normal Variances (One Population) 303
9.7 Testing Normal Variances (Two Populations) 304
9.8 Testing Normal Means (Two Populations with Equal Variances) 306
9.9 Testing Normal Means (Two Populations with Unequal Variances; The Behrens-Fisher Problem) 308
Problems 311
Honors Problems 318

10 Interval Estimation 323

10.1 Confidence Sets and Tests of Hypotheses 323
10.2 Confidence Intervals for One Normal Mean 326
10.3 Confidence Intervals for One Normal Variance 330
10.4 Confidence Intervals for a Success Probability 332
Problems 335
Honors Problems 338

11 Ranking and Selection Procedures 341

11.1 The General Ranking Problem 341
11.2 Selection of the Best (Largest Mean) Normal Population 343
11.3 Selection of the Best (Smallest Variance) Normal Population 349
11.4 Extensions and Generalizations 352
Problems 355
Honors Problems 359

12 Decision Theory 363

12.1 The General Problem of Statistics 363
12.2 Hypothesis Testing 379
12.3 Interval Estimation 382
12.4 Ranking and Selection 383
12.5 General Decision Theory 384
12.6 An Overview 388
Problems 391
Honors Problems 400

13 Nonparametric Statistical Inference 403

13.1 Nonparametric Confidence Intervals 404
13.2 Nonparametric Testing of Quantiles 408

13.3 Nonparametric Tests for Ordinal Data 408
Problems 413
Honors Problems 414

14 Regression and Linear Statistical Inference 416

14.1 Estimation of Regression Parameters; Linear Models 417
14.2 Testing Hypotheses about Regression Parameters 423
14.3 Prediction and Confidence Intervals in Regression 425
14.4 Selection of the Best Regression 428
Problems 429
Honors Problems 433

Statistical Tables 435

Table 1 Auxiliary Function $\sqrt{x}$ 435
Table 2 Auxiliary Function n 436
Table 3 Auxiliary Function $\binom{n}{r}$ 437
Table 4 Auxiliary Function e^{-x} 438
Table 5 The Binomial Distribution 440
Table 6 The Hypergeometric Distribution 444
Table 7 The Poisson Distribution 447
Table 8 The Beta Distribution 451
Table 9 The Gamma Distribution 454
Table 10 The Chi-Square Distribution 457
Table 11 The Normal Distribution 460
Table 12 The F Distribution 463
Table 13 The t Distribution 467
Selected References 469
Answers to Selected Problems 471
Glossary of Symbols 481
Author Index 491
Subject Index 495

MATHEMATICAL NOTATIONS

$\{x: \mathscr{A}(x)\}$	the set of all x such that $\mathscr{A}(x)$ holds; for example, $\{x: x \text{ is an integer}\} = \{\ldots, -3, -2, -1, 0, 1, 2, 3, \ldots\}$
$\in$	is a member of (belongs to)
$\notin$	is not a member of (does not belong to)
$\Rightarrow$	implies
$\not\Rightarrow$	does not imply
$\ni$	such that
iff	if and only if
$\Leftrightarrow$	if and only if
$\geq$	greater than or equal to
$>$	greater than
$\leq$	less than or equal to
$<$	less than
$\gg$	much larger than
$\equiv$	identically equal to
$\emptyset$	the empty set

GREEK ALPHABET

A	α	alpha	N	ν	nu
B	β	beta	Ξ	ξ	xi
Γ	γ	gamma	O	o	omicron
Δ	δ	delta	Π	π	pi
E	ε	epsilon	P	ρ	rho
Z	ζ	zeta	Σ	σ	sigma
H	η	eta	T	τ	tau
Θ, Θ	θ	theta	Υ	υ	upsilon
I	ι	iota	Φ	ϕ, φ	phi
K	κ	kappa	X	χ	chi
Λ	λ	lambda	Ψ	ψ	psi
M	μ	mu	Ω	ω	omega

1

NAIVE SET THEORY

In this chapter we introduce the basic notions and notations needed for dealing with *sets* (collections of objects of various kinds, such as numbers or pairs of numbers, called *points*). "Set" is an undefined notion; it is assumed that at least one set exists, and the treatment is not axiomatic. Attention is restricted to a *universe*, or *universal set*, Ω and all operations are with respect to Ω in order to avoid paradoxical situations (see the problems). For further details see S. Feferman, *The Number Systems: Foundations of Algebra and Analysis*, Addison-Wesley Publishing Co., Reading, Massachusetts, 1964.

How sets arise in the context of probability and statistics is discussed in Section 1.4 and a simple example is given.[1] [As we note at the beginning of Section 1.4, the example used there is simply a convenient probability-related illustration of set concepts and operations. Realistic probabilistic problems utilizing these concepts, such as the Post Office operations problem discussed following Definition 2.2.11 and Problem 2.8H, could also be given now but are deferred since their proper discussion entails many probabilistic concepts introduced in Chapters 2ff; some well-prepared students may wish to glance ahead to Chapter 2 in order to obtain an earlier view of the extensive applications of sets in probability and statistics.]

---

[1] Some teachers may wish to cover the example of Figure 1.4–1 (with the resulting universe Ω given on p. 8) at this point and to draw illustrations of the definitions and theorems of Sections 1.1–1.3 from the resulting universe Ω of ordered pairs. (This may be particularly appropriate for groups of students to whom the notions of Chapter 1 are largely new.) Others (especially those with groups of students for whom Chapter 1 is largely a review) will often prefer the crisp coverage of Sections 1.1–1.3 alone.

1.1. BASIC DEFINITIONS

Let A and B be sets[2]. We will now define four set operations and illustrate their meanings with diagrams called **Venn diagrams**. These diagrams are slightly unreliable, as are arguments by diagram in geometry (for example, the "typical" cases drawn often do not show any relations of inclusion or disjointness; a specific instance is given in Example 1.2.9).

Definition 1.1.1.[3] **Union** (*or* **join**, *or* **sum**) *of sets A and B* (see Figure 1.1–1):

$$A \cup B \equiv A + B \equiv \{\omega : \omega \in A \text{ or } \omega \in B\}. \qquad (1.1.2)$$

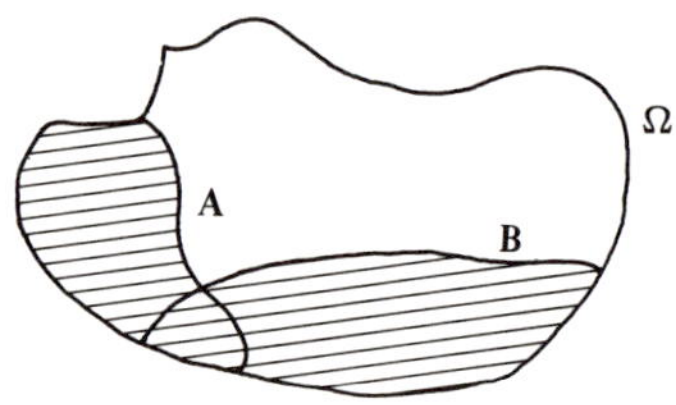

Figure 1.1–1. The shaded region is **A ∪ B**

Definition 1.1.3. **Intersection** (*or* **meet**, *or* **product**) *of sets A and B* (see Figure 1.1–2):

$$A \cap B \equiv AB \equiv A \cdot B \equiv \{\omega : \omega \in A \text{ and } \omega \in B\}. \qquad (1.1.4)$$

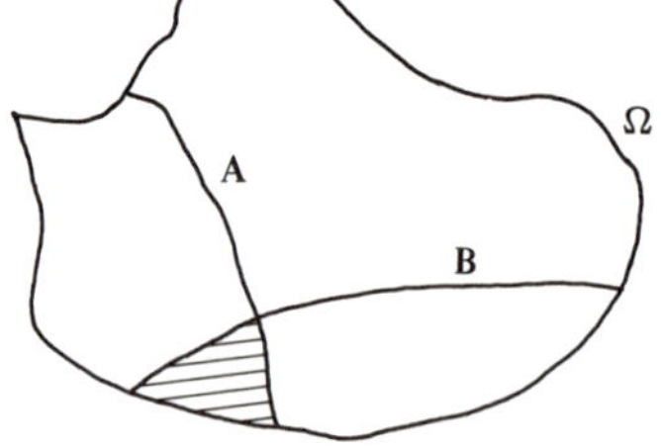

Figure 1.1–2. The shaded region is **AB**

[2] We will generally denote *points* by lower case Greek letters, *sets* by capital italic letters, collections of sets by capital script letters, and the *universe* by capital omega.

[3] In this text "or" is used in the usual mathematical sense; that is, "$\omega \in A$ or $\omega \in B$" means "either $\omega \in A$, or $\omega \in B$, *or both.*" As used in Equation (1.1.2), $\equiv$ defines the symbols $A \cup B$ and $A + B$; they are by definition shorthand for $\{\omega : \omega \in A \text{ or } \omega \in B\}$.

Definition 1.1.5. **Complement** *of a set A* (see Figure 1.1–3):

$$A^c \equiv \bar{A} \equiv A' \equiv \{\omega : \omega \in \Omega \text{ and } \omega \notin A\}. \tag{1.1.6}$$

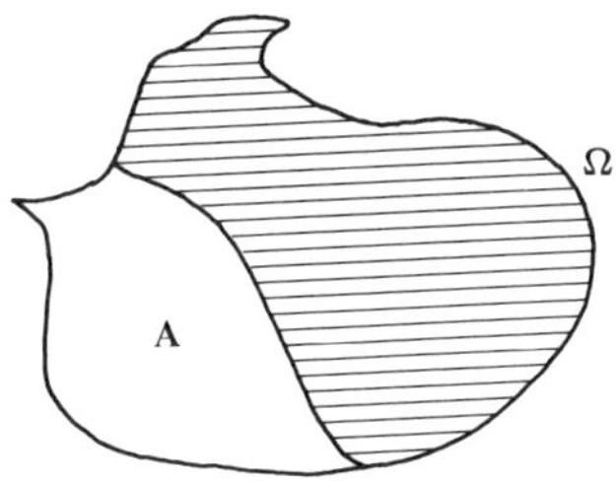

Figure 1.1–3. The shaded region is $\bar{\mathbf{A}}$

Note that we have given two notations for set unions ($A \cup B$ and $A + B$), three for set intersections ($A \cap B$, AB, and $A \cdot B$), and three for set complements (A^c, $\bar{A}$, and A'). Each notation is widely used in the fields of statistics and probability and each seems more convenient in certain contexts. We shall try below to use the first or second notations given most of the time.

Definition 1.1.7. **Difference** *of sets A and B* (see Figure 1.1–4):

$$A - B \equiv A \cap B^c. \tag{1.1.8}$$

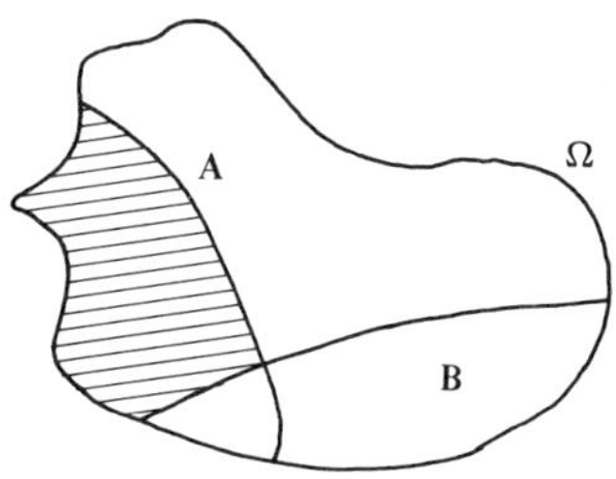

Figure 1.1–4. The shaded region is **A − B**

Example 1.1.9. As an illustration suppose that Ω is the set of all positive integers, that A is the set of even positive integers, that B is the set of odd positive integers, and that C is the set of positive integers which are powers of 2. Then

$$\begin{cases} A = \{2, 4, 6, 8, 10, 12, \ldots\}, \\ B = \{1, 3, 5, 7, 9, 11, \ldots\}, \\ C = \{1, 2, 4, 8, 16, \ldots\}, \end{cases} \tag{1.1.10}$$

are each subsets of $\Omega = \{1, 2, 3, 4, 5, 6, 7, 8, 9, 10, 11, 12, \ldots\}$. Simple calculations from the definitions given above (Definitions 1.1.1, 1.1.3, 1.1.5, and 1.1.7) yield

$$\begin{cases} A \cup B = \Omega, \\ A \cap B \quad \text{contains no elements,} \\ A \cap C = \{2, 4, 8, 16, \ldots\}, \\ B \cap C = \{1\}, \\ \quad \bar{A} = B, \\ \quad \bar{B} = A, \\ \quad \bar{C} = \{3, 5, 6, 7, 9, 10, 11, 12, 13, 14, 15, 17, \ldots\} = \Omega - C, \text{ and} \\ A - C = \{6, 10, 12, 14, 18, \ldots\}. \end{cases} \tag{1.1.11}$$

1.2. ADDITIONAL DEFINITIONS

Definition 1.2.1. **Identity** *of sets A and B:*

$$A = B \text{ iff (for each } \omega \in \Omega) \ \omega \in A \text{ iff } \omega \in B. \tag{1.2.2}$$

Definition 1.2.3. *A is a* **subset** *of B, written $A \subseteq B$, if $\omega \in A \Rightarrow \omega \in B$.*

Definition 1.2.4. *A is a* **proper subset** *of B, written $A \subsetneqq B$, if $A \subseteq B$ and* $A \neq B$ [where, of course, "$A \neq B$" means "not $A = B$"].

Note that $A \subseteq B$ will sometimes be shortened to $A \subset B$; some authors, however, use the latter notation for $A \subsetneqq B$.

Definition 1.2.5. *The* **empty set** (*or* **null set**), *usually written $\emptyset$ or 0, is the unique set such that for all $\omega \in \Omega$, $\omega \notin \emptyset$.* [In other words, the empty set is the set containing no points at all. Such a set, $A \cap B$, was encountered in Example 1.1.9.]

Definition 1.2.6. *$A \supseteq B$, $A \supset B$, and $A \supsetneqq B$ mean, respectively, $B \subseteq A$,* $B \subset A$, and $B \subsetneqq A$.

Definition 1.2.7. *Sets A and B are said to be* **mutually exclusive** (or **disjoint**) *if $A \cap B = \emptyset$.*

Definition 1.2.8. *Three* (*or more*) *sets A, B, C are said to be* **pairwise disjoint** (*or* **mutually exclusive**) *if every two of them are disjoint.*

Example 1.2.9. At the beginning of Section 1.1 we noted that Venn diagrams are not considered sufficient proof of a theorem of set theory since their "typical" cases may be misleading. Figure 1.2–1 could, for example, be used as a proof of the statement "$A \cup B$ is *not* a subset of B" *if* Venn diagrams were considered to be sufficient proof. As we see in Figure 1.2–2 the statement can be false. We therefore employ Venn diagrams solely as convenient aids to our intuition, recognizing that proofs of theorems about sets cannot rely solely on Venn diagrams.

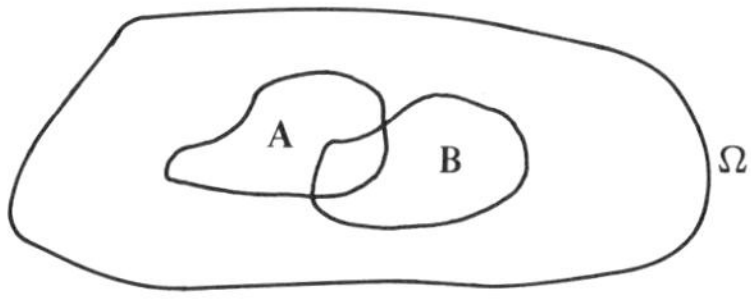

Figure 1.2–1. Is **A** ∪ **B** a subset of **B**?

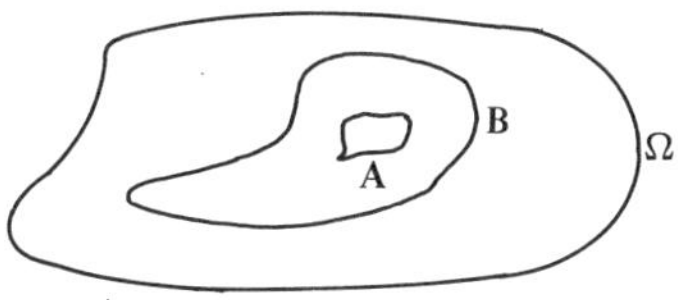

Figure 1.2–2. **A** ∪ **B** may be a subset of **B**

1.3. SIMPLE LAWS AND SOME EXTENSIONS

In the following theorem, we state some simple laws dealing with operations (such as union, intersection, and so on) on sets and prove one in order to show how a formal proof (as contrasted with an intuitive Venn diagram proof) proceeds. It is known that the set of statements 1–9 is **complete** in the sense that any equation formed using $\emptyset, \Omega, \cap, \cup$, and $\subset$ and any number of variables $A, B, C, \ldots$ which is true for every set Ω and subsets $A, B, C, \ldots$ of Ω can be deduced from 1–9.

Theorem 1.3.1. *Let Ω be a set and let A, B, C be subsets of Ω. Then:*

1. **Idempotency Laws**

$$A \cap A = A,$$
$$A \cup A = A.$$

2. **Commutative Laws**

$$A \cap B = B \cap A,$$
$$A \cup B = B \cup A.$$

3. Associative Laws

$$A \cap (B \cap C) = (A \cap B) \cap C,$$
$$A \cup (B \cup C) = (A \cup B) \cup C.$$

4. Distributive Laws

$$A \cap (B \cup C) = (A \cap B) \cup (A \cap C),$$
$$A \cup (B \cap C) = (A \cup B) \cap (A \cup C).$$

5. $A \cap \emptyset = \emptyset$, *and* $A \cup \emptyset = A$.

6. Ω *and* $\emptyset$ *act as* **identity elements** *for* $\cap$ *and* $\cup$.

$$A \cap \Omega = A,$$
$$A \cup \emptyset = A.$$

7.

$$A \cap A^c = \emptyset,$$
$$A \cup A^c = \Omega.$$

8. De Morgan's Laws

$$(A \cap B)^c = A^c \cup B^c,$$
$$(A \cup B)^c = A^c \cap B^c.$$

9.

$$(A^c)^c = A.$$

10.

$$B \subseteq A \text{ iff } A \cap B = B.$$

11.

$$B \subseteq A \text{ iff } A \cup B = A.$$

Proof of the commutative law $A \cap B = B \cap A$. We will prove that $A \cap B \subseteq B \cap A$. Since the proof that $B \cap A \subseteq A \cap B$ is similar, it then follows from Definitions 1.2.1 and 1.2.3 that $A \cap B = B \cap A$. Proof that $A \cap B \subseteq B \cap A$:

$$\begin{aligned} \omega \in A \cap B &\Rightarrow \omega \in A \text{ and } \omega \in B \\ &\Rightarrow \omega \in B \text{ and } \omega \in A \\ &\Rightarrow \omega \in B \cap A. \end{aligned}$$

Extended intersections and unions. Let $\mathscr{M}$ be any nonempty collection of subsets of Ω. Define

$$\bigcap_{A \in \mathscr{M}} A = \{\omega : \omega \in A \text{ for every } A \in \mathscr{M}\}, \tag{1.3.2}$$

$$\bigcup_{A \in \mathscr{M}} A = \{\omega : \omega \in A \text{ for some } A \in \mathscr{M}\}. \tag{1.3.3}$$

[Thus, (1.3.2) is the collection of those points ω which are in every set $A \in \mathscr{M}$, while (1.3.3) is the collection of those points ω which are in at least one set $A \in \mathscr{M}$. This notation allows us to talk with ease of intersections and unions of infinitely many sets.]

Many of the laws of Theorem 1.3.1 can be generalized to arbitrary unions and intersections. For example, De Morgan's Laws become

$$\left(\bigcap_{A\in\mathcal{M}} A\right)^c = \bigcup_{A\in\mathcal{M}} A^c, \tag{1.3.4}$$

$$\left(\bigcup_{A\in\mathcal{M}} A\right)^c = \bigcap_{A\in\mathcal{M}} A^c. \tag{1.3.5}$$

Example 1.3.6. Let Ω be a set and let A, B, C be subsets of Ω. Prove that

$$A \cap (B \cup C) \subseteq (A \cap B) \cup (A \cap C).$$

We shall in fact prove the two sets are equal, as follows:

$$\begin{aligned} \omega \in A \cap (B \cup C) &\Leftrightarrow \omega \in A \text{ and } \omega \in B \cup C \\ &\Leftrightarrow (\omega \in A \text{ and } \omega \in B) \text{ or } (\omega \in A \text{ and } \omega \in C) \\ &\Leftrightarrow (\omega \in A \cap B) \text{ or } (\omega \in A \cap C) \\ &\Leftrightarrow \omega \in (A \cap B) \cup (A \cap C) \end{aligned}$$

Example 1.3.7. For the Ω, A, B specified in Example 1.1.9,

$$\begin{cases} A \cap B = \emptyset, \\ A \cup B = \Omega, \\ \bar{A} = B, \\ \bar{B} = A. \end{cases} \tag{1.3.8}$$

Hence,

$$\begin{cases} (A \cap B)^c = \Omega = A \cup B = A^c \cup B^c, \\ (A \cup B)^c = \emptyset = A \cap B = A^c \cap B^c, \end{cases} \tag{1.3.9}$$

which verifies some special cases of De Morgan's Laws (part 8 of Theorem 1.3.1).

1.4. SETS IN PROBABILITY AND STATISTICS

In probability and statistics we usually study phenomena that are *random* (as opposed to phenomena that are *deterministic*); that is, we wish to consider some *experiment's results* and the experiment does not always produce the same results. We then become involved with the collection of *all possible outcomes* of the experiment, and this functions as our universal set Ω. Subsets A, B, C, . . . of Ω will then represent possible occurrences, and we will be interested in the "chances" (or probability) of their occurrence; in calculating those "chances" it will often be desirable (and sometimes necessary) to use the manipulations of set theory covered above. Example 1.4.1 details some sets of interest when a pair of dice is rolled. (This example is simply a convenient illustration of the concepts involved; more realistic examples will be covered in due course as in Problem 1.8.)

Example 1.4.1. Suppose that two 6-sided dice are rolled. If each has its sides marked with 1 through 6 dots in the usual manner, one way of looking at the experiment is as resulting in a pair of numbers (i, j) with $1 \leqslant i, j \leqslant 6$, where i is the number which "shows" on die 1 and j is the number which "shows" on die 2. (See Figure 1.4–1.) We may then take our universal set Ω as

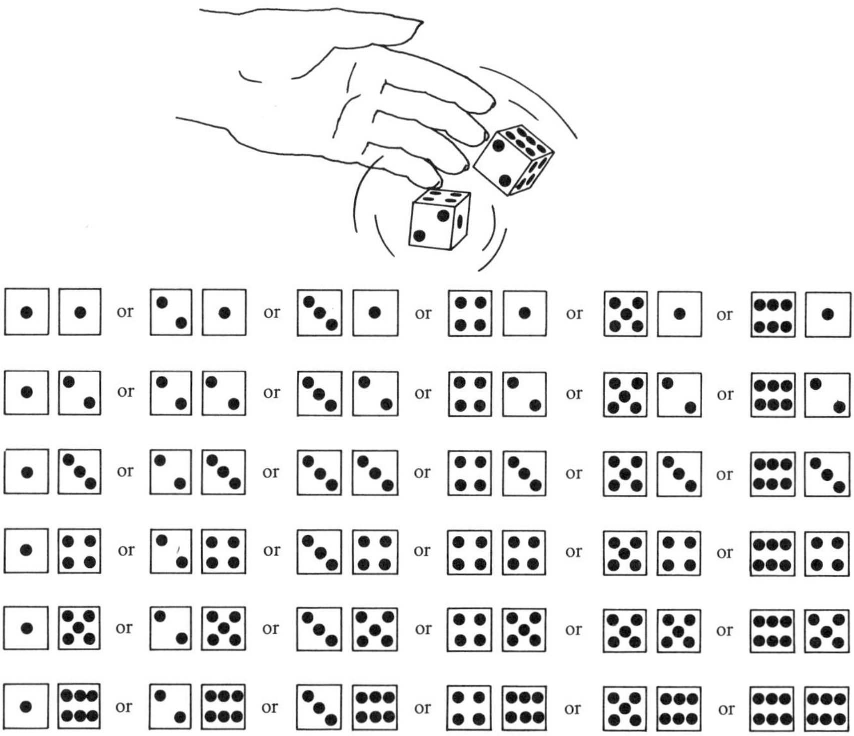

Figure 1.4–1. Results of dice-rolling experiment (with only tops of dice shown in results)

$$\Omega = \left\{ \begin{matrix} (1, 1), (2, 1), (3, 1), (4, 1), (5, 1), (6, 1), \\ (1, 2), (2, 2), (3, 2), (4, 2), (5, 2), (6, 2), \\ (1, 3), (2, 3), (3, 3), (4, 3), (5, 3), (6, 3), \\ (1, 4), (2, 4), (3, 4), (4, 4), (5, 4), (6, 4), \\ (1, 5), (2, 5), (3, 5), (4, 5), (5, 5), (6, 5), \\ (1, 6), (2, 6), (3, 6), (4, 6), (5, 6), (6, 6) \end{matrix} \right\}.$$

The "sum is 7" when and only when the result of the experiment is a point in

$$A = \{(1, 6), (2, 5), (3, 4), (4, 3), (5, 2), (6, 1)\},$$

which is a subset of Ω. "Neither die shows a 1" when and only when the result of the experiment is a point in

$$B = \begin{Bmatrix} (2,2), (3,2), (4,2), (5,2), (6,2), \\ (2,3), (3,3), (4,3), (5,3), (6,3), \\ (2,4), (3,4), (4,4), (5,4), (6,4), \\ (2,5), (3,5), (4,5), (5,5), (6,5), \\ (2,6), (3,6), (4,6), (5,6), (6,6) \end{Bmatrix}.$$

The "sum is 7, and neither die shows a 1" when and only when the result of the experiment is a point in $A \cap B = \{(2,5), (3,4), (4,3), (5,2)\}$. (See Figure 1.4–2.)

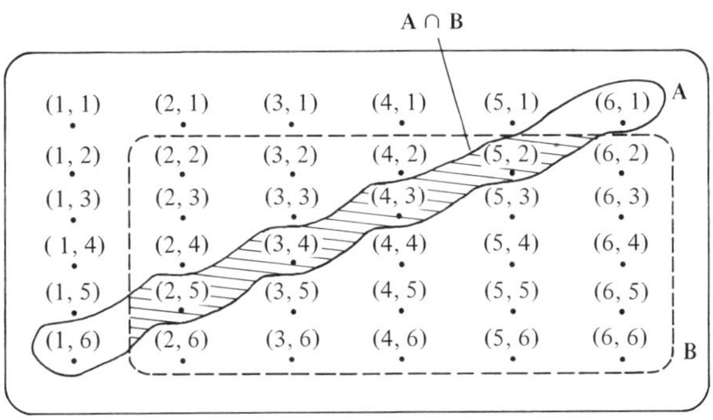

Figure 1.4–2. Ω, **A**, **B**, **A** $\cap$ **B** for Example 1.4.1

PROBLEMS[4]

1.1 Sets A, B, C, D are defined by $A = \{w_1, w_2, w_3\}$, $B = \{w_3, w_4, w_5\}$, $C = \{w_6\}$, $D = \{w_1, w_2, w_4, w_5\}$. Find $A \cup B$, $A \cap B$, $S(A, B)$ (see Problem 1.6 for the definition of this set), $A \cap B \cap C \cap D$, and $(B \cap C)^c$.

1.2 Suppose that the Librarian of Congress compiles, for inclusion in the Library of Congress, a bibliography of those bibliographies in the Library of Congress which do not list themselves. Should this bibliography list itself? How does this show that "not every property determines a set"? (Be explicit; for example, what property?)

[4] Problems 1.2–1.4 relate to the paradoxes of set theory mentioned on p. 1. For some recent work, see H. Nakano, "Russell's paradox in set spaces (axiomatic set theory is not set theory)," Abstract, *Notices of the American Mathematical Society*, Vol. 16 (1969), p. 764, and pp. 9–13 of G. Takeuti and W. M. Zaring, *Introduction to Axiomatic Set Theory*, Springer-Verlag, New York, 1971.

1.3 B. Russell discovered that unrestricted formation of sets can lead to a contradiction which, in consequence, has come to be known as *Russell's paradox*. Namely, form the set $A = \{B: B \text{ is a set, and } B \notin B\}$. Show that

(a) $A \in A \Rightarrow A \notin A$ (a contradiction);

(b) $A \notin A \Rightarrow A \in A$ (a contradiction).

How was such a contradiction avoided in the formulation of set theory in Chapter 1? (Prove that it was avoided.) [In "axiomatic set theory," there exists a systematization in which one explicitly lists all assumptions needed for a development consistently extending our conception of sets to the realm of infinite sets.]

1.4 The Barber of Seville shaves all men who don't shave themselves. Who shaves the Barber of Seville?

1.5 **(Poretsky's Law)** Let X and T be given sets. Prove that $X = \emptyset$ iff

$$T = (X \cap T^c) \cup (X^c \cap T).$$

1.6 Let A and B be given sets. Define a new set $S(A, B) = (A - B) \cup (B - A)$, called the **symmetric difference** of A and B. Prove that

$$S(A_1 \cup A_2, B_1 \cup B_2) \subseteq S(A_1, B_1) \cup S(A_2, B_2).$$

(Note that the result of Problem 1.5 says $X = \emptyset$ iff $T = S(T, X)$. Thus, introduction of the concept of the symmetric difference of two sets allows for a simplified notation.)

1.7 (*Suggested by P. Zuckerman*) In the definitions of extended intersections and unions, (1.3.2) and (1.3.3), we required $\mathscr{M} \neq \emptyset$. If we make the same definitions but allow $\mathscr{M} = \emptyset$. what sets will $\bigcap_{A \in \mathscr{M}} A$ and $\bigcup_{A \in \mathscr{M}} A$ be identical to when in fact $\mathscr{M} = \emptyset$? Check to see that your answer satisfies DeMorgan's Laws, (1.3.4) and (1.3.5). (A reference is P. R. Halmos, *Measure Theory*, D. Van Nostrand Company, Inc., Princeton, New Jersey, 1950, pp. 13–14.)

1.8 Suppose that one part of an experiment is to consist of observing the random phenomena (such as time of birth) that arise in a human birth. We may regard these phenomena in many different ways, depending on our experiment's purpose. (For example, we could regard the random phenomena as simply "year of birth," or more complexly as "year of birth, month of birth, day of birth, hour of birth, place of birth, weight at birth, sex at birth.")

(a) If the purpose of the experiment is to answer the question "Do more births occur from 1 a.m. to 6 a.m. than from 6 a.m. to 1 a.m.?" what minimal information should the statement of results contain? In that case what is the universal set Ω? How many points does Ω contain?

(b) If the purpose of the experiment is as stated in (a), it may still be desirable to have a more detailed statement of results. Why? Give an example.

1.9 Suppose that three 6-sided dice are rolled. If each has its sides marked with 1 through 6 dots in the usual manner, what universal set Ω would you, in analogy with Example 1.4.1, choose to represent the outcomes of the experiment? For what subset of Ω is the sum equal to 10? 11?

HONORS PROBLEMS

1.1H Let $\Omega = \{1, 2, 3, 4, 5, \ldots\}$ be the set of all positive integers. Let $\mathscr{F}$ be the set whose members are: Ω; $\emptyset$; all subsets of Ω which contain only a finite number of positive integers; and all subsets of Ω whose complements contain only a finite number of positive integers. Prove that $A \in \mathscr{F} \Rightarrow \bar{A} \in \mathscr{F}$, and that $A, B \in \mathscr{F} \Rightarrow A \cup B \in \mathscr{F}$. Prove that if $A_n \in \mathscr{F}$ $(n = 1, 2, \ldots)$, it is not necessarily the case that

$$\bigcup_{n=1}^{\infty} A_n \in \mathscr{F}.$$

[*Hint*: Consider the set $C = \{1, 3, 5, 7, \ldots\}$ of odd positive integers.] (In terms of the terminology of Problem 2.2, $\mathscr{F}$ is an algebra of sets, but is not a σ-algebra of sets.)

1.2H Let $A_1, \ldots, A_n$ be sets each of which contains a finite number of elements. Let

$$\Omega = A_1 \cup A_2 \cup \cdots \cup A_n.$$

Suppose that for some fixed integer k $(1 \leqslant k \leqslant n)$,

$$A_{\lambda_1} \cup \cdots \cup A_{\lambda_k} = \Omega$$

for every specification $(\lambda_1, \ldots, \lambda_k)$ of k of the integers $(1, 2, 3, \ldots, k, k+1, \ldots, n)$, *but* that for $j = 1, 2, \ldots, k-1$,

$$A_{\lambda_1} \cup \cdots \cup A_{\lambda_j} \subsetneqq \Omega$$

for every specification $(\lambda_1, \ldots, \lambda_j)$ of j of the integers $(1, 2, 3, \ldots, k, k+1, \ldots, n)$. *Find* formulas (involving k and n) for: (a) the smallest possible number of elements in Ω; (b) the number of elements in each of $A_1, \ldots, A_n$ when Ω contains the smallest possible number of elements; and (c) the number of elements in $A_{\lambda_1} \cap \cdots \cap A_{\lambda_j}$ when Ω contains the smallest possible number of elements. Illustrate your answers for the case $n = 5$, $k = 4$. [*Hint*: See Problem E 2322 by H. Lass and P. Gottlieb and its solution by D. M. Bloom, "Unions of finite sets of subsets," *American Mathematical Monthly*, Vol. 79 (1972), pp. 1039–1040.]

2

PROBABILITY

In Section 1.4 we have already discussed informally and briefly the idea of the "chance" of obtaining a certain type of outcome from an experiment which deals with random phenomena (that is, the "chance" of obtaining an outcome which belongs to some $A \subseteq \Omega$). In this chapter we introduce the idea of "probability," which formalizes our intuitive notion of "chance," and define a "probability function." We then illustrate basic definitions and rules for computing probabilities with a number of examples. (For simplicity, almost all of these examples deal with finite, discrete sample spaces and equally–likely outcomes; more complex situations are covered in Chapter 3.)

2.1. AXIOMS OF PROBABILITY; ELEMENTARY PROPERTIES

Let Ω be a set (called the **sample space**). The points $\omega \in \Omega$ are called **elementary events**. Ω corresponds to the collection of distinct outcomes which may result when we perform some particular experiment of interest. Each $A \subseteq \Omega$ is called an **event**. (For instance, in Example 1.4.1 Ω consisted of 36 possible outcomes resulting from a roll of dice, and any subset of Ω was an event.) We now wish to formalize the notion of the "chance" of obtaining an outcome lying in a specified subset A of Ω into a definition of probability; we will wish to talk of the probability of each event $A \subseteq \Omega$, and will hence need a *function* which assigns a unique number (called **probability**) to each such A. (For instance, in Example 1.4.1 the probability of a sum equal to 7 resulting from a roll of the dice is the probability of the set A of outcomes defined in that example.) After giving the formal definition, we will relate it to the "classical definition of probability" and to the frequency interpretation of probability. The latter [see (2.1.5)ff] provides the intuitive basis for Definition 2.1.1.

Definition 2.1.1. *$P(\cdot)$ is a* **probability function**[1] *if to each event $A \subseteq \Omega$ it assigns a number $P(A)$ such that*

$$P(A) \geqslant 0, \tag{2.1.2}$$

$$P(\Omega) = 1, \tag{2.1.3}$$

and

(Countable Additivity) *If $A_1, A_2, \ldots$ are events with $A_i \cap A_j = \emptyset\ (i \neq j)$, then*

$$P\left(\bigcup_{n=1}^{\infty} A_n\right) = \sum_{n=1}^{\infty} P(A_n). \tag{2.1.4}$$

The so-called **classical definition of probability** is actually a rule for assigning such a function $P(\cdot)$ to certain types of Ω's. It says: *If an experiment can lead to n mutually exclusive and equally likely outcomes, and if $n(A)$ of these outcomes have an attribute A, then*

$$\text{The probability of } A = \frac{n(A)}{n}. \tag{2.1.5}$$

This definition applies in many simple games of chance (cards, dice, balls in cells) but not usually in more complex situations. In Section 2.2 we study precisely those cases where this assignment of $P(\cdot)$ is appropriate.

We should ask, "How is such a probability function $P(\cdot)$ obtained in general?" One general method of conceptualizing obtaining such a probability function $P(\cdot)$ (which is really a conceptual connection between mathematical probability theory and the real world) is **frequency probability**. This says that if an event A occurs $n(A)$ times in n (unrelated) trials of an appropriate experiment, then we should define a frequency function (which depends on the number of trials n) by

$$R(A) = \frac{n(A)}{n}$$

and define a probability function by

$$P(A) = \lim_{n \to \infty} R(A) = \lim_{n \to \infty} \frac{n(A)}{n}. \tag{2.1.6}$$

In practice, n is finite, which is but one of the practical difficulties with such empirical estimates of probability (others including the problem of *having*

[1] This is sometimes also called a "probability measure."

unrelated trials). One often simply takes $P(A) = R(A)$ for all A, and it is easily verified that this *is* a probability function [that is, $P(\cdot)$ so defined satisfies Definition 2.1.1]. Therefore, of course, all of the properties that we will prove below for probability functions are also true for frequency functions. In particular, it is easy to see directly that (for any two events A and B)

$$0 \leqslant R(A) \leqslant 1, \tag{2.1.7}$$

$$R(A^c) = 1 - R(A), \tag{2.1.8}$$

$$R(A + B) = R(A) + R(B) - R(AB), \tag{2.1.9}$$

and that (2.1.2), (2.1.3), and (2.1.4) hold for any frequency function. Hence, if relative frequency furnishes a reasonable way of thinking about the "chances" of events, probability functions also incorporate some of this reasonableness [since they, by definition, must satisfy properties (2.1.2), (2.1.3), and (2.1.4)]. We shall now prove that probability functions, which we have, to this point, only assumed satisfy the three axioms (2.1.2), (2.1.3), and (2.1.4), possess properties corresponding to (2.1.7), (2.1.8), (2.1.9)[2] and other properties of relative frequencies; hence, they furnish a reasonable abstraction of our notion of "chance."

Theorem 2.1.10. $$P(\emptyset) = 0.$$

Proof: Let $A_n = \emptyset$ $(n = 1, 2, \ldots)$. Then $\bigcup_{n=1}^{\infty} A_n = \emptyset$ and, by countable additivity,

$$P(\emptyset) = P\left(\bigcup_{n=1}^{\infty} A_n\right) = \sum_{n=1}^{\infty} P(A_n) = \sum_{n=1}^{\infty} P(\emptyset).$$

Since $P(\cdot)$ is a real-valued function, $P(\emptyset) = 0$.

Theorem 2.1.11. **(Finite Additivity)** *Let $A_1, \ldots, A_n$ be disjoint events. Then*

$$P(A_1 \cup \cdots \cup A_n) = P(A_1) + \cdots + P(A_n).$$

Proof: Let $A_{n+1} = A_{n+2} = \cdots = \emptyset$. Then

$$P(A_1 \cup \cdots \cup A_n) = P\left(\bigcup_{n=1}^{\infty} A_n\right) = \sum_{k=1}^{\infty} P(A_k)$$
$$= \sum_{k=1}^{n} P(A_k) + 0 = P(A_1) + \cdots + P(A_n).$$

[2] The proof of the property corresponding to (2.1.9) is deferred to Corollary 2.3.8, where it follows easily from a more general result about probability functions.

Theorem 2.1.12. *If A and B are events and $A \subseteq B$, then $P(A) \leqslant P(B)$.*

Proof: Since $B = A \cup \bar{A}B$ and $A \cap \bar{A}B = \emptyset$, then by Theorem 2.1.11, $P(B) = P(A) + P(\bar{A}B)$. This is $\geqslant P(A)$ since $P(\bar{A}B) \geqslant 0$.

Corollary 2.1.13. *For every A, $P(A) \leqslant 1$.*

Proof: $A \subseteq \Omega$, so $P(A) \leqslant P(\Omega) = 1$.

Theorem 2.1.14. **(Boole's Inequality)** *If $A_1, A_2, \ldots$ are events, then*

$$P\left(\bigcup_{n=1}^{\infty} A_n\right) \leqslant \sum_{n=1}^{\infty} P(A_n).$$

Proof: Let $B_1 = A_1$ and (for $n \geqslant 2$) $B_n = \bar{A}_1 \cdots \bar{A}_{n-1}A_n$. Then $B_i \cap B_j = \emptyset$ $(i \neq j)$ and $\bigcup_{n=1}^{\infty} B_n = \bigcup_{n=1}^{\infty} A_n$, so

$$P\left(\bigcup_{n=1}^{\infty} A_n\right) = P\left(\bigcup_{n=1}^{\infty} B_n\right) = \sum_{n=1}^{\infty} P(B_n).$$

But $B_n \subseteq A_n$ $(n \geqslant 1)$, so $P(B_n) \leqslant P(A_n)$ and the theorem follows.

Theorem 2.1.15. $P(A^c) = 1 - P(A)$.

Proof: Since $A \cap \bar{A} = \emptyset$,

$$1 = P(\Omega) = P(A \cup \bar{A}) = P(A) + P(\bar{A}).$$

Example 2.1.16. *A sample space, the set of possible outcomes of a certain experiment or observation, may (in certain cases) be chosen in several ways.* We must choose the one which is appropriate for our aims. If a fair coin is tossed 100 times, we have many possible sample spaces. Let's consider two simple ones. First, the set of 100-tuples with T or H in each position (the possible results on each toss):

$$\Omega_1 = \{(x_1, x_2, \ldots, x_{100}): \ x_i = T \text{ or } H, \quad i = 1, 2, \ldots, 100\}. \tag{2.1.17}$$

Second, the set of 2-tuples where the first entry is the number of tails and the second entry is the number of heads:

$$\Omega_2 = \{(x_1, 100 - x_1): \ x_1 = 0, 1, \ldots, 100\}. \tag{2.1.18}$$

An event is a subset of the points in the sample space. Thus, "The 5th and 99th tosses result in H" would be an event in Ω_1, but not in Ω_2.

2.2. FINITE DISCRETE SAMPLE SPACE AND COMBINATORICS

A sample space Ω is called **discrete** if it contains only finitely many points or infinitely many points which can be arranged in a simple sequence $\omega_1, \omega_2, \ldots$; in the second case Ω is called **denumerably infinite**. With a *finite discrete sample space*, if the points are *equally probable* [that is, $P(\omega_1) = P(\omega_2)$ for every $\omega_1, \omega_2 \in \Omega$], then we can find $P(A)$ by finding the number $n(A)$ of points where A occurs, since then

$$P(A) = \frac{n(A)}{n}, \tag{2.2.1}$$

where n is the total number of points in Ω. In such a case, we find a need for **combinatorics** or **combinatorial analysis** (permutations, combinations, and so on) to find n and $n(A)$. The following theorem states a basic counting principle.

Theorem 2.2.2. (Multiplets) *Given n_1 elements $a_1, \ldots, a_{n_1}$; and n_2 elements $b_1, \ldots, b_{n_2}$; and ...; and n_r elements $x_1, \ldots, x_{n_r}$, it is possible to form precisely $n_1 \cdot n_2 \cdot \ldots \cdot n_r$ distinct ordered r-tuplets $(a_{j_1}, b_{j_2}, \ldots, x_{j_r})$ containing one element of each kind.*

Proof: This is easy for $r = 2$ (matrix of pairs) and in general follows by induction on r (a 3-tuple is a 2-tuple and a 1-tuple, and so on).

We will now introduce the concepts of sampling from a finite population, since probability problems with a finite discrete sample space with equally–likely outcomes have a simple interpretation in this context. Suppose we have a **population** (set) of n distinct elements (points) $a_1, \ldots, a_n$. An ordered arrangement $a_{j_1}, a_{j_2}, \ldots, a_{j_r}$ of r symbols is called an **ordered sample of size r**. If elements are selected from the population one by one, two simple (plus more complex) selection procedures are available. Under **simple random sampling (SRS) with replacement**, each selection is made from the entire population (so the same element may be drawn more than once) and each item is equally as likely to be drawn as any other item at any stage. Under **SRS without replacement**, an element once chosen is removed from the population (so there cannot be repetitions, and we must have $r \leqslant n$) and each item remaining is equally as likely to be drawn as any other item remaining at any stage. In Theorem 2.2.5 (which follows easily from the Multiplets Theorem 2.2.2) we note the number of possible samples under these two sampling procedures. (A "sample" is a "possible outcome" of the sampling experiment; how many such possible outcomes are there?)

Definition 2.2.3. $(n)_0 \equiv 1$. *If n, r are positive integers,*

$$(n)_r = \begin{cases} n(n-1)\cdots(n-r+1), & \text{if } r \leqslant n, \\ 0, & \text{if } r > n. \end{cases}$$

Definition 2.2.4. **(n-factorial)** $0! \equiv 1$. *If n is a positive integer,*

$$n! = n(n-1)\cdot\ldots\cdot 2\cdot 1.$$

[See Table 2 for some values.]

Theorem 2.2.5. *For a population of n elements and a prescribed sample size r, there exist*

n^r samples under SRS with replacement,
$(n)_r$ samples under SRS without replacement.

Corollary 2.2.6. *The number of different orderings of n elements is n!.* [These are called the **permutations** of the n elements.]

Proof: This is the number of samples under SRS without replacement when $r = n$.

We will also often be interested in unordered sets of elements from a given population. An unordered collection $\{a_{j_1}, a_{j_2}, \ldots, a_{j_r}\}$ of r distinct symbols is called an **unordered sample of size** r, or a **subpopulation of size** r. The number of subpopulations is specified in Theorem 2.2.9; it will be useful to first introduce a special notation (binomial coefficients).

Definition 2.2.7. **(Binomial Coefficients)** *If n and r are nonnegative integers,*

$$\binom{n}{r} \equiv {}^nC_r \equiv \frac{(n)_r}{r!}.$$

(*Thus*, $\binom{n}{0} \equiv 1$.) [See Table 3 for some values.]

Corollary 2.2.8. $\binom{n}{r} = \binom{n}{n-r}$.

Theorem 2.2.9. *A population of n elements possesses $\binom{n}{r}$ different subpopulations of size r ($\leqslant n$).*

Proof: $(n)_r$ = No. of ordered samples (of size r) under SRS without replacement;
= $r! \cdot$ (number of distinct subpopulations of size r).

Theorem 2.2.9 may be considered appropriate, for example, for selecting a committee of r people from a legislature of n people. Theorem 2.2.10 answers the question of how many ways we can form k disjoint committees having $r_1, \ldots, r_k$ members, respectively.

Theorem 2.2.10. (Multinomial Coefficients) *Let $r_1, \ldots, r_k$ be integers such that $r_1 + \cdots + r_k = n (r_i \geqslant 0)$. The number of ways in which a population of n elements can be partitioned into k subpopulations of which*

the 1st contains r_1 elements
the 2nd contains r_2 elements
.
.
.
the kth contains r_k elements

is

$$\frac{n!}{r_1!\, r_2! \cdots r_k!} = \binom{n}{r_1}\binom{n - r_1}{r_2}\binom{n - r_1 - r_2}{r_3}\cdots\binom{n - r_1 - \cdots - r_{k-2}}{r_{k-1}}.$$

Proof: To effect the desired partition, we first select r_1 elements out of the given n, then select r_2 of the $n - r_1$ remaining elements, and so on. After forming the $(k - 1)$st group, there remain $n - r_1 - \cdots - r_{k-1} = r_k$ elements, and these form the last group. The expression $n!/(r_1!\, r_2! \cdots r_k!)$ follows upon writing the binomial coefficients in terms of factorials and canceling equal terms. The product of binomial coefficients arose via use of the multiplets theorem (Theorem 2.2.2). (The present theorem may also be proven by induction.)

Many problems occur naturally (or are conveniently stated) in the form of **ball and cell models**. We will now explore some relationships between these models of tossing balls into cells, and sampling.

Definition 2.2.11. *If r balls are placed at random successively into n cells and each cell can hold 1, 2, . . . , i balls we shall say we have* **model** $(BC)_i$ $(1 \leqslant i \leqslant \infty)$. **At random** *means that, when any ball is to be tossed, all cells remaining eligible to receive it are equally likely to receive it.*

Thus, under model $(BC)_1$ each cell can hold only one ball, under model $(BC)_\infty$ each cell can hold unlimited balls, and in general under model $(BC)_m$ each cell can hold up to m balls.

Corollary[3] 2.2.12. *$(BC)_1$ is equivalent[4] to SRS sampling r times from a population of n objects, without replacement. $(BC)_\infty$ is equivalent[4] to SRS sampling r times from a population of n objects, with replacement.*

Due to the correspondence to random samples noted above (see Corollary 2.2.12), examples of $(BC)_1$ and of $(BC)_\infty$ models are easily given. For example, often when cargo is being unloaded from a ship and the total cargo weight is to be determined, only some of the cargo is weighed and observations of its weight are used to estimate the weight of the whole cargo. In such an instance we have a $(BC)_1$ model where n (the number of cells) is the number of cargo units and r (the number of balls) is the number of cargo units actually weighed.[5] Sampling with replacement here would yield a $(BC)_\infty$ model, but this is seldom used except in special circumstances since there is usually no gain from examining (for example, weighing) the same unit more than once.

Examples of $(BC)_m$ models can also be given. For example, a *post office* with n P.O. boxes with a capacity of m letters each might be considered, as a first approximation, to be a $(BC)_m$ situation. In such a situation, how many boxes could we expect to overflow each day? We are not in a position to study this question yet. The question could arise in a Post Office attempt to select box size m. (See Problems 2.8H and 5.5H.)

We now wish to distinguish two ways of specifying the outcomes of tossing r balls into n cells. The first way (assignment vectors) specifies which cell each ball entered. The second way (occupancy vectors) specifies how many balls entered each cell. For example, if 2 balls are tossed into 3 cells we have (under $(BC)_\infty$):

Possible Assignment Vectors (Ω_A)	*Possible Occupancy Vectors* (Ω_0)
(1, 1)	(2, 0, 0)
(2, 2)	(0, 2, 0)
(3, 3)	(0, 0, 2)
(1, 2) or (2, 1)	(1, 1, 0)
(1, 3) or (3, 1)	(1, 0, 1)
(2, 3) or (3, 2)	(0, 1, 1)

Thus, there are 9 possible assignment vectors, but only 6 possible occupancy vectors; hence, in calculating probabilities it will matter which items ($\omega \in \Omega_A$, or $\omega \in \Omega_0$) we consider to be equally–likely outcomes. If $\omega \in \Omega_A$ are

[3] This is a "corollary" to our discussion.

[4] To be more explicit, the equivalences arise from the fact that (considering the "cells" to be members of the populations) the cells which are "balled" may be considered as the members of the sample. We then have, for example, order or not depending on whether the balls carry numbers or not.

[5] An example involving $n = 1267$ bales of tobacco, of which some number r are to be weighed, is discussed by W. E. Deming, *Some Theory of Sampling*, Dover Publications, Inc., New York, 1966, p. 108. In that example r is set so as to obtain a desired accuracy of the estimate of the total cargo weight, resulting in $r = 93$. Note that weighing such a sample will, in general, be much less costly than weighing the whole cargo.

equally likely, $P[\text{Both balls land in the same cell}] = \frac{3}{9} = \frac{1}{3}$; if $\omega \in \Omega_0$ are equally likely, we obtain $P[\text{Both balls land in the same cell}] = \frac{3}{6} = \frac{1}{2}$. [Under $(BC)_1$ yet another result is found: $P[\text{Both balls land in the same cell}] = 0$.] As noted in detail in Feller (1968), pp. 38–42, in physics three possibilities are studied:

	Ω_A	Ω_0
$(BC)_\infty$	**Maxwell-Boltzmann statistical mechanics**	**Bose-Einstein statistical mechanics**
$(BC)_1$		**Fermi-Dirac statistical mechanics**

Definition 2.2.13. $\Omega_A = \{(x_1, \ldots, x_r)$: x_i is the number of the cell ball i goes into$\}$ *is called the set of* **assignment vectors**. $\Omega_0 = \{(r_1, \ldots, r_n)$: r_j is the number of balls in cell $j\}$ *is called the set of* **occupancy vectors**. *Let* $\#(\Omega)$ *denote the number of points in* Ω.

Theorem 2.2.14.

(1) *Under* $(BC)_1$, $\#(\Omega_A) = (n)_r$.
Under $(BC)_\infty$, $\#(\Omega_A) = n^r$.

(2) *Under* $(BC)_1$, $\#(\Omega_0) = \binom{n}{r}$.
Under $(BC)_\infty$, $\#(\Omega_0) = \binom{n+r-1}{r}$.

Proof: **(1)** These results follow from the correspondences with SRS sampling and Theorem 2.2.5.

(2) Under $(BC)_1$, $\Omega_0 = \{(r_1, \ldots, r_n)$: r_j is 0 or 1 and $r_1 + \cdots + r_n = r\}$. Thus, $\#(\Omega_0)$ is the number of ways of choosing r places out of n for 1's.

Under $(BC)_\infty$, $\Omega_0 = \{(r_1, \ldots, r_n)$: r_j is the number of balls in cell j, $r_1 + \cdots + r_n = r, r_j \geqslant 0\}$. If we consider that we have r balls (represented by stars) and n cells, thus $n + 1$ cell walls (represented by bars) (Figure 2.2–1),

Figure 2.2–1

then we must put one cell wall on the outside on each end in any distribution of balls into cells (distribution of stars and bars in a line); the $n - 1$ walls and r balls can be mixed up in any way, the number of such ways being the number of ways of choosing, out of $n + r - 1$ places, r for balls.

Example 2.2.15. $r = 7$ balls are distributed at random into $n = 7$ cells of unlimited capacity. [Thus, model $(BC)_\infty$ holds and each of the n^r possible assignment vectors has probability n^{-r}.] What is the probability that the occupancy numbers are 2, 2, 1, 1, 1, 0, 0 in some order?

The assignment vectors are $\Omega_A = \{(x_1, \ldots, x_7): x_i$ is the number of the cell ball i goes into$\}$. How many assignment vectors lead to occupancy numbers 2, 2, 1, 1, 1, 0, 0 *in that order*? (This order implies 2 "1's," 2 "2's," 1 "3," 1 "4," 1 "5," 0 "6's," 0 "7's" in some order.) The number of such assignment vectors is

$$\frac{7!}{2!2!1!1!1!0!0!} = \frac{7!}{2!2!}$$

by Theorem 2.2.10. How many *different* occupancy vectors are there with 2 doubly occupied, 3 singly occupied, and 2 unoccupied cells? (This eliminates the order on the occupancy vector.) This is the number of ways of subdividing 7 objects into 3 groups, of 2, 3, 2:

$$\frac{7!}{2!3!2!}.$$

Since there are 7^7 assignment vectors, the probability of occupancy numbers 2, 2, 1, 1, 1, 0, 0 in some order is

$$\frac{\dfrac{7!}{2!3!2!} \times \dfrac{7!}{2!2!}}{7^7} = \frac{45 \times 5!}{7^5}.$$

This number can be calculated with little trouble (for example, on a desk calculator). For cases where we need $n!$ and n is large, **Stirling's formula** yields a useful approximate evaluation:

$$n! \sim \sqrt{2\pi n}\left(\frac{n}{e}\right)^n. \tag{2.2.16}$$

Here the "$\sim$" means that

$$\lim_{n\to\infty} \frac{n!}{\sqrt{2\pi n}\left(\dfrac{n}{e}\right)^n} = 1. \tag{2.2.17}$$

Example 2.2.18. Assume that a year has 365 days, and that all birthdays are equally likely. What is the probability that, in a group of 60 people (whose birthdays are assumed unrelated), *some* 5 people will have the same birthday and all the other people will have different birthdays none of which falls on the birthday of the 5?

Here, $r = 60$ balls are distributed at random into $n = 365$ cells of unlimited capacity. [Thus, model $(BC)_\infty$ holds and each of the n^r possible assignment vectors has probability n^{-r}.] We desire

$$P[\text{Occupancy nos. are } \overbrace{5}^{1}, \overbrace{1, 1, \ldots, 1}^{55}, \overbrace{0, \ldots, 0}^{309} \text{ in some order}].$$

Here $\Omega_A = \{(x_1, \ldots, x_{60}): x_i \text{ is the number of the cell ball } i \text{ goes into}\}$. Now

$$\begin{bmatrix} \text{No. of assignment} \\ \text{vectors leading to} \\ \text{occupancy nos.} \\ 5, 1, \ldots, 1, 0, \ldots, 0 \\ \textit{in that order} \end{bmatrix} = \begin{bmatrix} \text{No. of assignment} \\ \text{vectors with} \\ 5 \text{ “1's”} \\ 1 \text{ “2”} \\ \cdot \\ \cdot \\ \cdot \\ 1 \text{ “56”} \\ 0 \text{ “57's”} \\ \cdot \\ \cdot \\ \cdot \\ 0 \text{ “365's”} \\ \text{in some order} \end{bmatrix} = \frac{60!}{5!1! \cdots 1!0! \cdots 0!},$$

and

$$\begin{bmatrix} \text{No. of occupancy} \\ \text{vectors with} \\ 1 \text{ cell w.5 occupants} \\ 55 \text{ cells w.1 occupants} \\ 309 \text{ cells w.0 occupants} \end{bmatrix} = \begin{bmatrix} \text{No. of ways to} \\ \text{subdivide 365} \\ \text{objects into} \\ \text{3 groups of} \\ 1, 55, 309 \end{bmatrix} = \frac{365!}{1!55!309!}.$$

Thus, the desired probability is

$$\frac{\dfrac{60!}{5!1!1! \cdots 1!0! \cdots 0!} \times \dfrac{365!}{1!55!309!}}{365^{60}} = \frac{\dbinom{60}{5}(365)_{56}}{365^{60}}.$$

In this last form, the result can be reasoned to as follows. Select 5 people to have the same birthday. This can be done in $\binom{60}{5}$ ways. For each of these ways, there are $(365) \cdots (310) = (365)_{56}$ different choices of 56 birthdays which are distinct, while with no restrictions there are 365^{60}.

Example 2.2.19. Seven passengers together enter an elevator on the first floor of an 11-floor building. Assume that the “egress pattern” to floors 2, 3, ..., 11 is the same. What is the probability of event A: all get off at different floors?

Again we use assignment vectors, with $\Omega_A = \{(x_1, \ldots, x_7)$: passenger i gets off at floor $x_i, 2 \leqslant x_i \leqslant 11\}$. $\#(\Omega_A) = 10^7$, $\#(A) = (10)_7$, so

$$P(A) = \frac{\#(A)}{\#(\Omega_A)} = \frac{10!}{10^7(10-7)!} = \frac{189}{3125} \approx .065.$$

Example 2.2.20. (Alcohol Tasting) As reported in a UPI story (for example, see *The Times-Union*, Rochester, New York, October 26, 1971, p. 9A), the Internal Revenue Service's argument that neutral grain alcohol tastes the same to everybody (and that special claims for various brands could, hence, not be permitted) suffered a blow when the quality control director of Joseph E. Seagram & Sons correctly identified the distilleries from which 14 such alcohol samples had come, by tasting and smelling the samples. What is the probability that a person with no such organoleptic abilities would obtain the same result? (Ignore the possibility of trickery.)

Suppose we are told that one sample came from each of 14 distilleries. There are 14! ways to assign distilleries to samples and we have probability 1/14! of selecting the correct one by guessing. Our chances of getting at least one distillery correctly paired with a sample by random guessing are studied in Example 2.3.9.

2.3. CONDITIONAL PROBABILITY; BAYES' RULE; ADDITIONAL PROPERTIES

It is often the case that we have partial information about a certain phenomenon, and wish to know how this affects the probabilities of outcomes of interest to us (for example, what is the probability of rain tomorrow given rain today). For this reason we will introduce conditional probability and rules for dealing with it. We then study additional properties of probability and give some examples.

Definition 2.3.1. *If A and B are events in a sample space Ω, and if $P(B) > 0$, then the* **conditional probability of** A **given** B *(say $P(A|B)$) is defined by*

$$P(A|B) \equiv \frac{P(AB)}{P(B)}.$$

Informally, Definition 2.3.1 can be interpreted as follows using frequency probability [see Equation (2.1.6)]. $R(A)$ is the relative frequency of event A in n trials. Let $R(A|B)$ denote the relative frequency of A in those of the n trials in which B occurred. Then

$$R(A|B) = \frac{n(AB)}{n(B)} = \frac{n(AB)/n}{n(B)/n} = \frac{R(AB)}{R(B)};$$

that is, $R(A|B)$ is by definition the ratio of the number of trials $n(AB)$ on which A and B both occurred to the number of trials $n(B)$ on which B occurred, and this equals $R(AB)/R(B)$. As $n \to \infty$ we obtain our definition of conditional probability from that of conditional frequency (which answers the question "How often does A occur on trials where B occurs?").

In many problems involving conditional and unconditional probability, the possible values of some variable (of possible interest) **stratify** the sample space into mutually exclusive and exhaustive subsets whose probabilities are either given or are easy to compute. Theorems 2.3.2, 2.3.4, and 2.3.6 consider how we can use such stratification. For example, suppose that in a certain species a family chosen at random has exactly j children with probability p_j ($j = 0, 1, 2, \ldots$) where $0 \leqslant p_0 \leqslant 1, 0 \leqslant p_1 \leqslant 1, \ldots$ and $p_0 + p_1 + p_2 + \cdots = 1$. Also, suppose that for any given number i of children, the 2^i possible sex distributions each have the same probability $1/2^i$. Let C_m denote the event "a family chosen at random has m children," for which $P(C_m) = p_m$ ($m = 0, 1, 2, \ldots$). Let D be the event "a family chosen at random has either all boys or all girls," and let E be the event "a family chosen at random has at least one child but no girls." Then $P(D)$ (the probability of an unmixed family) and $P(C_1|E)$ (the probability of an only child given at least one child and no girls) are of interest, and their calculation requires use of Theorems 2.3.4 and 2.3.6, respectively. The calculation is left as an exercise (see Problem 2.12H).

Theorem 2.3.2. *If B is an event and $P(B) > 0$, then $P(\cdot|B)$ is a probability function; that is,*

(i) $P(A|B) \geqslant 0$, *for all events A*;

(ii) $P(\Omega|B) = 1$;

(iii) $P\left(\bigcup_{n=1}^{\infty} A_n \middle| B\right) = \sum_{n=1}^{\infty} P(A_n|B)$, if $A_i \cap A_j = \emptyset \ (i \neq j)$.

Proof: Use Definition 2.3.1 and the properties of probability.

Theorem 2.3.3. (Multiplication Rule) *For each $n + 1$ events $A_0, A_1, \ldots, A_n$ for which $P(A_0 A_1 \cdots A_n) > 0$, we have*

$$P(A_0 A_1 \cdots A_n) = P(A_0)P(A_1|A_0)P(A_2|A_0 A_1) \cdots P(A_n|A_0 A_1 \cdots A_{n-1}).$$

Proof: Since

$$A_0 A_1 \cdots A_{n-1} \subseteq A_0 A_1 \cdots A_{n-2} \subseteq \cdots \subseteq A_0,$$

we have

$$0 < P(A_0 A_1 \cdots A_{n-1}) \leqslant \cdots \leqslant P(A_0),$$

so all conditional probabilities involved are defined. [Recall (see Definition 2.3.1) that $P(A|B)$ is defined only when $P(B) > 0$.] It is easy to show the result for $n = 1$, and the general result follows by mathematical induction.

Theorem 2.3.4. (Theorem of Total Probabilities) *If*

$$P\left(\bigcup_{n=1}^{\infty} B_n\right) = 1,$$
$$B_i \cap B_j = \emptyset \quad (i \neq j),$$
$$P(B_n) > 0, \quad \textit{for each } n,$$

then (for any event A)

$$P(A) = \sum_{n=1}^{\infty} P(A|B_n)P(B_n).$$

Proof[6]**:** If we denote $A \cap \left(\bigcup_{n=1}^{\infty} B_n\right)$ by X and $A \cap \left(\bigcup_{n=1}^{\infty} B_n\right)^c$ by Y, then $X \cap Y = \emptyset$, $X \cup Y = A$, and therefore $P(A) = P(X) + P(Y)$. Thus,

$$\begin{aligned} P(A) &= P\left(A \cap \bigcup_{n=1}^{\infty} B_n\right) + P\left(A \cap \left(\bigcup_{n=1}^{\infty} B_n\right)^c\right) \\ &= P\left(\bigcup_{n=1}^{\infty} AB_n\right) = \sum_{n=1}^{\infty} P(AB_n) = \sum_{n=1}^{\infty} P(A|B_n)P(B_n). \end{aligned}$$

Corollary 2.3.5. *Theorem 2.3.4 holds if ∞ is replaced by some finite number N.*

Theorem 2.3.6. (Bayes' Rule) *If*

$$P\left(\bigcup_{n=1}^{\infty} B_n\right) = 1,$$
$$B_i \cap B_j = \emptyset \quad (i \neq j),$$
$$P(B_n) > 0, \quad \textit{for each } n,$$

and A is an event with $P(A) > 0$, then for each k,

$$P(B_k|A) = \frac{P(A|B_k)P(B_k)}{\sum_{n=1}^{\infty} P(A|B_n)P(B_n)}.$$

---

[6] It may help the student to note that, since $P\left(\bigcup_{n=1}^{\infty} B_n\right) = 1$, $P\left(\overline{\bigcup_{n=1}^{\infty} B_n}\right) = 0$ and hence $P(Y) = 0$.

Proof: By the definition of conditional probability and the Theorem of Total Probabilities,

$$P(B_k|A) = \frac{P(AB_k)}{P(A)} = \frac{P(A|B_k)P(B_k)}{\sum_{n=1}^{\infty} P(A|B_n)P(B_n)}.$$

The hypotheses of the Theorem of Total Probabilities and of Bayes' Rule can be visualized as follows. Ω is essentially decomposed into disjoint sets $B_1, B_2, \ldots$. Then (Theorem of Total Probabilities) we can find $P(A)$ by adding up the probabilities of those parts of A which lie in $B_1, B_2, \ldots$. A diagram may help (Figure 2.3–1). [The final result is stated somewhat

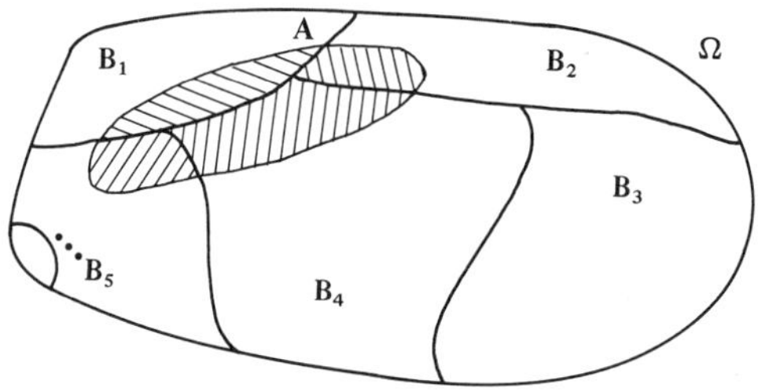

Figure 2.3–1

differently, since $P(AB_n) = P(A|B_n)P(B_n)$.] In Bayes' Rule we look at an inverse problem: "Given that an outcome has occurred which is in A, what is the probability that it is in B_k?" (At this point, examine Example 2.3.10 briefly.)

We already know how to find $P(A_1 \cup \cdots \cup A_n)$ if $A_1, \ldots, A_n$ are disjoint events (Theorem 2.1.11). If $A_1, \ldots, A_n$ are not disjoint, however, in general we have only the bound of Boole's Inequality 2.1.14. We now wish to remedy this lack (Theorem 2.3.7) and to give examples of the use of our new result (Examples 2.3.9 and 2.3.11).

Theorem 2.3.7. *Let $A_1, \ldots, A_n$ be events. Then*

$$P(A_1 \cup A_2 \cup \cdots \cup A_n) = \sum_{i=1}^{n} P(A_i) - \sum_{i<j} P(A_i \cap A_j)$$
$$+ \sum_{i<j<k} P(A_i \cap A_j \cap A_k) - \cdots + (-1)^{n+1} P(A_1 \cap \cdots \cap A_n).$$

Proof: Suppose a point ω is in $A_1 \cup \cdots \cup A_n$. Then its probability is counted once on the left-hand side. How many times is its probability counted on the right-hand side? Suppose, without loss of generality, that ω

is in exactly l of $A_1, \ldots, A_n$ ($1 \leqslant l \leqslant n$). Without loss of generality we may assume ω is in $A_1, \ldots, A_l$. Then

$$\begin{pmatrix}\text{No. of times } \omega\text{'s probability is} \\ \text{counted on the right-hand side}\end{pmatrix} = l - \binom{l}{2} + \binom{l}{3} - \cdots + (-1)^{l+1}\binom{l}{l}.$$

But, by the binomial theorem, $(a + b)^n = \sum_{k=0}^{n} \binom{n}{k} a^k b^{n-k}$. Thus,

$$\begin{aligned} 0 = (1-1)^l = \sum_{k=0}^{l}\binom{l}{k}(-1)^k &= 1 - \binom{l}{1} + \binom{l}{2} - \cdots + (-1)^l\binom{l}{l} \\ &= 1 - \left\{l - \binom{l}{2} + \binom{l}{3} - \cdots + (-1)^{l+1}\binom{l}{l}\right\}, \end{aligned}$$

so ω has its probability counted exactly once on the right-hand side also. On the other hand, if ω is not in $A_1 \cup \cdots \cup A_n$, its probability is counted zero times on each side (why?), so the proof is complete. (Note that this proof only covers the case of a finite or denumerable sample space Ω. Other cases are covered in the Honors Problems.)

Corollary 2.3.8. *Let A, B be events. Then*

$$P(A \cup B) = P(A) + P(B) - P(AB).$$

Proof: This is the result of Theorem 2.3.7 with $n = 2$. A Venn diagram may be intuitively helpful. See Figure 2.3–2.

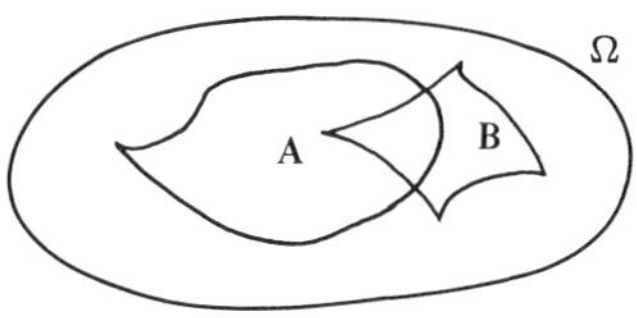

Figure 2.3–2

An alternative proof can be given as follows. First $A = (AB) \cup (A\bar{B})$ and $A \cup B = B \cup (A\bar{B})$ are each disjoint unions [that is, $(AB) \cap (A\bar{B}) = \emptyset$ and $B \cap (A\bar{B}) = \emptyset$]. Hence,

$$\begin{aligned} P(A) &= P(AB) + P(A\bar{B}), \\ P(A \cup B) &= P(B) + P(A\bar{B}), \end{aligned}$$

and

$$P(A) - P(A \cup B) = P(AB) - P(B),$$

or

$$P(A \cup B) = P(A) + P(B) - P(AB).$$

Example 2.3.9. (Probleme des Rencontres) Suppose that $N(N \geqslant 1)$ cards (numbered $1, 2, \ldots, N$) are placed at random onto N numbered places on a table (the places being numbered $1, 2, \ldots, N$), each place receiving one card. We say a match or rencontre occurs at the ith place if the card numbered i is placed there. What is the probability of at least one match?

Let A_i be the event "rencontre at the ith place" $(i = 1, \ldots, N)$. Then $P(A_1 \cup \cdots \cup A_N)$ is what we wish to find. Since our Ω consists of $N!$ orderings of N objects and each has probability $1/N!$, and since event $A_1 A_2 \cdots A_j$ occurs *iff* the cards numbered $1, 2, \ldots, j$ fall on the places numbered $1, 2, \ldots, j$, respectively $(1 \leqslant j \leqslant N)$, we find

$$P(A_1) = \frac{(N-1)!}{N!},$$

$$P(A_1 A_2) = \frac{(N-2)!}{N!},$$

$$\vdots$$

$$P(A_1 A_2 \cdots A_N) = \frac{1}{N!}.$$

However, the probability of the event $A_{i_1} \cdots A_{i_j}$ (where $i_1, \ldots, i_j$ are j distinct members of the set $\{1, \ldots, N\}$) is the same as the probability of the event $A_1 \cdots A_j$ $(1 \leqslant j \leqslant N)$. Thus, by Theorem 2.3.6,

$$\begin{aligned} P(A_1 \cup \cdots \cup A_N) &= N\frac{(N-1)!}{N!} - \binom{N}{2}\frac{(N-2)!}{N!} + \binom{N}{3}\frac{(N-3)!}{N!} - \cdots \\ &= N\frac{(N-1)!}{N!} - \frac{N!}{2!(N-2)!}\frac{(N-2)!}{N!} + \frac{N!}{3!(N-3)!}\frac{(N-3)!}{N!} - \cdots \\ &= 1 - \frac{1}{2!} + \frac{1}{3!} - \frac{1}{4!} + \cdots + (-1)^{N+1}\frac{1}{N!}. \end{aligned}$$

Since (for any real number x) $e^x = 1 + x + \frac{x^2}{2!} + \frac{x^3}{3!} + \cdots$, for N very large

$$P(A_1 \cup \cdots \cup A_N) = 1 - \left(1 - 1 + \frac{1}{2!} - \frac{1}{3!} + \cdots + (-1)^N \frac{1}{N!}\right) \approx 1 - e^{-1}.$$

Example 2.3.10. (How not to run a Presidential Poll) For the 1936 election, the *Literary Digest* picked names at random out of telephone books in some cities and sent these people sample ballots, attempting to predict the election results, Roosevelt versus Landon, by the returns. Now, even if 100 percent returned the ballots, even if all told how they really felt, even if all would vote, even if none would change their minds by election day, still this method could be (and was!) in trouble: They estimated a conditional probability in that part of the American population which had telephones, and (assuming their estimate in this sector was close to the true value) showed that that part was not typical of the total population in this regard (Roosevelt versus Landon) in 1936. (For a discussion of many such interesting examples of applications of statistics, see the book *The Nature of Statistics* by Wallis and Roberts listed in the references.)

Example 2.3.11. (The late Commuter) Ten pairs of shoes (20 individual shoes) are in a closet. If 4 shoes are selected at random, what is the probability of obtaining at least one pair?

Let A_i be the event "pair i is included in the sample" ($i = 1, \ldots, 10$). Then

$$P(A_i) = \frac{\binom{2}{2}\binom{18}{2}}{\binom{20}{4}}, \qquad P(A_1A_2) = \frac{1}{\binom{20}{4}},$$

$$P(A_1A_2A_3) = P(A_1A_2A_3A_4) = \cdots = P(A_1A_2A_3A_4 \cdots A_{10}) = 0,$$

so by Theorem 2.3.7,

$$P(A_1 \cup \cdots \cup A_{10}) = 10\frac{\binom{2}{2}\binom{18}{2}}{\binom{20}{4}} - \binom{10}{2}\frac{1}{\binom{20}{4}} + 0 = \frac{99}{323} \approx \frac{1}{3}.$$

For the reader's interest, we state a further result along the lines of Theorem 2.3.7.

Theorem 2.3.12. *Let $A_1, \ldots, A_N$ be events. Then*

$$P[\text{Exactly } m \text{ of } A_1, \ldots, A_N \text{ occur}]$$
$$= \sum_{i_1 < \cdots < i_m} P(A_{i_1} \cdots A_{i_m}) - \binom{m+1}{m} \sum_{i_1 < \cdots < i_{m+1}} P(A_{i_1} \cdots A_{i_{m+1}})$$
$$+ \binom{m+2}{m} \sum_{i_1 < \cdots < i_{m+2}} P(A_{i_1} \cdots A_{i_{m+2}}) - \cdots \pm \binom{N}{m} P(A_1 \cdots A_N).$$

It may be helpful to recall, at this point, **how we got involved with counting problems (combinatorics)**: If Ω is finite and discrete, for example $\Omega = \{a_1, \ldots, a_n\}$, and if $P(a_i) = P(a_j)$ for all i, j ($1 \leqslant i, j \leqslant n$), then for any $A \subseteq \Omega$

$$P(A) = \frac{n(A)}{n}.$$

Example 2.3.13. In Problem 1.6 we defined the symmetric difference of two sets A and B as

$$S(A, B) = (A - B) \cup (B - A) = (A\bar{B}) \cup (\bar{A}B).$$

If A and B are events (subsets of some sample space Ω), then prove that

$$P(S(A, B)) = P(A) + P(B) - 2P(AB). \tag{2.3.14}$$

First, since $(A\bar{B}) \cap (\bar{A}B) = \emptyset$, it follows that

$$P(S(A, B)) = P(A\bar{B}) + P(\bar{A}B). \tag{2.3.15}$$

Now $A = (AB) \cup (A\bar{B})$ and $B = (AB) \cup (\bar{A}B)$ are also disjoint unions; hence, $P(A) = P(AB) + P(A\bar{B})$ and $P(B) = P(AB) + P(\bar{A}B)$, or $P(A\bar{B}) = P(A) - P(AB)$ and $P(\bar{A}B) = P(B) - P(AB)$. Using these last two relations in (2.3.15) proves (2.3.14). We will now essentially leave this pursuit and go on to more general studies. Note that the results of this section (conditional probability, multiplication rule, Theorem of Total Probabilities, Bayes' Rule, 2.3.7, 2.3.8, 2.3.12), although illustrated by examples involving such Ω's and equally likely outcomes, are entirely general.

2.4. INDEPENDENT EVENTS

In this section we discuss briefly the notion of independent events, formulating precisely what it means for one event to have no influence on another, and indicating how this precise formulation fits our intuitive idea of "independence."

Definition 2.4.1. *Two events A and B are said to be* **independent** *if $P(AB) = P(A)P(B)$.*

Property 2.4.2. *If A and B are independent events, then A and B^c are independent.*

Property 2.4.3. *If A and B are independent events, and if $P(B) > 0$, then $P(A|B) = P(A)$.*

These properties are easy to prove, but we shall leave that as an exercise since it is only our purpose here to briefly introduce these notions. Property 2.4.3 seems to some, including the present author, to be a better expression of the intuitive idea of independence: events A and B are independent if knowing that B has occurred tells us nothing about the chances of A occurring. (The only reason we have not taken this as our definition is to avoid the requirement $P(B) > 0$ in our definition.) If independence is viewed through the conditional probability viewpoint of Property 2.4.3, then the concepts of positive, negative and no information (see Example 2.4.7) clarify the content of the "independence" concept immediately. Before considering that example, we note that (since often more than two events will be of interest to us) definitions and properties corresponding to 2.4.1, 2.4.2, and 2.4.3 are needed for arbitrary numbers of events. We will now state these.

Definition 2.4.4. *Let $\mathscr{B} = \{B_\lambda: \lambda \in \Lambda\}$ be a set of events. These events are said to be* **independent** *if for every positive integer n and each n distinct elements $\lambda_1, \ldots, \lambda_n$ in the indexing set Λ we have*

$$P[B_{\lambda_1} \cdots B_{\lambda_n}] = \prod_{j=1}^{n} P(B_{\lambda_j}).$$

Property 2.4.5. *If $\mathscr{C}$ is a collection of independent events, if $A_1, \ldots, A_n$, $B_1, \ldots, B_m$ are $m + n$ distinct events in $\mathscr{C}$, and if $P(B_1 \cdots B_m) > 0$, then*

$$P(A_1 \cdots A_n | B_1 \cdots B_m) = P(A_1 \cdots A_n).$$

Property 2.4.6. *If $\mathscr{C}$ is a class of independent events, and if each event in some subset of $\mathscr{C}$ is replaced by its complement, then the new class of events is also a class of independent events.*

Recall (Definition 1.2.7) that events A and B are said to be **mutually exclusive** if $AB = \emptyset$. This is often confused with the concept of independence, and we should be careful to see that we understand the difference. (Essentially, mutually exclusive events and independent events are mutually exclusive. See Problem 2.26.)

Example 2.4.7. (Negative Information) Suppose $P(B) > 0$ and $P(A) > 0$. We sometimes say that event B carries **negative information** about event A if $P(A|B) < P(A)$. [**Positive information** may be defined as the case when $P(A|B) > P(A)$. No information, or $P(A|B) = P(A)$, corresponds to independence.] Then $P(A|B) < P(A)$ implies $P(B|A) < P(B)$, for

$$P(B|A) = \frac{P(AB)}{P(A)} = \frac{P(A|B)P(B)}{P(A)}$$
$$< \frac{P(A)P(B)}{P(A)} = P(B).$$

Example 2.4.8. A certain water-supply system consists of a Source, 5 pumping stations, and a Destination. If each pumping station has probability p $(0 \leqslant p \leqslant 1)$ of being operable at a specified time t_0, if the stations are operable or failed independently of one another, and if the stations are connected as in Figure 2.4–1, find the probability $R(p)$ that water is available to the Destination at time t_0.

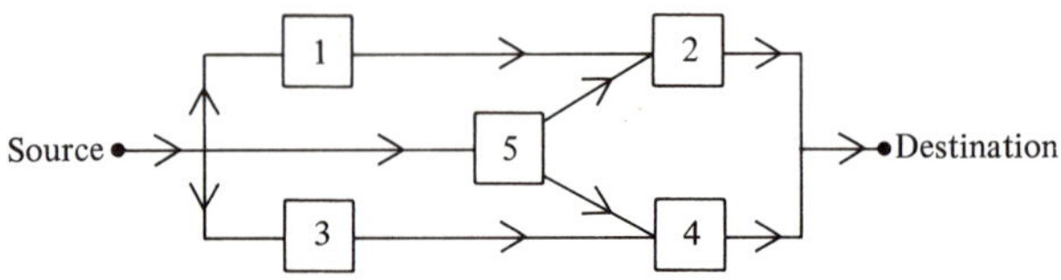

Figure 2.4–1. A water-supply system

Let A_i denote the event that pumping station i is operable at time t_0 $(1 \leqslant i \leqslant 5)$. Then $P(A_1) = P(A_2) = \cdots = P(A_5) = p$. Let B denote the event that water is available to the Destination at time t_0. Then $P(B) = R(p)$. Now

$$\begin{aligned} P(B) &= P(BA_5) + P(B\bar{A}_5) \\ &= P(B|A_5)P(A_5) + P(B|\bar{A}_5)P(\bar{A}_5) \\ &= p\,P(B|A_5) + (1 - p)P(B|\bar{A}_5). \end{aligned}$$

But (using Property 2.4.5 in the third equality)

$$\begin{aligned} P(B|\bar{A}_5) &= P((A_1A_2) \cup (A_3A_4)|\bar{A}_5) \\ &= P(A_1A_2|\bar{A}_5) + P(A_3A_4|\bar{A}_5) - P(A_1A_2A_3A_4|\bar{A}_5) \\ &= p^2 + p^2 - p^4 = 2p^2 - p^4, \end{aligned}$$

and

$$\begin{aligned} P(B|A_5) &= P(A_2 \cup A_4|A_5) \\ &= P(A_2|A_5) + P(A_4|A_5) - P(A_2A_4|A_5) \\ &= p + p - p^2 = 2p - p^2. \end{aligned}$$

Hence,

$$\begin{aligned} P(B) &= p(2p - p^2) + (1 - p)(2p^2 - p^4) \\ &= 4p^2 - 3p^3 - p^4 + p^5 \end{aligned}$$

Alternatively, we may express B as

$$B = (A_1A_2) \cup (A_3A_4) \cup (A_5A_2) \cup (A_5A_4)$$

and use Theorem 2.3.7:

$$\begin{aligned} P(B) &= 4p^2 - 3(p^4 + p^3) + 2(p^4 + p^5) - p^5 \\ &= 4p^2 - 3p^3 - p^4 + p^5. \end{aligned}$$

Can you devise a better way of utilizing 5 pumping stations [that is, a way such that, if the *sole* goal is to deliver water to the Destination at time t_0, the new way yields a larger $R(p)$]? [*Hint*: Pipe water through each station directly to the Destination.] Can you do better if water must (because of the distances involved) pass through at least 2 pumping stations and must (because of the pipe strengths involved) pass through at most 2 pumping stations (that is, if water must pass through exactly 2 pumping stations)?

Example 2.4.9. (Simpson's Paradox) Two companies manufacture a certain type of sophisticated electronic equipment for the government; to avoid lawsuits let us call them company C and company $\bar{C}$. In the past company C has had 5 percent good output, while company $\bar{C}$ has had 50 percent good output (that is, 95 percent of C's output and 50 percent of $\bar{C}$'s output is not of acceptable quality). The government has just ordered 21,100 of these electronic devices, 10,100 from company $\bar{C}$ and 11,000 from company C. [Company C often does receive part of the order, for political reasons, or because company $\bar{C}$ does not have a capacity large enough to allow it to take the whole order, and so on.]

Since the electronic devices are critically needed for an important government program, before production of the 21,100 devices starts, government scientists develop a new manufacturing method which they believe will almost double the percentage of good devices received. Companies C and $\bar{C}$ are given this information but (because of limited lead-time) its use is largely optional: they must each use the new method for at least 100 of their devices, but use of the new method beyond that point is left to their discretion.

When the devices are received and tested some time later, analysis of the 21,100 items shows a breakdown as in Table 2.4–1, which horrifies the government: 46 percent of items produced by the usual method were good, but only 11 percent of items produced by the new method were good. This shocks the responsible government officials, who begin to berate (and worse!) their scientists, and companies C and $\bar{C}$, for the "lousy new method" and for producing so many of the items with the clearly inferior new method, respectively. Yet the scientists and the companies insist that the new method *has* almost doubled the percentage of good items. *Can they possibly be right*?

Table 2.4–1. Analysis of Total Production

		Production Method	
		Standard	*New*
Results	Bad	5950	9005
	Good	5050 (46%)	1095 (11%)

The answer is (amazingly) *yes*, as we see in Table 2.4–2. The new method has nearly doubled the percent good in each case (10 percent versus 5 percent for company C and 95 percent versus 50 percent for company $\bar{C}$). The totals put into the experimental (new) method by each company are also quite reasonable: company $\bar{C}$ (which knew its implementation of the standard method was already quite good) put the minimum allowed (100), while company C (which knew its implementation of the standard method was inferior) put 10,000 of its 11,000 through the new method.

Table 2.4–2. Production Analysis by Company

		Company			
		C		$\bar{C}$	
		Production Method		Production Method	
		Standard	New	Standard	New
Results	Bad	950	9000	5000	5
	Good	50 (5%)	1000 (10%)	5000 (50%)	95 (95%)

The fact that results like those displayed above can occur is, when stated in terms of probabilities, called **Simpson's Paradox**[7]: *It is possible to have* $P(A|B) < P(A|\bar{B})$ *even though both*

$$
\begin{aligned}
P(A|BC) &\geqslant P(A|\bar{B}C),\\
P(A|B\bar{C}) &\geqslant P(A|\bar{B}\bar{C}).
\end{aligned}
\tag{2.4.10}
$$

PROBLEMS

2.1 A pair of dice is rolled once. Construct a sample space for this experiment and list its elements in full detail. [*Hint*: Number the sides of each die 1, 2, 3, 4, 5, 6 and consider 2-tuples.] On your diagram, circle and mark the following events:

A: faces are equal,
B: sum of faces is $\geqslant 10$,
C: faces sum to 7.

2.2 An **algebra** (or **field**) **of sets** is a collection $\mathscr{B}$ of subsets of a given set Ω such that

(a) $\Omega \in \mathscr{B}$,
(b) $A \in \mathscr{B} \Rightarrow \bar{A} \in \mathscr{B}$,
(c) $A, B \in \mathscr{B} \Rightarrow A \cup B \in \mathscr{B}$.

An algebra of sets $\mathscr{B}$ is called a **σ-algebra** (or **Borel field**) if also

[7] A good reference is C. R. Blyth, "On Simpson's paradox and the sure-thing principle," *Journal of the American Statistical Association*, Vol. 67 (1972), pp. 364–366.

$$A_n \in \mathscr{B}\ (n = 1, 2, \ldots) \Rightarrow \bigcup_{n=1}^{\infty} A_n \in \mathscr{B}.$$

Prove that (for a given set Ω) the collection $\mathscr{B}_0$ of all subsets of Ω is a σ-algebra.

2.3 Given that $P(\bar{A}) = .3$, $P(B) = .4$, and $P(A\bar{B}) = .5$, find
(a) $P(A)$,
(b) $P(AB)$,
(c) $P(A \cup B)$.

2.4 A certain City Council consists of 4 Democrats and 6 Republicans. What is the probability that a Councillor drawn at random will be a Democrat?

2.5 If 2 space vehicles (for example, one from the U.S.A. and one from the U.S.S.R.) attempt to land on the moon, what is the probability that exactly 1 succeeds? (Correct formulation of a model is part of this problem.)

2.6 If the 3 buttons on a push-button combination lock are pushed in random order, what is the probability that they will be in the correct order to unlock the lock? (Assume that each button is pushed exactly once, and that some one permutation unlocks the device.)

2.7 As in Problem 2.6, but a 24-button lock.

2.8 As in Problem 2.6, but an n-button lock ($n \geqslant 1$).

2.9 As in Problem 2.6, but a 24-button lock which opens no matter which button is used first as long as the successive pushes follow according to one basic sequence (for example, if 312 is basic, then 123 and 231 also work on a comparable 3-button lock).

2.10 **(a)** Prove Theorem 2.2.2 (the multiplets theorem).
(b) Prove Theorem 2.2.10 (multinomial coefficients) by induction.

2.11 How many 3-digit numbers can be formed with the integers 1, 2, 3, 4, 5 if duplication of the integers is not allowed? If duplication is allowed?

2.12 A kindergarten teacher has 10 chocolate bars, 5 bags of popcorn, and 5 ice cream bars to distribute to her class of 20 pupils. (Each pupil is to receive 1 item.) The chocolate bars are all alike; so are the bags of popcorn and the ice cream bars.
(a) In how many different ways can the teacher distribute these items?
(b) If the items are distributed at random, what is the probability that a given pair of boys in the class receive the same item?

2.13 If n persons, including Art and Bob, are randomly arranged in a straight line, what is the probability that there are exactly r persons in the line between Art and Bob?

2.14 Given that $P(A) = .7$ and $P(A \cup B) = .8$, find $P(B)$
(a) if A and B are mutually exclusive;
(b) if A and B are independent;
(c) if $P(A|B) = .6$.

2.15 Given that $P(\bar{A}) = .3$, $P(B) = .4$, and $P(A\bar{B}) = .5$, find $P(B|A \cup \bar{B})$.

2.16 Calculate (as a decimal) the probability in Example 2.2.18. [*Hint*: This can *either* be done by using tables and/or a desk calculator, *or* by using a digital computer.[8]]

2.17 Twelve dice are cast. What is the probability that each of the 6 faces will appear at least once? [*Hint*: If we regard this experiment as one of tossing 12 balls at random into 6 cells, each of which has unlimited capacity, then we are seeking $1 - P(A_1 \cup \cdots \cup A_6)$ where A_i is the event "cell i is empty" $(i = 1, \ldots, 6)$.]

2.18 Let $\mathscr{B} = \{B_\lambda : \lambda \in \Lambda\}$ be a set of events. These events are said to be **pairwise independent** if for every 2 distinct elements λ_1, λ_2 in the indexing set Λ, we have

$$P[B_{\lambda_1}B_{\lambda_2}] = P[B_{\lambda_1}]P[B_{\lambda_2}].$$

Does pairwise independence imply independence? (Prove your assertion.) [*Hint*: Let $\Omega = \{\omega_1, \omega_2, \omega_3, \omega_4\}$ with $P(\omega_i) = \frac{1}{4}$ $(i = 1, 2, 3, 4)$ and choose

$$A = \{\omega_1, \omega_2\}, B = \{\omega_1, \omega_3\}, C = \{\omega_1, \omega_4\}.]$$

2.19 Let k be a positive integer. Let $X_1, X_2, \ldots, X_k$ be a sample of size k, drawn at random with replacement, from the population of digits 0, 1, 2, 3, . . . , 9. (Each digit has probability $p = \frac{1}{10}$ of being drawn at each drawing, and drawings are unrelated.)

Let A be the event "neither 0 nor 1 appears in the sample," and let B be the event "1 does not appear but 2 does." Find

(a) $P(A)$,
(b) $P(B)$,
(c) $P(AB)$,
(d) $P(A \cup B)$.

2.20 A pair of fair dice is rolled as long as necessary for a score of 7 to turn up. Assuming that 7 does not appear on the first throw, what is the probability that more than two throws will be necessary?

2.21 **(a)** If 3 distinguishable balls are placed at random into 2 numbered cells, find the probability that all cells are occupied.
(b) If $n + 1$ distinguishable balls are placed at random into n numbered cells, find the probability that all cells are occupied. (n is a positive integer.)

2.22 An urn contains 7 red balls and 3 black balls. One ball is drawn at random and replaced by a ball of the opposite color; then a second ball is drawn at random. Given that the second ball is red, what is the conditional probability that the first ball was also red?

2.23 Balls are randomly drawn one at a time from an urn in which there are initially

---

[8] It is worthwhile considering making computer time available to students taking courses based on this text. The number of problems the students are required to do using the computer can be small; the number they are allowed to do can be much larger. This introduces the students to the computer and its uses in probability and statistics, and helps them grasp the elementary principles of its proper use. Students can be allowed to use any computer language they may know (for example, often FORTRAN or PL/I); or an official course language such as CUPL, which a student can master and write programs in after only a few hours' study (especially if a sample program or two are handed out and computer time is made available to the student), can be made available.

5 red and 10 black balls. After each draw, the ball is replaced and one additional ball of the same color is added.

(a) Given that the first n balls drawn are black, calculate the probability α_n that the $(n+1)$st is black, and find $\lim_{n\to\infty} \alpha_n$.

(b) Given that the 2nd through $(n+1)$st balls are black, calculate the probability β_n that the first ball drawn was black, and find $\lim_{n\to\infty} \beta_n$.

2.24 Let A, B, C be events (subsets of some sample space Ω). Assume $P(C) > 0$ and $P(BC) > 0$. Prove or disprove each of the following:

(a) $P(AB|C) = P(A|BC)P(B|C)$;

(b) $P(AB|C) = P(A|C)P(B|C)$ if A, B are independent.

If you disprove an equation, give 2 or more conditions under which it is true.

2.25 Imagine a population of $N+1$ urns. Urn number k contains $\{k \text{ red}, N-k \text{ white}\}$ balls $(k = 0, 1, \ldots, N)$. An urn is chosen at random and n random drawings are made from it, the ball drawn being replaced each time. Define

Event A: all n balls turn out to be red;
Event B: the $(n+1)$st draw yields a red ball.

(a) Find $P[A|\text{Urn } k \text{ is chosen}]$ $(k = 0, 1, \ldots, N)$.

(b) Find $P(A)$.

(c) Find $P(AB)$.

(d) Find $P(B|A)$.

(e) Find an approximation to $P(B|A)$, using the fact that if N is large,

$$\frac{1}{N}\sum_{k=1}^{N}\left(\frac{k}{N}\right)^n \sim \int_0^1 x^n\,dx = \frac{1}{n+1}.$$

2.26 Let A and B be events. Assume $P(A) > 0$ and $P(B) > 0$. Prove: If A, B are mutually exclusive, then A, B are not independent; if A, B are independent, then A, B are not mutually exclusive. Thus, "mutually exclusive events and independent events are mutually exclusive."

2.27 Let A_n, B_n $(n \geqslant 1)$ be events (subsets of some sample space Ω). Prove that if $\lim_{n\to\infty} P(B_n) = 1$, then $\lim_{n\to\infty} \{P(A_n|B_n) - P(A_n)\} = 0$. (You must prove that the latter limit exists.) Hence, show that if $\lim_{n\to\infty} P(A_n|B_n)$ or $\lim_{n\to\infty} P(A_n)$ exists, both exist and

$$\lim_{n\to\infty} P(A_n|B_n) = \lim_{n\to\infty} P(A_n).$$

2.28 Show that negative information (see Example 2.4.7) is not transitive; that is, $P(A|B) < P(A)$ and $P(B|C) < P(B)$ do not imply $P(A|C) < P(A)$. [*Hint*: Let $\Omega = \{\omega_1, \omega_2, \omega_3, \omega_4\}$ with $P(\omega_i) = \frac{1}{4}$ $(i = 1, 2, 3, 4)$ and choose

$$A = \{\omega_1, \omega_2\}, \qquad B = \{\omega_3, \omega_4\}, \qquad C = \{\omega_1\}.]$$

Find another example which also shows that negative information is not transitive, but in which all probabilities of interest are positive.

2.29 The confidential-data computing facility of a corporation contains 15 locked computer consoles. Each console is openable by a different key. Suppose an individual has obtained the 15 keys, but they are coded and the individual does not know the code; also, if a key is inserted into a wrong console lock, it is frozen in place electronically (it cannot be removed and the console cannot be operated).

If the individual (for example, an executive whose assistants are unavailable) goes down the line inserting keys into consoles at random, what is the probability that he unlocks at least one console?

2.30 Suppose we have a **network** of cities and roads. Each **segment** of road (that is, each piece of road on which one can go from one city to another without encountering a third city) has probability p of being open, independently of the other segments. The probability $R(p)$ that one can now go from city A to city B depends on the network. For the network

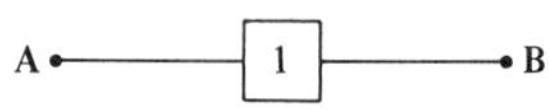

(which has one segment) $R(p) = p$. For the network

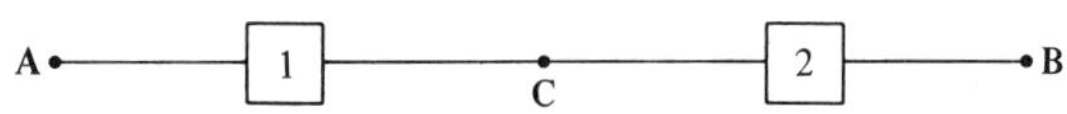

(which has two segments) $R(p) = p^2$. Not all two-segment networks have the same $R(p)$, for example, for the network

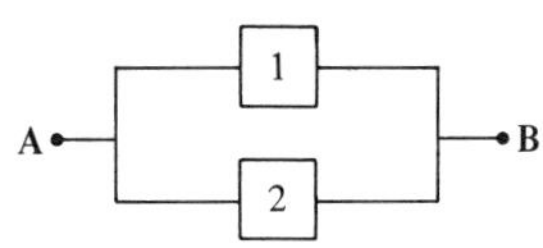

we find $R(p) = 2p - p^2$. Find $R(p)$ for the network

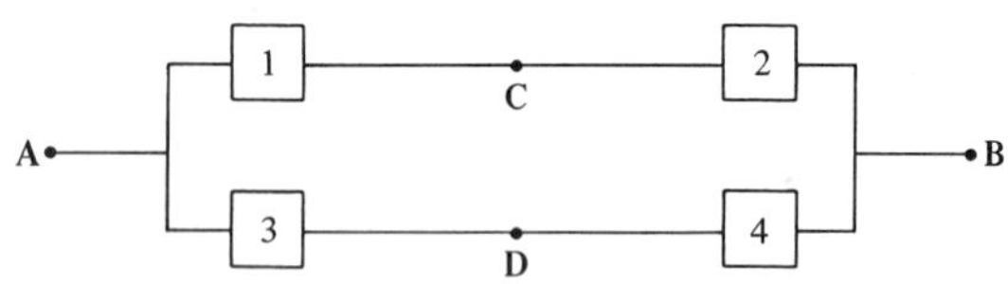

2.31 A **system** is made up of n parts. At any time t each part is either functional (1) or nonfunctional (0). We may regard the basic sample space Ω as consisting of all 2^n possible assignments of 1, 0 to each part. The sample space Ω is partitioned into two sets $\Omega = F \cup W$ (with $F \cap W = \emptyset$); if an $\omega \in F$ occurs at time t, the system will not work (has failed); while if an $\omega \in W$ occurs at time t, the system will work (has not failed).

Suppose that each part has probability p of being functional at time t, and that parts fail or operate independently of one another. $P(W)$ is a function of p, say $R(p)$, and is called the **reliability of the system**.

(a) A **series system** fails iff at least one of its parts fails. This situation may be diagrammed as

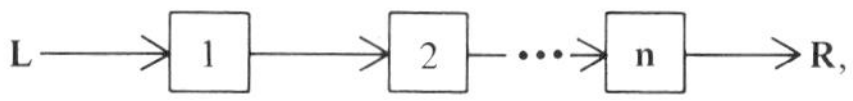

and the system works iff there is a *path* (connected sequence of functional parts) from L to R. Find $R(p)$ for such a system.

(b) A **parallel system** fails iff all of its parts fail. The situation may be diagrammed as

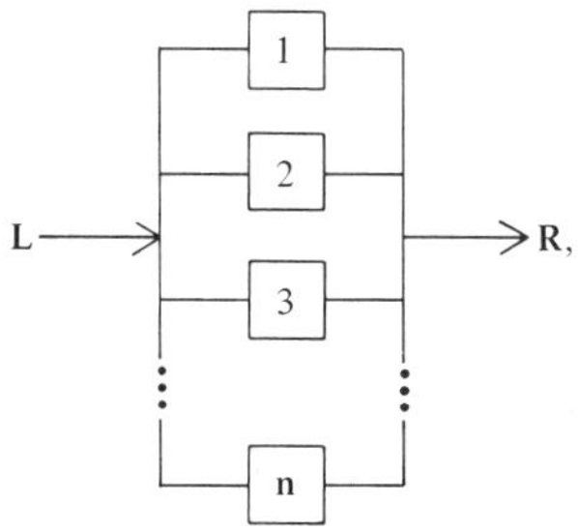

and the system works iff there is a path from L to R. Find $R(p)$ for such a system.

2.32 Use the example of Simpson's Paradox involving manufacture of electronic devices (see Example 2.4.9) to show that it may be the case that B gives positive information[9] about A, both when C obtains and when C does not obtain, but that B gives negative information about A. [*Hint*: Identify A with good output, B with the new method, and C, $\bar{C}$ with companies.] Also show that the **sure-thing principle** ("*If you would definitely prefer g to f, either knowing that the event C obtained, or knowing that the event C did not obtain, then you definitely prefer g to f*") is false. [*Hint*: Identify g, f with standard and new methods.]

2.33 Suppose that Democrat D and Republican R are running for the U.S. Presidency. Is it possible that a poll taken before the election could show *that*, in each of the 50 states, nearly twice the percent of people under 30 prefer D to R as do people over 30, *but* that overall (that is, for the 50 states combined) a higher percent of people over 30 prefer D to R than do people under 30? If so, give a numerical example demonstrating your contention. How could a pollster prevent such paradoxical situations from arising in his polls?

2.34 In cricket[10] one's bowling average is (number of runs) divided by (number of

---

[9] See Example 2.4.7 for the concepts of positive and negative information, which were apparently first introduced and studied by K. -L. Chung, "On mutually favorable events," *Annals of Mathematical Statistics*, Vol. 13 (1942), pp. 338–349. The author is grateful to Professor Colin R. Blyth for bringing this reference to his attention.

[10] This example was brought to the author's attention by M. C. Mitchelmore, and is contained (in a slightly different form) in *Joint Schools Project, Mathematics, Textbook 3*, edited by M. C. Mitchelmore and B. Raynor, Longmans Group Ltd., London, 1969, as Example 7 on p. 192.

wickets taken). If bowlers A and B compete in matches I and II with runs and wickets taken as follows:

	Match			
	I		*II*	
	Runs	Wickets	Runs	Wickets
A	185	31	99	9
B	186	31	38	2

then A beats B in Match I (his average of $5\frac{30}{31}$ runs per wicket is less than B's average of 6 runs per wicket) and A beats B in Match II (11 runs per wicket *versus* 19 runs per wicket). If Matches I and II comprise a Tournament, who wins the Tournament (that is, has the lowest runs-per-wicket overall)? Why is this so? How is this related to the sure-thing principle? (See Problem 2.32.) Could such a result happen in golf? Did A deserve to lose? (He did badly on much more play in Match II than did B, who played only $\approx\frac{1}{3}$ as much.)

HONORS PROBLEMS

2.1H Our proof of Theorem 2.3.7 was valid only for a finite or denumerable sample space Ω. Prove the result in general. [*Hint*: Most books which consider the general case only prove the result for $n = 2$ or $n = 3$ (for example, H. G. Tucker, *An Introduction to Probability and Mathematical Statistics*, Academic Press Inc., New York, 1962, p. 12). Such proofs may furnish you with a hint for the general case (for example, use induction).]

2.2H Prove Theorem 2.3.12, first for the denumerable case, then in the general case.

2.3H Given that events A and B are independent and that $P(A - B) = p_1$, $P(B - A) = p_2$ $(0 \leq p_1, p_2 \leq p_1 + p_2 \leq 1)$, find $r = P(AB)$.

What restrictions must be placed on p_1, p_2 besides those stated?

2.4H *A* **probability space** *is a triplet* $(\Omega, \mathscr{B}, P)$ *where*

(a) Ω *is a set called the* **sample space** *whose points* ω *are called* **elementary events**;

(b) $\mathscr{B}$ *is a* σ*-algebra of subsets of* Ω;

(c) $P(\cdot)$ *is a function called a* **probability function** *which assigns to each* **event** $A \in \mathscr{B}$ *a real number* $P(A)$ *such that*

(i) $P(A) \geq 0$ *for every* $A \in \mathscr{B}$;

(ii) $P(\Omega) = 1$;

(iii) **(Countable Additivity)** *If* $A_1, A_2, \ldots \in \mathscr{B}$ *and*

$$A_i \cap A_j = \emptyset \ (i \neq j), \text{ then } P\left(\bigcup_{n=1}^{\infty} A_n\right) = \sum_{n=1}^{\infty} P(A_n).$$

1. It is sometimes implied [for example, in Mood and Graybill (1963), p. 17, lines 16–17; this amounts to our Definition 2.1.1 without any compensating Problem 2.4H, and is not unreasonable in most undergraduate course treatments] that $\mathscr{B}_0$ = collection of all subsets of Ω is always our choice, in which case we could dispense with $\mathscr{B}$ and just talk of (Ω, P). Show that this is not so. [*Hint*: Exhibit a well-known $(\Omega, \mathscr{B}, P)$ such that it is a probability space but $(\Omega, \mathscr{B}_0, P)$ is not. (Why not?)]

2. **Complete additivity** means that if $A_\lambda \in \mathscr{B}$ for all $\lambda \in \Lambda$ (here Λ is some index set) and if $A_{\lambda_1} \cap A_{\lambda_2} = \emptyset$ for all $\lambda_1 \neq \lambda_2$, λ_1 and $\lambda_2 \in \Lambda$, then

$$P\left(\bigcup_{\lambda \in \Lambda} A_\lambda\right) = \sum_{\lambda \in \Lambda} P(A_\lambda).$$

Give an example of the sort of contradiction which can arise if we attempt to substitute complete additivity for countable additivity in our definition of a probability space.

2.5H Research the history of the "problem des rencontres" (matching problem) of Example 2.3.9, perhaps starting at Feller, Vol. I, pp. 90–91 (Second Edition) and write a well-documented paper on it.

Try to use the field of information theory to explain the fluctuations (with N) of P (At least one match), which goes up when N becomes odd and goes down when N becomes even (fluctuating about $1 - e^{-1}$).

No. of cards N	*P(at least one match)*
1	1.00000
2	.50000
3	.66667
4	.62500
5	.63333
6	.63196
.	
.	
.	
∞	.63212

Suppose that the matching problem is modified as follows: $N (N \geqslant 1)$ cards (numbered $1, 2, \ldots, N$) are shuffled thoroughly. Then N successive draws are made at random from the deck, with replacement. We say a match occurs at the ith draw if the card numbered i is drawn on that draw. Find the probability of at least one match, and show that it $\to 1 - e^{-1}$ as $N \to \infty$.

2.6H Magazines, newspapers, and other printed media have been known to print a collection of N pictures of individuals, followed by a list (in scrambled order) of N identifying characteristics (for example, names), one of which belongs to each individual.

(1) Suppose a movie magazine prints $N = 4$ baby pictures of movie stars, along with the 4 stars' names in scrambled order. Find, by direct enumeration, the probability that a person matching names with pictures purely at random would get at least 2 correct matchings.

(2) Suppose a newspaper, not to be outdone, prints $N = 52$ pictures, along with 52 names in scrambled order. We wish to find the probability that a person matching names with pictures purely at random would get at least 2 correct matchings.

There are $\binom{N}{2}$ distinct pairs of pictures. Mimicing the use of Theorem 2.3.7 in Example 2.3.9, let B_i be the event that pair i of pictures is correctly matched with names $\left(i = 1, 2, \ldots, \binom{N}{2}\right)$. Then $P(B_1 \cup B_2 \cup \cdots \cup B_{\binom{N}{2}})$ is what we wish to find. Why won't this method of Theorem 2.3.7 work here? [If you think it *will* work, do the calculation for $N = 4$ and compare with the answer found in part (1) by direct enumeration.] Finally, use Theorem 2.3.12 to solve this problem.

2.7H r ($\geqslant 1$) balls are placed at random into n cells, which consist of the following types:

n_1 which can hold 1 ball
n_2 which can hold 2 balls
.
.
.
n_{l-1} which can hold $l - 1$ balls
n_l which can hold l balls
.
.
.
n_∞ which can hold unlimited balls,

where $n_1, n_2, \ldots, n_\infty$ are nonnegative integers which sum to n. The set of assignment vectors is then

$$\Omega = \{(x_1, \ldots, x_r): x_i \text{ is the cell ball } i \text{ falls into } (1 \leqslant i \leqslant r)\}.$$

If we denote the

n_1 cells by $c_1, \ldots, c_{n_1}$,
n_2 cells by $c_{n_1+1}, \ldots, c_{n_1+n_2}$,
.
.
.
n_∞ cells by $c_{1+n_1+n_2+\ldots}, \ldots, c_{n_\infty+n_1+n_2+\ldots}$,

then we note that the cell types impose restrictions on the x_i in $(x_1, \ldots, x_r)$; for example, at most one may be c_1. (In the "usual" problems either $n_\infty = n$ or $n_1 = n$.) Let

$$\#_r(\Omega) = \text{The number of distinct assignment vectors.}$$

(1) Prove that $\dfrac{n!}{(n-r)!} \leqslant \#_r(\Omega) \leqslant n^r$.

(2) Suppose that from a population of n elements we draw a random sample of size r with replacement, but that n_i elements are so fragile that after being drawn i times they cannot be returned again ($i = 1, 2, \ldots, \infty$). How many samples are possible?

(3) Find $\#_1(\Omega)$, $\#_2(\Omega)$, $\#_3(\Omega)$. Find $\#_r(\Omega)$ $(r \geqslant 1)$. [*Note*: Tossing a ball into a cell reduces, for the next toss, the number of cells of that type and increases the number of cells of another type (unless the cell is of n_1 or n_∞ type).]

(4) Find a recursion relation between $\#_r(\Omega)$ and $\#_{r+1}(\Omega)$, or between $\#_1(\Omega), \ldots, \#_r(\Omega)$ and $\#_{r+1}(\Omega)$. Then solve it.

2.8H Following the definition of ball and cell model $(BC)_i$ (Definition 2.2.11), we noted an example dealing with post office boxes: a post office has n post office boxes, each with a capacity of m letters. Let X represent the number of boxes which overflow when N letters are distributed to the boxes at random.

If overflow letters must be handled by inserting special notices into boxes, then the post office could well wish to choose m so that X is suitably small; for example, so that $P[X \geqslant 1]$ is very small ($P[X \geqslant 1]$ is the probability that one or more boxes overflow). The problem can be approached in the following way: Consider a $(BC)_\infty$ model; then X is the number of cells which receive $\geqslant m$ balls when N balls are tossed into n cells. Find $P[X \geqslant 1]$, which is a function of n, m, N. For $n = 100$ and $N = 200$ compute $P[X \geqslant 1]$ for $m = 1, 2, 3, \ldots, 100$.

2.9H Consideration of P[All cells are occupied] when $n + 1$ balls are distributed at random into n cells (each of which has unlimited capacity) by two methods leads (show this) to two expressions, which must therefore be equal (see Problem 2.21):

$$P[\text{All cells are occupied}] = \sum_{k=0}^{n} \binom{n}{k} (-1)^k \frac{(n-k)^{n+1}}{n^{n+1}} = \frac{(n+1)!}{2n^n}.$$

Thus, it follows that

$$\sum_{k=0}^{n} \binom{n}{k} (-1)^k (n-k)^{n+1} = \frac{n}{2} \cdot (n+1)!. \qquad (*)$$

Calculate each side of $(*)$ for $n = 1, 2, 3, 4$ in order to check its veracity in those cases. Then prove $(*)$ in general. [In Feller, Vol. I, Second Edition, p. 58, 7, (11.6), upon setting $r = n + 1$ we find the left-hand side of $(*)$. On p. 63, (12.17) and its environs are related to this problem.]

2.10H (This problem is analogous to Example 2.2.18, but is more realistic.) Assume that there are 366 possible birthdays, the first 365 of which are equally likely, but the last of which is only one-fourth as likely as each of the others. What is the probability that, in a group of 60 people (whose birthdays are assumed unrelated), *some* 5 people will have the same birthday and all the other people will have different birthdays (none of which falls on the birthday of the 5)?

Calculate this probability as a decimal and compare it with the answer of Problem 2.16.

2.11H Suppose that a machine is built which takes a deck of 52 cards and shuffles them. If it always uses the same shuffle (for example, one simple shuffle would be to put the top card on the bottom of the deck), what is the largest number of shuffles that can possibly occur before the deck returns to its original order? For which shuffle can this occur? (We conceive of *exactly* the same operation being repeatedly applied to the deck. The deck then passes through a well-defined succession of orders, eventually returning to the original order. It cannot pass through all 52! possible orders. How many can it pass through at most? What rule achieves this?) [*References*: Feller (1968), pp. 406–407 and 422–423; and Problem E 2318 by T. Hughes and its solution by C. V. Heuer, *American Mathematical Monthly*, Vol. 79 (1972), p. 912.]

2.12H Immediately preceding Theorem 2.3.2 we stated a problem of sex composition of families which (for its solution) requires use of Theorems 2.3.4 and 2.3.6, namely, find $P(D)$ (the probability of an unmixed family) and $P(C_1|E)$ (the probability of an only child given at least one child and no girls).

(a) Find $P(D)$ and $P(C_1|E)$ as explicit formulas.

(b) For human families in the United States (or some other specific country) find $p_0, p_1, p_2, \ldots$ (the family size distribution) from statistical abstracts of the country for recent years. Then evaluate $P(D)$ and $P(C_1|E)$ as explicit numbers.

3

RANDOM VARIABLES: DISCRETE CASE

The outcome of an experiment (for example, the outcome of playing a game of chance) is completely described by a sample space Ω with probability function $P(\cdot)$ on its events, and by observing an $\omega \in \Omega$ selected according to $P(\cdot)$ one can answer questions concerning the experiment. However, usually we need[1] only observe some *function* of ω (for historical reasons such functions are called **random variables**) in order to answer the questions of interest. [For example, we may toss a pair of dice and, instead of observing the pair (No. on first die, No. on second die), simply report the resulting sum; this sum is a *function* of (No. on first die, No. on second die), namely the sum of the two numbers, and is often sufficient to answer the questions of practical import.] By the *discrete* case, we refer to random variables arising from discrete sample spaces Ω (and certain others to be specified below).

3.1. DISCRETE PROBABILITY FUNCTIONS AND DISTRIBUTION FUNCTIONS

For the reasons specified above (which will become clearer as we progress to Chapter 6ff) we now make the following:

Definition 3.1.1. *A* **random variable** *X is a real-valued function with domain* Ω [*that is, for each* $\omega \in \Omega$, $X(\omega) \in \mathscr{R} = \{y: -\infty < y < +\infty\}$]. *An n-***dimensional**

[1] Indeed, often we only can observe some function of ω (for example, whether a patient who is administered a new treatment is cured in a specified time period or not) and not the basic ω (which may belong to an Ω much too complex to even attempt to specify). For example, refer back to Problem 1.8.

random variable (*or n-dimensional random vector, or vector random variable*), $\mathbf{X} = (X_1, \ldots, X_n)$, *is a function with domain* Ω *and range in Euclidean n-space* $\mathscr{R}^n$ [*that is, for each* $\omega \in \Omega$, $\mathbf{X}(\omega) \in \mathscr{R}^n = \{(y_1, \ldots, y_n)\colon -\infty < y_i < +\infty \ (1 \leqslant i \leqslant n)\}$].

We may picture the situation as in Figure 3.1–1.

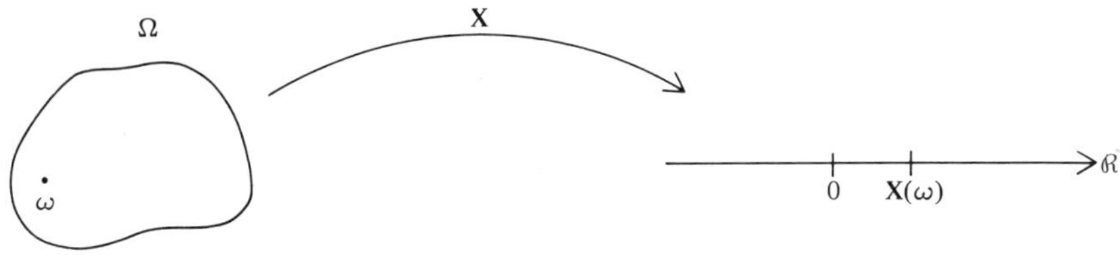

Figure 3.1–1

We can associate probabilities with values of X (or $\mathbf{X}$) because probabilities are already associated with points $\omega \in \Omega$ by the probability function $P(\omega)$. For example, suppose we have the situation as shown in Figure 3.1–2. Then $P[X = 2] = P(\{\omega\colon \omega \in \Omega, X(\omega) = 2\}) = P(\{\omega_1, \omega_2\}) = .4$, $P[X = 0] = .1$, $P[X = -2] = .5$. Our discussion will now be for a one-dimensional random variable X.

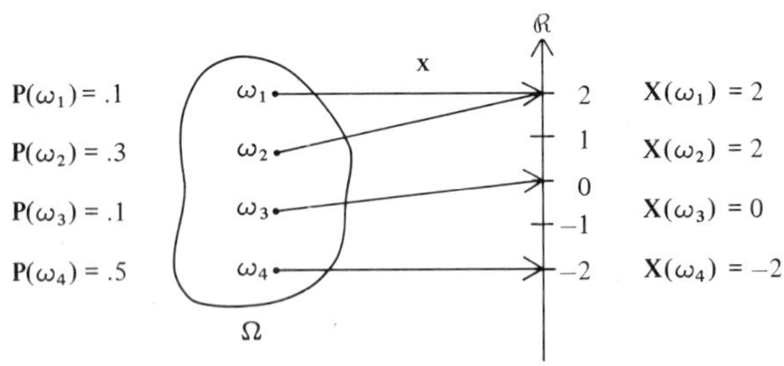

Figure 3.1–2

Definition 3.1.2. *If X is a random variable, its* **distribution function** (*or* **cumulative distribution function**, *or* **d.f.**, *or* **c.d.f.**) *is defined by*

$$F_X(x) = P[X \leqslant x], \quad \text{for all } x \in (-\infty, +\infty).$$

Theorem 3.1.3. *If X is a random variable* (**r.v.**), *then its d.f.* $F_X(x)$ *has the following properties:*

(a) $F_X(x)$ *is nondecreasing* [*that is,* $F_X(x) \leqslant F_X(y)$ *if* $x \leqslant y$];

(b) $F_X(-\infty) \equiv \lim_{x \to -\infty} F_X(x) = 0$, $F_X(+\infty) \equiv \lim_{x \to +\infty} F_X(x) = 1$;

(c) *$F_X(x)$ is continuous from the right* [*that is,* $\lim_{h \downarrow 0} F_X(x+h) = F_X(x)$ *for all* x].

We will not formally prove Theorem 3.1.3. A heuristic proof follows easily from the fact that $F_X(x) = P[X \leqslant x]$. The student should graph the d.f. of some very simple r.v. X. A typical $F_X(x)$ with the properties (a), (b), (c) of Theorem 3.1.3 is shown in Figure 3.1–3.

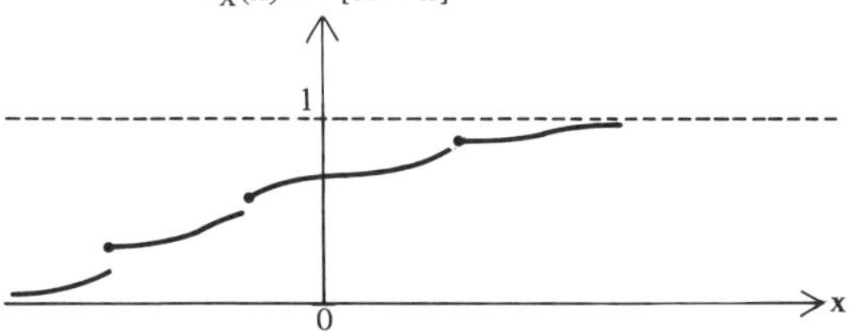

Figure 3.1–3

Theorem 3.1.4. *$F_X(x)$ has at most a countable number of discontinuities, and they are all jumps.* [For a discussion of discontinuities other than jumps, see W. Rudin, *Principles of Mathematical Analysis* (Second Edition), McGraw-Hill Book Company, New York, 1964, pp. 81–82.]

Theorem 3.1.5. *If $F(x)$ is a function defined over $(-\infty, +\infty)$ which satisfies (a), (b), (c) of Theorem 3.1.3, then there exists a sample space Ω with a probability function $P(\cdot)$, and a r.v. X defined over Ω, such that $F_X(x) = F(x)$ for all x.*

Theorem 3.1.5, whose proof we will omit, says that a real-valued function is a d.f. iff it satisfies (a), (b), (c) of Theorem 3.1.3. Note that, *up to this point, our results have been completely general* (in particular, not restricted to the discrete case). We now consider the discrete case. (The discrete case excludes such examples as picking a number at random from all numbers between 0 and 1. These are covered in Chapter 4.)

Definition 3.1.6. *X is called a (one-dimensional)* **discrete random variable** *if it is a random variable which assumes only a finite or denumerable number of values.*

Definition 3.1.7. *The* **probability function** *of a discrete r.v. X which assumes values $x_1, x_2, \ldots$ is $p(x_1), p(x_2), \ldots,$ where $p(x_i) = P[X = x_i]$ $(i = 1, 2, \ldots)$.* [Of course $P[X = y] = 0$ unless $y \in \{x_1, x_2, \ldots\}$, since $x_1, x_2, \ldots$ are all of the possible values of X. The "probability function" of Definition 2.1.1 should not be confused with the present "probability function" (sometimes called a "point probability function") which (see the proof of Theorem 3.1.8) can be

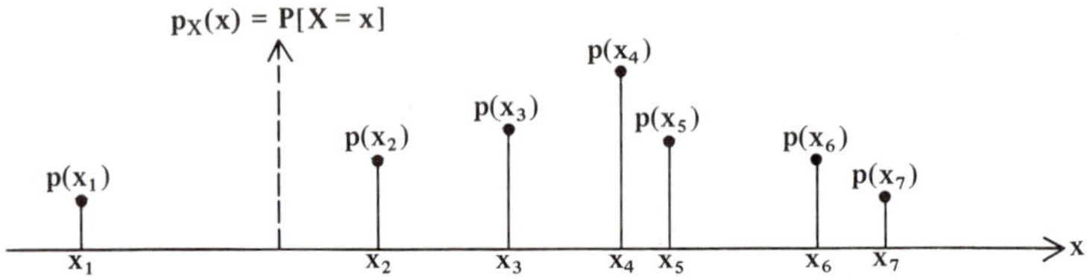

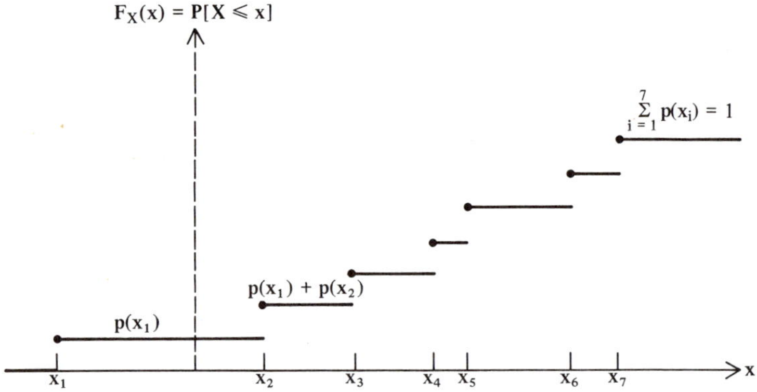

Figure 3.1–4

regarded as a special case of the one in Definition 2.1.1. In general, the context will make it clear which one is under consideration.]

For example, a typical discrete probability function and distribution function are shown in Figure 3.1–4. In the particular case $P(\omega_1) = .1$, $P(\omega_2) = .3$, $P(\omega_3) = .1$, $P(\omega_4) = .5$ with $X(\omega_1) = 2$, $X(\omega_2) = 2$, $X(\omega_3) = 0$, $X(\omega_4) = -2$, we obtain the discrete probability function and distribution function shown in Figures 3.1–5 and 3.1–6.

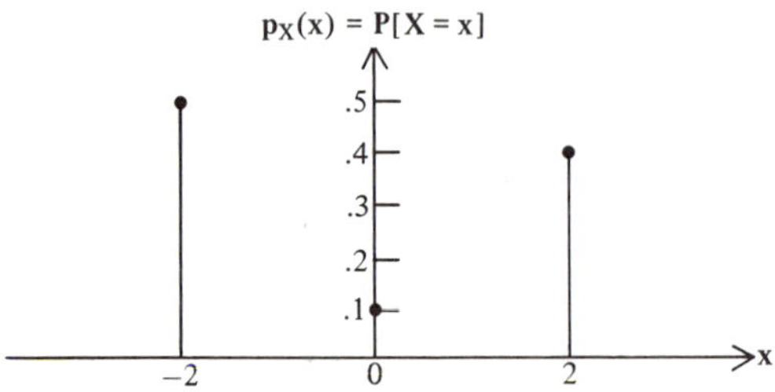

Figure 3.1—5

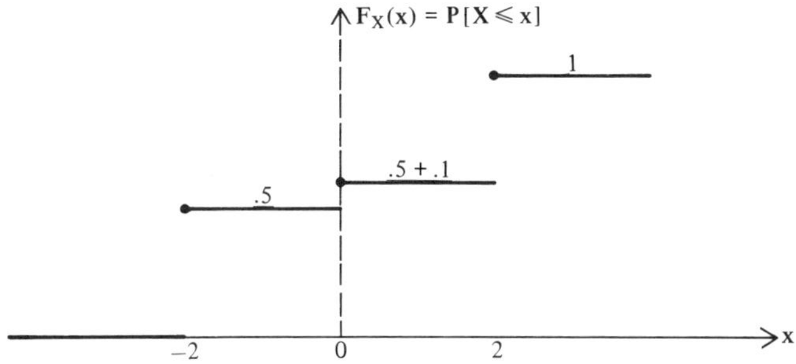

Figure 3.1–6

The following theorem says that all functions $p(x)$ with two very simple properties [(3.1.9) and (3.1.10)] are probability functions, and that all probability functions have those two properties. Thus, those two properties *characterize* probability functions. (Theorem 3.1.5 stated a similar result for general distribution functions.)

Theorem 3.1.8. *A function $p(x)$ is a probability function iff*

$$p(x) \geqslant 0, \quad \textit{for all } x \tag{3.1.9}$$

$$\sum_{i=1}^{\infty} p(x_i) = 1, \ \textit{and } p(x) = 0 \quad \text{if } x \notin \{x_1, x_2, \ldots\}.^{2} \tag{3.1.10}$$

Proof: First, suppose that $p(\cdot)$ is a probability function. Then there is some r.v. X such that $P[X = y] = p(y)$ for all values of y. (See Definition 3.1.7.) But $P[X = y]$ is the probability of $\{\omega: X(\omega) = y\} \subseteq \Omega$, and hence is $\geqslant 0$, proving (3.1.9). Now $P[X = y] = 0$ unless y is one of (a finite or denumerable number of values) $x_1, x_2, \ldots$. Hence,

$$\begin{aligned} \sum_x p(x) &= \sum_{i=1}^{\infty} p(x_i) = \sum_{i=1}^{\infty} P[X = x_i] \\ &= \sum_{i=1}^{\infty} P(\{\omega: \omega \in \Omega, X(\omega) = x_i\}) \qquad (3.1.11) \\ &= P\left(\bigcup_{i=1}^{\infty} \{\omega: \omega \in \Omega, X(\omega) = x_i\}\right) = P(\Omega) = 1 \end{aligned}$$

and (3.1.10) holds. [In (3.1.11) we used the facts that $\{\omega: \omega \in \Omega, X(\omega) = x_i\}$

[2] If only a finite number of distinct x's have $p(x) > 0$, we could replace ∞ by the number n of x's with $p(x) > 0$. Instead we adjoin x's with probability $p(x) = 0$ to $x_1, \ldots, x_n$. *The point* is that at most a countable number of distinct x's may have $p(x) > 0$, and we wish to make this point in a simple notation.

and $\{\omega: \omega \in \Omega, X(\omega) = x_j\}$ are disjoint if $i \neq j$, and that

$$\bigcup_{i=1}^{\infty} \{\omega: \omega \in \Omega, X(\omega) = x_i\} = \Omega$$

since each ω is mapped by X into either x_1, or x_2, or . . . .]

Second, suppose you are given a function $p(y)$ such that $p(y) \geqslant 0$, $p(x_1) + p(x_2) + \cdots = 1$, and $p(y) = 0$ if $y \notin \{x_1, x_2, \ldots\}$. To show that $p(\cdot)$ is a probability function we need to exhibit a r.v. X whose probability function *is* $p(\cdot)$. Define a sample space $\Omega = \{\omega_1, \omega_2, \ldots\}$ with probabilities $P(\omega_1) = p(x_1)$, $P(\omega_2) = p(x_2), \ldots$. (You should check to see that Definition 2.1.1 is satisfied by this definition.) Define a r.v. X by $X(\omega_1) = x_1$, $X(\omega_2) = x_2, \ldots$. Then (check this) $p(\cdot)$ *is* the probability function of X, which completes our proof.

The following example discusses a nonstandard probability function of a discrete r.v. which arose in practice.[3]

Example 3.1.12. The number X of microorganisms occurring on a surface is important in the design of procedures for cleansing surgical instruments, space vehicles, and so on. One study in this area[3] indicated that such organisms occur in "clumps," and that one basic distribution of interest would be that which allows some probability of one microorganism occurring in a clump and equal probability of having 2, 3, . . . , or k occur; namely,

$$P[X = x] = p_X(x) = \begin{cases} \theta, & \text{if } x = 1 \\ (1 - \theta)/k, & \text{if } x = 2, \ldots, k \\ 0, & \text{otherwise,} \end{cases}$$

for some θ with $0 \leqslant \theta \leqslant 1$. This example will be studied later (see Example 8.2.20) with regard to such questions as "How can I estimate θ?" (Usually in practice θ is unknown.)

3.2. SPECIAL DISCRETE DISTRIBUTIONS

In this section we will define several very important "discrete distributions" (that is, distribution functions and/or probability functions of discrete r.v.'s), and show by examples how they arise. Recall that a r.v. X has a discrete

---

[3] See the paper "Verification of a sampling distribution for microbiological assays" by Bruce A. Nelson of the Martin Marietta Corporation, Denver, Colorado, which was presented at the Ninth Annual Symposium on Biomathematics and Computer Science in the Life Sciences, Houston, Texas, March 22–24, 1971. (An abstract was contained in an abstracts booklet issued by the University of Texas' Graduate School of Biomedical Sciences at Houston.)

distribution if (with $p_i \geqslant 0, \sum_{i=1}^{\infty} p_i = 1$)

$$P[X = x] = p_X(x) = \begin{cases} p_i, & \text{if } x = x_i \ (i = 1, 2, \ldots) \\ 0, & \text{otherwise.} \end{cases}$$

Definition 3.2.1. *A r.v. X has the* **Bernoulli** *distribution if (for some p, $0 \leqslant p \leqslant 1$)*

$$P[X = x] = p_X(x) = \begin{cases} p^x(1 - p)^{1-x}, & x = 0, 1 \\ 0, & \text{otherwise.} \end{cases}$$

[A table is not needed due to the simplicity of $p_X(x)$.]

To prove that this is a probability function, we need only show that $p(x_i) \geqslant 0 \ (i = 1, 2, \ldots)$ and $\sum_{i=1}^{\infty} p(x_i) = 1$. For the Bernoulli this is clear; note that a Bernoulli r.v. assumes but two values (0 and 1) with respective probabilities $1 - p$ and p, and that $1 - p + p = 1$. (In more complicated, non-Bernoulli, cases it will sometimes not be at all clear, and the proof will be difficult.) Note that the values 0 and 1 are usually associated with "failure" and "success" in some context (for example, see Example 3.2.2).

Example 3.2.2. The green light at an intersection is on for 15 seconds at a time, the yellow for 5, and the red for 55. Assuming that traffic conditions induce random variations in arrival times at the light, so that "making the green light" is a chance event (called a "success") and we are equally likely to arrive at any time in the light cycle, *find* the distribution of X, the number of successes in one trial trip to the light.

Let G denote green, Y yellow, R red; the complete cycle of lights is G, Y, R. The total cycle length is $T = 15 + 5 + 55 = 75$ seconds and the probability that we arrive on that part of T which is G is $15/75 = .2$ (since we arrive at random within T and 15/75 of T is G). So, $P[X = 0] = .8$, $P[X = 1] = .2$; thus, X is *Bernoulli* with $p = .2$. (We can always interpret a Bernoulli r.v. Y as the number of successes in one trial of an experiment where the probability of success is p.)

Definition 3.2.3. *A r.v. X has the* **Binomial** *distribution if (for some positive integer n, and some p with $0 \leqslant p \leqslant 1$)*

$$P[X = x] = p_X(x) = \begin{cases} \binom{n}{x} p^x(1 - p)^{n-x}, & x = 0, 1, \ldots, n \\ 0, & \text{otherwise.} \end{cases}$$

[See Table 5 for a table of $p_X(x)$.]

Example 3.2.4. The distribution of X, the number of successes in 25 unrelated trials at the light of Example 3.2.2 is (show this)

$$P[X = x] = \binom{25}{x}(.2)^x(.8)^{25-x}, \quad x = 0, 1, \ldots, 25;$$

that is, X is *Binomial* with $n = 25$, $p = .2$. [In general, if the occurrence of an event E is subject to chance, the number of occurrences of E in n independent trials is a discrete r.v. X with possible values $0, 1, \ldots, n$. If the probability of occurrence of E on a trial is constant at some value p $(0 \leqslant p \leqslant 1)$ from trial to trial, X is (show this) a binomial r.v.]

Definition 3.2.5. *A r.v. X has the* **hypergeometric** *distribution if (for some integers n, a, N with $1 \leqslant n \leqslant N$ and $0 \leqslant a \leqslant N$)*

$$P[X = x] = p_X(x) = \begin{cases} \dfrac{\dbinom{a}{x}\dbinom{N-a}{n-x}}{\dbinom{N}{n}}, & x = \max(0, n-(N-a)), \ldots, \min(n, a) \\ 0, & \text{otherwise.} \end{cases}$$

[See Table 6 for a table of $p_X(x)$.]

As we noted after Definition 3.2.1, a direct proof that this is a probability function is difficult (see the Honors Problems). The next example gives an indirect, probabilistic proof.

Example 3.2.6. A population of N objects $(N \geqslant 1)$ contains a of type A $(0 \leqslant a \leqslant N)$. A random sample of size n $(1 \leqslant n \leqslant N)$ is drawn under SRS without replacement. Let X be the number in the sample which are of type A. Then $X \geqslant 0$ and $X \geqslant n - (N - a)$ [since we cannot obtain a negative number of items of type A in our sample, and since $n - (N - a)$ is the difference between the sample size and the number of non-type A items, respectively]; thus, $X \geqslant \max(0, n - (N - a))$. Also $X \leqslant n$ (it cannot exceed the sample size) and $X \leqslant a$ (it cannot exceed the number of elements of type A); thus, $X \leqslant \min(n, a)$. Now there are $\binom{N}{n}$ possible unordered samples, of which $\binom{a}{x}\binom{N-a}{n-x}$ contain x elements of type A. (To obtain ordered samples, we multiply each of these by $n!$.) Thus, X is *hypergeometric* with n, a, N and therefore the probabilities of Definition 3.2.5 sum to 1.

Definition 3.2.7. *A r.v. X has the* **Poisson** *distribution if (for some $\mu > 0$, called a* **parameter** *of the distribution)*

$$P[X = x] = p_X(x) = \begin{cases} e^{-\mu}\dfrac{\mu^x}{x!}, & x = 0, 1, 2, \ldots \\ 0, & \text{otherwise.} \end{cases}$$

[See Table 7 for a table of $p_X(x)$.]

Example 3.2.8. Suppose events occur randomly in time (and can therefore be represented as points along the time axis). Consider a time interval of length T anywhere on the time axis (Figure 3.2–1). We will find the probability of exactly n of these events occurring in the time interval T, say $P_n(T)$ (Assumption 3.2.9 justifies this notation) under the following assumptions.

Assumption 3.2.9. **(Time homogeneity)** *$P_n(T)$ is independent of where T is on the time axis.* [So we need not consider $P_n(t_1, t_2)$ with $t_2 - t_1 = T$.]

Assumption 3.2.10. *As $t \to 0$, $[1 - P_0(t)/t] \to \lambda$ (for some $\lambda > 0$).* This is equivalent to $1 - P_0(h) = \lambda h + o(h)$ where $\lim_{h \to 0} \dfrac{o(h)}{h} = 0$. [Restriction on the probability of one or more events in a small time interval of length h.]

Assumption 3.2.11. *$P_1(t) = \lambda t + o(t)$ and $P_2(t) + P_3(t) + \cdots = o(t)$.* [Whatever the number of events during $(0, t)$, the probability that during $(t, t + h)$ an event occurs is $\lambda h + o(h)$, *and* the probability that more than one event occurs is $o(h)$.]

Now we will find a system of differential equations for $P_n(t)$. Suppose $n \geqslant 1$. Then[4]

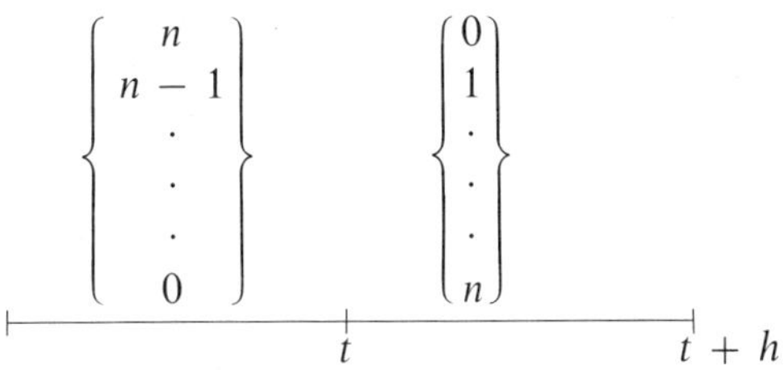

$$P_n(t + h) = P_n(t)(1 - \lambda h - o(h)) + P_{n-1}(t)(\lambda h + o(h)) + (P_{n-2}(t) + \cdots + P_0(t))(o(h))$$

[4] To obtain n events in time $t + h$, we can obtain n in time t and 0 in time h; or obtain $n - 1$ in time t and 1 in time h; and so on.

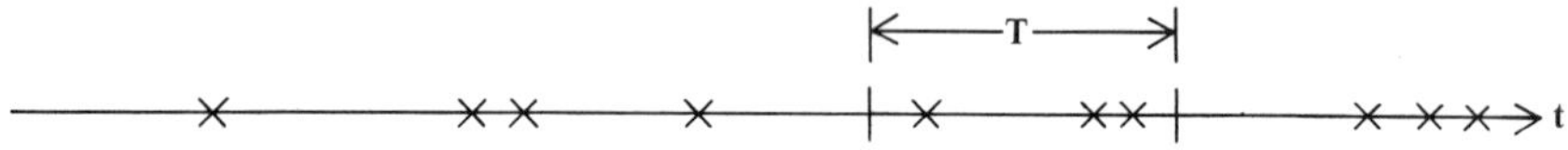

Figure 3.2–1

or

$$\frac{P_n(t+h) - P_n(t)}{h} = -\lambda P_n(t) + \lambda P_{n-1}(t) + \frac{o(h)}{h}.$$

Taking the limit as $h \to 0$ [also using $P_n(t-h)$ to prove a derivative exists] we find

$$P_n'(t) = -\lambda P_n(t) + \lambda P_{n-1}(t) \qquad (n \geqslant 1).$$

Suppose $n = 0$. Then $P_0(t+h) = P_0(t)(1 - \lambda h - o(h))$, which leads to $P_0'(t) = -\lambda P_0(t)$.

Since $P_0(0) = 1$, from $P_0'(t) = -\lambda P_0(t)$ [which has as its solution $P_0(t) = ce^{-\lambda t}$] we find

$$P_0(t) = e^{-\lambda t}.$$

Since $P_1(0) = 0$, from $P_1'(t) = -\lambda P_1(t) + \lambda P_0(t)$, or $P_1'(t) + \lambda P_1(t) = \lambda e^{-\lambda t}$, we find the solution

$$P_1(t) = \frac{(\lambda t)e^{-\lambda t}}{1!}.$$

Proceeding,[5] we find

$$P_n(t) = \frac{e^{-\lambda t}(\lambda t)^n}{n!}.$$

(This derivation uses some ideas of differential equations which the student is *not* expected to know. The important thing is to see how to obtain the differential equations; once obtained, reference to a book on the subject will allow solution. For example, see W. T. Martin and E. Reissner, *Elementary Differential Equations*, Addison-Wesley Publishing Co., Inc., Reading, Massachusetts, 1956, p. 44.) Thus X, the number of events in a specific interval of length T, is *Poisson* with $\mu = \lambda T$. (We will see later that λT is the "average" number of occurrences, so we may interpret the parameter μ in the Poisson distribution as the "average" number of occurrences.)

Definition 3.2.12. *A r.v. X has the (discrete)* **uniform** *distribution if (for some*

---

[5] This method of derivation is stated as Complement 1A on p. 123 of E. Parzen, *Stochastic Processes*, Holden-Day, Inc., San Francisco, California, 1962. Another derivation of $P_0(t) = e^{-\lambda t}$ is possible using the result that the only bounded functions such that $f(t_1 + t_2) = f(t_1)f(t_2)$ for all $t_1, t_2 > 0$ are (1) the function which is always zero and (2) $f(t) = e^{-\nu t}$ for $t > 0$ (for some $\nu > 0$); here since $P_0(t_1 + t_2) = P_0(t_1)P_0(t_2)$, we see immediately that $P_0(t) = e^{-\nu t}$ for some $\nu > 0$. Since $P_0(t) = 1 - \lambda t + o(t)$, it follows that $\nu = \lambda$.

integer $n \geq 1$, *and some* n *distinct numbers* $x_1, \ldots, x_n$)

$$P[X = x] = p_X(x) = \begin{cases} \dfrac{1}{n}, & x = x_1, x_2, \ldots, x_n \\ 0, & \text{otherwise.} \end{cases}$$

[A table is not needed due to the simplicity of $p_X(x)$.]

Example 3.2.13. If Ω is a finite discrete sample space with equally likely points $\{\omega_1, \ldots, \omega_n\}$, then let $X(\omega_i) = x_i$ and [if $x_i \neq x_j (i \neq j)$] X will be *uniform* on $x_1, \ldots, x_n$.

Definition 3.2.14. *A r.v. X has the* **geometric** *distribution if* (*for some* p, $0 \leq p \leq 1$)

$$P[X = x] = p_X(x) = \begin{cases} (1 - p)^{x-1} p, & x = 1, 2, 3, \ldots \\ 0, & \text{otherwise.} \end{cases}$$

[A table is not needed due to the simplicity of $p_X(x)$.]

Example 3.2.15. If an event E has constant probability p of occurring on any given trial, and if trials are independent, let X be the number of trials needed for the first occurrence of E. Then X is *geometrically* distributed.

Definition 3.2.16. *A r.v. X has the* **negative binomial** *distribution if* (*for some integer* $r \geq 1$, *and some* p *with* $0 \leq p \leq 1$)

$$P[X = x] = p_X(x) = \begin{cases} p^r \dbinom{x + r - 1}{r - 1} q^x, & x = 0, 1, 2, \ldots \\ 0, & \text{otherwise.} \end{cases}$$

(Here we have used the standard notation $q = 1 - p$.) [A table is not given below.]

Example 3.2.17. A given event occurs (success) with probability p (and therefore fails to occur with probability $q = 1 - p$), and independent trials are conducted. Let X be r fewer than the number of trials needed to produce r successes; so $X = x$ means we needed $x + r$ trials ($x = 0, 1, 2, \ldots$). Then

$$P[X = x] = p \binom{x + r - 1}{r - 1} p^{r-1} q^x$$

since: the last trial must be a success (p), there are $\dbinom{x + r - 1}{r - 1}$ ways to order with $r - 1$ successes, and $p^{r-1}q^x$ is the probability of each such order. Thus, X is *negative binomial* with r. The *geometric* distribution is the case $r = 1$, with one added to X (that is, in the geometric case we considered the number of trials needed to produce the first success, while in the negative binomial case we consider the number of failures encountered on the way to the first r successes).

3.3. RELATIONS AMONG CERTAIN DISCRETE DISTRIBUTIONS

We will now show that certain of the distributions already discussed "tend to" (in senses yet to be specified) certain others as indicated in the diagram of Figure 3.3–1.

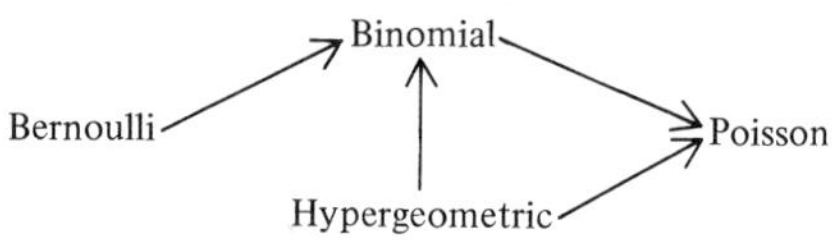

Figure 3.3–1

Bernoulli → Binomial. Suppose that $X_1, \ldots, X_n$ are r.v.'s with the common Bernoulli distribution (for some p, $0 \leqslant p \leqslant 1$)

$$P[X_i = x] = p_{X_i}(x) = \begin{cases} p^x(1-p)^{1-x}, & x = 0, 1 \\ 0, & \text{otherwise,} \end{cases}$$

and that they are based on independent events. Let $Y = X_1 + \cdots + X_n$. Then the distribution of Y is[6]

$$P[Y = y] = p_Y(y) = \begin{cases} \binom{n}{y} p^y(1-p)^{n-y}, & y = 0, 1, \ldots, n \\ 0, & \text{otherwise;} \end{cases}$$

that is, Y is binomial with parameters n and p. (Intuitively, the number of successes in n trials is a sum, in which we count 1 for success and 0 for failure on each of the trials.)

Binomial → Poisson. Suppose that X is binomial with parameters n and p (that is, for example, the number of successes in n trials taken independently with probability of success p, $0 \leqslant p \leqslant 1$, on each). Then X is approximately Poisson with $\lambda = np$. (The word "approximately" will be clarified in the derivation.)

---

[6] See Example 3.2.4.

Fix $k = 0, 1, 2, \ldots$ and $\lambda > 0$. Then since

$$\lim_{n\to\infty}\left(1 + \frac{a}{n}\right)^n = e^a,$$

so that

$$\lim_{n\to\infty}\left(1 + \frac{-\lambda}{n}\right)^n = e^{-\lambda},$$

$$\lim_{n\to\infty}\binom{n}{k}\left(\frac{\lambda}{n}\right)^k\left(1 - \frac{\lambda}{n}\right)^{n-k}$$
$$= \lim_{n\to\infty}\frac{1}{k!}\lambda^k\left(1 - \frac{\lambda}{n}\right)^{n-k}\frac{n(n-1)\cdots(n-k+1)}{n^k} = e^{-\lambda}\frac{\lambda^k}{k!}.$$

Since this holds as a limit, it is true that for large n (and thus small p, so that $\lambda = np$ is moderate)

$$\binom{n}{k}p^k(1-p)^{n-k} \approx e^{-np}\frac{(np)^k}{k!}.$$

Hypergeometric → Binomial. A population of N objects contains a of type A. A random sample of size n is drawn without replacement. ($1 \leqslant n \leqslant N$, $0 \leqslant a \leqslant N$.) Let X be the number of type A in the sample, so that for $x = \max(0, n-(N-a)), \ldots, \min(n, a)$ we have (letting $p = a/N$, the proportion of items of type A in the original population)

$$P_X(x) = \frac{\binom{a}{x}\binom{N-a}{n-x}}{\binom{N}{n}} = \frac{a!}{x!(a-x)!}\frac{(N-a)!}{(n-x)!(N-a-n+x)!}\frac{n!(N-n)!}{N!}$$
$$= \binom{n}{x}a(a-1)\cdots(a-x+1)\frac{(N-a)(N-a-1)\cdots(N-a-n+x+1)}{N(N-1)\cdots(N-n+1)}$$
$$= \binom{n}{x}\frac{a^x}{N^n}(N-a)^{n-x}1\left(1-\frac{1}{a}\right)\cdots\left(1-\frac{x-1}{a}\right)$$
$$\cdot\frac{1\left(1-\frac{1}{N-a}\right)\cdots\left(1-\frac{n-x-1}{N-a}\right)}{1\left(1-\frac{1}{N}\right)\cdots\left(1-\frac{n-1}{N}\right)}$$
$$= \binom{n}{x}p^x(1-p)^{n-x}\left(1-\frac{1}{a}\right)\left(1-\frac{2}{a}\right)\cdots\left(1-\frac{x-1}{a}\right)$$
$$\cdot\frac{\left(1-\frac{1}{N-a}\right)\left(1-\frac{2}{N-a}\right)\cdots\left(1-\frac{n-x-1}{N-a}\right)}{\left(1-\frac{1}{N}\right)\left(1-\frac{2}{N}\right)\cdots\left(1-\frac{n-1}{N}\right)}.$$

Thus, it is clear that if

$$\frac{x}{a}, \qquad \frac{n-x}{N-a}, \qquad \frac{n}{N}$$

are small (say, $\leqslant .1$), then the probability that $X = x$ is approximately the same in sampling without replacement as it is in sampling with replacement. $\left(\frac{n}{N}, \frac{n-x}{N-a}, \frac{x}{a}\right.$ are, respectively, the proportion of the population which we sample, the proportion which the non–type A items in the sample are of the total non–type A items in the population, and the proportion which the type A items in the sample are of the total type A items in the population.)

Hypergeometric → Poisson. If $\frac{n}{N}, \frac{x}{a}, \frac{n-x}{N-a}$ are small and $\lambda = np = n\frac{a}{N}$ is moderate while n is large, then the above results imply

Hypergeometric → Binomial → Poisson.

This is easy to verify directly also.

Further results along these lines are noted in the book by Feller (Volume I) listed in the references.

Example 3.3.1. Suppose that a certain small fixed volume of blood contains, on the average, 20 red blood cells for normal people. If a specimen of blood of this volume is taken from a normal person, what is the probability that it will contain fewer than 15 red cells? (By thus examining a blood specimen under a microscope, red blood cell deficiency can be determined.)

Suppose that N blood cells (red) are distributed at random in a volume of blood V, and that the small sample volume is D. Then, for $x = 0, \ldots, N$,

$$P[\text{No. in volume } D, \text{ say } X, = x] = \binom{N}{x}\left(\frac{D}{V}\right)^x\left(1 - \frac{D}{V}\right)^{N-x}$$

$$\approx e^{-\lambda}\frac{\lambda^x}{x!}$$

with $\lambda = Np = N\frac{D}{V} = 20$ (for "large" N). So,

$$P[X < 15] \approx \sum_{x=0}^{14} e^{-20}\frac{20^x}{x!} = .105.$$

Probabilities such as this can be found in many sources, for example, H. Arkin and R. R. Colton, *Tables for Statisticians* (Second Edition), Barnes & Noble, Inc., New York, 1963, pp. 129–139. See also Table 7. Note that in our example, since N and V are unknown, the exact binomial probability *cannot* be calculated.

3.4. MULTIVARIATE DISTRIBUTIONS; INDEPENDENCE

In Sections 3.1 through 3.3 we studied r.v.'s X which could arise as follows. You go to a casino with \$100 and, following a predetermined strategy, bet on roulette until your strategy tells you to stop. Then a typical outcome $\omega \in \Omega$ would describe each outcome of each play, of which there could easily be hundreds. However, your only interest might be your winnings when outcomes specified by ω occurred [which is a function of ω, say $X_1(\omega)$]; this is a one-dimensional r.v.

In other instances, your interests might include not only $X_1(\omega)$ = winnings, but also $X_2(\omega)$ = number of plays completed before stopping. Our interest in this situation[7] is in a r.v. $X(\omega) = (X_1(\omega), X_2(\omega))$ which for each outcome $\omega \in \Omega$ specifies two real numbers $X_1(\omega)$ and $X_2(\omega)$ (winnings and time to completion). If we ask "What is the probability our winnings are at most \$1000 and we play at most 5000 plays," we desire $P[X_1 \leqslant 1000, X_2 \leqslant 5000]$; this is exactly the sort of information provided by a d.f. A d.f. for these situations is formally defined in Definition 3.4.1 (previously we only considered this concept for one r.v., in Definition 3.1.2).

Definition 3.4.1. *If* $\mathbf{X} = (X_1, \ldots, X_n)$ *is an n-dimensional r.v.* $(n \geqslant 1)$, *its* **(joint) distribution function** *is defined by*

$$F_{\mathbf{X}}(x_1, \ldots, x_n) = P[X_1 \leqslant x_1, \ldots, X_n \leqslant x_n]$$

for $-\infty < x_i < +\infty$ $(i = 1, \ldots, n)$. [This extends Definition 3.1.2 to the case of n-dimensional random variables.]

Definition 3.4.2. *The joint distribution of any subset of the random variables* $X_1, \ldots, X_n$ *is called a* **marginal distribution function** [*of* $F_{X_1, \ldots, X_n}(x_1, \ldots, x_n)$]. [For example, $F_{X_1, \ldots, X_{n-1}}(x_1, \ldots, x_{n-1})$ is a marginal d.f.]

Lemma 3.4.3. *If* $\mathbf{X} = (X_1, \ldots, X_n)$ *is a vector r.v., then (for* $n \geqslant 2$)

$$\lim_{x_n \to \infty} F_{X_1, \ldots, X_n}(x_1, \ldots, x_n) = F_{X_1, \ldots, X_{n-1}}(x_1, \ldots, x_{n-1}).$$

Theorem 3.4.4. *If* $\mathbf{X} = (X_1, \ldots, X_n)$ *is an n-dimensional r.v., then:*

(1) $\lim_{\min x_i \to +\infty} F_{\mathbf{X}}(x_1, \ldots, x_n) = 1.$

(2) *For each* i $(1 \leqslant i \leqslant n)$ $\lim_{x_i \to -\infty} F_{\mathbf{X}}(x_1, \ldots, x_n) = 0.$

(3) $F_{\mathbf{X}}(x_1, \ldots, x_n)$ *is continuous from above in each argument.*

(4) *For each* **cell** $(\mathbf{a}, \mathbf{b}] = \{\mathbf{x} = (x_1, \ldots, x_n): a_i < x_i \leqslant b_i, 1 \leqslant i \leqslant n\}$ *in*

---

[7] Some instructors may wish to carry this example through the section (perhaps generalizing it) to illustrate the various concepts. We have not done so in order not to obscure the development in a cloud of examples.

$\mathscr{R}^n$, letting $\Delta_{k,n}$ denote the set of $\binom{n}{k}$ n-tuples $(z_1, \ldots, z_n)$ where each z_i is a_i or b_i and such that exactly k of the z_i's are a_i's,

$$\sum_{k=0}^{n} (-1)^k \sum_{\boldsymbol{\delta} \in \Delta_{k,n}} F_{\mathbf{X}}(\boldsymbol{\delta}) \geqslant 0.$$

We will, as in Section 3.1, not prove these or most other results. Note that $\Delta \equiv \bigcup_{k=0}^{n} \Delta_{k,n}$ is the set of the 2^n vertices of the cell $(\mathbf{a}, \mathbf{b}]$ in $\mathscr{R}^n$. Also, examine the result of Theorem 3.4.4 in the cases $n = 1$ and $n = 2$ to see especially what (4) means; it is the generalization of Theorem 3.1.3 needed for n-dimensions in order that Theorem 3.4.5 hold:

Theorem 3.4.5. *If a real-valued function F on $\mathscr{R}^n$ satisfies* (1), (2), (3), (4) *of Theorem 3.4.4, then there exists a sample space Ω with a probability function $P(\cdot)$, and r.v.'s $X_1, \ldots, X_n$ such that*

$$F_{X_1, \ldots, X_n}(x_1, \ldots, x_n) = F(x_1, \ldots, x_n).$$

Up to this point, our results in this section *have been completely general* (in particular, not restricted to the discrete case) and also generalizations of results of Section 3.1 to n-dimensional r.v.'s. We will continue this, introducing *conditional d.f.'s* and *independent r.v.'s*; these will of course also have meaning for the one-dimensional case, but it seems most natural to introduce them at this point (rather than essentially twice).

Definition 3.4.6. *The* **conditional d.f.** *of* $\mathbf{X} = (X_1, \ldots, X_n)$ *given that* $\mathbf{X} \in \mathbf{G}$, $\mathbf{G}$ *some set in* $\mathscr{R}^n$ *(that is, given that* $\mathbf{X}$ *lies in a specified subset* $\mathbf{G}$ *of n-space* $\mathscr{R}^n$*), is defined for* $P(\mathbf{G}) > 0$ *as*

$$F_{\mathbf{X}|\mathbf{G}}(\mathbf{x}|\mathbf{G}) = \frac{P[X_1 \leqslant x_1, \ldots, X_n \leqslant x_n; \mathbf{X} \in \mathbf{G}]}{P[\mathbf{X} \in \mathbf{G}]}.$$

[It may be verified that $F_{\mathbf{X}|\mathbf{G}}(\mathbf{x}|\mathbf{G})$ is a d.f.]

Consider the case $\mathbf{X} = (X_1, X_2)$, and let $\mathbf{G}$ be the set in $\mathscr{R}^2$ where $x_1'' < x_1 \leqslant x_1'$ with $P(G) > 0$. Then $F_{X_1, X_2|\mathbf{G}}(x_1, x_2|\mathbf{G}) = P[X_1 \leqslant x_1, X_2 \leqslant x_2, \mathbf{X} \in \mathbf{G}]/P[\mathbf{X} \in \mathbf{G}]$, and

$$\begin{aligned} F_{X_1, X_2|\mathbf{G}}(x_1', x_2|\mathbf{G}) &= \frac{P[X_1 \leqslant x_1', X_2 \leqslant x_2, x_1'' < X_1 \leqslant x_1']}{P[x_1'' < X_1 \leqslant x_1']} \\ &= \frac{F_{X_1, X_2}(x_1', x_2) - F_{X_1, X_2}(x_1'', x_2)}{F_{X_1}(x_1') - F_{X_1}(x_1'')}. \end{aligned}$$

Note that $F_{X_1, X_2|\mathbf{G}}(x_1', x_2|\mathbf{G})$ is (show this) a d.f. as a function of x_2. If the limit exists as $x_1'' \to x_1'$, let

$$F_{X_2|X_1=x_1'}(x_2|x_1') \equiv \lim_{x_1'' \to x_1'} F_{X_1, X_2|\mathbf{G}}(x_1, x_2|\mathbf{G}).$$

If $F_{X_2|X_1=x_1'}(x_2|x_1')$ exists for every x_2, it can be shown that it is a d.f. as a function of x_2. It is called the **conditional d.f. of X_2 given $X_1 = x_1'$**.

Definition 3.4.7. *Let $\{X_\lambda: \lambda \in \Lambda\}$ be a family of r.v.'s. They are said to be* **independent** *if for every integer $n \geqslant 1$ and every n distinct $\lambda_1, \ldots, \lambda_n \in \Lambda$,*

$$F_{X_{\lambda_1}, \ldots, X_{\lambda_n}}(x_1, \ldots, x_n) = \prod_{j=1}^{n} F_{X_{\lambda_j}}(x_j) \qquad \text{all } \mathbf{x} \in \mathscr{R}^n.$$

Suppose we have a sample space Ω and r.v.'s $X_1, X_2, \ldots, X_n$. Then $A_1, A_2, \ldots, A_n$ defined by

$$A_1 = \{\omega: \omega \in \Omega, X_i(\omega) \leqslant x_i\}$$

are events (that is, subsets of Ω); for example A_2 is the event that $X_2 \leqslant x_2$. Previously (Section 2.4) we called $A_1, A_2, \ldots, A_n$ *independent* events iff

$$P(A_1 A_2 \cdots A_n) = P(A_1)P(A_2) \cdots P(A_n)$$

$$\vdots$$

$$P(A_1 A_2) = P(A_1)P(A_2).$$

This was a reasonable definition intuitively in terms of conditional probability and of positive and negative information. In Definition 3.4.7 we have defined independence of r.v.'s $X_1, \ldots, X_n$ as meaning that (for all $x_1, x_2, \ldots, x_n$) the events $A_1, A_2, \ldots, A_n$ are independent. (This is the case where $\Lambda = \{1, 2, \ldots, n\}$; our definition covered much more complex situations also; for example, $\Lambda = \{1, 2, 3, \ldots\}$, which corresponds to a definition of independence of $X_1, X_2, X_3, \ldots$.) Note that if r.v.'s are independent, their joint (multivariate) d.f. may be trivially obtained from their individual (univariate) d.f.'s. However, often (for instance, in the example of X_1 = winnings, X_2 = number of plays, discussed at the beginning of this section) we have nonindependent r.v.'s.

Lemma 3.4.8. *The r.v.'s $\{X_\lambda: \lambda \in \Lambda\}$ are independent iff those of every finite subset are independent.*

Lemma 3.4.9. *$X_1, \ldots, X_n$ are independent iff $F_{X_1, \ldots, X_n}(x_1, \ldots, x_n) = \prod_{j=1}^{n} F_{X_j}(x_j)$, all $\mathbf{x} \in \mathscr{R}^n$.* [By our results on marginal d.f.'s, this follows directly: there is no need to check for subsets.]

We now consider the discrete case; Definitions 3.4.10 and 3.4.11 are generalizations of Definitions 3.1.6 and 3.1.7, while the other concepts are first introduced now. (Here $i_1, \ldots, i_t$, for example, are some $t \geqslant 1$ of the numbers $1, \ldots, n$. Similarly, for $i_{t+1}, \ldots, i_s$.)

Definition 3.4.10. $(X_1, \ldots, X_n)$ *is called a* **discrete** n**-dimensional r.v.** *if it assumes as values only a finite or denumerable number of points* $(x_1, \ldots, x_n)$ *in* $\mathscr{R}^n$.

Definition 3.4.11. *The* **probability function** *of a discrete n-dimensional r.v.* $\mathbf{X}$ *which assumes values* $\mathbf{x}_1, \mathbf{x}_2, \ldots$ *is* $p(\mathbf{x}_1), p(\mathbf{x}_2), \ldots$ *where* $p(\mathbf{x}_i) = P[\mathbf{X} = \mathbf{x}_i]$ $(i = 1, 2, \ldots)$. [Of course $P[\mathbf{X} = \mathbf{y}] = 0$ unless $\mathbf{y} \in \{\mathbf{x}_1, \mathbf{x}_2, \ldots\}$.]

Definition 3.4.12. *The* **marginal probability function** *of* $X_{i_1}, \ldots, X_{i_t}$ *is the name given to (for* $i_l \neq i_k$ $(l \neq k)$*)*

$$P[X_{i_1} = x_{i_1}, \ldots, X_{i_t} = x_{i_t}] = \sum_{x_{i_{t+1}}, \ldots, x_{i_n}} P[X_1 = x_1, \ldots, X_n = x_n].$$

Definition 3.4.13. *The* **conditional probability function** *of* $X_{i_1}, \ldots, X_{i_t}$ *given* $X_{i_{t+1}} = x_{i_{t+1}}, \ldots, X_{i_s} = x_{i_s}$ *is, if* $P[X_{i_{t+1}} = x_{i_{t+1}}, \ldots, X_{i_s} = x_{i_s}] > 0$ *(for disjoint sets of variables)*

$$\begin{aligned} P[X_{i_1} &= x_{i_1}, \ldots, X_{i_t} = x_{i_t} | X_{i_{t+1}} = x_{i_{t+1}}, \ldots, X_{i_s} = x_{i_s}] \\ &= \frac{P[X_{i_1} = x_{i_1}, \ldots, X_{i_t} = x_{i_t}, X_{i_{t+1}} = x_{i_{t+1}}, \ldots, X_{i_s} = x_{i_s}]}{P[X_{i_{t+1}} = x_{i_{t+1}}, \ldots, X_{i_s} = x_{i_s}]}. \end{aligned}$$

Theorem 3.4.14. *The discrete r.v.'s* $X_1, \ldots, X_n$ *are independent iff (for all* $x_1, \ldots, x_n$*)*

$$p_{X_1, \ldots, X_n}(x_1, \ldots, x_n) = p_{X_1}(x_1) \cdots p_{X_n}(x_n).$$

Proof: First suppose that (for all $x_1, \ldots, x_n$)

$$p_{X_1, \ldots, X_n}(x_1, \ldots, x_n) = p_{X_1}(x_1) \cdots p_{X_n}(x_n).$$

Then for any $y_1, \ldots, y_n$ $(-\infty < y_1, \ldots, y_n < +\infty)$

$$\begin{aligned} F_{X_1, \ldots, X_n}(y_1, \ldots, y_n) &= \sum_{x_1 < y_1} \cdots \sum_{x_n < y_n} p_{X_1, \ldots, X_n}(x_1, \ldots, x_n) \\ &= \sum_{x_1 < y_1} \cdots \sum_{x_n < y_n} p_{X_1}(x_1) \cdots p_{X_n}(x_n) \\ &= \left(\sum_{x_1 \leqslant y_1} p_{X_1}(x_1) \right) \cdots \left(\sum_{x_n \leqslant y_n} p_{X_n}(x_n) \right) \\ &= F_{X_1}(y_1) \cdots F_{X_n}(y_n), \end{aligned}$$

and (by Lemma 3.4.9) $X_1, \ldots, X_n$ are independent r.v.'s.

Second, suppose that $X_1, \ldots, X_n$ are independent discrete r.v.'s. Consider first the case $n = 2$. Then (for all x_1, x_2)

$$
\begin{aligned}
p_{X_1, X_2}(x_1, x_2) \\
&= \lim_{\delta \downarrow 0} P[x_1 - \delta < X_1 \leqslant x_1, x_2 - \delta < X_2 \leqslant x_2] \\
&= \lim_{\delta \downarrow 0} \{P[X_1 \leqslant x_1, X_2 \leqslant x_2] - P[X_1 \leqslant x_1 - \delta, X_2 \leqslant x_2] \\
&\quad - P[X_1 \leqslant x_1, X_2 \leqslant x_2 - \delta] + P[X_1 \leqslant x_1 - \delta, X_2 \leqslant x_2 - \delta]\} \\
&= \lim_{\delta \downarrow 0} \{F_{X_1}(x_1)F_{X_2}(x_2) - F_{X_1}(x_1 - \delta)F_{X_2}(x_2) \\
&\qquad - F_{X_1}(x_1)F_{X_2}(x_2 - \delta) + F_{X_1}(x_1 - \delta)F_{X_2}(x_2 - \delta)\} \\
&= F_{X_1}(x_1)F_{X_2}(x_2) - P[X_1 < x_1]F_{X_2}(x_2) - F_{X_1}(x_1)P[X_2 < x_2] \\
&\qquad + P[X_1 < x_1]P[X_2 < x_2] \\
&= F_{X_2}(x_2)(F_{X_1}(x_1) - P[X_1 < x_1]) \\
&\qquad - P[X_2 < x_2](F_{X_1}(x_1) - P[X_1 < x_1]) \\
&= F_{X_2}(x_2)p_{X_1}(x_1) - P[X_2 < x_2]p_{X_1}(x_1) \\
&= p_{X_1}(x_1)(F_{X_2}(x_2) - P[X_2 < x_2]) \\
&= p_{X_1}(x_1)p_{X_2}(x_2),
\end{aligned}
$$

which proves the desired result. The general case follows easily by induction.

Example 3.4.15. Let X_1 and X_2 be two jointly discrete r.v.'s whose joint probability function is given by the following table:

	$p_{X_1, X_2}(x_1, x_2)$	
x_2 \ x_1	-1	1
0	$\frac{1}{6}$	$\frac{1}{6}$
.5	$\frac{1}{3}$	$\frac{1}{3}$

Also, let Y_1 and Y_2 be two jointly discrete r.v.'s whose joint probability function is given by

	$p_{Y_1, Y_2}(y_1, y_2)$	
y_2 \ y_1	-1	1
0	$\frac{1}{3}$	0
.5	$\frac{1}{6}$	$\frac{1}{2}$

Show that X_1 and Y_1 are identically distributed [that is, $p_{X_1}(z) = p_{Y_1}(z)$ for all z], and that X_2 and Y_2 are identically distributed. Are the r.v.'s X_1 and X_2 independent? Are the r.v.'s Y_1 and Y_2 independent? Find the conditional probability function of Y_1 given $Y_2 = 0$. Find the conditional probability function of X_1 given $X_2 = 0$.

First (see Definition 3.4.12 of the marginal probability function),

$$p_{X_1}(-1) = p_{X_1, X_2}(-1, 0) + p_{X_1, X_2}(-1, .5) = \tfrac{1}{6} + \tfrac{1}{3} = \tfrac{1}{2},$$
$$p_{X_1}(1) = p_{X_1, X_2}(1, 0) + p_{X_1, X_2}(1, .5) = \tfrac{1}{6} + \tfrac{1}{3} = \tfrac{1}{2},$$

and

$$p_{Y_1}(-1) = p_{Y_1, Y_2}(-1, 0) + p_{Y_1, Y_2}(-1, .5) = \tfrac{1}{3} + \tfrac{1}{6} = \tfrac{1}{2},$$
$$p_{Y_1}(1) = p_{Y_1, Y_2}(1, 0) + p_{Y_1, Y_2}(1, .5) = 0 + \tfrac{1}{2} = \tfrac{1}{2},$$

so $p_{X_1}(z) = p_{Y_1}(z)$ for all z. Similarly, we find $p_{X_2}(0) = p_{Y_2}(0) = \frac{1}{3}$ and $p_{X_2}(.5) = p_{Y_2}(.5) = \frac{2}{3}$. Hence, X_2 and Y_2 (as well as X_1 and Y_1) are identically distributed.

X_1 and X_2 are independent (using the criterion of Theorem 3.4.14) since $p_{X_1, X_2}(x_1, x_2) = p_{X_1}(x_1)p_{X_2}(x_2)$ for all x_1, x_2. However, Y_1 and Y_2 are not independent since $p_{Y_1, Y_2}(y_1, y_2)$ does not equal $p_{Y_1}(y_1)p_{Y_2}(y_2)$ for all y_1, y_2; for example, $p_{Y_1, Y_2}(1, 0) = 0 \neq \frac{1}{2} \cdot \frac{1}{3} = p_{Y_1}(1)p_{Y_2}(0)$.

The conditional probability function of Y_1 given $Y_2 = 0$ (see Definition 3.4.13) is

$$P[Y_1 = -1 | Y_2 = 0] = \frac{P[Y_1 = -1, Y_2 = 0]}{P[Y_2 = 0]} = \frac{\frac{1}{3}}{\frac{1}{3}} = 1,$$
$$P[Y_1 = y | Y_2 = 0] = 0, \quad \text{if } y \neq -1.$$

The conditional probability function of X_1 given $X_2 = 0$ is

$$P[X_1 = -1 | X_2 = 0] = \frac{P[X_1 = -1, X_2 = 0]}{P[X_2 = 0]} = \frac{\frac{1}{6}}{\frac{2}{6}} = \tfrac{1}{2},$$
$$P[X_1 = 1 | X_2 = 0] = \frac{P[X_1 = 1, X_2 = 0]}{P[X_2 = 0]} = \frac{\frac{1}{6}}{\frac{2}{6}} = \tfrac{1}{2}.$$

This could be obtained directly since (by independence of X_1 and X_2) $P[X_1 = x_1 | X_2 = 0] = P[X_1 = x_1]$ for all x_1.

3.5. SPECIAL MULTIVARIATE DISCRETE DISTRIBUTIONS

Definition 3.5.1. *A r.v.* $\mathbf{X} = (X_1, \ldots, X_k)$ *has the* **multinomial distribution** *if* [*for some integers* $k \geqslant 2$, $n \geqslant 1$ *and some* $0 \leqslant p_i \leqslant 1$ $(i = 1, \ldots, k)$ *with* $p_1 + \cdots + p_k = 1$]

$$P[X_1 = x_1, \ldots, X_k = x_k] = \begin{cases} \dfrac{n!}{x_1! \cdots x_k!} p_1^{x_1} \cdots p_k^{x_k}, & x_i = 0, \ldots, n (1 \leqslant i \leqslant k) \\ & x_1 + \cdots + x_k = n \\ 0, & \text{otherwise.} \end{cases}$$

Example 3.5.2. Suppose an experiment can result in exactly k possible outcomes, say $o_1, \ldots, o_k$, and that the probabilities of these outcomes are, respectively, $p_1, \ldots, p_k$ [where $0 \leqslant p_i \leqslant 1$ $(i = 1, \ldots, k)$ and $p_1 + \cdots + p_k = 1$]. Suppose the experiment is repeated n times independently, and let $X_i =$ no. of outcomes of type o_i $(i = 1, \ldots, k)$. Then (show this) $(X_1, \ldots, X_k)$ has the *multinomial distribution* as given in Definition 3.5.1. Note that we could write simply $P[X_1 = x_1, \ldots, X_{k-1} = x_{k-1}]$ since $X_1, \ldots, X_{k-1}$ (or, any $k - 1$ of $X_1, \ldots, X_k$) determine X_k (or, determine the other X_i).

Example 3.5.3. At a certain hour, people may be watching any of k (for example, $k = 3$) television shows. If the probabilities are $p_1, \ldots, p_3$ that a person is watching shows 1, 2, 3 (respectively), and if n people are sampled independently as to show being watched, let $X_i =$ no. watching show i $(i = 1, 2, 3)$. Then

$$P[X_1 = x_1, X_2 = x_2, X_3 = x_3] = \begin{cases} \dfrac{n!}{x_1!x_2!x_3!} p_1^{x_1} p_2^{x_2} p_3^{x_3}, & \begin{array}{l} x_i = 0, \ldots, n \\ x_1 + x_2 + x_3 = n \end{array} \\ 0, & \text{otherwise.} \end{cases}$$

Example 3.5.4. Let X_1 and X_2 be two jointly discrete r.v.'s whose joint probability function is given by the following table in which $h = \frac{1}{60}$:

	$p_{X_1, X_2}(x_1, x_2)$			
x_2 \ x_1	0	1	2	$p_{X_2}(x_2)$
0	h	$4h$	$9h$	$14h$
1	$2h$	$6h$	$12h$	$20h$
2	$3h$	$8h$	$3h$	$14h$
3	$4h$	$2h$	$6h$	$12h$
$p_{X_1}(x_1)$	$10h$	$20h$	$30h$	

Are the r.v.'s X_1 and X_2 independent or dependent? What is $P[X_1 \geqslant X_2]$? What is $P[X_1 - X_2 = 1 | X_1 + X_2 = 3]$?

For independence of X_1 and X_2, we must have $p_{X_1, X_2}(x_1, x_2) = p_{X_1}(x_1)p_{X_2}(x_2)$ for all x_1, x_2. But $h = p_{X_1, X_2}(0, 0) \neq 2\frac{1}{3}h = \frac{140}{60}h = (10h)(14h) = p_{X_1}(0) \cdot p_{X_2}(0)$. Next, $P[X_1 \geqslant X_2] = \sum_{x_1 \geqslant x_2} p_{X_1, X_2}(x_1, x_2) = 35h = \frac{7}{12}$. Finally,

$$P[X_1 - X_2 = 1 | X_1 + X_2 = 3] = \frac{P[X_1 - X_2 = 1, X_1 + X_2 = 3]}{P[X_1 + X_2 = 3]}$$

$$= \frac{P[X_1 = 2, X_2 = 1]}{P[X_1 + X_2 = 3]} = \frac{12h}{24h} = \frac{1}{2}.$$

3.6. DISTRIBUTIONS OF FUNCTIONS OF DISCRETE RANDOM VARIABLES

Often we will observe a random variable but real interest will center, not about the distribution of the observed r.v., but about the distribution of some function of the observed r.v. (for example, not sales but profit). We here state the (obvious) results for one- and n-dimensional r.v.'s.

Theorem 3.6.1. *Let X be a discrete r.v. Define $Y = g(X)$ for some real-valued function $g(x)$. Then*

$$F_Y(y) = P[Y \leqslant y] = P[g(X) \leqslant y] = \sum_{x:g(x) \leqslant y} p_X(x),$$

$$p_Y(y) = P[Y = y] = P[g(X) = y] = \sum_{x:g(x) = y} p_X(x).$$

Theorem 3.6.2. *Let $\mathbf{X} = (X_1, \ldots, X_n)$ be an n-dimensional discrete r.v. Define $\mathbf{Y} = (Y_1, \ldots, Y_r) = (g_1(\mathbf{X}), \ldots, g_r(\mathbf{X}))$. Then*

$$F_{\mathbf{Y}}(\mathbf{y}) = P[Y_1 \leqslant y_1, \ldots, Y_r \leqslant y_r] = P[g_1(\mathbf{X}) \leqslant y_1, \ldots, g_r(\mathbf{X}) \leqslant y_r]$$
$$= \sum_{\substack{x_1, \ldots, x_n \text{ such that} \\ g_1(\mathbf{x}) < y_1, \ldots, g_r(\mathbf{x}) < y_r}} p_{X_1, \ldots, X_n}(x_1, \ldots, x_n)$$

$$p_{\mathbf{Y}}(\mathbf{y}) = \sum_{\substack{x_1, \ldots, x_n \text{ such that} \\ g_1(\mathbf{x}) = y_1, \ldots, g_r(\mathbf{x}) = y_r}} p_{X_1, \ldots, X_n}(x_1, \ldots, x_n)$$

Example 3.6.3. For X_1 and X_2 as given in Example 3.4.15, find the distribution of $Y = |X_1 - X_2|$.

By Theorem 3.6.2, $p_Y(y) = P[Y = y] = P[|X_1 - X_2| = y]$ is the sum of $p_{X_1, X_2}(x_1, x_2)$ for all x_1, x_2 such that $|x_1 - x_2| = y$. We have

(x_1, x_2)	$p_{X_1, X_2}(x_1, x_2)$	$\lvert x_1 - x_2\rvert$
$(-1, 0)$	$\frac{1}{6}$	1
$(-1, .5)$	$\frac{1}{3}$	1.5
$(1, 0)$	$\frac{1}{6}$	1
$(1, .5)$	$\frac{1}{3}$	.5

Hence,

$$p_Y(y) = P[Y = y] = \begin{cases} \frac{1}{3}, & \text{if } y = 1.5 \\ \frac{2}{6} = \frac{1}{3}, & \text{if } y = 1 \\ \frac{1}{3}, & \text{if } y = .5. \end{cases}$$

Example 3.6.4. Suppose that (X, Y) is a two-dimensional discrete r.v. Then

(Theorem 3.6.2)

$$P[X + Y = z] = \sum_{\substack{x,\, y \text{ such that} \\ x+y=z}} p_{X,Y}(x, y) \\ = \sum_x p_{X,Y}(x, z - x). \tag{3.6.5}$$

If, in addition, X and Y are independent [so that by Theorem 3.4.14, $p_{X,Y}(x, y) = p_X(x)p_Y(y)$],

$$P[X + Y = z] = \sum_x p_X(x)p_Y(z - x). \tag{3.6.6}$$

The probability distribution $p(z)$ defined in (3.6.6) is called the **convolution** of the probability distributions $p_X(\cdot)$ and $p_Y(\cdot)$. Hence, we have Theorem 3.6.7.

Theorem 3.6.7. *If X and Y are independent discrete r.v.'s, then the distribution of $X + Y$ is the convolution of the distributions of X and of Y.*

Example 3.6.8. (Due to W. J. Hall) Suppose that the distribution of $X + Y$ is the convolution of the distributions of X and of Y. Does it follow that X and Y are independent r.v.'s?

Suppose that X and Y have the joint probability function given by the following table (where θ is fixed, $|\theta| \leqslant \frac{1}{9}$):

	$p_{X,Y}(x, y)$			
$y \backslash x$	-1	0	1	$p_Y(y)$
-1	$\frac{1}{9}$	$\frac{1}{9} + \theta$	$\frac{1}{9} - \theta$	$\frac{1}{3}$
0	$\frac{1}{9} - \theta$	$\frac{1}{9}$	$\frac{1}{9} + \theta$	$\frac{1}{3}$
1	$\frac{1}{9} + \theta$	$\frac{1}{9} - \theta$	$\frac{1}{9}$	$\frac{1}{3}$
$p_X(x)$	$\frac{1}{3}$	$\frac{1}{3}$	$\frac{1}{3}$	

Then $P[X + Y = z] = \frac{1}{9}, \frac{2}{9}, \frac{3}{9}$ for $z = \pm 2, \pm 1, 0$ for all θ. Although this is the convolution of the distributions of X and of Y, X and Y are independent r.v.'s iff $\theta = 0$. Hence, the proposition that, under the stated conditions, X and Y are independent r.v.'s does not follow (that is, the converse of Theorem 3.6.7 is false).

PROBLEMS

3.1 Let $\Omega = \{H, T\}$ with $P(H) = P(T) = \frac{1}{2}$. Define real-valued random variables X and Y by $X(H) = 1$, $X(T) = 0$, and $Y(H) = 0$, $Y(T) = 1$. Find and graph $F_X(x)$ and $F_Y(y)$.

3.2 Can two different random variables have the same d.f.? Prove your assertion.

3.3 Prove that (for $0 \leqslant p \leqslant 1$, n an integer $\geqslant 1$)

$$p_X(x) = \begin{cases} \binom{n}{x} p^x(1-p)^{n-x}, & x = 0, 1, \ldots, n \\ 0, & \text{otherwise} \end{cases}$$

is a probability function (called the *binomial probability function*).

3.4 Prove that (for any fixed $\lambda > 0$)

$$p_X(x) = \begin{cases} e^{-\lambda}\dfrac{\lambda^x}{x!}, & x = 0, 1, 2, \ldots \\ 0, & \text{otherwise} \end{cases}$$

is a probability function (called the *Poisson probability function*).

3.5 Ten questions are directed at random to 4 students so that each question is equally likely to be directed to any student. What is the probability function for the number of questions directed to the first student? What is the d.f.?

3.6 A silver dollar is tossed repeatedly and independently until a head appears. What is the probability function for the number of tosses required?

3.7 Cards are drawn from an ordinary 52-card shuffled deck, without replacement, until a club appears. What is the probability function for the number of draws required?

3.8 A machine pits prunes with an average of 1 percent of its output being defective. What is the probability function of the number of defectives in a sample of 60 consecutive prunes if

(1) each "pitted" prune has probability 1/100 of being defective and is defective or good independently of the other prunes;

(2) the machine turns out cycle after cycle of prunes with 99 good followed by 1 bad, and the "time" we started to sample in a cycle was random.

3.9 Suppose that teams B and C are competing in the World Series, that the (at most) 7 games are independent, and that

$$P[B \text{ beats } C \text{ in any game}] = p$$

$(0 \leqslant p \leqslant 1)$. Let

$$X = \begin{cases} 1, & \text{if } B \text{ wins the series} \\ 0, & \text{if } C \text{ wins the series.} \end{cases}$$

Find $P[X = 1]$.

3.10 In problem 3.9, $P[X = 1]$ is a function of p $(0 \leqslant p \leqslant 1)$. Plot a graph of $P[X = 1]$ against p, and include specifically the values of $P[X = 1]$ for $p = 0, .1, .2, .3, .4, .5, .6, .7, .8, .9, 1$.

3.11 Plot the probability function and d.f. of X (see Problem 3.10) for $p = .6$.

3.12 Suppose that teams B and C (Problem 3.9) played not 7 but $2n + 1$ games, for n some positive integer. In that case what would $P[X = 1]$ be?

3.13 Suppose that a woman, in her lifetime, is equally likely to have 1, 2, or 3 female offspring, and suppose that these second-generation women are in turn each equally likely to have 1, 2, or 3 third-generation female offspring. What is the probability function of the number of third-generation women?

3.14 A telephone switchboard handles 600 calls, on the average, during an hour. The board can make a maximum of 20 connections per minute. Use the Poisson distribution to evaluate the probability that the board will be overtaxed during a given minute. What is the probability that 10 seconds elapse with no calls?

3.15 If X and Y are independent random variables, each having the Poisson distribution (with respective parameters λ_1 and λ_2) show that the conditional distribution of X, given $X + Y$, is binomial.

3.16 Let A and B be events such that $P(A) = \frac{1}{4}$, $P(B|A) = \frac{1}{2}$, and $P(A|B) = \frac{1}{4}$. Define random variables X and Y by

$$X = \begin{cases} 1, & \text{if } A \text{ occurs} \\ 0, & \text{if } A \text{ does not occur;} \end{cases}$$

$$Y = \begin{cases} 1, & \text{if } B \text{ occurs} \\ 0, & \text{if } B \text{ does not occur.} \end{cases}$$

Determine whether and why each of the following statements is true or false.

(i) X and Y are independent.
(ii) $P[X^2 + Y^2 = 1] = \frac{1}{4}$.
(iii) $P[XY = X^2Y^2] = 1$.
(iv) X is uniformly distributed on the interval 0 to 1.
(v) X and Y are identically distributed.

3.17 The number X of calculators that a certain store sells in a week obeys the Poisson probability law with parameter $\lambda = 10$. The profit on each calculator sold is \$2. If at the beginning of the week 10 calculators are in stock, the profit Y from sale of calculators during the week is

$$Y = 2 \cdot \text{minimum}\,(X, 10).$$

Find the probability law of Y.

3.18 Prove that it is impossible for two independent identically distributed random variables X_1 and X_2, each taking values 1 to 6, to have the property that

$$P[X_1 + X_2 = k] = 1/11, \quad \text{for } k = 2, 3, \ldots, 12.$$

Is it possible to weight a pair of dice so that the probability of occurrence of every sum from 2 to 12 will be the same?

3.19 Let X, Y be independent discrete r.v.'s with Poisson distributions of parameters λ_1 and λ_2 respectively. Show that the distribution of $X + Y$ is Poisson. What is the parameter of the Poisson distribution of $X + Y$?

3.20 If X is a random variable with distribution function $F_X(x)$ $(-\infty < x < +\infty)$, and if $Y = aX + b$, where $a > 0$, express $P[Y \leqslant z]$ using $F_X(\cdot)$.

3.21 **(a)** Let X be a binomial r.v. based on n trials each with probability p of success. Table 5 contains numerical values of $P[X = x]$ for various n, for $x = 0$, $1, \ldots, n$, and for $p = .05\,(.05).50$. How could we use this table to find

$P[Y = y]$ if Y were a binomial r.v. based on n trials, each with probability of success $q = .75$?

(b) Let X be a Poisson r.v. with parameter $\mu > 0$. If $P[X = y + 1] = P[X = y] > 0$, what value must μ have? Give an example of values of μ and y for which this equality $P[X = y + 1] = P[X = y]$ is exemplified in Table 7.

3.22 Suppose that S_n is a random variable whose probability distribution is binomial for some $p\,(0 \leqslant p \leqslant 1)$. If we observe $S_n = x_0$ (where x_0 is one of $0, 1, \ldots, n$), then the **Maximum-Likelihood Estimator (MLE)** of p based on S_n is that $\hat{p}\,(0 \leqslant \hat{p} \leqslant 1)$ which maximizes

$$P[S_n = x_0] = \binom{n}{x_0} p^{x_0}(1 - p)^{n - x_0}.$$

Find it.

[*The following material is used in Questions* 3.23, 3.24, 3.25.] It is sometimes the case that a sequence (say $Y_1, Y_2, \ldots$) of random variables cannot be observed, although we can observe results of the values of $Y_1, Y_2, \ldots$ (say $X_1, X_2, \ldots$).

Let $Y_1, Y_2, \ldots$ be independent and identically distributed random variables. Let a be a fixed number $(-\infty < a < +\infty)$. For $i = 1, 2, \ldots$ define

$$X_i = \begin{cases} 1, & \text{if } Y_i \geqslant a \\ 0, & \text{if } Y_i < a. \end{cases}$$

Define $p = P[Y_1 \geqslant a]$, $q = 1 - p$, $S_n = X_1 + \cdots + X_n$. (Note that S_n is the number of $Y_1, \ldots, Y_n$ which exceed the threshold a.)

3.23 Let n be a fixed positive integer. Let $A_i(n)$ be the event "At least i of $Y_1, \ldots, Y_n$ are $\geqslant a$" $(i = 1, \ldots, n)$. Find (in terms of p, q, n)

(a) $P\left[\bigcap_{i=1}^{n} A_i(n)\right]$;

(b) $P\left[\bigcup_{i=1}^{n} A_i(n)\right]$;

(c) $P\left[\bigcap_{i=1}^{j} A_i(n)\right]$, where $1 \leqslant j \leqslant n$.

3.24 Let n be a fixed positive integer. Let $B_i(n)$ be the event "Exactly i of $Y_1, \ldots, Y_n$ are $\geqslant a$" $(i = 0, 1, \ldots, n)$. Find (in terms of p, q, n)

$$P[B_0(n) \cup B_1(n) \cup \cdots \cup B_n(n)].$$

3.25 **(a)** How many different values can the random vector $(X_1, \ldots, X_n)$ take on with positive probability? (Assume that $0 < p < 1$.)

(b) How many different values can $S_n = X_1 + \cdots + X_n$ take on with positive probability? (Assume that $0 < p < 1$.)

3.26 Show that if a random variable X is geometric, that is, for some $p(0 \leqslant p < 1)$

$$P[X = x] = \begin{cases} (1 - p)p^x, & \text{for } x = 0, 1, \ldots \\ 0, & \text{otherwise,} \end{cases}$$

then it has the **lack of memory property**: for all positive integers j and k,

$$P[X \geqslant j + k | X \geqslant k] = P[X \geqslant j].$$

3.27 Suppose $X_1, X_2, \ldots, X_n$ are independent random variables and that $P[X_i = 0] = 1 - p$, $P[X_i = 1] = p$ for some p with $0 \leqslant p \leqslant 1$ for $i = 1, 2, \ldots, n$. Let X denote the number of $X_1, \ldots, X_n$ which are at least as large as X_1. Find $P[X = x]$ for all x. [*Hint*: First find $P[X = n]$.]

3.28 A fair coin is tossed 3 times independently. Define random variables X and Y by

X = the number of heads occurring in the 3 tosses,
Y = the number of tails occurring before the first head.

Tabulate (see Example 3.5.4) the joint probability function $p_{X,Y}(x, y) = P[X = x, Y = y]$. Determine whether or not X and Y are independent.

3.29 An unknown number N of rats live in a certain neighborhood. In order to design an effective rat control program, we need to have an estimate of N. Suppose the following experiment is performed. a of the N rats are caught, tagged, and released. After they have had time to mix thoroughly with the total population, we catch (at random without replacement) n of the rats. If x of the rats sampled have tags, what can we say about N? Specifically,

(a) What is the probability, say $p(N)$, that x of the rats sampled have tags?

(b) Regarding $p(N)$ as a function of N, for what value of N is $p(N)$ largest? [*Hint*: Find $p(N)/p(N - 1)$. Now $p(N)$ increases if this ratio is >1 and decreases if it is <1.]

3.30 For a discrete random variable X the **conditional failure probability function** (or **discrete failure rate**) is defined as

$$q(x) = P[X = x | X \geqslant x]$$

for x such that $P[X \geqslant x] > 0$. If the r.v. X represents the life of some item, $q(x)$ can be interpreted as the conditional probability of death at x given that death has not occurred before x. Find and plot the discrete failure rate $q(x)$ if X has the geometric distribution (see Definition 3.2.14); the uniform distribution (see Definition 3.2.12).

3.31 Let X be a discrete r.v. such that

$$\sum_{i=-\infty}^{\infty} p_i = 1,$$

where $p_i = P[X = i]$ (that is, X assumes only integers as values). The distribution $\{p_i\}$ is said to be **unimodal** if there is at least one integer M such that

$$\begin{cases} p_i \geqslant p_{i-1}, & \text{whenever } i \leqslant M \\ p_{i+1} \leqslant p_i, & \text{whenever } i \geqslant M. \end{cases}$$

The distribution $\{p_i\}$ is said to be **log-concave** if

$$p_i^2 \geqslant p_{i+1} p_{i-1}, \quad \text{for all } i.$$

Prove directly[8] that the Poisson distribution is both unimodal and log-concave. Prove directly that the binomial distribution is both unimodal and log-concave.

HONORS PROBLEMS

3.1H In a rigorous theoretical development of probability theory, certain measurability restrictions are required on a "random variable" (and a "random n-vector"). What are they? Why are they needed?

3.2H Prove that the hypergeometric distribution for sampling without replacement n times at random from a population of N elements (a of which are of type A) is a probability distribution. Note that, if X = No. of type A in the sample, then

$$P[X = x] = p_X(x) = \begin{cases} \dfrac{\dbinom{a}{x}\dbinom{N-a}{n-x}}{\dbinom{N}{n}}, & x = \max(0, n-(N-a)), \ldots, \min(n, a) \\ 0, & \text{otherwise.} \end{cases}$$

3.3H Suppose given n items of type A and n items of type B ($n \geqslant 1$), as well as n cells (each with two positions, each position being capable of holding an item of type A or one of type B). Let the $2n$ items be distributed at random into the $2n$ cell positions, and define

$$X = \begin{pmatrix} \text{number of cells containing one item} \\ \text{of type } A \text{ and one item of type } B \end{pmatrix}.$$

Find $p_X(x) = P[X = x]$, the probability function of X. [*Hint*: See Feller, Vol. I (Second Edition), p. 54, No. 12, for a simpler problem in this set-up. See the journal article referenced there. Tossing the n of type A determines X, since the places to be occupied by the items of type B are then determined.]

3.4H Give a complete derivation of the Poisson distribution as the solution to a problem of random events, along the lines indicated in Example 3.2.8. (In particular, show in detail the general solution of the differential equation

$$\frac{dy}{dx} + P(x)y = Q(x),$$

and indicate explicitly how you use it.) Give a second derivation along the lines of footnote 5: "Another derivation . . . ."

3.5H **Theorem 1:** *If* $X = (X_1, X_2)$ *is a two-dimensional random variable, then*

(a) $\lim_{\min(x_1, x_2) \to +\infty} F_{X_1, X_2}(x_1, x_2) = 1;$

---

[8] An indirect proof is also possible since J. Keilson and H. Gerber ["Some results for discrete unimodality," *Journal of the American Statistical Association*, Vol. 66 (1971), pp. 386–389] have proven that log-concavity implies unimodality.

(b) $\lim_{x_i \to -\infty} F_{X_1, X_2}(x_1, x_2) = 0 \quad (i = 1, 2)$;

(c) $F_{X_1, X_2}(x_1, x_2)$ *is continuous from above in each argument*;

(d) *for all a, b, c, d (with* $a < b, c < d$*),*
$F_{X_1, X_2}(b, d) - F_{X_1, X_2}(a, d) - F_{X_1, X_2}(b, c) + F_{X_1, X_2}(a, c) \geqslant 0$;

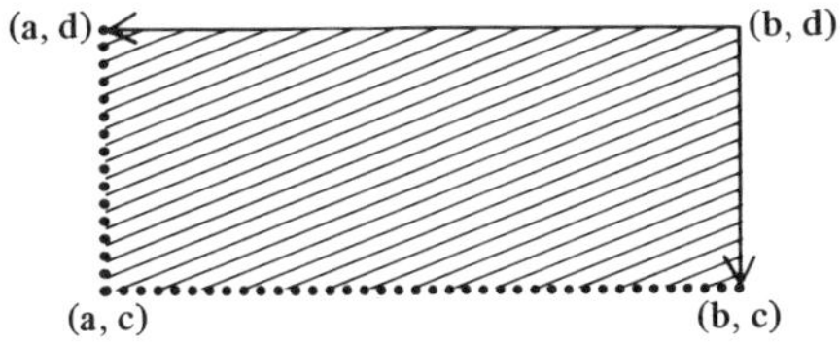

that is, $F_{X_1, X_2}(\cdot, \cdot)$ *assigns positive mass to the indicated rectangle.*

Theorem 2: *If a real-valued function F on* $\mathscr{R}^2$ *satisfies* (a), (b), (c), *and* (d) *of Theorem* 1, *then there is a probability space and r.v.'s* X_1, X_2 *such that* $F_{X_1, X_2}(x_1, x_2) = F(x_1, x_2)$.

Show that if (d) in Theorem 2 is replaced by (d'): $F_{X_1, X_2}(x_1, x_2) \leqslant F_{X_1, X_2}(y_1, y_2)$ for $x_1 \leqslant y_1$, $x_2 \leqslant y_2$, then the result is false. [*Hint*: Find an $F(\cdot, \cdot)$ which satisfies (a), (b), (c), (d') but violates (d).] Prove Theorem 1. Prove Theorem 2.

3.6H Let X_1 and X_2 be two independent discrete random variables. Denote $P[X_1 = i] = p_i \geqslant 0\,(1 \leqslant i \leqslant n)$, $P[X_2 = j] = q_j \geqslant 0\,(1 \leqslant j \leqslant n)$, where $n \geqslant 2$ is a fixed positive integer, $\sum p_i = \sum q_j = 1$. Thus, each variable takes on only values in the set $\{1, 2, \ldots, n\}$. Prove that there do not exist $p_1, \ldots, p_n; q_1, \ldots, q_n$ such that $P[X_1 + X_2 = k] = 1/(2n - 1)$ for $k = 2, \ldots, 2n$. [*Note*: This is easily shown for $n = 2$. With the further restriction that $p_i = q_i$ $(1 \leqslant i \leqslant n)$, the problem is well-known (Problem 3.18 covers the case $n = 6$) and we easily show that it is not possible to weight a pair of n-sided dice so that the probability of occurrence of every sum from 2 to $2n$ will be the same. We seek here to remove the requirement of identical weightings.]

3.7H Prove or find a counterexample: If $F(x_1, \ldots, x_n)$ is the distribution function of an n-dimensional random variable, then its points of discontinuity are contained in an (at most countable) union of subsets of dimension $n - 1$. (The result for $n = 1$ is well-known—see Theorem 3.1.4.)

3.8H Let X and Y be two discrete r.v.'s. Given $P[X = z]$, $P[Y = z]$, and $P[\max(X, Y) = z]$, find $P[\min(X, Y) = z]$. [*Reference*: N. D. Kazarinoff, L. Moser, and A. Wilansky, "Problem B–1 and Solution," *American Mathematical Monthly*, Vol. 76 (1969), pp. 912, 914.] Generalize this in a meaningful way to n discrete r.v.'s $X_1, \ldots, X_n$.

3.9H Investigate the discrete failure rate (see Problem 3.30) of the Poisson distribution.

3.10H The following table (extracted from p. 150 of *Vital Statistics of the United States 1950, Volume I* by the National Office of Vital Statistics, Washington, D.C., 1954) gives the proportion of persons (in each of four groups) who would die in a specified age interval *if* these persons were to die in the same proportions as observed for their age and group in 1950 and were alive at the beginning of that interval. Thus, the tabulation essentially gives $P[X = x|X \geqslant x]$, the discrete failure rate (see Problem 3.30), for 19 values of x (age of death intervals).

White Male		*White Female*		*Nonwhite Male*		*Nonwhite Female*	
0–1	0.0302	0–1	0.0231	0–1	0.0492	0–1	0.0402
1–5	.0054	1–5	.0044	1–5	.0110	1–5	.0094
5–10	.0033	5–10	.0024	5–10	.0048	5–10	.0039
10–15	.0034	10–15	.0021	10–15	.0048	10–15	.0036
15–20	.0065	15–20	.0031	15–20	.0108	15–20	.0086
20–25	.0086	20–25	.0040	20–25	.0179	20–25	.0126
25–30	.0085	25–30	.0049	25–30	.0212	25–30	.0163
30–35	.0100	30–35	.0064	30–35	.0276	30–35	.0222
35–40	.0145	35–40	.0093	35–40	.0350	35–40	.0307
40–45	.0234	40–45	.0143	40–45	.0501	40–45	.0435
45–50	.0377	45–50	.0217	45–50	.0706	45–50	.0619
50–55	.0590	50–55	.0322	50–55	.1059	50–55	.0881
55–60	.0896	55–60	.0493	55–60	.1495	55–60	.1211
60–65	.1306	60–65	.0769	60–65	.1946	60–65	.1643
65–70	.1835	65–70	.1172	65–70	.2075	65–70	.1536
70–75	.2593	70–75	.1887	70–75	.2669	70–75	.2204
75–80	.3587	75–80	.2896	75–80	.3566	75–80	.2926
80–85	.4730	80–85	.4084	80–85	.3494	80–85	.2929
85 and over	1.0000	85 and over	1.0000	85 and over	1.0000	85 and over	1.0000

(a) Graph the discrete failure rate for all four groups. (Put all four curves on the same graph, and connect the 19 points by a smooth curve for each. Use semilog graph paper so the whole curve can be fitted onto one graph.) Compare white and nonwhite males; white and nonwhite females; white males and females; and nonwhite males and females, each with regard to discrete failure rate (as estimated). What observations can you make about comparative failure rates which are true over a broad range of ages? What observations are true for only a narrow range of ages?

(b) For the similar recent (1966) data given on page 75 (extracted from p. 5–4 of *Vital Statistics of the United States 1966, Volume II—Mortality, Part A* by the Division of Vital Statistics of the National Center for Health Statistics, Washington, D.C., 1968) plot comparable failure rate curves. Compare these with each other, and with the 1950 curves. What progress has occurred? How are the comparisons changed (if at all)?

(c) Investigate the degree of confidence we can have in our above comparisons (essentially the precision with which $P[X = x|\, X \geqslant x]$ has been estimated from the data for all x simultaneously).

White Male		*White Female*		*Nonwhite Male*		*Nonwhite Female*	
0–1	0.0234	0–1	0.0176	0–1	0.0422	0–1	0.0349
1–5	.0036	1–5	.0029	1–5	.0066	1–5	.0057
5–10	.0024	5–10	.0017	5–10	.0034	5–10	.0025
10–15	.0024	10–15	.0014	10–15	.0035	10–15	.0021
15–20	.0070	15–20	.0028	15–20	.0090	15–20	.0040
20–25	.0091	20–25	.0031	20–25	.0157	20–25	.0067
25–30	.0080	25–30	.0036	25–30	.0199	25–30	.0096
30–35	.0090	30–35	.0051	30–35	.0257	30–35	.0151
35–40	.0129	35–40	.0073	35–40	.0343	35–40	.0219
40–45	.0206	40–45	.0118	40–45	.0485	40–45	.0309
45–50	.0339	45–50	.0183	45–50	.0669	45–50	.0440
50–55	.0560	50–55	.0277	50–55	.0913	50–55	.0583
55–60	.0880	55–60	.0408	55–60	.1307	55–60	.0845
60–65	.1287	60–65	.0600	60–65	.1671	60–65	.1220
65–70	.1877	65–70	.0982	65–70	.2698	65–70	.1995
70–75	.2639	70–75	.1569	70–75	.3081	70–75	.2082
75–80	.3524	75–80	.2452	75–80	.3286	75–80	.2501
80–85	.4763	80–85	.3853	80–85	.3576	80–85	.3006
85 and over	1.0000	85 and over	1.0000	85 and over	1.0000	85 and over	1.0000

3.11H In Section 3.3 we proved that the binomial distribution "tends to" the Poisson in the sense that, if $n \to \infty$ and np ($= \lambda$, say) is held fixed, then

$$\binom{n}{k} p^k(1 - p)^{n-k} \to e^{-\lambda}\lambda^k/k!. \tag{$*$}$$

Let $h(0), h(1), h(2), \ldots$ be $\geqslant 0$. Prove that

$$\lim_{n\to\infty} \sum_{k=0}^{\infty} h(k) \left| \binom{n}{k} p^k(1 - p)^{n-k} - e^{-\lambda} \frac{\lambda^k}{k!} \right| = 0 \tag{$**$}$$

iff

$$\sum_{k=0}^{\infty} h(k)e^{-\lambda} \frac{\lambda^k}{k!} < \infty. \tag{$***$}$$

[*Reference*: G. Simons and N. L. Johnson, "On the convergence of binomial to Poisson distributions," *Annals of Mathematical Statistics*, Vol. 42 (1971), pp. 1735–1736.]

4

RANDOM VARIABLES: CONTINUOUS AND MIXED CASES

In Chapter 3, we introduced the notions of random variables and n-dimensional random variables, and discussed (for a general n-dimensional random variable)

1. multivariate distribution functions,
2. marginal distributions,
3. conditional distributions, and
4. independence.

For a *discrete* n-dimensional random variable we discussed

1. probability functions,
2. marginal probability functions,
3. conditional probability functions,
4. independence, and
5. distributions of functions of r.v.'s.

In this chapter we discuss, for an (*absolutely*) *continuous* n-dimensional r.v.,

1. multivariate density functions,
2. marginal density functions,
3. conditional density functions,
4. independence, and
5. distributions of functions of r.v.'s.

A simple example which illustrates how continuous r.v.'s may arise is given in Section 4.1.

4.1. CLASSIFICATION OF DISTRIBUTION FUNCTIONS

In Section 3.1 we defined the distribution function of any one-dimensional r.v. X as $F_X(x) = P[X \leqslant x]$ for all $x \in (-\infty, +\infty)$. Such a r.v. X was called *discrete* (see Definitions 3.1.6 and 3.1.7 for details) essentially iff (for all x)

$$F_X(x) = \sum_{\substack{x' \leqslant x \\ \text{with } p_X(x')>0}} p_X(x').$$

Definition 4.1.1. *X is called a (one-dimensional)* **(absolutely) continuous random variable** *if its d.f. $F_X(x)$ may be represented as*

$$F_X(x) = \int_{-\infty}^{x} f_X(y)\,dy \qquad (-\infty < x < +\infty),$$

where $f_X(y)$ is a **probability density function (p.d.f.)**; *that is by definition,*

$$f_X(y) \geqslant 0 \quad \text{for all } y, \text{ and}$$
$$\int_{-\infty}^{\infty} f_X(y)\,dy = 1.$$

In this case $F_X(x)$ is called an **(absolutely) continuous d.f.**

Example 4.1.2. Suppose that we pick a number "at random" between 0 and 1, say X (a random variable). What is the d.f. of X, and is X absolutely continuous or discrete (or as is sometimes possible, neither)?

Since the number is picked at random between 0 and 1, it should have equal probability of being in the interval $(0, \frac{1}{2})$ and the interval $(\frac{1}{2}, 1)$, while its probability of being less than 0 is 0, and its probability of being greater than 1 is 0. Thus,

$$F_X(x) = 0, \quad \text{if } x \leqslant 0,$$
$$F_X(\tfrac{1}{2}) = \tfrac{1}{2},$$
$$F_X(x) = 1, \quad \text{if } x \geqslant 1.$$

However, since the point is chosen at random between 0 and 1, the probability that it lies in the interval $(0, x)$ should be the proportion that the length of the interval $(0, x)$ is of the interval $(0, 1)$; that is, $x/1 = x$ if $0 \leqslant x \leqslant 1$. Thus, $F_X(x) = x$ if $0 \leqslant x \leqslant 1$, and $F_X(x)$ is as given in Figure 4.1–1. Now let us define $f(x) = 1$ if $0 \leqslant x \leqslant 1$ and $f(x) = 0$ otherwise. Then $f(y) \geqslant 0$ for all y and $\int_{-\infty}^{\infty} f(y)\,dy = 1$, so by Definition 4.1.1 $f(y)$ is a p.d.f. But, for all x,

$$F_X(x) = \int_{-\infty}^{x} f(y)\,dy,$$

so X is an (absolutely) continuous r.v. with p.d.f. $f(x)$.

We now note two results on d.f.'s.

Theorem 4.1.3. *If $F(x)$ is an absolutely continuous d.f., then [at all values of*

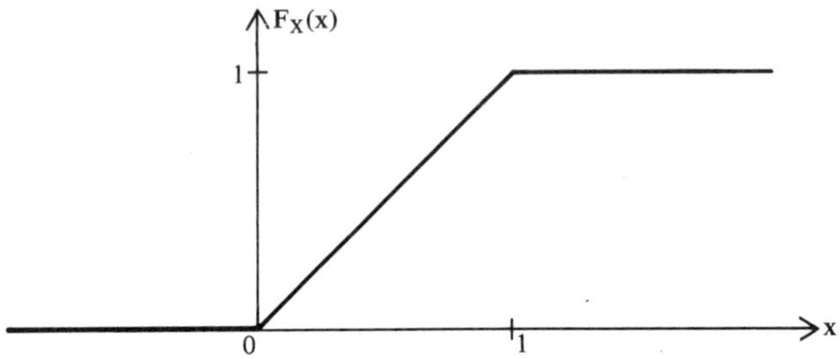

Figure 4.1–1. $F_X(x)$ for Example 4.1.2

x where f(x) is continuous]

$$F'(x) \equiv \frac{d}{dx} F(x) = f(x),$$

where f(x) is the p.d.f. of F(x).

Theorem 4.1.4. *Every d.f. F(x) may be represented as*

$$F(x) = c_1 F^d(x) + c_2 F^{ac}(x) + c_3 F^{sc}(x)$$

($c_1, c_2, c_3 \geqslant 0$, $c_1 + c_2 + c_3 = 1$) *where $F^d(x)$ is a discrete d.f., $F^{ac}(x)$ is an (absolutely) continuous d.f., and $F^{sc}(x)$ is a* **singularly continuous** *d.f.* [the latter being defined as a d.f. which is a continuous function of x but has $F^{sc'}(x) = 0$ at almost all points x].

With regard to singularly continuous d.f.'s, the statement "in most practical cases $c_3 = 0$" has been made so often that it is a cliché. [Actually, Feller (1966) on p. 139 suggests that this cliché obscures the true nature of things: singularly continuous d.f.'s are important, but their importance is hidden.] An example of a singularly continuous d.f. will be given in Section 4.2.

We will find the following special functions useful.

Definition 4.1.5. **(Beta Functions)** *For $m > 0$, $n > 0$ (not necessarily integers)*

$$\begin{aligned}\beta(m, n) &\equiv \int_0^1 x^{m-1}(1-x)^{n-1}\,dx \\ &= 2\int_0^{\pi/2} (\sin\theta)^{2m-1}(\cos\theta)^{2n-1}\,d\theta.\end{aligned}$$

[The transformation used was $x = \sin^2\theta$.]

Definition 4.1.6. **(Gamma Functions)** *For $n > 0$ (not necessarily an integer),*

$$\Gamma(n) \equiv \int_0^\infty x^{n-1}e^{-x}\,dx = 2\int_0^\infty y^{2n-1}e^{-y^2}\,dy.$$

[The transformation used was $x = y^2$.]

Theorem 4.1.7. $\Gamma(1) = 1$. *If n is a positive integer, then* $\Gamma(n) = (n-1)!$. $\Gamma(\frac{1}{2}) = \sqrt{\pi}$.

Proof:

$$\begin{aligned}
\Gamma(1) &= \int_0^\infty x^{1-1}e^{-x}\,dx = \int_0^\infty e^{-x}\,dx = -e^{-x}\Big]_0^\infty = 1.\\
\Gamma(n) &= \int_0^\infty x^{n-1}e^{-x}\,dx = -x^{n-1}e^{-x}\Big]_0^\infty + \int_0^\infty e^{-x}(n-1)x^{n-2}\,dx\\
&= (n-1)\Gamma(n-1).
\end{aligned}$$

Thus, we proved directly that $\Gamma(1) = 1$, and used integration by parts to show that for $n > 1$, $\Gamma(n) = (n-1)\Gamma(n-1)$. It follows by mathematical induction that $\Gamma(n) = (n-1)!$ when n is a positive integer. For a proof of $\Gamma(\frac{1}{2}) = \sqrt{\pi}$, see, for example, Wadsworth and Bryan (1960), pp. 3–4; this proof uses the material on Jacobians of transformations which is discussed in our Theorem 4.6.5 below.

Theorem 4.1.8.[1] $\beta(m, n) = \dfrac{\Gamma(m)\Gamma(n)}{\Gamma(m+n)}$.

Corollary 4.1.9. $\beta(m, n) = \beta(n, m)$.

Example 4.1.10. Let X be an (absolutely) continuous random variable with density function given by

$$f_X(t) = \begin{cases} A(t^2+1), & \text{if } 0 \leqslant t \leqslant 1\\ 0, & \text{otherwise.}\end{cases}$$

Determine the value of the constant A and sketch the graph of $f_X(t)$. Determine the distribution function $F(t)$ corresponding to the density function $f_X(t)$ and sketch its graph (see Figure 4.1–2).

$$1 = \int_{-\infty}^\infty f_X(t)\,dt = \int_0^1 A(t^2+1)\,dt = A\left[\frac{t^3}{3} + t\right]_0^1 = \tfrac{4}{3}A \Rightarrow A = \tfrac{3}{4}.$$

$$F(t) = \int_{-\infty}^t f_X(t)\,dt = \begin{cases} 1, & \text{if } t \geqslant 1\\ \int_0^t A(t^2+1)\,dt = \dfrac{3}{4}\left(\dfrac{t^3}{3} + t\right), & \text{if } 0 < t < 1\\ 0, & \text{if } t \leqslant 0.\end{cases}$$

[1] The proof uses a transformation to polar coordinates and is omitted. [See, for example, Wadsworth and Bryan (1960), p. 5, for the details.]

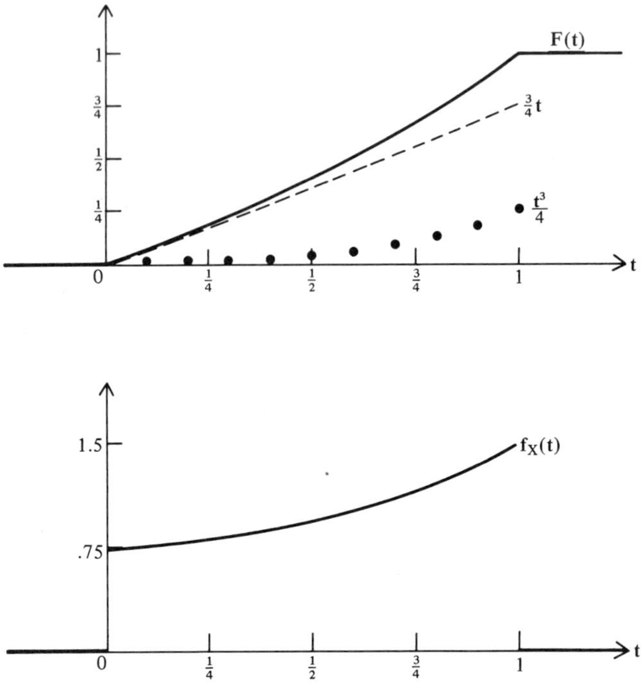

Figure 4.1–2

4.2. SPECIAL CONTINUOUS DISTRIBUTIONS

In this section we will define several[2] very important continuous distributions. Few examples will be given at this point, since these are more natural in Chapter 6 and the chapters on statistics (where continuous distributions are indispensable in the solution of practical problems).

Definition 4.2.1. *A r.v. X has the* **beta** *distribution* $B(x|\alpha, \lambda)$ *if (for some* α, $\lambda > -1$)

$$f_X(x) = \begin{cases} \dfrac{1}{\beta(\alpha + 1, \lambda + 1)} x^{\alpha}(1 - x)^{\lambda}, & 0 \leqslant x \leqslant 1 \\ 0, & \text{otherwise.} \end{cases}$$

It is easy to show that $f_X(x) \geqslant 0$ and (using Definition 4.1.5) $\int_{-\infty}^{\infty} f_X(x)\,dx = 1$ so that $f_X(x)$ is a p.d.f. by Definition 4.1.1. The fact that if $f(x)$ is a p.d.f., then

[2] At this point the student should read the note which follows Definition 4.2.17. [It should also be reread after that definition.]

$F(x) = \int_{-\infty}^{x} f(y)\,dy$ is a d.f. follows from Theorem 3.1.5. Of course (by Definition 4.1.1), $F(x)$ is then an absolutely continuous d.f. [See Table 8 for graphs and a table of x such that $F_X(x) = .05$.]

Definition 4.2.2. *A r.v. X has the* **uniform** *distribution on* (0, 1) *if*

$$f_X(x) = \begin{cases} 1, & 0 \leqslant x \leqslant 1 \\ 0, & \text{otherwise.} \end{cases}$$

[Here $F_X(x)$ is easy to calculate, so a table is not needed.]

Example 4.2.3. The beta distribution $B(x|0, 0)$ is the uniform distribution on (0, 1).

Definition 4.2.4. *A r.v. X has the* **uniform** *distribution on* (b, c) *(for* $b < c$ *fixed) if*

$$f_X(x) = \begin{cases} \dfrac{1}{c - b}, & b \leqslant x \leqslant c \\ 0, & \text{otherwise.} \end{cases}$$

[Here $F_X(x)$ is easy to calculate, so a table is not needed.]

Example 4.2.5. If Y is uniform on (0, 1), then (for $b < c$ *fixed*) $X = (c - b)Y + b$ is uniform on (b, c).

Definition 4.2.6. *A r.v. X has the* **gamma** *distribution* $G(x|\alpha, \beta, A)$ *if (for some* $\alpha > -1, \beta > 0, -\infty < A < +\infty$*)*

$$f_X(x) = \begin{cases} \dfrac{1}{\beta^{\alpha+1}\Gamma(\alpha + 1)}(x - A)^{\alpha} e^{-(x-A)/\beta}, & A \leqslant x < \infty \\ 0, & \text{otherwise.} \end{cases}$$

[See Table 9 for graphs and a table of $F_X(x)$.]

Definition 4.2.7. *A r.v. X has the* **exponential** *distribution if (for some* $\beta > 0$, $-\infty < A < +\infty$*)*

$$f_X(x) = \begin{cases} \dfrac{1}{\beta} e^{-(x-A)/\beta}, & A \leqslant x < \infty \\ 0, & \text{otherwise.} \end{cases}$$

[Here $F_X(x)$ is easy to calculate, so a table is not needed.]

Example 4.2.8. The gamma distribution $G(x|0, \beta, A)$ is the exponential distribution.

Definition 4.2.9. *A r.v. X has the* **chi-square** *distribution with n degrees of freedom*, say $\chi_n^2(0)$, if (for some integer $n > 0$)

$$f_X(x) = \begin{cases} \dfrac{1}{2^{n/2}\Gamma(n/2)} x^{(n/2)-1} e^{-x/2}, & 0 \leqslant x < \infty \\ 0, & \text{otherwise.} \end{cases}$$

[See Table 10 for graphs and a table of $F_X(x)$.]

Example 4.2.10. The gamma distribution $G(x|(n/2) - 1, 2, 0)$ (n an integer, $n > 0$) is the $\chi_n^2(0)$ distribution.

Definition 4.2.11. *A r.v. X has the* **noncentral chi-square** *distribution with n degrees of freedom and noncentrality λ, say $\chi_n^2(\lambda)$, if (for some integer $n > 0$, $\lambda \geqslant 0$)*

$$f_X(x) = \begin{cases} e^{-\lambda} \displaystyle\sum_{i=0}^{\infty} \frac{\lambda^i x^{(1/2)(n+2i)-1} e^{-x/2}}{i!\,2^{(1/2)(n+2i)}\Gamma\left(\dfrac{n+2i}{2}\right)}, & 0 \leqslant x < \infty \\ 0, & \text{otherwise.} \end{cases}$$

Example 4.2.12. The noncentral chi-square distribution with n degrees of freedom and noncentrality $\lambda = 0$ *is* the chi-square distribution with n degrees of freedom (as our notation in Definitions 4.2.9 and 4.2.11 suggests).

Definition 4.2.13. *A r.v. X has the* **normal** *distribution $N(\mu, \sigma^2)$ if (for some*[3] *$\sigma^2 > 0$ and $-\infty < \mu < +\infty$)*

$$f_X(x) = \frac{1}{\sqrt{2\pi}\,\sigma} e^{-\frac{1}{2}\left(\frac{x-\mu}{\sigma}\right)^2}, \qquad -\infty < x < +\infty.$$

The $N(0, 1)$ distribution is called the **standard normal** *distribution.* [See Table 11 for graphs and a table of $F_X(x)$ which assumes $\mu = 0$ and $\sigma^2 = 1$.]

Definition 4.2.14. *A r.v. X has the* **F-distribution** *$F(n_1, n_2)$ if (for some positive integers n_1, n_2)*

$$f_X(x) = \begin{cases} \dfrac{\Gamma\left(\dfrac{n_1+n_2}{2}\right)\left(\dfrac{n_1}{n_2}\right)^{n_1/2}}{\Gamma\left(\dfrac{n_1}{2}\right)\Gamma\left(\dfrac{n_2}{2}\right)} \dfrac{x^{(1/2)(n_1-2)}}{\left(1 + \dfrac{n_1}{n_2}x\right)^{(1/2)(n_1+n_2)}}, & 0 < x < \infty \\ 0, & \text{otherwise.} \end{cases}$$

[See Table 12 for graphs and a table of $F_X(x)$.]

[3] σ is used to denote the positive square root of σ^2.

Definition 4.2.15. *A r.v. X has the* **noncentral F-distribution** $F(n_1, n_2, \lambda)$ *with noncentrality* λ *if (for some positive integers* n_1 *and* n_2, $\lambda \geqslant 0$)

$$f_X(x) = \begin{cases} e^{-\lambda} \sum_{i=0}^{\infty} \dfrac{\lambda^i x^{(1/2)(n_1+2i)-1}}{i!\left(1+\dfrac{n_1}{n_2}x\right)^{(1/2)(2i+n_1+n_2)}} \dfrac{\Gamma\left(\dfrac{2i+n_1+n_2}{2}\right)\left(\dfrac{n_1}{n_2}\right)^{(1/2)(2i+n_1)}}{\Gamma\left(\dfrac{n_2}{2}\right)\Gamma\left(\dfrac{2i+n_1}{2}\right)}, & x \geqslant 0 \\ 0, & \text{otherwise.} \end{cases}$$

Example 4.2.16. $F(n_1, n_2, 0)$ is $F(n_1, n_2)$. Note that, in each case, n_1 and n_2 are called **degrees of freedom**. One should compare $F(n_1, n_2, \lambda)$ with $\chi_n^2(\lambda)$.

Definition 4.2.17. *A r.v. X has* **Student's t-distribution** *with n degrees of freedom if (for some integer* $n > 0$)

$$f_X(x) = \frac{\Gamma\left(\dfrac{n+1}{2}\right)}{\sqrt{n\pi}\,\Gamma\left(\dfrac{n}{2}\right)} \frac{1}{\left(1+\dfrac{x^2}{n}\right)^{(n+1)/2}}, \qquad -\infty < x < +\infty.$$

[See Table 13 for graphs and a table of $F_X(x)$.]

Note that at this point the student is *not* expected to memorize these densities and is *not* expected to realize how or why they are reasonable or useful. (In many cases it is difficult even to show that they are p.d.f.'s!) This is simply the logical place to list them for use later in Chapter 6 and in the chapters on statistics. They should give the student some idea of what an actual p.d.f. may look like.

Also note that various authors will define the parameters of a distribution differently; for example, α will be called $\alpha + 1$, and so on. (Also, the "starting point" A is often taken as 0 without a statement to that effect.) You have now been warned!

Example 4.2.18.[4] **(Cantor-type Distributions)** A new casino has an automatic coin flipper containing a fair coin. If we deposit \$.75, the machine operates as follows. The machine flips the coin and if heads turns up it pays \$3/4; it then flips again and if heads turns up it pays \$3/16; it then flips again and if heads turns up it pays \$3/64; and so on. (Successive flips are assumed to be independent.) That is, on flip i ($i \geqslant 1$) it pays $\$3/4^i$ for heads and \$0 otherwise, so that our total gross (that is, ignoring the \$.75 fee) winnings are

$$W = \sum_{i=1}^{\infty} \frac{3}{4^i} X_i, \tag{4.2.19}$$

[4] This example is not essential to an understanding of other material in this book.

where $X_1, X_2, \ldots$ are independent r.v.'s and $P[X_i = 0] = P[X_i = 1] = .5$ $(i = 1, 2, \ldots)$. *We now ask*: What is the d.f. of W? This will turn out to be *singularly continuous.*

Let us consider $F(x) = P[W \leqslant x]$. First let us note that

$$0 \leqslant W \leqslant \frac{3}{4} + \frac{3}{4^2} + \frac{3}{4^3} + \cdots = 1 \tag{4.2.20}$$

[where we used the formula $1 + x + x^2 + \cdots = (1-x)^{-1}$ for $0 < x < 1$], so that $F(x) = P[W \leqslant x] = 0$ for $x < 0$ and $F(x) = P[W \leqslant x] = 1$ for $x \geqslant 1$. Also $P[W \leqslant 0] = P[W < 0] + P[W = 0] = 0 + P[W = 0] = 0$ since (for every integer $m > 0$)

$$\begin{aligned} P[W = 0] &= P[X_1 = 0, X_2 = 0, X_3 = 0, \ldots] \\ &\leqslant P[X_1 = 0, X_2 = 0, \ldots, X_m = 0] \\ &= P[X_1 = 0]P[X_2 = 0] \cdots P[X_m = 0] = 1/2^m. \end{aligned} \tag{4.2.21}$$

Now

$$\begin{aligned} X_1 &= 1 \Rightarrow W \geqslant \tfrac{3}{4}, \\ X_1 &= 0 \Rightarrow W \leqslant \frac{3}{4^2} + \frac{3}{4^3} + \cdots = \frac{1}{4}, \end{aligned} \tag{4.2.22}$$

so that $P[\frac{1}{4} \leqslant W \leqslant \frac{3}{4}] = 0$ and, for $\frac{1}{4} \leqslant x \leqslant \frac{3}{4}$, $F(x) = P[W \leqslant \frac{1}{4}] = P[X_1 = 0] = \frac{1}{2}$. [Any items like $P[W = y]$ which arise, for example $P[W = \frac{1}{4}]$, are handled as at (4.2.21).] Thus, $F(x)$ has no jumps greater than $\frac{1}{2}$. (See Figure 4.2–1 for our information about $F(x)$ to this point.)

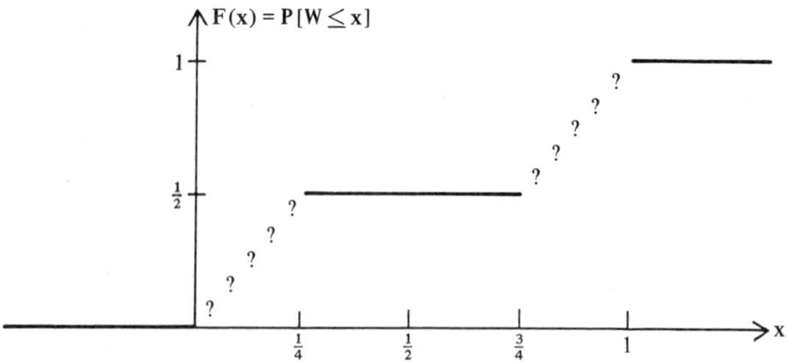

Figure 4.2–1

Now, the winnings on tosses 2, 3, ... are

$$3 \sum_{i=2}^{\infty} \frac{1}{4^i} X_i = \tfrac{1}{4} \cdot 3 \sum_{i=2}^{\infty} \frac{1}{4^{i-1}} X_i = \tfrac{1}{4} \cdot \left(3 \sum_{i=1}^{\infty} \frac{1}{4^i} X_{i+1} \right)$$

which has the same distribution as $\frac{1}{4}W$. Hence, $F(x)$ in $[0, \frac{1}{4}]$ can be related to $F(x)$ in $[0, 1]$ as follows: For $0 < x < \frac{1}{4}$,

$$\begin{aligned} F(x) = P[W \leqslant x] &= P[W \leqslant x \text{ and } X_1 = 0] + P[W \leqslant x \text{ and } X_1 = 1] \\ &= P\left[3\sum_{i=2}^{\infty}\frac{1}{4^i}X_i \leqslant x \text{ and } X_1 = 0\right] + 0 \\ &= P\left[3\sum_{i=2}^{\infty}\frac{1}{4^i}X_i \leqslant x\right]P[X_1 = 0] \\ &= \tfrac{1}{2}P\left[\frac{1}{4}\cdot\left(3\sum_{i=1}^{\infty}\frac{1}{4^i}X_{i+1}\right) \leqslant x\right] \\ &= \tfrac{1}{2}P\left[\left(3\sum_{i=1}^{\infty}\frac{1}{4^i}X_{i+1}\right) \leqslant 4x\right] \\ &= \tfrac{1}{2}F(4x). \end{aligned} \tag{4.2.23}$$

Thus, $F(x) = \frac{1}{4}$ for $\frac{1}{16} \leqslant x \leqslant \frac{3}{16}$. Similarly, $F(x)$ in $[\frac{3}{4}, 1]$ can be related to $F(x)$ in $[0, 1]$ as follows: For $\frac{3}{4} < x < 1$,

$$\begin{aligned} F(x) &= P[W \leqslant x] = P[W \leqslant x \text{ and } X_1 = 0] + P[W \leqslant x \text{ and } X_1 = 1] \\ &= \tfrac{1}{2} + P\left[\frac{3}{4} + 3\sum_{i=2}^{\infty}\frac{1}{4^i}X_i \leqslant x \text{ and } X_1 = 1\right] \\ &= \tfrac{1}{2} + P\left[3\sum_{i=2}^{\infty}\frac{1}{4^i}X_i \leqslant x - \frac{3}{4}\right]P[X_1 = 1] \\ &= \tfrac{1}{2} + \tfrac{1}{2}P\left[\frac{1}{4}\cdot\left(3\sum_{i=1}^{\infty}\frac{1}{4^i}X_{i+1}\right) \leqslant x - \frac{3}{4}\right] \\ &= \tfrac{1}{2} + \tfrac{1}{2}P\left[\left(3\sum_{i=1}^{\infty}\frac{1}{4^i}X_{i+1}\right) \leqslant 4(x - \tfrac{3}{4})\right] \\ &= \tfrac{1}{2} + \tfrac{1}{2}F(4(x - \tfrac{3}{4})). \end{aligned} \tag{4.2.24}$$

Thus, $F(x) = \frac{3}{4}$ for $\frac{13}{16} \leqslant x \leqslant \frac{15}{16}$. [See Figure 4.2–2 for our information about $F(x)$ to this point.]

We have now shown that in *each* of three intervals $F(x)$ is constant (at values $\frac{1}{4}$, $\frac{1}{2}$, $\frac{3}{4}$); these intervals have a total length $\frac{1}{8} + \frac{1}{2} + \frac{1}{8} = \frac{1}{2} + \frac{2}{8} = \frac{3}{4}$.

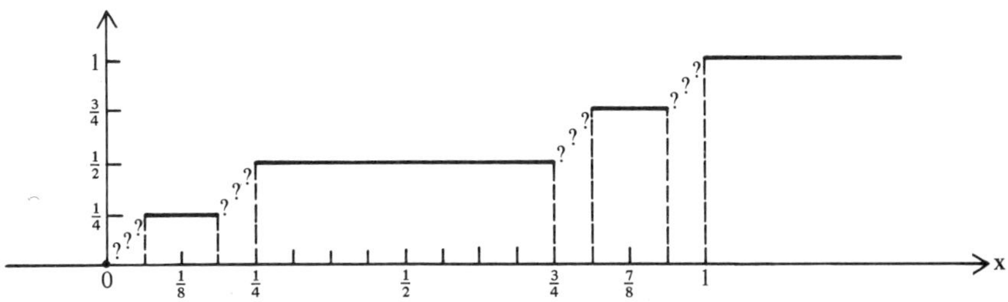

Figure 4.2–2

It follows that, if $F(x)$ has a jump, that jump cannot be greater than $\frac{1}{4}$ (see Figure 4.2–2). In the remaining 4 intervals (each of which has length $\frac{1}{16}$) we can use the approach used at (4.2.23) to get the graph related to that of $F(x)$; hence the respective 4 intervals will contain subintervals of length $\frac{1}{32}$ in which $F(x)$ is constant at values $\frac{1}{8}$, $\frac{3}{8}$, $\frac{5}{8}$, and $\frac{7}{8}$.

Using mathematical induction, after m steps we will find $1 + 2 + 2^2 + 2^3 + \cdots + 2^{m-1}$ intervals with total length $2^{-1} + 2^{-2} + 2^{-3} + 2^{-4} + \cdots + 2^{-m} = 1 - 1/2^m$ in *each* of which $F(x)$ is constant. *It follows that*: $F(x)$ is a continuous nondecreasing function; $F(0) = 0$; $F(1) = 1$; and the subintervals of $[0, 1]$ where $F(x)$ is a constant have total length 1. Since the derivative of a constant is zero, $F'(x) = 0$ at almost all x, so $F(x)$ is *singularly continuous.*

[Note that $F(x)$ is constant at most of the interval $[0, 1]$, essentially because most values x with $0 \leqslant x \leqslant 1$ are not possible values of W. The possible values of W (for example, $0, \frac{1}{4}, \frac{3}{4}, 1$) all have probability 0 as shown at (4.2.21); we have shown $F'(x) = 0$ except at such values.]

Example 4.2.25. (A Life Distribution) Suppose that a certain company manufactures light bulbs which, while their life X is a random number of hours, have an exponential distribution for life with $\beta = 10$, $A = 0$. For what number of hours t is it the case that the probability is .9 that the bulb will burn out before t?

We wish to find t such that $P[X < t] = .9$. Since X has the exponential distribution with $\beta = 10$ and $A = 0$ (see Definition 4.2.7), we set

$$\int_0^t \tfrac{1}{10} e^{-x/10}\, dx = .9,$$

$$-e^{-x/10}\Big]_0^t = 1 - e^{-t/10} = .9.$$

Hence, $e^{-t/10} = .1$ and $t = -10 \ln(.1) \approx 23$.

Example 4.2.26. Let W be a r.v. uniformly distributed over the interval $[1, 3]$. Determine the density function $f_W(x)$ of W and sketch its graph. If we define a random variable G by $G = 2W - 1$, determine the density function $f_G(x)$ of G and sketch its graph.

The height of $f_W(x)$ is constant on $[1, 3]$ and the integral of $f_W(x)$ from 1 to 3 is 1, so

$$f_W(x) = \begin{cases} \frac{1}{2}, & \text{if } 1 \leqslant x \leqslant 3 \\ 0, & \text{otherwise.} \end{cases}$$

(This could also be obtained directly from Definition 4.2.4.) Hence, the graph is as shown in Figure 4.2–3. The d.f. of W, say $F(t)$, is

$$F(t) = P[W \leqslant t] = \begin{cases} 1, & \text{if } t > 3 \\ \dfrac{t-1}{2}, & \text{if } 1 \leqslant t \leqslant 3 \\ 0, & \text{if } t < 1. \end{cases}$$

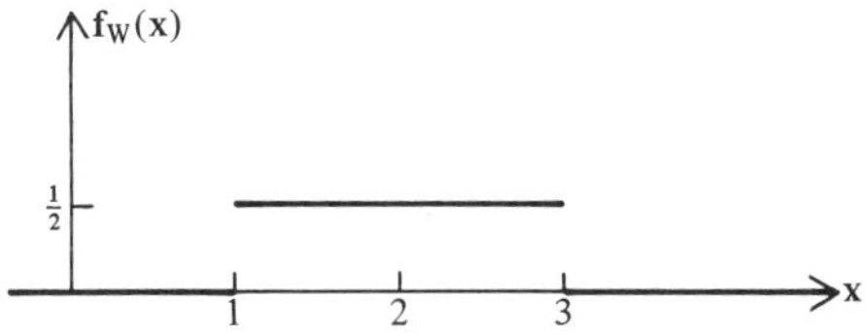

Figure 4.2–3

We will now seek to directly find $F_G(x)$, the d.f. of G, and differentiate it to obtain $f_G(x)$ (see Theorem 4.1.3). Since $G = 2W - 1$,

$$F_G(x) = P[G \leqslant x] = P[2W - 1 \leqslant x]$$

$$= P\left[W \leqslant \frac{x+1}{2}\right] = \begin{cases} 1 & \text{if } \dfrac{x+1}{2} > 3 \\ \dfrac{\dfrac{x+1}{2} - 1}{2}, & \text{if } 1 \leqslant \dfrac{x+1}{2} \leqslant 3 \\ 0, & \text{if } \dfrac{x+1}{2} < 1 \end{cases}$$

$$= \begin{cases} 1 & \text{if } x > 5 \\ \dfrac{x-1}{4} & \text{if } 1 \leqslant x \leqslant 5 \\ 0, & \text{if } x < 1. \end{cases}$$

Differentiating, we find

$$f_G(x) = \begin{cases} \frac{1}{4}, & \text{if } 1 \leqslant x \leqslant 5 \\ 0, & \text{otherwise,} \end{cases}$$

which shows that G is uniform on (1, 5). (This is a special case of Example 4.2.5; the proof in that case, which was omitted, is similar to the one we have just given.) The graph is as shown in Figure 4.2–4.

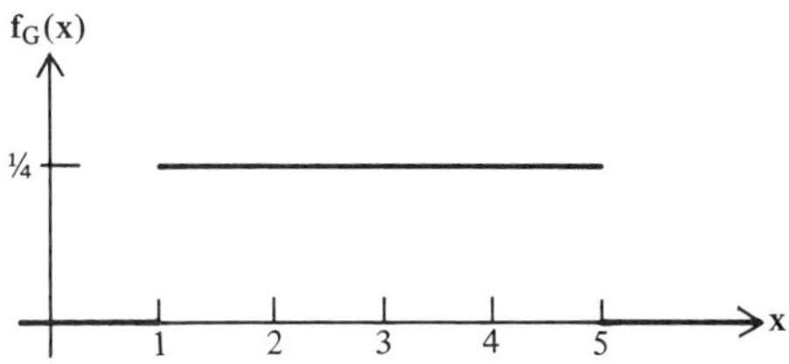

Figure 4.2–4

Example 4.2.27. If a d.f. $F(x)$ has a p.d.f. $f(x)$, the ratio

$$q(x) = \frac{f(x)}{1 - F(x)} \tag{4.2.28}$$

is defined for $F(x) < 1$ and is called the **failure rate** (or **hazard rate**, or **force of mortality**). If $F(x)$ is the d.f. of a r.v. Y which represents the life of some item, $q(x)\,dx$ can be interpreted as the conditional probability of death in $(x, x + dx)$ given that death has not occurred before x. We say $F(x)$ has an *increasing (decreasing) failure rate*—denoted **IFR (DFR)**—if $q(x)$ is increasing (decreasing). IFR intuitively represents "wear-out," while DFR represents "work-hardening."

For example, if an item has a random life length Y with d.f. $F(x)$, then the probability $Y \leqslant 80$ given $Y > 70$ is

$$\begin{aligned} P[Y \leqslant 80 | Y > 70] &= \frac{P[70 < Y \leqslant 80]}{P[Y > 70]} \\ &= \frac{\int_{70}^{80} f(x)\,dx}{1 - F(70)} = \frac{F(80) - F(70)}{1 - F(70)}. \end{aligned}$$

Find and plot the failure rate if Y has the exponential distribution (see Definition 4.2.7) with $A = 0$, $\beta = 1/\alpha$, namely,

$$f(t) = \begin{cases} \alpha e^{-\alpha t}, & t \geqslant 0 \\ 0, & t < 0. \end{cases} \tag{4.2.29}$$

Also find and plot the failure rate if Y has the so-called **Weibull distribution** with $\alpha > 0$ and $\beta > 0$

$$f(t) = \begin{cases} \alpha\beta t^{\beta-1} e^{-\alpha t^{\beta}}, & t \geqslant 0 \\ 0, & t < 0. \end{cases} \tag{4.2.30}$$

First, for the exponential distribution of (4.2.29) we find (for $x \geqslant 0$)

$$\begin{aligned} F(x) &= \int_0^x \alpha e^{-\alpha t}\,dt = 1 - e^{-\alpha x} \Rightarrow 1 - F(x) = e^{-\alpha x}, \\ q(x) &= \frac{f(x)}{1 - F(x)} = \frac{\alpha e^{-\alpha x}}{e^{-\alpha x}} = \alpha. \end{aligned}$$

In light of Problem 4.16 (the "lack of memory property") this is not surprising. (See Figure 4.2–5.)

Second, for the Weibull distribution of (4.2.30) we find (for $x \geqslant 0$)

$$\begin{aligned} F(x) &= \int_0^x \alpha\beta t^{\beta-1} e^{-\alpha t^{\beta}}\,dt = 1 - e^{-\alpha x^{\beta}} \Rightarrow 1 - F(x) = e^{-\alpha x^{\beta}}, \\ q(x) &= \frac{f(x)}{1 - F(x)} = \frac{\alpha\beta x^{\beta-1} e^{-\alpha x^{\beta}}}{e^{-\alpha x^{\beta}}} = \alpha\beta x^{\beta-1}. \end{aligned}$$

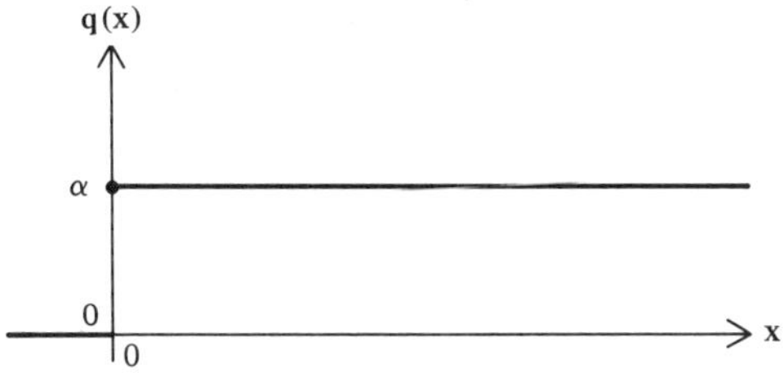

Figure 4.2–5. Failure rate of an exponential life

Here we find constant failure rate if $\beta = 1$ [in which case $F(x)$ is just the exponential distribution previously discussed], while $\alpha\beta x^{\beta-1}$ is an increasing function of x for $\beta > 1$ and a decreasing function of x for $\beta < 1$. (See Figure 4.2–6.)

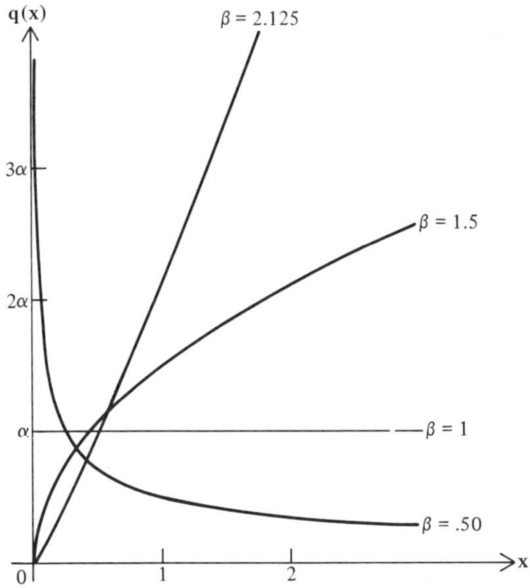

Figure 4.2–6. Failure rate of a Weibull life

The Weibull distribution is important in applications for several reasons. First, by simply making an appropriate choice of β it can be used to model both increasing and decreasing failure rates. Second, it appropriately models cases where the failure rate $q(x)$ is proportional to powers of x. Third, if Y is exponential, then (as we shall see in Example 4.6.3) $Y^{1/\beta}$ is Weibull.

4.3. RELATIONS AMONG CERTAIN CONTINUOUS DISTRIBUTIONS

It is most natural to postpone relations among the distributions of Section 4.2 until later, since they are derived using results of Section 4.6 on distributions of functions of continuous random variables and results on characteristic functions (Chapter 5). Certain of these results go by special names, for example, the Central Limit Theorem (Chapter 6). Some relations of the sort "the A distribution is a special case of the B distribution" were given in Section 4.2.

4.4. MULTIVARIATE CONTINUOUS DISTRIBUTIONS; INDEPENDENCE

In Section 3.4 we defined the distribution function of any n-dimensional r.v. $\mathbf{X} = (X_1, \ldots, X_n)$ as $F_{\mathbf{X}}(x_1, \ldots, x_n) = P[X_1 \leqslant x_1, \ldots, X_n \leqslant x_n]$ for $-\infty < x_i < +\infty (i = 1, \ldots, n)$. Such a r.v. $\mathbf{X}$ was called *discrete* (see Definitions 3.4.10 and 3.4.11 for details) essentially iff (for all $x_1, \ldots, x_n$)

$$F_{\mathbf{X}}(x_1, \ldots, x_n) = \sum_{\substack{x_1' \leqslant x_1, \ldots, x_n' \leqslant x_n \\ \text{with } p_x(x_1', \ldots, x_n') > 0}} p_{\mathbf{X}}(x_1', \ldots, x_n').$$

Definition 4.4.1. $\mathbf{X}$ *is called an* (*n-dimensional*) **(absolutely) continuous random variable** *if its d.f.* $F_{\mathbf{X}}(x_1, \ldots, x_n)$ *may be represented as*

$$F_{\mathbf{X}}(x_1, \ldots, x_n) = \int_{-\infty}^{x_n} \int_{-\infty}^{x_{n-1}} \cdots \int_{-\infty}^{x_1} f_{\mathbf{X}}(y_1, \ldots, y_n)\, dy_1 \cdots dy_{n-1}\, dy_n,$$

where $f_{\mathbf{X}}(x_1, \ldots, x_n)$ *is a* **probability density function (p.d.f.)**; *that is, by definition*

$$\begin{cases} f_{\mathbf{X}}(y_1, \ldots, y_n) \geqslant 0 & \text{for all } y_1, \ldots, y_n, \text{ and} \\ \int_{-\infty}^{\infty} \cdots \int_{-\infty}^{\infty} f_{\mathbf{X}}(y_1, \ldots, y_n)\, dy_1 \cdots dy_n = 1. \end{cases}$$

In this case $F_{\mathbf{X}}(x_1, \ldots, x_n)$ *is called an* **(absolutely) continuous d.f.**

Note that Theorems 4.1.3 and 4.1.4. generalize to n-dimensional random variables.

Definition 4.4.2. *If* $\mathbf{X} = (X_1, \ldots, X_n)$ *is an (absolutely) continuous n-dimensional r.v., the d.f. of any subset of* $X_1, \ldots, X_n$ *is (absolutely) continuous and its p.d.f. is called a* **marginal density function**.

Example 4.4.3. If $\mathbf{X} = (X_1, \ldots, X_n)$ is an (absolutely) continuous n-dimensional r.v., the d.f. of $(X_1, \ldots, X_{n-1})$ is (absolutely) continuous. We may calculate (by Lemma 3.4.3)

$$\begin{aligned}F_{X_1,\ldots,X_{n-1}}(x_1,\ldots,x_{n-1}) &= F_{X_1,\ldots,X_{n-1},X_n}(x_1,\ldots,x_{n-1},+\infty)\\ &= \int_{-\infty}^{x_1}\cdots\int_{-\infty}^{x_{n-1}}\int_{-\infty}^{\infty} f_{X_1,\ldots,X_n}(y_1,\ldots,y_n)\,dy_1\cdots dy_{n-1}\,dy_n\\ &= \int_{-\infty}^{x_1}\cdots\int_{-\infty}^{x_{n-1}}\left[\int_{-\infty}^{\infty} f_{X_1,\ldots,X_n}(y_1,\ldots,y_n)\,dy_n\right]dy_1\cdots dy_{n-1}.\end{aligned}$$

Hence,

$$f_{X_1,\ldots,X_{n-1}}(x_1,\ldots,x_{n-1}) = \int_{-\infty}^{\infty} f_{X_1,\ldots,X_n}(x_1,\ldots,x_{n-1},y)\,dy,$$

the marginal p.d.f. of $(X_1,\ldots,X_{n-1})$.

If $\mathbf{X} = (X_1, X_2)$ is an (absolutely) continuous r.v., then (under certain assumptions; see Definition 3.4.6 ff) the conditional d.f. of X_2 given $X_1 = x_1'$ is

$$\lim_{x_1''\to x_1'} \frac{\int_{x_1''}^{x_1'}\int_{-\infty}^{x_2} f_{X_1,X_2}(x_1,x_2)\,dx_2\,dx_1}{\int_{x_1''}^{x_1'} f_{X_1}(x_1)\,dx_1} = \frac{\int_{-\infty}^{x_2} f_{X_1,X_2}(x_1',x_2)\,dx_2}{f_{X_1}(x_1')}. \tag{4.4.4}$$

[This requires that $f_{X_1}(x_1)$ and $\int_{-\infty}^{x_2} f_{X_1,X_2}(x_1,x_2)\,dx_2$ are continuous functions of x_1 at x_1' and that $f_{X_1}(x_1') > 0$.] This suggests that we define the **conditional density of X_2 given $X_1 = x_1'$** to be

$$f_{X_2|X_1=x_1'}(x_2|x_1') \equiv \frac{f_{X_1,X_2}(x_1',x_2)}{f_{X_1}(x_1')} \tag{4.4.5}$$

[or zero if $f_{X_1}(x_1') = 0$], since integrating on x_2 will then yield the conditional d.f. of X_2 given $X_1 = x_1'$. We will not formalize this definition, but will note that it extends to the case of n-dimensional r.v.'s.

Theorem 4.4.6. *The (absolutely) continuous r.v.'s $X_1, \ldots, X_n$ are independent iff (for all $x_1, \ldots, x_n$)*

$$f_{X_1,\ldots,X_n}(x_1,\ldots,x_n) = f_{X_1}(x_1)\cdots f_{X_n}(x_n).$$

The proof of this theorem uses differentiation to prove the "only if" and integration to prove the "if." (See Definition 3.4.7.)

4.5. SPECIAL MULTIVARIATE CONTINUOUS DISTRIBUTIONS

In this section, we will introduce the single most important multivariate continuous distribution, the multivariate normal distribution $N(\boldsymbol{\mu}, \boldsymbol{\Sigma})$. Its properties and uses will be clarified in the chapters on statistics.[5] We will

---

[5] This section from Definition 4.5.3 on can be omitted if desired. See Remark 8.4.16.

proceed as follows: Recall the definition of the univariate normal distribution, define the bivariate normal distribution, and define the multivariate normal distribution (of which the two previous are special cases).

Definition 4.5.1. *A r.v. X has the* **(univariate) normal** *distribution $N(\mu, \sigma^2)$ if (for some*[6] *$\sigma^2 > 0$ and $-\infty < \mu < +\infty$)*

$$f_X(x) = \frac{1}{\sqrt{2\pi}\,\sigma} e^{-\frac{1}{2}\left(\frac{x-\mu}{\sigma}\right)^2}, \qquad -\infty < x < +\infty.$$

Definition 4.5.2. *A r.v. $\mathbf{X} = (X_1, X_2)$ has the* **(bivariate) normal** *distribution $N(\mu_1, \mu_2; \sigma_1^2, \sigma_2^2, \rho)$ if (for some*[6] *$\sigma_1^2 > 0$, $\sigma_2^2 > 0$, $-1 < \rho < +1$, $-\infty < \mu_1 < +\infty$, $-\infty < \mu_2 < +\infty$)*

$$f_{X_1, X_2}(x_1, x_2) = \frac{1}{2\pi\,\sigma_1\,\sigma_2\sqrt{1-\rho^2}} e^{-\frac{1}{2(1-\rho^2)}\left\{\left(\frac{x_1-\mu_1}{\sigma_1}\right)^2 - 2\rho\left(\frac{x_1-\mu_1}{\sigma_1}\right)\left(\frac{x_2-\mu_2}{\sigma_2}\right) + \left(\frac{x_2-\mu_2}{\sigma_2}\right)^2\right\}}$$
$$-\infty < x_1, x_2 < +\infty.$$

Definition 4.5.3. *A r.v. $\mathbf{X} = (X_1, \ldots, X_n)$ has the* **(multivariate) normal** *distribution $N(\boldsymbol{\mu}, \boldsymbol{\Sigma})$ where*

$$\boldsymbol{\mu} \equiv \begin{bmatrix} \mu_1 \\ \cdot \\ \cdot \\ \cdot \\ \mu_n \end{bmatrix}$$

is a vector of n real numbers and

$$\boldsymbol{\Sigma} = \begin{bmatrix} \sigma_{11} & \sigma_{12} & \cdots & \sigma_{1n} \\ \sigma_{21} & \sigma_{22} & \cdots & \sigma_{2n} \\ \cdot & & & \\ \cdot & & & \\ \cdot & & & \\ \sigma_{n1} & \sigma_{n2} & \cdots & \sigma_{nn} \end{bmatrix}$$

is any $n \times n$ symmetric positive definite matrix of real numbers if [defining $\mathbf{x} = (x_1, \ldots, x_n)'$]

$$f_{\mathbf{X}}(\mathbf{x}) = \frac{1}{(2\pi)^{n/2}\sqrt{|\boldsymbol{\Sigma}|}} e^{-\frac{1}{2}(\mathbf{x}-\boldsymbol{\mu})'\boldsymbol{\Sigma}^{-1}(\mathbf{x}-\boldsymbol{\mu})}$$

for all vectors $\mathbf{x}$ of real numbers.

---

[6] $\sigma, \sigma_1, \sigma_2$ are used (respectively) to denote the positive square roots of $\sigma^2, \sigma_1^2, \sigma_2^2$.

After a review of vector and matrix algebra (for example, Appendix II, pp. 387–405, of H. Scheffé, *The Analysis of Variance*, John Wiley & Sons, Inc., New York, 1959), the student should be able to verify that (if $n = 2$) Definitions 4.5.2 and 4.5.3 are equivalent.

For example, if $\mathbf{X}$ has the multivariate normal distribution according to Definition 4.5.3 and $n = 2$, then

$$|\boldsymbol{\Sigma}| = \begin{vmatrix} \sigma_{11} & \sigma_{12} \\ \sigma_{21} & \sigma_{22} \end{vmatrix} = \sigma_{11}\sigma_{22} - \sigma_{12}\sigma_{21};$$

$$\boldsymbol{\Sigma}^{-1} = \begin{bmatrix} \dfrac{\sigma_{22}}{\sigma_{11}\sigma_{22} - \sigma_{12}\sigma_{21}} & \dfrac{-\sigma_{12}}{\sigma_{11}\sigma_{22} - \sigma_{12}\sigma_{21}} \\ \dfrac{-\sigma_{21}}{\sigma_{11}\sigma_{22} - \sigma_{12}\sigma_{21}} & \dfrac{\sigma_{11}}{\sigma_{11}\sigma_{22} - \sigma_{12}\sigma_{21}} \end{bmatrix}$$

$$= \frac{1}{\sigma_{11}\sigma_{22} - \sigma_{12}\sigma_{21}} \begin{bmatrix} \sigma_{22} & -\sigma_{12} \\ -\sigma_{21} & \sigma_{11} \end{bmatrix};$$

$$f_{\mathbf{X}}(x) = \frac{1}{2\pi\sqrt{|\boldsymbol{\Sigma}|}} e^{-\frac{1}{2}(\mathbf{x}-\boldsymbol{\mu})'\boldsymbol{\Sigma}^{-1}(\mathbf{x}-\boldsymbol{\mu})} \tag{4.5.4}$$

$$= \frac{1}{2\pi\sqrt{\sigma_{11}\sigma_{22} - \sigma_{12}\sigma_{21}}} e^{-\frac{1}{2(\sigma_{11}\sigma_{22}-\sigma_{12}\sigma_{21})}\{\sigma_{22}(x_1-\mu_1)^2 - (\sigma_{12}+\sigma_{21})(x_1-\mu_1)(x_2-\mu_2) + \sigma_{11}(x_2-\mu_2)^2\}}$$

$$= \frac{1}{2\pi\sqrt{\sigma_{11}}\sqrt{\sigma_{22}}\sqrt{1-\dfrac{\sigma_{12}\sigma_{21}}{\sigma_{11}\sigma_{22}}}} e^{-\frac{1}{2\left(1-\frac{\sigma_{12}\sigma_{21}}{\sigma_{11}\sigma_{22}}\right)}\left\{\left(\frac{x_1-\mu_1}{\sqrt{\sigma_{11}}}\right)^2 - \frac{\sigma_{12}+\sigma_{21}}{\sqrt{\sigma_{11}}\sqrt{\sigma_{22}}}\left(\frac{x_1-\mu_1}{\sqrt{\sigma_{11}}}\right)\left(\frac{x_2-\mu_2}{\sqrt{\sigma_{22}}}\right) + \left(\frac{x_2-\mu_2}{\sqrt{\sigma_{22}}}\right)^2\right\}}.$$

If in Definition 4.5.2 we choose our $\sigma_1^2 > 0$ as σ_{11}, our $\sigma_2^2 > 0$ as σ_{22}, our μ_1 as μ_1, our μ_2 as μ_2, and our $\rho(-1 < \rho < +1)$ as $\sigma_{12}/\sqrt{\sigma_{11}\sigma_{22}}$, then (if $\sigma_{11} > 0$, $\sigma_{22} > 0$, $\sigma_{12} = \sigma_{21}$) the above $f_{\mathbf{X}}(x)$ (4.5.4) becomes

$$\frac{1}{2\pi\sigma_1\sigma_2\sqrt{1-\rho^2}} e^{-\frac{1}{2(1-\rho^2)}\left\{\left(\frac{x_1-\mu_1}{\sigma_1}\right)^2 - 2\rho\left(\frac{x_1-\mu_1}{\sigma_1}\right)\left(\frac{x_2-\mu_2}{\sigma_2}\right) + \left(\frac{x_2-\mu_2}{\sigma_2}\right)^2\right\}},$$

as was to be shown. The restriction $\sigma_{12} = \sigma_{21}$ follows since $\boldsymbol{\Sigma}$ was assumed symmetric; the restrictions $\sigma_{11} > 0$ and $\sigma_{22} > 0$ are necessary for $\boldsymbol{\Sigma}$ to be positive definite. (Why?) We know $-1 < \rho < +1$. (Why?)

4.6. DISTRIBUTIONS OF FUNCTIONS OF CONTINUOUS RANDOM VARIABLES

In Section 3.6, we considered the distribution of a function of a discrete r.v., and obtained a comprehensive formula. In the case of continuous r.v.'s the situation is more complicated; below some of the results are stated and motivated (and formulas for the comparable discrete case are given in a few instances to illustrate the difference).

Theorem 4.6.1. *If $Y = aX + b$ and $a > 0$, then $F_Y(y) = F_X\left(\frac{y-b}{a}\right)$. If X is (absolutely) continuous, so is Y and*

$$f_{aX+b}(y) = \frac{1}{a} f_X\left(\frac{y-b}{a}\right).$$

If X is discrete, so is Y and

$$p_{aX+b}(y) = p_X\left(\frac{y-b}{a}\right).$$

[Note that these results are for the case $a > 0$; if $a < 0$, similar results can be obtained.]

Proof: (Assuming $a > 0$.) First,

$$F_Y(y) = F_{aX+b}(y) = P[aX + b \leqslant y] = P\left[X \leqslant \frac{y-b}{a}\right] = F_X\left(\frac{y-b}{a}\right).$$

Differentiating this equation with respect to y, we find (in the case of absolutely continuous X)

$$\frac{d}{dy} F_Y(y) = \frac{d}{dy} F_X\left(\frac{y-b}{a}\right) = \frac{1}{a} f_X\left(\frac{y-b}{a}\right).$$

Let us call this function $f(y)$. The $f(y) \geqslant 0$ [since $f_X(x) \geqslant 0$ and $a > 0$] and

$$\int_{-\infty}^{\infty} f(y)\, dy = \int_{-\infty}^{\infty} \frac{1}{a} f_X\left(\frac{y-b}{a}\right) dy = \int_{-\infty}^{\infty} f_X(y)\, dy = 1,$$

so (by Definition 4.1.1) $f(y)$ is a p.d.f. However,

$$\int_{-\infty}^{z} f(y)\, dy = \int_{-\infty}^{z} \frac{1}{a} f_X\left(\frac{y-b}{a}\right) dy = \int_{-\infty}^{(z-b)/a} f_X(x)\, dx$$

$$= P\left[X \leqslant \frac{z-b}{a}\right] = P[aX + b \leqslant z] = P[Y \leqslant z] = F_Y(z)$$

for all z. Hence (by Definition 4.1.1), Y is an absolutely continuous r.v. with p.d.f. $f(y)$.

In this discrete case, it is easy to see that

$$p_{aX+b}(y) = P[aX + b = y] = P\left[X = \frac{y-b}{a}\right] = p_X\left(\frac{y-b}{a}\right).$$

Theorem 4.6.2. *If $y = g(x)$ is differentiable for all x, and either $g'(x) > 0$ for all x or $g'(x) < 0$ for all x, and if X is (absolutely) continuous, then $Y = g(X)$ is (absolutely) continuous and*

$$f_Y(y) = \begin{cases} f_X(g^{-1}(y)) \left| \dfrac{d}{dy} g^{-1}(y) \right|, & \text{if } \min(\lim_{x\to-\infty} g(x), \lim_{x\to+\infty} g(x)) \\ & \quad < y < \max(\lim_{x\to-\infty} g(x), \lim_{x\to+\infty} g(x)) \\ 0, & \text{otherwise.} \end{cases}$$

Proof: As motivation, consider the fact that $F_Y(y) = P[Y \leqslant y] = P[g(X) \leqslant y] = P[X \in \{x: g(x) \leqslant y\}]$.

Case I: $g'(x) > 0$. Then

$$F_Y(y) = P[X \leqslant g^{-1}(y)] = F_X(g^{-1}(y))$$

and by differentiation

$$f_Y(y) = f_X(g^{-1}(y)) \frac{d}{dy} g^{-1}(y).$$

[As motivation, consider Figure 4.6–1.]

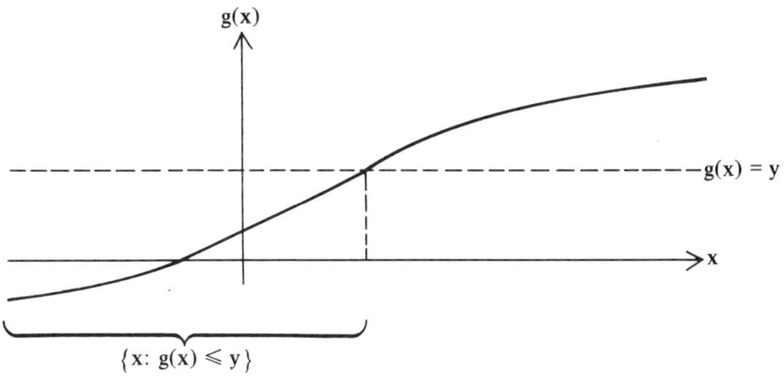

Figure 4.6–1

Case II: $g'(x) < 0$. Then

$$\begin{aligned} F_Y(y) &= P[X \geqslant g^{-1}(y)] = 1 - P[X < g^{-1}(y)] \\ &= 1 - F_X(g^{-1}(y)) + P[X = g^{-1}(y)] = 1 - F_X(g^{-1}(y)) \end{aligned}$$

and by differentiation

$$f_Y(y) = -f_X(g^{-1}(y)) \frac{d}{dy} g^{-1}(y).$$

As motivation, consider Figure 4.6–2.

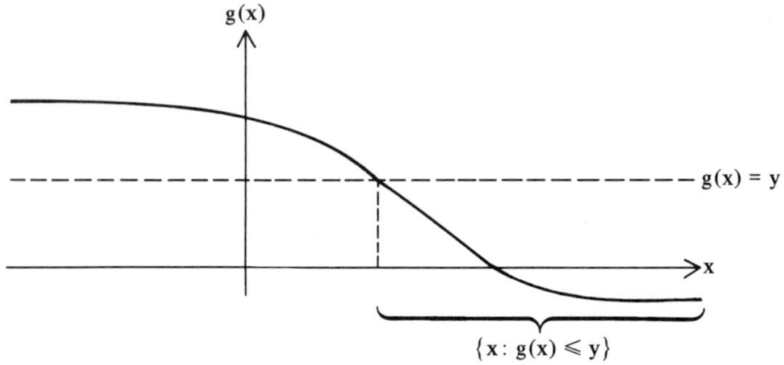

Figure 4.6–2

Since $y = g(x)$, $x = g^{-1}(y)$ and

$$\frac{d}{dy} g^{-1}(y) = \frac{d}{d[g(x)]} x = \frac{1}{\frac{dg(x)}{dx}} = \frac{1}{g'(x)|_{x=g^{-1}(y)}} = \frac{1}{g'(x)},$$

which is >0 if $g'(x) > 0$, and is <0 if $g'(x) < 0$. Thus, we may combine cases I and II to obtain

$$f_Y(y) = f_X(g^{-1}(y)) \left| \frac{d}{dy} g^{-1}(y) \right|.$$

The range of positivity and validity of this formula is determined by consideration of where $F_Y(y)$ is 0 and 1.

Example 4.6.3. Suppose a r.v. X has the Weibull distribution with $\alpha > 0$ and $\beta > 0$ [see (4.2.30)]

$$f_X(x) = \begin{cases} \alpha\beta x^{\beta-1} e^{-\alpha x^\beta}, & x \geq 0 \\ 0, & x < 0. \end{cases}$$

Find the density of $Y = X^\beta$.

Here $y = g(x) = x^\beta$ is differentiable for all x, and $g'(x) = \beta x^{\beta-1} > 0$ for all $x > 0$. Also $g^{-1}(y) = x = y^{1/\beta}$, so

$$f_X(g^{-1}(y)) = f_X(y^{1/\beta}) = \alpha\beta y^{1-\frac{1}{\beta}} e^{-\alpha y}$$

and

$$\frac{d}{dy} g^{-1}(y) = \frac{d}{dy} y^{1/\beta} = \frac{1}{\beta} y^{\frac{1}{\beta}-1};$$

hence (by Theorem 4.6.2),

$$f_Y(y) = \begin{cases} \alpha e^{-\alpha y}, & \text{if } 0 < y < \infty \\ 0, & \text{otherwise.} \end{cases}$$

[Note that we have actually used a slight extension of Theorem 4.6.2, wherein now we talk only of x for which $f_X(x) > 0$; for example, $g'(x) > 0$ for all x for which $f_X(x) > 0$, that is, for all $x > 0$. Only a slight modification of the proof of Theorem 4.6.2 is needed in order to restrict attention throughout to $\{x: f_X(x) > 0\}$. We then also take limits of $g(x)$ with x in this set, of course.]

Thus, *if* X *is Weibull,* X^β *is exponential. (And, if* Y *is exponential,* $Y^{1/\beta}$ *is Weibull.)*

Theorem 4.6.4. *Let* $Y = X_1 + X_2$. *If* (X_1, X_2) *is (absolutely) continuous,*

$$\begin{aligned} F_Y(y) = P[X_1 + X_2 \leqslant y] &= \iint_{\{(x_1, x_2):\, x_1 + x_2 \leqslant y\}} f_{X_1, X_2}(x_1, x_2)\, dx_1\, dx_2 \\ &= \int_{-\infty}^{\infty} \left[\int_{-\infty}^{y - x_1} f_{X_1, X_2}(x_1, x_2)\, dx_2\right] dx_1 \\ &= \int_{-\infty}^{\infty} \left[\int_{-\infty}^{y} f_{X_1, X_2}(x_1, x_2' - x_1)\, dx_2'\right] dx_1. \end{aligned}$$

Differentiating,

$$f_Y(y) = f_{X_1+X_2}(y) = \int_{-\infty}^{\infty} f_{X_1, X_2}(x_1, y - x_1)\, dx_1 = \int_{-\infty}^{\infty} f_{X_1, X_2}(y - x_2, x_2)\, dx_2.$$

If X_1 *and* X_2 *are independent,*

$$f_{X_1+X_2}(y) = \int_{-\infty}^{\infty} f_{X_1}(x) f_{X_2}(y - x)\, dx = \int_{-\infty}^{\infty} f_{X_1}(y - x) f_{X_2}(x)\, dx.$$

This theorem's statement actually contains its proof. Similarly, we may derive the following theorem.

Theorem 4.6.5. *If* (X_1, X_2) *is (absolutely) continuous,*

(1) $f_{X_1 - X_2}(y) = \int_{-\infty}^{\infty} f_{X_1, X_2}(y + x, x)\, dx = \int_{-\infty}^{\infty} f_{X_1, X_2}(x, x - y)\, dx;$

(2) $f_{X_1 X_2}(y) = \int_{-\infty}^{\infty} \frac{1}{|x|} f_{X_1, X_2}\left(\frac{y}{x}, x\right) dx = \int_{-\infty}^{\infty} \frac{1}{|x|} f_{X_1, X_2}\left(x, \frac{y}{x}\right) dx;$

(3) $f_{X_1/X_2}(y) = \int_{-\infty}^{\infty} |x|\, f_{X_1, X_2}(yx, x)\, dx.$

Theorem 4.6.6[7]**.** *We wish to find the density of* $Y_1 = g_1(X_1, \ldots, X_n), \ldots, Y_m = g_m(X_1, \ldots, X_n)$ *where* $m \leqslant n$ *and* $(X_1, \ldots, X_n)$ *is an (absolutely)*

---

[7] A simplification of the conditions of this theorem results if we use the simple conditions of D. E. Varberg, "Changes of variables in multiple integrals," *American Mathematical Monthly*, Vol. 78 (1971), pp. 42–45. Since a substantial amount of mathematics is needed to understand this simplification, we leave it as an exercise for the mathematically well-prepared student.

continuous r.v. Suppose we consider $Y_1, \ldots, Y_{m+1}, \ldots, Y_n$ *where*

$$\begin{aligned} Y_{m+1} &= g_{m+1}(X_1, \ldots, X_n), \\ &\vdots \\ Y_n &= g_n(X_1, \ldots, X_n). \end{aligned}$$

Suppose the functions $g_1(x_1, \ldots, x_n), \ldots, g_n(x_1, \ldots, x_n)$ *have continuous first partial derivatives at all* $(x_1, \ldots, x_n)$ *and are such that at all* $(x_1, \ldots, x_n)$ *the* **Jacobian**

$$J\left(\frac{y_1, \ldots, y_n}{x_1, \ldots, x_n}\right) \equiv \begin{vmatrix} \frac{\partial y_1}{\partial x_1} & \cdots & \frac{\partial y_1}{\partial x_n} \\ \frac{\partial y_2}{\partial x_1} & \cdots & \frac{\partial y_2}{\partial x_n} \\ \vdots & & \\ \frac{\partial y_n}{\partial x_1} & \cdots & \frac{\partial y_n}{\partial x_n} \end{vmatrix} \neq 0.$$

If the p.d.f. of $(X_1, \ldots, X_n)$ *is continuous at all but a finite number of points* $(x_1, \ldots, x_n)$ *then* $(Y_1, \ldots, Y_n)$ *is (absolutely) continuous with*

$$f_{Y_1, \ldots, Y_n}(y_1, \ldots, y_n) = \begin{cases} f_{X_1, \ldots, X_n}(x_1, \ldots, x_n) \dfrac{1}{\left| J\left(\dfrac{y_1, \ldots, y_n}{x_1, \ldots, x_n}\right)\right|}, & \mathbf{y} \in C \\ 0, & \mathbf{y} \notin C, \end{cases}$$

where C is the set of $\mathbf{y} = (y_1, \ldots, y_n)'$ *such that the n equations*

$$\begin{cases} y_1 = g_1(x_1, \ldots, x_n), \\ \vdots \\ y_n = g_n(x_1, \ldots, x_n), \end{cases}$$

possess at least one solution $(x_1, \ldots, x_n)$; *the* $(x_1, \ldots, x_n)$ *are the unique solution, say*

$$\begin{cases} x_1 = g_1^{-1}(y_1, \ldots, y_n), \\ \vdots \\ x_n = g_n^{-1}(y_1, \ldots, y_n). \end{cases}$$

Finally,

$$f_{Y_1,\ldots,Y_m}(y_1,\ldots,y_m)$$
$$= \int_{-\infty}^{\infty}\cdots\int_{-\infty}^{\infty} f_{Y_1,\ldots,Y_m,Y_{m+1},\ldots,Y_n}(y_1,\ldots,y_m,y_{m+1},\ldots,y_n)\,dy_{m+1}\cdots dy_n.$$

Note 4.6.7. $$\left|J\left(\frac{x_1,\ldots,x_n}{y_1,\ldots,y_n}\right)\right| = \left|J\left(\frac{y_1,\ldots,y_n}{x_1,\ldots,x_n}\right)\right|^{-1}$$

but the right-hand side is often easy to obtain, rather than solving for $x_1,\ldots,x_n$ in terms of $y_1,\ldots,y_n$ *or* differentiating implicitly.

Example 4.6.8. Let X_1, X_2 be independent r.v.'s, each with the exponential distribution with $\beta^{-1} = \lambda$, $A = 0$ (see Definition 4.2.7). Are the r.v.'s $X_1 + X_2$ and X_1/X_2 independent?

We know that

$$f_{X_1}(x) = f_{X_2}(x) = \begin{cases}\lambda e^{-\lambda x}, & x > 0\\ 0, & x \leqslant 0,\end{cases}$$

so that

$$f_{X_1,X_2}(x_1,x_2) = \begin{cases}\lambda^2 e^{-\lambda(x_1+x_2)}, & x_1 > 0 \text{ and } x_2 > 0\\ 0, & \text{otherwise.}\end{cases}$$

Let $Y_1 = X_1 + X_2$, $Y_2 = X_1/X_2$. If we find $f_{Y_1,Y_2}(y_1,y_2)$ and show it equals $f_{Y_1}(y_1)\cdot f_{Y_2}(y_2)$ for all y_1, y_2, then Y_1 and Y_2 are independent r.v.'s by Theorem 4.4.6. Now (since the conditions of Theorem 4.6.6 are satisfied)

$$f_{Y_1,Y_2}(y_1,y_2) = \begin{cases} f_{X_1,X_2}(x_1,x_2)\dfrac{1}{\left|J\left(\dfrac{y_1,y_2}{x_1,x_2}\right)\right|}, & \mathbf{y}\in C\\[2ex] 0, & \mathbf{y}\notin C.\end{cases}$$

(Here it is understood that x_1, x_2 are to be eliminated from the right-hand side.) Since $y_1 = x_1 + x_2$ and $y_2 x_2 = x_1$,

$$\begin{cases} x_1 = \dfrac{y_1 y_2}{1+y_2},\\[2ex] x_2 = \dfrac{y_1}{1+y_2}.\end{cases}$$

Also,

$$J\left(\frac{y_1,y_2}{x_1,x_2}\right) = \begin{vmatrix}\dfrac{\partial y_1}{\partial x_1} & \dfrac{\partial y_1}{\partial x_2}\\[2ex] \dfrac{\partial y_2}{\partial x_1} & \dfrac{\partial y_2}{\partial x_2}\end{vmatrix} = \begin{vmatrix} 1 & 1\\[2ex] \dfrac{1}{x_2} & \dfrac{-x_1}{x_2^2}\end{vmatrix} = \frac{-x_1}{x_2^2} - \frac{1}{x_2}.$$

Thus, for $\mathbf{y} \in C$,

$$f_{Y_1, Y_2}(y_1, y_2) = \lambda^2 e^{-\lambda(x_1 + x_2)} \frac{1}{\left| -\dfrac{x_1}{x_2^2} - \dfrac{1}{x_2} \right|}$$

$$= \lambda^2 e^{-\lambda y_1} \frac{x_2^2}{x_1 + x_2} = \lambda^2 e^{-\lambda y_1} \frac{\dfrac{y_1^2}{(1 + y_2)^2}}{y_1}$$

$$= \lambda^2 y_1 e^{-\lambda y_1} \cdot \frac{1}{(1 + y_2)^2}.$$

Since $C = \{(y_1, y_2): y_1 > 0, y_2 > 0\}$, it follows easily that

$$f_{Y_1, Y_2}(y_1, y_2) = f_{Y_1}(y_1) f_{Y_2}(y_2), \qquad \text{all } y_1, y_2.$$

Example 4.6.9. Suppose that $X_1, \ldots, X_k$ are independent r.v.'s, each distributed as chi-square with n degrees of freedom $\chi_n^2(0)$ (see Definition 4.2.9). Define new random variables

$$Y_1 = \frac{X_2}{X_1}, \qquad Y_2 = \frac{X_3}{X_1}, \qquad \ldots, \qquad Y_{k-1} = \frac{X_k}{X_1}. \tag{4.6.10}$$

Find the joint density of $Y_1, \ldots, Y_{k-1}$.

First, since $X_1, \ldots, X_k$ are independent r.v.'s we have (using Definition 4.2.9)

$$\begin{aligned} f_{X_1, \ldots, X_k}(x_1, \ldots, x_k) &= f_{X_1}(x_1) \cdots f_{X_k}(x_k) \\ &= \prod_{i=1}^{k} \frac{1}{\Gamma\left(\dfrac{n}{2}\right) 2^{n/2}} x_i^{(n-2)/2} e^{-x_i/2} \\ &= \frac{1}{\left(\Gamma\left(\dfrac{n}{2}\right)\right)^k 2^{nk/2}} x_1^{(n-2)/2} \cdots x_k^{(n-2)/2} e^{-\frac{1}{2}(x_1 + \cdots + x_k)} \end{aligned} \tag{4.6.11}$$

when $x_1 > 0$ and ... and $x_k > 0$, and zero otherwise. We now wish to use Theorem 4.6.6; to do so we must define a Y_k (of no interest to us) in such a way that all k functions of $X_1, \ldots, X_k$ have continuous partial derivatives and $J \neq 0$. Let us try $Y_k = X_1 + \cdots + X_k$. Then our functions are

$$\begin{cases} y_1 = g_1(x_1, \ldots, x_k) = \dfrac{x_2}{x_1}, \\ y_2 = g_2(x_1, \ldots, x_k) = \dfrac{x_3}{x_1}, \\ \cdot \\ \cdot \\ \cdot \end{cases} \tag{4.6.12}$$

$$\begin{cases} y_{k-1} = g_{k-1}(x_1, \ldots, x_k) = \dfrac{x_k}{x_1}, \\ y_k = g_k(x_1, \ldots, x_k) = x_1 + \cdots + x_k, \end{cases}$$

all of which have continuous first partial derivatives at all $(x_1, \ldots, x_k)$ [as in Example 4.6.3, we need only talk of $(x_1, \ldots, x_k)$ for which $f_{X_1, \ldots, X_k}(x_1, \ldots, x_k) > 0$]. We now find the Jacobian

$$J\left(\frac{y_1, \ldots, y_k}{x_1, \ldots, x_k}\right) = \begin{vmatrix} -\dfrac{x_2}{x_1^2} & \dfrac{1}{x_1} & 0 & 0 & \cdots & 0 & 0 \\ -\dfrac{x_3}{x_1^2} & 0 & \dfrac{1}{x_1} & 0 & \cdots & 0 & 0 \\ -\dfrac{x_4}{x_1^2} & 0 & 0 & \dfrac{1}{x_1} & \cdots & 0 & 0 \\ \vdots & & & & & & \\ -\dfrac{x_{k-1}}{x_1^2} & 0 & 0 & 0 & \cdots & \dfrac{1}{x_1} & 0 \\ -\dfrac{x_k}{x_1^2} & 0 & 0 & 0 & \cdots & 0 & \dfrac{1}{x_1} \\ 1 & 1 & 1 & 1 & \cdots & 1 & 1 \end{vmatrix}$$

$$= \begin{vmatrix} 0 & \dfrac{1}{x_1} & 0 & 0 & \cdots & 0 & 0 \\ 0 & 0 & \dfrac{1}{x_1} & 0 & \cdots & 0 & 0 \\ 0 & 0 & 0 & \dfrac{1}{x_1} & \cdots & 0 & 0 \\ \vdots & & & & & & \\ 0 & 0 & 0 & 0 & \cdots & \dfrac{1}{x_1} & 0 \\ 0 & 0 & 0 & 0 & \cdots & 0 & \dfrac{1}{x_1} \\ 1 + \dfrac{x_2}{x_1} + \cdots + \dfrac{x_k}{x_1} & 1 & 1 & 1 & \cdots & 1 & 1 \end{vmatrix} \tag{4.6.13}$$

$$= \left(\frac{1}{x_1}\right)^{k-1} \left(1 + \frac{x_2}{x_1} + \cdots + \frac{x_k}{x_1}\right)$$

$$= \frac{x_1 + x_2 + \cdots + x_k}{x_1^k}$$

(where we multiplied columns 2, 3, ..., k by $x_2/x_1, x_3/x_1, \ldots, x_k/x_1$, respectively, and added them to the first column). Hence,

$$f_{Y_1,\ldots,Y_k}(y_1,\ldots,y_k) = f_{X_1,\ldots,X_k}(x_1,\ldots,x_k)\left|J\left(\frac{y_1,\ldots,y_k}{x_1,\ldots,x_k}\right)\right|^{-1} = \frac{1}{\left(\Gamma\left(\frac{n}{2}\right)\right)^k 2^{nk/2}} \frac{x_1^{(n-2)/2}\cdots x_k^{(n-2)/2}x_1^{\ k}}{x_1+x_2+\cdots+x_k} e^{-(x_1+\cdots+x_k)/2} \tag{4.6.14}$$

where $x_1, \ldots, x_k$ are to be replaced by their unique solutions [from system (4.6.12)] in terms of $y_1, \ldots, y_k$ [and where $f_{Y_1,\ldots,Y_k}(y_1,\ldots,y_k) = 0$ otherwise; that is, unless $(y_1,\ldots,y_k)$ can arise from some $(x_1,\ldots,x_k)$]. Doing this replacement, we find

$$f_{Y_1,\ldots,Y_k}(y_1,\ldots,y_k) = \frac{1}{\left(\Gamma\left(\frac{n}{2}\right)\right)^k 2^{nk/2}} \frac{\prod_{i=1}^{k-1} y_i^{(n-2)/2}}{(1+y_1+\cdots+y_{k-1})^{nk/2}} y_k^{nk/2-1}e^{-y_k/2}, \tag{4.6.15}$$

where $0 < y_i < \infty$ $(i = 1, 2, \ldots, k)$. [Since $y_1 = x_2/x_1, \ldots, y_{k-1} = x_{k-1}/x_1$, we can clearly fix $y_1, \ldots, y_{k-1}$ (independently of each other) to any positive values. Can we then do similarly for y_k? Yes, for $y_k = x_1 + \cdots + x_k = \left(1 + \frac{x_2}{x_1} + \cdots + \frac{x_k}{x_1}\right)x_1 = (1 + y_1 + \cdots + y_{k-1})x_1$ and x_1 is still under our control.]

Now, however, recall that Y_k was an extraneous variable introduced for mathematical reasons. We complete this example by eliminating Y_k as follows.

$$f_{Y_1,\ldots,Y_{k-1}}(y_1,\ldots,y_{k-1}) = \int_0^\infty f_{Y_1,\ldots,Y_{k-1},Y_k}(y_1,\ldots,y_{k-1},y_k)\,dy_k = \frac{1}{\left(\Gamma\left(\frac{n}{2}\right)\right)^k 2^{nk/2}} \frac{\prod_{i=1}^{k-1} y_i^{(n-2)/2}}{(1+y_1+\cdots+y_{k-1})^{nk/2}} \int_0^\infty y_k^{nk/2-1}e^{-y_k/2}\,dy_k. \tag{4.6.16}$$

Making the substitution $z = y_k/2$, the last integral becomes

$$2^{nk/2}\int_0^\infty z^{nk/2-1}e^{-z}\,dz = 2^{nk/2}\,\Gamma\left(\frac{nk}{2}\right)$$

[where we used Definition 4.1.6 of $\Gamma(n)$]. Hence,

$$f_{Y_1,\ldots,Y_{k-1}}(y_1,\ldots,y_{k-1}) = \begin{cases} \dfrac{\Gamma\left(\dfrac{kn}{2}\right)}{\left(\Gamma\left(\dfrac{n}{2}\right)\right)^k} \dfrac{\prod_{i=1}^{k-1} y_i^{(n-2)/2}}{(1+y_1+\cdots+y_{k-1})^{nk/2}}, & \text{if } 0 < y_1, \ldots, 0 < y_{k-1} \\ 0, & \text{otherwise.} \end{cases} \tag{4.6.17}$$

(The problem of finding the density of $Y_1, \ldots, Y_{k-1}$ as specified in Example 4.6.9 arose in a paper "A single-sample multiple decision procedure for ranking variances of normal populations" by R. E. Bechhofer and M. Sobel, *Annals of Mathematical Statistics*, Vol. 25 (1954), pp. 273–289. In particular, our final result (4.6.17) is a special case of Equation (13) on their p. 277 (the case $t = 1$). One practical example of the class of problems they were considering would be the selection of that one of several effective drugs which has the least variability in its effects. Problems of this sort are called problems of "ranking and selection" and are studied in Chapter 11.)

PROBLEMS

4.1 If a r.v. X has density

$$f(x) = \begin{cases} 2x, & \text{if } 0 < x < 1 \\ 0, & \text{otherwise,} \end{cases}$$

find the probability that: **(a)** $X < \frac{1}{2}$; **(b)** $\frac{1}{4} < X < \frac{1}{2}$; **(c)** $X > \frac{3}{4}$, given that $X > \frac{1}{2}$.

4.2 Define a density using $x(2 - x)$ over the set $0 < x < 2$. Find the probability that $a < X < b$ if: **(i)** $0 < a < b < 2$; **(ii)** $a < 0, 2 < b$.

4.3 If a r.v. X has density

$$f(x) = \begin{cases} 4x^3, & \text{if } 0 < x < 1 \\ 0, & \text{otherwise,} \end{cases}$$

find the number a such that X is equally likely to be greater than a or less than a. Find the number b such that $P[X > b] = .05$.

4.4 A r.v. X has density

$$f(x) = \begin{cases} 1, & \text{if } 0 < x < 1 \\ 0, & \text{otherwise.} \end{cases}$$

Determine the number a such that the probability will be .9 that at least one of 4 values of X drawn at random (and hence independent r.v.'s) will exceed a.

4.5 A certain house is supplied with fuel oil on the first day of each month. If its monthly need for fuel oil during the "worst" month of the year is, in 1000's of gallons, distributed as

$$f(x) = \begin{cases} 5(1-x)^4, & 0 < x < 1 \\ 0, & \text{otherwise,} \end{cases}$$

what capacity tank is needed in order that the probability that its supply will be exhausted in a typical "worst" month will be: **(a)** .01; **(b)** 0.

4.6 Suppose that a r.v. X has the $N(\mu, \sigma^2)$ distribution (see Definition 4.2.13). Express $F(z) = P[X \leqslant z]$ in terms of $\Phi(\cdot)$ where (for all x, $-\infty < x < +\infty$)

$$\Phi(x) = \int_{-\infty}^{x} \frac{1}{\sqrt{2\pi}} e^{-\frac{1}{2}y^2}\, dy.$$

[Note that $\Phi(\cdot)$ is tabled in Table 11.]

4.7 If

$$f_{X,Y}(x, y) = \begin{cases} e^{-(x+y)}, & x > 0 \text{ and } y > 0 \\ 0, & \text{otherwise,} \end{cases}$$

find $P(X > 1)$; $P(a < X + Y < b)$ if $0 < a < b$; and $P(X < Y | X < 2Y)$.

4.8 Suppose that the total length in minutes of intercontinental satellite long distance telephone calls is found to have d.f.

$$F(x) = \begin{cases} 0, & x \leqslant 0 \\ 1 - \frac{2}{3}e^{-x/3} - \frac{1}{3}e^{-[x/3]}, & x > 0, \end{cases}$$

where $[y]$ means the largest integer which is $\leqslant y$ (for example, $[5.3] = 5$ and $[6] = 6$).

(a) Sketch the d.f.

(b) Is the d.f. absolutely continuous, discrete, singularly continuous or a combination of these? (If a combination, specify c_1, c_2, c_3 as noted in Theorem 4.1.4.)

(c) What is the probability that the total length in minutes of a telephone call of this type will be > 6 minutes; < 4 minutes; $= 3$ minutes; between 4 and 7 minutes?

(d) What is the conditional probability that the total length in minutes of a telephone call of this type will be < 9 minutes, given that it has lasted more than 5 minutes?

4.9 The choice of a random point Y between 0 and 1 can be described as follows in terms of discrete r.v.'s. Let $Y_k(y)$ denote the kth decimal of y and, to avoid ambiguities, use terminating expansions when possible. Let the r.v. Y_k take values $0, 1, \ldots, 9$ with probability 1/10 each. Suppose that $Y_1, Y_2, \ldots$ are independent r.v.'s. Then

$$Y = \sum_{i=1}^{\infty} \frac{1}{10^i} Y_i,$$

which reduces the random choice of a point Y to choice of its successive decimals:

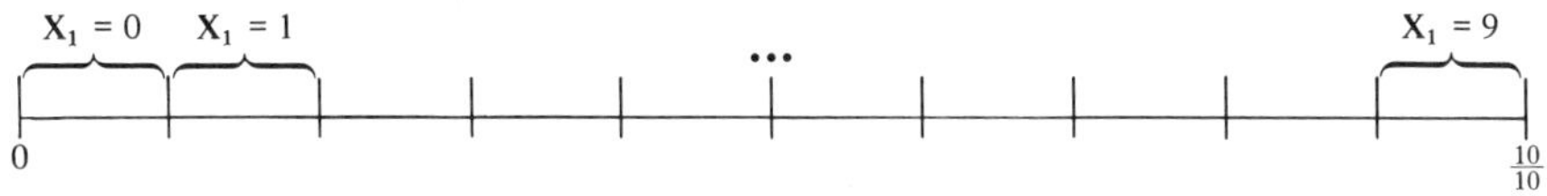

Find the d.f. of $99Z$, where

$$Z = \sum_{k=2,4,\ldots} \frac{1}{10^k} Y_k = \sum_{v=1}^{\infty} \frac{1}{100^v} Y_{2v}.$$

[*Hint*: See Example 4.2.18.]

4.10 We define a function $\mathbf{x}^+$, called the **positive part of x**, as

$$x^+ = \begin{cases} x, & \text{if } x \geqslant 0 \\ 0, & \text{if } x < 0. \end{cases}$$

Similarly the **negative part of x, x⁻**, is

$$x^- = \begin{cases} 0, & x \geqslant 0 \\ -x, & x < 0. \end{cases}$$

(Note that $x = x^+ - x^-$ and $|x| = x^+ + x^-$.) If X is a random variable with d.f. $F_X(x)$, find the d.f. of $Y = X^+$.

4.11 Let X_1, X_2 be independent random variables, each with a uniform distribution on $(0, 1)$. Find and sketch the density functions of: $X_1 + X_2$; $X_1 - X_2$; $|X_1 - X_2|$; $\max(X_1, X_2)$; $\min(X_1, X_2)$; X_1X_2; X_1/X_2.

4.12 Prove the **theorem**: *Let $(X_1, \ldots, X_n)$ be an n-dimensional (absolutely) continuous random variable with density function $f_{X_1, \ldots, X_n}(x_1, \ldots, x_n)$. Then, $X_1, \ldots, X_n$ are stochastically independent iff, for some functions $g_1(\cdot), \ldots, g_n(\cdot)$,*

$$f_{X_1, \ldots, X_n}(x_1, \ldots, x_n) = g_1(x_1) \cdots g_n(x_n), \quad \textit{for all } \mathbf{x} \in \mathscr{R}^n. \qquad (*)$$

4.13 If $(*)$ holds in Problem 4.12, find $f_{X_1}(x_1)$ in terms of $g_1(\cdot), \ldots, g_n(\cdot)$.

4.14 Let $\mathbf{X} = (X_1, X_2)$ have the joint density function

$$f_{X_1, X_2}(x_1, x_2) = f_1(x_1)f_2(x_2)\{1 + \alpha[2F_1(x_1) - 1][2F_2(x_2) - 1]\}, \qquad (*)$$

$|\alpha| \leqslant 1$, where

$$f_i(x_i) = \frac{1}{\sqrt{2\pi}} e^{-\frac{1}{2}x_i^2} \qquad (i = 1, 2)$$

$$F_i(x_i) = \int_{-\infty}^{x_i} \frac{1}{\sqrt{2\pi}} e^{-\frac{1}{2}t^2}\, dt \qquad (i = 1, 2).$$

(a) Verify that $\int_{-\infty}^{\infty} \int_{-\infty}^{\infty} f_{X_1, X_2}(x_1, x_2)\, dx_1\, dx_2 = 1$.

(b) Find $f_{X_1}(x_1)$ and $f_{X_2}(x_2)$. What are these particular densities called?

(c) Is (X_1, X_2) jointly normal? [*Hint*: Consider two cases, $\alpha = 0$, $\alpha \neq 0$. In the case $\alpha \neq 0$, for (X_1, X_2) to be jointly normal it must be the case that $(*)$ equals the normal bivariate density with parameters $\mu_1 = \mu_2 = 0$, $\sigma_1{}^2 = \sigma_2{}^2 = 1$, and ρ. (Why?) This equality must hold for all (x_1, x_2), in particular for $(x_1, x_2) = (0, 0)$. Thus, $\rho = 0$. (Why?) Thus, (X_1, X_2) is not jointly normal. (Why?)]

(d) (Optional.) Show that $W = X_1 + X_2$ is not normally distributed unless $\alpha = 0$.

[*Reference*: L. Rosenberg, "Nonnormality of linear combinations of normally distributed random variables," *American Mathematical Monthly*, Vol. 72 (1965), pp. 888–890.]

4.15 Let U be a random variable uniform on (0, 1). Find the density of $X = \sin(2\pi U)$.

4.16 Suppose that X and Y are jointly distributed random variables with joint density function

$$f_{X,Y}(x, y) = \begin{cases} e^{-(x+y)}, & \text{if } x \geqslant 0 \text{ and } y \geqslant 0 \\ 0, & \text{otherwise.} \end{cases}$$

Find the density function of $Z = XY$.

4.17 Show that if a random variable X is exponential, that is, for some $\lambda(\lambda > 0)$

$$f_X(x) = \begin{cases} \lambda e^{-\lambda x}, & \text{for } x > 0 \\ 0, & \text{otherwise,} \end{cases}$$

then it has the "lack of memory property." That is, for all positive numbers x_1 and x_2,

$$P[X \geqslant x_1 + x_2 | X \geqslant x_1] = P[X \geqslant x_2].$$

4.18 Prove (see Example 4.2.5) that if Y is uniform on (0, 1), then (for $b < c$ fixed) $X = (c - b)Y + b$ is uniform on (b, c).

4.19 Supply the remaining details in the verification that (if $n = 2$) Definitions 4.5.2 and 4.5.3 are equivalent.

4.20 $X_1, \ldots, X_n$ are independent random variables distributed according to the uniform density on the interval from 0 to 1. Let $Z_n = \max(X_1, \ldots, X_n)$. Derive an expression for $P[Z_n \leqslant x]$, and use this to find how large n must be in order that $P[Z_n > .9] \geqslant .99$. (Your answer may be left in terms of logarithms, and so on.)

4.21 Suppose that X and Y are jointly distributed random variables with joint density function

$$f_{X,Y}(x, y) = \begin{cases} \frac{1}{4}(1 + xy(x^2 - y^2)), & \text{if } -1 \leqslant x, y \leqslant 1 \\ 0, & \text{otherwise.} \end{cases}$$

Find the density of X; of Y; of $X + Y$. Find

$$f(z) = \int_{-\infty}^{\infty} f_X(x) f_Y(z - x)\, dx$$

[called the **convolution** of $f_X(\cdot)$ and $f_Y(\cdot)$]. Interpret the results as in Example 3.6.8.

4.22 If X and Y have the joint density function

$$f_{X,Y}(x, y) = \begin{cases} \dfrac{1}{2\pi} e^{-\frac{1}{4\pi}x^2 - y}, & \text{if } -\infty < x < +\infty, y \geqslant 0 \\ 0, & \text{otherwise,} \end{cases}$$

find $f_X(x)$ and $f_Y(y)$. Are X and Y independent?

4.23 Suppose that X is a random variable with d.f. $F(x)$ $(-\infty < x < +\infty)$. Find the d.f. of $Z = \sqrt{|X|}$. [*Hint*: First find the d.f. of $Y = |X|$; then find the d.f. of $Z = \sqrt{Y}$.]

4.24 Suppose that $X_1, \ldots, X_n$ are independent r.v.'s and that X_i has the exponential distribution

$$f_i(t) = \begin{cases} \alpha_i e^{-\alpha_i t}, & t \geqslant 0 \\ 0, & t < 0. \end{cases}$$

Find the distribution function and density function of $Y_1 = \min(X_1, \ldots, X_n)$. Find and plot the failure rate[8] (see Example 4.2.21) for Y_1 (which corresponds to a series system). Find the distribution function and density function of $Y_2 = \max(X_1, \ldots, X_n)$. What can you say about the failure rate[9] for Y_2 (which corresponds to a parallel system)?

4.25 Suppose that X and Y are independent r.v.'s, that X has the $\chi_m^2(0)$ distribution, and that Y has the $\chi_n^2(0)$ distribution (see Definition 4.2.9). Find the density of $U = X/(X + Y)$. Find the density of $V = X/Y$. (These will turn out to be distributions we have already discussed; find out which they are.)

4.26 Let $F(x, y)$ be the d.f. of some two-dimensional r.v. (X, Y), and let $F_1(x)$ and $F_2(y)$, respectively, be the marginal d.f.'s of X and of Y. Define functions

$$U(x, y) = \min\{F_1(x), F_2(y)\},$$
$$L(x, y) = \max\{F_1(x) + F_2(y) - 1, 0\}.$$

(a) Prove that $L(x, y)$ and $U(x, y)$ are each distribution functions, and that their marginal distributions are the same as those of $F(x, y)$.

(b) Prove that $L(x, y) \leqslant F(x, y) \leqslant U(x, y)$. [This bound for a bivariate probability in terms of given marginals was attributed to a 1951 paper of Fréchet by A. W. Marshall and I. Olkin, "A multivariate exponential distribution," *Journal of the American Statistical Association*, Vol. 62 (1967), pp. 30–40; Professor Ingram Olkin has recently called attention to the fact that W. Hoeffding had this result in 1940.]

(c) Suppose that X and Y each are $N(0, 1)$ r.v.'s, but their joint distribution function is otherwise unknown. Graph (as a function of x) the bounds $L(x, y_0)$ and $U(x, y_0)$ on $F(x, y_0)$ of (b), for $y_0 = 0$. Also for $y_0 = -1$. Also for $y_0 = 1$.

4.27 Let X be a r.v. with d.f. $F_1(x)$. Let $F_2(y)$ denote the d.f. of $Y = g(X)$. Let $F(x, y)$ denote the d.f. of the two-dimensional r.v. (X, Y).

(a) Show that if $g(\cdot)$ is monotone increasing, then

$$F(x, y) = \min\{F_1(x), F_2(y)\}.$$

---

[8] This failure rate is still the subject of research. For example, see T. K. Sarkar, "An exact lower confidence bound for the reliability of a series system where each component has an exponential time to failure distribution," *Technometrics*, Vol. 13 (1971), pp. 535–546, and the references therein.

[9] This failure rate exhibits unusual properties. For example, see M. C. Bryson, "A Gibbs-type phenomenon in the failure rates of parallel systems," *Technometrics*, Vol. 13 (1971), pp. 692–695.

(b) Show that if $g(\cdot)$ is monotone decreasing, then

$$F(x, y) = \max\{F_1(x) + F_2(y) - 1, 0\}.$$

4.28 In Example 4.6.9 we were given the joint distribution of $X_1, X_2, \ldots, X_k$ and desired to find the joint density of $Y_1, \ldots, Y_{k-1}$ where

$$Y_1 = \frac{X_2}{X_1}, \qquad Y_2 = \frac{X_3}{X_1}, \qquad \ldots, \qquad Y_{k-1} = \frac{X_k}{X_1}.$$

In order to use Theorem 4.6.6 we defined another r.v. $Y_k = X_1 + \cdots + X_k$, found the joint density of $Y_1, \ldots, Y_{k-1}, Y_k$, and then (integrating out Y_k) found the joint density of $Y_1, \ldots, Y_{k-1}$ as desired.

Many other possible choices of Y_k can be made, some of which will simplify (and some of which will complicate) the calculation of the joint density of $Y_1, \ldots, Y_{k-1}$ when the above method is used. Solve the problem of finding the joint density of $Y_1, \ldots, Y_{k-1}$ by this method when Y_k is defined as $Y_k = X_1$ [hence, $y_k = g_k(x_1, \ldots, x_k) = x_1$ in (4.6.12)]. Is it better to choose Y_k to be $X_1 + \cdots + X_k$ or to be X_1?

HONORS PROBLEMS

4.1H Supply the details noted at the end of the example on a singularly continuous d.f. in Section 4.2.

4.2H Give an example of a two-dimensional singularly continuous d.f. (It is not necessary to show a random variable which leads to this d.f., but it is necessary to prove that it is a d.f. and that it is singularly continuous.) [*Note*: If we construct a random variable X which has a singularly continuous d.f., we may be able to construct a two-dimensional analog.]

Give an example of a random variable which has a singularly continuous two-dimensional d.f. Try to do this so that, if $\mathbf{X} = (X_1, X_2)$ is the random variable, X_1 and X_2 are not independent. [*Reference*: L. H. Koopmans, "Some simple singular and mixed probability distributions," *American Mathematical Monthly*, Vol. 76 (1969), pp. 297–299. Additionally, see A. Brown, "An elementary example of a continuous singular function," *American Mathematical Monthly*, Vol. 76 (1969), pp. 295–297.]

4.3H An example of a singularly continuous $F(x)$ was given as Equation (10) by J. Lamperti and W. Rudin, "Annihilators of some function algebras," *American Mathematical Monthly*, Vol. 74 (1967), pp. 1204–1208. Consult that article and supply the details.

4.4H Fix $n \geqslant 3$. Show that there exist r.v.'s $X_1, \ldots, X_n$ such that any two are pairwise independent (in fact any proper subset is a set of independent r.v.'s) but the set of n r.v.'s are not independent. [*Reference*: R. L. Dykstra and D. A. Pierce, "Independence and the normal distribution," *The American Statistician*, Vol. 23 (1969), p. 39.]

4.5H Find and plot (using a computer plotter if possible) the failure rate $q(x)$ [see (4.2.28)] for the $N(0, 1)$ distribution [see Definition 4.2.13].

Note that the failure rate for the $N(\mu, \sigma^2)$ distribution, say $q_{\mu,\sigma^2}(x)$, has the property that

$$
\begin{aligned}
q_{\mu,\sigma^2}(x) &= \frac{\dfrac{1}{\sqrt{2\pi}\,\sigma}\, e^{-\frac{1}{2}\left(\frac{x-\mu}{\sigma}\right)^2}}{\int_x^{\infty} \dfrac{1}{\sqrt{2\pi}\,\sigma}\, e^{-\frac{1}{2}\left(\frac{y-\mu}{\sigma}\right)^2}\, dy} \\
&= \frac{1}{\sigma}\, \frac{\dfrac{1}{\sqrt{2\pi}}\, e^{-\frac{1}{2}\left(\frac{x-\mu}{\sigma}\right)^2}}{\int_{(x-\mu)/\sigma}^{\infty} \dfrac{1}{\sqrt{2\pi}}\, e^{-\frac{1}{2}z^2}\, dz} \\
&= \frac{1}{\sigma}\, q_{0,1}\left(\frac{x-\mu}{\sigma}\right) = \frac{1}{\sigma}\, q\left(\frac{x-\mu}{\sigma}\right).
\end{aligned}
$$

Thus, the failure rate for the $N(\mu, \sigma^2)$ distribution can be easily obtained from the $N(0, 1)$ distribution's failure rate. (An illustration of how this works, for an arbitrary function, is given below.)

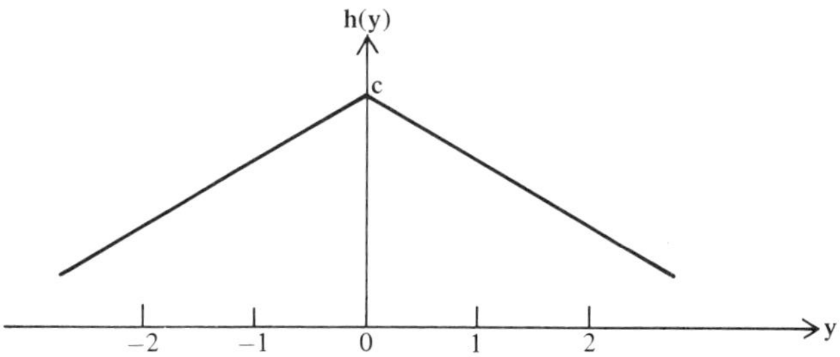

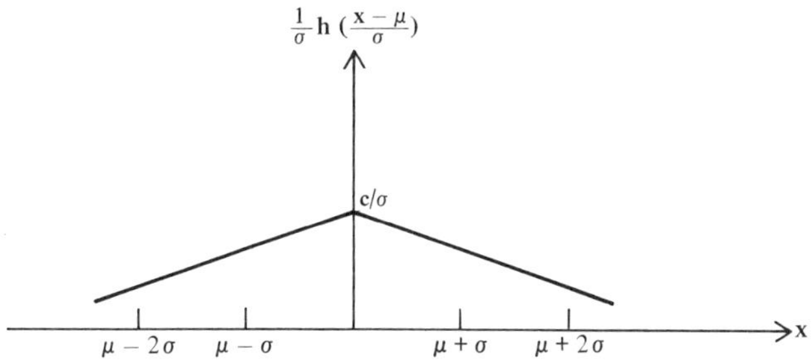

4.6H Prove that the $N(\mu, \sigma^2)$ distribution is an IFR distribution. [*Hint*: From Problem 4.5H we see that $q'_{\mu,\sigma^2}(x)$ is simply related to $q'_{0,1}(y)$. Then use Equation (1.8) of Feller (1968), p. 175.]

5

MOMENTS

Many frequently used r.v.'s can be both characterized, and dealt with effectively for practical purposes, by consideration of quantities called their **moments**. (For example, a gambler might be interested in his average winnings at a game, a business in its average profits on a product, a physicist in the average charge of a particle, and so on. The "average" value of a random phenomenon is one of the "moments.") In this chapter we define and study these quantities, which will be used extensively in Chapter 6 (where they play a vital role in our first simple statistical problems).

5.1. EXPECTATION AND OTHER MOMENTS; MEDIAN

The definitions and results of this section are stated only for two cases: X a discrete r.v. and X an (absolutely) continuous r.v. These are the cases of most practical interest. They can be unified [and the definitions and results extended to a r.v. X with an arbitrary d.f. $F(x)$] by use of the Stieltjes (or Lebesgue-Stieltjes) integral, but this is not central to our theme and will be omitted. In any case it will be transparent how we should deal with mixed discrete and (absolutely) continuous d.f.'s. [For details of these notions for arbitrary d.f.'s $F(x)$ see the book by Tucker (1967).]

Definition 5.1.1. *For a r.v. X we define its* **expectation** [**EX** *or* **E(X)**] *as* $EX = \int_{-\infty}^{\infty} x f_X(x)\,dx$ [$EX = \sum x_i p_X(x_i)$] *if X is absolutely continuous with density function $f_X(x)$* [*if X is discrete with probability function $p_X(x)$*], *provided this integral (sum) exists*[1] *and is finite.*

[1] This requires absolute convergence.

Theorem 5.1.2. *If X is a r.v., then the expectation of the r.v. $g(X)$ exists iff $\int_{-\infty}^{\infty} |g(x)| f_X(x)\,dx < \infty$ when X is absolutely continuous $[\sum |g(x_i)| p_X(x_i) < \infty$ when X is discrete], in which case $Eg(X) = \int_{-\infty}^{\infty} g(x) f_X(x)\,dx$ $[Eg(X) = \sum g(x_i) p_X(x_i)]$.*

Corollary 5.1.3. *If X is a r.v., then EX exists iff $\int_0^{\infty} x f_X(x)\,dx$ and $\int_{-\infty}^{0} x f_X(x)\,dx$ are finite when X is absolutely continuous $\left[\sum_{x_i \geqslant 0} x_i p_X(x_i)\right.$ and $\sum_{x_i < 0} x_i p_X(x_i)$ are finite when X is discrete$\left.\right]$, in which case $EX = \int_{-\infty}^{\infty} x f_X(x)\,dx$ $[EX = \sum x_i p_X(x_i)]$.*

Corollary 5.1.4. *If X is a r.v. with d.f. $F(x)$, then [for any d.f. $F(x)$] EX exists iff $\int_0^{\infty} (1 - F(x))\,dx$ and $\int_{-\infty}^{0} F(x)\,dx$ are finite, in which case*

$$EX = \int_0^{\infty} (1 - F(x))\,dx - \int_{-\infty}^{0} F(x)\,dx.$$

We shall not prove these results, although we note that Corollary 5.1.4 follows from Corollary 5.1.3 and integration by parts.

Suppose that X and $g(X)$ are both (absolutely) continuous r.v.'s. Then by Definition 5.1.1, $Eg(X) = \int_{-\infty}^{\infty} y f_{g(X)}(y)\,dy$, so that we will need to find $f_{g(X)}(y)$ (an onerous task, as we saw in Chapter 4). But Theorem 5.1.2 tells us that $Eg(X) = \int_{-\infty}^{\infty} g(x) f_X(x)\,dx$, a very useful result which allows us to bypass $f_{g(X)}(y)$. In Parzen (1960) this result is shown to be due to the fact that we define expectation only when our integrals are absolutely convergent (pp. 344–345), and an example is given where this result fails to hold if absolute convergence is not required in the definition of expectation (pp. 350–351). Parzen (p. 347) also notes the extension that, if $Y = g_1(X)$, then $Eg(Y) = \int_{-\infty}^{\infty} g(y) f_{g_1(X)}(y)\,dy = \int_{-\infty}^{\infty} g(g_1(x)) f_X(x)\,dx$. [So, we can find $Eg(Y)$ without finding the probability distribution of Y, where $Y = g_1(X)$.]

Theorem 5.1.5. (Properties of Expectation) *If c is a constant and $g(X)$, $g_1(X)$, and $g_2(X)$ are functions whose expectations exist, then*

(1) $E(c) = c$;
(2) $E(cg(X)) = cEg(X)$;
(3) $E(g_1(X) + g_2(X)) = Eg_1(X) + Eg_2(X)$;
(4) $Eg_1(X) \leqslant Eg_2(X)$ *if* $g_1(x) \leqslant g_2(x)$ *for all* x;
(5) $|Eg(X)| \leqslant E|g(X)|$.

Proof: (Part (3) for the case where X is (absolutely) continuous.)

$$\begin{aligned} E(g_1(X) + g_2(X)) &= \int_{-\infty}^{\infty} (g_1(x) + g_2(x)) f_X(x)\,dx \\ &= \int_{-\infty}^{\infty} g_1(x) f_X(x)\,dx + \int_{-\infty}^{\infty} g_2(x) f_X(x)\,dx = Eg_1(X) + Eg_2(X). \end{aligned}$$

Definition 5.1.6. *Let X be a r.v. with d.f. $F(x)$. The* **n^{th} (noncentral) moment** *of X is (if this expectation exists)* $\mu_n' \equiv EX^n$.

Definition 5.1.7. *Let X be a r.v. with d.f. $F(x)$. The* **n^{th} central moment** *of X is (if this expectation exists)* $\mu_n \equiv E(X - EX)^n$. *The* **variance** *of X, denoted* **Var(X)** or **σ^2(X)**, *is μ_2 (the second central moment of X).*

The *mean* (expectation) and variance of a r.v. X are usefully thought of as **measures of location** and **dispersion** (respectively) of the distribution of the r.v. X. The **mean** of an absolutely continuous r.v. X corresponds to the center of gravity of a bar with density $f_X(x)$. Then "the" **median** of X (if X is absolutely continuous) is any point m where

$$\int_{-\infty}^{m} f_X(x)\,dx = \frac{1}{2} = \int_{m}^{\infty} f_X(x)\,dx.$$

(For discrete and other distributions this concept is not as easy to formulate, but may, nevertheless, be usefully thought of in this way.) The **mode** of X (if X is absolutely continuous) is the point where $f_X(x)$ is maximized; it may not be unique, and in fact may not exist. We now wish to investigate relations among the mean, median, and mode, all of which are in a sense measures of the location of a distribution. The following concepts will be useful in our discussion.

Definition 5.1.8. *A density function $f_X(x)$ is* **symmetric about μ** *if (for all x)* $f_X(\mu + x) = f_X(\mu - x)$. (See Figure 5.1–1.)

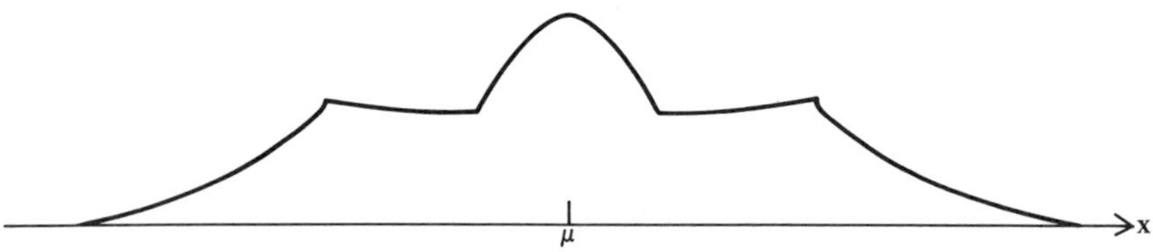

Figure 5.1–1. A typical symmetric density

Definition 5.1.9. *The* 3rd *moment about the mean,* $\mu_3 = E(X - EX)^3$, *is called the* **skewness** *of the distribution of X.* [This is a measure (one of many measures) of asymmetry.]

Corollary 5.1.10. *If the distribution of X is symmetric, then*

$$\mu_3 = E(X - EX)^3 = 0.$$

(See Figure 5.1–2.)

Lemma 5.1.11. *If a r.v. X has a symmetric density, then* Mean (X) = Median (X).

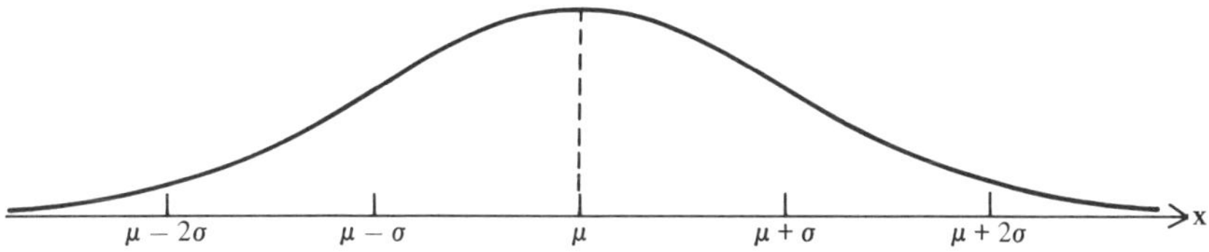

Figure 5.1–2. Measures of location for the $N(\mu, \sigma^2)$ distribution (symmetric, so $\mu_3 = 0$)

Proof: Suppose X is symmetric about a, so that [letting $f_X(x)$ denote the density of X] $f_X(a + x) = f_X(a - x)$ for all x. We first show Median $(X) = a$:

$$\begin{aligned} P[X > a] &= \int_a^\infty f_X(x)\,dx = \int_0^\infty f_X(a + y)\,dy; \\ P[X < a] &= \int_{-\infty}^a f_X(x)\,dx = \int_{-\infty}^0 f_X(y + a)\,dy \\ &= -\int_\infty^0 f_X(a - z)\,dz = \int_0^\infty f_X(a - z)\,dz. \end{aligned}$$

We now prove Mean $(X) = a$, which completes our proof of the lemma:

$$\begin{aligned} EX &= \int_{-\infty}^\infty x f_X(x)\,dx = \int_{-\infty}^\infty (y + a) f_X(a + y)\,dy \\ &= a + \int_{-\infty}^\infty y f_X(a + y)\,dy \\ &= a + \int_0^\infty y f_X(a + y)\,dy + \int_{-\infty}^0 y f_X(a + y)\,dy = a. \end{aligned}$$

Many common fallacies about the mean, median, mode, and symmetry exist even in high quality reference works. For example, in M. G. Kendall and A. Stuart, *The Advanced Theory of Statistics, Volume 1, Distribution Theory* (Second Edition), Hafner Publishing Company, New York, 1963, pp. 40, 85, we find the statements: "It is a useful mnemonic to observe that the mean, median and mode occur in the same order (or the reverse order) as in the dictionary; and that the median is nearer to the mean than to the mode, just as the corresponding words are nearer together in the dictionary" (shown totally false by the example of Figure 5.1–3) and "In a symmetrical population, mean, median and mode coincide" (true as to mean and median, but shown false as to mode by the example of Figure 5.1–4).

Definition 5.1.12. *Let us denote $\sigma^2(X)$ as, simply, σ^2. Then σ (the positive square root of σ^2) is called the* **standard deviation** *of X, and is often written* **$\sigma(X)$**.

Lemma 5.1.13. $\sigma^2(X) \geqslant 0$ *with equality iff X is a constant with probability* 1.

Lemma 5.1.14. $\mathrm{Var}(X) = EX^2 - (EX)^2$.

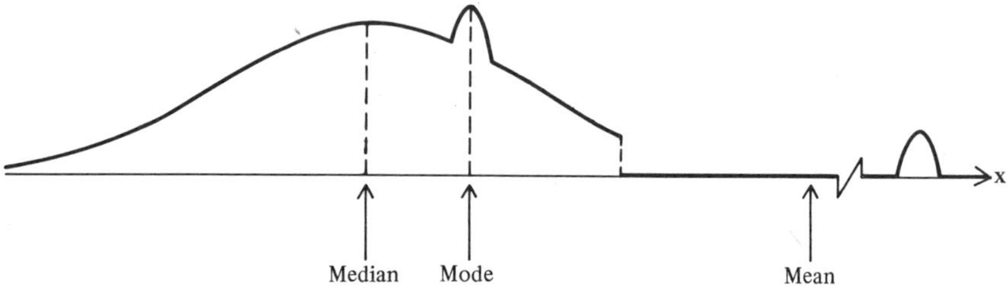

Figure 5.1–3. Measures of location for an altered $N(\mu, \sigma^2)$

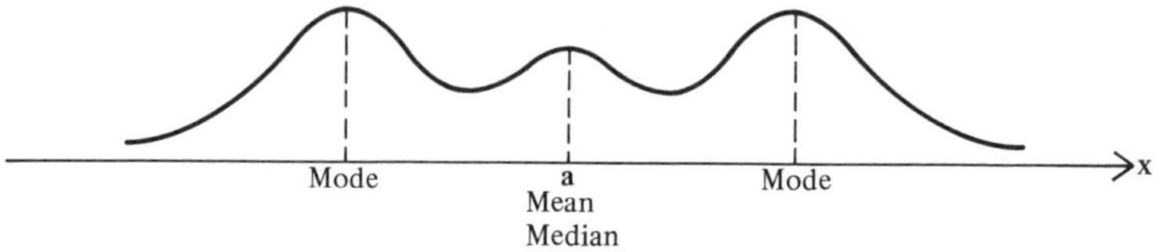

Figure 5.1–4

Proof:

$$\begin{aligned}\text{Var}(X) &= E(X - EX)^2 \\ &= E\{X^2 - 2XEX + (EX)^2\} \\ &= EX^2 - 2(EX)^2 + (EX)^2 \\ &= EX^2 - (EX)^2.\end{aligned}$$

Theorem 5.1.15. $\text{Var}(aX + b) = a^2 \text{Var}(X)$.

Proof:

$$\begin{aligned}\text{Var}(aX + b) &= E[(aX + b) - E(aX + b)]^2 \\ &= E[a(X - EX)]^2 \\ &= a^2 E(X - EX)^2 \\ &= a^2 \text{Var}(X).\end{aligned}$$

Theorem 5.1.16. $X^* = (X - EX)/\sigma(X)$ *has* $EX^* = 0$, $\text{Var}(X^*) = 1$. (*Note that* X^* *is called the* **standardized X**.)

We wish to note briefly **the moment problem**, which considers the question: Suppose a r.v. X has d.f. $F(x)$ and that all moments $EX, EX^2, EX^3, EX^4, \ldots$ (which we called $\mu_1', \mu_2', \mu_3', \mu_4', \ldots$) exist—when do $\mu_1', \mu_2', \mu_3', \ldots$ uniquely determine $F(x)$? [For a reference, see Mood and Graybill (1963), p. 110 (paragraph 2) and pp. 116–117, for a brief treatment; and Feller (1966) for an extensive treatment.] The following results are available.

Theorem 5.1.17. **(Carleman's Theorem)** *A d.f. $F(x)$ is uniquely determined by its moments μ_1', μ_2', μ_3', ... (which are assumed to exist) iff*

$$\sum_{n=1}^{\infty} (M_n)^{-1/n} = \infty,$$

where $M_n = E|X|^n$.

Theorem 5.1.18. **(Hausdorff's Theorem)** *A sequence of numbers μ_0', μ_1', ... represents the moments of some d.f. $F(x)$ concentrated on $[0, 1]$ [that is, such that $F(1) = 1$ and $F(0) = 0 + P[X = 0]$] iff*

$$E(X^k(1 - X)^r) \geqslant 0, \qquad \mu_0' = 1.$$

In such a case the d.f. $F(x)$ is unique, and at each point t where $F(x)$ is a continuous function,

$$F(t) = \lim_{n \to \infty} \sum_{k \leqslant nt} \binom{n}{k} E\{X^k(1 - X)^{n-k}\}.$$

Lemma 5.1.19. *The following criterion may be substituted in Carleman's Theorem (Theorem 5.1.17):*

$$\sum_{n=1}^{\infty} (\mu_{2n}')^{-1/2n} = \sum_{n=1}^{\infty} \frac{1}{(\mu_{2n}')^{1/2n}} = \infty.$$

We now give an example showing how r.v.'s with expectations which do not exist may arise.

Example 5.1.20. **(The persistence of bad luck)** Let X_0 denote my financial loss (or waiting time) at some chance event. Suppose friends of mine encounter the same type of chance event, with results $X_1, X_2, \ldots$. Assume $X_0, X_1, X_2, \ldots$ are independent r.v.'s, each with the same distribution, and are (absolutely) continuous.

Let N = first subscript n such that $X_n > X_0$ (that is, a random quantity, the time it takes until a friend experiences worse luck than mine).

The event $\{N > n - 1\}$ occurs iff the largest of $X_0, X_1, \ldots, X_{n-1}$ is X_0. $P[N > n - 1] = 1/n$ by symmetry. For $n = 1, 2, \ldots,$

$$P[N = n] = P[N > n - 1] - P[N > n] = \frac{1}{n} - \frac{1}{n + 1} = \frac{1}{n(n + 1)}.$$

Now,

$$EN = \sum_{n=1}^{\infty} nP[N = n] = \sum_{n=1}^{\infty} \frac{1}{n + 1} = +\infty.$$

Thus (although $EN = 10{,}000$ would be bad) the waiting time in fact has infinite expectation. However, the probability is only, for example, $\frac{1}{5}$ that you will have to wait longer than 4 friends until someone has worse luck than you. [This should give insight into the nature (and deficiencies) of the mean as a central value.]

As a final word on measures of location, we note the following concept of concentration of a r.v. at a point, which will prove useful in the chapters on statistics.

Definition 5.1.21. *If X and Y are r.v.'s, X is said to be* **more concentrated** *about a than Y if (for all $r > 0$)*

$$P[|X - a| \leqslant r] \geqslant P[|Y - a| \leqslant r].$$

Lemma 5.1.22[2]**.** *If X is more concentrated than Y about μ, and if $EX = EY = \mu$, then $\operatorname{Var}(X) \leqslant \operatorname{Var}(Y)$.*

5.2. MEAN AND VARIANCE OF SPECIAL RANDOM VARIABLES

Theorem 5.2.1. *If X has the Bernoulli distribution (see Definition 3.2.1)*

$$p_X(x) = \begin{cases} p^x(1 - p)^{1-x}, & x = 0, 1 \\ 0, & \text{otherwise,} \end{cases}$$

then $EX = p$, $\operatorname{Var}(X) = pq$.

Proof:

$$\begin{aligned} EX &= 0 \cdot (1 - p) + 1 \cdot p = p. \\ EX^2 &= 0^2 \cdot (1 - p) + 1^2 \cdot p = p. \\ \operatorname{Var}(X) &= EX^2 - (EX)^2 = p - p^2 = p(1 - p) = pq. \end{aligned}$$

Theorem 5.2.2. *If X has the Binomial distribution (see Definition 3.2.3)*

$$p_X(x) = \begin{cases} \dbinom{n}{x} p^x(1 - p)^{n-x}, & x = 0, 1, \ldots, n \\ 0, & \text{otherwise,} \end{cases}$$

then $EX = np$, $\operatorname{Var}(X) = npq$.

Proof:

$$\begin{aligned} EX &= \sum_{x=0}^{n} x \binom{n}{x} p^x(1 - p)^{n-x} = \sum_{x=1}^{n} n \binom{n - 1}{x - 1} p^x q^{n-x} \\ &= np \sum_{x=1}^{n} \binom{n - 1}{x - 1} p^{x-1} q^{(n-1)-(x-1)} \\ &= np \sum_{k=0}^{n-1} \binom{n - 1}{k} p^k q^{(n-1)-k} \\ &= np(p + q)^{n-1} = np. \end{aligned}$$

---

[2] This is easy to see intuitively. A rigorous proof is straightforward (but messy) and is omitted.

$$\begin{aligned}
EX^2 &= \sum_{x=0}^{n} x^2 \binom{n}{x} p^x(1-p)^{n-x} = \sum_{x=0}^{n} [x(x-1)+x]\binom{n}{x} p^x(1-p)^{n-x} \\
&= \sum_{x=2}^{n} x(x-1)\binom{n}{x} p^x(1-p)^{n-x} + EX \\
&= n(n-1)\sum_{x=2}^{n}\binom{n-2}{x-2} p^x q^{n-x} + EX \\
&= n(n-1)p^2 \sum_{k=0}^{n-2}\binom{n-2}{k} p^k q^{(n-2)-k} + EX \\
&= n(n-1)p^2(p+q)^{n-2} + np = n^2p^2 + npq. \\
\operatorname{Var}(X) &= EX^2 - (EX)^2 = npq.
\end{aligned}$$

Theorem 5.2.3. *If X has the Poisson distribution (see Definition 3.2.7)*

$$p_X(x) = \begin{cases} \dfrac{e^{-\mu}\mu^x}{x!}, & x = 0, 1, 2, \ldots \\ 0, & \text{otherwise,} \end{cases}$$

then $EX = \mu$, $\operatorname{Var}(X) = \mu$.

Proof:

$$\begin{aligned}
EX &= \sum_{x=0}^{\infty} x e^{-\mu}\frac{\mu^x}{x!} = e^{-\mu}\sum_{x=1}^{\infty}\frac{\mu^x}{(x-1)!} \\
&= e^{-\mu}\mu \sum_{x=1}^{\infty}\frac{\mu^{x-1}}{(x-1)!} = e^{-\mu}\mu e^{\mu} = \mu. \\
EX^2 &= \sum_{x=0}^{\infty} x^2 e^{-\mu}\frac{\mu^x}{x!} = \sum_{x=0}^{\infty}\{x(x-1)+x\}e^{-\mu}\frac{\mu^x}{x!} \\
&= \sum_{x=2}^{\infty} e^{-\mu}\frac{\mu^x}{(x-2)!} + \mu = \mu^2 + \mu. \\
\operatorname{Var}(X) &= EX^2 - (EX)^2 = \mu.
\end{aligned}$$

Theorem 5.2.4. *If X has the uniform distribution (see Definition 4.2.4)*

$$f_X(x) = \begin{cases} \dfrac{1}{c-b}, & b \leqslant x \leqslant c \\ 0, & \text{otherwise,} \end{cases}$$

then $EX = (b+c)/2$, $\operatorname{Var}(X) = (c-b)^2/12$.

Proof:

$$EX = \int_b^c x \frac{1}{c-b}\,dx = \frac{1}{2}\frac{1}{c-b}x^2\bigg|_b^c = \frac{b+c}{2}.$$

$$EX^2 = \int_b^c x^2 \frac{1}{c-b}\,dx = \frac{1}{3}\frac{1}{c-b}(c^3 - b^3) = \frac{c^2 + bc + b^2}{3}.$$

$$\operatorname{Var}(X) = EX^2 - (EX)^2 = \frac{(c-b)^2}{12}.$$

Theorem 5.2.5. *If X has the exponential distribution (see Definition 4.2.7)*

$$f_X(x) = \begin{cases} \dfrac{1}{\lambda}e^{-x/\lambda}, & 0 \leqslant x < \infty \\ 0, & \text{otherwise,} \end{cases}$$

then $EX = \lambda$, $\operatorname{Var}(X) = \lambda^2$.

Proof:

$$\begin{aligned} EX &= \int_0^\infty x \frac{1}{\lambda} e^{-x/\lambda}\,dx = \lambda \int_0^\infty ye^{-y}\,dy \\ &= \lambda\Gamma(2) = \lambda(1!) = \lambda. \\ EX^2 &= \int_0^\infty x^2 \frac{1}{\lambda} e^{-x/\lambda}\,dx = \lambda^2 \int_0^\infty y^2 e^{-y}\,dy = \lambda^2\Gamma(3) = 2\lambda^2. \\ \operatorname{Var}(X) &= EX^2 - (EX)^2 = \lambda^2. \end{aligned}$$

Theorem 5.2.6. *If X has the normal distribution (see Definition 4.2.13)*

$$f_X(x) = \frac{1}{\sqrt{2\pi}\,\sigma} e^{-\frac{1}{2}\left(\frac{x-\mu}{\sigma}\right)^2}, \qquad -\infty < x < +\infty,$$

then $EX = \mu$, $\operatorname{Var}(X) = \sigma^2$.

Proof:

$$\begin{aligned} EX &= \int_{-\infty}^\infty \frac{x}{\sqrt{2\pi}\,\sigma} e^{-\frac{1}{2}\left(\frac{x-\mu}{\sigma}\right)^2}\,dx = \int_{-\infty}^\infty \frac{\sigma y + \mu}{\sqrt{2\pi}\,\sigma} e^{-\frac{1}{2}y^2}\sigma\,dy \\ &= \frac{\sigma}{\sqrt{2\pi}} \int_{-\infty}^\infty ye^{-\frac{1}{2}y^2}\,dy + \mu \int_{-\infty}^\infty \frac{1}{\sqrt{2\pi}} e^{-\frac{1}{2}y^2}\,dy \\ &= 0 + \mu \cdot 1 = \mu. \end{aligned}$$

$$EX^2 = \int_{-\infty}^{\infty} \frac{x^2}{\sqrt{2\pi}\,\sigma} e^{-\frac{1}{2}\left(\frac{x-\mu}{\sigma}\right)^2} dx = \int_{-\infty}^{\infty} \frac{(\sigma y + \mu)^2}{\sqrt{2\pi}\,\sigma} e^{-\frac{1}{2}y^2}\sigma\, dy$$

$$= \sigma^2 \int_{-\infty}^{\infty} \frac{y^2}{\sqrt{2\pi}} e^{-\frac{1}{2}y^2} dy + \frac{2\sigma\mu}{\sqrt{2\pi}} \int_{-\infty}^{\infty} y e^{-\frac{1}{2}y^2} dy + \mu^2 \int_{-\infty}^{\infty} \frac{1}{\sqrt{2\pi}} e^{-\frac{1}{2}y^2} dy$$

$$= \sigma^2\, 2 \int_0^{\infty} \frac{y^2}{\sqrt{2\pi}} e^{-\frac{1}{2}y^2} dy + 0 + \mu^2 \cdot 1$$

$$= \sigma^2 \frac{1}{\sqrt{2\pi}}\, 2 \int_0^{\infty} z^2 \frac{2}{1} e^{-z^2} \sqrt{2}\, dz + \mu^2$$

$$= \sigma^2 \frac{2}{\sqrt{\pi}} \Gamma\left(\frac{3}{2}\right) + \mu^2 = \sigma^2 \frac{2}{\sqrt{\pi}} \frac{1}{2} \sqrt{\pi} + \mu^2 = \sigma^2 + \mu^2.$$

$$\mathrm{Var}(X) = EX^2 - (EX)^2 = \sigma^2.$$

5.3. MULTIVARIATE MOMENTS

While Section 5.1 considered moments for a single (that is, one-dimensional) random variable X, in this section we deal with moments of several random variables (which, in general, are not independent—thus, they "affect" each other).

If $(X_1, \ldots, X_n)$ is an n-dimensional random variable, then $g(X_1, \ldots, X_n)$ is itself a random variable with some d.f. and hence its expectation has been defined (Definition 5.1.1) and various theorems relative to it have been stated in Section 5.1, for example, Corollary 5.1.3 and Corollary 5.1.4. We now wish to state without proof a result analogous to Theorem 5.1.2.

Theorem 5.3.1. *If $(X_1, \ldots, X_n)$ is an n-dimensional r.v. and if $Eg(X_1, \ldots, X_n)$ exists, then*

$$Eg(X_1, \ldots, X_n) = \int_{-\infty}^{\infty} \cdots \int_{-\infty}^{\infty} g(x_1, \ldots, x_n) f_{X_1, \ldots, X_n}(x_1, \ldots, x_n)\, dx_1 \cdots dx_n$$

when $(X_1, \ldots, X_n)$ is absolutely continuous, and

$$Eg(X_1, \ldots, X_n) = \Sigma \cdots \Sigma\, g(x_1, \ldots, x_n) p_{X_1, \ldots, X_n}(x_1, \ldots, x_n)$$

when $(X_1, \ldots, X_n)$ is discrete.

Definition 5.3.2. *Let (X_1, X_2) be a two-dimensional r.v. For any n_1, n_2 (non-negative integers) we define $\boldsymbol{\mu}'_{n_1, n_2} = E(X_1^{n_1} X_2^{n_2})$ (if this expectation exists). This is called a* **joint moment of $(\mathbf{X}_1, \mathbf{X}_2)$ of order $\mathbf{n}_1 + \mathbf{n}_2$**.

Example 5.3.3. Let (X_1, X_2) be a two-dimensional r.v. Then $\mu'_{1,0} = EX_1$,

$\mu'_{0,1} = EX_2$, $\mu'_{2,0} = EX_1^{\,2}$, $\mu'_{0,2} = EX_2^{\,2}$, $\mu'_{1,1} = E(X_1X_2)$. Note that $\mu'_{1,1}$ is called the **product moment**.

Definition 5.3.4. *Let (X_1, X_2) be a two-dimensional r.v. For any n_1, n_2 (nonnegative integers) we define* $\boldsymbol{\mu}_{n_1,n_2} = E\{(X_1 - EX_1)^{n_1}(X_2 - EX_2)^{n_2}\}$ *(if this expectation exists). This is called a* **joint central moment of $(\mathbf{X}_1, \mathbf{X}_2)$ of order $\mathbf{n}_1 + \mathbf{n}_2$.**

Example 5.3.5. Let (X_1, X_2) be a two-dimensional r.v. Then $\mu_{1,0} = \mu_{0,1} = 0$, $\mu_{2,0} = \text{Var}(X_1)$, $\mu_{0,2} = \text{Var}(X_2)$, $\mu_{1,1} = E\{(X_1 - EX_1)(X_2 - EX_2)\}$. *Note that* $\boldsymbol{\mu}_{1,1}$ *is called the* **covariance of $\mathbf{X}_1$ and $\mathbf{X}_2$**, *denoted* **$\text{Cov}(\mathbf{X}_1, \mathbf{X}_2)$.**

Lemma 5.3.6. $\text{Cov}(X_1, X_2) = E(X_1X_2) - EX_1EX_2.$

Proof:

$$\begin{aligned}\text{Cov}(X_1, X_2) &\equiv E\{(X_1 - EX_1)(X_2 - EX_2)\}\\ &= E\{X_1X_2 - X_2EX_1 - X_1EX_2 + EX_1EX_2\}\\ &= E(X_1X_2) - 2EX_2EX_1 + EX_1EX_2\\ &= E(X_1X_2) - EX_1EX_2.\end{aligned}$$

Lemma 5.3.7. $\text{Var}(X_1 + X_2) = \text{Var}(X_1) + \text{Var}(X_2) + 2\,\text{Cov}(X_1, X_2).$

Proof:

$$\begin{aligned}&\text{Var}(X_1 + X_2)\\ &\quad= E(X_1 + X_2)^2 - (E(X_1 + X_2))^2\\ &\quad= EX_1^{\,2} - (EX_1)^2 + EX_2^{\,2} - (EX_2)^2 + 2\{EX_1X_2 - EX_1EX_2\}.\end{aligned}$$

Thus, the variance of the sum of two r.v.'s is the sum of their variances plus two times the quantity called their "covariance" (a measure of how they are interrelated). If the r.v.'s are unrelated in the sense of zero covariance (or correlation), are they independent? This question will be studied below, with the final results being given in Theorem 5.3.11; related questions will also be considered.

Theorem 5.3.8. *If X_1 and X_2 are independent, then [for any two functions $g_1(x)$ and $g_2(x)$]*

$$Eg_1(X_1)g_2(X_2) = Eg_1(X_1)Eg_2(X_2)$$

if the expectations on the right-hand side exist. (The obvious generalization to n r.v.'s $X_1, \ldots, X_n$ is also true.)

The proof of Theorem 5.3.8 may be easily shown, for example, for $n = 2$ in the absolutely continuous case. There, $f_{X_1,X_2}(x_1, x_2) = f_{X_1}(x_1)f_{X_2}(x_2)$ by independence and Theorem 4.4.6, and the integral defining $Eg_1(X_1)g_2(X_2)$ can easily be written as the product of the integrals defining $Eg_1(X_1)$ and $Eg_2(X_2)$. Now, two random variables X_1 and X_2 are called **uncorrelated** if $EX_1X_2 = EX_1EX_2$ [that is, if the result of Theorem 5.3.8 holds for $g_1(x) = g_2(x) = x$]. We now wish to first study correlation and then its relation to independence. (Rephrasing the question asked following the proof of Lemma 5.3.7, we ask "Can uncorrelated r.v.'s fail to be independent?")

Lemma 5.3.9. *X_1 and X_2 are uncorrelated iff* $\text{Cov}(X_1, X_2) = 0$.

Proof: X_1 and X_2 are uncorrelated iff $EX_1X_2 = EX_1EX_2$,

iff $EX_1X_2 - EX_1EX_2 = 0$,

iff $\text{Cov}(X_1, X_2) = 0$, by Lemma 5.3.6.

Theorem 5.3.10. $\text{Var}(X_1 + X_2) = \text{Var}(X_1) + \text{Var}(X_2)$ *iff X_1 and X_2 are uncorrelated.*

Proof:

$$\text{Var}(X_1 + X_2) = \text{Var}(X_1) + \text{Var}(X_2) + 2\,\text{Cov}(X_1, X_2).$$

Theorem 5.3.11.

(1) *If X_1 and X_2 are independent, they are uncorrelated.*
(2) *If X_1 and X_2 are uncorrelated, they are not necessarily independent.*

Proof:

(1) This follows by Theorem 5.3.8 with $g_1(x) = g_2(x) = x$.
(2) Let $X_1 = \sin(2\pi U)$, $X_2 = \cos(2\pi U)$ with U uniform on (0, 1). We can show that X_1 and X_2 are then uncorrelated, but since $F_{X_1, X_2}(x_1, x_2) \neq F_{X_1}(x_1)F_{X_2}(x_2)$ for all x_1 and x_2, they are not independent.

The previous theorem answers the question, raised following the proof of Lemma 5.3.7, as to whether covariance zero implies independence, while the following theorem gives a precise and complete statement of the relationship between correlation and independence. (Feller says: "As frequently happens, the history of the problem and the luster of partial results easily obscured the fact that the solution is extremely simple by modern methods. The following theorem contains various results proved in the literature by laborious methods.")

Theorem 5.3.12. *X_1 and X_2 are independent iff $Eg_1(X_1)g_2(X_2) = Eg_1(X_1)Eg_2(X_2)$ for all continuous functions $g_1(x)$ and $g_2(x)$ which vanish outside a finite interval.* (See Figure 5.3–1.)

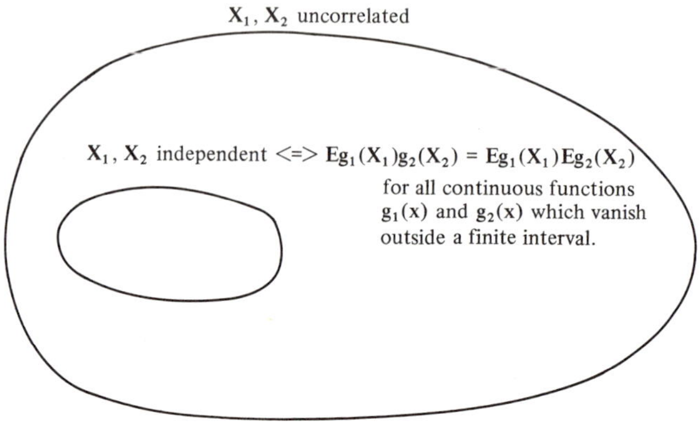

Figure 5.3–1. Relation between independence and correlation

Definition 5.3.13. *The* **correlation coefficient** *of two jointly distributed r.v.'s* X_1 *and* X_2 *is* [*if* $\sigma^2(X_1) > 0$, $\sigma^2(X_2) > 0$ *and* $\sigma^2(X_1)$, $\sigma^2(X_2)$ *are finite*]

$$\rho(X_1, X_2) = \frac{\operatorname{Cov}(X_1, X_2)}{\sigma(X_1)\sigma(X_2)}.$$

Lemma 5.3.14. *Suppose that* $\rho(X_1, X_2)$ *is defined.*[3] *Then* X_1 *and* X_2 *are uncorrelated iff* $\rho(X_1, X_2) = 0$.

The correlation coefficient provides a measure of the **linear relationship** between two r.v.'s, that is, a measure of how good a prediction of the value of one of the r.v.'s can be formed by a linear function of an observation on the other. But suppose $X_1 = X$ and $X_2 = X^2$; then $X_2 = X_1{}^2$ and we have an exact **quadratic relationship** between X_1 and X_2; then $\operatorname{Cov}(X_1, X_2) = EX^3 - EXEX^2$ and if $EX^3 = EXEX^2$ [for example, if $X = -1, 1$ with probability $\frac{1}{2}$ each, or if X has any density symmetric about zero, for example, the $N(0, \sigma^2)$ density], then $\operatorname{Cov}(X_1, X_2) = 0$. However, given X_1 we can predict X_2 very well (if we know that $X_2 = X_1{}^2$). This shows that while the correlation coefficient may measure linear relationships fairly well, it is not good at measuring nonlinear relationships.

Theorem 5.3.15. (Schwarz' Inequality) *For r.v.'s* X *and* Y, *if the expectations exist,* $(EXY)^2 \leqslant EX^2EY^2$. *Equality holds iff* (*for some constant* a) $Y = aX$.

---

[3] We see from Definition 5.3.13 that $\rho(X_1, X_2)$ fails to be defined iff at least one of X_1, X_2 has a constant value (variance zero).

Proof: Assume $EY^2 > 0$ (otherwise $Y = 0$ with probability 1 and the result is trivial). Then

$$E(X + tY)^2 = EX^2 + 2tEXY + t^2EY^2 \geqslant 0$$

is a quadratic polynomial in t. Since it is nonnegative, it has either (1) two complex roots [when > 0] or (2) a double real root λ [when $= 0$]. These are (or this is)

$$t = \frac{-2EXY \pm \sqrt{4(EXY)^2 - 4EX^2EY^2}}{2EY^2}.$$

So, we have case (1) iff $4(EXY)^2 - 4EX^2EY^2 < 0$, that is, iff $(EXY)^2 < EX^2EY^2$. We have case (2) iff $(EXY)^2 = EX^2EY^2$; and this is the case $E(X + t_0Y)^2 = 0$, so $X + t_0Y = 0$ with probability 1 and $t_0 = -EXY/EY^2$.

Theorem 5.3.16.

(1) $-1 \leqslant \rho(X_1, X_2) \leqslant 1$;

(2) $\rho(X_1, X_2) = 1$ *iff* $\dfrac{X_2 - EX_2}{\sigma(X_2)} = \dfrac{X_1 - EX_1}{\sigma(X_1)}$;

and

(3) $\rho(X_1, X_2) = -1$ *iff* $\dfrac{X_2 - EX_2}{\sigma(X_2)} = -\dfrac{X_1 - EX_1}{\sigma(X_1)}$.

Proof:

(1) Take (in Theorem 5.3.15) $X = X_1 - EX_1$, $Y = X_2 - EX_2$. Then

$$\begin{aligned} (EXY)^2 &\leqslant EX^2EY^2; \\ [\mathrm{Cov}(X_1, X_2)]^2 &\leqslant \mathrm{Var}(X_1)\,\mathrm{Var}(X_2); \\ |\mathrm{Cov}(X_1, X_2)| &\leqslant \sigma(X_1)\,\sigma(X_2); \\ |\rho(X_1, X_2)| &\leqslant 1; \\ -1 \leqslant \rho(X_1, X_2) &\leqslant 1. \end{aligned}$$

(2) *and* **(3)**

$$(EXY)^2 = EX^2EY^2 \text{ iff } X = \frac{EXY}{EY^2}\,Y.$$

So,

$$\begin{aligned} |\rho(X_1, X_2)| = 1 \text{ iff } X_1 - EX_1 &= \frac{E(X_1 - EX_1)(X_2 - EX_2)}{\sigma^2(X_2)}\,(X_2 - EX_2) \\ \text{iff } \frac{X_1 - EX_1}{\sigma(X_1)} &= \frac{E(X_1 - EX_1)(X_2 - EX_2)}{\sigma(X_1)\sigma(X_2)}\,\frac{X_2 - EX_2}{\sigma(X_2)} \\ \text{iff } \frac{X_1 - EX_1}{\sigma(X_1)} &= \rho(X_1, X_2)\,\frac{X_2 - EX_2}{\sigma(X_2)}. \end{aligned}$$

As an example, recall (Definition 4.5.2) that a r.v. $\mathbf{X} = (X_1, X_2)$ has the bivariate normal distribution $N(\mu_1, \mu_2; \sigma_1^2, \sigma_2^2, \rho)$ if $(-\infty < x_1, x_2 < +\infty)$

$$f_{X_1, X_2}(x_1, x_2) = \frac{1}{2\pi\sigma_1\sigma_2\sqrt{1-\rho^2}} e^{-\frac{1}{2(1-\rho^2)}\left\{\left(\frac{x_1-\mu_1}{\sigma_1}\right)^2 - 2\rho\left(\frac{x_1-\mu_1}{\sigma_1}\right)\left(\frac{x_2-\mu_2}{\sigma_2}\right) + \left(\frac{x_2-\mu_2}{\sigma_2}\right)^2\right\}}.$$

Then, we may prove the following:

Theorem 5.3.17. *Let X_1, X_2 be jointly normally distributed. Then $EX_1 = \mu_1$, $\operatorname{Var}(X_1) = \sigma_1^2$, $EX_2 = \mu_2$, $\operatorname{Var}(X_2) = \sigma_2^2$, $\rho(X_1, X_2) = \rho$. Also, X_1 and X_2 are uncorrelated iff they are independent.*

Proof: We first note that if X_1, X_2 has the bivariate normal distribution, then

$$\begin{aligned}
f_{X_1}(x_1) &= \int_{-\infty}^{\infty} f_{X_1, X_2}(x_1, x_2)\, dx_2 \\
&= \frac{1}{2\pi\sigma_1\sigma_2\sqrt{1-\rho^2}} \int_{-\infty}^{\infty} e^{-\frac{1}{2(1-\rho^2)}\left\{\left(\frac{x_1-\mu_1}{\sigma_1}\right)^2 - 2\rho\left(\frac{x_1-\mu_1}{\sigma_1}\right)\left(\frac{x_2-\mu_2}{\sigma_2}\right) + \left(\frac{x_2-\mu_2}{\sigma_2}\right)^2\right\}} dx_2 \\
&= \frac{1}{2\pi\sigma_1\sqrt{1-\rho^2}} \int_{-\infty}^{\infty} e^{-\frac{1}{2(1-\rho^2)}\left\{\left(\frac{x_1-\mu_1}{\sigma_1}\right)^2 - 2\rho\left(\frac{x_1-\mu_1}{\sigma_1}\right)y + y^2\right\}} dy \\
&= \frac{1}{2\pi\sigma_1\sqrt{1-\rho^2}} \int_{-\infty}^{\infty} e^{-\frac{1}{2(1-\rho^2)}\left\{\left(\frac{x_1-\mu_1}{\sigma_1}\right)^2 - \rho^2\left(\frac{x_1-\mu_1}{\sigma_1}\right)^2 + \left(y-\rho\left(\frac{x_1-\mu_1}{\sigma_1}\right)\right)^2\right\}} dy \\
&= \frac{1}{\sqrt{2}\pi\sigma_1} \int_{-\infty}^{\infty} e^{-\frac{1}{2}\left(\frac{x_1-\mu_1}{\sigma_1}\right)^2} e^{-z^2}\, dz \\
&= \frac{1}{\sqrt{2}\pi\sigma_1} e^{-\frac{1}{2}\left(\frac{x_1-\mu_1}{\sigma_1}\right)^2} \Gamma(\tfrac{1}{2}) = \frac{1}{\sqrt{2\pi}\sigma_1} e^{-\frac{1}{2}\left(\frac{x_1-\mu_1}{\sigma_1}\right)^2},
\end{aligned}$$

where we made the substitution $y = (x_2 - \mu_2)/\sigma_2$, completed the square via

$$-2\rho\left(\frac{x_1-\mu_1}{\sigma_1}\right)y + y^2 = \left(y - \rho\left(\frac{x_1-\mu_1}{\sigma_1}\right)\right)^2 - \rho^2\left(\frac{x_1-\mu_1}{\sigma_1}\right)^2,$$

and made the substitution

$$z = \frac{y - \rho\left(\dfrac{x_1-\mu_1}{\sigma_1}\right)}{\sqrt{2(1-\rho^2)}}.$$

Thus, if X_1, X_2 has the bivariate normal distribution, then X_1 has the (univariate) normal distribution which (see Theorem 5.2.6) shows that $EX_1 = \mu_1$ and $\operatorname{Var}(X_1) = \sigma_1^2$. Similarly, we find $EX_2 = \mu_2$, $\operatorname{Var}(X_2) = \sigma_2^2$. The remainder of this theorem's proof is left to Problem 5.4. Theorem 5.3.17 may be extended to the multivariate normal distribution of Definition 4.5.3; in particular, $X_1, \ldots, X_n$ has an n-dimensional normal distribution iff every linear combination of $X_1, \ldots, X_n$ is normally distributed. Note, however, that X_1 and X_2 must be *jointly* normal for their "normality" to imply independence iff $\rho(X_1, X_2) = 0$, as the following example shows.

Example 5.3.18. Let X be normal with mean 0. Let Y be independent of X and $P[Y = 1] = P[Y = -1] = \frac{1}{2}$. Define $Z = XY$. *Then* X and Z have marginal normal distributions, $\rho(X, Z) = 0$, *but* X and Z are dependent.

The crux of the argument is that the normal distribution is symmetric about its mean, so that $P[X < a] = P[X > -a]$ for all a. Then, since X and Y are independent,

$$\begin{aligned} P[Z < a] &= P[XY < a] \\ &= P[Y = 1]P[X < a] + P[Y = -1]P[X > -a] \\ &= P[X < a], \end{aligned}$$

that is, Z (as well as X) has a marginal normal distribution. Now $\rho(X, Z) = 0$ since $E(XZ) = E(X^2Y) = EX^2EY = 0$. But $Z = \pm X$, so Z and X are quite dependent.

As a final note, we make a definition of **conditional expectation**. (For example, if we know $X = x$, what is then EY?)

Definition 5.3.19. *If $F_{Y|X}(y|x)$ is absolutely continuous or discrete, then (respectively) the* **conditional mean of Y, given X** *is*

$$\begin{aligned} E(Y|X = x) &= \textstyle\int_{-\infty}^{\infty} y f_{Y|X}(y|x)\,dy \\ E(Y|X = x) &= \textstyle\sum y p_{Y|X}(y|x). \end{aligned}$$

Theorem 5.3.20[4]**.** *If the quantities indicated exist, if $F_X(x)$ is absolutely continuous or discrete, then (respectively)*

$$\begin{aligned} EY &= \textstyle\int_{-\infty}^{\infty} E(Y|X = x) f_X(x)\,dx, \\ EY &= \textstyle\sum E(Y|X = x) p_X(x). \end{aligned}$$

This can be quite useful in the calculation of expected values. We note that it is possible to define not only $E(Y|X = x)$, but also a r.v. $E(Y|X)$ (at least if EY exists). Then, our last result says that

$$E(E(Y|X)) = EY.$$

A useful extension of Theorem 5.3.20 *says* (in this last notation) *that*

$$E(g(Y)h(X)) = E(h(X)E(g(Y)|X)).$$

---

[4] The proof follows easily using Definition 5.3.19 and properties of conditional probability and conditional densities.

Example 5.3.21. Suppose that (X, Y) is a two-dimensional r.v. with density function

$$f_{X,Y}(x, y) = \begin{cases} 6, & \text{if } x^2 \leqslant y \leqslant x, 0 \leqslant x \leqslant 1 \\ 0, & \text{otherwise;} \end{cases}$$

that is, (X, Y) is uniformly distributed over the region shown in Figure 5.3–2. [Check that $f_{X,Y}(\cdot, \cdot)$ is a density function, that is, it is $\geqslant 0$ and $\int_{-\infty}^{\infty} \int_{-\infty}^{\infty} f_{X,Y}(x, y)\,dx\,dy = 1$.] What is the average of Y in those cases in which X has value x_0 $(0 \leqslant x_0 \leqslant 1)$?

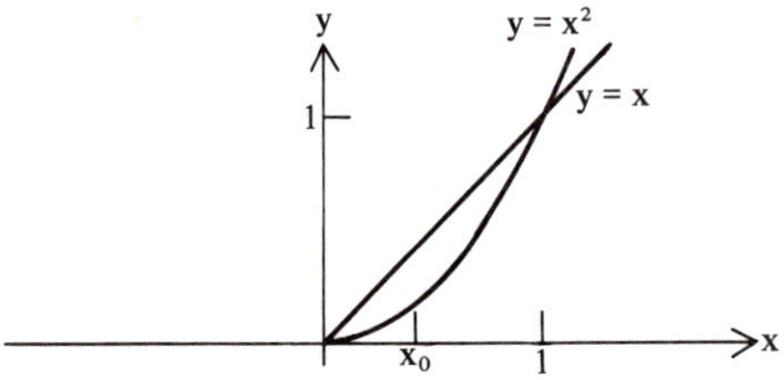

Figure 5.3–2. Region $\mathbf{x}^2 \leqslant \mathbf{y} \leqslant \mathbf{x}, 0 \leqslant \mathbf{x} \leqslant 1$

The desired quantity is that made precise by Definition 5.3.19; namely, $E(Y|X = x_0)$. Now [see Equation (4.4.5)]

$$E(Y|X = x_0) = \int_{-\infty}^{\infty} y f_{Y|X}(y|x_0)\,dy = \int_{-\infty}^{\infty} y \frac{f_{X,Y}(x_0, y)}{f_X(x_0)}\,dy. \quad (5.3.22)$$

We now pause to calculate $f_X(x_0)$:

$$f_X(x) = \int_{-\infty}^{\infty} f_{X,Y}(x, y)\,dy = \begin{cases} \int_{x^2}^{x} 6\,dy, & \text{if } 0 \leqslant x \leqslant 1 \\ 0, & \text{otherwise} \end{cases}$$

$$= \begin{cases} 6(x - x^2), & \text{if } 0 \leqslant x \leqslant 1 \\ 0, & \text{otherwise.} \end{cases}$$

Thus,

$$f_{Y|X}(y|x) = \frac{f_{X,Y}(x, y)}{f_X(x)} = \begin{cases} \dfrac{1}{x - x^2}, & \text{if } x^2 \leqslant y \leqslant x, 0 \leqslant x \leqslant 1 \\ 0, & \text{otherwise} \end{cases}$$

and

$$E(Y|X = x_0) = \int_{x_0^2}^{x_0} y \frac{1}{x_0 - x_0^2}\,dy = \frac{x_0(x_0 + 1)}{2}.$$

Note that $f_{Y|X}(y|x)$ is the uniform distribution on (x^2, x), so it follows immediately from Theorem 5.2.4 that $E(Y|X = x_0) = (x_0^2 + x_0)/2 = x_0(x_0 + 1)/2$.

Also note that, having found $E(Y|X = x_0)$, we can easily find $E(Y)$ via Theorem 5.3.20:

$$\begin{aligned} E(Y) &= \int_{-\infty}^{\infty} E(Y|X = x_0) f_X(x_0)\, dx_0 \\ &= \int_0^1 \frac{x^2 + x}{2} 6(x - x^2)\, dx \\ &= \tfrac{2}{5}. \end{aligned}$$

5.4. MOMENT-GENERATING FUNCTIONS AND CHARACTERISTIC FUNCTIONS

In this section, we introduce and study moment-generating functions (m.g.f.'s) and characteristic functions (ch.f.'s). These will allow us to prove theorems of fundamental importance in statistics (for example, in Chapter 6, the Central Limit Theorem). We first introduce the conceptually simpler but less elegant m.g.f., then only going to the ch.f.

Definition 5.4.1. *The* **moment-generating function** *of a r.v. X is defined for every real number t by* $\boldsymbol{\Psi_X(t) = Ee^{tX}}$. [Thus, the **m.g.f.** is a function of a real variable t. Of course, it is only defined for values t for which Ee^{tX} exists. Recall that we spoke previously of expected values of functions of random variables, $Eg(X)$. Here $g(X) = e^{tX}$.]

Theorem 5.4.2. *If the m.g.f. $\psi_X(t)$ of the r.v. X exists for $|t| \leqslant T$ (for some $T > 0$), then EX^n exists ($n = 1, 2, 3, \ldots$) and*

$$EX^n = \psi_X{}^{(n)}(0) \equiv \left. \frac{d^n}{dt^n} \psi_X(t) \right|_{t=0}.$$

Proof: We know that (by definition) $\psi_X(t) = Ee^{tX}$, and for the function e^y we have the series expansion $e^y = 1 + y + \dfrac{y^2}{2!} + \dfrac{y^3}{3!} + \cdots$. Thus, for $|t| < T$ (where Ee^{tX} exists)

$$\psi_X(t) = 1 + (EX)t + (EX^2)\frac{t^2}{2!} + \cdots + (EX^n)\frac{t^n}{n!} + \cdots.$$

(We have omitted the proof that the expected value, of this infinite sum of terms, equals the sum of the expected values.) The result follows by differentiation (where we also omit the proof that the derivative, of this infinite sum of terms, equals the sum of the derivatives of the individual terms).

Note that we can therefore readily obtain EX^n for any $n = 1, 2, \ldots$ if we can obtain a power series expansion of $\psi_X(t)$; EX^n will simply be the coefficient of $t^n/n!$ in the power series expansion.

Alternate Proof: Since $\psi_X(t)$ exists for $|t| < T$, we may (we omit the proof) form its successive derivatives by successively differentiating "under the integral sign." Thus,

$$\psi_X^{(n)}(t) = \frac{d^n}{dt^n}\psi_X(t) = E\left[\frac{\partial}{\partial t}\frac{d^{n-1}}{dt^{n-1}}e^{tX}\right] = E[X^n e^{tX}].$$

Setting $t = 0$, $\psi_X^{(n)}(0) = EX^n$.

Definition 5.4.3. *The* **joint m.g.f.** *of $X_1, \ldots, X_n$ is defined for real numbers $t_1, \ldots, t_n$ by*

$$\psi_{X_1,\ldots,X_n}(t_1, \ldots, t_n) = Ee^{t_1X_1 + \cdots + t_nX_n}.$$

We will simply note informally that as in Theorem 5.4.2, we may obtain expressions like

$$\psi_{X_1,X_2}(t_1, t_2) = \sum_{n_1=0}^{\infty}\sum_{n_2=0}^{\infty}\frac{t_1^{n_1}}{n_1!}\frac{t_2^{n_2}}{n_2!}E(X_1^{n_1}X_2^{n_2}).$$

These can then be used to prove results such as the following:

$$E(X_1) = \frac{\partial}{\partial t_1}\psi_{X_1,X_2}(0, 0),$$

$$E(X_1^2) = \frac{\partial^2}{\partial t_1^2}\psi_{X_1,X_2}(0, 0),$$

$$E(X_1X_2) = \frac{\partial^2}{\partial t_1\partial t_2}\psi_{X_1,X_2}(0, 0),$$

$$\text{Var}(X_1) = \frac{\partial^2}{\partial t_1^2}\psi_{X_1-EX_1,X_2-EX_2}(0, 0),$$

$$\text{Cov}(X_1, X_2) = \frac{\partial^2}{\partial t_1\partial t_2}\psi_{X_1-EX_1,X_2-EX_2}(0, 0).$$

If we have a r.v. X, then its probability distribution is completely specified by its d.f. $F_X(x)$ $(-\infty < x < +\infty)$ and this always exists. Alternatively, its probability distribution is completely specified by $P[X \in S]$ (all $S \subseteq \{x: -\infty < x < +\infty\}$), and this always exists. [In the absolutely continuous and discrete cases, respectively, we may give its density function $f_X(x)$ $(-\infty < x < +\infty)$ and its probability function $p_X(x)$ $(-\infty < x < +\infty)$.] The m.g.f., of course, does not have this property, since it does not always exist. We will now show that $F_X(x)$ $(-\infty < x < +\infty)$ and $P[X \in S]$ $(S \subseteq \{x: -\infty < x < +\infty\})$ share a common property. After this, the *characteristic function*, which also serves to specify the probability distribution of a r.v., will be introduced.

Lemma 5.4.4. *Both the probability function $P[X \in S]$ for sets S, and the d.f. $F_X(x)$ $(-\infty < x < +\infty)$, can be regarded as expectations.*

Proof: Let the **indicator function** of the set S be defined by

$$I_S(x) = \begin{cases} 1, & \text{if } x \in S \\ 0, & \text{if } x \notin S. \end{cases} \tag{5.4.5}$$

Then $P[X \in S] = 1 \cdot P[X \in S] + 0 \cdot P[X \notin S] = E\{I_S(X)\}$, proving our first result.

Secondly, define a function $I_y(x)$ of a real variable x $(-\infty < x < +\infty)$ by

$$I_y(x) = \begin{cases} 1, & \text{if } x \leqslant y \\ 0, & \text{if } x > y. \end{cases} \tag{5.4.6}$$

Then $F_X(y) = E\{I_y(X)\}$, proving our second result. [Another proof could be given by simply noting that $F_X(y) = P[X \in S]$, itself an expectation, when $S = (-\infty, x)$.]

Question: Do other families of functions on the real line, besides those of form (5.4.5) and (5.4.6), exist such that (when evaluated at X) knowledge of their expectations is sufficient to determine the probability distribution of X? *Answer*: Yes, the **complex exponential functions**.

Definition 5.4.7. *For a function $g(x)$ which (for $-\infty < x < +\infty$) takes values from among the complex numbers, we define*

$$Eg(X) = E\{\text{Re}\, g(X)\} + i\, E\{\text{Im}\, g(X)\},$$

where "Re z" *means* "real part of z" *and* "Im z" *means* "imaginary part of z." [Any complex number z may be written as $z = x + iy$ where x and y are real numbers. Then Re $z = x$ and Im $z = y$.] *We say $Eg(X)$ exists if $E|g(X)| < \infty$.* [Of course, $i = \sqrt{-1}$ and $i^2 = -1$.]

Corollary 5.4.8. *All the usual properties of expectation hold for complex-valued functions whose expectations exist (under Definition 5.4.7).*

Theorem 5.4.9. $|Eg(X)| \leqslant E|g(X)|$; *that is,*

$$\sqrt{(E\,\text{Re}\, g(X))^2 + (E\,\text{Im}\, g(X))^2} \leqslant E\sqrt{(\text{Re}\, g(X))^2 + (\text{Im}\, g(X))^2}.$$

Definition 5.4.10. *The* **characteristic function (ch. f.)** *of a r.v. X is given by (for all real numbers t)*

$$\boldsymbol{\phi_X(t) = Ee^{itX}} = E\{\cos(tX) + i\sin(tX)\} = E\cos(tX) + i\,E\sin(tX).$$

Theorem 5.4.11.

(1) *The ch.f. of a r.v. X always exists;*
(2) $|\phi_X(t)| \leqslant 1$ *for all real numbers* t;
(3) $\phi_X(0) = 1$ *and* $\phi_X(t) = \overline{\phi_X(-t)}$;
(4) $\phi_X(t)$ *is continuous (in fact uniformly continuous) as a function of* $t(-\infty < t < +\infty)$.

Proof:

(1) $E|e^{itX}| = E(1) = 1 < \infty$ since $|e^{itx}|^2 = (\cos(tx))^2 + (\sin(tx))^2 = 1$.
(2) $|Ee^{itX}| \leqslant E|e^{itX}| = 1$.
(3) $\phi_X(0) = Ee^0 = 1$; $Ee^{itX} = \overline{(Ee^{-itX})}$.
(4) This is beyond the scope of this book; see, for example, E. Lukacs, *Characteristic Functions*, Charles Griffin & Co., London, 1960.

Theorem 5.4.12. $\phi_{aX+b}(t) = e^{ibt}\phi_X(at)$.

Proof:

$$\phi_{aX+b}(t) = Ee^{it(aX+b)} = e^{itb}Ee^{i(at)X}.$$

Theorem 5.4.13. *If* EX^k *exists, then* $\phi_X(t)$ *can be expanded in a neighborhood of* $t = 0$ *as follows:*

$$\phi_X(t) = 1 + \frac{i\,EX}{1!}t + \frac{i^2\,EX^2}{2!}t^2 + \cdots + \frac{i^kEX^k}{k!}t^k + o(t^k),$$

where $\lim_{t\to 0} \frac{o(t^k)}{t^k} = 0$.

Proof: If $h(x)$ is k-times differentiable, then by Taylor's Theorem with Remainder,

$$h(x) = h(0) + \frac{h^{(1)}(0)}{1!}x + \frac{h^{(2)}(0)}{2!}x^2 + \cdots + \frac{h^{(k-1)}(0)}{(k-1)!}x^{k-1} + \frac{h^{(k)}(\xi)}{k!}x^k,$$

with $0 < \xi < x$ (or, if $x < 0$, $x < \xi < 0$). In our case, since EX^k exists, the differentiations may be taken under the integral sign (yielding our coefficients). The error term follows by continuity of $\phi_X{}^{(k)}(t)$ for all real t (which we shall not prove).

Theorem 5.4.14. (Uniqueness Theorem)

(1) *Two ch.f.'s which are equal at all (except a countable number of) points are the ch.f.'s of the same d.f.*
(2) *Two d.f.'s which are equal at all (except a countable number of) points give rise to the same ch.f.*

Corollary 5.4.15. *If EX^k exists, then $EX^k = \phi_X^{(k)}(0)/i^k$.*

Proof: Use Theorem 5.4.13.

Example 5.4.16. If X has the Binomial distribution (see Definition 3.2.3)

$$p_X(x) = \begin{cases} \binom{n}{x} p^x(1-p)^{n-x}, & x = 0, 1, \ldots, n \\ 0, & \text{otherwise,} \end{cases}$$

then

$$\begin{aligned} \phi_X(t) = Ee^{itX} &= \sum_{k=0}^{n} \binom{n}{k} p^k q^{n-k} e^{itk} \\ &= \sum_{k=0}^{n} \binom{n}{k} (pe^{it})^k q^{n-k} \\ &= (pe^{it} + q)^n. \end{aligned}$$

Example 5.4.17. If X has the Poisson distribution (see Definition 3.2.7)

$$p_X(x) = \begin{cases} \dfrac{e^{-\mu}\mu^x}{x!}, & x = 0, 1, 2, \ldots \\ 0, & \text{otherwise,} \end{cases}$$

then

$$\begin{aligned} \phi_X(t) = Ee^{itX} &= \sum_{k=0}^{\infty} e^{-\mu} \frac{\mu^k}{k!} e^{itk} \\ &= e^{-\mu} \sum_{k=0}^{\infty} \frac{(\mu e^{it})^k}{k!} = e^{-\mu} e^{\mu e^{it}} \\ &= e^{\mu(e^{it}-1)}. \end{aligned}$$

Example 5.4.18. If $P[X = c] = 1$ for some constant c, then

$$\phi_X(t) = Ee^{itX} = e^{itc}.$$

We will now state, without proof, several more ch.f.'s of important distributions of r.v.'s.

Example 5.4.19. If X has the normal distribution (see Definition 4.2.13)

$$f_X(x) = \frac{1}{\sqrt{2\pi}\,\sigma} e^{-\frac{1}{2}\left(\frac{x-\mu}{\sigma}\right)^2}, \qquad -\infty < x < +\infty,$$

then

$$\phi_X(t) = e^{\mu i t - \frac{1}{2}\sigma^2 t^2}.$$

[So, if X is standard normal with $\mu = 0$ and $\sigma^2 = 1$, then $\phi_X(t) = e^{-\frac{1}{2}t^2}$.

Letting $Y = aX + b$ and using Theorem 5.4.12, the general result follows from this special one.]

Example 5.4.20. If X has the uniform distribution (see Definition 4.2.4)

$$f_X(x) = \begin{cases} \dfrac{1}{b - a}, & a \leqslant x \leqslant b \\ 0, & \text{otherwise,} \end{cases}$$

then

$$\phi_X(t) = \frac{e^{itb} - e^{ita}}{it(b - a)}.$$

Example 5.4.21. If X has the exponential distribution (see Definition 4.2.7)

$$f_X(x) = \begin{cases} \lambda e^{-\lambda x}, & 0 \leqslant x < \infty \\ 0, & \text{otherwise,} \end{cases}$$

then

$$\phi_X(t) = \left(1 - \frac{it}{\lambda}\right)^{-1}.$$

Example 5.4.22. If X has the gamma distribution (see Definition 4.2.6) $G(x|\alpha = r - 1, \beta = 1/\lambda, A = 0)$, that is,

$$f_X(x) = \begin{cases} \dfrac{\lambda}{\Gamma(r)} (\lambda x)^{r-1} e^{-\lambda x}, & x \geqslant 0 \\ 0, & \text{otherwise,} \end{cases}$$

then

$$\phi_X(t) = \left(1 - \frac{it}{\lambda}\right)^{-r}.$$

Example 5.4.23. (Generating moments) Suppose X has the binomial distribution of Example 5.4.14. Then, by Corollary 5.4.15,

$$\begin{aligned} EX &= \frac{\phi_X{}^{(1)}(0)}{i^1} = \frac{1}{i} \frac{d}{dt} \{(pe^{it} + q)^n\}\big|_{t=0} \\ &= \frac{1}{i} n(pe^{it} + q)^{n-1} ipe^{it}\big|_{t=0} \\ &= \frac{1}{i} n(p + q)^{n-1} ip1 = np. \end{aligned}$$

Example 5.4.24. (Determining distributions) Suppose that $X_1, \ldots, X_n$ are independent r.v.'s, each with the exponential distribution of Example 5.4.21. What is the distribution of $Y = X_1 + \cdots + X_n$?

To find the distribution of Y, we could use the methods of Chapter 4. However, this can also be found as follows. The characteristic function of Y is

$$\begin{aligned}\phi_Y(t) &= Ee^{itY} = Ee^{it(X_1 + \cdots + X_n)}\\ &= E(e^{itX_1} \cdots e^{itX_n}), \quad \text{and by Theorem 5.3.8}\\ &= (Ee^{itX_1}) \cdots (Ee^{itX_n})\\ &= \phi_{X_1}(t) \cdots \phi_{X_n}(t), \quad \text{and by Example 5.4.19}\\ &= \left(1 - \frac{it}{\lambda}\right)^{-1} \cdots \left(1 - \frac{it}{\lambda}\right)^{-1}\\ &= \left(1 - \frac{it}{\lambda}\right)^{-n},\end{aligned}$$

and by Theorem 5.4.14 and Example 5.4.22 we know Y has the gamma distribution $G(x|\alpha = n - 1, \beta = 1/\lambda, A = 0)$. (Note that Theorem 5.3.8's use needed the independence of $X_1, \ldots, X_n$.) Thus, use of characteristic functions can yield simple determinations of the distributions of functions of r.v.'s (*if* we can recognize the relevant characteristic functions). This is studied in more detail in Section 6.1.

Example 5.4.25. Suppose that X is a discrete r.v. with distribution given by $P[X = -1] = P[X = +1] = \frac{1}{3}$, $P[X = 0] = P[X = 6] = \frac{1}{6}$. Find $E(X)$, $E(X^2)$, $\operatorname{Var}(X)$, $E(e^{itX})$.

$$\begin{aligned}E(X) &= (-1) \cdot \tfrac{1}{3} + (1) \cdot \tfrac{1}{3} + (0) \cdot \tfrac{1}{6} + (6) \cdot \tfrac{1}{6} = 1.\\ E(X^2) &= (-1)^2 \cdot \tfrac{1}{3} + (1)^2 \cdot \tfrac{1}{3} + (0)^2 \cdot \tfrac{1}{6} + (6)^2 \cdot \tfrac{1}{6} = 6\tfrac{2}{3}.\\ \operatorname{Var}(X) &= E(X^2) - (EX)^2 = 6\tfrac{2}{3} - 1 = 5\tfrac{2}{3}.\\ E(e^{itX}) &= \tfrac{1}{6}\{2e^{-it} + 2e^{it} + e^{6it} + 1\}.\end{aligned}$$

PROBLEMS

5.1 **(a)** Let $X_0, X_1, X_2, \ldots$ be independent and identically distributed absolutely continuous random variables. Let

$$\begin{aligned}N_1 &= \text{The first subscript } n \text{ such that } X_n > X_0;\\ N_2 &= \text{The first subscript } n \text{ such that } X_n < X_0.\end{aligned}$$

[Intuitively, we may think of N_1 as the number of the first person to experience worse luck, and N_2 as the number of the first person to experience better luck.] Find $E(N_1)$; $E(N_2)$; $E\min(N_1, N_2)$; $E[\min(N_1, N_2)]^2$; $\operatorname{Var}[\min(N_1, N_2)]$.

(b) If $P[X = 1] = 1$, find $E(X)$ and $\operatorname{Var}(X)$.

(c) Suppose $P[X = n] = 1/(n^2 + n)$ for $n = 1, 2, \ldots$. Find $E(X)$.

5.2 Show that random variables X and Y may be uncorrelated and yet non-independent.

5.3 Consider the following gambling game. A player bets \$1 on a number i_0 ($1 \leqslant i_0 \leqslant 6$), where i_0 is an integer. Two fair dice are then tossed. If two i_0's show, the player receives \$3; if one i_0 shows, the player receives \$2; and if no i_0's show, the player receives \$0.

(a) Let X_1 = the no. of i_0's that appear in a single play of this game. Find the probability distribution of X_1 and $E(X_1)$.

(b) Suppose each *winning* payoff is increased by \$1. Then what is $E(X_1)$?

(c) Let W_1 = the player's winnings on a single play of the original game. Find $E(W_1)$.

(d) Suppose that 100 plays of the game are made, but that rather than being independent they are such that

$$\rho(W_i, W_{i+1}) = 1 - \frac{1}{2^i} \qquad (i = 1, 2, \ldots)$$

$$\rho(W_i, W_j) = 0 \qquad (j > i + 1).$$

Find $E(W_1 + \cdots + W_{100})$.

5.4 Prove that, if X_1, X_2 has the bivariate normal distribution, then $\rho(X_1, X_2) = \rho$. Deduce that X_1 and X_2 are then uncorrelated iff they are independent. (This completes the proof of Theorem 5.3.17.)

5.5 Show that, if X is a r.v. which has the $\chi_n^2(0)$ distribution (see Definition 4.2.9) and $n \geqslant 2$, then $EX = n$ and the mode of the distribution of X is $n - 2$. Find EX and the mode when $n = 1$.

5.6 **(a)** Prove that $e^{5it - 10t^2 + 6(e^{it} - 1)}$ is a characteristic function.

(b) Prove that $e^{4it + t^2}$ is not a characteristic function.

5.7 If random variables $X_1, X_2, \ldots, X_n$ are independent and identically distributed with common density function $f(x)$, and if

$$Q = \frac{1}{n} \sum_{i=1}^{n} \sin X_i,$$

$$R = (\sin X_1)(\sin X_2) \cdots (\sin X_n),$$

show that: **(a)** $E(Q) = \int_{-\infty}^{\infty} (\sin x) f(x)\, dx$;

(b) $E(R) = \{\int_{-\infty}^{\infty} (\sin x) f(x)\, dx\}^n$.

5.8 Prove that

$$(\tfrac{1}{10} e^{it} + 1) \cdot e^{5(e^{it} - 1)} \cdot e^{5it - t^2}$$

is not a characteristic function.

5.9 Let $X_1, \ldots, X_n$ be independent random variables with $E(X_i) = \mu_i$ and $\operatorname{Var}(X_i) = a_i \sigma^2$ ($i = 1, \ldots, n$). Find $E(S_n{}^2)$, where $S_n = X_1 + \cdots + X_n$.

5.10 It is sometimes the case that a sequence (say $Y_1, Y_2, \ldots$) of random variables cannot be observed, although we can observe results of the values of $Y_1, Y_2, \ldots$ (say $X_1, X_2, \ldots$).

Let $Y_1, Y_2, \ldots$ be independent and identically distributed random variables.

Let a be a fixed number $(-\infty < a < +\infty)$. For $i = 1, 2, \ldots$ define

$$X_i = \begin{cases} 1, & \text{if } Y_i \geqslant a \\ 0, & \text{if } Y_i < a. \end{cases}$$

Define $p = P[Y_1 \geqslant a]$, $q = 1 - p$, $S_n = X_1 + \cdots + X_n$. (Note that S_n is the number of $Y_1, \ldots, Y_n$ which exceed a.)

(a) Define the characteristic function, say $\phi_X(t)$, of a random variable X.

(b) Find $\phi_{X_n}(t)$.

(c) Find $\phi_{S_n}(t)$.

(d) Show that $P[S_n = j] = \binom{n}{j} p^j q^{n-j}$ for $j = 0, 1, \ldots, n$.

5.11 Lemma 5.3.7 stated that

$$\mathrm{Var}(X_1 + X_2) = \mathrm{Var}(X_1) + \mathrm{Var}(X_2) + 2\,\mathrm{Cov}(X_1, X_2).$$

Extending this to n r.v.'s, show that

$$\mathrm{Var}(X_1 + \cdots + X_n) = \sum_{i=1}^{n} \mathrm{Var}(X_i) + 2 \sum_{i<j} \mathrm{Cov}(X_i, X_j).$$

[*Hint*: Use mathematical induction on n.]

5.12 (*Due to J. Keilson*) Let $X_1, X_2, \ldots$ be a sequence of r.v.'s and let $p_i \geqslant 0$ $(i = 1, 2, \ldots)$ with $\sum_{i=1}^{\infty} p_i = 1$. Prove that

$$\sigma\left(\sum_{i=1}^{\infty} p_i X_i\right) \leqslant \sum_{i=1}^{\infty} p_i\, \sigma(X_i) \leqslant \sqrt{\sum_{i=1}^{\infty} p_i\, \sigma^2(X_i)}.$$

5.13 Let $X_1, X_2, \ldots, X_n$ be n r.v.'s. Prove that

$$\mathrm{Var}\left(\frac{X_1 + \cdots + X_n}{n}\right) \leqslant \frac{\sum_{i=1}^{n} \mathrm{Var}(X_i)}{n}.$$

5.14 Let X, Y, Z be r.v.'s, and let a, b be constants. Prove that

(a) $\mathrm{Cov}(X, aY + b) = a\,\mathrm{Cov}(X, Y)$;

(b) $\mathrm{Cov}(X, Y + Z) = \mathrm{Cov}(X, Y) + \mathrm{Cov}(X, Z)$;

(c) $\rho(X, aY + b) = \rho(X, Y)$ for $a > 0$;

(d) $\rho(X, Y + Z) = \dfrac{\sigma(Y)}{\sigma(Y + Z)}\rho(X, Y) + \dfrac{\sigma(Z)}{\sigma(Y + Z)}\rho(X, Z)$.

5.15 Let X_1, X_2, X_3 be three independent r.v.'s, each with variance σ^2. If we define new r.v.'s W_1, W_2, W_3 by

$$W_1 = X_1,$$

$$W_2 = \frac{\sqrt{3} - 1}{2} X_1 + \frac{3 - \sqrt{3}}{2} X_2,$$

$$W_3 = (\sqrt{2} - 1)X_2 + (2 - \sqrt{2})X_3,$$

show that

$$\rho(W_i, W_{i+1}) = \tfrac{1}{2} \qquad (i = 1, 2)$$
$$\rho(W_i, W_j) = 0 \qquad (j > i + 1).$$

5.16 Suppose that Z_1, Z_2, Z_3 are independent random variables with $\text{Var}(Z_i) = \sigma_i^2 > 0$ $(i = 1, 2, 3)$. Let $X = Z_1 + Z_2$, $Y = Z_1 + Z_3$. Find the correlation coefficient $\rho(X, Y)$. What is the value of $\rho(X, Y)$ when $\sigma_1^2 = \sigma_2^2 = \sigma_3^2 > 0$? What is the range of possible values of $\rho(X, Y)$ as $\sigma_1^2, \sigma_2^2, \sigma_3^2$ vary?

5.17 Let X_1 and X_2 be two jointly discrete r.v.'s whose joint probability function assigns probability 1/17 to each of the following 17 pairs of values:

(86, 100), (75, 98), (93, 98), (92, 95),
(95, 93), (87, 91), (81, 83), (81, 79),
(66, 66), (67, 64), (78, 63), (94, 58),
(95, 56), (88, 53), (98, 33), (87, 31),
(54, 5).

Find EX_1, EX_2, $E|X_1 - EX_1|$, $\rho(X_1, X_2)$, and the smallest d such that $P[|X_1 - EX_1| \leqslant d] \geqslant .75$. [The 17 pairs of values actually represent the scores of 17 students on two examinations. As discussed briefly following Lemma 5.3.14, $\rho(X_1, X_2)$ is a measure of how well a quantity like $aX_1 + b$ can predict the score X_2 on the second test.]

5.18 Suppose a given event occurs (success) with probability p (and therefore fails to occur with probability $q = 1 - p$), and independent trials are conducted. Let X denote the number of trials needed to obtain a total of k successes. Find $\text{Var}(X)$. [*Hint*: Show that we can express X as $X = Y_1 + \cdots + Y_k$ where $Y_1, \ldots, Y_k$ are independent r.v.'s and Y_i has the geometric distribution of Definition 3.2.14. Then use Problem 5.10 and Theorem 5.3.11.]

5.19 Suppose that X is an (absolutely) continuous random variable with density function given by

$$f_X(y) = \begin{cases} 5e^{-5y}, & \text{if } 0 \leqslant y < \infty \\ 0, & \text{otherwise.} \end{cases}$$

Find $E(X)$, $E(X^2)$, $\text{Var}(X)$, Mode of X, Median of X.

5.20 A die is tossed $n + 2$ times. After each toss a "+" is recorded if a 4, 5, 6 occurs; and a "−" is recorded if a 1, 2, 3 occurs. We thus form an ordered sequence of signs. Let (for $i = 2, \ldots, n + 1$)

$$X_{i-1} = \begin{cases} 1, & \text{if sign } i - 1 \text{ and sign } i + 1 \text{ both differ from sign } i \\ 0, & \text{otherwise.} \end{cases}$$

We thus define random variables $X_1, X_2, \ldots, X_n$. (For example, if $n = 3$ and we observe dice outcomes 4, 3, 1, 6, 5, then the sign sequence is $+ - - + +$ and $X_1 = 0$, $X_2 = 0$, $X_3 = 0$.) Let $Y = X_1 + \cdots + X_n$. Find $E(Y)$ and $\text{Var}(Y)$.

5.21 Assume r balls ($r > 1$, r an integer) are distributed at random into n cells ($n \geqslant 1$);

thus, each assignment vector has probability $1/n^r$. Let

$$p_m(r, n) = \text{probability of finding exactly } m \text{ cells empty.}$$

In Feller [Vol. I (Second Edition), p. 92] we can find that

$$p_m(r, n) = \binom{n}{m} \sum_{\nu=0}^{n-m} (-1)^\nu \binom{n-m}{\nu} \left(1 - \frac{m+\nu}{n}\right)^r; \tag{1}$$

direct numerical evaluation of (1) is limited to the case of relatively small n and r. However, the following is true:

Limit Theorem: *If $n \to \infty$ and $r \to \infty$ in such a way that $\lambda = ne^{-r/n}$ remains bounded, then for each fixed m*

$$p_m(r, n) - e^{-\lambda} \frac{\lambda^m}{m!} \to 0. \tag{2}$$

Recall that if a random variable $\tilde{M}$ is such that $P[\tilde{M} = m] = e^{-\lambda}\lambda^m/m!$ $(m = 0, 1, 2, \ldots)$, then $E(\tilde{M}) = \lambda$. (Poisson distribution.)

Suppose that r balls are placed at random into n cells; r is known, but n is unknown and we desire to estimate it. We observe the random variable $n - M$, the number of cells with at least 1 ball, and find value c (say).

Because of Limit Theorem (2), we might (for "large" r and n, and "moderate" $\lambda = ne^{-r/n}$) hope that

$$E(M) \approx E(\tilde{M}) = \lambda = ne^{-r/n}, \tag{3}$$

or

$$E(n - M) \approx E(n - \tilde{M}) = n(1 - e^{-r/n}). \tag{4}$$

If we also assume that the observed value c of $n - M$ is "close" to $E(n - M)$, from (4) we obtain

$$n(1 - e^{-r/n}) \approx c. \tag{5}$$

We could solve (5) numerically for an estimate of the (unknown) number n of cells, obtaining (say)

$$\hat{n}_1(c, r). \tag{6}$$

Alternatively, we could also assume (as a first approximation)

$$n \approx c \tag{7}$$

and, using (7) in (5), obtain a second more easily computed estimate of n, say

$$\hat{n}_2(c, r) = \frac{c}{1 - e^{-r/c}}. \tag{8}$$

(a) In general, $E\hat{n}_2(c, r)$ will be difficult to calculate. For the case $r = 1$, find $\{p_m(r, n)\}$: $p_0(1, n)$, $p_1(1, n)$, $p_2(1, n), \ldots, p_{n-1}(1, n)$, $p_n(1, n)$. Then calculate $E\hat{n}_2(c, 1)$.

(b) Show that

$$\hat{n}_2(c, r) \leqslant \frac{\min(r, n)}{1 - e^{-r/\min(r, n)}}.$$

(c) Show that, if $r \leqslant n$, $\hat{n}_2(c, r) \leqslant (1 - e^{-1})^{-1}r$. Does this mean that we should try to have $r \geqslant (1 - e^{-1})n$ when using $\hat{n}_2$ as an estimator? Why or why not?

(d) If $r = 9000$, $n = 1000$, we have (by the Poisson approximation) the approximate probabilities

$p_m(9000, 1000)$	.8869	.1064	.0064	.0003	...
m	0	1	2	3	...

Find, using these approximate probabilities, $E\hat{n}_2(c, 9000)$. What is an "ideal" value for $\hat{n}_2(c, r)$?

x	4.50	4.51	4.52	4.53	4.54
e^{-x}	.011	.011	.011	.011	.011

x	9.00	9.05	9.10	9.15	9.20
e^{-x}	.0001	.0001	.0001	.0001	.0001

5.22 Let $X_1, X_2, \ldots, X_n$ be independent random variables such that $P[X_i = 1] = P[X_i = -1] = \frac{1}{2}$ $(i = 1, \ldots, n)$. Let $a_1, a_2, \ldots, a_n$ be constants. Find the characteristic function of X_1; of $Y_n = a_1X_1 + \cdots + a_nX_n$. Let $a_1 = 1/2$, $a_2 = 1/2^2$, $a_3 = 1/2^3, \ldots$. What is $\phi_n(t)$, the characteristic function of Y_n, in this case? Show that $\lim_{n\to\infty} \phi_n(t) = \frac{\sin(t)}{t}$. [*Hint*: $\sin(x) = (e^{ix} - e^{-ix})/2i$ and $\cos(x) = (e^{ix} + e^{-ix})/2$.]

5.23 Suppose that (X, Y) is a two-dimensional r.v. with density function

$$f_{X,Y}(x, y) = \begin{cases} 1, & \text{if } 0 \leqslant y \leqslant x, 0 \leqslant x \leqslant \sqrt{2} \\ 0, & \text{otherwise;} \end{cases}$$

that is, (X, Y) is uniformly distributed over a certain triangle. Find $E(Y|X = x_0)$ $(0 \leqslant x_0 \leqslant \sqrt{2})$ and $E(Y)$.

5.24 Suppose that (for $i = 1, 2$) the r.v. X_i has the exponential density

$$f_{X_i}(x) = \begin{cases} (1/\beta_i)e^{-x/\beta_i}, & x > 0 \\ 0, & x \leqslant 0 \end{cases}$$

where $\beta_i > 0$, and that X_1 and X_2 are independent r.v.'s. Show that the density function of $Y = X_1/X_2$ is

$$f_Y(y) = \begin{cases} \alpha/(y + \alpha)^2, & y > 0 \\ 0, & y \leqslant 0 \end{cases}$$

where $\alpha = \beta_1/\beta_2$. Show that, for $\delta \geqslant 1$, EY^δ does not exist. [This distribution was discussed by P. A. Lachenbruch and D. R. Brogan, "Some distributions on the positive real line which have no moments," *The American Statistician*, Vol. 25 (1971), pp. 46–47.]

HONORS PROBLEMS

5.1H Prove Corollary 5.1.4.

5.2H Let α be fixed, $0 < \alpha < 1$. Show that

$$f(x) = \begin{cases} \frac{1}{24} e^{-\sqrt[4]{x}} (1 - \alpha \sin \sqrt[4]{x}), & 0 \leqslant x \\ 0, & \text{otherwise} \end{cases}$$

is a density function. Show that the moments $\mu_1' = EX, \mu_2' = EX^2, \mu_3' = EX^3, \ldots$ do not depend on $\alpha\,(0 < \alpha < 1)$. What does this imply (via Carleman's Theorem) about the value of

$$\sum_{n=1}^{\infty} \left\{ \int_{-\infty}^{\infty} |x|^n f(x)\, dx \right\}^{-1/n} ? \tag{$*$}$$

Evaluate $(*)$ explicitly.

5.3H Let $X_1, X_2, X_3, \ldots$ be a sequence of independent random variables, each with d.f. $F(x)$. Can you construct a sequence of functions of $X_1, X_2, X_3, \ldots$ (say $W_1, W_2, W_3, \ldots$) such that

$$\rho(W_i, W_{i+1}) = 1 - \frac{1}{2^i} \qquad (i = 1, 2, \ldots)$$

$$\rho(W_i, W_j) = 0 \qquad (j > i + 1)?$$

[*Hint*: See Problems 5.15 and 5.4H.]

5.4H Let $-1 \leqslant \rho_{ij} \leqslant 1$ $(1 \leqslant i, j \leqslant n)$ be n^2 numbers and let (ρ_{ij}) be an $n \times n$ matrix whose entry in row i and column j is ρ_{ij}. Give a necessary and sufficient condition for (ρ_{ij}) to be the matrix of correlation coefficients of some n random variables $Z_1, \ldots, Z_n$.

5.5H For the random variable X defined in Problem 2.8H, find $E(X)$ as a general formula. Evaluate this formula for $n = 100$, $N = 200$ and $m = 1, 2, 3, \ldots, 100$. [If you are unable to solve this problem analytically, you may try a **simulation** approach. Namely, write a computer program which—via use of pseudo-random numbers—distributes $N = 200$ balls into $n = 100$ cells and records the number of balls in each cell. Then let $S(i)$ be the number of cells which have more than m balls in them if $m = i$; record $S(1), S(2), \ldots, S(100)$. Then do the experiment over again and add the new results $S(i)$ to the old ones, and so on, performing the experiment $t = 1000$ times. Then $S(i)/1000$ is an estimate of $E(X)$ when $m = i$.]

Even if you were able to solve the problem analytically, try the simulation approach (which is often used when analytic solutions cannot be obtained). How should t be chosen? Why do you expect $E(X) \approx S(i)/1000$? (Recall the relative frequency interpretation of probabilities.) A reference is T. H. Naylor, J. L. Balintfy, D. S. Burdick, and K. Chu, *Computer Simulation Techniques*, John Wiley & Sons, Inc., New York, 1966.

5.6H Construct a probability distribution for a random variable $X \geqslant 0$ such that:

(a) EX does not exist;

(b) EX^δ does not exist for any $\delta > 0$;

(c) EX^δ does not exist for $\delta = -1$;

(d) EX^δ does not exist for any $\delta < 0$;

(e) EX^δ exists iff $\delta = 0$.

5.7H Let $A_1, \ldots, A_n$ be $n \geqslant 2$ events and define (for $i = 1, \ldots, n$) indicator r.v.'s by

$$I_{A_i} = \begin{cases} 1, & \text{if } A_i \text{ occurs} \\ 0, & \text{otherwise.} \end{cases}$$

We know (see Boole's Inequality, Theorem 2.1.14) that it is always the case that

$$P(A_1 \cup \cdots \cup A_n) \leqslant P(A_1) + \cdots + P(A_n).$$

What can be said about $P(A_1 \cap \cdots \cap A_n)$ [usually written $P(A_1 \cdots A_n)$]? More specifically:

(a) Prove that $P(A_1 \cdots A_n) = P(A_1) \cdots P(A_n)$ if $A_1, \ldots, A_n$ are independent events.

(b) Prove that $P(A_1 A_2) \geqslant P(A_1)P(A_2)$ iff $\operatorname{Cov}(I_{A_1}, I_{A_2}) \geqslant 0$.

(c) Prove that $P(A_1 \cdots A_n) \geqslant P(A_1) \cdots P(A_n)$ if $\operatorname{Cov}(I_{A_j}, I_{A_1 \cdots A_{j-1}}) \geqslant 0$ for $j = 2, \ldots, n$.

(d) R.v.'s $X_1, \ldots, X_n$ (for example, $I_{A_1}, \ldots, I_{A_n}$) are said to be **associated** if, for all nondecreasing functions f and g, $\operatorname{Cov}(f(X_1, \ldots, X_n), g(X_1, \ldots, X_n)) \geqslant 0$. [See J. D. Esary, F. Proschan, and D. W. Walkup, "Association of random variables, with applications," *Annals of Mathematical Statistics*, Vol. 38 (1967), pp. 1466–1474 for further details.] Try to use this concept to provide a theorem giving a condition C such that it is true that "$P(A_1 \cdots A_n) \geqslant P(A_1) \cdots P(A_n)$ iff C."

6

SUMS OF RANDOM VARIABLES AND LIMIT LAWS

In many statistical problems, we have random variables $X_1, \ldots, X_n$ which are assumed to be independent, and we desire to make probability statements about their sum $X_1 + \cdots + X_n$. If the distribution function of $X_1, \ldots, X_n$ is known, then we can use the techniques of Section 4.6 to determine the distribution function of $X_1 + \cdots + X_n$, but this is a hard task. Also, we may not know the d.f. of $X_1, \ldots, X_n$.

In Section 6.1 we study the use of characteristic functions in determining the distribution of $X_1 + \cdots + X_n$. In Section 6.2 we study statements we can make with regard to bounding the probability that, for example, $X_1 + \cdots + X_n$ is "large." In Section 6.3 we consider what happens to the distribution of $X_1 + \cdots + X_n$ as n grows large. This is continued in Section 6.4, where we prove the Central Limit Theorem *and* give some statistical applications.

6.1. CHARACTERISTIC FUNCTIONS AND THE DISTRIBUTIONS OF SUMS

In the Uniqueness Theorem (Theorem 5.4.14) we saw that if we could identify the ch.f. of a r.v. Y, we would know its d.f. We show below how this may be done in some special cases, and we then give some examples.

Note that *if* X_1 *and* X_2 *are independent r.v.'s, then the d.f. of* $Z = X_1 + X_2$ *is called the* **convolution** *of the d.f.'s* $F_{X_1}(x)$ *and* $F_{X_2}(x)$:

$$F_{X_1+X_2}(z) = \int_{-\infty}^{\infty} F_{X_1}(z - x) f_{X_2}(x)\, dx$$

[*if* $F_{X_2}(x)$ *has a density*]. We then write $F_{X_1+X_2} = F_{X_1} * F_{X_2}$, $F^{2*} = F * F$, and F convoluted with itself $n-1$ times is called F^{n*}.

Theorem 6.1.1. *Let* $X_1, \ldots, X_n$ *be r.v.'s which are independent. Then*

$$\phi_{X_1+\cdots+X_n}(t) = \phi_{X_1}(t)\cdots\phi_{X_n}(t).$$

Proof:

$$\begin{aligned}\phi_{X_1+\cdots+X_n}(t) &= Ee^{it(X_1+\cdots+X_n)}\\ &= E(e^{itX_1}\cdots e^{itX_n}), \text{ and by independence}^1\\ &= Ee^{itX_1}\cdots Ee^{itX_n} = \phi_{X_1}(t)\cdots\phi_{X_n}(t).\end{aligned}$$

Theorem 6.1.2. *Let* $X_1, \ldots, X_n$ *be r.v.'s which are independent and identically distributed. Then*

$$\phi_{X_1+\cdots+X_n}(t) = [\phi_{X_1}(t)]^n.$$

Proof: By Theorem 6.1.1,

$$\phi_{X_1+\cdots+X_n}(t) = \phi_{X_1}(t)\cdots\phi_{X_n}(t),$$

and by identical distribution

$$\begin{aligned}&= \phi_{X_1}(t)\cdots\phi_{X_1}(t)\\ &= [\phi_{X_1}(t)]^n.\end{aligned}$$

Theorem 6.1.3. *Let* $X_1,\ldots,X_n$ *be independent r.v.'s. Let* $S_n = X_1 + \cdots + X_n$.

(1) *If* X_k *is* $N(\mu_k, \sigma_k^2)$ [Definition 4.2.13 and Example 5.4.19] (*for* $k = 1, \ldots, n$), *then* S_n *is* $N(\mu_1 + \cdots + \mu_n, \sigma_1^2 + \cdots + \sigma_n^2)$.

(2) *If* X_k *is binomial* [Definition 3.2.3 and Example 5.4.16] *with* n_k, p (*for* $k = 1, \ldots, n$), *then* S_n *is binomial with* $n_1 + \cdots + n_n$, p.

(3) *If* X_k *is Poisson* [Definition 3.2.7 and Example 5.4.17] *with* λ_k (*for* $k = 1, \ldots, n$), *then* S_n *is Poisson with* $\lambda_1 + \cdots + \lambda_n$.

(4) *If* X_k *is* $\chi^2_{n_k}(0)$ [Definition 4.2.9] (*for* $k = 1, \ldots, n$), *then* S_n *is* $\chi^2_{n_1+\cdots+n_k}(0)$.

(5) *X is said to be* **Cauchy** *with a* ($-\infty < a < +\infty$) *and* $b > 0$ *if*

$$f_X(x) = \frac{1}{\pi b}\left[1 + \left(\frac{x-a}{b}\right)^2\right]^{-1}, \qquad -\infty < x < +\infty.$$

[Note that if $a = 0$, $b = 1$ then X is Student's t with $n = 1$ degrees of freedom; see Definition 4.2.17.] *If* X_k *is Cauchy with* a_k *and* b_k (*for* $k = 1,\ldots,n$), *then* S_n *is Cauchy with* $a_1 + \cdots + a_n$, $b_1 + \cdots + b_n$.

---

[1] A slight strengthening of Theorem 5.3.8 is needed here, since we are using complex-valued functions.

Proof: We will prove only part (1) since parts (2), (3), (4), and (5) are proven similarly.

$$\begin{aligned}\phi_{S_n}(t) &= \phi_{X_1}(t)\cdots\phi_{X_n}(t)\\ &= e^{it\mu_1-\frac{1}{2}\sigma_1{}^2t^2}\cdots e^{it\mu_n-\frac{1}{2}\sigma_n{}^2t^2}\\ &= e^{it(\mu_1+\cdots+\mu_n)-\frac{1}{2}(\sigma_1{}^2+\cdots+\sigma_n{}^2)t^2}.\end{aligned}$$

By the Uniqueness Theorem, S_n is $N(\mu_1 + \cdots + \mu_n, \sigma_1{}^2 + \cdots + \sigma_n{}^2)$.

6.2. CHEBYSHEV AND MARKOV INEQUALITIES

Theorem 6.2.1. (Chebyshev's Inequality) *Let $\delta > 0$ be fixed. Let X be a r.v. with mean μ and variance σ^2. Then*

$$P[|X-\mu| \geqslant \delta] \leqslant \frac{\sigma^2}{\delta^2}.$$

Proof: Consider the diagram of Figure 6.2–1. Now $\sigma^2 = E(X - EX)^2 \geqslant \delta^2 P[|X - EX| \geqslant \delta]$, which proves the theorem. In this inequality, we have first neglected the integral from $\mu - \delta$ to $\mu + \delta$, then replaced the integrand on the remaining integral (which ranges from $-\infty$ to $\mu - \delta$, and from $\mu + \delta$ to $+\infty$) by its minimal value δ^2. Note that $P[|X - EX| \geqslant \delta]$ is the probability mass outside the interval from $\mu - \delta$ to $\mu + \delta$: $P[X \in S] = \int_{x\in S} f(x)\,dx$ where $S = \{x: |x - \mu| \geqslant \delta\}$.

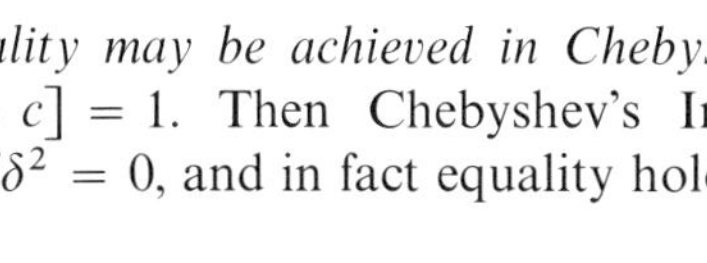

Figure 6.2–1

Example 6.2.2. *Equality may be achieved in Chebyshev's Inequality.* For example, let $P[X = c] = 1$. Then Chebyshev's Inequality states that $P[|X - c| \geqslant \delta] \leqslant 0/\delta^2 = 0$, and in fact equality holds for all $\delta > 0$.

Corollary 6.2.3. *Let $\delta > 0$ be fixed. Let X be a r.v. with mean μ and variance $\sigma^2 > 0$. Then*

$$P\left[\left|\frac{X-\mu}{\sigma}\right| \geqslant \delta\right] \leqslant \frac{1}{\delta^2}.$$

Proof: By Chebyshev's Inequality, for any $t > 0$, $P[|X - \mu| \geqslant t] \leqslant \sigma^2/t^2$.

Let $t = \delta\sigma$, which is > 0. Then

$$P[|X - \mu| \geqslant \delta\sigma] \leqslant \frac{\sigma^2}{(\delta\sigma)^2},$$

that is,

$$P\left[\left|\frac{X - \mu}{\sigma}\right| \geqslant \delta\right] \leqslant \frac{1}{\delta^2}.$$

Example 6.2.4. *Equality may be achieved in Chebyshev's Inequality with* $\sigma^2 > 0$. For example, let $P[X = -1] = P[X = 1] = \frac{1}{2}$. Then $\mu = 0$, $\sigma^2 = 1$, and Chebyshev's Inequality states that $P[|X| \geqslant \delta] \leqslant 1/\delta^2$. This is an equality when $\delta = 1$.

Theorem 6.2.5. **(Chebyshev's Inequality)** *Let* $t > 0$ *be fixed. Let* Y *be a r.v. such that* $P[Y \geqslant 0] = 1$. *Then* $P[Y \geqslant t] \leqslant EY^2/t^2$. [This implies Chebyshev's Inequality in the form of Theorem 6.2.1 upon taking $Y = |X - \mu|$.]

Proof: Essentially as in Theorem 6.2.1; that is, $EY^2 \geqslant t^2 P[Y \geqslant t]$, q.e.d.

Theorem 6.2.6. **(Markov's Inequality)** *Let* $t > 0$, $a \geqslant 0$ *be fixed. Let* Y *be a r.v. such that* $P[Y \geqslant 0] = 1$. *Then* $P[Y \geqslant t] \leqslant EY^a/t^a$.

Proof: As in Theorem 6.2.5.

Note: A reference to these (and more general) inequalities is Loève (1963), p. 158. An interesting and readable application is made by P. A. Samuelson, "How deviant can you be?," *Journal of the American Statistical Association*, Vol. 63 (1968), pp. 1522–1525.]

Example 6.2.7. A newspaper stand has observed that, over the course of 10 days, the numbers of newspapers demanded were $X_1, X_2, \ldots, X_{10}$, respectively. If we assume that these are independent and identically distributed observations of demand, what bound can be put on the probability that their average $\bar{X} \equiv (X_1 + \cdots + X_{10})/10$ is more than 3σ newspapers from the (unknown) average demand μ? We have $E\bar{X} = \mu$, $\text{Var}(\bar{X}) = \sigma^2/10$, so by Corollary 6.2.3,

$$P[|\bar{X} - \mu| \geqslant 3\sigma] = P\left[\left|\frac{\bar{X} - \mu}{\sigma/\sqrt{10}}\right| \geqslant 3\sqrt{10}\right] \leqslant \frac{1}{(3\sqrt{10})^2} = \frac{1}{90}.$$

6.3. MODES OF CONVERGENCE; LAWS OF LARGE NUMBERS

Let $Z_1, \ldots, Z_n$ be a sequence of jointly distributed r.v.'s (for $n \geqslant 1$) defined on the same sample space Ω. Let Z be another r.v. defined on this same sample space. We shall consider now what can be meant by Z_n "tending (or converging) to" Z.

Definition 6.3.1. *We say* **Z_n converges to Z with probability one (w.p. 1)** *if* $P[\lim_{n \to \infty} Z_n = Z] = 1$. (*This is also called* **almost sure (a.s.) convergence**, *or* **almost everywhere (a.e. or p.p.) convergence**, *or* **strong convergence**.)

Definition 6.3.2. *We say* **Z_n converges to Z in probability ($Z_n \xrightarrow{P} Z$)** *if for every* $\varepsilon > 0$

$$\lim_{n \to \infty} P[|Z_n - Z| > \varepsilon] = 0.$$

(*This is also called* **stochastic convergence**, *or* **convergence in measure**, *or* **weak convergence**.)

Definition 6.3.3. *We say* **Z_n converges to Z in mean square** *if*

$$\lim_{n \to \infty} E(Z_n - Z)^2 = 0.$$

(*We also say* **$Z_n \to Z$ in $\mathscr{L}_2$**, *or* **$Z_n \xrightarrow{2} Z$**.)

The relationships among these various types of convergence are given in the following theorem. For our purposes, which are statistical, convergence in probability is the central concept.

Theorem 6.3.4.

$$\begin{array}{l} (Z_n \to Z \text{ w.p. } 1) \\ \qquad\qquad \searrow \\ \qquad\qquad\qquad (Z_n \xrightarrow{P} Z) \Rightarrow \left(\begin{array}{l}\text{There is a subsequence} \\ n_k \text{ such that} \\ \qquad Z_{n_k} \to Z \text{ w.p. } 1\end{array}\right) \\ \qquad\qquad \nearrow \\ (Z_n \to Z \text{ in } \mathscr{L}_2) \end{array}$$

The following theorem, called *a* Weak Law of Large Numbers, asserts that in many cases the sample mean $(X_1 + \cdots + X_n)/n$ converges in probability to $E(X_1 + \cdots + X_n)/n$.

Theorem 6.3.5. **[Weak Law of Large Numbers (WLLN)]** *Let $X_1, \ldots, X_n$ be independent random variables such that $EX_i = a$, $\sigma^2(X_i) = b^2$ $(i = 1, \ldots, n)$. Let $\delta > 0$ and $\varepsilon > 0$ be arbitrary. Then there exists $N(\varepsilon, \delta)$ such that for all*

$n > N(\varepsilon, \delta)$

$$P\left[\left|\frac{X_1 + \cdots + X_n}{n} - a\right| > \delta\right] < \varepsilon.$$

$\left(\textit{More concisely, } \dfrac{X_1 + \cdots + X_n}{n} \xrightarrow{P} a.\right)$

Proof: Let $Z_n = (X_1 + \cdots + X_n)/n$. Then

$$EZ_n = \frac{1}{n} na = a, \quad \sigma^2(Z_n) = \frac{1}{n^2} n\sigma^2(X_1) = \frac{b^2}{n},$$

and by Chebyshev's Inequality

$$P\left[\left|\frac{X_1 + \cdots + X_n}{n} - a\right| > \delta\right] \leqslant \frac{\sigma^2(Z_n)}{\delta^2} = \frac{b^2}{n\delta^2}.$$

Let $N(\varepsilon, \delta)$ be the smallest positive integer n such that $b^2/n\delta^2 < \varepsilon$ and the theorem follows.

Note that the proof of Theorem 6.3.5 *illustrates* **a general method for proving stochastic convergence**. Let $X_1, \ldots, X_n$ be any r.v.'s such that their means and variances exist. Let $S_n/n = (X_1 + \cdots + X_n)/n$. We say $X_1, \ldots, X_n$ *obey the WLLN* if $S_n/n - E(S_n/n) \xrightarrow{P} 0$. By Chebyshev's Inequality,

$$P\left[\left|\frac{S_n}{n} - E\left(\frac{S_n}{n}\right)\right| > \delta\right] \leqslant \frac{\sigma^2(S_n/n)}{\delta^2}. \tag{6.3.6}$$

Thus, $\sigma^2(S_n/n) \to 0$ will imply the WLLN for $X_1, \ldots, X_n$.

Example 6.3.7. Suppose a coin has probability p $(0 < p < 1)$ of turning up heads when flipped in a certain manner, and we flip it independently n times. Let (for $i = 1, \ldots, n$)

$$X_i = \begin{cases} 1, & \text{if the } i\text{th toss yields a head} \\ 0, & \text{otherwise.} \end{cases}$$

Then we may decide to estimate p by

$$\frac{S_n}{n} = \frac{X_1 + \cdots + X_n}{n},$$

and we desire to know *in what sense is this close to p*. The WLLN provides one answer. Also, our general method will allow us to consider the question: If our trials are correlated, how "large" can the correlation be and yet have $S_n/n \xrightarrow{P} p$? [We are interested in convergence as $n \to \infty$ because, even though the actual n is "small" or "moderate" in many cases, it is helpful to know your procedure would behave well for n very large ($n \to \infty$). We intuitively feel that if a procedure cannot do well with lots of observations it probably is not very good with just a few.]

6.4. CHARACTERISTIC FUNCTION METHODS FOR LIMITS; CENTRAL LIMIT THEOREM

In Section 6.3 we discussed what we could mean by the statement "the sequence of r.v.'s $Z_1, Z_2, Z_3, \ldots$ tends to the r.v. Z" and studied and applied some Laws of Large Numbers. In this section we discuss and apply convergence in terms of distribution functions of r.v.'s.

Definition 6.4.1. *We say a sequence of d.f.'s $F_1(\cdot), F_2(\cdot), F_3(\cdot), \ldots$ (or simply $F_1, F_2, F_3, \ldots$) converges[2] to a d.f. $L(\cdot)$ (or simply L), and write $F_n \to L$, if $F_n(x) \to L(x)$ at each point x where L is continuous.*

Theorem 6.4.2. (Lévy's Continuity Theorem) *Let $Z_1, Z_2, \ldots$ be a sequence of r.v.'s; let the d.f. of Z_n be F_n and let the ch.f. of Z_n be ϕ_n. Then $F_n \to F$ (a d.f.) iff $\phi_n(t) \to \phi(t)$ for every point t and $\phi(t)$ is continuous at $t = 0$. (In either case ϕ is the ch.f. of F, and the convergence $\phi_n \to \phi$ is uniform in every finite interval.)*

As an application of Lévy's Continuity Theorem we now obtain a second general method for proving stochastic convergence of a r.v. to a constant (the first having been given in Section 6.3). As an illustration we prove Khinchin's Theorem.

Corollary 6.4.3. *A r.v. $X_n \xrightarrow{P} c$ (a constant) iff the d.f. of X_n, $F_n \to F$ where $F(x) = 0$ for $x < c$ and $F(x) = 1$ for $x \geqslant c$; this occurs iff $\phi_{X_n}(t) \to e^{ict}$ for all t.*

Proof: (1a) Suppose that $X_n \xrightarrow{P} c$ (a constant). Then for any $\delta > 0$, $P[|X_n - c| \geqslant \delta] \to 0$. Hence, $F_n(x) = P[X_n \leqslant x] \to 0$ for $x < c$, and $F_n(x) = P[X_n \leqslant x] \to 1$ for $x > c$. Thus, $F_n \to F$ as stated.

(1b) Suppose $F_n \to F$ with $F(x) = 0$ for $x < c$ and $F(x) = 1$ for $x \geqslant c$. Then for any $\delta > 0$, we have

$$\begin{aligned} P[|X_n - c| \geqslant \delta] &= P[X_n \leqslant c - \delta] + P[X_n \geqslant c + \delta] \\ &= P[X_n \leqslant c - \delta] + 1 - P[X_n < c + \delta] \\ &\to 0 + 1 - 1 = 0, \end{aligned}$$

so $X_n \xrightarrow{P} c$.

---

[2] This is often called "weak" or "complete" convergence.

(2) The specified d.f., $F(x)$ with $F(x) = 0$ for $x < c$ and $F(x) = 1$ for $x \geqslant c$ (see Figure 6.4–1), corresponds to a r.v. X such that $P[X = c] = 1$. Such a r.v. X has characteristic function $\phi_X(t) = Ee^{itX} = e^{ict}$, which is continuous at $t = 0$. By Lévy's Continuity Theorem, $F_n \to F$ iff $\phi_{X_n}(t) \to \phi_X(t)$ for all t and $\phi_X(t)$ is continuous at $t = 0$. Hence, the second "iff" of the theorem follows.

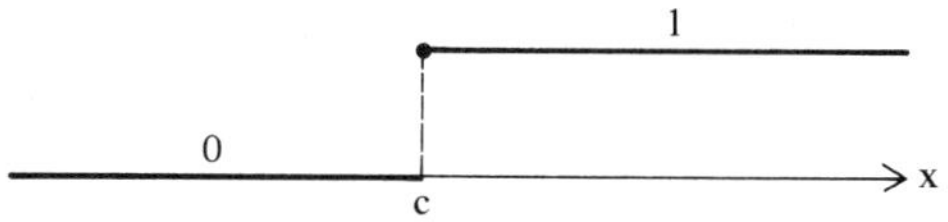

Figure 6.4–1. $F(x)$ for Corollary 6.4.3

Theorem 6.4.4. (Khinchin's Theorem) *If $X_1, X_2, \ldots$ are independent and identically distributed r.v.'s and EX_1 exists ($EX_1 = a$, say), then*

$$\frac{X_1 + \cdots + X_n}{n} \xrightarrow{p} a.$$

Proof:

$$\begin{aligned}
\phi_{\frac{X_1 + \cdots + X_n}{n}}(t) &= \phi_{\frac{X_1}{n} + \cdots + \frac{X_n}{n}}(t) \\
&= [\phi_{X_1/n}(t)]^n, \quad \text{by Theorem 6.1.2} \\
&= \left[\phi_{X_1}\left(\frac{t}{n}\right)\right]^n, \quad \text{by Theorem 5.4.12} \\
&= \left[1 + \frac{iEX_1}{1!}\frac{t}{n} + o\left(\frac{t}{n}\right)\right]^n, \quad \text{by Theorem 5.4.13} \\
&= \left[1 + \frac{iat + no(t/n)}{n}\right]^n \\
&\to e^{iat}, \quad \text{as } n \to \infty.
\end{aligned}$$

The result then follows by Corollary 6.4.3. Note that we have not used the well-known result

$$\lim_{n\to\infty}\left(1 + \frac{a}{n}\right)^n = e^a,$$

but rather its slight generalization (see Problem 6.3H)

$$\lim_{n\to\infty}\left(1 + \frac{a_n}{n}\right)^n = e^{\lim_{n\to\infty} a_n}.$$

Definition 6.4.5. *A sequence of jointly distributed r.v.'s $X_1, \ldots, X_n$ with finite means and variances is said to obey the (classical)* **central limit theorem**

(CLT) *iff* Z_n, *defined by*

$$Z_n = \frac{S_n - E(S_n)}{\sigma(S_n)}, \quad S_n = X_1 + \cdots + X_n,$$

converges in distribution to a r.v. which is normally distributed with mean 0 *and variance* 1. [That is, by Lévy's Continuity Theorem, $\{X_n\}$ obeys the CLT iff (for all t)

$$\phi_{Z_n}(t) \to e^{-\frac{1}{2}t^2}.]$$

We now prove a result which indicates that this is a common (or "normal") situation.

Theorem 6.4.6. (Central Limit Theorem) *Let* $X_1, X_2, \ldots$ *be independent and identically distributed r.v.'s with* $EX_1 = \mu$ *and* $\operatorname{Var}(X_1) = \sigma^2 > 0$ *(both finite). Then (for all z,* $-\infty < z < +\infty$*) as* $n \to \infty$,

$$P\left[\frac{(X_1 - \mu) + \cdots + (X_n - \mu)}{\sqrt{n}\sigma} < z\right] \to \frac{1}{\sqrt{2\pi}} \int_{-\infty}^{z} e^{-\frac{1}{2}y^2}\, dy.$$

Proof:

$$\begin{aligned}
\phi_{\frac{(X_1-\mu)+\cdots+(X_n-\mu)}{\sqrt{n}\sigma}}(t) &= \phi_{\frac{(X_1-\mu)}{\sqrt{n}\sigma}+\cdots+\frac{(X_n-\mu)}{\sqrt{n}\sigma}}(t) \\
&= \left[\phi_{\frac{X_1-\mu}{\sqrt{n}\sigma}}(t)\right]^n, \quad \text{by Theorem 6.1.2} \\
&= \left[\phi_{X_1-\mu}\left(\frac{t}{\sqrt{n}\sigma}\right)\right]^n, \quad \text{by Theorem 5.4.12} \\
&= \left[1 + \frac{iE(X_1-\mu)}{1!}\frac{t}{\sqrt{n}\sigma} + \frac{i^2E(X_1-\mu)^2}{2!}\frac{t^2}{n\sigma^2}\right. \\
&\qquad \left. + o\left(\frac{t^2}{n\sigma^2}\right)\right]^n, \quad \text{by Theorem 5.4.13} \\
&= \left[1 - \frac{1}{2}\frac{t^2}{n} + o\left(\frac{t^2}{n\sigma^2}\right)\right]^n \\
&= \left[1 + \frac{-\frac{1}{2}t^2 + no\left(\frac{t^2}{n\sigma^2}\right)}{n}\right]^n \\
&\to e^{-\frac{1}{2}t^2}, \quad \text{as } n \to \infty.
\end{aligned}$$

[**Note:** Since $F_n \to F$, F being a d.f. which is continuous everywhere, it follows that F_n converges uniformly to F. (This is called **Polya's Theorem**; for example, see C. R. Rao (1965), p. 100.) This means that we may say "*the distribution of* Z_n *looks more and more like the normal distribution.*"]

Example 6.4.7. A casino has a coin and wishes to estimate p, the probability of a Head on any toss, in such a way that they can be 99 percent confident that their estimate ($\hat{p}$, say) is within .01 of p.

They take n independent tosses, which yield $X_1,\ldots,X_n$ with (for $i = 1,\ldots,n$)

$$X_i = \begin{cases} 1, & \text{if a head occurs on the } i\text{th toss} \\ 0, & \text{otherwise.} \end{cases}$$

If they decide to use $\hat{p} = \bar{X} \equiv (X_1 + \cdots + X_n)/n$ to estimate p, then *how large* must n be? *Approximately how large* must n be?

By the WLLN, Theorem 6.3.5, $\bar{X} \xrightarrow{P} p$, so there is a smallest sample size n_0 which will "do the job," that is, such that for $n = n_0$, $P[|\bar{X} - p| < .01] \geqslant .99$. (In fact, there is an n_0 such that for $n \geqslant n_0$ we know $P[|\bar{X} - p| < .01] \geqslant .99$.) Now

$$\begin{aligned} P[|\bar{X} - p| < .01] &= P[-.01 < \bar{X} - p < .01] \\ &= P[p - .01 < \bar{X} < p + .01] \\ &= P[n(p - .01) < X_1 + \cdots + X_n < n(p + .01)] \\ &= \sum_{n \in J(n,\, p)} \binom{n}{j} p^j (1 - p)^{n-j} \end{aligned}$$

where the summation runs from

$$j = \max([n(p - .01)] + (1 - \delta_{[n(p-.01)],\, n(p-.01)}), 0)$$

to $j = \min([n(p + .01)], n)$ and δ_{xy} is 0 if $x \neq y$ and 1 if $x = y$. We have called this set of values $J(n, p)$. If we show that, whatever value n may have, $P[|\bar{X} - p| < .01]$ is a minimum when $p = .5$, then we simply calculate $P_{p=.5}[|\bar{X} - p| < .01]$ for $n = 1, 2, 3, \ldots$ and choose n_0 as the first n for which this probability is at least .99. (See Honors Problem 6.4H.)

We now seek an approximation to n_0. We have

$$\begin{aligned} P[|\bar{X} - p| < .01] &= P[n(p - .01) < X_1 + \cdots + X_n < n(p + .01)] \\ &= P[-.01n < X_1 + \cdots + X_n - np < .01n] \\ &= P\left[\frac{-.01n}{\sqrt{npq}} < \frac{X_1 + \cdots + X_n - np}{\sqrt{npq}} < \frac{.01n}{\sqrt{npq}}\right] \\ &= P\left[-.01\sqrt{\frac{n}{pq}} < \frac{X_1 + \cdots + X_n - np}{\sqrt{npq}} < .01\sqrt{\frac{n}{pq}}\right] \\ &\approx P\left[-.01\sqrt{\frac{n}{pq}} < Y < .01\sqrt{\frac{n}{pq}}\right], \end{aligned}$$

where Y is a $N(0, 1)$ r.v. [This follows by the CLT, Theorem 6.4.6; here $EX_1 = p$, $\text{Var}(X_1) = pq$.] Now this approximation is a function of p, and is clearly minimized when $pq = p(1 - p)$ is maximized, which occurs at $p = \frac{1}{2}$ (so

$pq = \frac{1}{4}$). We will now find n such that

$$P[-.02\sqrt{n} < Y < .02\sqrt{n}] \approx .99.$$

Let $\Phi(z) \equiv P[Y \leqslant z]$, $-\infty < z < +\infty$ (the area under the normal density to the left of z). Of course $\Phi(0) = \frac{1}{2}$. Then we wish to set n by

$$\begin{aligned}
.99 &= P[-.02\sqrt{n} < Y < .02\sqrt{n}] \\
&= \Phi(.02\sqrt{n}) - \Phi(-.02\sqrt{n}) \\
&= \Phi(.02\sqrt{n}) - [1 - \Phi(.02\sqrt{n})] \\
&= 2\Phi(.02\sqrt{n}) - 1
\end{aligned}$$

so $\Phi(.02\sqrt{n}) = .995$ and $.02\sqrt{n} = 2.58$, or $n = (129)^2 \approx (130)^2 = 16{,}900$. So our approximation to n_0 is $n_0 = 16{,}900$.

Example 6.4.8. A standard process is known to yield about 80 percent acceptable items. A new process has been found to yield 85 acceptable items out of its first 100. Is the superiority of the new process well established?

If the new process' probability of producing an acceptable item (p_2, say) is less than or equal to the old process' probability of producing an acceptable item (p_1, say), how large can the probability of obtaining 85 acceptables in a sample of 100 from the new process be? Denote the 100 items by $X_1, \ldots, X_{100}$, where

$$X_i = \begin{cases} 1, & \text{if the } i\text{th unit is acceptable} \\ 0, & \text{otherwise.} \end{cases}$$

Then $X_1, \ldots, X_{100}$ will be assumed independent with $P[X_i = 1] = p_2$ ($i = 1, \ldots, 100$). We wish to study

$$\begin{aligned}
&P_{p_2}[X_1 + \cdots + X_{100} \geqslant 85] \\
&\qquad = P_{p_2}\left[\frac{(X_1 + \cdots + X_{100}) - 100p_2}{\sqrt{100}\sqrt{p_2(1 - p_2)}} \geqslant \frac{85 - 100p_2}{\sqrt{100}\sqrt{p_2(1 - p_2)}}\right] \\
&\qquad \approx 1 - \Phi\left(\frac{85 - 100p_2}{10\sqrt{p_2(1 - p_2)}}\right), \quad \text{by the CLT.}
\end{aligned}$$

For $p_2 \leqslant p_1 = .80$, this is maximized when $\Phi\left(\frac{85 - 100p_2}{10\sqrt{p_2(1 - p_2)}}\right)$ is a minimum, that is, when $\frac{85 - 100p_2}{10\sqrt{p_2(1 - p_2)}}$ is a minimum. Since the derivative of this expression with respect to p_2 is negative for $0 \leqslant p_2 \leqslant 1$, the minimum for $p_2 \leqslant .80$ occurs at $p_2 = .80$. (Alternatively, looking at the exact binomial probability sum for $P_{p_2}[X_1 + \cdots + X_{100} \geqslant 85]$, we see that the derivative with respect to p_2 is $\geqslant 0$ if $p_2 \leqslant .85$, so the maximum for $p_2 \leqslant .80$ is at

$p_2 = .80$.) Now

$$1 - \Phi\left(\frac{85 - 80}{10\sqrt{(.8)(.2)}}\right) = 1 - \Phi(1.25)$$
$$= 1 - .8944 = .1056 \approx .11.$$

So, there is at most 1 chance in 10 of obtaining 85 or more acceptable out of 100 if, in fact, $p_2 \leqslant p_1 = .80$.

We could decide to switch to the new process based on this evidence, or we could decide to wait for more evidence, or we could decide to reject the new process. (For example, we might in advance have decided to reject upon observing 75 or fewer acceptable out of 100, wait for more evidence upon observing 76, . . . , 90 acceptable out of 100, and accept upon observing 91 or more acceptable out of 100.) Reaching a realistic decision rule involves consideration of cost of observations, loss suffered by not switching to a better process, loss suffered by switching to an inferior process, and so on, and is studied in Statistical Decision Theory [see, for example, Weiss (1961) and Chapter 12].

Note that

$$\frac{S_n - ES_n}{\sigma(S_n)} = \frac{\bar{X} - \mu}{\sigma/\sqrt{n}} = \frac{\sqrt{n}}{\sigma}(\bar{X} - \mu),$$

so that approximate normality of the former implies approximate normality of the latter. The latter form is often encountered.

PROBLEMS

6.1 Let Y be a random variable which is always $\geqslant 0$. Prove that (for any $t > 0$)

$$P[Y \geqslant t] \leqslant \min\left(\frac{EY^2}{t^2}, 1\right).$$

6.2 Let Y be a random variable such that $Y \geqslant 0$. Show that (for any $t > 0$)

$$P[Y \geqslant t] \leqslant \frac{EY}{t}.$$

Prove that the Chebyshev bound EY^2/t^2 is better iff

$$\frac{EY^2}{EY} \leqslant t.$$

6.3 Let X be a discrete random variable taking on values $0, 1, \ldots, n$. Assume that $E(X) = 1$ and $\operatorname{Var}(X) = 1$. For $k = 2, 3, \ldots, n$ prove that

$$P[X \geqslant k] \leqslant \frac{1}{(k-1)^2}.$$

6.4 Suppose given a sequence of n independent tosses of a coin with probability p $(0 \leqslant p \leqslant 1)$ of heads. Let T_n = no. of times in n tosses that a head is followed by a tail. Prove that

$$\frac{T_n}{n} \xrightarrow{P} c,$$

where c is some constant, and find c. [*Hint*: Use Chebyshev's Inequality.]

6.5 Approximately how many independent flips n of a fair coin are needed in order that

$$P[.49 \leqslant (\text{Proportion of heads in } n \text{ tosses}) \leqslant .51] \geqslant .99?$$

6.6 A country has estimated that the probability that any particular nuclear weapon will explode unintentionally in a given decade is .00001. If this country has 100,000 nuclear weapons, let us consider the probability that exactly j or more explode unintentionally between 1976 and 1986. Give: **(a)** an exact expression; **(b)** the Poisson approximation; and **(c)** the normal approximation. Evaluate (b) and (c) for $j = 1$; for $j = 2$.

6.7 A fair die is thrown 1000 times independently. Let X = the number of 6's which occur. Find an exact expression for $P[X \geqslant 170]$. Use the normal approximation to evaluate $P[X \geqslant 170]$.

6.8 It costs \$1 to play a certain slot machine in Las Vegas. The machine is set by the house to pay \$2 with probability .45 (and to pay nothing with probability .55). Let X_i be the house's net winnings on the ith play of the machine. Then $S_n = \sum_{i=1}^{n} X_i$ equals the house's winnings after n plays of the machine. Assuming that successive plays are independent, find:

(a) $E(S_n)$;

(b) $\text{Var}(S_n)$;

(c) The approximate probability that after 10,000 plays of the machine the house's winnings are between \$800 and \$1100. Express your answer as a decimal.

6.9 Let $X_1, \ldots, X_n$ be n independent and identically distributed random variables with $E(X_1) = \mu(-\infty < \mu < +\infty)$, $\text{Var}(X_1) = \sigma^2(0 \leqslant \sigma^2 < +\infty)$, and $E(X_1 - \mu)^4 = \sigma^4 + 1$.

(a) State the Weak Law of Large Numbers.

(b) It is the case that, as $n \to \infty$,

$$\frac{X_1^2 + X_2^2 + \cdots + X_n^2}{n} \xrightarrow{P} c$$

for some constant $c(0 \leqslant c < +\infty)$. Find c.

(c) State the Central Limit Theorem.

(d) Find $\lim_{n\to\infty} P\left[\sigma^2 - \frac{1}{\sqrt{n}} \leqslant \frac{(X_1 - \mu)^2 + \cdots + (X_n - \mu)^2}{n} \leqslant \sigma^2 + \frac{1}{\sqrt{n}}\right]$.

6.10 (Refer to Problem 5.3.)

(a) Suppose that 100 plays of this game are made independently, yielding winnings $W_1, \ldots, W_{100}$. What does Chebyshev's Inequality say about

$$P\left[\left|\frac{W_1 + \cdots + W_{100}}{100} + \frac{13}{36}\right| > \frac{1}{2}\sqrt{\frac{1235}{1296}}\right]?$$

Can you improve on this statement?

(b) Prove that $\dfrac{W_1 + \cdots + W_n}{n} \xrightarrow{P} -\dfrac{13}{36}$ by using:

(1) Chebyshev's Inequality;
(2) Lévy's Continuity Theorem;
(3) The Weak Law of Large Numbers;
(4) Khinchin's Theorem.

6.11 The random variables $X_1, X_2, \ldots$ are such that $EX_k = 0$, $P[|X_k| < 4] = 1$ ($k = 1, 2, \ldots$), and

$$\operatorname{Var}\left(\frac{X_1 + \cdots + X_n}{n}\right) \to \frac{1}{2}, \quad \text{as } n \to \infty.$$

($X_1, X_2, \ldots$ are not necessarily identically distributed.)

(a) Prove that the X's cannot all be independent.

(b) Prove that the Law of Large Numbers does not hold for the X's.

6.12 Let $X_1, \ldots, X_n$ be independent r.v.'s with $P[X_i = 1] = p$ and $P[X_i = 0] = 1 - p$ for $i = 1, \ldots, n$, where p is unknown and $0 \leqslant p \leqslant 1$. Let $S_n = X_1 + \cdots + X_n$. Fix $t > 0$.

(a) Using Chebyshev's Inequality, how large an n will guarantee

$$P\left[\left|\frac{S_n}{n} - p\right| \geqslant t\right] \leqslant .01$$

no matter what value the (unknown) p may have? (Note that your n cannot depend on the unknown p.)

(b) Using the Central Limit Theorem, the approximate n needed so that

$$\min_{0 \leqslant p \leqslant 1} P\left[-t \leqslant \frac{S_n}{n} - p \leqslant t\right] \geqslant \frac{99}{100}$$

is what?

(c) Compare the answers to (a) and (b) when $t = .01$.

6.13 Let $X_1, \ldots, X_n$ be independent random variables, each $N(\mu, \sigma^2)$ where μ is unknown ($-\infty < \mu < +\infty$) and $\sigma^2 > 0$ is known. Let $\bar{X} \equiv (X_1 + \cdots + X_n)/n$.

Only some pairs of real numbers (d_1, d_2) will be such that (irrespective of the value of μ)

$$P[\bar{X} - d_1 \leqslant \mu \leqslant \bar{X} + d_2] = .99; \qquad (*)$$

if (d_1, d_2) is such a pair then the interval of real numbers $(\bar{X} - d_1, \bar{X} + d_2)$ is called a **confidence interval** for μ with **confidence coefficient** .99.

(a) Specify precisely which pairs (d_1, d_2) satisfy $(*)$.

(b) Among all pairs (d_1, d_2) which satisfy $(*)$, which one minimizes $d_1 + d_2$? (That is, which pair minimizes the length of the confidence interval?)

6.14 Show that in Problem 5.22 when $a_i = 1/2^i\ (i = 1, 2, 3, \ldots)$ the r.v. $a_1X_1 + a_2X_2 + \cdots$ has a uniform distribution on $(-1, 1)$. [*Hint*: Use the results of Problem 5.22 and Lévy's Continuity Theorem (Theorem 6.4.2).]

HONORS PROBLEMS

6.1H In the literature of **"ESP"** and of **"Parapsychology"**, much reference is made to probability and statistics: it is used and misused to justify conclusions. [See, for example, Feller, Vol. I (Second Edition), p. 54, footnote 17; p. 368, footnote 10; and the articles referred to in those footnotes.]

(a) Read the articles referred to in Feller's book. Summarize them, with your analysis, in a well-written essay.

(b) Read "Can your hunches make you rich" by E. Cudlipp, which appeared on October 29, 1967, in *This Week* (The National Newspaper Magazine for a Better America). Write a critique, including: coverage of misuse of probability and statistics; clarification of poorly stated or misstated probability and statistics (for example, what does the table titled "Law of Averages" mean); and unscientific reasoning.

(c) Read one recent article in the *Journal of Parapsychology* and write a report on it. Read and report on C. E. M. Hansel, *ESP: A Scientific Evaluation*, Charles Scribner's Sons, New York, 1966; and Milbourne Christopher, *ESP, Seers & Psychics*, Thomas Y. Crowell Company, New York, 1970.

6.2H Suppose $F_n \to L$ in the sense of Definition 6.4.1, but without any restrictions on L. When must L be $\geqslant 0$ and monotonic? Show by example that L need not be a d.f.

6.3H Prove that

$$\lim_{n\to\infty}\left(1 + \frac{a_n}{n}\right)^n = e^{\lim_{n\to\infty} a_n}.$$

6.4H Investigate whether the exact probability in Example 6.4.7 is a minimum when $p = .5$.

6.5H Let $\Phi(x) = P[Z \leqslant x]$ where Z is a $N(0, 1)$ r.v. Let X_1 be a $N(0, 1/n)$ r.v., let X_2 be a $N(1, 1/n)$ r.v., and suppose that X_1 and X_2 are independent.

(a) If a new r.v. Y is a result of choosing from the values $\{X_1, X_2\}$ at random with probability $1 - 1/n$ of choosing X_1 and probability $1/n$ of choosing X_2, show that

$$P[Y \leqslant x] = \left(1 - \frac{1}{n}\right)\Phi(\sqrt{n}x) + \frac{1}{n}\Phi(\sqrt{n}(x - 1)).$$

Find $\phi_Y(t) = Ee^{itY}$.

(b) Suppose that $Y_1, \ldots, Y_n$ are independent r.v.'s with the same distribution as Y in (a). Show that

$$\lim_{n\to\infty} \phi_{Y_1+\cdots+Y_n}(t) = e^{-\frac{1}{2}t^2+e^{it}-1}.$$

(Note that the common d.f. of $Y_1, \ldots, Y_n$ changes as n changes.) How could we obtain a r.v. with this characteristic function? [*Hint*: If T_1 is a $N(0, 1)$ r.v. and T_2 is a Poisson r.v., what is the characteristic function of $T_1 + T_2$ if T_1 and T_2 are independent r.v.'s?] Express

$$\lim_{n\to\infty} P[Y_1 + \cdots + Y_n \leqslant x]$$

as an infinite series.

6.6H Suppose that $X_1, X_2, \ldots$ is a sequence of r.v.'s such that

$$X_n \xrightarrow{P} X(\theta) \qquad (*)$$

for a specified r.v. $X(\theta)$ for all θ in a specified set Θ. (The r.v.'s $X_1, X_2, \ldots$ are not necessarily either independent or identically distributed, and in general their d.f. depends on $\theta \in \Theta$.) *We then ask*: Does there always exist a subsequence n_k such that for all $\theta \in \Theta$

$$X_{n_k} \to X(\theta) \text{ w.p. } 1? \qquad (**)$$

[From Theorem 6.3.4 it does not follow that (**) holds, but only that there is a subsequence $n_k(\theta)$ such that $X_{n_k(\theta)} \to X(\theta)$ w.p. 1. (**) asserts we can find a subsequence which "works" for all θ simultaneously.] A reference is G. Simons, "Identifying probability limits," *Annals of Mathematical Statistics*, Vol. 42 (1971), pp. 1429–1433.

7

AN EXAMPLE OF STATISTICAL ANALYSIS

In this chapter we first (Section 7.1) study the problem of finding the root of an equation $M(x) = \alpha$ where $M(\cdot)$ is a known function and α is a fixed number. This problem is nonstochastic (that is, contains no random elements or random variables in its formulation) and is called a problem of **(nonstochastic) approximation**. However, in many real-world situations (one such situation from pharmacology is noted in Section 7.3) random variables arise when we attempt to evaluate $M(\cdot)$ at a specified point θ. [Such evaluations are needed to solve $M(x) = \alpha$ for the root θ.] In Section 7.3 we study one approach to this problem, called **stochastic approximation**. The problem of Section 7.3 gives us an example of statistical analysis in the area of *point estimation*; hence, some basic terminology and concepts of this area are given in Section 7.2.

7.1. (NONSTOCHASTIC) APPROXIMATION

We now formulate and study a simple problem which involves no stochastic (random) elements: Given an increasing function $M(x)$ (where $a \leqslant x \leqslant b$ with $a < b$) for which there is a unique value (root) θ such that $M(\theta) = \alpha$, find (more precisely, approximate) that value θ. Since $M(x)$ is an increasing function, if we evaluate $M(\cdot)$ at x_1 and find $M(x_1) < \alpha$ we know that $\theta > x_1$. Similarly, if we evaluate $M(\cdot)$ at x_2 and find $M(x_2) > \alpha$, we know that $\theta < x_2$. Since we start off with the knowledge that $a \leqslant \theta \leqslant b$, observing that $M(x_1) < \alpha$ increases our knowledge; we then know that $x_1 \leqslant \theta \leqslant b$. [Similarly, observing $M(x_2) > \alpha$ increases our knowledge; we then know that $a \leqslant \theta \leqslant x_2$.] But if we are allowed to evaluate[1] $M(\cdot)$ at only N points x, how should we choose those points x in order that the smallest interval $L \leqslant \theta \leqslant U$ in which we can be *sure* [after evaluating $M(\cdot)$ at the N points x] θ lies has minimal length?

We will find it convenient to reparametrize the problem by defining a function $f(x)$ with $0 \leqslant x \leqslant 1$ by

$$f(x) = M(bx + a(1 - x)) - \alpha; \tag{7.1.1}$$

we are then seeking that unique x $(0 \leqslant x \leqslant 1)$ for which the increasing function $f(x)$ is 0.

Let $\mathscr{F}$ be the class of all $f(x)$ which are monotone nondecreasing on $I = \{x\colon 0 \leqslant x \leqslant 1\}$ with $f(x) = 0$ for exactly one x with $0 \leqslant x \leqslant 1$. Let D^* be the collection of all subintervals of I. Let S_N be the class of all procedures s [for estimating the point $g(f)$ where $f(x) = 0$] which take N observations, and which state as their final decision $g(f) \in d$ for a specific $d \in D^*$. Let $\bar{S}_N$ be those procedures s in S_N for which it is *always*[2] true that $g(f) \in d$. Then the length of d depends on $f \in \mathscr{F}$ and $s \in \bar{S}_N$; call it $L(f, s)$. We will find a procedure $s_N^* \in \bar{S}_N$ for which[3]

$$\sup_{f \in \mathscr{F}} L(f, s_N^*) = \inf_{s \in \bar{S}_N} \sup_{f \in \mathscr{F}} L(f, s); \tag{7.1.2}$$

such a procedure s_N^* is called a (nonrandomized) **minimax decision procedure**. Such procedures are important in the general theory of statistical decisions (Chapter 12).

In the following theorem we show that the "best" way of choosing each point x at which to evaluate $M(\cdot)$ is: as the midpoint of the smallest interval in which previous evaluations of $M(\cdot)$ allow us to be *certain* θ lies.

Theorem 7.1.3. *The following procedure s_N^* is minimax [that is, satisfies (7.1.2)]. Let $x_1 = \frac{1}{2}$. Let $x_k = (\alpha_{k-1} + \beta_{k-1})/2$, where α_{k-1} is the largest previously observed x with $f(x) < 0$ (or 0 if there is no such x), and β_{k-1} is the smallest for which $f(x) > 0$ (or 1 if there is none). Take interval $d = [\alpha_N, \beta_N]$ (or a single point x_r if $f(x_r) = 0$ for some r).*

Also, $L(f, s_N^*) = 2^{-N}$ unless $g(f) = q/2^N$ for some integer $q = 1, 2, \ldots, N - 1$ (then $L = 0$).

---

[1] These N evaluations are all we can observe about $M(\cdot)$.

[2] "Always" means for all $f \in \mathscr{F}$ and in all applications of the procedure.

[3] The expressions "**sup**" and "**inf**" (abbreviations for "**supremum**" and "**infimum**," respectively) will be used throughout the following chapters. They are essentially "max" and "min" (for "maximum" and "minimum," respectively), with the following proviso. The maximum of a specified set of numbers may not exist (for example, $\max_{0<x<1} x$ does not, since there is *no* largest number x between 0 and 1), but there will be a least number $\geqslant$ all numbers in the set (for example, 1 in the case of $\max_{0<x<1} x$). We therefore usually write $\sup_{0<x<1} x$ (which must exist) instead of $\max_{0<x<1} x$ (which often, as in this case, does not exist). (Similarly for inf.) The student not familiar with these notions can simply read "max" for "sup" and "min" for "inf" with little loss.

Proof: If $N = 1$, it is easy to see that any procedure $s \in \bar{S}_N$ which takes $x_1 \neq \frac{1}{2}$ will have a larger $\sup\{L(f, s): f \in \mathscr{F}\}$. Now suppose we have proven the theorem for $n = 1, 2, \ldots, N - 1$. We now show the theorem is then true for $n = N$. For, any procedure which has taken $n - 1$ evaluations (at say $x_1, \ldots, x_{n-1}$) does best to take x_n midway between $\alpha_{n-1}(x_1, \ldots, x_{n-1})$ and $\beta_{n-1}(x_1, \ldots, x_{n-1})$, and then achieves minimax length $(\beta_{n-1} - \alpha_{n-1})/2$. But taking $x_1, \ldots, x_{n-1}$ in accord with s_{n-1}^* minimizes $\beta_{n-1} - \alpha_{n-1}$.

7.2. POINT ESTIMATION

The basic situation in "point estimation" (of which we will encounter a special case in Section 7.3, where random variables are added to the approximation problem of Section 7.1) is as follows. We observe r.v.'s $X_1, \ldots, X_n$ (not necessarily independent or identically distributed) and the d.f. of $X_1, \ldots, X_n$ depends on an unknown parameter θ (for example, the mean of a normal distribution can be unknown) which is known to be in some given set Θ. The problem of estimating θ is a problem of **point estimation**. [We have a basic sample space Ω, the experiment essentially chooses an $\omega \in \Omega$, the observations are $X_1(\omega) = x_1, \ldots, X_n(\omega) = x_n$, and the d.f. of $X_1, \ldots, X_n$ is, say, $F_{X_1,\ldots,X_n}(y_1, \ldots, y_n|\theta)$. We then wish to estimate θ.] Of course, we estimate θ by some (appropriate) function of the observations $X_1, \ldots, X_n$; such a function being called a "statistic" or an "estimator":

Definition 7.2.1. *Any function of the random variables which are being observed, say $t_n(X_1, \ldots, X_n)$, is called a* **statistic**. *It is also called an* **estimator of** θ; *since $X_1, \ldots, X_n$ are r.v.'s, it is a r.v. A particular value of the estimator, say $t_n(x_1, \ldots, x_n)$, is called an* **estimate of** θ.

We would like our estimator of θ to be exactly θ on the average, and if it is we call it an "unbiased" estimator of θ:

Definition 7.2.2. *We say $t_n(X_1, \ldots, X_n)$ is an* **unbiased estimator** *of θ if*

$$E_\theta t_n(X_1, \ldots, X_n) \stackrel{\theta}{\equiv} \theta, \quad \text{all } \theta \in \Theta.$$

[The subscript θ on the expectation sign indicates which d.f. $F_{X_1,\ldots,X_n}(y_1, \ldots, y_n|\theta)$ is to be utilized in the evaluation.]

However, many estimators are not unbiased, so we also define a measure of the difference between our estimator and θ:

Definition 7.2.3. (**Bias** of t_n) $= E_\theta t_n(X_1, \ldots, X_n) - \theta \equiv \mathbf{b_n(\theta)}$.

Another property we would like to have in estimators of θ is that they converge in probability to θ (that is, their probability "collects" close to θ) as the number of observations $n \to \infty$:

Definition 7.2.4. *If $X_1, \ldots, X_n$ are r.v.'s to be observed and they have d.f. $F_{X_1,\ldots,X_n}(y_1, \ldots, y_n|\theta)$ where the unknown $\theta \in \Theta$, then a sequence $t_1, t_2, \ldots$ is called* **a consistent sequence of estimators of $\boldsymbol{\theta}$** *iff (as $n \to \infty$)*

$$t_n(X_1, \ldots, X_n) \xrightarrow{P} \theta, \quad \text{all } \theta \in \Theta.$$

7.3. STOCHASTIC APPROXIMATION

Consider the pharmacological situation in which a dosage x of a drug is administered to animals, and a certain (random) fraction of the animals die. On the average $M(x)$ die at dosage x, and usually $M(x)$ increases as x increases. The problem in pharmacology is to determine LD50 (Lethal Dose 50 percent), the θ such that $M(\theta) = \frac{1}{2}$. In general, the problem is that for the unknown function $M(x)$ we wish to determine "the" point θ such that $M(\theta) = \alpha$ for some specified α. (See Figure 7.3–1.) Examples of other situations with the same goal are plentiful. For example, the author once consulted with a medical researcher who was going to remove a specified

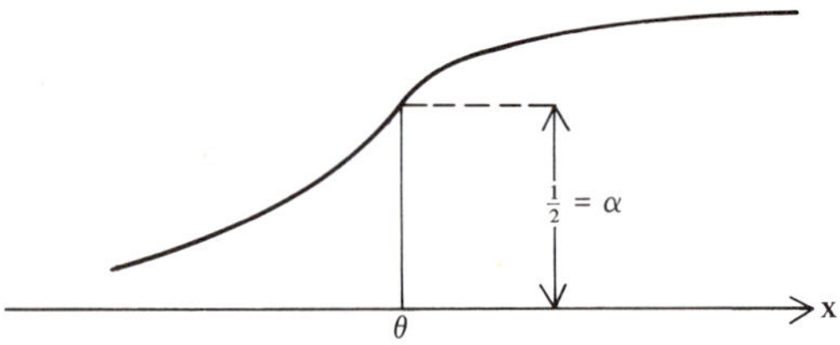

Figure 7.3–1. **M(x)** for LD50

proportion x of the blood volume from certain animals (it happened to be pigs) for 2 hours. This would deliver a shock to the pigs, and proportion $M(x)$ on the average would die at specified proportion (dosage) x. The goal of the study was to find LD50, that is, the dosage x at which (on the average) half of the pigs die from the shock and the other half survive. (The purpose of the experimental determination of LD50 was as follows. It is well-established that in certain respects pigs are similar to humans. The shock of blood removal for 2 hours is thought to be similar to the shock of a heart attack. Now if we know LD50, we can later administer not only LD50 to a pig *but* also a drug which may help people survive a heart attack. We would then be in a position to see if the drug seemed to increase the chances of surviving a heart attack, since we know about 50 percent would die without the drug. A similar experiment on humans would be highly unethical, and we could not otherwise control "heart attack severity" in them.)

The distribution of responses obtained at dosage x, represented by its d.f. say $H(y|x)$ *by definition*, varies as the real-valued parameter x varies. If Y is a r.v. with d.f. $H(y|x)$, the expected value of Y at x [say $M(x)$, called for historical reasons a *regression function*] is[4]

$$M(x) = \int_{-\infty}^{\infty} y \, d_y H(y|x).$$

We will assume (**Assumption A–1**) that *$M(x)$ is a uniformly bounded function* [$|M(x)| \leqslant C < \infty$ for some number C], *and* that *the variance of Y at dose x is uniformly bounded* [(Variance of Y at x) $\leqslant \sigma^2 < \infty$ for some number σ^2]; that is,

$$\operatorname{Var}(Y|x) = \int_{-\infty}^{\infty} (y - M(x))^2 \, d_y H(y|x) \leqslant \sigma^2 < \infty.$$

In addition, we will assume either that (**Assumption A–2**) *for some $\delta > 0$, $M(x) \leqslant \alpha - \delta$ for $x < \theta$, and $M(x) \geqslant \alpha + \delta$ for $x > \theta$* [this is the case of a jump of $M(\cdot)$ at θ; see Figure 7.3–2] or (**Assumption A–3**) that: (a) *$M(x) < \alpha$ for $x < \theta$, $M(\theta) = \alpha$, $M(x) > \alpha$ for $x > \theta$*; (b) *For some $\delta > 0$, $M(x)$ is strictly increasing if $|x - \theta| < \delta$; and* (c) $\inf\{|M(x) - \alpha| : |x - \theta| \geqslant \delta\} > 0$. [Thus, we have, for example, an $M(x)$ as in Figure 7.3–3. Note that (c)

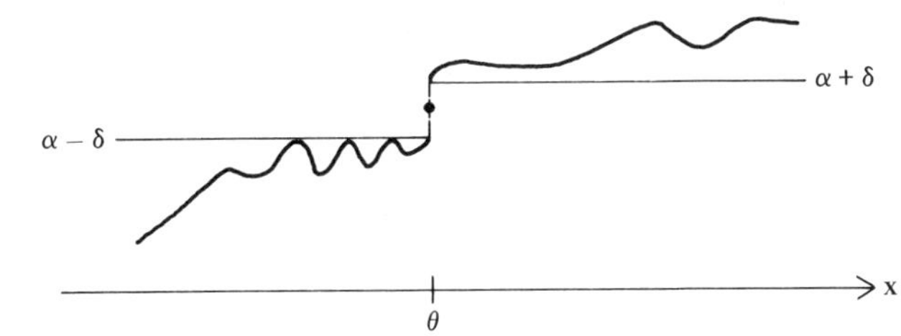

Figure 7.3–2. **M(x)** Meeting A–2

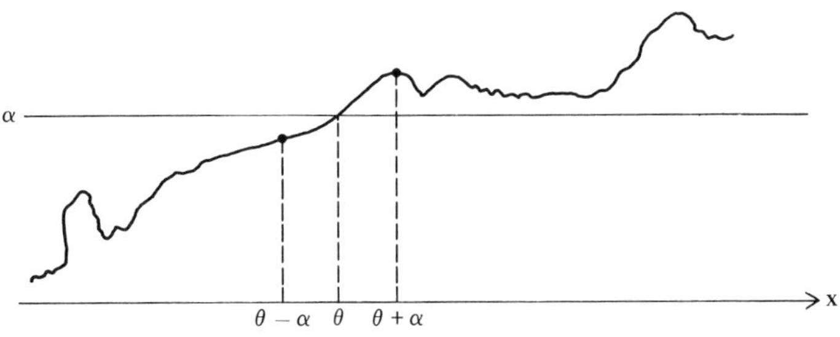

Figure 7.3–3. **M(x)** Meeting A–3

[4] By the notation "$d_y H(y|x)$" we mean "$h(y|x)\, dy$" in the absolutely continuous case, and "$P_x[Y = y]$" (with $\int$ replaced by $\sum$) in the discrete case. The expression actually has meaning in the general case as well, and is called a Stieltjes integral.

requires that the $M(x)$ not dip arbitrarily closer and closer to height α outside of the strictly increasing range of values specified in (b).] [For our purposes it will suffice below if either A–1 and A–2 hold, or A–1 and A–3 hold. In either case $M(x)$ will be restricted so that it does not hit height α several times, and so that it does not get closer and closer to α.]

We would now like to find a method which, based on an observation taken at dosage $\mathbf{x_n}$, tells us a new dosage $\mathbf{x_{n+1}}$ at which to take our next observation. Heuristically, we can reason as follows: If x_n is too small ($x_n < \theta$), we expect to observe a $Y_n < \alpha$ (so $\alpha - Y_n > 0$) and we would therefore like to choose $x_{n+1} > x_n$. [Similarly, if x_n is too large ($x_n > \theta$), we expect to observe a $Y_n > \alpha$ (so $\alpha - Y_n < 0$) and we would then like to choose $x_{n+1} < x_n$.] *If* we choose

$$x_{n+1} = x_n + a(\alpha - Y_n)$$

for some constant $a > 0$, then we will have $x_{n+1} > x_n$ iff $Y_n < \alpha$ (and $x_{n+1} < x_n$ iff $Y_n > \alpha$). *However*, we would not want the same value $a > 0$ to be used each time we take an observation, for then our sequence of dosages $x_1, x_2, x_3, \ldots$ would not converge to θ (since it would not converge). So we need a sequence of a's $a_1, a_2, a_3, \ldots$ which are not so large (for example, all equal) that we "bounce around" instead of converging, yet not so small that we cannot move from x_n to θ. Following this line of reasoning we now let, *by definition*, $a_1, a_2, a_3, \ldots$ be any sequence of positive numbers such that $\sum_{n=1}^{\infty} a_n = \infty$ and $\sum_{n=1}^{\infty} a_n^2 < \infty$ (for example, $a_n = 1/n$), and consider the following **method of generating the nth estimator (dosage)** X_n:

Let X_1 be an arbitrary number; take an observation at dosage X_1, obtaining a random variable Y_1 whose d.f. is $H(\cdot|X_1)$. Next let $X_2 = X_1 + a_1(\alpha - Y_1)$; take an observation at dosage X_2, obtaining a r.v. Y_2 (independent[5] of Y_1) whose d.f. is $H(\cdot|X_2)$. Next let $X_3 = X_2 + a_2(\alpha - Y_2)$; take an observation at dosage X_3, obtaining a r.v. Y_3 (independent[5] of Y_1, Y_2) whose d.f. is $H(\cdot|X_3)$. Next let $X_4 = X_3 + a_3(\alpha - Y_3) \ldots$. [Of course after stage n, we let $X_{n+1} = X_n + a_n(\alpha - Y_n) \ldots$.]

We will now prove that (with this method of moving from one dosage to another) **$\mathbf{X_n}$ converges to $\boldsymbol{\theta}$ in probability ($\mathbf{X_n} \xrightarrow{P} \boldsymbol{\theta}$).** Hence (see Definition 7.2.4), $\{X_n\}$ is a consistent sequence of estimators of θ regardless of our choice of $a_1, a_2, a_3, \ldots$ as long[6] as $a_1 + a_2 + \cdots = \infty$, $a_1^2 + a_2^2 + \cdots < \infty$, and $a_i > 0$ $(i = 1, 2, \ldots)$.

---

[5] This independence is conditional on having fixed the dosage; since X_{n+1} depends on $Y_1, \ldots, Y_n$ we do not have real independence.

[6] A natural question is "Which sequence $a_1, a_2, a_3, \ldots$ is best?" For this and related considerations see M. T. Wasan, *Stochastic Approximation*, Cambridge University Press, London, 1969.

Theorem 7.3.1. *Under* A–1 *and* A–2, *or* A–1 *and* A–3, $X_n \xrightarrow{P} \theta$ *(that is,* $X_1, X_2, \ldots$ *is a consistent sequence of estimators of* θ*).*

We will prove Theorem 7.3.1 in the following manner. A sequence of lemmas (Lemmas 7.3.2, 7.3.5, 7.3.7, and 7.3.11) shows that for some subsequence $n_1, n_2, \ldots$ of $1, 2, 3, \ldots$ we have $X_{n_j} \xrightarrow{P} \theta$. Using this we show that $X_n \xrightarrow{P} \theta$.

Lemma 7.3.2. *Define*

$$e_n = E\left[\int_{-\infty}^{\infty} (y - \alpha)^2 \, d_y H(y|X_n)\right].$$

Then

$$0 \leqslant e_n \leqslant \alpha^2 + (C + |\alpha|)^2 = h \text{ (say)}. \tag{7.3.3}$$

[This says that the average value of the square of the difference $Y_n - \alpha$, which we call e_n, is at most $\alpha^2 + (C + |\alpha|)^2$.]

Proof: From A–1 we know that

$$0 \leqslant \int_{-\infty}^{\infty} (y - M(X_n))^2 \, d_y H(y|X_n) \leqslant \alpha^2 < \infty;$$

but

$$\begin{aligned}
&\int_{-\infty}^{\infty} (y - M(X_n))^2 \, d_y H(y|X_n) \\
&\quad = \int_{-\infty}^{\infty} ((y - \alpha) + (\alpha - M(X_n)))^2 \, d_y H(y|X_n) \\
&\quad = \int_{-\infty}^{\infty} (y - \alpha)^2 \, d_y H(y|X_n) + 2(\alpha - M(X_n)) \int_{-\infty}^{\infty} (y - \alpha) \, d_y H(y|X_n) \\
&\qquad\qquad + (\alpha - M(X_n))^2 \int_{-\infty}^{\infty} d_y H(y|X_n) \\
&\quad = \int_{-\infty}^{\infty} (y - \alpha)^2 \, d_y H(y|X_n) - 2(\alpha - M(X_n))^2 + (\alpha - M(X_n))^2 \\
&\quad = \int_{-\infty}^{\infty} (y - \alpha)^2 \, d_y H(y|X_n) - (\alpha - M(X_n))^2,
\end{aligned}$$

and hence

$$0 \leqslant \int_{-\infty}^{\infty} (y - \alpha)^2 \, d_y H(y|X_n) \leqslant \sigma^2 + (\alpha - M(X_n))^2 \leqslant \sigma^2 + (|\alpha| + C)^2. \tag{7.3.4}$$

Result (7.3.3) follows upon taking the expectation of (7.3.4) and referring to the definition of e_n.

Lemma 7.3.5. *Define*

$$\begin{aligned}
b_n &= E(X_n - \theta)^2, \\
d_n &= E[(X_n - \theta)(M(X_n) - \alpha)].
\end{aligned}$$

Then

$$b_{n+1} = b_1 + \sum_{i=1}^{n} a_i^2 e_i - 2 \sum_{i=1}^{n} a_i \, d_i. \tag{7.3.6}$$

[This relates b_n (the average value of the square of the difference $X_n - \theta$) to e_n and d_n by a direct formula.]

Proof: First, $b_1 = b_1$. For b_2, we have

$$\begin{aligned} b_2 &= E(X_2 - \theta)^2 = E[X_1 + a_1(\alpha - Y_1) - \theta]^2 \\ &= E[(X_1 - \theta) + a_1(\alpha - Y_1)]^2 \\ &= E(X_1 - \theta)^2 + a_1{}^2E(\alpha - Y_1)^2 - 2a_1E(X_1 - \theta)(Y_1 - \alpha) \\ &= b_1 + a_1{}^2e_1 - 2a_1\, d_1 \end{aligned}$$

[where we used the fact that $E(X_1 - \theta)(Y_1 - \alpha) = (X_1 - \theta)E(Y_1 - \alpha) = (X_1 - \theta)(M(X_1) - \alpha) = E(X_1 - \theta)(M(X_1) - \alpha)$, since X_1 is a constant]. Proceeding by induction, for b_{n+1}

$$\begin{aligned} b_{n+1} &= E(X_{n+1} - \theta)^2 \\ &= E(X_n - \theta)^2 + a_n{}^2E(\alpha - Y_n)^2 - 2a_nE(X_n - \theta)(Y_n - \alpha) \\ &= b_n + a_n{}^2e_n - 2a_n\, d_n \end{aligned}$$

[where we used the fact that $E(X_n - \theta)(Y_n - \alpha) = E(X_n - \theta)[(M(X_n) - \alpha) + (Y_n - M(X_n))] = d_n + E(X_n - \theta)(Y_n - M(X_n)) = d_n$, since[7] $E(X_n - \theta)\cdot(Y_n - M(X_n)) = E((X_n - \theta)E(Y_n - M(X_n)|X_n)) = 0$]. Hence, we find (7.3.6).

Lemma 7.3.7.

$$\lim_{n\to\infty} \inf\{d_j : j \geqslant n\} = 0. \tag{7.3.8}$$

[This implies that for some positive integers $n_1 < n_2 < \cdots$ the subsequence $d_{n1}, d_{n2}, \ldots$ of the sequence $d_1, d_2, \ldots$ is such that $\lim d_{n_j} = 0$.]

Proof: Since $b_{n+1} \geqslant 0$ by definition, from (7.3.6) and (7.3.3) we find

$$0 \leqslant b_{n+1} \leqslant b_1 + h\sum_{i=1}^{n} a_i^2 - 2\sum_{i=1}^{n} a_i\, d_i \leqslant b_1 + h\sum_{i=1}^{\infty} a_i^2 - 2\sum_{i=1}^{n} a_i\, d_i,$$

where $\sum_{i=1}^{\infty} a_i^2 < \infty$; hence,

$$\sum_{i=1}^{\infty} a_i\, d_i < \infty. \tag{7.3.9}$$

[Note that (7.3.9) follows since $a_i\, d_i$ is a sequence of positive numbers. By

[7] Here we use the fact (see Theorem 5.3.20 and the succeeding discussion) that $E(g(Y)) = E(E(g(Y)|X))$.

definition $a_n \geqslant 0$. We now show $d_n = E(X_n - \theta)(M(X_n) - \alpha) \geqslant 0$. First, if A–2 holds,

$$d_n = E[(X_n - \theta)(M(X_n) - \alpha)|X_n \geqslant \theta]P[X_n \geqslant \theta] + E[(X_n - \theta)(M(X_n) - \alpha)|X_n < \theta]P[X_n < \theta] \geqslant 0,$$

since both terms are $\geqslant 0$. The proof is similar for the case where A–3 holds.] Note that $b_{n+1} \to$ a limit b (say), since (7.3.6) holds, (7.3.9) holds, $a_i\, d_i \geqslant 0$, and $0 \leqslant e_i \leqslant h$. But now (since $\sum_{n=1}^{\infty} a_n = \infty$, $\sum_{n=1}^{\infty} a_n\, d_n < \infty$, and $d_n \geqslant 0$)

$$\liminf d_n \equiv \lim_{n\to\infty} \inf\{d_j : j \geqslant n\} = 0; \tag{7.3.10}$$

for if not, then $\liminf d_n = c > 0$, and hence $d_i > c/2$ for $i = N, N + 1, \ldots$ (for some N). Thus, $\sum_{n=1}^{\infty} a_n\, d_n \geqslant (c/2) \sum_{N}^{\infty} a_i = \infty$, a contradiction.

Lemma 7.3.11. *Let* $n_1 < n_2 < \cdots$ *be a sequence of positive integers such that* $\lim d_{n_j} = 0$. *Then*

$$X_{n_j} \xrightarrow{P} \theta. \tag{7.3.12}$$

Proof: Suppose (7.3.12) fails to hold. Then there is a sequence $t_1 < t_2 < \cdots$ (a subsequence of $n_1 < n_2 < \cdots$) and some $\varepsilon > 0$, $\eta > 0$ such that for all j

$$P[|X_{t_j} - \theta| > \eta] > \varepsilon. \tag{7.3.13}$$

But then for all t_j we would have[8]

$$\begin{aligned} d_{t_j} &= E(X_{t_j} - \theta)(M(X_{t_j}) - \alpha) \\ &= E|X_{t_j} - \theta|\,|M(X_{t_j}) - \alpha| \\ &\geqslant E[|X_{t_j} - \theta|\,|M(X_{t_j}) - \alpha|\,\big|\,|X_{t_j} - \theta| > \eta]P[|X_{t_j} - \theta| > \eta] \\ &\geqslant \eta \inf\{|M(X_{t_j}) - \alpha| : |X_{t_j} - \theta| > \eta\}\,\varepsilon = \eta \inf\{|M(x) - \alpha| : |x - \theta| > \eta\}\,\varepsilon > 0; \end{aligned}$$

here we used the facts that (1) $X_{t_j} - \theta$ and $M(X_{t_j}) - \alpha$ are either both positive or both negative [shown following (7.3.9)], and that the inf of $|M(x) - \alpha|$ over $|x - \theta| > \eta$ is (by A–2 or A–3) a fixed positive number. This contradicts $n_1, n_2, \ldots$ being such that $\lim d_{n_j} = 0$; hence, (7.3.12) holds.

Proof of Theorem 7.3.1. We will show that $X_n \xrightarrow{P} \theta$ by finding (for arbitrary $\varepsilon > 0$ and $\eta > 0$) an $N(\eta, \varepsilon)$ such that

$$n > N(\eta, \varepsilon) \Rightarrow P[|X_n - \theta| > \eta] \leqslant \varepsilon. \tag{7.3.14}$$

---

[8] We use the fact that $E(Y) = E(Y|A)P(A) + E(Y|\bar{A})P(\bar{A})$, which is (show this) a special case of $E(XY) = E(XE(Y|X))$ (see Theorem 5.3.20 ff).

Let $s > 0$ be such that $(s^2 + s)/\eta^2 < \varepsilon/2$. Since $X_{n_j} \overset{P}{\to} \theta$ [(7.3.12)] and since $\sum_{n=1}^{\infty} a_n^{\,2} < \infty$, there is an integer N_0 such that

$$P[|X_{N_0} - \theta| \geqslant s] < \frac{\varepsilon}{2} \quad \text{and} \quad \sum_{n=N_0}^{\infty} a_n^2 < \frac{s}{2h}. \tag{7.3.15}$$

Now we define, for $n \geqslant N_0$,

$$\begin{aligned} b_n'(z) &= E[(X_n - \theta)^2 | X_{N_0} = z] \\ &= b_{N_0}'(z) + \sum_{i=N_0}^{n} a_i^2 e_i - 2 \sum_{i=N_0}^{n} a_i\, d_i \\ &\leqslant b_{N_0}'(z) + \sum_{i=N_0}^{n} a_i^2 e_i \\ &\leqslant (z - \theta)^2 + h \sum_{i=N_0}^{\infty} a_i^2 \\ &\leqslant (z - \theta)^2 + h \frac{s}{2h} \\ &= (z - \theta)^2 + \frac{s}{2}, \end{aligned} \tag{7.3.16}$$

where we used (7.3.6), the definition of b_n, the facts $a_i \geqslant 0$ and $d_i \geqslant 0$, and (7.3.3). Hence, using (7.3.16),

$$\begin{aligned} &E[(X_n - \theta)^2 |\, |X_{N_0} - \theta| < s] \\ &\qquad \leqslant \sup\{(z - \theta)^2 + \frac{s}{2} : \theta - s \leqslant z \leqslant \theta + s\} \\ &\qquad = s^2 + \frac{s}{2} < s^2 + s. \end{aligned} \tag{7.3.17}$$

Now, when $n > N_0$, by Chebyshev's Inequality (Theorem 6.2.5) and (7.3.17), we have

$$\begin{aligned} P[|X_n - \theta| \geqslant \eta \,|\, |X_{N_0} - \theta| < s] &\leqslant \frac{E[(X_n - \theta)^2 \,|\, |X_{N_0} - \theta| < s]}{\eta^2} \\ &< \frac{s^2 + s}{\eta^2} < \frac{\varepsilon}{2}, \end{aligned} \tag{7.3.18}$$

where the last step used the definition of s [just above (7.3.15)]. Now using the definition of conditional probability [(2.3.1)], the Theorem of Total

Probabilities (Theorem 2.3.4), and (7.3.15), for $n > N_0$

$$\begin{aligned}
&P[|X_n - \theta| \geqslant \eta] \\
&= P[|X_n - \theta| \geqslant \eta, |X_{N_0} - \theta| < s] + P[|X_n - \theta| \geqslant \eta, |X_{N_0} - \theta| \geqslant s] \\
&\leqslant P[|X_n - \theta| \geqslant \eta, |X_{N_0} - \theta| < s] + P[|X_{N_0} - \theta| \geqslant s] \\
&\leqslant P[|X_n - \theta| \geqslant \eta \mid |X_{N_0} - \theta| < s]P[|X_{N_0} - \theta| < s] + \frac{\varepsilon}{2} \\
&\leqslant P[|X_n - \theta| \geqslant \eta \mid |X_{N_0} - \theta| < s] + \frac{\varepsilon}{2} \leqslant \varepsilon.
\end{aligned} \tag{7.3.19}$$

This demonstrates (7.3.14) and thus concludes our proof.

Note that the method of choosing successive points X_n at which to make observations Y_n studied in this section (Theorem 7.3.1) has the property that $X_n \not\xrightarrow{P} \theta$. In Section 7.1, we clearly satisfy assumptions A–1 and A–3; hence, the present method will provide a consistent sequence of estimators of θ for the context studied in Section 7.1. However, with that sequence, one could not stop after $N(\varepsilon)$ stages and be assured of, say, $|X_n - \theta| \leqslant \varepsilon$ for a specified $\varepsilon > 0$. In contrast, in Section 7.1 we found a "best" sequence.

In the following example we detail the explicit results of an experiment in biological science where stochastic approximation was used.

Example 7.3.20. It has been found that (in *Paramecium caudatum*) increased rates of cell division are induced by treatment with the substance kinetin. In particular, suppose that one group of culture slides of paramecia is treated with kinetin for x hours, and that another similar group of culture slides of paramecia is kept untreated for the same period. It is assumed that the average $M(x)$ of the ratio

$$\frac{\text{(No. of divisions in paramecia treated with kinetin)}}{\text{(No. of divisions in paramecia untreated)}}$$

is an increasing function of the length of exposure x. A problem of biological significance[9] was to estimate the x for which $M(x) = 1.10$, that is, the length of exposure x for which on the average 10 percent more divisions would be noted in the group treated with kinetin.

At initial exposure $X_1 = 30$ hours, the observed ratio of divisions was $Y_1 = 1.067$. Since the a_n's chosen were $a_n = 20/n$ ($n = 1, 2, 3, \ldots$), the second exposure was taken to be (see the discussion preceding Theorem 7.3.1)

$$X_2 = X_1 + a_1(\alpha - Y_1) = 30 + 20(1.10 - 1.067) = 30.66 \doteq 30.7.$$

[9] For a complete discussion, see L. Guttman and R. Guttman, "An illustration of the use of stochastic approximation," *Biometrics*, Vol. 15 (1959), pp. 551–559.

The experimentation then proceeded sequentially as in Table 7.3–1; since $X_6, X_7, \ldots, X_{13}$ were all approximately 25 hours, Guttman and Guttman concluded that this was the approximate value of x for which $M(x) = 1.10$.

Table 7.3–1. Paramecia Experiment Data.

n	X_n	Y_n
1	30	1.067
2	30.7	1.300
3	28.7	1.131
4	27.3	1.223
5	26.6	1.577
6	24.8	1.133
7	24.6	.890
8	25.2	1.000
9	25.5	.810
10	25.6	1.310
11	25.1	1.210
12	24.8	1.030
13	24.9	—

For a discussion of the choice $a_n = 20/n$ (which affects the speed of convergence of X_n) see p. 555 of the previously cited article by Guttman and Guttman. [Theoretical considerations related to choice of a_n have been given in the work of Wasan cited just before Theorem 7.3.1.]

PROBLEMS

7.1 In the proof of Theorem 7.1.3 we stated that (with $N = 1$) the $s \in \bar{S}_1$, which minimizes

$$\sup\{L(f, s): f \in \mathscr{F}\} \qquad (*)$$

is the one which takes $x_1 = \frac{1}{2}$. Evaluate $(*)$ for any $s \in \bar{S}_1$ which takes $x_1 = x$ for some $0 \leqslant x \leqslant 1$.

7.2 In the approximation problem of Section 7.1, s_N^* as specified in Theorem 7.1.3 gives an interval of length at most 2^{-N}, regardless of which $f \in \mathscr{F}$ underlies the problem. Show that if $f(x) = x^b$ $(b > 0)$, Newton-Raphson iteration yields loss $x_1(1 - b^{-1})^{N-1}$ after n steps. How does this compare to 2^{-N} if $b > 2$ and N is large? [This example was noted by J. Kiefer, "Optimum sequential search and approximation methods under minimum regularity assumptions," *Journal of the Society for Industrial and Applied Mathematics*, Vol. 5 (1957), pp. 105–136.]

7.3 It is sometimes the case that a sequence (say $Y_1, Y_2, \ldots$) of random variables cannot be observed, although we can observe results of the values of $Y_1, Y_2, \ldots$ (say $X_1, X_2, \ldots$). Let $Y_1, Y_2, \ldots$ be independent and identically distributed random variables. Let a be a fixed number ($-\infty < a < +\infty$). For $i = 1, 2, \ldots$ define

$$X_i = \begin{cases} 1, & \text{if } Y_i \geqslant a \\ 0, & \text{if } Y_i < a. \end{cases}$$

Define $p = P[Y_1 \geqslant a]$, $q = 1 - p$, $S_n = X_1 + \cdots + X_n$. (Note that S_n is the number of $Y_1, \ldots, Y_n$ which exceed a.)

(a) Find an expression for $E(X_1^2)$ in terms of $\text{Var}(X_1)$ and $E(X_1)$.
(b) Find $E(S_n)$.
(c) Find $E(S_n^2)$.
(d) Find $E(S_n/n)$.
(e) Find $E\{(S_n/n)^2\}$.
(f) State precisely what we would mean by the statement "$t(X_1, \ldots, X_n)$ is an unbiased estimator of $f(p)$ for p in $\Theta = \{z: 0 \leqslant z \leqslant 1\}$" where $f(\cdot)$ is some fixed function.
(g) Find an unbiased estimator of p for p in Θ.
(h) Find an unbiased estimator of q for p in Θ.
(i) Show that $\dfrac{S_n}{n}\left(1 - \dfrac{S_n}{n}\right)$ is not an unbiased estimator of pq for p in Θ.
(j) Show that $\dfrac{S_n}{n}\left(1 - \dfrac{S_n}{n}\right)$ is an unbiased estimator of pq for p in $\Theta' = \{0, 1\}$.

7.4 A certain coin has probability θ of showing heads, and probability $1 - \theta$ of showing tails, when it is flipped in a certain specified manner. (Here θ is unknown, $0 \leqslant \theta \leqslant 1$.) If we perform n independent tosses, the results may be represented as n independent random variables $X_1, \ldots, X_n$ with (for $i = 1, \ldots, n$)

$$X_i = \begin{cases} 1, & \text{with probability } \theta \\ 0, & \text{with probability } 1 - \theta. \end{cases}$$

Consider the following five estimators of θ. For each, determine whether it is unbiased, and whether it is consistent.

$$\hat{\theta}_1 = \tfrac{1}{2},$$
$$\hat{\theta}_2 = X_1,$$
$$\hat{\theta}_3 = \frac{X_{n-1} + X_n}{2},$$
$$\hat{\theta}_4 = \frac{X_1 + \cdots + X_{n-1}}{n - 1},$$
$$\hat{\theta}_5 = \frac{X_1 + \cdots + X_n}{n}.$$

Find (see Problem 8.8) the BLUE of θ.

7.5 Let $X_1, X_2, \ldots, X_n$ be independent random variables with (for $i = 1, 2, \ldots, n$)

$$P[X_i = z] = \begin{cases} \theta, & \text{if } z = 1 \\ 1 - \theta, & \text{if } z = 0 \\ 0, & \text{otherwise,} \end{cases}$$

where θ is unknown, $\theta \in \Theta = \{x : 0 \leqslant x \leqslant 1\}$.

(a) Give an example of an unbiased estimator of θ. Prove that your estimator is unbiased.

(b) Give an example of a consistent estimator of θ. Prove that your estimator is consistent.

7.6 For the stochastic approximation procedure of Section 7.3, prove that (for $n \geqslant 1$)

$$X_{n+1} = X_1 + \sum_{i=1}^{n} a_i(\alpha - Y_i).$$

7.7 If an experimenter uses the stochastic approximation procedure of Section 7.3 to estimate θ, starts with $X_1 = 5$, uses $a_n = 1/n$, and happens to observe $(Y_1, Y_2, Y_3, Y_4, Y_5, Y_6) = (0, 0, 1, 0, 0, 1)$, what will be the value of the sequence $(X_1, X_2, X_3, X_4, X_5, X_6)$?

7.8 We proved that under conditions A–1 and A–3 $X_n \xrightarrow{P} \theta$ (Theorem 7.3.1). Suppose we define condition A–3′ as: (a) $M(x) < \alpha$ for $x < \theta$, $M(\theta) = \alpha$, $M(x) > \alpha$ for $x > \theta$. Show that under conditions A–1 and A–3′ it is not always true that $X_n \xrightarrow{P} \theta$. [*Hint*: Construct a specific example where A–1 and A–3′ hold but it is not true that $X_n \xrightarrow{P} \theta$. For example, take $a_n = 1/(4n)$, $\theta = \alpha = 0$, and let $H(\cdot|x)$ be the d.f. which corresponds to always obtaining value $M(x)$ at dosage x. If we define $M(0) = 0$, $M(1) = 1$, and let $x_1 = 1$, we find $x_2 = \frac{3}{4}$. Let $M(x_n) = 1/4n$ for $n \geqslant 2$. Then x_n is a decreasing sequence, but $P[x_n > \frac{1}{4}] = 1$, and $M(x)$ can be defined at points x not covered above in such a manner that A–3′ is satisfied. This example is due to J. Wolfowitz, "On the stochastic approximation method of Robbins and Monro," *Annals of Mathematical Statistics*, Vol. 23 (1952), pp. 457–461.]

7.9 The field of **bio-assay** concerns itself with the assessment of poisons and other stimuli. There, if we apply a stimulus of strength x to an animal, we observe a response Y such that

$$P[Y = y] = \begin{cases} F(x), & \text{if } y = 1 \\ 1 - F(x), & \text{if } y = 0; \end{cases}$$

that is, the animal reacts or responds with probability $F(x)$. Suppose that $F(x)$ is a d.f. Then we often wish to determine that point (which we assume is unique) θ such that

$$F(\theta) = \alpha$$

where $0 < \alpha < 1$ [θ is called the αth quantile of the d.f. $F(\cdot)$]; for example, if $\alpha = .95$, we wish to determine the stimulus strength which elicits 95 percent response on the average. Using the results of Section 7.3, give a consistent sequence of estimators of θ. [This example was first given by H. Robbins and

S. Monro, "A stochastic approximation method," *Annals of Mathematical Statistics*, Vol. 22 (1951), pp. 400–407. Bio-assay is discussed by, for example, B. L. van der Waerden, *Mathematical Statistics*, 1969, Springer-Verlag, New York, pp. 212–224.]

HONORS PROBLEMS

7.1H Suppose that we wish to test the quality of a quantity of a substance, for example, a quantity of a microbial insecticide. Frequently such insecticides cannot easily be judged in quality via chemical assay, and in some cases (for example, where the active ingredient is the organism itself) no chemical assay is even possible. Hence, a bio-assay (see Problem 7.9) may be run to assess the quality of the specific lot of the substance; we will then run experiments at various doses with this lot and report out a single number which measures the quality of the lot. If that number is an estimate of LDXX[10] (where XX is some percentage . . . see Section 7.3 where LD50 was defined), then it has been claimed[11] that *the error of the estimate is least when* XX = 50, *that is, when we are estimating* LD50. Justify this claim.

[10] We could as easily talk of ED50 in other cases, namely the quantity of a drug resulting in a cure rate of 50 percent.

[11] See p. 188 of H. T. Dulmage, "Assay and standardization of microbial insecticides," *Annals of the New York Academy of Sciences*, Vol. 217 (*Regulation of Insect Populations by Microorganisms*) (1973), pp. 187–199.

8

POINT ESTIMATION AND STATISTICAL INFERENCE PRINCIPLES

In this chapter we study point estimation extensively. This study, begun in Section 7.2 (and a special example of which—stochastic approximation—was considered in Section 7.3), involves methods for estimating an unknown value θ after observing random variables $X_1, \ldots, X_n$ whose distribution is (usually) different for different values of θ. In the course of this study, we cover thoroughly many concepts and methods of statistical inference which also find use in later chapters (on hypothesis testing, interval estimation, ranking and selection, decision theory and so on); thus, those later chapters proceed efficiently (and the student can refer back to this chapter if and as the need arises).

8.1. BIAS AND CONSISTENCY WITH EXAMPLES

In Section 7.2 we introduced the problem where we observe random variables $X_1, \ldots, X_n$ (not necessarily independent or identically distributed) and the d.f. of $X_1, \ldots, X_n$ depends on an unknown value θ which is known to be in a given set Θ. **Point estimation** is the study of how to estimate θ after an experiment yields $X_1, \ldots, X_n$.

Suppose $X_1, \ldots, X_n$ have d.f. $F_{X_1, \cdots, X_n}(y_1, \ldots, y_n|\theta)$ and our sample space is Ω. The experiment essentially yields an $\omega \in \Omega$ and values (observations) $X_1(\omega) = x_1, \ldots, X_n(\omega) = x_n$. Then (see Definition 7.2.1) any function $t_n(X_1, \ldots, X_n)$ is called a *statistic* or an *estimator*, and $t_n(x_1, \ldots, x_n)$ is called an *estimate*. Also (Definitions 7.2.3, 7.2.2) the *bias* of $t_n = t_n(X_1, \ldots, X_n)$ is $b_n(\theta) = E_\theta t_n - \theta$, and t_n is called *unbiased* if $b_n(\theta) \overset{\theta}{\equiv} 0$. Often an estimator t_n

will not be unbiased, but its bias will go to 0 as n increases (for example, if $t_n = s_n + 1/n$ where s_n is unbiased, then t_n is biased but asymptotically unbiased); we now formalize this notion:

Definition 8.1.1. *We say $t_n(X_1, \ldots, X_n)$ is an* **asymptotically unbiased** *estimator of θ if*

$$\lim_{n\to\infty} b_n(\theta) = 0, \qquad \textit{all } \theta \in \Theta.$$

Finally, recall (Definition 7.2.4) that a sequence of estimators $t_1, t_2, t_3, \ldots$ is called *consistent* for θ if it converges in probability to θ (no matter which $\theta \in \Theta$ is the actual value).

In Definitions 5.1.6 and 5.1.7, for any r.v. X for which these quantities exist, we talked about the *population moments* $\mu'_k = \mu'_k(X) = EX^k$ (k^{th} noncentral[1] moment) and $\mu_k = \mu_k(X) = E(X - \mu'_1)^k$ (k^{th} central moment), and their importance became clear in Chapters 5 and 6. Often they are unknown and are estimated by *sample moments*, which we will now define.

Definition 8.1.2. *Let $X_1, \ldots, X_n$ be n r.v.'s (not necessarily independent or identically distributed). Their* **kth (noncentral) sample moment** *is*

$$m'_k = \frac{X_1^k + \cdots + X_n^k}{n};$$

their **kth central sample moment** *is*

$$m_k = \frac{(X_1 - m'_1)^k + \cdots + (X_n - m'_1)^k}{n}.$$

Example 8.1.3. The $k = 1^{st}$ (noncentral) sample moment is

$$m'_1 = \frac{X_1 + \cdots + X_n}{n} \equiv \bar{X}(n) \equiv \bar{X},$$

called the **sample mean**, and we shall utilize the notations $\bar{X}(n)$ and $\bar{X}$ (instead of m'_1) quite extensively.

Note that we want to consider the bias and consistency of sample moments as estimators of population moments (some special cases were covered by Theorem 6.4.4 and Problem 6.9) but we first consider why the m'_k and m_k are called *sample moments*:

Example 8.1.4. Suppose we perform an experiment and obtain results $X_1 = x_1, \ldots, X_n = x_n$. Let a r.v. Z be such that $P[Z = x_i] = 1/n$ ($i = 1, \ldots, n$).

---

[1] Of course $E(X - a)^k$ is a "noncentral" moment as long as $a \neq \mu'_1$. However, the term is usually reserved for the case $a = 0$.

Then

$$EZ = m_1', \quad EZ^k = m_k', \quad E(Z - EZ)^k = \frac{(x_1 - \bar{x})^k + \cdots + (x_n - \bar{x})^k}{n},$$

which justifies the name sample moments (moments of the sample). (Problem 5.17 involved the calculation of some sample moments.)

Theorem 8.1.5. *Let $X_1, \ldots, X_n$ be independent and identically distributed r.v.'s, each with d.f. $F(x|\theta)$ for some fixed $\theta \in \Theta$. (Suppose that any moment mentioned exists.) Then, by definition,*

$$\mu_k' = EX_1^k, \qquad \mu_k = E(X_1 - \mu_1')^k,$$

$$m_k' = \frac{X_1^k + \cdots + X_n^k}{n}, \qquad m_k = \frac{(X_1 - \bar{X})^k + \cdots + (X_n - \bar{X})^k}{n}.$$

We shall prove that:

(1) $Em_k' \equiv \mu_k'$, *but* $Em_k \not\equiv \mu_k$; *and*

(2) $m_k' \xrightarrow{p} \mu_k'$, *and* $m_k \xrightarrow{p} \mu_k$.

Proof of (1): First, we easily show $Em_k' \equiv \mu_k'$ as follows:

$$Em_k' = E\left(\frac{X_1^k + \cdots + X_n^k}{n}\right) = \frac{EX_1^k + \cdots + EX_n^k}{n} = \frac{n\mu_k'}{n} = \mu_k';$$

we used identical distribution, but not independence.

Second, we show with some algebra that $Em_k \not\equiv \mu_k$. Consider the case $k = 2$. Then $\mu_k = E(X_1 - \mu_1')^2 = \sigma^2$ (say) is called the variance of the distribution of the X_i's. Here

$$m_2 = \frac{\sum_{i=1}^{n} (X_i - \bar{X})^2}{n},$$

and

$$\begin{aligned} Enm_2 &= E \sum_{i=1}^{n} [(X_i - \mu_1') - (\bar{X} - \mu_1')]^2 \\ &= E[\textstyle\sum(X_i - \mu_1')^2 + n(\bar{X} - \mu_1')^2 - 2\sum(\bar{X} - \mu_1')(X_i - \mu_1')] \\ &= n\sigma^2 + \sigma^2 - 2\sigma^2 = (n - 1)\sigma^2, \end{aligned}$$

where: (because it is obvious) we have written $\sum$ for $\sum_{i=1}^{n}$ (this will be done below without further mention); the second term gave

$$En\left(\frac{(X_1 - \mu_1') + \cdots + (X_n - \mu_1')}{n}\right)^2 = 0 + \frac{n}{n^2} n\sigma^2 = \sigma^2$$

(cross-products had expected value 0, using independence); a typical last term gave

$$E\left(\frac{(X_1 - \mu_1') + \cdots + (X_n - \mu_1')}{n}\right)(X_1 - \mu_1') = \frac{\sigma^2}{n}$$

(cross-products had expected value 0, except when they had the same index), so the last term yielded $-2\sigma^2$; and we introduced μ_1' since $E(X_i - \mu_1') = 0$ [hence, $E(\bar{X} - \mu_1') = 0$]. Thus,

$$Em_2 = \frac{n-1}{n}\sigma^2 = \frac{n-1}{n}\mu_2 \not\equiv \mu_2.$$

However, $E\frac{n}{n-1}m_2 = \sigma^2 = \mu_2$, so

$$E\frac{\sum_{i=1}^{n}(X_i - \bar{X})^2}{n-1} = \sigma^2. \tag{8.1.6}$$

The quantity $\sum_{i=1}^{n}(X_i - \bar{X})^2/(n-1)$, an unbiased estimator of σ^2, is usually used instead of m_2. [Of course, if we knew $EX_1 = \mu$ (say), we could use $\sum_{i=1}^{n}(X_i - \mu)^2/n$ as an unbiased estimator of σ^2. Since we do not know μ, we in effect replace it by $\bar{X}$, and this requires division by $n - 1$ instead of n in order to achieve unbiasedness; this is called a *loss of 1 degree of freedom*, for reasons which will become apparent later.]

Proof of (2): First, the fact that $m_k' \xrightarrow{P} \mu_k'$ follows on applying Khinchin's Theorem (Theorem 6.4.4) (a law of large numbers) to $X_1^k, \ldots, X_n^k$.

Second, we wish to prove that $m_k \xrightarrow{P} \mu_k$. Now $\mu_k = E(X_1 - \mu_1')^k$; if we expand $(X_1 - \mu_1')^k$ by the Binomial Theorem and use properties of expectation (for example, the expected value of a sum equals the sum of the expected values—see Theorem 5.1.5),

$$\mu_k = \Pi(\mu_1', \ldots, \mu_k'),$$

that is, μ_k is a polynomial in $\mu_1', \ldots, \mu_k'$. We now claim that

$$m_k = \Pi(m_1', \ldots, m_k'), \tag{8.1.7}$$

that is, m_k is *the same* polynomial in $m_1', \ldots, m_k'$. For, consider any set of observations $X_1 = x_1, \ldots, X_n = x_n$ and let D be a r.v. such that $P[D = x_i] = 1/n\ (i = 1, \ldots, n)$. Then $\mu_k(D) = m_k$; hence, $m_k = \Pi(m_1', \ldots, m_k')$ for any observations $x_1, \ldots, x_n$. Since (8.1.7) holds for any fixed numbers $x_1, \ldots, x_n$, it also holds for random numbers (r.v.'s) $X_1, \ldots, X_n$. But we already proved that

$$m_k' \xrightarrow{P} \mu_k',$$

and a polynomial Π is a jointly continuous function of its arguments. Hence,

$$m_k = \Pi(m_1', \ldots, m_k') \xrightarrow{p} \Pi(\mu_1', \ldots, \mu_k') = \mu_k.$$

A problem often encountered (for example, see Example 8.1.13 below) is that of estimating (based on a number of observations) an unknown d.f. One standard estimator is given in the following definition.

Definition 8.1.8. *Suppose that r.v.'s $Y_1, \ldots, Y_n$ are independent and identically distributed with a d.f. $F(\cdot)$. The* **empiric**[2] **d.f.** *(based on $Y_1, \ldots, Y_n$) is defined for all x ($-\infty < x < +\infty$) by*

$$F_n(x|Y_1, \ldots, Y_n) = \frac{(\text{No. of } Y_1, \ldots, Y_n \text{ which are} \leqslant x)}{n}.$$

[This is a chance d.f. (chance functional); that is, a function of $\omega \in \Omega$.]

Example 8.1.9. If we observe $Y_1 = y_1, \ldots, Y_n = y_n$, let us denote the ordered[3] $y_1, \ldots, y_n$ by $z_n = \max(y_1, \ldots, y_n), \ldots, z_1 = \min(y_1, \ldots, y_n)$. Then $F_n(\cdot|y_1, \ldots, y_n)$ is as shown in Figure 8.1–1.

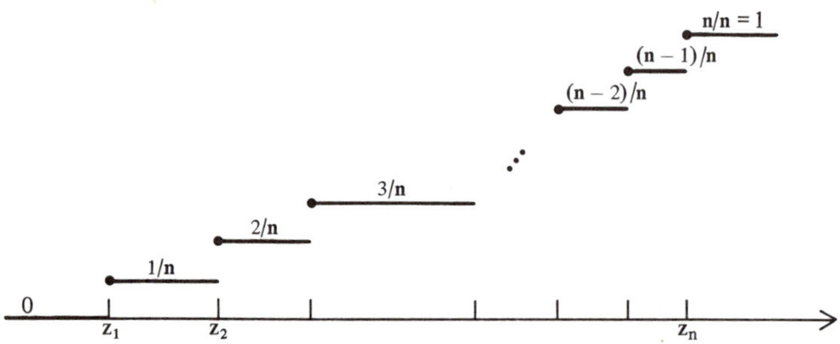

Figure 8.1–1. A typical empiric d.f.

We will now prove that the empiric d.f. furnishes a consistent estimator of $F(x)$. Then we will state an even stronger result (sometimes called the **Fundamental Theorem of Statistics** in certain circles) and give a simple example of use of the empiric d.f.

Theorem 8.1.10. *Suppose that independent r.v.'s $Y_1, \ldots, Y_n$ each have d.f.*

---

[2] Some authors call this the *empirical d.f.*

[3] The ordered $Y_1, \ldots, Y_n$ (say $Z_1, \ldots, Z_n$) are called the **order statistics**.

$F(\cdot)$. *Then (for any x)*

$$F_n(x|Y_1, \ldots, Y_n) \xrightarrow{P} F(x).$$

[*That is, for any* $\varepsilon > 0$, $P[|F_n(x|Y_1, \ldots, Y_n) - F(x)| < \varepsilon] \to 1$ *as* $n \to \infty$.]

Proof: Let x be fixed but arbitrary. Let $W_i = 1$ if $Y_i \leqslant x$ (and zero otherwise) for $i = 1, \ldots, n$. Then $W_1, \ldots, W_n$ are independent r.v.'s, each binomial with success probability $P[Y_i \leqslant x] = F(x)$ (see Definition 3.2.3). Hence, $EW_1 = F(x)$ and (by Khinchin's Theorem 6.4.4),

$$\frac{W_1 + \cdots + W_n}{n} \xrightarrow{P} F(x).$$

But

$$\frac{W_1 + \cdots + W_n}{n} = \frac{(\text{No. of } Y_1, \ldots, Y_n \text{ which are} \leqslant x)}{n}$$
$$= F_n(x|Y_1, \ldots, Y_n),$$

so the theorem follows.

Theorem 8.1.11. **(Glivenko-Cantelli Theorem)** *Suppose that independent r.v.'s* $Y_1, \ldots, Y_n$ *each have d.f.* $F(\cdot)$. *Then for any* $\varepsilon > 0$, *as* $n \to \infty$,

$$P[\sup_x |F_n(x|Y_1, \ldots, Y_n) - F(x)| < \varepsilon] \to 1. \tag{8.1.12}$$

[*That is,* $\sup_x |F_n(x|Y_1, \ldots, Y_n) - F(x)| \xrightarrow{P} 0$.] *In fact, the weak convergence of (8.1.12) can be replaced by strong convergence* [see Definitions 6.3.1 and 6.3.2 for details of these two notions].

Example 8.1.13. A newsboy wishes to determine how many newspapers to purchase from a distributor. He is allowed to purchase $0, 1, 2, \ldots, U$ papers on a specified day, and while each one costs him \$$p$, each will fetch \$$r$ from a customer (if it sells) and will fetch \$$s$ from a scrappaper dealer (if it fails to sell). Now our newsboy does not know how many newspapers (say X) he will be able to sell today, but on the past n days he could have sold respectively $X_1, X_2, \ldots, X_n$ copies. He is willing to assume that $X_1, X_2, \ldots, X_n$ are independent r.v.'s with some common d.f. (distribution function of demand) $F(\cdot)$, and that today's demand will be independent of past demands, but will have the same d.f. as did past demands.

The boy's father, a statistician, reasons as follows. If my son buys K papers from the distributor, his profit will be (on the average) $E\{P(K, X)\}$ where $P(K, X)$ (the profit if K papers are bought and demand is X) is

$$P(K, X) = \begin{cases} -pK + rX + s(K - X), & \text{if } K \quad X \\ -pK + rK, & \text{if } K < X. \end{cases}$$

Hence, the expected profit is [denoting $P[X = x]$ by $f(x)$ and assuming $\sum_{x=0}^{\infty} f(x) = 1$]

$$\begin{aligned}
\bar{P}_K \equiv E\{P(K, X)\} &= \sum_{x=0}^{\infty} P(K, x)f(x) \\
&= \sum_{x=0}^{K} \{(s - p)K + (r - s)x\}f(x) + \sum_{x=K+1}^{\infty} (r - p)Kf(x) \\
&= (s - p)K \sum_{x=0}^{K} f(x) + (r - s) \sum_{x=0}^{K} xf(x) + (r - p)K \sum_{x=K+1}^{\infty} f(x) \\
&= (s - p)KF(K) + (r - s) \sum_{x=0}^{K} xf(x) + (r - p)K[1 - F(K)].
\end{aligned} \tag{8.1.14}$$

Now we may wish to choose K so as to maximize $\bar{P}_K$; then we simply use (8.1.14) to find the appropriate value of K *if* we know $F(\cdot)$. Here we do not know $F(\cdot)$, but we do have information about it in the form of $X_1, \ldots, X_n$, independent observations with the same d.f. $F(\cdot)$. One approach [suggested by Weiss (1961), pp. 134–135] is to replace $F(\cdot)$ by $F_n(\cdot|X_1, \ldots, X_n)$. Other approaches have also been suggested [for example, see R. H. Hayes' paper "Efficiency of simple order statistic estimates when losses are piecewise linear," *Journal of the American Statistical Association*, Vol. 66 (1971), pp. 127–135]. The newsboy problem is a special case of a statistical decision problem; we will study such problems in detail in Chapter 12.

Note that logic requires $s < p < r$ (if $s \geqslant p$, he can buy all U papers and either make a profit by selling them for scrap ($s > p$) or at least break even on unsold papers by selling them for scrap ($s = p$); if $p \geqslant r$, selling to customers is a losing proposition and he should either go out of business or sell papers for scrap). Of course, logic does not always prevail; thus, during the 1960s in the United States, certain coins cost, for example, \$1 at the bank but could be melted down in large quantities and sold as metal for \$1.25.

We will now give two examples which show that if we have more than one unbiased estimator, then we have infinitely many unbiased estimators, but that we need not have even one unbiased estimator. [Two[4] questions arise. First, if there are many unbiased estimators, how should we choose one of them to use? Some possible answers (maximum likelihood[5], minimum variance, and so on) are studied in this chapter. Second, if there are

[4] A third question could arise: "Does an estimator possess any advantage simply by virtue of being unbiased?" For us lack of bias is appealing in itself. (However, biased estimators can sometimes possess desirable properties which unbiased ones lack.)

[5] This is not the only (or even the most important) context in which maximum likelihood is important.

no unbiased estimators, which sorts of biased estimators should we utilize? This question is only treated implicitly (maximum likelihood, and so on).]

Example 8.1.15. Suppose that we observe random variables $X_1, \ldots, X_n$ (not necessarily independent or identically distributed) and that their joint d.f. depends on an unknown $\theta \in \Theta$. Further suppose that we find functions of $X_1, \ldots, X_n$ (statistics) $t_1(X_1, \ldots, X_n), \ldots, t_k(X_1, \ldots, X_n)$ such that

$$E_\theta t_i(X_1, \ldots, X_n) \overset{\theta}{\equiv} \theta, \quad \text{all } \theta \in \Theta \tag{8.1.16}$$

(that is, $t_1, \ldots, t_k$ are each unbiased estimators of θ). Then suppose $w_1, \ldots, w_k$ are k fixed constants ($-\infty < w_i < +\infty$). We have

$$E_\theta\{w_1 t_1(X_1,\ldots,X_n) + \cdots + w_k t_k(X_1,\ldots,X_n)\} \overset{\theta}{\equiv} (w_1 + \cdots + w_k)\theta \ (\theta \in \Theta), \tag{8.1.17}$$

so $\sum_{i=1}^{k} w_i t_i(X_1, \ldots, X_n)$ is an unbiased estimator of θ iff $w_1 + \cdots + w_k = 1$. There are infinitely many such $w_1, \ldots, w_k$. Also, if t_1 is an unbiased estimator of θ, then so is $t_1 + s_1$ if $E_\theta s_1 \overset{\theta}{\equiv} 0$.

Example 8.1.18. Suppose that $X_1, X_2, \ldots$ are independent r.v.'s, each with $P[X_i = 1] = p$ and $P[X_i = 0] = q \equiv 1 - p$ for some unknown p ($0 \leqslant p \leqslant 1$) (this is the Bernoulli distribution of Definition 3.2.1). Let N be the number of observations we need to make to obtain our first success (then $N - 1$ has the negative binomial distribution of Definition 3.2.16 with $r = 1$ success). Hence, $X_1 = X_2 = \cdots = X_{N-1} = 0$ and $X_N = 1$. Also, $P[N = l] = q^{l-1}p$ for $l = 1, 2, 3, \ldots$.

We now wish to find all functions ζ of N alone which are unbiased estimators of p [so, such functions as $\zeta(N) + X_1 - X_2$ are excluded]. Our requirement is

$$E_p\zeta(N) \overset{p}{\equiv} p; \tag{8.1.19}$$

that is [taking the expected value in (8.1.19)],

$$\sum_{j=1}^{\infty} \zeta(j)q^{j-1}p \overset{p}{\equiv} p. \tag{8.1.20}$$

If $q = 1$ (so $p = 0$), Equation (8.1.20) becomes $0 = 0$. If $q \neq 1$ (so $p \neq 0$), upon division by p (8.1.20) becomes

$$\sum_{j=1}^{\infty} \zeta(j)q^{j-1} \overset{p}{\equiv} 1. \tag{8.1.21}$$

Now it is well-known that two power series $\sum_0^\infty a_j x^j$ and $\sum_0^\infty b_j x^j$ have the same value for every x in some interval iff $a_0 = b_0, a_1 = b_1, \ldots$. Thus, from (8.1.21), $\zeta(1) = 1$ and $\zeta(j) = 0$ for $j > 1$. Hence, exactly one unbiased estimator $\zeta(N)$ exists, namely

$$\zeta(N) = \begin{cases} 1, & \text{if } N = 1 \\ 0, & \text{if } N > 1. \end{cases} \tag{8.1.22}$$

Note that this absurd estimator need only observe X_1; in fact, $P[\zeta(N) = X_1] = 1$. If we require that $P[\zeta(N) = X_1] \neq 1$, no unbiased estimators exist.

We now wish to introduce (in addition to bias and consistency) a third concept and to study its relation to the other two.

Definition 8.1.23. *We say $t_n(X_1, \ldots, X_n)$ is* **consistent in mean square** (*or* **consistent in quadratic mean**) *if*

$$E_\theta(t_n - \theta)^2 \to 0 \quad \textit{as } n \to \infty \quad (\textit{all } \theta \in \Theta). \tag{8.1.24}$$

Theorem 8.1.25. *$t_n(X_1,\ldots,X_n)$ is consistent in mean square iff (1) $Var_\theta(t_n) \to 0$ as $n \to \infty$, and (2) t_n is asymptotically unbiased.*

Proof: Since $EX^2 = \text{Var}(X) + (EX)^2$, we have

$$\begin{aligned} E_\theta(t_n - \theta)^2 &= \text{Var}_\theta(t_n - \theta) + [E_\theta(t_n - \theta)]^2 \\ &= \text{Var}_\theta(t_n) + [b_n(\theta)]^2, \end{aligned}$$

which will $\to 0$ iff both of its (positive) terms $\to 0$.

Theorem 8.1.26. *If $E_\theta(t_n - \theta)^2 \to 0$ (all θ), then $t_n \xrightarrow{P} \theta$ (all θ).*

Proof: For any $\varepsilon > 0$, Chebyshev's Inequality (Theorem 6.2.5) states that

$$P[|t_n - \theta| \leqslant \varepsilon] \geqslant 1 - \frac{E(t_n - \theta)^2}{\varepsilon^2},$$

which (by the hypotheses of our theorem) $\to 1$ as $n \to \infty$.

We have previously (Example 8.1.3) defined the sample mean of $X_1,\ldots,X_n$; we now wish to define the sample median and then to compare them as estimators (of what?). First, however, we must define the population median [in Chapter 5, following Definition 5.1.7, we defined the population median for the special case of a r.v. X with a density $f(x)$, in which case "it" was any m such that $\int_{-\infty}^{m} f(x)\,dx = \frac{1}{2} = \int_{m}^{\infty} f(x)\,dx$].

Definition 8.1.27. *Let X be a r.v. with d.f. $F(x)$. "The"* **median** *of X is: (1) v if there is a unique v such that $F(v) = \frac{1}{2}$; (2) v if there is a v such that $F(v) \geqslant \frac{1}{2}$ and $\lim_{\substack{x \to v \\ x < v}} F(x) < \frac{1}{2}$; (3) $\{v\colon F(v) = \frac{1}{2}\}$ if there is more than one point v such that $F(v) = \frac{1}{2}$.* [These three cases are illustrated in Figure 8.1–2.]

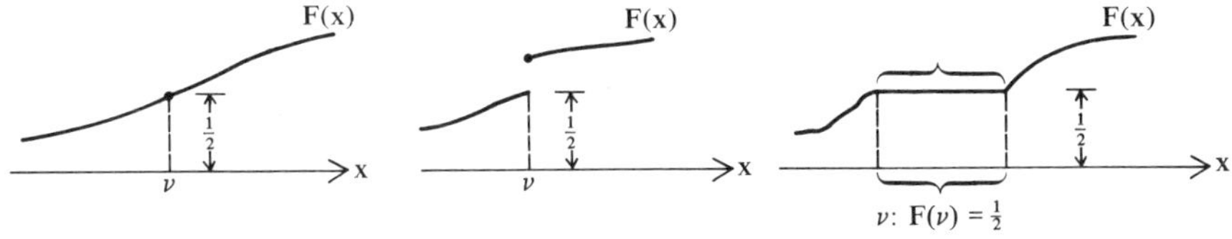

(a) (b) (c)

Figure 8.1–2. Population medians

Definition 8.1.28. *For n r.v.'s* $X_1, \ldots, X_n$ *let* $Y_1, \ldots, Y_n$ *(called the* **order statistics***) denote the ordered* X_i*'s (that is, the* X_i*'s ordered so that* $Y_1 \leqslant \cdots \leqslant Y_n$*; if several* X_i*'s are equal, we can choose one at random for a place* Y_i*). Then the* **sample median** $\mathbf{M_n}$ *of* $X_1, \ldots, X_n$ *is (usually) taken to be*

$$M_n = \begin{cases} Y_{\frac{n+1}{2}}, & \text{if } n \text{ is odd,} \\ \dfrac{Y_{\frac{n}{2}} + Y_{\frac{n}{2}+1}}{2}, & \text{if } n \text{ is even.} \end{cases}$$

[When $X_1, \ldots, X_n$ are lined up in order of value, M_n is the "middle" one if there is one (n odd) or the midpoint of the two values in the "middle" (n even). Thus, if we have the three observations 4.2, 6.7, 8.9, their median is 6.7; while if we have the four observations 4.2, 6.7, 8.9, 10.2, their median is $(6.7 + 8.9)/2 = 7.8$.]

Lemma 8.1.29. *For any r.v.'s* $X_1, \ldots, X_n$ *let* $Y_1 \leqslant \cdots \leqslant Y_n$ *be the ordered* X_i*'s. Then* $Y_1, \ldots, Y_n$ *are not independent r.v.'s (unless for some* $c_1, \ldots, c_n$ $P[Y_1 = c_1] = \cdots = P[Y_n = c_n] = 1$*).*

Proof: We have $P[Y_1 \leqslant x | Y_2 \leqslant x] = 1$; but then

$$P[Y_2 \leqslant x] = P[Y_1 \leqslant x | Y_2 \leqslant x] P[Y_2 \leqslant x] = P[Y_1 \leqslant x, Y_2 \leqslant x].$$

However, by Lemma 3.4.9 and Lemma 3.4.8 we need

$$P[Y_1 \leqslant x, Y_2 \leqslant x] = P[Y_1 \leqslant x] P[Y_2 \leqslant x]$$

for all x if $Y_1, \ldots, Y_n$ are to have any chance of being independent.

Theorem 8.1.30. *Let* $X_1, \ldots, X_n$ *be independent r.v.'s, each with d.f.* $F(x|\theta)$, $\theta \in \Theta$. *Then (supposing* $E_\theta X_1$ *exists):*

(1) $\bar{X}$ *is an unbiased and consistent estimator of* $E_\theta X_1$*; and*

(2) M_n *is a consistent (generally not unbiased) estimator of the population median* $\nu_\theta = \nu$ *(say).*

Proof: Part (1) follows from Theorem 8.1.5, since $\bar{X}$ is a sample (noncentral) moment. We now prove that $M_n \xrightarrow{p} \nu$. So, for any fixed $\delta > 0$ and $\varepsilon > 0$, we must show there is an integer $N(\delta, \varepsilon)$ such that, for $n > N(\delta, \varepsilon)$, $P[|M_n - \nu| < \delta] > 1 - \varepsilon$.

For convenience[6], assume ν is unique and is a point of continuity of $F(x|\theta)$ [case (a) of Figure 8.1–2]. Then there are points A and B with $A \geqslant \nu - \delta$ and $B \leqslant \nu + \delta$ (at which $F(x|\theta)$ is continuous) with $A < \nu < B$ such that $|F(A|\theta) - \frac{1}{2}| < \delta/4$ and $|F(B|\theta) - \frac{1}{2}| < \delta/4$. (See Figure 8.1–3.)

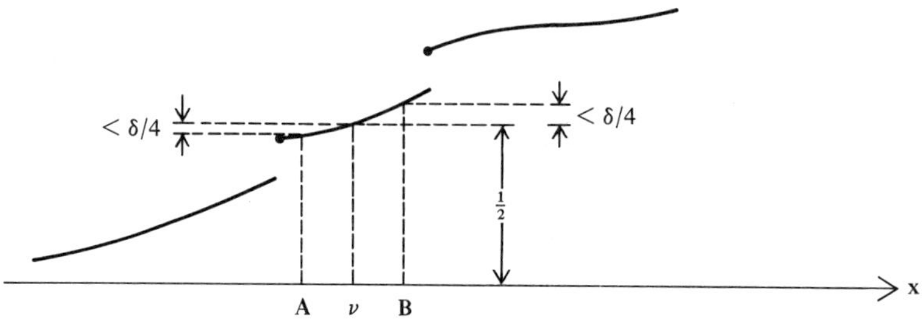

Figure 8.1–3. Points **A**, **B**, and ν

Now define a r.v. $\alpha(X)$ to be 1 if $X < A$ and 0 if $X \geqslant A$ (a Bernoulli r.v.). Then $\alpha(X_1), \ldots, \alpha(X_n)$ are independent Bernoulli r.v.'s, and (for $i = 1, \ldots, n$)

$$P[\alpha(X_i) = 1] = P[X_i < A] > \frac{1}{2} - \frac{\delta}{4}.$$

Now by Khinchin's Theorem (Theorem 6.4.4) or by the WLLN (Theorem 6.3.5),

$$\frac{\alpha(X_1) + \cdots + \alpha(X_n)}{n} \xrightarrow{p} P[X_1 < A] > \frac{1}{2} - \frac{\delta}{4},$$

so for all $n > N_1(\delta, \varepsilon)$ (since $P[X_1 < A] < \frac{1}{2}$)

$$P\left[(\text{No. of } X_1, \ldots, X_n < A) < \frac{n}{2}\right] > 1 - \frac{\varepsilon}{2}.$$

[The details involved in this last step are as follows. By the convergence in probability, there is an integer $N_1^*(\delta, \varepsilon)$ such that for all $n > N_1^*(\delta, \varepsilon)$ we

---

[6] This is not totally a matter of convenience, since if the distribution has an interval of medians we cannot expect any one estimator sequence to estimate all consistently. However (as we note at the end of the proof), we can identify all medians with each other in the general case and the result then follows.

have

$$P\left[\left|\frac{\alpha(X_1) + \cdots + \alpha(X_n)}{n} - P[X_1 < A]\right| < \delta\right] > 1 - \frac{\varepsilon}{2}.$$

We now relate this to the desired quantity as follows:

$$\begin{aligned} P&\left[(\text{No. of } X_1, \ldots, X_n < A) < \frac{n}{2}\right] \\ &= P\left[\frac{(\text{No. of } X_1, \ldots, X_n < A)}{n} < \frac{1}{2}\right] \\ &= P\left[\frac{\alpha(X_1) + \cdots + \alpha(X_n)}{n} - P[X_1 < A] < \frac{1}{2} - P[X_1 < A]\right] \\ &\geqslant P\left[\left|\frac{\alpha(X_1) + \cdots + \alpha(X_n)}{n} - P[X_1 < A]\right| < \frac{1}{2} - P[X_1 < A]\right]. \end{aligned}$$

Taking $N_1(\delta, \varepsilon)$ to be the $N_1^*(\delta, \varepsilon)$ for $\delta = \frac{1}{2} - P[X_1 < A]$ completes the argument.] Similarly, for all $n > N_2(\delta, \varepsilon)$,

$$P\left[(\text{No. of } X_1, \ldots, X_n > B) < \frac{n}{2}\right] > 1 - \frac{\varepsilon}{2}.$$

Thus, for $n > \max(N_1(\delta, \varepsilon), N_2(\delta, \varepsilon))$

$$P\left[(\text{No. of } X_1, \ldots, X_n < A) < \frac{n}{2}, (\text{No. of } X_1, \ldots, X_n > B) < \frac{n}{2}\right] > 1 - \varepsilon;$$

for let $E = \{(\text{No. of } X_1, \ldots, X_n < A) < n/2\}$ and $F = \{(\text{No. of } X_1, \ldots, X_n > B) < n/2\}$. Then for $n > \max(N_1(\delta, \varepsilon), N_2(\delta, \varepsilon))$,

$$\begin{aligned} P&\left[(\text{No. of } X_1, \ldots, X_n < A) < \frac{n}{2}, (\text{No. of } X_1, \ldots, X_n > B) < \frac{n}{2}\right] \\ &= P(EF) = 1 - P(E\bar{F}) - P(F\bar{E}) - P(\overline{EF}) \\ &= 1 - P(E\bar{F}) - P(\bar{E}) \\ &\geqslant 1 - P(\bar{F}) - P(\bar{E}) \\ &> 1 - \frac{\varepsilon}{2} - \frac{\varepsilon}{2} = 1 - \varepsilon. \end{aligned}$$

Thus, w.p. $\geqslant 1 - \varepsilon$ fewer than $n/2$ of $X_1, \ldots, X_n$ are $< A < v$ *and* fewer than $n/2$ of $X_1, \ldots, X_n$ are $> B > v$; thus, w.p. $\geqslant 1 - \varepsilon$ the sample median is between A and B. Hence, $P[|M_n - v| < \delta] = P[-\delta < M_n - v < \delta] = P[v - \delta < M_n < v + \delta] \geqslant P[A < M_n < B] \geqslant 1 - \varepsilon$, as was to be shown. The result for cases (b) and (c) of Figure 8.1–2 follows similarly. [In case (c), we consider $\{v: F(v) = \frac{1}{2}\}$ as a single point.] $M_n \xrightarrow{P} v$ also follows from the Glivenko-Cantelli Theorem (Theorem 8.1.11). It is easy to see that in general

$EM_n \neq \nu$; for, both densities of Figure 8.1–4 have the same median, but $P[M_n \leqslant x]$ is smaller (at most equal) in case (b). Hence, $EM_n \neq \nu$ for at least one of these d.f.'s.

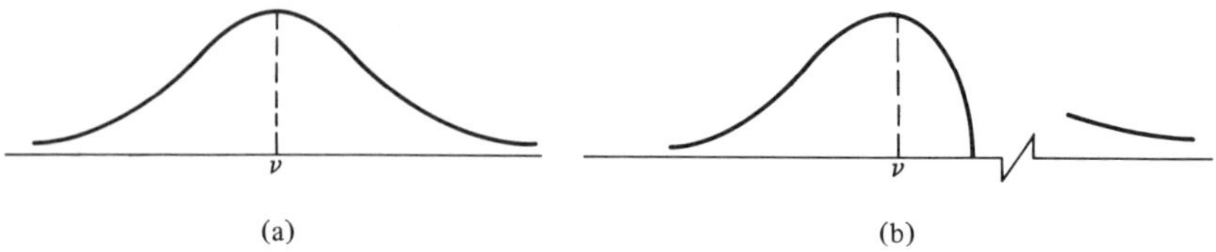

Figure 8.1–4. Densities with the same median

8.2 MAXIMUM-LIKELIHOOD ESTIMATION

In Section 8.1, we noted several properties (unbiasedness, consistency, and their variants) which seem desirable in the context of point estimation. Thus, we would like to check a proposed estimator to see if it is consistent, unbiased, and so on. However, faced with a new point-estimation problem, where can we start to look for estimators? It would be convenient to have one (or several) intuitively reasonable methods of generating possibly-good estimators to study for our problem; in particular, a method which often gives good results (that is, generates "good" estimators often) would be desirable. In this section we introduce one such method, that of "maximum likelihood," which will be studied in or related to much of the rest of Chapter 8.

Definition 8.2.1. *Let $X_1, \ldots, X_n$ be n r.v.'s (not necessarily independent or identically distributed) with d.f. $F(x_1, \ldots, x_n|\theta)$ where $\theta \in \Theta$ is unknown. The* **likelihood function** *is*

$$L(\theta) = \begin{cases} f(X_1, \ldots, X_n|\theta), & \text{if } F \text{ has a density } f \\ p(X_1, \ldots, X_n|\theta), & \text{if } F \text{ has a probability function } p. \end{cases} \tag{8.2.2}$$

[Note that the density f or probability function p is evaluated at the random point provided by the observations $X_1, \ldots, X_n$ and is considered as a function of θ (which may be a vector).] *Any $\hat{\theta} = \hat{\theta}_n(X_1, \ldots, X_n) \in \Theta$ such that*

$$L(\hat{\theta}) = \sup\{L(\theta): \theta \in \Theta\} \tag{8.2.3}$$

is called a **maximum-likelihood estimator (MLE)** *of θ.* [There may be none, one, or many such $\hat{\theta}$'s for any given problem. Fortunately, there is often one with "good" properties.]

In the past, we usually held θ fixed and let $x_1, \ldots, x_n$ vary in a density

$f(x_1, \ldots, x_n|\theta)$; here we hold $x_1, \ldots, x_n$ fixed and let θ vary [in which case $f(x_1, \ldots, x_n|\theta)$ is called the likelihood function].

Note that (1) if $X_1, \ldots, X_n$ are independent r.v.'s and X_i has density $f_i(x|\theta)$ $(i = 1, \ldots, n)$, then

$$L(\theta) = f_1(X_1|\theta) \cdots f_n(X_n|\theta), \quad \theta \in \Theta; \tag{8.2.4}$$

(2) if in addition $f_1 = f_2 = \cdots = f_n = f$ (say), of course the subscripts on $f_1, \ldots, f_n$ in (8.2.4) can be deleted; and (3):

Lemma 8.2.5. *Those* $\hat{\theta}$'s *which maximize* $L(\theta)$ *are precisely those which maximize* $\ln L(\theta)$ [this latter form is often easier to deal with].

Proof: This follows since $L(\theta_1) < L(\theta_2)$ iff $\ln L(\theta_1) < \ln L(\theta_2)$ (for all $\theta_1, \theta_2 \in \Theta$).

The following example should help clarify why, intuitively, we might think an MLE would be a "good" estimator of θ.

Example 8.2.6. A certain lion has three possible states of activity each night; they are "very active" (denoted by θ_1), "moderately active" (denoted by θ_2), and "lethargic" (denoted by θ_3). Also, each night this lion eats people; it eats i people with probability $p(i|\theta)$, $\theta \in \Theta = \{\theta_1, \theta_2, \theta_3\}$. (Of course, the probability distribution of the number of people eaten depends on the lion's activity state $\theta \in \Theta$.) The numerical values are given in Table 8.2–1.

Table 8.2–1. Lion's Appetite Distribution

i	0	1	2	3	4
$p(i\|\theta_1)$	.00	.05	.05	.80	.10
$p(i\|\theta_2)$	.05	.05	.80	.10	.00
$p(i\|\theta_3)$	.90	.08	.02	.00	.00

If we are told $X = x_0$ people were eaten last night, how should we estimate the lion's activity state (θ_1, θ_2, or θ_3)? One seemingly reasonable method is to estimate θ as *that* θ *in* Θ *for which* $p(x_0|\theta)$ *is largest* (the $\theta \in \Theta$ which provides the largest probability of observing what we did observe). That is, $\hat{\theta} = \hat{\theta}(X)$, the MLE of θ based on X. We have, of course, $\hat{\theta}(0) = \theta_3$, $\hat{\theta}(1) = \theta_3$, $\hat{\theta}(2) = \theta_2$, $\hat{\theta}(3) = \theta_1$, $\hat{\theta}(4) = \theta_1$. (Note that we *must* have $\theta = \theta_1$ in order to observe $X = 4$.)

Suppose, as a slight modification, that we know $\theta \in \Theta' = \{\theta_1, \theta_2\}$. Then the MLE is no longer unique, for although $\hat{\theta}(0) = \theta_2$, $\hat{\theta}(2) = \theta_2$, $\hat{\theta}(3) = \theta_1$, $\hat{\theta}(4) = \theta_1$, still $\hat{\theta}(1)$ can be *either* θ_1 *or* θ_2 [since $p(1|\theta_1) = p(1|\theta_2) = .05$].

We now wish to illustrate the workings of the method in a number of problems, and to compare the resulting estimators with previously developed ones. We first consider the problem of Example 8.1.18, where we found that the only unbiased estimator was absurd.

Example 8.2.7. Suppose we observe N and[7] that $P_p[N = l] = q^{l-1}p$ $(l = 1, 2, 3, \ldots)$ for some $p \in \Theta = \{x\colon 0 \leqslant x \leqslant 1\}$. What is "the" MLE of p? Our likelihood function is $L(p) = q^{N-1}p$, which has a derivative

$$\frac{dL(p)}{dp} = q^{N-1} - (N - 1)q^{N-2}p;$$

this is $\geqslant 0$ iff $p \leqslant 1/N$. Hence, $L(p)$ is maximized uniquely by $\hat{p} = \hat{p}(N) = 1/N$. This estimator says "If it took you N trials to observe 1 success, estimate the probability of a success as $1/N$." The absurd estimator of Example 8.1.18, which was unbiased, said "If it took you N trials to observe 1 success, estimate the probability of a success as 1 if $N = 1$ and as 0 if $N > 1$."

Example 8.2.8. Suppose we observe $X_1, \ldots, X_n$, independent r.v.'s each $N(\theta, 1)$ (see Definition 4.2.13) with θ unknown. What is "the" MLE of θ? (Here, since it is not stated what Θ is, it is to be taken as all possible θ, that is, since $-\infty < \theta < +\infty$, $\Theta = \{x\colon -\infty < x < +\infty\}$.) Here (see (8.2.4))

$$\begin{aligned} L(\theta) &= \prod_{i=1}^{n} f(X_i|\theta) = \prod_{i=1}^{n} \left\{\frac{1}{\sqrt{2\pi}} e^{-.5(X_i-\theta)^2}\right\} \\ &= \left(\frac{1}{\sqrt{2\pi}}\right)^n e^{-.5\sum_{i=1}^{n}(X_i-\theta)^2} \end{aligned} \tag{8.2.9}$$

and it is convenient (see Lemma 8.2.5) to take

$$\ln L(\theta) = -n\ln(\sqrt{2\pi}) - .5\sum_{i=1}^{n}(X_i - \theta)^2. \tag{8.2.10}$$

Now we wish to maximize $\ln L(\theta)$ in (8.2.10) with respect to θ (see Definition 8.2.1); this is equivalent to minimizing $g(\theta) = \sum_{i=1}^{n}(X_i - \theta)^2$. Now

$$g'(\theta) \equiv \frac{dg(\theta)}{d\theta} = -2\sum_{i=1}^{n}(X_i - \theta),$$

$$g''(\theta) = 2\sum_{i=1}^{n} 1 = 2n.$$

[7] Recall that $q = 1 - p$.

Hence, solving $g'(\theta) = 0$, we find $\theta = \left(\sum_{i=1}^{n} X_i\right)\Big/ n$. We denote this MLE by $\hat{\theta}$; $\sum_{i=1}^{n} X_i/n$ is simply the sample mean $\bar{X}$, so $\hat{\theta} = \bar{X}$. Since $g''(\theta) > 0$ for all θ, in particular $g''(\hat{\theta}) > 0$, which proves (see Theorem 8.2.11) that $g(\theta)$ has a relative minimum [and hence $L(\theta)$ has a relative maximum] at $\hat{\theta}$. Since $\theta \in \Theta = \{x: -\infty < x < +\infty\}$ and $\lim_{|\theta| \to \infty} g(\theta) = +\infty$, $\hat{\theta}$ yields the absolute minimum of $g(\theta)$ [and hence the absolute maximum of $L(\theta)$]. We previously studied the sample mean as an estimator of the population mean (Theorem 8.1.5), and in this case it is both unbiased and consistent.

Theorem 8.2.11. *Let $g(x)$ be continuous and twice differentiable in some open interval I containing a. If $g'(a) = 0$, then $g(x)$ has a relative maximum at $x = a$ if $g''(a) < 0$, and has a relative minimum at $x = a$ if $g''(a) > 0$. The set $\{a: g'(a) = 0\}$ is called the set of* **critical points** *of $g(x)$. The only possible maxima and minima of $g(x)$ for $c \leqslant x \leqslant d$ are at the critical points between c* and *d, and at c and at d.*

Proof: See any calculus book; for example, G. B. Thomas, Jr., *Calculus and Analytic Geometry* (Third Edition), Addison-Wesley Publishing Company, Inc., Reading, Massachusetts, 1960, Chapter 3.

Example 8.2.12. We obtain the same MLE $\hat{\theta} = \bar{X}$ if, in Example 8.2.8, each of the r.v.'s is $N(\theta, \sigma^2)$ where $\sigma^2 > 0$ is known.

Example 8.2.13. Suppose we observe $X_1, \ldots, X_n$, independent r.v.'s each $N(\mu, \sigma^2)$ with μ known, and σ^2 unknown. What is the MLE of $\theta = \sigma^2$? (Here $\Theta = \{x: x > 0\}$ since $0 < \sigma^2 < +\infty$.) Now

$$L(\sigma^2) = \prod_{i=1}^{n} \left\{\frac{1}{\sqrt{2\pi}\sigma} e^{-.5((X_i-\mu)/\sigma)^2}\right\}$$
$$= \left(\frac{1}{\sqrt{2\pi}\sigma}\right)^n e^{-.5\sum_{i=1}^{n}((X_i-\mu)/\sigma)^2}$$

and

$$\ln L(\sigma^2) = -n \ln(\sqrt{2\pi}) - n\ln(\sigma) - .5\sum_{i=1}^{n}\left(\frac{X_i-\mu}{\sigma}\right)^2.$$

Choosing σ^2 to maximize $\ln L(\sigma^2)$ is equivalent to choosing σ^2 to minimize

$$g(\sigma^2) = n\ln(\sigma) + .5\sum_{i=1}^{n}\left(\frac{X_i-\mu}{\sigma}\right)^2. \tag{8.2.14}$$

Now, it would be easier to differentiate with respect to σ in (8.2.14) (instead of with respect to σ^2). We shall do this [and thus essentially seek the MLE of σ, say $\hat{\sigma}$, instead of the MLE of σ^2, say $\widehat{(\sigma^2)}$; we will prove below that, under conditions satisfied here, $(\hat{\sigma})^2 = \widehat{(\sigma^2)}$ ... an *invariance* property of MLE's (see Theorem 8.2.15)]. Hence, we find

$$\frac{dg(\sigma^2)}{d\sigma} = \frac{n}{\sigma} - \frac{\sum_{i=1}^{n}(X_i - \mu)^2}{\sigma^3},$$

$$\frac{d^2g(\sigma^2)}{d\sigma^2} = -\frac{n}{\sigma^2} + 3\frac{\sum_{i=1}^{n}(X_i - \mu)^2}{\sigma^4}.$$

Setting the first derivative to zero, we find the unique solution

$$\hat{\sigma} = \sqrt{\frac{\sum_{i=1}^{n}(X_i - \mu)^2}{n}},$$

and

$$(\hat{\sigma})^2 = \frac{\sum_{i=1}^{n}(X_i - \mu)^2}{n}$$

(which is an unbiased and consistent estimator of σ^2, as may be easily verified). At $\hat{\sigma}$, the second derivative is

$$\left.\frac{d^2g(\sigma^2)}{d\sigma^2}\right|_{\sigma=\hat{\sigma}} = \left(\frac{1}{\hat{\sigma}}\right)^2 \{-n + 3n\} > 0,$$

so we have minimized $g(\sigma^2)$ (and hence maximized the likelihood $L(\sigma^2)$). The case of μ and σ^2 both unknown will be covered in the problems.

Theorem 8.2.15. *Suppose we have likelihood function* $L(\theta) = f(X_1, \ldots, X_n|\theta)$, $\theta \in \Theta$. *Suppose* $\hat{\theta}$ *is an MLE of* θ. *If* $g(\theta)$ *is a one-to-one function* [*that is, if* $g(\theta_1) = g(\theta_2) \Leftrightarrow \theta_1 = \theta_2$], *then* $g(\hat{\theta})$ *is an MLE of* $g(\theta)$ [*that is,* $g(\hat{\theta})$ *is a value of* $g(\theta)$ *at which* $L(\theta)$ *is maximized*].

Proof: By $\widehat{g(\theta)}$ we mean a value of $g(\theta)$ at which $L(\theta)$ is maximized. However, in general several θ's (some of which do *not* maximize $L(\theta)$) correspond to one value of $g(\theta)$. With a one-to-one function $g(\cdot)$ this cannot occur, so $\widehat{g(\theta)} = g(\hat{\theta})$.

Formally, $L(\theta) = L(g^{-1}(g(\theta)))$ are both maximized by "the" same $\hat{\theta}$, so $\hat{\theta} = g^{-1}(\widehat{g(\theta)})$, or $g(\hat{\theta}) = \widehat{g(\theta)}$.

Example 8.2.16. Suppose that $X_1, \ldots, X_n$ are independent r.v.'s, each with the uniform distribution on $[-d, d]$, where d is unknown, $d \in \Theta = \{x:$

$x > 0\}$. Find the MLE of d. Here, for $i = 1, \ldots, n$,

$$f(X_i|d) = \begin{cases} \dfrac{1}{2d}, & \text{if } -d \leqslant X_i \leqslant d \\ 0, & \text{otherwise.} \end{cases}$$

Hence,

$$L(d) = \begin{cases} \left(\dfrac{1}{2d}\right)^n, & \text{if}^8 \ -d \leqslant X_1, \ldots, X_n \leqslant d \\ 0, & \text{otherwise.} \end{cases}$$

This $L(d)$ will be maximized by the smallest possible d which keeps $L(d) > 0$. But[9] $d \geqslant (-X_1, \ldots, -X_n, X_1, \ldots, X_n)$ iff $d \geqslant (|X_1|, \ldots, |X_n|)$, so

$$\hat{d} = \max(|X_1|, \ldots, |X_n|).$$

[This $\hat{d}$ is the "smallest" d which can explain the data (that is, for which the data can occur).] Note that if we had a uniform distribution centered at a (a known) of width $2d$, that is, uniform on $[a - d, a + d]$, the $X_i - a$ would be uniform on $[-d, +d]$ and an analysis like that above (with center $a = 0$) would apply, yielding

$$\hat{d} = \max(|X_1 - a|, \ldots, |X_n - a|).$$

Example 8.2.17. Suppose that $X_1, \ldots, X_n$ are independent r.v.'s, each with the uniform distribution on $[c - d, c + d]$, where d is known ($d > 0$), c is unknown, and $c \in \Theta = \{x: -\infty < x < +\infty\}$. Find the MLE of c. Here,

$$L(c) = \begin{cases} \left(\dfrac{1}{2d}\right)^n, & \text{if } c - d \leqslant X_1, \ldots, X_n \leqslant c + d \\ 0, & \text{otherwise.} \end{cases}$$

Hence any c, such that $L(c) > 0$, is an MLE. The restriction is

$$c - d \leqslant X_1, \ldots, X_n \leqslant c + d, \text{ or } X_1 - d, \ldots, X_n - d \leqslant c \leqslant d + X_1, \ldots, d + X_n,$$
$$\text{or } \max(X_1, \ldots, X_n) - d \leqslant \hat{c} \leqslant \min(X_1, \ldots, X_n) + d.$$

We thus have another example of a nonunique MLE (the other being in Example 8.2.6).

Example 8.2.18. We now complete the trilogy whose first two parts were Examples 8.2.16 and 8.2.17. Let $X_1, \ldots, X_n$ be as in Example 8.2.17, but now with (c, d) unknown, $(c, d) \in \Theta = \{(x, y): -\infty < x < +\infty, y > 0\}$.

---

[8] "$-d \leqslant X_1, \ldots, X_n \leqslant d$" means "$-d \leqslant X_1 \leqslant d, \ldots, -d \leqslant X_n \leqslant d$."

[9] "$d \geqslant (a, b, c, \ldots, x)$" means "$d \geqslant a, d \geqslant b, d \geqslant c, \ldots, d \geqslant x$." Similar notation occurs later.

Find the MLE of (c, d). Since

$$L(c, d) = \begin{cases} \left(\dfrac{1}{2d}\right)^n, & \text{if } c - d \leqslant X_1, \ldots, X_n \leqslant c + d \\ 0, & \text{otherwise,} \end{cases}$$

for any fixed c, $L(c, d)$ is maximized by the smallest d which yields $L(c, d) > 0$, that is, the smallest d such that

$$\begin{gathered} c - d \leqslant X_1, \ldots, X_n \leqslant c + d \\ \max(X_1, \ldots, X_n) - d \leqslant c \leqslant \min(X_1, \ldots, X_n) + d \\ c - \min(X_1, \ldots, X_n) \leqslant d, \text{ and } \max(X_1, \ldots, X_n) - c \leqslant d \\ d \geqslant \max(-\min(X_1 - c, \ldots, X_n - c), \max(X_1 - c, \ldots, X_n - c)) \end{gathered}$$

that is,

$$d = \begin{cases} \max(X_1 - c, \ldots, X_n - c), \\ \qquad \text{if } \max(X_1 - c, \ldots, X_n - c) \geqslant -\min(X_1 - c, \ldots, X_n - c) \\ -\min(X_1 - c, \ldots, X_n - c), \\ \qquad \text{if } \max(X_1 - c, \ldots, X_n - c) \leqslant -\min(X_1 - c, \ldots, X_n - c) \end{cases}$$

$$= \begin{cases} \max(X_1, \ldots, X_n) - c, \\ \qquad \text{if } c \leqslant (\max(X_1, \ldots, X_n) + \min(X_1, \ldots, X_n))/2 \\ -\min(X_1, \ldots, X_n) + c, \\ \qquad \text{if } c \geqslant (\max(X_1, \ldots, X_n) + \min(X_1, \ldots, X_n))/2. \end{cases}$$

In either case, it is

$$c = \frac{\max(X_1, \ldots, X_n) + \min(X_1, \ldots, X_n)}{2}$$

which allows d to be minimized. Hence, we obtain

$$(\hat{c}, \hat{d}) = \left(\frac{\max(X_1,\ldots,X_n) + \min(X_1,\ldots,X_n)}{2}, \frac{\max(X_1,\ldots,X_n) - \min(X_1,\ldots,X_n)}{2}\right).$$

In Definition 8.2.1, we defined an MLE as any $\hat{\theta} \in \Theta$ such that $L(\hat{\theta}) = \sup\{L(\theta): \theta \in \Theta\}$. In Examples 8.2.6 and 8.2.17 we saw that such a $\hat{\theta}$ need not be unique. It may also not exist, as will be noted in Problems 8.5 and 8.6.

Example 8.2.19. Suppose that $X_1, \ldots, X_n$ are independent r.v.'s, each with the exponential distribution

$$f(x) = \begin{cases} \dfrac{1}{\theta_1} e^{-(x-\theta_2)/\theta_1}, & x \geqslant \theta_2 \\ 0, & x < \theta_2 \end{cases}$$

where $-\infty < \theta_2 < +\infty$ and $\theta_1 > 0$ are both unknown (see Definition 4.2.7; a slight generalization of Theorem 5.2.5 shows that $EX_1 = \theta_1 + \theta_2$). Find the MLE of (θ_1, θ_2), say $(\hat{\theta}_1, \hat{\theta}_2)$. Find $E\hat{\theta}_1$ and $E\hat{\theta}_2$.

Here

$$L(\theta_1, \theta_2) = f(X_1)\cdots f(X_n) = \begin{cases} \dfrac{1}{\theta_1^n} e^{-\sum_{i=1}^{n}(X_i-\theta_2)/\theta_1}, & \min(X_1, \ldots, X_n) \geqslant \theta_2 \\ 0, & \min(X_1, \ldots, X_n) < \theta_2. \end{cases}$$

Now for any fixed $\theta_1 > 0$, $L(\theta_1, \theta_2)$ is maximized by choosing θ_2 as large as possible such that $L(\theta_1, \theta_2) > 0$, so

$$\hat{\theta}_2 = \min(X_1, \ldots, X_n).$$

Next, if we set $\dfrac{\partial \ln L(\theta_1, \hat{\theta}_2)}{\partial \theta_1}$ to zero, we find

$$\frac{\partial \ln L(\theta_1, \hat{\theta}_2)}{\partial \theta_1} = \frac{\partial}{\partial \theta_1}\left\{-n \ln \theta_1 - \frac{\sum_{i=1}^{n}(X_i - \hat{\theta}_2)}{\theta_1}\right\} = 0,$$

$$-\frac{n}{\theta_1} + \frac{\sum_{i=1}^{n}(X_i - \hat{\theta}_2)}{\theta_1^2} = 0,$$

$$\hat{\theta}_1 = \bar{X} - \min(X_1, \ldots, X_n).$$

This yields a maximum since

$$\frac{\partial^2}{\partial \theta_1^2} \ln L(\theta_1, \hat{\theta}_2) = \frac{n}{\theta_1^2} - 2\frac{\sum_{i=1}^{n}(X_i - \hat{\theta}_2)}{\theta_1^3}$$

is <0 iff $\theta_1 < 2\hat{\theta}_1$. [To complete the proof we should note that $L(\theta_1, \theta_2) \to 0$ as $\theta_1 \to 0$ and also as $\theta_1 \to +\infty$.]

Now to find $E\hat{\theta}_1$ and $E\hat{\theta}_2$ we need to find $E\bar{X}$ and $E\min(X_1, \ldots, X_n)$. First, $E(\bar{X}) = E(X_1) = \theta_1 + \theta_2$. Next, the distribution function of $\min(X_1, \ldots, X_n)$ $(x \geqslant \theta_2)$ is

$$\begin{aligned} P[\min(X_1, \ldots, X_n) \leqslant x] &= 1 - P[\min(X_1, \ldots, X_n) > x] \\ &= 1 - P[X_1 > x]\cdots P[X_n > x] = 1 - (P[X_1 > x])^n \\ &= 1 - (1 - P[X_1 \leqslant x])^n = 1 - e^{-n\left(\frac{x-\theta_2}{\theta_1}\right)}; \end{aligned}$$

hence the density function of $\min(X_1, \ldots, X_n)$ is (via differentiation)

$$g(x) = \begin{cases} \dfrac{n}{\theta_1} e^{-n\left(\frac{x-\theta_2}{\theta_1}\right)}, & x \geqslant \theta_2 \\ 0, & x < \theta_2. \end{cases}$$

Thus,

$$\begin{aligned}
E\min(X_1, \ldots, X_n) &= \int_{\theta_2}^{\infty} y g(y)\, dy \\
&= \int_{\theta_2}^{\infty} y \frac{n}{\theta_1} e^{-n\left(\frac{y-\theta_2}{\theta_1}\right)}\, dy = \int_0^{\infty} (\theta_2 + x) \frac{n}{\theta_1} e^{-nx/\theta_1}\, dx \\
&= \theta_2 + \int_0^{\infty} x \frac{n}{\theta_1} e^{-nx/\theta_1}\, dx = \theta_2 + \frac{\theta_1}{n} \int_0^{\infty} z e^{-z}\, dz \\
&= \theta_2 + \frac{\theta_1}{n}.
\end{aligned}$$

Hence, $E\hat{\theta}_1 = (\theta_1 + \theta_2) - (\theta_2 + \theta_1/n) = \theta_1(1 - 1/n)$, and $E\hat{\theta}_2 = \theta_2 + \theta_1/n$.

Example 8.2.20. Suppose that $X_1, \ldots, X_n$ are independent r.v.'s, each with the distribution

$$p(x) = \begin{cases} \theta, & \text{if } x = 1 \\ (1 - \theta)/k, & \text{if } x = 2, \ldots, k \\ 0, & \text{otherwise,} \end{cases}$$

where θ $(0 \leqslant \theta \leqslant 1)$ and k $(k = 1, 2, \ldots)$ are unknown. What is the MLE of (θ, k)? [This distribution arose in Example 3.1.12 as the distribution of the number of microorganisms per clump on a surface. Thus, if n such clumps have been examined independently and respective numbers of organisms $X_1, \ldots, X_n$ found in each, what is the MLE of (θ, k), where θ is the probability a clump consists of one microorganism and k is the maximum possible number of microorganisms?]

Here, if we let Z represent the number of $X_1, \ldots, X_n$ which are one, then

$$L(\theta, k) = \begin{cases} \dfrac{\theta^Z (1 - \theta)^{n-Z}}{k^{n-Z}}, & \text{if } 0 \leqslant \theta \leqslant 1, k \geqslant X_1, \ldots, k \geqslant X_n \\ 0, & \text{otherwise.} \end{cases}$$

Since (subject to $k \geqslant X_1, \ldots, k \geqslant X_n$) k^{n-Z} is minimized by $k = \max(X_1, \ldots, X_n)$ [unless $Z = n$, in which case k^{n-Z} is constant for all k] and since $\theta^Z(1 - \theta)^{n-Z}$ is maximized by $\theta = Z/n$, the MLE of (θ, k) is

$$(\hat{\theta}, \hat{k}) = \begin{cases} \left(\dfrac{Z}{n}, \max(X_1, \ldots, X_n)\right), & \text{if } 0 \leqslant Z < n \\ (1, 1) \text{ or } (1, 2) \text{ or } \ldots & \text{if } Z = n. \end{cases}$$

That is, θ is estimated by the proportion of $X_1, \ldots, X_n$ which were zero, and k is (except when $Z = n$, where the MLE is not unique) estimated by the largest number of microorganisms observed.

8.3. CONSISTENCY OF MLE'S[10]

It follows from the results of Problem 8.6 and Part (1) of Theorem 8.1.5 that MLE's need not be unbiased estimators. However, it is the case that MLE's, the likelihood function (see (8.2.2)) and related quantities are central to a large part of statistics. While not unbiased in general, MLE's are usually consistent, and a proof of this develops much geometry and methods of proof which students with mathematical inclination should find exciting (and a part of which will be used later). Thus, this section is devoted to developing certain geometric concepts and utilizing them in a proof of consistency of MLE's.

Recall that the set of all **vectors x** of k real numbers $x_1, \ldots, x_k$ is denoted by $\mathscr{R}^k$, and that we write vectors as columns

$$\mathbf{x} = \begin{pmatrix} x_1 \\ x_2 \\ \vdots \\ x_k \end{pmatrix}.$$

The **transpose** of $\mathbf{x}$, say $\mathbf{x}'$, is a row vector $(x_1, \ldots, x_k)$. The **inner product** is $\mathbf{x}'\mathbf{y} = x_1 y_1 + \cdots + x_k y_k = \sum_{i=1}^{k} x_i y_i$. The **origin** is the zero vector $\mathbf{0} = (0, 0, \ldots, 0)'$.

Also, a subset S of the set of vectors $\mathscr{R}^k$ is **closed** if every limit point of S is a point in S. The **distance** from $\mathbf{x}$ to $\mathbf{y}$ (if $\mathbf{x}, \mathbf{y} \in \mathscr{R}^k$) is

$$\sqrt{(x_1 - y_1)^2 + (x_2 - y_2)^2 + \cdots + (x_n - y_n)^2} = \sqrt{(\mathbf{x} - \mathbf{y})'(\mathbf{x} - \mathbf{y})}.$$

The **sphere**[11] of radius $\alpha > 0$ about $\mathbf{a}$ is $\{\mathbf{x} : \mathbf{x} \in \mathscr{R}^k, (\mathbf{x} - \mathbf{a})'(\mathbf{x} - \mathbf{a}) \leqslant \alpha^2\}$.

Definition 8.3.1. *A subset S of $\mathscr{R}^k$ is called* **convex** *if $\mathbf{x}, \mathbf{y} \in S$ implies $\lambda\mathbf{x} + (1 - \lambda)\mathbf{y} \in S$ for all λ with $0 < \lambda < 1$.* [This means S is convex if all of the line joining each two points of S, is itself in S. See Figure 8.3–1.]

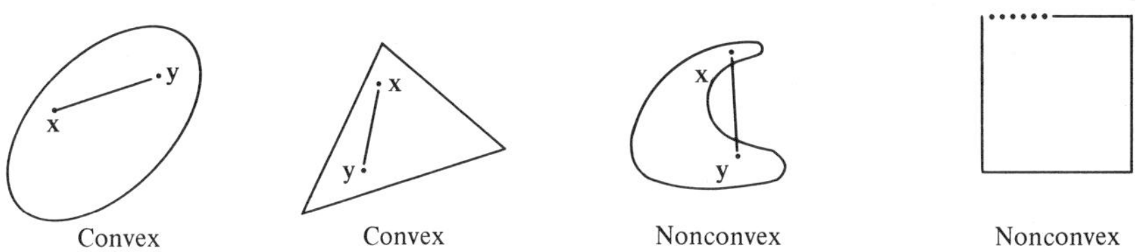

Figure 8.3–1. Some convex and nonconvex sets

[10] The material of this section may be omitted at this time. Those results needed in later chapters' sections may be covered at that time, as they comprise only a modest part of this section's material.

[11] Or "ball," or "solid sphere," or "spherical cell."

Our goal now is to prove what is called the Separating Hyperplane Theorem (which essentially says two disjoint convex subsets of $\mathscr{R}^k$ may be separated by a plane (in $\mathscr{R}^k$ a "plane" is usually called a hyperplane; in $\mathscr{R}^2$ it is called a line, while in $\mathscr{R}^3$ it is called a plane)). We need one other result in order to prove the Separating Hyperplane Theorem, and several related ones (especially the Supporting Hyperplane Theorem) will be useful in later chapters (especially in Chapter 12 on Decision Theory). These shall be simply stated; for detailed proofs, see, for example, T. S. Ferguson: *Mathematical Statistics: A Decision Theoretic Approach*, Academic Press Inc., New York, 1967, pp. 70–73.

Lemma 8.3.2. *If $S \subseteq \mathscr{R}^k$ is closed and convex, and if $\mathbf{0} \notin S$, then there is a $\mathbf{p} \in \mathscr{R}^k$ such that* [see Figure 8.3–2]

$$\mathbf{p}'\mathbf{x} > 0 \quad (\textit{all } \mathbf{x} \in S).$$

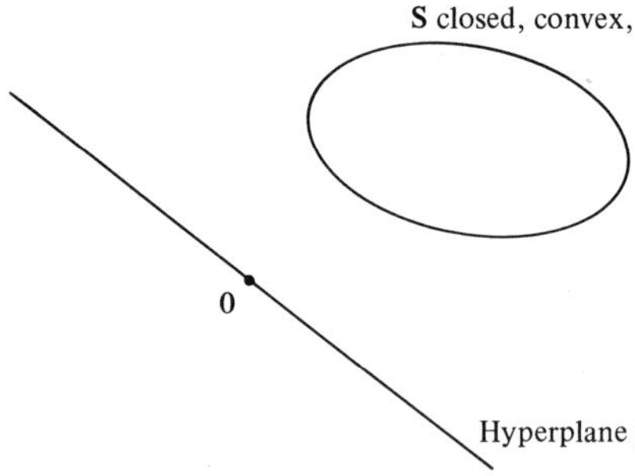

Figure 8.3–2. Illustration of Lemma 8.3.2

Note that the origin $\mathbf{0}$ lies in every **hyperplane**[12] $\mathbf{p}'\mathbf{x} = 0$; Lemma 8.3.2 says we can choose a $\mathbf{p}$ so that $\mathbf{p}'\mathbf{x} > 0$ for all $\mathbf{x} \in S$. Since (by definition and by analogy with $\mathscr{R}$, $\mathscr{R}^2$, and $\mathscr{R}^3$) $\{\mathbf{x}: \mathbf{p}'\mathbf{x} > 0\}$ and $\{\mathbf{x}: \mathbf{p}'\mathbf{x} < 0\}$ are called the **two sides of the hyperplane** $\mathbf{p}'\mathbf{x} = 0$, this result says we can choose a hyperplane $\mathbf{p}'\mathbf{x} = 0$ so that S is kept on one side of the hyperplane (that is, the hyperplane does not intersect S).

Lemma 8.3.3. *If $S \subseteq \mathscr{R}^k$ is convex, and if $A \subseteq \mathscr{R}^k$ is open, and if $A \subseteq \text{cl}(S) =$ closure of S, then $A \subseteq S$.*

Lemma 8.3.3 means essentially that any open set contained in the closure of a convex set, is already contained in the convex set which (as can be seen by deleting the boundary from the first two figures of Figure 8.3–1) need not be closed itself.

[12] Formally, a **hyperplane** consists of all $\mathbf{x}$ such that $\mathbf{p}'\mathbf{x} = c$, where $\mathbf{p} \neq \mathbf{0}$ and c are constants.

Theorem 8.3.4. *If $S \subseteq \mathscr{R}^k$ is convex and if $\mathbf{x}_0$ is either $\notin S$ or is a boundary point of S, then there is a $\mathbf{p} \in \mathscr{R}^k$ such that*

$$\mathbf{p} \neq \mathbf{0}, \quad \mathbf{p}'\mathbf{x} \geqslant \mathbf{p}'\mathbf{x}_0 \quad (\textit{all } \mathbf{x} \in S).$$

The situation described in Theorem 8.3.4 is depicted in Figure 8.3–3, where in case (a) $\mathbf{x}_0 \notin S$ and in cases (b) and (c) $\mathbf{x}_0$ is a boundary point of S. Note

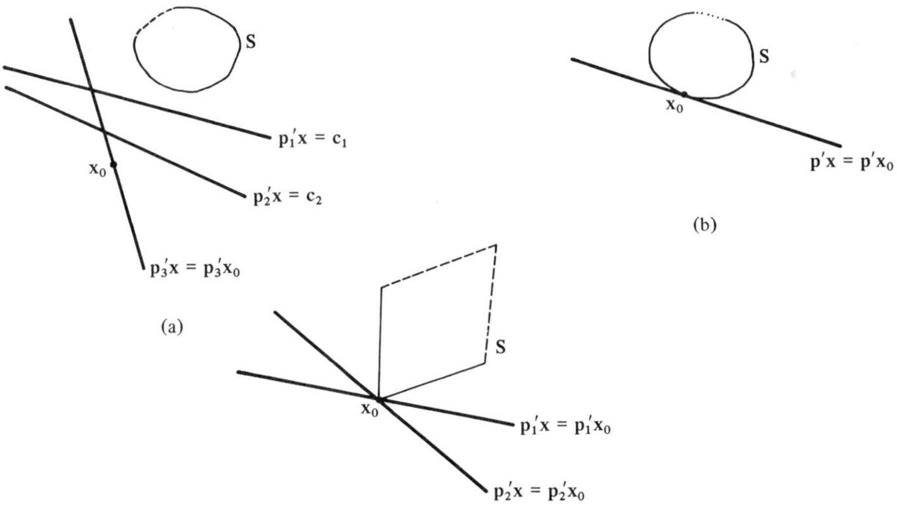

Figure 8.3–3. Illustration of Theorem 8.3.4

that $\mathbf{p}$ is not always unique. The part of Theorem 8.3.4 relating to $\mathbf{x}_0$ in the boundary of S is called the **Supporting Hyperplane Theorem** because the hyperplane $\mathbf{p}'\mathbf{x} = \mathbf{p}'\mathbf{x}_0$ is tangent to S and keeps S on one side [essentially by the definition of *sides* of the hyperplane, as noted after Figure 8.3–2]. As we saw in Figure 8.3–3, there may be infinitely many supporting hyperplanes at $\mathbf{x}_0$.

Theorem 8.3.5. **(Separating Hyperplane Theorem)** *If S_1 and S_2 are disjoint, convex subsets of $\mathscr{R}^k$, then there is a vector $\mathbf{p} \neq \mathbf{0}$ such that*

$$\mathbf{p}'\mathbf{y} \leqslant \mathbf{p}'\mathbf{x} \quad (\textit{all } \mathbf{x} \in S_1, \mathbf{y} \in S_2).$$

Proof: We will proceed by defining a new set

$$S = \{\mathbf{z} : \mathbf{z} = \mathbf{x} - \mathbf{y} \text{ for some } \mathbf{x} \in S_1 \text{ and } \mathbf{y} \in S_2\}.$$

Then $\mathbf{0} \in S$, for if $\mathbf{0}$ were in S, there would be an $\mathbf{x}$ in S_1 and a $\mathbf{y}$ in S_2 such that $\mathbf{x} - \mathbf{y} = \mathbf{0}$. It would follow that $\mathbf{x} = \mathbf{y}$ with $\mathbf{x} \in S_1$, and $\mathbf{y} \in S_2$, which would be a contradiction since S_1 and S_2 are disjoint. (This proof of $\mathbf{0} \notin S$ fails only if at least one of S_1, S_2 is empty; but in that case $S = \emptyset$, and the theorem is obvious anyway. Hence, we may assume $S_1 \neq \emptyset$, $S_2 \neq \emptyset$.)

Secondly, S is a convex set. For suppose $\mathbf{z}_1$ and $\mathbf{z}_2$ are in S; then (for some $\mathbf{x}_1, \mathbf{x}_2 \in S_1$ and $\mathbf{y}_1, \mathbf{y}_2 \in S_2$) $\mathbf{z}_1 = \mathbf{x}_1 - \mathbf{y}_1$ and $\mathbf{z}_2 = \mathbf{x}_2 - \mathbf{y}_2$. Hence,

$$\begin{aligned}\beta\mathbf{z}_1 + (1-\beta)\mathbf{z}_2 &= \beta(\mathbf{x}_1 - \mathbf{y}_1) + (1-\beta)(\mathbf{x}_2 - \mathbf{y}_2)\\ &= \{\beta\mathbf{x}_1 + (1-\beta)\mathbf{x}_2\} - \{\beta\mathbf{y}_1 + (1-\beta)\mathbf{y}_2\},\end{aligned}$$

which (for $0 < \beta < 1$) is in S since (for $0 < \beta < 1$) $\beta\mathbf{x}_1 + (1-\beta)\mathbf{x}_2 \in S_1$ and $\beta\mathbf{y}_1 + (1-\beta)\mathbf{y}_2 \in S_2$. Hence (see Definition 8.3.1) S is a convex set.

Finally, let us apply Theorem 8.3.4 to the convex subset S of $\mathscr{R}^k$ and $\mathbf{x}_0 = \mathbf{0} \notin S$. This assures us there is a $\mathbf{p} \in \mathscr{R}^k$ $(\mathbf{p} \neq \mathbf{0})$ such that

$$\mathbf{p}'\mathbf{z} \geqslant \mathbf{p}'\mathbf{0} = 0 \quad (\text{all } \mathbf{z} \in S);$$

hence,

$$\mathbf{p}'\mathbf{x} \geqslant \mathbf{p}'\mathbf{y} \qquad (\text{all } \mathbf{x} \in S_1,\ \mathbf{y} \in S_2).$$

(See Figure 8.3–4.)

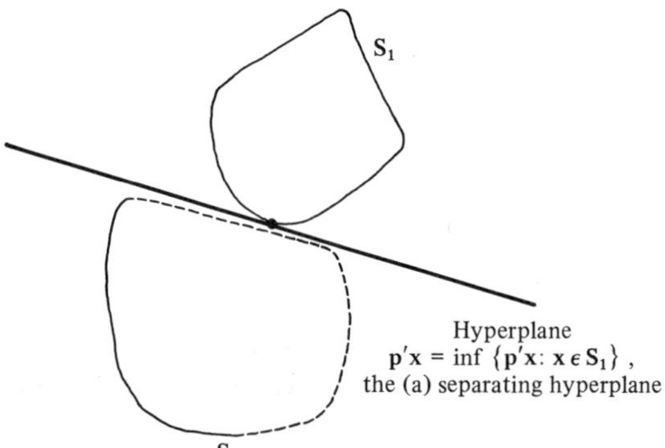

Figure 8.3–4. Illustration of Theorem 8.3.6

As a final preparation for the proof of consistency we develop some properties of convex functions and Jensen's Inequality (which generalizes the inequality between geometric and arithmetic means).

Definition 8.3.6. *A function $f(\mathbf{x})$, which has values in $\mathscr{R}$ and is defined for all $\mathbf{x}$ in some convex subset S of $\mathscr{R}^k$, is called* **convex** *if (for all $0 < \beta < 1$, $\mathbf{x} \in S$, $\mathbf{y} \in S$).*

$$f(\beta\mathbf{x} + (1-\beta)\mathbf{y}) \leqslant \beta f(\mathbf{x}) + (1-\beta)f(\mathbf{y}).$$

Figure 8.3–5 illustrates the convex function $f(x) = 1/x$ on $S = \{x : x \in \mathscr{R}, x > 0\}$. Of course, $f(x)$ is not defined for $x = 0$ and is not convex if points

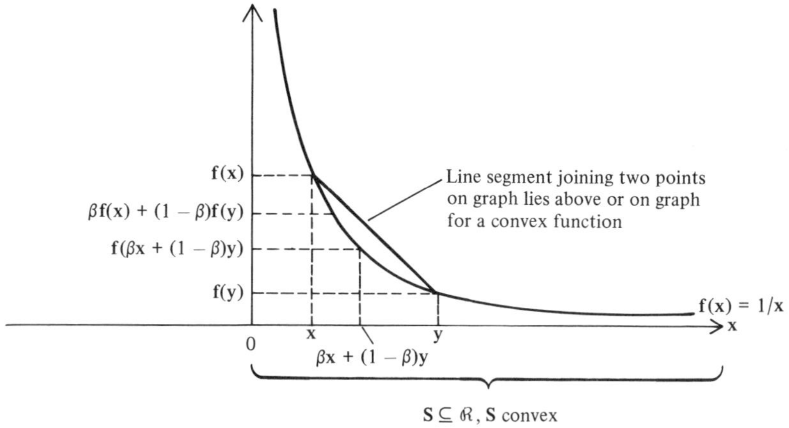

Figure 8.3–5. Convex function $\mathbf{f(x) = 1/x}$ on $\mathbf{S} = \{\mathbf{x}: \mathbf{x} \in \mathscr{R}, \mathbf{x} > 0\}$

$x < 0$ are allowed in S; but it is only claimed that, on this particular $S = \{x: x \in \mathscr{R}, x > 0\}$, $f(x) = 1/x$ is a convex function.

Theorem 8.3.7. (Jensen's Inequality) *If $f(\mathbf{x})$ is a convex function defined on $S \neq \emptyset$, $S \subseteq \mathscr{R}^k$, and if $\mathbf{Z}$ is a k-dimensional r.v. such that $E\mathbf{Z}$ exists*[13] *and $P[\mathbf{Z} \in S] = 1$, then $E\mathbf{Z} \in S$ and (provided $Ef(\mathbf{Z})$ exists)*

$$f(E\mathbf{Z}) \leqslant Ef(\mathbf{Z}).$$

The proof we shall give that the MLE is consistent in general is due to A. Wald ["Note on the consistency of the maximum likelihood estimate," *Annals of Mathematical Statistics*, Vol. 20 (1949), pp. 595–601] and J. Wolfowitz ["On Wald's proof of the consistency of the maximum likelihood estimate," *Annals of Mathematical Statistics*, Vol. 20 (1949), pp. 601–602]. In order to find the $\hat{\theta}$ which maximizes the likelihood $L(\theta)$ [see (8.2.2)], we often set $L'(\hat{\theta})$ (the derivative of $L(\hat{\theta})$ with respect to θ) to 0 and solve for $\hat{\theta}$, the MLE. Of course, the result is not necessarily a unique $\hat{\theta}$, and this is important. Now the equation

$$L'(\theta) = 0 \tag{8.3.8}$$

is usually called the **likelihood equation**. H. Cramér proved that some root of the likelihood equation is consistent, and V. S. Huzurbazar proved (under conditions) that (8.3.8) has at most one consistent solution and that $L(\theta)$ has a relative maximum at such a solution. (For references, see Wald's paper.) All this, however, still does not prove that a $\hat{\theta}$ which achieves the

[13] In Chapter 5, moments were discussed with regard to only a univariate r.v. X (which was allowed to be a function of, say, $\mathbf{Y} = (Y_1, \ldots, Y_n)$). By such a notation as $E\mathbf{Z}$, we simply mean $(EZ_1, \ldots, EZ_k)$; similarly, $\text{Var}(\mathbf{Z})$ simply means $(\text{Var}(Z_1), \ldots, \text{Var}(Z_k))$, and so on.

absolute maximum of $L(\theta)$ is consistent. Another approach, also with its problems, was that of J. L. Doob (again, see Wald's paper for details). Recently M. D. Perlman ["The limiting behavior of multiple roots of the likelihood equation," Abstract, *Annals of Mathematical Statistics*, Vol. 40 (1969), p. 1864] has announced[14] results which seem to partially salvage the Cramér-Huzurbazar proof. We are fortunate to have the Wald-Wolfowitz proof, which we now present.

Assumption 8.3.9. *Suppose that $X_1, \ldots, X_n$ are independent r.v.'s, each with d.f. $F(x|\boldsymbol{\theta}) = P_{\boldsymbol{\theta}}[X_1 \leqslant x]$ where $\boldsymbol{\theta} \in \Theta \subseteq \mathscr{R}^k$, and Θ is closed. Assume $F(x|\boldsymbol{\theta})$ is either absolutely continuous [in which case let $f(x|\boldsymbol{\theta})$ denote its density at x] or discrete [in which case let $f(x|\boldsymbol{\theta})$ denote $P_{\boldsymbol{\theta}}[X_1 = x]$]. Now simply write θ for $\boldsymbol{\theta}$.*

Definition 8.3.10. *Let $|\theta|$ denote $\sqrt{\theta_1^2 + \cdots + \theta_k^2}$, the distance from θ to $\mathbf{0}$. For any θ and any $\rho > 0$, define*

$$f(x, \theta, \rho) = \sup\{f(x|\theta'): \theta' \ \text{such that}\ |\theta' - \theta| \leqslant \rho\},$$

and for any $r > 0$ define

$$\varphi(x, r) = \sup\{f(x|\theta'): \theta' \ \text{such that}\ |\theta'| > r\}.$$

[These regions of θ are illustrated in Figure 8.3–6.] *Finally, define*

$$f^*(x, \theta, \rho) = \begin{cases} f(x, \theta, \rho), & \text{if } f(x, \theta, \rho) > 1 \\ 1, & \text{otherwise,} \end{cases}$$

$$\varphi^*(x, r) = \begin{cases} \varphi(x, r), & \text{if } \varphi(x, r) > 1 \\ 1, & \text{otherwise.} \end{cases}$$

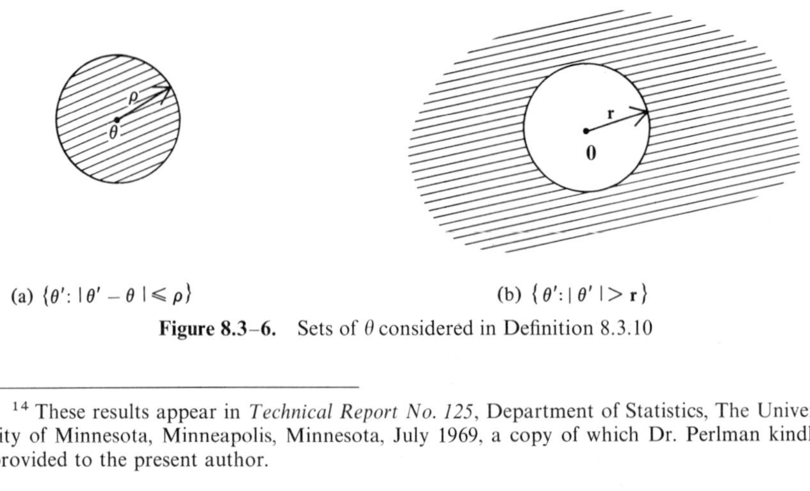

Figure 8.3–6. Sets of θ considered in Definition 8.3.10

[14] These results appear in *Technical Report No. 125*, Department of Statistics, The University of Minnesota, Minneapolis, Minnesota, July 1969, a copy of which Dr. Perlman kindly provided to the present author.

Assumption 8.3.11. (**1**) *$F(x|\theta)$ is either discrete for all θ, or absolutely continuous for all θ.* (**2**) *For ρ sufficiently small and r sufficiently large, both of*

$$E_{\theta_0} \ln f^*(X, \theta, \rho),\ E_{\theta_0} \ln \varphi^*(X, r)$$

exist (where θ_0 denotes the true parameter point). (**3**) *If $\lim_{i\to\infty} \theta_i = \theta$, then $\lim_{i\to\infty} f(x|\theta_i) = f(x|\theta)$ for all x [it is sufficient if $P_{\theta_0}(E_\theta) = 0$, where E_θ is the set of x, which may depend on θ but not on the sequence θ_i, for which $\lim_{i\to\infty} f(x|\theta_i) \neq f(x|\theta)$; the probability is calculated using $F(x|\theta_0)$].* (**4**) *If $\theta_1 \neq \theta_0$ (the true parameter point) and $\theta_1 \in \Theta$, then $F(x|\theta_1) \neq F(x|\theta_0)$ for at least one value of x* [otherwise, how could θ_1 and θ_0 be distinguished from each other?]. (**5**) *For the true parameter point θ_0,*

$$E_{\theta_0}|\ln f(X|\theta_0)|$$

exists. (**6**) *Agree to say, for any r.v. U, $E_{\theta_0}U$ "exists" if $E_{\theta_0} \max(U, 0) < \infty$, and "$E_{\theta_0}U = -\infty$" if $E_{\theta_0} \max(U, 0) < \infty$ but $E_{\theta_0} \min(U, 0)$ does not exist (where θ_0 is the true parameter point).*

Lemma 8.3.12. *Under assumptions* 8.3.9 *and* 8.3.11, *for any $\theta \neq \theta_0$*

$$E_{\theta_0} \ln f(X|\theta) < E_{\theta_0} \ln f(X|\theta_0).$$

[On the average, the value of the density is smaller with a false value θ than with the true θ_0.]

Proof: By part (2) of Assumption 8.3.11, both expected values "exist" in the sense of part (6) of Assumption 8.3.11. By part (5) of Assumption 8.3.11, $E_{\theta_0} \ln f(X|\theta_0)$ exists. Now if "$E_{\theta_0} \ln f(X|\theta) = -\infty$" the conclusion is obvious. Suppose, on the other hand, that $E_{\theta_0}|\ln f(X|\theta)|$ exists. Then let

$$U = \ln f(X|\theta) - \ln f(X|\theta_0).$$

Since $E_{\theta_0}|\ln f(X|\theta)|$ and $E_{\theta_0} \ln f(X|\theta_0)$ exist, $P_{\theta_0}[f(X|\theta) = 0 \text{ or } f(X|\theta_0) = 0] = 0$, and $E|U|$ exists. Now by part (4) of Assumption 8.3.11, U is *not* {constant w.p. 1}. Thus, by Jensen's Inequality 8.3.7 (also see Problem 8.16)

$$E_{\theta_0}U < \ln E_{\theta_0}e^U.$$

But

$$E_{\theta_0}e^U = E_{\theta_0}\frac{f(X|\theta)}{f(X|\theta_0)} = \int_{\{x:\, f(x|\theta_0)\neq 0\}} \frac{f(x|\theta)}{f(x|\theta_0)} f(x|\theta_0)\, dx = \int_{\{x:\, f(x|\theta_0)\,\neq\, 0\}} f(x|\theta)\, dx \leqslant 1,$$

so $E_{\theta_0}U < \ln 1 = 0$, which proves the lemma.

Lemma 8.3.13. *Under Assumptions* 8.3.9 *and* 8.3.11, $\lim_{\rho\to 0} E_{\theta_0} \ln f(X, \theta, \rho) = E_{\theta_0} \ln f(X|\theta)$, *and* $\lim_{r\to\infty} E_{\theta_0} \ln \varphi(X, r) = -\infty$.

[Note, for example, that the latter result says what happens, on the average, to the largest value of $f(x|\theta)$ for θ outside a sphere of radius $r \to \infty$ about $\mathbf{0}$.]

Theorem 8.3.14. *Under Assumptions* 8.3.9 *and* 8.3.11, *the MLE of θ is consistent; that is, for any $\delta > 0$, $\varepsilon > 0$ there is an $N(\delta, \varepsilon)$ such that $n > N(\delta, \varepsilon)$ implies*

$$P_{\theta_0}[|\hat{\theta}_n(X_1, \ldots, X_n) - \theta_0| < \delta] > 1 - \varepsilon.$$

Proof: Let $r_0 > 0$ be chosen so that

$$E_{\theta_0} \ln \varphi(X, r_0) < E_{\theta_0} \ln f(X|\theta_0) \tag{8.3.15}$$

[by part (5) of Assumption 8.3.11 the right-hand side exists, and by Lemma 8.3.13 the left-hand side $\to -\infty$ as $r_0 \to \infty$, so this choice of r_0 is possible]. Let Θ be any closed subset of Θ such that $\theta_0 \notin \Theta$, and let $\Theta_1 = \{\theta: \theta \in \Theta, |\theta| \leq r_0\}$ (see Figure 8.3.7). With each[15] $\theta \in \Theta_1$ we may associate a number $\rho_\theta > 0$ such that

$$E_{\theta_0} \ln f(X, \theta, \rho_\theta) < E_{\theta_0} \ln f(X|\theta_0). \tag{8.3.16}$$

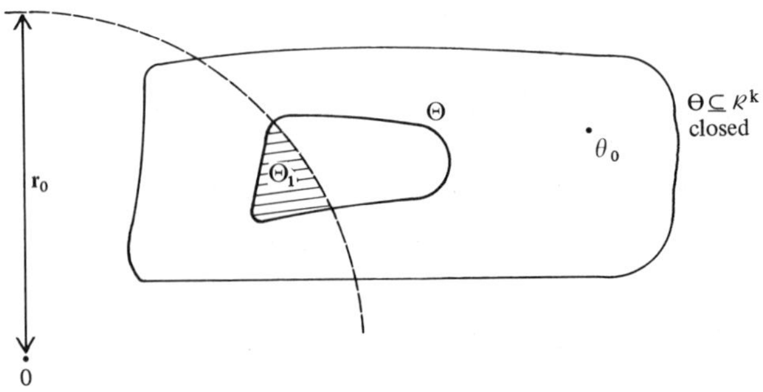

Figure 8.3–7. Sets Θ, Θ, and Θ_1

That there is such a ρ_θ we now prove. By Lemma 8.3.12, for any $\theta \neq \theta_0$ we have $E_{\theta_0} \ln f(X|\theta) < E_{\theta_0} \ln f(X|\theta_0)$; hence,

$$E_{\theta_0} \ln f(X|\theta_0) - E_{\theta_0} \ln f(X|\theta) \equiv \gamma(\theta) > 0.$$

[15] For r_0 "large enough," $\Theta_1 \neq \emptyset$.

Also, by Lemma 8.3.13 we have

$$\lim_{\rho \to 0} E_{\theta_0} \ln f(X, \theta, \rho) = E_{\theta_0} \ln f(X|\theta),$$

so there is a value $\rho_\theta > 0$ of ρ such that $|E_{\theta_0} \ln f(X, \theta, \rho_\theta) - E_{\theta_0} \ln f(X|\theta)| < \gamma(\theta)/2$. Thus [and this proves the assertion of (8.3.16)],

$$\begin{aligned} E_{\theta_0} \ln f(X, \theta, \rho_\theta) &< E_{\theta_0} \ln f(X|\theta) + \frac{\gamma(\theta)}{2} \\ &= E_{\theta_0} \ln f(X|\theta_0) - \frac{\gamma(\theta)}{2} \\ &< E_{\theta_0} \ln f(X|\theta_0). \end{aligned}$$

Next, let $S(\theta, \rho) = \{\theta' : |\theta' - \theta| < \rho\}$ (the collection of points with distance less th n ρ from θ, that is, the open sphere with center θ and radius ρ). Now since Θ_1 (being a closed and bounded, hence compact, subset of $\mathscr{R}^k$) is such that the union of all of the open sets $S(\theta, \rho_\theta)$ for $\theta \in \Theta_1$ contains Θ_1 [that is, $\{S(\theta, \rho_\theta) : \theta \in \Theta_1\}$ is an open cover of Θ_1], it follows from the Heine-Borel Theorem that a finite number (say m) of the open sets $S(\theta, \rho_\theta)$ contain Θ_1 in their union [these m sets are called a finite subcover of Θ_1]; call these sets $S(\theta_1, \rho_{\theta_1}), \ldots, S(\theta_m, \rho_{\theta_m})$. Define $T(\theta_i)$, for $i = 1, \ldots, m, m + 1$, as follows:

$$\begin{cases} -2T(\theta_i) = E_{\theta_0} \ln f(X, \theta_i, \rho_{\theta_i}) - E_{\theta_0} \ln f(X|\theta_0) \quad (i = 1, \ldots, m) \\ -2T(\theta_{m+1}) = E_{\theta_0} \ln \varphi(X, r_0) - E_{\theta_0} \ln f(X|\theta_0); \end{cases} \tag{8.3.17}$$

but if any of the right-hand side quantities are infinite, for that case let $T(\theta_i) = 1$ (say). Thus [see (8.3.15), (8.3.16), and (8.3.17)] $T(\theta_i) > 0$ for $i = 1, \ldots, m, m + 1$.

Now by Khinchin's Theorem (Theorem 6.4.4) or by the WLLN (Theorem 6.3.5), it follows that for any i $(1 \leqslant i \leqslant m + 1)$ there is an integer $N_i > 0$ such that, for $n > N_i$, we have

$$\begin{cases} P_{\theta_0}\left[\dfrac{\prod_{j=1}^{n} f(X_j, \theta_i, \rho_{\theta_i})}{\prod_{j=1}^{n} f(X_j|\theta_0)} > e^{-nT(\theta_i)}\right] < \dfrac{\varepsilon}{m+1} \quad (i = 1, \ldots, m) \\ P_{\theta_0}\left[\dfrac{\prod_{j=1}^{n} \varphi(X_j, r_0)}{\prod_{j=1}^{n} f(X_j|\theta_0)} > e^{-nT(\theta_{m+1})}\right] < \dfrac{\varepsilon}{m+1}. \end{cases} \tag{8.3.18}$$

We prove (8.3.18) as follows. Consider first the case $1 \leqslant i \leqslant m$; there, the following three statements are equivalent:

$$\frac{\prod_{j=1}^{n} f(X_j, \theta_i, \rho_{\theta_i})}{\prod_{j=1}^{n} f(X_j|\theta_0)} > e^{-nT(\theta_i)},$$

$$\sum_{j=1}^{n} \ln \frac{f(X_j, \theta_i, \rho_{\theta_i})}{f(X_j|\theta_0)} > -nT(\theta_i),$$

$$\frac{\sum_{j=1}^{n} \ln \frac{f(X_j, \theta_i, \rho_{\theta_i})}{f(X_j|\theta_0)}}{n} > -T(\theta_i).$$

Now

$$E_{\theta_0} \ln \frac{f(X_j, \theta_i, \rho_{\theta_i})}{f(X_j|\theta_0)} = E_{\theta_0}[\ln f(X_j, \theta_i, \rho_{\theta_i}) - \ln f(X_j|\theta_0)],$$

which is <0 by (8.3.16). Hence, by Khinchin's Theorem and (8.3.17), when θ_0 is true

$$\frac{\sum_{j=1}^{n} \ln \frac{f(X_j, \theta_i, \rho_{\theta_i})}{f(X_j|\theta_0)}}{n} \xrightarrow{P} -2T(\theta_i) < 0.$$

Hence, there is an N_i such that for $n > N_i$, we have [since $-2T(\theta_i) < -T(\theta_i) < 0$]

$$P_{\theta_0}\left[\frac{\sum_{j=1}^{n} \ln \frac{f(X_j, \theta_i, \rho_{\theta_i})}{f(X_j|\theta_0)}}{n} > -T(\theta_i)\right] < \frac{\varepsilon}{m+1}.$$

The proof of (8.3.18) for $i = m + 1$ follows similarly (see Honors Problem 8.7H).

We now "put it all together" to prove the theorem (whose statement the student should, at this point, review). Let $N(\varepsilon) = \max(N_1, \ldots, N_{m+1})$ and $h = \max(e^{-T(\theta_1)}, \ldots, e^{-T(\theta_{m+1})})$. Then for $n > N = N(\varepsilon)$, all $m + 1$ statements of (8.3.18) hold simultaneously; they continue to hold if $e^{-T(\theta_i)}$ is replaced by h (for $i = 1, \ldots, m, m + 1$), since this latter quantity is harder for each left-hand side in (8.3.18) to exceed. Thus, for $n > N$,

$$\left\{ P_{\theta_0}\left[\frac{\prod_{j=1}^{n} f(X_j, \theta_i, \rho_{\theta_i})}{\prod_{j=1}^{n} f(X_j|\theta_0)} > h^n\right] < \frac{\varepsilon}{m+1} \quad (i = 1, \ldots, m) \right.$$

$$\left\{ P_{\theta_0}\left[\frac{\prod_{j=1}^{n} \varphi(X_j, r_0)}{\prod_{j=1}^{n} f(X_j|\theta_0)} > h^n\right] < \frac{\varepsilon}{m+1}; \right.$$

that is, for $n > N$,

$$\left\{ \begin{aligned} &P_{\theta_0}\left[\frac{\prod_{j=1}^{n} \sup_{\{\theta': |\theta' - \theta_i| \leqslant \rho_{\theta_i}\}} f(X_j|\theta')}{\prod_{j=1}^{n} f(X_j|\theta_0)} > h^n\right] < \frac{\varepsilon}{m+1} \quad (i = 1, \ldots, m), \\ &P_{\theta_0}\left[\frac{\prod_{j=1}^{n} \sup_{|\theta'| > r_0} f(X_j|\theta')}{\prod_{j=1}^{n} f(X_j|\theta_0)} > h^n\right] < \frac{\varepsilon}{m+1}. \end{aligned} \right. \tag{8.3.19}$$

Recall that Θ was (at the beginning of this proof) defined to be any closed subset of θ such that $\theta_0 \notin \Theta$, while we took $\Theta_1 = \{\theta: \theta \in \Theta, |\theta| \leqslant r_0\}$. Now, for any $\eta > 0$, we take

$$\left\{ \begin{aligned} &\Theta = \Theta(\eta) = \theta - S(\theta_0, \eta), \\ &\Theta_1(\eta) = \{\theta: \theta \in \Theta(\eta), |\theta| \leqslant r_0\}. \end{aligned} \right.$$

Then [where the sets $S(\cdot, \cdot)$ depend on η since $h = h(\eta)$ and $N = N(\eta, \varepsilon)$]

$$\begin{aligned} \Theta(\eta) &= \Theta_1(\eta) \cup \{\theta: \theta \in \Theta(\eta), |\theta| > r_0\} \\ &\subseteq S(\theta_1, \rho_{\theta_1}) \cup \cdots \cup S(\theta_m, \rho_{\theta_m}) \cup \{\theta: |\theta| > r_0\} \\ &\subseteq \mathrm{cl}(S(\theta_1, \rho_{\theta_1})) \cup \cdots \cup \mathrm{cl}(S(\theta_m, \rho_{\theta_m})) \cup \{\theta: |\theta| > r_0\}. \end{aligned}$$

Now for $n > N(\eta, \varepsilon)$,

$$\begin{aligned} &P_{\theta_0}\left[\frac{\sup_{\theta \in \Theta(\eta)} \prod_{j=1}^{n} f(X_j|\theta)}{\prod_{j=1}^{n} f(X_j|\theta_0)} > h^n\right] \\ &\leqslant P_{\theta_0}\left[\max\left\{\sup_{\theta \in \mathrm{cl}(S(\theta_1, \rho_{\theta_1}))} \prod_{j=1}^{n} f(X_j|\theta), \ldots, \right.\right. \\ &\qquad \left.\left. \sup_{\theta \in \mathrm{cl}(S(\theta_m, \rho_{\theta_m}))} \prod_{j=1}^{n} f(X_j|\theta), \sup_{|\theta| > r_0} \prod_{j=1}^{n} f(X_j|\theta)\right\} \Big/ \prod_{j=1}^{n} f(X_j|\theta_0) > h^n\right] \end{aligned}$$

$$\leqslant P_{\theta_0}\left[\frac{\sup\limits_{\theta \in \mathrm{cl}(S(\theta_1, \rho_{\theta_1}))} \prod_{j=1}^{n} f(X_j|\theta)}{\prod_{j=1}^{n} f(X_j|\theta_0)} > h^n, \text{ or} \ldots, \text{ or } \frac{\sup\limits_{\theta \in \mathrm{cl}(S(\theta_m, \rho_{\theta_m}))} \prod_{j=1}^{n} f(X_j|\theta)}{\prod_{j=1}^{n} f(X_j|\theta_0)} > h^n, \right.$$

$$\left. \text{or } \frac{\sup\limits_{|\theta| > r_0} \prod_{j=1}^{n} f(X_j|\theta)}{\prod_{j=1}^{n} f(X_j|\theta_0)} > h^n \right]$$

$$\leqslant P_{\theta_0}\left[\frac{\sup\limits_{\theta \in \mathrm{cl}(S(\theta_1, \rho_{\theta_1}))} \prod_{j=1}^{n} f(X_j|\theta)}{\prod_{j=1}^{n} f(X_j|\theta_0)} > h^n\right] + (m \text{ similar terms})$$

$$\leqslant P_{\theta_0}\left[\frac{\prod_{j=1}^{n} \sup\limits_{\theta \in \mathrm{cl}(S(\theta_1, \rho_{\theta_1}))} f(X_j|\theta)}{\prod_{j=1}^{n} f(X_j|\theta_0)} > h^n\right] + (m \text{ similar terms})$$

$$< \frac{\varepsilon}{m+1} + (m \text{ similar terms}) = \frac{\varepsilon}{m+1}(m+1) = \varepsilon,$$

where we used Boole's Inequality (Theorem 2.1.14) and (8.3.19). Thus, for any $\eta > 0$ and $\varepsilon > 0$, there is an $h(\eta)$, $0 < h < 1$, such that $n > N(\eta, \varepsilon)$ implies

$$P_{\theta_0}\left[\frac{\sup\limits_{\theta \in \Theta - S(\theta_0, \eta)} \prod_{j=1}^{n} f(X_j|\theta)}{\prod_{j=1}^{n} f(X_j|\theta_0)} > h^n\right] < \varepsilon. \tag{8.3.20}$$

Let us now choose $\eta \leqslant \delta$ (where $\delta > 0$ is specified as in the statement of the theorem). Then we can take n so large [say $n > N(\delta, \varepsilon)$] that $n > N(\eta, \varepsilon)$ and $h^n \leqslant \delta$. Then for any $\delta > 0$ and $\varepsilon > 0$, $n > N(\delta, \varepsilon)$ implies

$$P_{\theta_0}\left[\frac{\sup\limits_{\theta \in \Theta - S(\theta_0, \delta)} \prod_{j=1}^{n} f(X_j|\theta)}{\prod_{j=1}^{n} f(X_j|\theta_0)} > \delta\right] < \varepsilon. \tag{8.3.21}$$

From (8.3.21), it follows that [for $n > N(\delta, \varepsilon)$] the sup of $\prod_{j=1}^{n} f(X_j|\theta)$ occurs (w.p. $> 1 - \varepsilon$) within δ of θ_0; that is, for any $\delta > 0$ and $\varepsilon > 0$, $n > N(\delta, \varepsilon)$

implies

$$P_{\theta_0}[|\hat{\theta}_n(X_1, \ldots, X_n) - \theta_0| < \delta] > 1 - \varepsilon, \tag{8.3.22}$$

which proves the theorem.

At this point, it may be helpful to give a brief **heuristic proof** of the consistency of the MLE; hopefully the pieces (and technical details) of the proof of Theorem 8.3.14 will then fall into place. In seeking a maximum-likelihood estimator of θ, under Assumption 8.3.9 we seek to maximize (over θ)

$$L(\theta) = \prod_{i=1}^{n} f(X_i|\theta);$$

this is equivalent to maximizing $1/n^{\text{th}}$ the logarithm of $L(\theta)$, namely

$$\frac{\sum_{i=1}^{n} \ln f(X_i|\theta)}{n},$$

which is of the form $\bar{Y} = (Y_1 + \cdots + Y_n)/n$ with $Y_1, \ldots, Y_n$ independent and identically distributed r.v.'s (where $Y_i = \ln f(X_i|\theta), i = 1, \ldots, n$). Hence, by the *WLLN* (for any θ)

$$\frac{\sum_{i=1}^{n} \ln f(X_i|\theta)}{n} \xrightarrow{p} E_{\theta_0} \ln f(X|\theta),$$

where θ_0 is the true parameter point. But $E_{\theta_0} \ln f(X|\theta)$ is *maximized* by $\theta = \theta_0$ (a precise proof was just Lemma 8.3.12); intuitively, this allows us to distinguish between θ_0 and any other value θ. However, by *continuity*, we expect a little change in θ to result in a little change in $E_{\theta_0} \ln f(X|\theta)$. Hence, since $\left(\sum_{i=1}^{n} \ln f(X_i|\theta)\right)\Big/n$ is "close" to $E_{\theta_0} \ln f(X|\theta)$, its maximum should be "close" to the true θ_0. The student should look over the proof of Theorem 8.3.14 and see that he understands how it embodies in a precise way this intuitive use of *WLLN*, *maximization at* θ_0, and *continuity* to prove consistency of the MLE.

8.4. ASYMPTOTIC NORMALITY; ASYMPTOTIC RELATIVE EFFICIENCY

In Chapter 7 we introduced the problem of point estimation of an unknown parameter θ, about which r.v.'s $X_1, \ldots, X_n$ (say) have something to tell us. In particular, we considered various criteria of "goodness" of an estimator $t_n(X_1, \ldots, X_n)$ of θ [that is, in what sense(s) $t_n(X_1, \ldots, X_n)$ might be "close" to θ]; special note was given to the criteria of *bias* and *consistency* and their variants in Section 8.1, while in Sections 8.2 and 8.3 respectively a method (maximum likelihood) of finding possibly good estimators was introduced and discussed.

Now, however, we have seen (for example, in Problem 7.4) that there may be more than one estimator which is both consistent and unbiased (although, as in Example 8.1.18, there may be none). How, then, do we select one out of these many? We could choose the MLE, but it need not always exist. What then can we do? An answer can be developed as follows.

First, we need a criterion of goodness for comparing two estimators. With an eye toward this need, in Definition 5.1.21 we introduced the notion of **concentration of a r.v. about a point**, namely that t_1 is said to be more concentrated about θ than t_2 if (for all $r > 0$)

$$P[|t_1 - \theta| \leqslant r] \geqslant P[|t_2 - \theta| \leqslant r]. \tag{8.4.1}$$

Now, if (8.4.1) holds for all $r > 0$, we ought [at least if in addition

$$P[|t_1 - \theta| \leqslant r] > P[|t_2 - \theta| \leqslant r]$$

for at least one $r > 0$] to prefer t_1 to t_2 as an estimator of θ, since in a quite reasonable and general sense it is "closer" to θ.

Now, in general, it will *not* be the case that t_1 is more concentrated than t_2, and also it will not be the case that t_2 is more concentrated than t_1 (the relation between $P[|t_1 - \theta| \leqslant r]$ and $P[|t_2 - \theta| \leqslant r]$ will vary depending on the value of r). However, in one special case the relation is simple:

Theorem 8.4.2. *If t_1 and t_2 are (respectively) $N(\theta, \sigma_1^2)$ and $N(\theta, \sigma_2^2)$ r.v.'s, then t_1 is more concentrated about θ than t_2 iff $\sigma_1^2 \leqslant \sigma_2^2$.*

Proof: If t_i is $N(\theta, \sigma_i^2)$ $(i = 1, 2)$, then for any $r > 0$,

$$\begin{aligned} P[|t_i - \theta| \leqslant r] &= P[-r \leqslant t_i - \theta \leqslant r] \\ &= P\left[-\frac{r}{\sigma_i} \leqslant \frac{t_i - \theta}{\sigma_i} \leqslant \frac{r}{\sigma_i}\right] = 2\Phi\left(\frac{r}{\sigma_i}\right) - 1, \end{aligned}$$

where $\Phi(x)$ is (review Example 6.4.7) the area under the $N(0, 1)$ density to the left of x. Hence,

$$\begin{aligned} & P[|t_1 - \theta| \leqslant r] \geqslant P[|t_2 - \theta| \leqslant r] \\ \Leftrightarrow \quad & 2\Phi\left(\frac{r}{\sigma_1}\right) - 1 \geqslant 2\Phi\left(\frac{r}{\sigma_2}\right) - 1 \\ \Leftrightarrow \quad & \Phi\left(\frac{r}{\sigma_1}\right) \geqslant \Phi\left(\frac{r}{\sigma_2}\right) \\ \Leftrightarrow \quad & \frac{r}{\sigma_1} \geqslant \frac{r}{\sigma_2} \\ \Leftrightarrow \quad & \sigma_1 \leqslant \sigma_2 \\ \Leftrightarrow \quad & \sigma_1^2 \leqslant \sigma_2^2. \end{aligned}$$

Thus, if our competing estimators of θ, t_1 and t_2 (say), both had normal distributions with mean θ it would be easy to choose between them. Generally life is not so simple; but $t_1 = t_1(X_1, \ldots, X_n)$ and $t_2 = t_2(X_1, \ldots, X_n)$ are functions of the number of observations n and we have already seen some cases (see the CLT, Theorem 6.4.6) where as $n \to \infty$ a r.v. has a distribution which is approximately normal (and in fact—see the note following the CLT—the distribution "looks more and more like the $N(0, 1)$ distribution" as $n \to \infty$).

We shall now develop precisely this method of comparing estimators, and see how the MLE fits into the scheme of things.

Recall (Definition 6.4.1) that we say a sequence of d.f.'s $F_1, F_2, \ldots$ converges to a d.f. F (written $F_n \to F$) if $F_n(x) \to F(x)$ at each point x at which $F(x)$ is a continuous function. As we noted after the CLT (Theorem 6.4.6), for large n $F_n(x)$ is within δ of $F(x)$ for all x, that is:

Theorem 8.4.3. (Polya's Theorem) *Let $F, F_1, F_2, \ldots$ be d.f.'s. If $F_n \to F$ and F is continuous, then the convergence is uniform; that is,*

$$\lim_{n\to\infty} \sup_x |F_n(x) - F(x)| = 0.$$

Hence, if we have two estimators t_1 and t_2 of θ (say), and can prove that the d.f.'s F_n of t_1 and G_n of t_2 satisfy $F_n \to F_1$ and $G_n \to F_2$ (with F_1 and F_2 each continuous d.f.'s), we will be able to compare t_1 and t_2 by comparing the concentrations of F_1 and F_2 about θ. Reality is slightly more complex, for if $t_1 \xrightarrow{p} \theta$ and $t_2 \xrightarrow{p} \theta$, then $F_1(x) = F_2(x)$ is 0 for $x < \theta$ and 1 if $x \geqslant \theta$. We will find it therefore necessary to "blow up" the difference between t_i and θ in order to obtain the continuous limiting distribution of t_i about θ ($i = 1, 2$). We now make the following definition:

Definition 8.4.4. *If an estimator φ_n is such that $\sqrt{n}(\varphi_n - \theta)$ is asymptotically $N(0, J^2(\theta))$ for some function $J^2(\theta)$, it is called a* **consistent asymptotically normal (CAN)** *estimator.*

In Definition 8.4.4, of course, $\varphi_n = \varphi_n(X_1, \ldots, X_n)$ where $X_1, \ldots, X_n$ are some observed (not necessarily independent or identically distributed) r.v.'s and θ is an unknown parameter with $\theta \in \Theta$ for some set $\Theta \subseteq \mathscr{R}$. By the phraseology "$\sqrt{n}(\varphi_n - \theta)$ is asymptotically $N(0, J^2(\theta))$" we mean that the d.f. F_n of $\sqrt{n}(\varphi_n - \theta)$ satisfies $F_n \to F$ where F is the d.f. of a $N(0, J^2(\theta))$ r.v., that is, of a normal r.v. with mean zero and variance which may depend on θ, $J^2(\theta)$.

Assumption 8.4.5. *Suppose that $X_1, \ldots, X_n$ are independent r.v.'s, each with d.f. $F(x|\theta) = P_\theta[X_1 \leqslant x]$ where $\theta \in \Theta \subseteq \mathscr{R}$, and Θ is an interval (with or without its endpoints). Let $\bar{\Theta}$ be any interval which contains the endpoints of Θ in its interior. Assume $F(x|\theta)$ is either absolutely continuous [in which case let*

$f(x|\theta)$ denote its density at x] or discrete [in which case let $f(x|\theta)$ denote $P_\theta[X_1 = x]$].

Note that Assumption 8.4.5 is essentially the same as Assumption 8.3.9, except that the latter allows θ to be a vector $\boldsymbol{\theta} \in \Theta \subseteq \mathscr{R}^k$ with Θ closed.

Assumption 8.4.6. *Suppose that Assumption* 8.4.5 *is satisfied and that, in addition, the following hold.*

(1) *For all x, the derivatives*

$$\frac{\partial \ln f(x|\theta)}{\partial \theta}, \quad \frac{\partial^2 \ln f(x|\theta)}{\partial \theta^2}, \quad \frac{\partial^3 \ln f(x|\theta)}{\partial \theta^3}$$

exist for each $\theta \in \bar{\Theta}$ [it is sufficient if $P_\theta(E) = 0$, where E is the set of x for which: at least one of these derivatives fails to exist for some value of θ].

(2) *For each $n \geqslant 1$ and all $(x_1, \ldots, x_n)$, the MLE $\hat{\theta}_n(x_1, \ldots, x_n)$ exists and is a root of the likelihood equation* (8.3.8) [it is sufficient if $P_\theta(E) = 0$ where E is the set of values $(x_1, \ldots, x_n)$ of $(X_1, \ldots, X_n)$ for which the MLE either does not exist or is not a root of the likelihood equation].

(3) *For each $\theta \in \bar{\Theta}$ and all x,*

$$\left|\frac{\partial f(x|\theta)}{\partial \theta}\right| < F_1(x), \quad \left|\frac{\partial^2 f(x|\theta)}{\partial \theta^2}\right| < F_2(x), \quad \left|\frac{\partial^3 \ln f(x|\theta)}{\partial \theta^3}\right| < H(x),$$

where $F_1(x)$, $F_2(x)$, $H(x)$ are functions of x such that

$$\int_{-\infty}^{\infty} F_1(x)\,dx < \infty, \quad \int_{-\infty}^{\infty} F_2(x)\,dx < \infty,$$
$$\int_{-\infty}^{\infty} H(x)f(x|\theta)\,dx < M \quad (M \text{ a constant not depending on } \theta)$$

[it is sufficient if $P_\theta(E) = 0$ where E is the set of x for which the condition fails to hold for at least one θ].

(4) *We have, for each $\theta \in \bar{\Theta}$,*

$$I(\theta) \equiv E_\theta\left(\frac{\partial \ln f(X|\theta)}{\partial \theta}\right)^2 \geqslant m > 0$$

(m a constant not depending on θ).

Theorem 8.4.7. *Under Assumption* 8.4.6, *the MLE $\hat{\theta}_n = \hat{\theta}_n(X_1, \ldots, X_n)$ of θ is a CAN estimator with $J^2(\theta) = 1/I(\theta)$ [that is, is such that $\sqrt{n}(\hat{\theta}_n - \theta)$ is asymptotically $N(0, I^{-1}(\theta))$].*

We shall not give the proof of Theorem 8.4.7; for the details (under essentially the same assumptions, which are now standard) we may refer to, for example, H. Cramér, *Mathematical Methods of Statistics*, Princeton University Press, Princeton, New Jersey, 1946 (Tenth Printing, 1963), pp. 500–504. (This proof is simpler than that of the consistency of the MLE given in Section 8.3.) For use below, we note explicitly at this point that

Assumption 8.4.6 involves essentially only the density function $f(x|\theta)$ and its properties.

We will have occasion to compare our MLE $\hat{\theta}_n$ to other estimators φ_n. We noted (at the beginning of this section) that which of two proposed estimators t_1 and t_2 is preferable is usually not clear. Now if an estimator, say t, is a CAN estimator, then (letting $r = r(n)$ and $r^* = \lim_{n\to\infty} rn^{1/2}$)

$$P_\theta[|t - \theta| \leqslant r] = P_\theta\left[\frac{\sqrt{n}|t - \theta|}{J(\theta)} \leqslant \frac{rn^{1/2}}{J(\theta)}\right] \to 2\Phi\left(\frac{r^*}{J(\theta)}\right) - 1. \quad (8.4.8)$$

Hence, our concentration would $\to 1$ as $n \to \infty$ for all fixed r, for any CAN estimator t [for then $rn^{1/2} \to +\infty$ and $\Phi(+\infty) = 1$]; yet if $r = r(n)$ tends to zero at the proper rate, we get a limit which is strictly between 0 and 1. Since this rate is independent of which CAN estimator is being considered, it makes sense to compare CAN estimators by using $r = kn^{-1/2}$ for $k > 0$ fixed. However, by Theorem 8.4.2, the better CAN estimator will be the one with the lower $J^2(\theta)$. Hence, it makes sense to give the following:

Definition 8.4.9. *If t_1 and t_2 are two CAN estimators of θ with respective $J^2(\theta)$'s of $J_1^2(\theta)$ and $J_2^2(\theta)$, t_1 is* **asymptotically more concentrated about θ** *if $J_1^2(\theta) \leqslant J_2^2(\theta)$ for all θ (with $<$ for at least one θ).*

Since by Theorem 8.4.7 we know that (for many densities $f(x|\theta)$) the MLE is a CAN estimator, $\sqrt{n}(\hat{\theta}_n - \theta)$ being asymptotically $N(0, I^{-1}(\theta))$, $\hat{\theta}_n$ will be asymptotically more concentrated about θ than another CAN estimator φ_n [for which, for example, $\sqrt{n}(\varphi_n - \theta)$ is asymptotically $N(0, J^2(\theta))$] iff

$$\frac{1}{I(\theta)} \leqslant J^2(\theta) \quad (8.4.10)$$

for all θ (with $<$ for at least one θ). Now a program of R. A. Fisher was to show that (for many densities $f(x|\theta)$) (8.4.10) holds, that is, the MLE is, for large n, the most concentrated about θ of all CAN estimators. If this were true, then we would have (for all θ)

$$\frac{1/I(\theta)}{J^2(\theta)} \leqslant 1;$$

this ratio might be called a measure of efficiency of $\hat{\theta}_n$ versus φ_n, as in the following:

Definition 8.4.11. *The* **asymptotic relative efficiency** *of two CAN estimators t_1 and t_2 (with respective $J^2(\theta)$'s of $J_1^2(\theta)$ and $J_2^2(\theta)$) is*

$$e_\theta(t_1, t_2) = \frac{J_1^2(\theta)}{J_2^2(\theta)}.$$

The conjecture about the MLE would then state that (for many densities $f(x|\theta)$) $e_\theta(\hat{\theta}_n, \varphi_n) \leqslant 1$ for all θ for any CAN estimator φ_n. We now wish to show that $e_\theta(t_1, t_2)$ is, in a very reasonable sense, a measure of the relative efficiency of t_1 and t_2 (both CAN estimators) for large n when θ is the true value of the unknown parameter.

Lemma 8.4.12. *If t_1 is $N(\theta, J_1^2(\theta)/n_1)$ and t_2 is $N(\theta, J_2^2(\theta)/n_2)$ with $J_1^2(\theta) < J_2^2(\theta)$, the number of observations n_2 which will make t_2's variance as small as the variance $J_1^2(\theta)/n_1$ of t_1 must satisfy*

$$\frac{J_2^2(\theta)}{n_2} = \frac{J_1^2(\theta)}{n_1},$$

or

$$\frac{J_1^2(\theta)}{J_2^2(\theta)} = \frac{n_1}{n_2}.$$

Lemma 8.4.12 is obvious. The lemma itself justifies defining an efficiency for normally distributed unbiased estimators of θ $e_\theta(t_1, t_2) = J_1^2(\theta)/J_2^2(\theta)$ and not, say, $J_1(\theta)/J_2(\theta)$ or some other quantity; for example, if $J_1^2(\theta)/J_2^2(\theta) = \frac{1}{2}$, then $n_2 = 2n_1$ observations are needed by t_2 to achieve the same variance which t_1 achieves with only n_1 observations. The same definition for CAN estimators (given in Definition 8.4.11) is also justified since for n_1 and n_2 large, we can act as though t_1 and t_2 are normally distributed (in which case, by Theorem 8.4.2, variance is *the* measure of concentration about the mean).

Note that in the above work, we refer to the variance of a limiting distribution of some statistic, and not to the limit of the variance; for example, in Definition 8.4.4 $J^2(\theta)$ is the variance of the asymptotic distribution of $\sqrt{n}(\varphi_n - \theta)$ and not the limit as $n \to \infty$ of $\mathrm{Var}_\theta(\sqrt{n}(\varphi_n - \theta))$. The latter limit is (if it exists) always at least as large as $J^2(\theta)$, as we will now prove. First, we will state a useful theorem on convergence of integrals [for a proof see, for example, M. Fisz, *Probability Theory and Mathematical Statistics* (Third Edition), John Wiley & Sons, Inc., New York, 1963, p. 239].

Theorem 8.4.13. (Helly-Bray Theorem) *Let $F, F_1, F_2, \ldots$ be d.f.'s. If $F_n \to F$ and $g(x)$ is a bounded and continuous function of x for all x, then*

$$\lim_{n\to\infty} \int_{-\infty}^{\infty} g(x)\,dF_n(x) = \int_{-\infty}^{\infty} g(x)\,dF(x).$$

Theorem 8.4.14. *Let X_n be a r.v. with d.f. F_n and $EX_n = 0$ $(n = 1, 2, \ldots)$, and suppose $F_n \to F$ where F is a d.f. with $\int_{-\infty}^{\infty} x^2\,dF(x) = \sigma^2 < +\infty$. Then*

$$\lim_{n\to\infty} \inf\{\mathrm{Var}(X_j): j \geqslant n\} \geqslant \sigma^2.$$

Proof: For any finite $a \leqslant b$, it follows from the Helly-Bray Theorem

(Theorem 8.4.13) that

$$\int_a^b x^2\,dF_n(x) \to \int_a^b x^2\,dF(x) = L(a, b) \quad \text{(say)}.$$

Now for any fixed $\varepsilon > 0$ we can take $a = a(\varepsilon)$ so small (that is, large negative) and $b = b(\varepsilon)$ so large that $\sigma^2 - \varepsilon \leqslant L(a, b) \leqslant \sigma^2$, so that (for any $\varepsilon > 0$)

$$\begin{aligned}\lim_{n\to\infty} \inf\{\operatorname{Var}(X_j): j \geqslant n\} &= \lim_{n\to\infty} \inf\{\textstyle\int_{-\infty}^{\infty} x^2\,dF_j(x): j \geqslant n\}\\ &\geqslant \lim_{n\to\infty} \inf\{\textstyle\int_a^b x^2\,dF_j(x): j \geqslant n\} = \lim_{n\to\infty} \int_a^b x^2\,dF_n(x) \geqslant \sigma^2 - \varepsilon;\end{aligned}$$

the theorem follows.

Corollary 8.4.15. *Let X_n be a r.v. with d.f. F_n and $EX_n = 0$ ($n = 1, 2, \ldots$), and suppose $F_n \to \Phi_{\sigma^2}$ where Φ_{σ^2} is the $N(0, \sigma^2)$ d.f. Then*

$$\lim_{n\to\infty} \inf\{\operatorname{Var}(X_j): j \geqslant n\} \geqslant \sigma^2.$$

Remark 8.4.16. Theorem 8.4.7 continues to be true for $\theta = \boldsymbol{\theta}$, a k-dimensional unknown parameter ($k \geqslant 2$), under similar conditions. The asymptotic distribution of $\hat{\boldsymbol{\theta}}_n$ (suitably standardized) is then the multivariate normal of Definition 4.5.3 with appropriate $\boldsymbol{\mu}$ and $\boldsymbol{\Sigma}$. [These considerations are contained in many standard books, and we omit them: it is, in general, in this book not our purpose to write an encyclopedia, but to cover the univariate in depth (especially with regard to why results are true, what value they have, and so on) and hint at the multivariate so that a student then needing the (usually messy but straightforward) multivariate analogs of the univariate can easily grasp them and will understand their value.] Similarly, the discrete case is entirely analogous, with integrals being replaced by summations. These two remarks hold in most cases below, but we shall no longer usually even mention them.

Having discussed asymptotic relative efficiency for CAN estimators, let us now return to the conjecture given following (8.4.10), which R. A. Fisher wished to demonstrate, that the MLE is (for large n) the most concentrated about θ of all CAN estimators. The following example, due to J. L. Hodges, shows the conjecture to be false; this is done by exhibiting an alternative estimator.

Example 8.4.17. (J. L. Hodges' Superefficienciation) Suppose that the MLE $\hat{\theta}_n$ of θ based on $X_1, \ldots, X_n$ is a CAN estimator, so that $\sqrt{n}(\hat{\theta}_n - \theta)$ is asymptotically $N(0, J^2(\theta))$ for some function $J^2(\theta)$ [for instance, in Example 8.2.8 $X_1, \ldots, X_n$ were independent $N(\theta, 1)$ r.v.'s with θ unknown and we had $\hat{\theta}_n = \bar{X} = \sum_{i=1}^{n} X_i/n$ and $J^2(\theta) = 1$; in Example 8.2.12 $X_1, \ldots, X_n$

were independent $N(\theta, \sigma^2)$ r.v.'s with θ unknown and σ^2 known and we had $\hat{\theta}_n = \bar{X}$ and $J^2(\theta) = \sigma^2$]. Let $\alpha \neq 0$ be any constant. Define a new estimator

$$t_n = \begin{cases} \alpha\hat{\theta}_n, & \text{if } |\hat{\theta}_n| < \dfrac{1}{\sqrt[4]{n}} \\ \hat{\theta}_n, & \text{otherwise.} \end{cases} \tag{8.4.18}$$

We now wish to determine the asymptotic distribution of $\sqrt{n}(t_n - \theta)$. First,

$$\begin{aligned} P_\theta[\sqrt{n}(t_n - \theta) \leqslant x] &= P_\theta\left[\sqrt{n}(t_n - \theta) \leqslant x \,\middle|\, |\hat{\theta}_n| < \frac{1}{\sqrt[4]{n}}\right] P_\theta\left[|\hat{\theta}_n| < \frac{1}{\sqrt[4]{n}}\right] \\ &\quad + P_\theta\left[\sqrt{n}(t_n - \theta) \leqslant x \,\middle|\, |\hat{\theta}_n| \geqslant \frac{1}{\sqrt[4]{n}}\right] P_\theta\left[|\hat{\theta}_n| \geqslant \frac{1}{\sqrt[4]{n}}\right]. \end{aligned} \tag{8.4.19}$$

Now (since $\sqrt{n}(\hat{\theta}_n - \theta)$ is asymptotically $N(0, J^2(\theta))$)

$$\begin{aligned} P_\theta\left[|\hat{\theta}_n| < \frac{1}{\sqrt[4]{n}}\right] &= P_\theta[-n^{-1/4} < \hat{\theta}_n < n^{-1/4}] \\ &= P_\theta[-n^{1/4} - \sqrt{n}\theta < \sqrt{n}(\hat{\theta}_n - \theta) < n^{1/4} - \sqrt{n}\theta] \\ &\to \begin{cases} 1, & \text{if } \theta = 0 \\ 0, & \text{if } \theta \neq 0, \end{cases} \end{aligned} \tag{8.4.20}$$

since $-n^{1/4} - \sqrt{n}\theta \to -\infty(+\infty)$ for $\theta \geqslant 0$ $(\theta < 0)$ and $n^{1/4} - \sqrt{n}\theta \to -\infty(+\infty)$ for $\theta > 0$ $(\theta \leqslant 0)$. From use of (8.4.20) in (8.4.19), we find [if the limits indicated do not exist, the result means that the difference of the two sides $\to 0$; we shall show below that the limits indicated do exist]

$$\begin{aligned} P_\theta[\sqrt{n}(t_n - \theta) \leqslant x] &\to \begin{cases} \lim\limits_{n\to\infty} P_\theta[\sqrt{n}(t_n - \theta) \leqslant x \,|\, |\hat{\theta}_n| < n^{-1/4}], & \text{if } \theta = 0 \\ \lim\limits_{n\to\infty} P_\theta[\sqrt{n}(t_n - \theta) \leqslant x \,|\, |\hat{\theta}_n| \geqslant n^{-1/4}], & \text{if } \theta \neq 0 \end{cases} \\ &= \begin{cases} \lim\limits_{n\to\infty} P_\theta[\sqrt{n}(\alpha\hat{\theta}_n - \theta) \leqslant x \,|\, |\hat{\theta}_n| < n^{-1/4}], & \text{if } \theta = 0 \\ \lim\limits_{n\to\infty} P_\theta[\sqrt{n}(\hat{\theta}_n - \theta) \leqslant x \,|\, |\hat{\theta}_n| \geqslant n^{-1/4}], & \text{if } \theta \neq 0. \end{cases} \end{aligned} \tag{8.4.21}$$

Now (see Problem 2.27) if A_n, B_n $(n \geqslant 1)$ are events in some sample space Ω, $\lim_{n\to\infty} P(B_n) = 1$ and $\lim_{n\to\infty} P(A_n)$ exists, then $\lim_{n\to\infty} P(A_n|B_n) = \lim_{n\to\infty} P(A_n)$. However, by (8.4.20) the conditioning events in (8.4.21) are (for their specific possible θ's) each B_n's such that $\lim_{n\to\infty} P(B_n) = 1$ and since the events $\sqrt{n}(\alpha\hat{\theta}_n - \theta) \leqslant x$ (for $\theta = 0$) and $\sqrt{n}(\hat{\theta}_n - \theta) \leqslant x$ (for $\theta \neq 0$) are A_n's such that $\lim_{n\to\infty} P(A_n)$ exists [this is shown below],

$$P_\theta[\sqrt{n}(t_n - \theta) \leqslant x] \to \begin{cases} \lim_{n\to\infty} P_\theta[\sqrt{n}(\alpha\hat{\theta}_n - \theta) \leqslant x], & \text{if } \theta = 0 \\ \lim_{n\to\infty} P_\theta[\sqrt{n}(\hat{\theta}_n - \theta) \leqslant x], & \text{if } \theta \neq 0 \end{cases}$$
$$= \begin{cases} \lim_{n\to\infty} P_\theta[\sqrt{n}\alpha\hat{\theta}_n \leqslant x], & \text{if } \theta = 0 \\ \lim_{n\to\infty} P_\theta[\sqrt{n}(\hat{\theta}_n - \theta) \leqslant x], & \text{if } \theta \neq 0. \end{cases} \tag{8.4.22}$$

Hence, $\sqrt{n}(t_n - \theta)$ is asymptotically $N(0, J_2^2(\theta))$ where

$$J_2^2(\theta) = \begin{cases} \alpha^2 J^2(\theta), & \text{if } \theta = 0 \\ J^2(\theta), & \text{if } \theta \neq 0. \end{cases} \tag{8.4.23}$$

Thus, for any choice α with $|\alpha| < 1$ (and $\alpha \neq 0$), the alternate estimator t_n is a CAN estimator which is at least as efficient as the MLE $\hat{\theta}_n$ and (at one point θ, $\theta = 0$) is more efficient.

We called the method of Example 8.4.17 Hodges' "Superefficienciation" Method since points θ where $e_\theta(\hat{\theta}_n, t_n) > 1$ are called points of **superefficiency** of the estimator t_n relative to the MLE $\hat{\theta}_n$ (when, as required by Definition 8.4.11, both are CAN estimators), and Hodges' method allows the construction of such superefficient CAN estimators t_n as alternatives to *any* MLE $\hat{\theta}_n$ which is a CAN estimator.

Note that in Example 8.5.34 we compare two CAN estimators in detail.

So, R. A. Fisher's idea [that maybe the MLE is (for large n) the most concentrated about θ of all CAN estimators] seems doomed by Hodges' example. More recent research has tried to save the idea of MLE supremacy via two tracks, which Wolfowitz[16] calls respectively the "deprecation of the evil" and the "exclusion of the evil" approaches. In the first approach ["deprecation of the evil"], Bahadur and Le Cam have shown that the (evil) points θ at which we have superefficiency cannot be a large set [in mathematical language, it is a set of measure zero]. In the second approach ["exclusion of the evil"], various authors have stated conditions *on the estimators we are allowed to compare the MLE to* which are sufficient to eliminate estimators which have points of superefficiency from the class of competitors to the MLE. In the paragraph following Theorem 8.4.7 we noted that the assumptions needed there [for an MLE to be a CAN] involved essentially only the

[16] See p. 269 of J. Wolfowitz, "Asymptotic efficiency of the maximum likelihood estimator," *Teorija Verojatnosteĭ i ee Primenenija*, Vol. 10 (1965), pp. 267–281.

density $f(x|\theta)$. This is a valid approach, for (among the class of densities which meet the assumptions) we can make a useful statement about the asymptotic character of the MLE. Wolfowitz (loc. cit.) makes it clear that this is a different matter than excluding competitors, since any estimator can be "best": simply declare all better estimators ineligible to compete.

The problem of providing estimators with good asymptotic concentration received a major step toward solution when Weiss and Wolfowitz[17] showed that, among all estimators which satisfy a condition U [similar to being CAN as in Definition 8.4.4, but much more general and defensible as a prerequisite to an estimator being considered at all] an estimator Z (which is called a "generalized maximum-likelihood estimator" since in simple cases it is the MLE) has the highest asymptotic concentration. Details of this important work are beyond the scope of this book. A number of papers utilizing and generalizing these ideas to prove striking results may be found in the recent literature.[18]

8.5. QUANTILES

If some random variable X of interest to us has an unknown distribution function $F(x)$, we are sometimes interested (for a given p) in estimating the (a) point ξ such that $P[X \leqslant \xi] = p$. (For example, if $p = .95$ and X is the level of methyl mercury in a certain type of fish, then estimating the point ξ such that $P[X \leqslant \xi] = .95$ means estimating the level of contamination which is exceeded by 5 percent of such fish.) Such a ξ is called a *quantile*, and in this section we wish to study its estimation. [In Section 8.1 we introduced the quantile corresponding to $p = .5$ (the median in Definition 8.1.27) and studied its estimation by the sample median (Definition 8.1.28 and Theorem 8.1.30).] We will begin by defining the population quantiles, and their usual estimators the sample quantiles.

Definition 8.5.1. *For any* p $(0 < p < 1)$, *the* (*a*) **quantile of order p** *of the*

---

[17] L. Weiss and J. Wolfowitz, "Generalized maximum likelihood estimators," *Teorija Verojatnosteĭ i ee Primenenija*, Vol. 11 (1966), pp. 68–93.

[18] For example, see L. Weiss and J. Wolfowitz, "Maximum probability estimators," *Annals of the Institute of Statistical Mathematics*, Vol. 19 (1967), pp. 193–206; L. Weiss and J. Wolfowitz, "Maximum probability estimators with a general loss function," *Probability and Information Theory*, edited by M. Behara, K. Krickeberg, and J. Wolfowitz, Springer-Verlag, Berlin, 1969, pp. 232–256; and L. Weiss and J. Wolfowitz, "Asymptotically efficient non-parametric estimators of location and scale parameters," *Zeitschrift für Wahrscheinlichkeitstheorie und verwandte Gebiete*, Vol. 16 (1970), pp. 134–150.

d.f. $F(x)$ is any root of the equation

$$F(\xi) = p;$$

we will often call such a root ξ_p, or simply ξ, and will usually consider only cases where it is unique.

Definition 8.5.2. *For n r.v.'s $X_1, \ldots, X_n$ let $Y_1, \ldots, Y_n$ denote the ordered X_i's (that is, the X_i's ordered so that $Y_1 \leqslant \cdots \leqslant Y_n$; if several X_i's are equal, we can choose one at random for a place Y_i). The* **sample quantile of order p** *$(0 < p < 1)$ is*

$$Z_p = Y_{[np]+1}$$

(where $[np]$ denotes the largest integer $<np$).

Note that (taking $p = .5$ in Definition 8.5.2) this definition takes, as the sample quantile of order .5, $Y_{(n+1)/2}$ if n is odd, and $Y_{n/2}$ if n is even. This differs slightly (when n is even) from Definition 8.1.28 of the sample median; the difference is of no consequence when n is "large." (See Problem 8.15H.)

We will now show that Z_p is a CAN estimator of ξ_p, that is (see Definition 8.4.4), $\sqrt{n}(Z_p - \xi_p)$ is asymptotically $N(0, J^2)$. It will be slightly simpler to show (which is an equivalent statement) that $\sqrt{n}(Z_p - \xi_p)/J$ is asymptotically $N(0, 1)$. (The CAN nature of Z_p will allow us to compare it readily with other estimators.) We shall give two proofs of this result; the first is traditional and messy, while the second is simpler (perhaps a bit more sophisticated) and extends to more general situations with ease.

Theorem 8.5.3. *Fix p $(0 < p < 1)$. Let $X_1, \ldots, X_n$ be independent r.v.'s with d.f. $F(x)$ and density $f(x)$. Suppose that $f(x)$ is continuous and positive at $x = \xi_p$. Then the following three (equivalent) results hold:*

(1) *Z_p is a CAN estimator of ξ_p with $J^2 = pq/f^2(\xi_p)$;*

(2) *$\sqrt{\dfrac{n}{pq}}\, f(\xi_p)(Z_p - \xi_p)$ is asymptotically $N(0, 1)$;*

(3) *For all $y_1 < y_2$,*

$$\lim_{n \to \infty} P\left[y_1 < \sqrt{\frac{n}{pq}}\, f(\xi_p)(Z_p - \xi_p) < y_2\right] = \frac{1}{\sqrt{2\pi}} \int_{y_1}^{y_2} e^{-y^2/2}\, dy.$$

Proof 1: First, let us call $[np] + 1$ by the name k. Then (see Problem 8.26) the density of $Z_p = Y_{[np]+1} = Y_k$ is

$$f_{Z_p}(x) = \frac{n!}{(k-1)!(n-k)!} [F(x)]^{k-1}[1 - F(x)]^{n-k} f(x). \qquad (8.5.4)$$

Now

$$P\left[\sqrt{\frac{n}{pq}}\,f(\xi_p)(Z_p - \xi_p) \leqslant y\right] = P\left[Z_p \leqslant \frac{y}{f(\xi_p)}\sqrt{\frac{pq}{n}} + \xi_p\right],$$

where the right-hand side is the d.f. of Z_p, so by differentiation and use of (8.5.4) we find the density function

$$\begin{aligned}
&g_{\sqrt{(n/pq)}f(\xi_p)(Z_p-\xi_p)}(y)\\
&\quad= \frac{1}{f(\xi_p)}\sqrt{\frac{pq}{n}}\,f_{Z_p}\left(\xi_p + \sqrt{\frac{pq}{n}}\frac{y}{f(\xi_p)}\right)\\
&\quad= \sqrt{\frac{pq}{n}}\,\frac{n!}{(k-1)!(n-k)!}\,p^{k-1}q^{n-k}\cdot\frac{f\left(\xi_p + \sqrt{\frac{pq}{n}}\frac{y}{f(\xi_p)}\right)}{f(\xi_p)}\\
&\qquad\cdot\left[\frac{F\left(\xi_p + \sqrt{\frac{pq}{n}}\frac{y}{f(\xi_p)}\right)}{p}\right]^{k-1}\left[\frac{1 - F\left(\xi_p + \sqrt{\frac{pq}{n}}\frac{y}{f(\xi_p)}\right)}{q}\right]^{n-k} \qquad (8.5.5)\\
&\quad= A_{n1}A_{n2}A_{n3},
\end{aligned}$$

where (by definition)

$$\begin{cases}
A_{n1} = \sqrt{\dfrac{pq}{n}}\,\dfrac{n!}{(k-1)!(n-k+1)!}\,p^{k-1}q^{n-k+1}\,\dfrac{n-k+1}{q},\\[2ex]
A_{n2} = f\left(\xi_p + \sqrt{\dfrac{pq}{n}}\dfrac{y}{f(\xi_p)}\right)\Big/ f(\xi_p),\\[2ex]
A_{n3} = \left[\dfrac{F\left(\xi_p + \sqrt{\frac{pq}{n}}\frac{y}{f(\xi_p)}\right)}{p}\right]^{k-1}\left[\dfrac{1 - F\left(\xi_p + \sqrt{\frac{pq}{n}}\frac{y}{f(\xi_p)}\right)}{q}\right]^{n-k}.
\end{cases} \qquad (8.5.6)$$

We will now find

$$\lim_{n\to\infty} g_{\sqrt{(n/pq)}f(\xi_p)(Z_p-\xi_p)}(y) = \lim_{n\to\infty} A_{n1}A_{n2}A_{n3} = (\lim_{n\to\infty} A_{n1})(\lim_{n\to\infty} A_{n2})(\lim_{n\to\infty} A_{n3}).$$

First, by Stirling's formula [(2.2.16)], $n! \sim \sqrt{2\pi n}(n/e)^n$ where "$\sim$" means the ratio of the two sides $\to 1$ as $n \to \infty$. Hence,

$$\begin{aligned}
&\frac{n!}{(k-1)!(n-k+1)!}\\
&\quad\sim \frac{n^n\sqrt{2\pi n}}{\sqrt{2\pi(k-1)}\sqrt{2\pi(n-k+1)}\,(k-1)^{k-1}(n-k+1)^{n-k+1}}.
\end{aligned} \qquad (8.5.7)$$

Now recalling that $k = [np] + 1$, hence that $np - 1 \leqslant k \leqslant np + 1$, we find that $\lim_{n\to\infty} k/n = p$, and thus $k - 1 \sim np$ and $n - k + 1 \sim n - np$.

Substituting these last two relations in (8.5.7) and simplifying, we find

$$\begin{aligned}\frac{n!}{(k-1)!(n-k+1)!} &\sim \frac{1}{\sqrt{2\pi pqn}}\cdot\frac{n^n}{(np)^{np}(n-np)^{n-np}}\\ &= \frac{1}{\sqrt{2\pi pqn}}\cdot\frac{1}{p^{np}q^{n-np}}\cdot\frac{n^n}{n^{np}n^{n-np}}\\ &= \frac{1}{\sqrt{2\pi pqn}}\cdot\frac{1}{p^{k-1}q^{n-k+1}}.\end{aligned} \tag{8.5.8}$$

Using (8.5.8) in the definition of A_{n1} [(8.5.6)], we find

$$A_{n1} \sim \sqrt{\frac{pq}{n}}\cdot\frac{1}{\sqrt{2\pi pqn}}\cdot\frac{n-k+1}{1-p} = \frac{1}{\sqrt{2\pi}}\cdot\frac{n-k+1}{n}\cdot\frac{1}{1-p}. \tag{8.5.9}$$

Since $\lim_{n\to\infty}\frac{n-k+1}{n} = 1 - p$, it follows from (8.5.9) that

$$\lim_{n\to\infty} A_{n1} = \frac{1}{\sqrt{2\pi}}. \tag{8.5.10}$$

Second, since (by assumption) $f(x)$ is continuous and positive at $x = \xi_p$, it follows directly that

$$\lim_{n\to\infty} A_{n2} = \lim_{n\to\infty}\frac{f\left(\xi_p + \sqrt{\frac{pq}{n}}\frac{y}{f(\xi_p)}\right)}{f(\xi_p)} = \lim_{h\to 0}\frac{f(\xi_p + h)}{f(\xi_p)} = 1. \tag{8.5.11}$$

Third, if a function $h(x)$ is k-times differentiable, then by Taylor's Theorem with Remainder,

$$h(x) = h(0) + \frac{h^{(1)}(0)}{1!}x + \frac{h^{(2)}(0)}{2!}x^2 + \cdots + \frac{h^{(k-1)}(0)}{(k-1)!}x^{k-1} + \frac{h^{(k)}(\xi)}{k!}x^k,$$

with $0 < \xi < x$ (or, if $x < 0$, $x < \xi < 0$); if $h^{(k)}(\cdot)$ is continuous, the last term is

$$\frac{h^{(k)}(0)}{k!}x^k + o(x^k).$$

(This was previously used in the proof of Theorem 5.4.13.) Taking as our function $h(x) = F(x)$ and expanding about ξ_p rather than about zero, we find

$$\begin{aligned}F(x) &= F(\xi_p) + \frac{F^{(1)}(\xi_p+\eta)}{1!}(x-\xi_p) \text{ with } \begin{cases}0 < \eta < x, & \text{if } x > 0\\ x < \eta < 0, & \text{if } x < 0\end{cases}\\ &= F(\xi_p) + F^{(1)}(\xi_p)(x-\xi_p) + o(x-\xi_p)\\ &= p + f(\xi_p)(x-\xi_p) + o(x-\xi_p).\end{aligned} \tag{8.5.12}$$

Hence,

$$F\left(\xi_p + \frac{y}{f(\xi_p)}\sqrt{\frac{pq}{n}}\right) = p + \sqrt{\frac{pq}{n}}\,y + o\left(\sqrt{\frac{pq}{n}}\frac{y}{f(\xi_p)}\right) = p + y\sqrt{\frac{pq}{n}}\{1 + Q_n(y)\}, \tag{8.5.13}$$

where

$$\lim_{n\to\infty} Q_n(y) = 0.$$

Thus, using (8.5.13) in (8.5.6),

$$A_{n3} = \left[1 + \frac{y\sqrt{\frac{pq}{n}}\{1 + Q_n(y)\}}{p}\right]^{k-1}\left[1 - \frac{y\sqrt{\frac{pq}{n}}\{1 + Q_n(y)\}}{q}\right]^{n-k} = \left(1 + \frac{cy}{\sqrt{n}}\right)^{k-1}\left(1 - \frac{dy}{\sqrt{n}}\right)^{n-k} \tag{8.5.14}$$

where

$$\begin{cases} c = \sqrt{\frac{q}{p}}\{1 + Q_n(y)\}, \\ d = \sqrt{\frac{p}{q}}\{1 + Q_n(y)\}. \end{cases}$$

Thus,

$$\ln A_{n3} = (k - 1)\ln\left(1 + \frac{cy}{\sqrt{n}}\right) + (n - k)\ln\left(1 - \frac{dy}{\sqrt{n}}\right). \tag{8.5.15}$$

However, using Taylor's Theorem for expanding $\ln(1 + x)$ about 1 we know that

$$\ln(1 + x) = \ln(1) + \frac{\ln^{(1)}(1)}{1!}x + \frac{\ln^{(2)}(1)}{2!}x^2 + o(x^2)$$
$$= 0 + x - \tfrac{1}{2}x^2 + o(x^2) = x - \tfrac{1}{2}x^2 + o(x^2).$$

Hence,

$$\ln\left(1 + \frac{sy}{\sqrt{n}}\right) = \frac{sy}{\sqrt{n}} - \frac{1}{2}\frac{s^2}{n}y + o\left(\frac{a^2y^2}{n}\right). \tag{8.5.16}$$

Using (8.5.16) in (8.5.15), we find

$$\begin{aligned} \ln A_{n3} &= (k - 1)\left[\frac{c}{\sqrt{n}}y - \frac{c^2}{2n}y^2 + o\left(\frac{1}{n}\right)\right] \\ &\quad - (n - k)\left[-\frac{d}{\sqrt{n}}y - \frac{d^2}{2n}y^2 + o\left(\frac{1}{n}\right)\right] \\ &= \frac{(k - 1)c - (n - k)d}{\sqrt{n}}y \\ &\quad - \frac{(k - 1)c^2 + (n - k)d^2}{n}\frac{y^2}{2} + no\left(\frac{1}{n}\right). \end{aligned} \tag{8.5.17}$$

But $\lim_{n\to\infty} no(1/n) = 0$ and (see Problem 8.16H)

$$\begin{cases} \lim_{n\to\infty} \dfrac{(k-1)c - (n-k)d}{\sqrt{n}} = 0, \\ \lim_{n\to\infty} \dfrac{(k-1)c^2 + (n-k)d^2}{n} = 1, \end{cases}$$

so

$$\lim_{n\to\infty} \ln A_{n3} = -\frac{y^2}{2} \tag{8.5.18}$$

or (see Problem 8.29)

$$\lim_{n\to\infty} A_{n3} = e^{-y^2/2}. \tag{8.5.19}$$

Now from (8.5.5) and {(8.5.10), (8.5.11), (8.5.19)} it follows that

$$\lim_{n\to\infty} g_{\sqrt{(n/pq)}f(\xi_p)(Z_p-\xi_p)}(y) = \frac{1}{\sqrt{2\pi}} e^{-y^2/2}. \tag{8.5.20}$$

Then

$$\begin{aligned} \lim_{n\to\infty} P\left[y_1 < \sqrt{\frac{n}{pq}} f(\xi_p)(Z_p - \xi_p) < y_2\right] \\ &= \lim_{n\to\infty} \int_{y_1}^{y_2} g_{\sqrt{(n/pq)}f(\xi_p)(Z_p-\xi_p)}(y)\,dy \\ &= \int_{y_1}^{y_2} \left\{\lim_{n\to\infty} g_{\sqrt{(n/pq)}f(\xi_p)(Z_p-\xi_p)}(y)\right\} dy \\ &= \frac{1}{\sqrt{2\pi}} \int_{y_1}^{y_2} e^{-y^2/2}\,dy, \end{aligned} \tag{8.5.21}$$

which proves the theorem. [For a justification of the interchange of the limit and the integral in (8.5.21) see Problem 8.17H.]

Next we shall give, as promised, a simpler and more sophisticated proof of Theorem 8.5.3. This proof extends with ease to more complex and non-standard situations (see Problem 8.19H).

Proof 2: For any y let $N_n(y)$ denote the number of $X_1, \ldots, X_n$ which are $> y$. Then, since

$$\begin{aligned} \{Z_p \leqslant y\} &\text{ iff } \{Y_{[np]+1} \leqslant y\} \\ &\text{ iff } \{\text{At most } Y_{[np]+2}, \ldots, Y_n \text{ exceed } y\} \\ &\text{ iff } \{N_n(y) \leqslant n - ([np] + 2) + 1\}, \end{aligned} \tag{8.5.22}$$

we have the relation

$$P[Z_p \leqslant y] = P[N_n(y) \leqslant n - [np] - 1]. \tag{8.5.23}$$

But $N_n(y)$ is a binomial r.v. based on n trials, each with probability $1 - F(y)$

of success. Thus (see Theorem 5.2.2),

$$\begin{cases} EN_n(y) = n(1 - F(y)), \\ \operatorname{Var}(N_n(y)) = nF(y)(1 - F(y)), \end{cases} \tag{8.5.24}$$

and

$$\begin{aligned} P[Z_p \leqslant y] &= P[N_n(y) \leqslant n - [np] - 1] \\ &= P\left[\frac{N_n(y) - n(1 - F(y))}{\sqrt{nF(y)(1 - F(y))}} \leqslant \frac{(n - [np] - 1) - n(1 - F(y))}{\sqrt{nF(y)(1 - F(y))}}\right] \\ &= P\left[N_n^*(y) \leqslant \frac{nF(y) - [np] - 1}{\sqrt{nF(y)(1 - F(y))}}\right], \end{aligned} \tag{8.5.25}$$

where $N_n^*(y)$ is the standardized $N_n(y)$ (see Theorem 5.1.16); hence, $EN_n^*(y) = 0$ and $\operatorname{Var}(N_n^*(y)) = 1$. Now the quantity of interest to us is

$$\begin{aligned} P\left[\sqrt{\frac{n}{pq}} f(\xi_p)(Z_p - \xi_p) \leqslant t\right] &= P\left[Z_p \leqslant \xi_p + \sqrt{\frac{pq}{n}} \frac{t}{f(\xi_p)}\right] \\ &= P\left[N_n^*(y_p) \leqslant \frac{nF(y_p) - [np] - 1}{\sqrt{nF(y_p)(1 - F(y_p))}}\right], \end{aligned} \tag{8.5.26}$$

where we made use of Equation (8.5.25) and defined

$$y_p = \xi_p + \sqrt{\frac{pq}{n}} \frac{t}{f(\xi_p)}.$$

But by (8.5.12) [which resulted from our use of Taylor's Theorem in Proof 1] we know $F(x) = p + f(\xi_p)(x - \xi_p) + o(x - \xi_p)$, so substituting we find

$$\begin{aligned} F(y_p) &= F\left(\xi_p + \sqrt{\frac{pq}{n}} \frac{t}{f(\xi_p)}\right) \\ &= p + f(\xi_p)\left(\left(\xi_p + \sqrt{\frac{pq}{n}} \frac{t}{f(\xi_p)}\right) - \xi_p\right) \\ &\qquad + o\left(\left(\xi_p + \sqrt{\frac{pq}{n}} \frac{t}{f(\xi_p)}\right) - \xi_p\right) \\ &= p + \sqrt{\frac{pq}{n}}\, t + o\left(\sqrt{\frac{pq}{n}} \frac{t}{f(\xi_p)}\right) \\ &= p + \sqrt{\frac{pq}{n}}\, t + o\left(\frac{1}{\sqrt{n}}\right). \end{aligned} \tag{8.5.27}$$

Substituting (8.5.27) into (8.5.26), we find

$$P\left[\sqrt{\frac{n}{pq}} f(\xi_p)(Z_p - \xi_p) \leqslant t\right]$$
$$= P\,[\,N_n^*(y_p)$$

$$\leqslant \frac{n\left(p + \sqrt{\frac{pq}{n}}\,t + o\left(\frac{1}{\sqrt{n}}\right)\right) - [np] - 1}{\sqrt{n}\sqrt{\left(p + \sqrt{\frac{pq}{n}}\,t + o\left(\frac{1}{\sqrt{n}}\right)\right)\left(q - \sqrt{\frac{pq}{n}}\,t + o\left(\frac{1}{\sqrt{n}}\right)\right)}}\Bigg] \tag{8.5.28}$$

$$= P[N_n^*(y_p) \leqslant a_n t + b_n],$$

where we have defined

$$a_n = 1\Bigg/\sqrt{\left(p + \sqrt{\frac{pq}{n}}\,t + o\left(\frac{1}{\sqrt{n}}\right)\right)\left(q - \sqrt{\frac{pq}{n}}\,t + o\left(\frac{1}{\sqrt{n}}\right)\right)},$$

$$b_n = \frac{\sqrt{np} + \sqrt{n}o\left(\frac{1}{\sqrt{n}}\right) - \frac{[np]}{\sqrt{n}} - \frac{1}{\sqrt{n}}}{\sqrt{\left(p + \sqrt{\frac{pq}{n}}\,t + o\left(\frac{1}{\sqrt{n}}\right)\right)\left(q - \sqrt{\frac{pq}{n}}\,t + o\left(\frac{1}{\sqrt{n}}\right)\right)}}. \tag{8.5.29}$$

Since $a_n \to 1$ and $b_n \to 0$ as $n \to \infty$, it follows (see Problem 8.30) that

$$\begin{aligned} \lim_{n\to\infty} P\left[\sqrt{\frac{n}{pq}}\, f(\xi_p)(Z_p - \xi_p) \leqslant t\right] &= \lim_{n\to\infty} P[N_n^*(y_p) \leqslant a_n t + b_n] \\ &= \lim_{n\to\infty} P[N_n^*(y_p) \leqslant t]. \end{aligned} \tag{8.5.30}$$

Now by the CLT (Theorem 6.4.6), $N_n^*(y)$ is asymptotically $N(0, 1)$, that is,

$$\lim_{n\to\infty} P[N_n^*(y) \leqslant t] = \frac{1}{\sqrt{2\pi}} \int_{-\infty}^{t} e^{-z^2/2}\, dz. \tag{8.5.31}$$

Hence, we expect to find (see Problem 8.31)

$$\lim_{n\to\infty} P[N_n^*(y_p) \leqslant t] = \frac{1}{\sqrt{2\pi}} \int_{-\infty}^{t} e^{-z^2/2}\, dz, \tag{8.5.32}$$

which [see (8.5.30)] proves the theorem since then

$$\lim_{n\to\infty} P\left[y_1 < \sqrt{\frac{n}{pq}}\, f(\xi_p)(Z_p - \xi_p) < y_2\right] = \frac{1}{\sqrt{2\pi}} \int_{y_1}^{y_2} e^{-z^2/2}\, dz. \tag{8.5.33}$$

Example 8.5.34. In Section 8.1 (see especially Theorem 8.1.30) we found that if $X_1, \ldots, X_n$ are independent r.v.'s, each with some d.f. $F(x)$, then $\bar{X}$ (the sample mean) is an unbiased and consistent estimator of EX_1 (the population mean), while M_n (the sample median) is a consistent estimator of v (the population median). If $F(x)$ is the $N(\mu, \sigma^2)$ d.f., then $EX_1 = \mu = v$; that is, $\bar{X}$ and M_n are estimating *the same* quantity. How do these estimators compare?

Both are consistent and unbiased estimators. (The proof that M_n is unbiased is Problem 8.32.) Now $\bar{X}$ is a CAN estimator of μ (see Problem 8.33), and M_n is a CAN estimator of μ (see Theorem 8.5.3 and Problem 8.15H), so (by Lemma 8.4.12) it makes sense to compare $\bar{X}$ and M_n via asymptotic relative efficiency (Definition 8.4.11). Now

$$\begin{cases} \sqrt{n}(\bar{X} - \mu) \text{ is asymptotically } N(0, \sigma^2) \\ \sqrt{n}(M_n - \mu) \text{ is asymptotically } N\left(0, \dfrac{pq}{f^2(\xi_p)}\right) \end{cases}$$

so

$$e_\mu(\bar{X}, M_n) = \frac{\sigma^2}{\dfrac{pq}{f^2(\xi_p)}} = \frac{\sigma^2}{\dfrac{(\frac{1}{2})(\frac{1}{2})}{f^2(\mu)}}$$

$$= \frac{4\sigma^2}{\left(\dfrac{1}{\sqrt{2\pi}\sigma} e^{-.5(\mu-\mu)^2/\sigma^2}\right)^{-2}} = \frac{4\sigma^2}{2\pi\sigma^2} = \frac{2}{\pi} < 1.$$

Since $2/\pi = .64$, $\bar{X}$ can (in large samples) estimate μ as well with $.64n$ observations as M_n can with n observations [assuming we have independent $N(\mu, \sigma^2)$ observations]. Estimation of μ by M_n nevertheless *is* performed in some cases, for the following reason. If one or several of $X_1, \ldots, X_n$ may contain gross errors (for example, in practice this occurs when gauges are misread, when decimal points are misplaced, when computers malfunction, and so on) $\bar{X}$ may be affected to a much larger degree (since it averages the gross observations with all the others) than is M_n (which depends only on which are the one or two middle observations and what are their values).

Another reason for use of M_n instead of $\bar{X}$ is that often we want to estimate not the mean EX_1, but rather $\xi_{.5}$ (the population median). In general, $\xi_{.5} \neq EX_1$, so M_n is an appropriate estimator and $\bar{X}$ is not. Theorem 8.5.3 tells us that under very general conditions M_n is a CAN estimator of $\xi_{.5}$.

Example 8.5.35. For the 88 observations of seasonal snowfall (say $X_1, X_2, \ldots, X_{88}$) given in Problem 8.6, we find $\bar{X} = 79.7$ and $M_n = M_{88} = (75.8 + 76.4)/2 = 76.1$.

8.6. CRAMÉR-RAO INEQUALITY

In Section 8.4 we discussed asymptotic normality and asymptotic efficiency in some detail, and saw that the MLE was conjectured to be the "best" CAN; that conjecture was false, so two ways of "salvaging" the MLE as best were considered (which Wolfowitz calls "deprecation of the evil" and "exclusion of the evil," respectively)

The Cramér-Rao inequality gives a lower bound on the variance of an unbiased estimator, and essentially "excludes evil." Nevertheless this is an important result. The fact that an unbiased estimator has a variance equal to this bound means it has the best (smallest) variance of a large class of estimators (not *all* estimators—and perhaps some important estimators have been "excluded"—but *many* estimators are beaten). Relations with the MLE are covered in Section 8.7.

We will assume that Assumption 8.4.5 holds, that is, $X_1, \ldots, X_n$ are independent r.v.'s, each with d.f. $F(x|\theta) = P_\theta[X_1 \leqslant x]$ where $\theta \in \Theta \subseteq \mathscr{R}$, Θ is an interval, and $f(x|\theta)$ denotes either a density [if $F(x|\theta)$ is absolutely continuous] or a probability function [if $F(x|\theta)$ is discrete]. Let $\varphi_n(X_1, \ldots, X_n)$ be any estimator of θ, and write

$$E_\theta \varphi_n(X_1, \ldots, X_n) = \theta + b_n(\theta), \tag{8.6.1}$$

so that $b_n(\theta)$ is (see Definition 7.2.3) the bias of φ_n.

Note that those points $(x_1, \ldots, x_n)$ such that $\varphi_n(x_1, \ldots, x_n) = c$ form a subset of $\mathscr{R}^n$, called a **hypersurface**, and as we vary c we obtain a **family of hypersurfaces**. It will sometimes be the case that

Assumption 8.6.2. *Assumption* 8.4.5 *holds and, in addition, the following hold.*

(1) *We are able to choose* $n - 1$ *functions* $Y_1 = Y_1(X_1, \ldots, X_n), \ldots, Y_{n-1} = Y_{n-1}(X_1, \ldots, X_n)$ [called **local coordinates**] *such that each point in* $\mathscr{R}^n$ *is uniquely determined by: (1) the value of* $\varphi_n(X_1, \ldots, X_n)$ *on the hypersurface to which it belongs and* (2) *the* $n - 1$ *"local coordinates"* $Y_1, \ldots, Y_{n-1}$ [which determine the position of the point on the hypersurface].

(2) $\varphi_n(X_1, \ldots, X_n)$ *and each of* $Y_1(X_1, \ldots, X_n), \ldots, Y_{n-1}(X_1, \ldots, X_n)$ *are continuous functions of* $X_1, \ldots, X_n$.

(3) *The transformation* $(X_1, \ldots, X_n) \to (\varphi_n, Y_1, \ldots, Y_{n-1})$ *is one-to-one* [that is, for any set of specified values of $\varphi_n, Y_1, \ldots, Y_{n-1}$ there is at most *one* set of values of $X_1, \ldots, X_n$ which yields the specified values of $\varphi_n, Y_1, \ldots, Y_{n-1}$].

(4) [This requirement is needed only if $F(x|\theta)$ is absolutely continuous.] *The partial derivatives*

$$\frac{\partial \varphi_n(x_1, \ldots, x_n)}{\partial x_i} \quad \textit{and} \quad \frac{\partial y_j(x_1, \ldots, x_n)}{\partial x_i} \quad (i = 1, \ldots, n; j = 1, \ldots, n - 1)$$

exist and are continuous at all $(x_1, \ldots, x_n)$. [It is permissible for this condition to fail to hold for some of the $(x_1, \ldots, x_n)$'s on a finite number of hypersurfaces $\varphi_n(x_1, \ldots, x_n) = c$.]

Lemma 8.6.3. *If Assumption* 8.6.2 *holds, then* [in the case $F(x|\theta)$ is absolutely continuous] *the density function of* $(\varphi_n, Y_1, \ldots, Y_{n-1})$ *is*

$$f(x_1|\theta)\cdots f(x_n|\theta)\frac{1}{\left|J\begin{pmatrix}\varphi_n, y_1, \ldots, y_{n-1}\\ x_1, x_2, \ldots, x_n\end{pmatrix}\right|} \tag{8.6.4}$$
$$= g(x|\theta)h(y_1, \ldots, y_{n-1}|x, \theta),$$

where $g(x|\theta)$ *denotes the density function of* $\varphi_n(X_1, \ldots, X_n)$ *and* $h(y_1, \ldots, y_{n-1}|x, \theta)$ *denotes the density function of the conditional d.f. of* $Y_1, \ldots, Y_{n-1}$ *given that* $\varphi_n(X_1, \ldots, X_n) = x$. *In the discrete case* [that is, $F(x|\theta)$ discrete], (8.6.4) *holds if* J *is omitted and* $g(x|\theta)$ *denotes* $P_\theta[\varphi_n = x]$ *while* $h(y_1, \ldots, y_{n-1}|x, \theta)$ *denotes* $P_\theta[Y_1 = y_1, \ldots, Y_{n-1} = y_{n-1}|\varphi_n = x]$. [Below we shall use the appropriate interpretation and not separate the absolutely continuous and discrete cases.]

Proof: This lemma essentially follows from use of Theorem 4.6.5 (actually from a slight generalization of Theorem 4.6.5). (Here J is the Jacobian introduced in Section 4.6. Conditional d.f.'s and densities were discussed in Section 4.4.) The discrete case follows easily from Theorem 3.6.2.

With some additional assumptions (see Assumption 8.6.5) we will be able to define precisely the problem ("regular" estimation) to be considered. When we do (Definition 8.6.6), note that the "regularity" conditions have restricted φ_n's as well as $F(x|\theta)$'s. ("Exclusion of the evil.")

Assumption 8.6.5. *Suppose Assumption* 8.6.2 *holds. In addition (using the notation of Lemma* 8.6.3), *suppose the following. For all values of* x, φ_n, $y_1, \ldots, y_{n-1}$ *the partial derivatives*

$$\frac{\partial f(x|\theta)}{\partial\theta}, \quad \frac{\partial g(\varphi_n|\theta)}{\partial\theta}, \quad \frac{\partial h(y_1, \ldots, y_{n-1}|\varphi_n, \theta)}{\partial\theta}$$

exist for every $\theta \in \Theta$ [it is sufficient if $P_\theta(E) \equiv 0$, where E is the set of $(x, \varphi_n, y_1, \ldots, y_{n-1})$ for which: at least one of these derivatives fails to exist for some value of θ], *and can be bounded as follows:*

$$\left|\frac{\partial f(x|\theta)}{\partial\theta}\right| < F_0(x), \quad \left|\frac{\partial g(\varphi_n|\theta)}{\partial\theta}\right| < G_0(\varphi_n),$$
$$\left|\frac{\partial h(y_1, \ldots, y_{n-1}|\varphi_n, \theta)}{\partial\theta}\right| < H_0(y_1, \ldots, y_{n-1}, \varphi_n)$$

where

$$\int_{-\infty}^{\infty} F_0(x)\,dx, \quad \int_{-\infty}^{\infty} G_0(x)\,dx, \quad \int_{-\infty}^{\infty} uG_0(u)\,du,$$

and

$$\int_{-\infty}^{\infty}\cdots\int_{-\infty}^{\infty} H_0(y_1, \ldots, y_{n-1}, u)\,dy_1\cdots dy_{n-1}$$

are finite. [For $F(x|\theta)$ discrete the integrals are replaced by summations.]

Definition 8.6.6. *If Assumption* 8.6.5 *holds, we shall say that we are dealing with a* **regular estimation case** (*which is either of the absolutely continuous type, or of the discrete type*) *and will call* φ_n *a* **regular estimator of θ**.

We are now almost ready to state and prove the Cramér-Rao Inequality. We shall do this after stating (without proof) a lemma. [The proof, which uses Schwarz' Inequality (Theorem 5.3.15), can be found on pp. 475–477 of Cramér (1946).]

Lemma 8.6.7. *Suppose that* $g(x|\theta)$ *is a density function, where* $\theta \in \Theta$ (Θ *an interval in* $\mathscr{R}$), $E_\theta X = \psi(\theta)$ *and* $E_\theta X^2$ *both exist if* X *has density function* $g(x|\theta)$, $\partial g(x|\theta)/\partial\theta$ *exists for each* $\theta \in \Theta$ [this can be weakened as in Assumption 8.6.5], *and for some* $G_0(x)$ (*such that* $\int_{-\infty}^{\infty} G_0(x)\,dx$ *and* $\int_{-\infty}^{\infty} xG_0(x)\,dx$ *are finite*) $\left|\dfrac{\partial g(x|\theta)}{\partial\theta}\right| < G_0(x)$. *Then* $d\psi(\theta)/d\theta$ *exists for all* $\theta \in \Theta$ *and*

$$E_\theta(X - \theta)^2 \cdot E\left(\frac{\partial \ln g(X|\theta)}{\partial\theta}\right)^2 \geqslant \left(\frac{d\psi(\theta)}{\partial\theta}\right)^2. \tag{8.6.8}$$

Equality holds in (8.6.8), *for a specified* θ, *iff*

$$\frac{\partial \ln g(x|\theta)}{\partial\theta} = k(\theta)(x - \theta) \tag{8.6.9}$$

for some $k(\theta)$ *and for all* x *where* $g(x|\theta) > 0$. [It is permissible for (8.6.9) to fail for a set E of x's with $P_\theta(E) = 0$.]

Theorem 8.6.10. **(Cramér-Rao Inequality)** *In any regular estimation case,*

$$E_\theta(\varphi_n - \theta)^2 \geqslant \frac{(1 + b_n'(\theta))^2}{nE_\theta\left(\dfrac{\partial \ln f(X|\theta)}{\partial\theta}\right)^2}. \tag{8.6.11}$$

Equality for all $\theta \in \Theta$ *holds iff* (*whenever* $g(x|\theta) > 0$) (1) $h(y_1, \ldots, y_{n-1}|x, \theta)$ *is independent of* θ *and* (2)

$$\frac{\partial \ln g(x|\theta)}{\partial\theta} = k(\theta)(x - \theta).$$

Proof: [We shall prove only the case where $F(x|\theta)$ is absolutely continuous. A similar proof, replacing integrals by summations, yields the result when $F(x|\theta)$ is discrete.] First, by the assumptions made on $f(x|\theta)$ and $h(y_1, \ldots, y_{n-1}|u, \theta)$ it follows (see Cramér (1946), p. 67 for a proof) that we may interchange integration and differentiation (with respect to θ) of $f(x|\theta)$ and of $h(y_1, \ldots, y_{n-1}|u, \theta)$. Then (recalling that $f(x|\theta)$ and $h(y_1, \ldots, y_{n-1}|u, \theta)$ are density functions)

$$\int_{-\infty}^{\infty} f(x|\theta)\,dx = \int_{-\infty}^{\infty} \cdots \int_{-\infty}^{\infty} h(y_1, \ldots, y_{n-1}|u, \theta)\,dy_1 \cdots dy_{n-1} = 1,$$

and

$$\frac{\partial}{\partial\theta}\int_{-\infty}^{\infty} f(x|\theta)\,dx = \frac{\partial}{\partial\theta}\int_{-\infty}^{\infty}\cdots\int_{-\infty}^{\infty} h(y_1,\ldots,y_{n-1}|u,\theta)\,dy_1\cdots dy_{n-1}$$
$$= \frac{\partial}{\partial\theta}(1) = 0.$$

Interchanging the differentiation and integration, we obtain

$$\int_{-\infty}^{\infty}\frac{\partial f(x|\theta)}{\partial\theta}\,dx = \int_{-\infty}^{\infty}\cdots\int_{-\infty}^{\infty}\frac{\partial h(y_1,\ldots,y_{n-1}|u,\theta)}{\partial\theta}\,dy_1\cdots dy_{n-1} = 0.$$

Finally, since for any positive function $w(\theta)$ we have the equation

$$\frac{dw(\theta)}{d\theta} = \frac{d\ln w(\theta)}{d\theta}\,w(\theta),$$

our result becomes

$$\int_{-\infty}^{\infty}\frac{\partial\ln f(x|\theta)}{\partial\theta}f(x|\theta)\,dx = \int_{-\infty}^{\infty}\cdots\int_{-\infty}^{\infty}\frac{\partial\ln h(y_1,\ldots,y_{n-1}|u,\theta)}{\partial\theta}$$
$$\cdot\, h(y_1,\ldots,y_{n-1}|u,\theta)\,dy_1\cdots dy_{n-1} \equiv 0. \tag{8.6.12}$$

This will be used below.

Now, by Lemma 8.6.3, $f(x_1|\theta)\cdots f(x_n|\theta)|J|^{-1} = g(u|\theta)h(y_1,\ldots,y_{n-1}|u,\theta)$. Taking the logarithm of this relation and differentiating it with respect to θ we find

$$\sum_{i=1}^{n}\frac{\partial\ln f(x_i|\theta)}{\partial\theta} = \frac{\partial\ln g(u|\theta)}{\partial\theta} + \frac{\partial\ln h(y_1,\ldots,y_{n-1}|u,\theta)}{\partial\theta}. \tag{8.6.13}$$

If we square each side of (8.6.13), multiply by $f(x_1|\theta)\cdots f(x_n|\theta)\,dx_1\cdots dx_n = g(u|\theta)h(y_1,\ldots,y_{n-1}|u,\theta)\,du\,dy_1\cdots dy_{n-1}$, and integrate, we obtain

$$\int_{-\infty}^{\infty}\cdots\int_{-\infty}^{\infty}\left\{\sum_{i=1}^{n}\left(\frac{\partial\ln f(x_i|\theta)}{\partial\theta}\right)^2 + \sum_{\substack{i,j=1\\ i\neq j}}^{n}\left(\frac{\partial\ln f(x_i|\theta)}{\partial\theta}\right)\left(\frac{\partial\ln f(x_j|\theta)}{\partial\theta}\right)\right\} f(x_1|\theta)\cdots f(x_n|\theta)\,dx_1\cdots dx_n$$
$$= \int_{-\infty}^{\infty}\int_{-\infty}^{\infty}\cdots\int_{-\infty}^{\infty}\left\{\left(\frac{\partial\ln g(u|\theta)}{\partial\theta}\right)^2 + \left(\frac{\partial\ln h(y_1,\ldots,y_{n-1}|u,\theta)}{\partial\theta}\right)^2 + \left(\frac{\partial\ln g(u|\theta)}{\partial\theta}\right)\left(\frac{\partial\ln h(y_1,\ldots,y_{n-1}|u,\theta)}{\partial\theta}\right)\right\}$$
$$\cdot\, g(u|\theta)h(y_1,\ldots,y_{n-1}|u,\theta)\,du\,dy_1\cdots dy_{n-1}$$

or (integrating out densities for which no variable occurs, and using (8.6.12) to show the cross-product integrals are zero)

$$
\begin{aligned}
n \int_{-\infty}^{\infty} & \left(\frac{\partial \ln f(x|\theta)}{\partial \theta}\right)^2 f(x|\theta)\, dx \\
&= \int_{-\infty}^{\infty} \left(\frac{\partial \ln g(u|\theta)}{\partial \theta}\right)^2 g(u|\theta)\, du \\
&\quad + \int_{-\infty}^{\infty} \int_{-\infty}^{\infty} \cdots \int_{-\infty}^{\infty} g(u|\theta) \left(\frac{\partial \ln h(y_1, \ldots, y_{n-1}|u, \theta)}{\partial \theta}\right)^2 \qquad (8.6.14) \\
&\qquad \cdot h(y_1, \ldots, y_{n-1}|u, \theta)\, dy_1 \cdots dy_{n-1}\, du \\
&\geqslant \int_{-\infty}^{\infty} \left(\frac{\partial \ln g(u|\theta)}{\partial \theta}\right)^2 g(u|\theta)\, d\theta,
\end{aligned}
$$

where the last inequality follows upon dropping a nonnegative term. Note that (8.6.14) will be an equality iff (whenever $g(u|\theta) > 0$) $h(y_1, \ldots, y_{n-1}|u, \theta)$ is independent of θ (so that $\partial h(y_1, \ldots, y_{n-1}|u, \theta)/\partial\theta = 0$).

Now since the density function $g(u|\theta)$ satisfies (by assumption) the conditions of Lemma 8.6.7, and since $\psi(\theta) = \theta + b_n(\theta)$ for which

$$\frac{d}{d\theta}\psi(\theta) = 1 + b_n'(\theta),$$

we obtain from Equation (8.6.8)

$$E_\theta(\varphi_n - \theta)^2 \geqslant \frac{(1 + b_n'(\theta))^2}{\int_{-\infty}^{\infty} \left(\frac{\partial \ln g(u|\theta)}{\partial \theta}\right)^2 g(u|\theta)\, du}, \qquad (8.6.15)$$

which has equality iff (8.6.9) holds. Using (8.6.14) in (8.6.15) we find

$$E_\theta(\varphi_n - \theta)^2 \geqslant \frac{(1 + b_n'(\theta))^2}{n \int_{-\infty}^{\infty} \left(\frac{\partial \ln f(x|\theta)}{\partial \theta}\right)^2 f(x|\theta)\, dx} = \frac{(1 + b_n'(\theta))^2}{n E_\theta \left(\frac{\partial \ln f(X|\theta)}{\partial \theta}\right)^2}. \qquad (8.6.16)$$

This proves the theorem, since equality can hold in (8.6.16) iff both (8.6.15) and (8.6.14) are equalities.

The result of the following lemma is sometimes of value in evaluating the Cramér-Rao bound.

Lemma 8.6.17.

$$E_\theta \left(\frac{\partial \ln f(X|\theta)}{\partial \theta}\right)^2 = E_\theta \left(\frac{f'(X|\theta)}{f(X|\theta)}\right)^2.$$

Proof: We have (in the absolutely continuous case, the discrete case being analogous with summations replacing integrals)

$$E_\theta\left(\frac{\partial \ln f(X|\theta)}{\partial\theta}\right)^2 = \int_{-\infty}^{\infty}\left(\frac{\partial \ln f(x|\theta)}{\partial\theta}\right)^2 f(x|\theta)\,dx$$
$$= \int_{-\infty}^{\infty}\left(\frac{1}{f(x|\theta)}\frac{\partial f(x|\theta)}{\partial\theta}\right)^2 f(x|\theta)\,dx$$
$$= \int_{-\infty}^{\infty}\left(\frac{f'(x|\theta)}{f(x|\theta)}\right)^2 f(x|\theta)\,dx = E_\theta\left(\frac{f'(X|\theta)}{f(X|\theta)}\right)^2.$$

[As near the beginning of the proof of the Cramér-Rao Inequality, we used the relation

$$\frac{\partial \ln f(x|\theta)}{\partial\theta} = \frac{1}{f(x|\theta)}\frac{\partial f(x|\theta)}{\partial\theta}.\Bigg]$$

Note that the Cramér-Rao Inequality remains valid if $X_1, \ldots, X_n$ are each multivariate r.v.'s (instead of univariate r.v.'s as above), the proof being essentially unchanged. The Cramér-Rao Inequality is also attributed (with some merit) to Fréchet and others; we have used the name by which it appears to be most widely known. It is of most use when it furnishes a bound which does not depend on the estimator $\varphi_n(X_1, \ldots, X_n)$ . . . the bound of (8.6.11) depends on $\varphi_n(X_1, \ldots, X_n)$ through the term $b_n'(\theta)$. But $b_n'(\theta)$ is zero for unbiased estimators. Hence, we obtain the following.

Corollary 8.6.18. (Cramér-Rao Inequality for unbiased estimators) *In any regular estimation case, if $\varphi_n(X_1, \ldots, X_n)$ is unbiased then*

$$\operatorname{Var}_\theta(\varphi_n) \geqslant \frac{1}{nE_\theta\left(\dfrac{\partial \ln f(X|\theta)}{\partial\theta}\right)^2}. \tag{8.6.19}$$

In form (8.6.19) the result is often simply called the Cramér-Rao Inequality. It furnishes a lower bound (which depends only on the density function or probability function $f(x|\theta)$] on the variance of any unbiased estimators for the problem. [Of course, we must have a regular case, which restricts the estimators φ_n and $f(x|\theta)$.] If an estimator achieves this lower bound, it is the best of the unbiased regular estimators (in the sense of having a small variance).

Example 8.6.20. Suppose that $X_1, \ldots, X_n$ are $n(\geqslant 2)$ independent r.v.'s, each $N(\mu, \sigma^2)$ where μ is known and $\sigma^2(\sigma^2 > 0)$ is unknown. Suppose that this is a regular estimation case, and find the lower bound (8.6.19) on the variance of any regular unbiased estimator φ_n of σ^2.

Here we have

$$
\begin{aligned}
E_{\sigma^2}\left(\frac{\partial \ln f(X|\sigma^2)}{\partial \sigma^2}\right)^2 &= E_{\sigma^2}\left(\frac{\partial}{\partial \sigma^2}\ln\left\{\frac{1}{\sqrt{2\pi}\sigma}e^{-\frac{1}{2}\left(\frac{X-\mu}{\sigma}\right)^2}\right\}\right)^2 \\
&= E_{\sigma^2}\left(\frac{\partial}{\partial \sigma^2}\left\{-.5\ln(2\pi\sigma^2) - \frac{(X-\mu)^2}{2\sigma^2}\right\}\right)^2 \\
&= E_{\sigma^2}\left(-\frac{1}{2\sigma^2} + \frac{(X-\mu)^2}{2\sigma^4}\right)^2 \\
&= E_{\sigma^2}\left(\frac{1}{4\sigma^4} - \frac{(X-\mu)^2}{2\sigma^6} + \frac{(X-\mu)^4}{4\sigma^8}\right) \\
&= \frac{1}{4\sigma^4} - \frac{1}{2\sigma^4}E(Y^2) + \frac{1}{4\sigma^4}EY^4 \\
&= \frac{1}{4\sigma^4} - \frac{1}{2\sigma^4}\cdot(1) + \frac{1}{4\sigma^4}\cdot(3) \\
&= \frac{1}{2\sigma^4},
\end{aligned}
$$

where Y is a $N(0, 1)$ r.v. (since $(X - \mu)/\sigma$ is $N(0, 1)$) and EY^4 was evaluated using the gamma function (see (4.1.6)). Thus, for any regular unbiased estimator φ_n,

$$
\operatorname{Var}_{\sigma^2}(\varphi_n) \geqslant \frac{2\sigma^4}{n}. \tag{8.6.21}
$$

Recall that the MLE of σ^2 was (see Example 8.2.13)

$$
\hat{\sigma}^2 = \frac{\sum_{i=1}^{n}(X_i - \mu)^2}{n} \tag{8.6.22}
$$

and was unbiased. We may easily show that $\operatorname{Var}_{\sigma^2}(\hat{\sigma}^2) = 2\sigma^4/n$, so we have an estimator with the minimal variance.

Definition 8.6.23. *In any regular estimation case, if $\varphi_n(X_1, \ldots, X_n)$ is unbiased and* (8.6.19) *is an equality, then $\varphi_n(X_1, \ldots, X_n)$ will be called a* **best regular unbiased estimator (BRUE)**.

Thus, in Example 8.6.20 we have (aside from showing that we have a regular estimation case) shown that $\hat{\sigma}^2$ is a BRUE of σ^2 [for independent $N(\mu, \sigma^2)$ observations with μ known].

8.7. EFFICIENCY

In Section 8.6 we derived (Corollary 8.6.18) a lower bound on the variance of certain unbiased estimators. Thus, it makes sense to evaluate unbiased estimators by comparing their variances to this bound. Below we do this, and also study when this bound is achieved.

Definition 8.7.1. *In any regular estimation case the* **efficiency** *of an unbiased regular estimator* $\varphi_n(X_1, \ldots, X_n)$ *is*

$$e_\theta(\varphi_n) = \frac{1\bigg/\left\{nE_\theta\left(\dfrac{\partial \ln f(X|\theta)}{\partial\theta}\right)^2\right\}}{\operatorname{Var}_\theta(\varphi_n)}. \tag{8.7.2}$$

If $e_\theta(\varphi_n) \equiv 1$, *then* φ_n *is called* **efficient**.

Corollary 8.7.3. *In any regular estimation case,* $0 \leqslant e_\theta(\varphi_n) \leqslant 1$. *We have* $e_\theta(\varphi_n) \equiv 1$ *iff* $\operatorname{Var}_\theta(\varphi_n)$ *achieves the lower bound for all* θ.

Definition 8.7.4. *In any regular estimation case the* **asymptotic efficiency** *of an unbiased regular estimator* $\varphi_n(X_1, \ldots, X_n)$ *is* $\lim_{n\to\infty} e_\theta(\varphi_n)$.

Note that efficiency and asymptotic efficiency (defined in this section) should not be confused with the asymptotic relative efficiency (of two CAN estimators) introduced in Definition 8.4.11. Also bear in mind that (1) it may be that *no* estimator achieves the lower bound for all θ, and that (2) there may be nonregular estimators which beat the bound.

Having considered evaluating unbiased estimators by comparing their variances to the lower bound, we now ask: "When is the lower bound achieved?" The answer is a simple corollary to Theorem 8.6.10 and Corollary 8.6.18.

Corollary 8.7.5. *In a regular estimation case, a regular unbiased estimator* φ_n *is efficient iff* [*whenever* $g(x|\theta) > 0$] *(1)* $h(y_1, \ldots, y_{n-1}|x, \theta)$ *is independent of* θ *and (2)*

$$\frac{\partial \ln g(x|\theta)}{\partial\theta} = k(\theta)(x - \theta).$$

Proof: See Theorem 8.6.10 and Corollary 8.6.18.

Let us consider condition (1) of Corollary 8.7.5 further. This states that (whenever $g(x|\theta) > 0$) we should have $h(y_1, \ldots, y_{n-1}|x, \theta)$ independent of θ; that is, given that $\varphi_n = x$, the conditional distribution of the $n - 1$ "local coordinates" $Y_1, \ldots, Y_{n-1}$ should be independent of θ. *We claim*: **This means we can then say the estimator $\varphi_n(X_1, \ldots, X_n)$ summarizes all the information relative to θ which is contained in the sample** $X_1, \ldots, X_n$; for, if we know $\varphi_n = x$, then the sample point $(x_1, \ldots, x_n)$ must lie on the hypersurface $\varphi_n = x$, and the conditional distribution on this hypersurface does not depend on θ (so that further specification of the position of $(x_1, \ldots, x_n)$

gives no new information about θ). In light of this it seems reasonable to call $\varphi_n(X_1, \ldots, X_n)$ a *sufficient statistic* (that is, one which has all the information which $X_1, \ldots, X_n$ can provide about θ).

Definition 8.7.6. *In a regular estimation case, a regular estimator φ_n is a* **sufficient statistic** *iff $h(y_1, \ldots, y_{n-1}|x, \theta)$ is independent of θ whenever $g(x|\theta) > 0$.*

Theorem 8.7.7. *In a regular estimation case, a regular estimator φ_n is a sufficient statistic iff*

$$f(x_1|\theta) \cdots f(x_n|\theta) = g(x|\theta)H(x_1, \ldots, x_n) \tag{8.7.8}$$

for all $x, x_1, \ldots, x_n, \theta$ (for some function $H(x_1, \ldots, x_n)$).

Proof: Since we have a regular estimation case and a regular estimator φ_n, by (8.6.4)

$$f(x_1|\theta) \cdots f(x_n|\theta)|J|^{-1} = g(x|\theta)h(y_1, \ldots, y_{n-1}|x, \theta),$$

where $|J|^{-1}$ is not a function of θ. Thus,

$$f(x_1|\theta) \cdots f(x_n|\theta) = g(x|\theta)\frac{h(y_1, \ldots, y_{n-1}|x, \theta)}{|J|^{-1}},$$

which is of form (8.7.8) iff $h(y_1, \ldots, y_{n-1}|x, \theta)$ is independent of θ. Using Definition 8.7.6, the theorem follows.

The concept of *sufficiency*, which arose in our study of regular cases of estimation, is valuable in general. It will be studied in general (regular or nonregular) estimation cases in Section 8.8. Finally, as promised at the beginning of Section 8.6, we study relations with the MLE. (This will be followed by an example.)

Theorem 8.7.9. **(1)** *In a regular estimation case, if a regular unbiased estimator φ_n is efficient, then the likelihood equation will have a unique solution equal to φ_n.*

(2) *In a regular estimation case, if a regular estimator φ_n is sufficient, then any solution of the likelihood equation will be a function of φ_n.*

Proof: **(1)** Since by assumption φ_n is efficient, by Corollary 8.7.5 the conditions for equality in the Cramér-Rao Inequality are satisfied. However, as Equation (8.6.13) in the proof of the Cramér-Rao Inequality we found

$$\sum_{i=1}^{n} \frac{\partial \ln f(x_i|\theta)}{\partial \theta} = \frac{\partial \ln g(u|\theta)}{\partial \theta} + \frac{\partial \ln h(y_1, \ldots, y_{n-1}|u, \theta)}{\partial \theta}. \tag{8.7.10}$$

The "equality" conditions yield

$$\frac{\partial \ln g(u|\theta)}{\partial\theta} = k(\theta)(u - \theta), \quad \frac{\partial \ln h(y_1, \ldots, y_{n-1}|u, \theta)}{\partial\theta} = 0. \quad (8.7.11)$$

Since

$$\frac{\partial}{\partial\theta} \ln \{f(x_1|\theta) \cdots f(x_n|\theta)\} = \sum_{i=1}^{n} \frac{\partial \ln f(x_i|\theta)}{\partial\theta},$$

using (8.7.11) in (8.7.10) we find

$$\frac{\partial}{\partial\theta} \ln \{f(x_1|\theta) \cdots f(x_n|\theta)\} = k(\theta)(u - \theta). \quad (8.7.12)$$

Hence (see (8.3.12)) the likelihood equation $L'(\theta) = 0$ is $k(\theta)(\varphi_n - \theta) = 0$, for which $\hat{\theta} = \varphi_n(X_1, \ldots, X_n)$ is the unique solution.

(2) Since by assumption φ_n is sufficient, by Theorem 8.7.7 the likelihood equation $L'(\theta) = 0$ is of the form

$$\frac{\partial g(x|\theta)}{\partial\theta} H(x_1, \ldots, x_n) = 0, \quad (8.7.13)$$

any solution of which depends on $x_1, \ldots, x_n$ only through φ_n; that is, any solution is a function of φ_n alone.

Note that Theorem 8.7.9 says that (1) any efficient estimator φ_n can be found as a solution of the likelihood equation, and that (2) any solution of the likelihood equation is based only on any sufficient statistic. There is a relation between *efficiency* (Definition 8.7.1) and *asymptotic efficiency* (Definition 8.7.4), on the one hand, and *asymptotic relative efficiency* of two CAN estimators (Definition 8.4.11) on the other hand, via the MLE. Namely, under conditions (we have already discussed the nature of such "regularity" conditions extensively) the MLE is a CAN estimator and is asymptotically efficient; hence, if in Definition 8.4.11 we compare with the MLE, we are looking at the same quantity as in Definition 8.7.4 [as long as φ_n is a CAN estimator]. For some further results and references, see, for example, M. Woodroofe, "Maximum likelihood estimation of a translation parameter of a truncated distribution," *Annals of Mathematical Statistics*, Vol. 43 (1972), pp. 113–122.

Example 8.7.14. Suppose that $X_1, \ldots, X_n$ are $n(\geqslant 1)$ independent r.v.'s, each Poisson with parameter $\lambda > 0$ unknown. Suppose that this is a regular estimation case, and find the lower bound (8.6.19) on the variance of any regular unbiased estimator φ_n of λ.

Here we have

$$E_\lambda\left(\frac{\partial \ln f(X|\lambda)}{\partial\lambda}\right)^2 = E_\lambda\left(\frac{\partial}{\partial\lambda}\ln\left\{\frac{e^{-\lambda}\lambda^X}{X!}\right\}\right)^2$$
$$= E_\lambda\left(\frac{\partial}{\partial\lambda}\{-\lambda + X\ln\lambda - \ln(X!)\}\right)^2$$
$$= E_\lambda\left(\frac{X}{\lambda} - 1\right)^2 = \frac{E_\lambda(X-\lambda)^2}{\lambda^2}$$
$$= \frac{1}{\lambda^2}\operatorname{Var}_\lambda(X) = \frac{\lambda}{\lambda^2} = \frac{1}{\lambda},$$

where we used the result of Theorem 5.2.3 that $\operatorname{Var}_\lambda(X) = \lambda$. Thus, for any regular unbiased estimator φ_n,

$$\operatorname{Var}_\lambda(\varphi_n) \geqslant \frac{\lambda}{n}. \tag{8.7.15}$$

Recall that (if Θ was slightly altered; see Problem 8.5) the MLE of λ was $\bar{X}$, which is unbiased. We may easily show that $\operatorname{Var}_\lambda(\bar{X}) = \lambda/n$, so we have an estimator with the minimal variance.

Details of the extensions of the results in Sections 8.6 and 8.7 to the case of several unknown parameters (which were referred to following the proof of Lemma 8.6.17) may be found in, for example, Cramér (1946), pp. 490–496.

8.8. SUFFICIENCY

In Section 8.7 (see Definition 8.7.6 and the note following Theorem 8.7.7) we dealt with the concept of a "sufficient statistic" in a special context, that of regular estimation cases. The concept is of importance not just in those cases, however, and can be defined in general. The idea is if we observe a r.v. X whose distribution depends on an unknown parameter θ [for example, often $X = (X_1, \ldots, X_n)$ where $X_1, \ldots, X_n$ are independent and identically distributed r.v.'s, each having density function $f(x|\theta)$ where θ is unknown, $\theta \in \Theta$), often X can be reduced via a function without losing any information about θ (for example, $t(X) = t(X_1, \ldots, X_n) = (X_1 + \cdots + X_n)/n$, the sample mean, may in some cases contain all relevant information about θ]; in that case $t(X)$ is called a sufficient statistic. [Note that $t(X)$ may also be a random vector; for example, the order statistics.] We can then simply base our inference about θ on $t(X)$, which can be considerably simpler than X (for example, one number $\bar{X}$ versus n numbers $X_1, \ldots, X_n$). The formal definition is as follows.

Definition 8.8.1. *A statistic $t(X_1, \ldots, X_n)$ is* **sufficient (for θ)** *iff*

$$P_\theta[(X_1, \ldots, X_n) \in \mathscr{U} | t(X_1, \ldots, X_n) = c] = l(\mathscr{U}|c) \tag{8.8.2}$$

for all θ and all $\mathscr{U} \subseteq \mathscr{R}^n$. [Here θ and c may be vectors.]

This definition of a "sufficient" statistic means that $t(X_1, \ldots, X_n)$ is a "sufficient" statistic if, once you are told the value of $t(\;)$ at $(X_1, \ldots, X_n)$, a more exact specification of where $(X_1, \ldots, X_n)$ lies in $\mathscr{R}^n$ has no information about θ [in the sense that the conditional distribution is independent of θ]. More precisely, $\{(x_1, \ldots, x_n): t(x_1, \ldots, x_n) = c\}$ is a subset of $\mathscr{R}^n$ called a hypersurface [see the paragraph following (8.6.1)]. If we are told the value of $t(\;)$, we only know that *some* point on this hypersurface was the result of our experiment. The left side of Equation (8.8.2) is the conditional probability of having obtained a value $(X_1, \ldots, X_n)$ in a certain subset of $t(\;) = c$, given a value in $t(\;) = c$ was obtained. (See Figure 8.8–1.) If this is not dependent on θ, then further specification of the point obtained contains no information about θ.

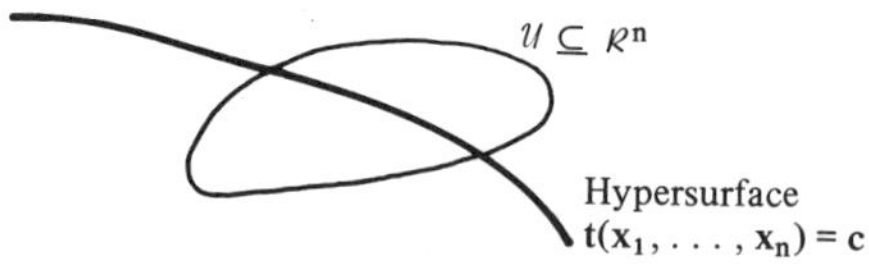

Figure 8.8–1

Although Definition 8.8.1 is easy to understand intuitively, it is usually difficult to verify, and a criterion more like that of Theorem 8.7.7 in the regular estimation case is often more useful in practice. Now, after stating an assumption slightly more general than Assumption 8.4.5, we will give a criterion more like that of Theorem 8.7.7.

Assumption 8.8.3. *Suppose that $(X_1, \ldots, X_n)$ is a r.v. with d.f. $F(x_1, \ldots, x_n|\theta) = P_\theta[X_1 \leqslant x_1, \ldots, X_n \leqslant x_n]$ where $\theta = (\theta_1, \ldots, \theta_r) \in \Theta \subseteq \mathscr{R}^r$. Assume $F(x_1, \ldots, x_n|\theta)$ is either absolutely continuous [in which case let $f(x_1, \ldots, x_n|\theta)$ denote its density at $(x_1, \ldots, x_n)$] or discrete [in which case let $f(x_1, \ldots, x_n|\theta)$ denote $P_\theta[X_1 = x_1, \ldots, X_n = x_n]$].*

Theorem 8.8.4. (Neyman Factorization Theorem) *Under Assumption* 8.8.3, *$t(X_1, \ldots, X_n)$ is a sufficient statistic (for θ) iff*

$$f(x_1, \ldots, x_n|\theta) = h(x_1, \ldots, x_n) g(t(x_1, \ldots, x_n)|\theta) \tag{8.8.5}$$

for all $x_1, \ldots, x_n, \theta$ for some functions $h(\;)$ and $g(\;)$.

Proof: [We will give the proof only for the case[19] of a r.v. $X = (X_1, \ldots, X_n)$ which assumes values $x = (x_1, \ldots, x_n)$ in a set containing at most denumerably many points; this is to be the same set for every $\theta \in \Theta$.]

First, suppose $t(X_1, \ldots, X_n)$ is a sufficient statistic (that is, Definition 8.8.1 holds). Let X denote $(X_1, \ldots, X_n)$, and let x denote $(x_1, \ldots, x_n)$. Let $S = \{x: f(x|\theta) > 0 \text{ for some } \theta \in \Theta\}$, the set of values which X can assume. We suppose S contains at most denumerably many points. Now (by (8.8.2))

$$P_\theta[X = x | t(X) = c] = l(\{x\}|c) \tag{8.8.6}$$

is not a function of $\theta \in \Theta$. Now let x_0 be an arbitrary member of S, and denote $t(x_0)$ by c_0. Then

$$\begin{aligned} f(x_0|\theta) &= P_\theta[X = x_0] = \sum_d P_\theta[t(X) = d]P_\theta[X = x_0|t(X) = d] \\ &= P_\theta[t(X) = c_0]P_\theta[X = x_0|t(X) = c_0] \\ &= P_\theta[t(X) = c_0]l(\{x_0\}|t(x_0)), \end{aligned} \tag{8.8.7}$$

where we used the Theorem of Total Probabilities (Theorem 2.3.4), the fact that $P_\theta[X = x_0|t(X) = d]$ is zero unless $d = c_0$, and (8.8.6). But now in (8.8.7) the factor $l(\{x_0\}|t(x_0))$ is a function of x_0 alone, say $h(x_0)$; and $P_\theta[t(X) = c_0]$ is a function of $t(x_0)$ and θ, say $g(t(x_0)|\theta)$. Hence, (8.8.5) follows and since $x_0 \in S$ was arbitrary this part of the theorem [sufficiency implies (8.8.5)] is proven.

Next, to prove the converse, suppose that (8.8.5) holds (that is, the density, or probability function, factors). Then, if x_0 is an arbitrary member of S, we have

$$f(x_0|\theta) = h(x_0)g(t(x_0)|\theta). \tag{8.8.8}$$

Hence (denoting $t(x_0)$ by c_0)

$$\begin{aligned} &P_\theta[X = x_0|t(X) = c_0] \\ &\quad = \frac{P_\theta[X = x_0, t(X) = c_0]}{P_\theta[t(X) = c_0]} = \frac{P_\theta[X = x_0]}{\sum_{d \ni t(d) = c_0} P_\theta[X = d]} \\ &\quad = \frac{h(x_0)g(c_0|\theta)}{\sum_{d \ni t(d) = c_0} h(d)g(c_0|\theta)} = \frac{h(x_0)}{\sum_{d \ni t(d) = c_0} h(d)}, \end{aligned} \tag{8.8.9}$$

which is a function only of x_0 and c_0, say $l(\{x_0\}|c_0)$. Since x_0 was arbitrary and, in a denumerable case, $P_\theta[X \in \mathscr{U}|t(X) = c]$ is simply a sum of such

[19] For the general case see, for example, E. L. Lehmann, *Testing Statistical Hypotheses*, John Wiley & Sons, Inc., New York, 1959, pp. 49–50. This treatment requires understanding of the concepts mentioned in Problems 2.4H and 3.1H, but the basic ideas involved are the same as for the discrete case covered by our proof.

terms (8.8.9) (each of which we have shown is independent of θ), the left-hand side of (8.8.2) is independent of θ, hence proving the theorem. [Note that in (8.8.9) we ignored the possibility $P_\theta[t(X) = c_0] = 0$ and the possibility of considering $P_\theta[X = x_0|t(X) = c]$ with $c \neq c_0$. This is easily taken care of, and comprises Problem 8.40.]

Note [see Definition 8.8.1 and Equation (8.8.2)] that what makes a statistic $t(X_1, \ldots, X_n)$ sufficient is a property of sets of points $\{(x_1, \ldots, x_n): t(x_1, \ldots, x_n) = c\}$, that is, of sets of points to which $t(\ )$ assigns the same value [and hence which $t(\ )$ chooses to "lump together" or not distinguish from each other]. Thus, any one-to-one function of t, say $k(t)$, is also a sufficient statistic (see Problem 8.41); for example, if $t(\ )$ is sufficient, then so is $10 + t(\ )$. [Hence, of course, $t(\ )$ itself may not be a proper (that is, good) estimator of θ; rather, any good estimator of θ needs only the information in $(X_1, \ldots, X_n)$ about θ which $t(X_1, \ldots, X_n)$ contains, and hence can be based solely on $t(X_1, \ldots, X_n)$.] It is useful to call the hypersurface of points $(x_1, \ldots, x_n)$ where a sufficient statistic has a specified value a **contour of** $t(\ )$. Then, all that matters about a sufficient statistic is not its specified values, but its contours; a function (which $t(\ )$ is) unites points onto contours, and any other statistic which unites the same points in the same way is essentially the same (see Figure 8.8–2).

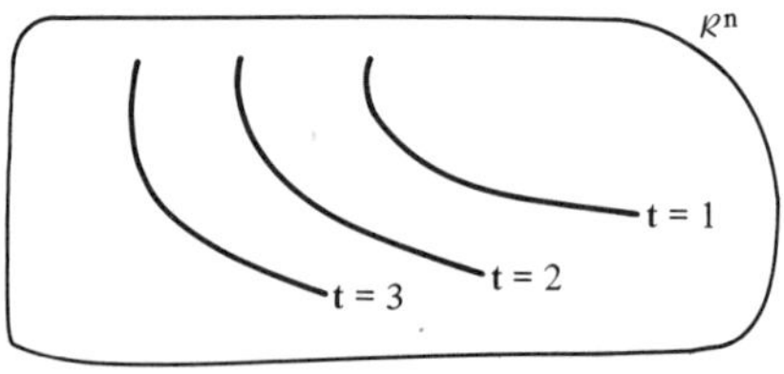

Figure 8.8–2. Contours of a sufficient statistic

To sum up, in an experiment we find a value of $X = (X_1, \ldots, X_n)$. If we are told the value of $t(X)$, that is, which contour of $t(\ )$ the observed point lies on, and further specification of the location of the observed point can shed no light on the value of θ (this was formalized in Definition 8.8.1, via independence of θ of the conditional distribution on the contour), then knowledge of the contour alone is *sufficient*. Intuitively, then, if there exists a sufficient statistic $t(\ )$, any estimator of θ should be a function of $t(\ )$ only. Hence, once we agree on what $f(x_1, \ldots, x_n|\theta)$ is, we need not report or record all our data $X = (X_1, \ldots, X_n)$; rather, $t(X_1, \ldots, X_n)$ is a sufficient summary of the data.

We have thus seen that it is desirable to have a sufficient statistic $t(X_1, \ldots, X_n)$. In fact **at least one sufficient statistic always exists, namely**

the identity statistic $\mathbf{I}(\mathbf{X}_1, \ldots, \mathbf{X}_n) = (\mathbf{X}_1, \ldots, \mathbf{X}_n)$; the proof of sufficiency is trivial from Definition 8.8.1, since the conditional distribution is independent of θ due to there being only one point on each contour, whose conditional probability is 1. But suppose there is more than one sufficient statistic: which should we choose? If there are two, say $t(\ )$ and $\varphi(\ )$, which are the same except as illustrated in Figure 8.8–3, then we should prefer $\varphi(\ )$

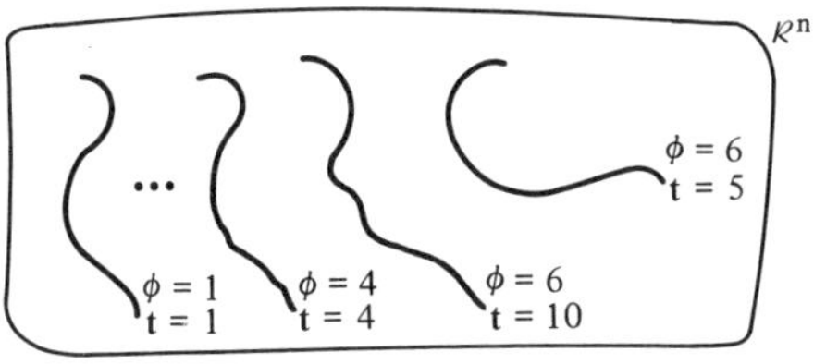

Figure 8.8–3. Two sufficient statistics' contours

because it unites more points (that is, summarizes the data more) but is still sufficient. Note that in fact $\varphi(\ ) = k(t(\ ))$, that is $\varphi(\ )$ is a function of $t(\ )$ (this is not true for every pair of sufficient statistics). Thus, if we are seeking a *best sufficient statistic*, it should be the one with the largest contours. We now formalize this notion.

Definition 8.8.10. *$T(X_1, \ldots, X_n)$ is a* **minimal sufficient statistic** *if $T(\ )$ is a function of every other sufficient statistic; that is,*

$$T(X_1, \ldots, X_n) = k_t(t(X_1, \ldots, X_n)) \tag{8.8.11}$$

for every sufficient statistic $t(\ )$.

Thus, a minimal sufficient statistic [the minimal information about $(X_1, \ldots, X_n)$ which must be preserved for good estimation of θ] has maximal contours. Such a statistic need not always exist (but often does).

We already showed (above Figure 8.8–3) that one sufficient statistic always exists, the identity statistic. Often another also does, and each of these is called a **trivial sufficient statistic**. Namely (see Problem 8.43 for the proof), the following.

Theorem 8.8.12. *If Assumption 8.8.3 holds and $X_1, \ldots, X_n$ are independent and identically distributed r.v.'s, then the order statistic $t(X_1, \ldots, X_n) = (Y_1, \ldots, Y_n)$* [see Definition 8.1.28] *is a sufficient statistic.*

Next we wish to show that sometimes only trivial sufficient statistics exist. For this we need the following lemma.

Lemma 8.8.13. *Suppose Assumption 8.8.3 holds. Then $f(x|\theta)/f(y|\theta)$ is independent of θ if $t(x) = t(y)$ for some sufficient statistic $t(\ )$ (that is, if*

$x = (x_1, \ldots, x_n)$ and $y = (y_1, \ldots, y_n)$ are on the same contour of some sufficient statistic).

Proof: By the Neyman Factorization Theorem, since $t(\ )$ is sufficient and $t(x) = t(y)$,

$$\frac{f(x|\theta)}{f(y|\theta)} = \frac{h(x)g(t(x)|\theta)}{h(y)g(t(y)|\theta)} = \frac{h(x)}{h(y)},$$

which is independent of θ, hence proving the lemma.

Example 8.8.14. Suppose we have $n = 2$ independent observations from the Cauchy distribution (see Theorem 6.1.3) with $a = \theta$ and $b = 1$, that is, from density

$$f_X(x) = \frac{1}{\pi}\frac{1}{1 + (x - \theta)^2}, \quad -\infty < x < +\infty. \tag{8.8.15}$$

Show that no nontrivial sufficient statistic exists. (Note that the Cauchy distribution is essentially the Student's t-distribution of Definition 4.2.17 with 1 degree of freedom.)

Suppose there were a nontrivial sufficient statistic $t(\ )$. Then for some two points $x = (x_1, x_2)$ and $y = (y_1, y_2)$ [which are not permutations of each other] we would have $t(x) = t(y)$. From Lemma 8.8.13 it would then follow that $f(x|\theta)/f(y|\theta)$ is independent of θ, that is,

$$\frac{f(x|\theta)}{f(y|\theta)} = \frac{\dfrac{1}{\pi}\dfrac{1}{1 + (x_1 - \theta)^2} \cdot \dfrac{1}{\pi}\dfrac{1}{1 + (x_2 - \theta)^2}}{\dfrac{1}{\pi}\dfrac{1}{1 + (y_1 - \theta)^2} \cdot \dfrac{1}{\pi}\dfrac{1}{1 + (y_2 - \theta)^2}} = Q(x_1, x_2, y_1, y_2)$$

or

$$\begin{aligned}(1 + (x_1 - \theta)^2)(1 + (x_2 - \theta)^2)Q(x_1, x_2, y_1, y_2)& \\ = (1 + (y_1 - \theta)^2)(1 + (y_2 - \theta)^2).&\end{aligned} \tag{8.8.16}$$

Now (8.8.16) must hold for all θ; but by multiplying out the terms in (8.8.16) and collecting them all on one side (8.8.16) becomes

$$P_4(\theta) \overset{\theta}{\equiv} 0, \tag{8.8.17}$$

where $P_4(\theta)$ is a polynomial of degree 4 (that is, $a + b\theta + c\theta^2 + d\theta^3 + e\theta^4$) in θ. Since such a polynomial can have only (at most) 4 real roots, (8.8.17) cannot be true for all θ; hence, there cannot be a nontrivial sufficient statistic $t(\ )$ in this case. (Also see Problem 8.44.)

We now wish to consider how to find a minimal sufficient statistic.

Theorem 8.8.18. *Suppose Assumption 8.8.3 holds. Define a statistic $t(\ )$ as follows: Only put x and y on the same contour iff $f(x|\theta)/f(y|\theta)$ is independent of θ. Then $t(\ )$ is a minimal sufficient statistic.*

Proof: [We will give the proof only for the case[20] of a r.v. $X = (X_1, \ldots, X_n)$ which assumes values $x = (x_1, \ldots, x_n)$ in a set containing at most denumerably many points; this is to be the same set for every $\theta \in \Theta$. This is the same case for which we proved the Neyman Factorization Theorem.] Note that (Lemma 8.8.13) we previously showed the ratio $f(x|\theta)/f(y|\theta)$ to be independent of θ if $t(x) = t(y)$ for some sufficient statistic $t(\ )$.

First suppose that the statistic $t(\ )$ defined in the statement of the theorem is sufficient. Then it is minimal sufficient. For, if $s(\ )$ is another sufficient statistic and $s(w) = s(z)$, then [by Lemma 8.8.13] $f(w|\theta)/f(z|\theta)$ is independent of θ, and (by the definition of $t(\ )$) $t(w) = t(z)$. Hence, $t(\ )$ is a function of $s(\ )$ and (by Definition 8.8.10) is minimal sufficient.

Next, we wish to show $t(\ )$ is sufficient. Now for an arbitrary x_0 [denoting $t(x_0)$ by c_0],

$$P_\theta[X = x_0|t(X) = c_0]$$
$$= \frac{P_\theta[X = x_0, t(X) = c_0]}{P[t(X) = c_0]} = \frac{P_\theta[X = x_0]}{\sum\limits_{d \ni t(d)=c_0} P_\theta[X = d]} \qquad (8.8.19)$$
$$= 1/\{\sum_{d \ni t(d)=c_0} (f(d|\theta)/f(x_0|\theta))\}.$$

But since $t(d) = c_0$ and $t(x_0) = c_0$, d and x_0 have been put on the same contour; hence, it must have been the case that $f(d|\theta)/f(x_0|\theta)$ was independent of θ. Hence, (8.8.19) is independent of θ. As following (8.8.9), this proves $P_\theta[X \in \mathscr{U}|t(X) = c]$ is independent of θ; hence, $t(\ )$ is a sufficient statistic.

Example 8.8.20. Suppose $X_1, \ldots, X_n$ are independent r.v.'s each $N(\theta, \sigma^2)$ where θ is unknown $(-\infty < \theta < +\infty)$ and $\sigma^2 > 0$ is known. Find a sufficient statistic for θ.

Here, we have

$$f(x_1, \ldots, x_n|\theta) = \prod_{i=1}^{n} f(x_i|\theta) = \prod_{i=1}^{n} \left\{\frac{1}{\sqrt{2\pi}\sigma} e^{-\frac{1}{2\sigma^2}(x_i-\theta)^2}\right\}$$
$$= \left(\frac{1}{\sqrt{2\pi}\sigma}\right)^n e^{-\frac{1}{2\sigma^2}\sum_{i=1}^{n}(x_i-\theta)^2} \qquad (8.8.21)$$
$$= \left(\frac{1}{\sqrt{2\pi}\sigma}\right)^n e^{-\frac{1}{2\sigma^2}\sum_{i=1}^{n}x_i^2} e^{-\frac{1}{2\sigma^2}\left\{-2\theta\sum_{i=1}^{n}x_i + n\theta^2\right\}},$$

from which it follows via the Neyman Factorization Theorem (Theorem

[20] For the general case see pp. 327–337 of E. L. Lehmann and H. Scheffé, "Completeness, similar regions, and unbiased estimation—Part I," *Sankhyā*, Vol. 10 (1950), pp. 305–340. The basic ideas (but not the mathematical details) are the same as for the discrete case covered by our proof.

8.8.4) taking

$$\begin{cases} h(x_1, \ldots, x_n) = \left(\dfrac{1}{\sqrt{2\pi}\sigma}\right)^n e^{-\frac{1}{2\sigma^2}\sum_{i=1}^n x_i^2}, \\ g(x_1 + \cdots + x_n|\theta) = e^{-\frac{1}{2\sigma^2}\{-2\theta(x_1+\cdots+x_n)+n\theta^2\}}, \end{cases}$$

that $t(X_1, \ldots, X_n) = X_1 + \cdots + X_n$ is a sufficient statistic for θ. Hence (see the discussion following the proof of the Neyman Factorization Theorem), so is any one-to-one function of $X_1 + \cdots + X_n$; for example, $\bar{X} = (X_1 + \cdots + X_n)/n$.

Example 8.8.22. Suppose $X_1, \ldots, X_n$ are independent r.v.'s, each uniform on $(0, \theta)$ where $\theta > 0$ is unknown. Find a sufficient statistic for θ.

We have

$$\begin{aligned} f(x_1, \ldots, x_n|\theta) &= \prod_{i=1}^n f(x_i|\theta) = \begin{cases} (1/\theta)^n, & \text{if } 0 \leqslant x_1, \ldots, x_n \leqslant \theta \\ 0, & \text{otherwise} \end{cases} \\ &= \begin{cases} 1/\theta^n, & \text{if } 0 \leqslant \min(x_1, \ldots, x_n), \max(x_1, \ldots, x_n) \leqslant \theta \\ 0, & \text{otherwise,} \end{cases} \end{aligned} \tag{8.8.23}$$

from which it follows via the Neyman Factorization Theorem taking

$$\begin{cases} h(x_1, \ldots, x_n) = \begin{cases} 1, & \text{if } \min(x_1, \ldots, x_n) \geqslant 0 \\ 0, & \text{otherwise,} \end{cases} \\ g(\max(x_1, \ldots, x_n)|\theta) = \begin{cases} 1/\theta^n, & \text{if } \max(x_1, \ldots, x_n) \leqslant \theta \\ 0, & \text{otherwise,} \end{cases} \end{cases}$$

that $t(X_1, \ldots, X_n) = \max(X_1, \ldots, X_n)$ is a sufficient statistic for θ.

Example 8.8.24. Suppose $X_1, \ldots, X_n$ are independent r.v.'s, each $N(\mu, \sigma^2)$ where $\mu(-\infty < \mu < +\infty)$ and $\sigma^2(\sigma^2 > 0)$ are both unknown. Find a minimal sufficient statistic for $\theta = (\mu, \sigma^2)$.

Here, proceeding as in Example 8.8.20, we find that

$$f(x_1, \ldots, x_n|\mu, \sigma^2) = \left(\frac{1}{\sqrt{2\pi}\sigma}\right)^n e^{-\frac{1}{2\sigma^2}\left\{\sum_{i=1}^n x_i^2 - 2\mu\sum_{i=1}^n x_i + n\mu^2\right\}}. \tag{8.8.25}$$

Now (by Theorem 8.8.18) if we set $t(x) = t(y)$ iff $f(x|\theta)/f(y|\theta)$ is independent of θ, then $t(\ )$ will be a minimal sufficient statistic. But using (8.8.25) we find

$$\frac{f(x|\theta)}{f(y|\theta)} = e^{-\frac{1}{2\sigma^2}\left\{\left(\sum_{i=1}^n x_i^2 - \sum_{i=1}^n y_i^2\right) - 2\mu\left(\sum_{i=1}^n x_i - \sum_{i=1}^n y_i\right)\right\}}, \tag{8.8.26}$$

which is independent of (μ, σ^2) iff $\left\{\sum_{i=1}^{n} x_i = \sum_{i=1}^{n} y_i \text{ and } \sum_{i=1}^{n} x_i^2 = \sum_{i=1}^{n} y_i^2\right\}$. Hence, $t(X_1, \ldots, X_n) = \left(\sum_{i=1}^{n} X_i, \sum_{i=1}^{n} X_i^2\right)$ is a minimal sufficient statistic for (μ, σ^2). [Note that this $t(\ )$ puts x and y on the same contour iff $\left\{\sum_{i=1}^{n} x_i = \sum_{i=1}^{n} y_i \text{ and } \sum_{i=1}^{n} x_i^2 = \sum_{i=1}^{n} y_i^2\right\}$.]

In the following theorem we give a statistic which is minimal sufficient in general.

Theorem 8.8.27. (Dynkin's Theorem) *Suppose Assumption* 8.8.3 *holds. Let* θ_0 *be any fixed possible value of* θ. *Then*

$$\left\{\left(\theta, \frac{f(X|\theta)}{f(X|\theta_0)}\right) : \theta \in \Theta\right\} = t(X) \ \ (say) \tag{8.8.28}$$

is a minimal sufficient statistic for θ.

Proof: First, if we define

$$\begin{cases} h(x_1, \ldots, x_n) = f(x_1, \ldots, x_n|\theta_0), \\ g(t(x_1, \ldots, x_n)|\theta) = \dfrac{f(x_1, \ldots, x_n|\theta)}{f(x_1, \ldots, x_n|\theta_0)} \end{cases} \tag{8.8.29}$$

then sufficiency follows by the Neyman Factorization Theorem.

Next we will show that $t(X)$ is a minimal sufficient statistic, by showing that if $s(X)$ is any other sufficient statistic, then $t(X)$ is a function of $s(X)$. To show this it suffices to show that, whenever $s(x) = s(y)$, also $t(x) = t(y)$. Note that $t(x)$ is a function; for example, if $\Theta = \mathscr{R}$, a typical value of $t(x)$ is an entire function (see Figure 8.8–4). Now, since the other statistic $s(X)$ is sufficient, by the Neyman Factorization Theorem

$$f(x|\theta) = h^*(x)g^*(s(x)|\theta) \tag{8.8.30}$$

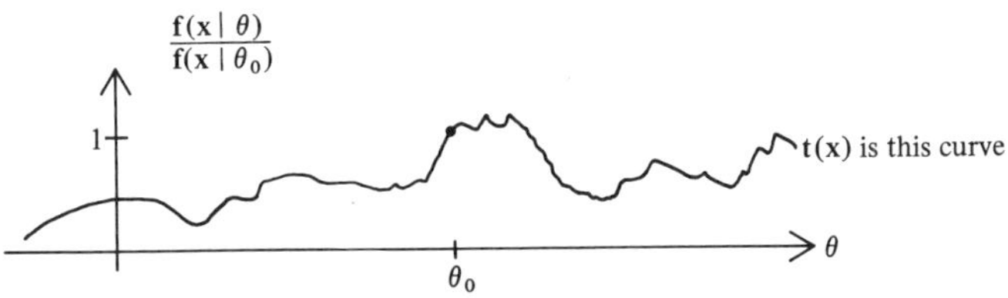

Figure 8.8–4. A typical **t(x)** if $\Theta = \mathscr{R}$

for some functions $h^*(\ )$ and $g^*(\ )$. Now suppose $s(x) = s(y)$; then (by Lemma 8.8.13) $f(x|\theta)/f(y|\theta)$ is independent of θ. We wish to show $t(x) = t(y)$; by (8.8.28), this will follow iff we can show (for all θ)

$$\frac{f(x|\theta)}{f(x|\theta_0)} = \frac{f(y|\theta)}{f(y|\theta_0)},$$

that is, (using (8.8.30)) iff for all θ

$$\frac{h^*(x)g^*(s(x)|\theta)}{h^*(x)g^*(s(x)|\theta_0)} = \frac{h^*(y)g^*(s(y)|\theta)}{h^*(y)g^*(s(y)|\theta_0)},$$

that is, iff for all θ

$$g^*(s(x)|\theta)g^*(s(y)|\theta_0) = g^*(s(x)|\theta_0)g^*(s(y)|\theta).$$

This is true since $s(x) = s(y)$.

Example 8.8.31. Suppose Assumption 8.8.3 holds and $\Theta = \{\theta_0, \theta_1\}$ where $\theta_0 \neq \theta_1$. What then is the sufficient statistic provided by Dynkin's Theorem (Theorem 8.8.27)?

Here we need to evaluate and specify

$$\begin{aligned} t(X) &= \left\{\left(\theta, \frac{f(X|\theta)}{f(X|\theta_0)}\right) : \theta \in \Theta\right\} \\ &= \left\{\left(\theta_0, \frac{f(X|\theta_0)}{f(X|\theta_0)}\right), \left(\theta_1, \frac{f(X|\theta_1)}{f(X|\theta_0)}\right)\right\}; \end{aligned}$$

but this is equivalent to specifying the quantity $f(X|\theta_1)/f(X|\theta_0)$, called the **likelihood-ratio statistic**.

In this chapter considerable attention has been paid to maximum-likelihood estimators, and to the method of maximum likelihood as a way of generating possibly good estimators. Since we have seen in this section that a good estimator need only be based on any sufficient statistic, the question arises: how are MLE's related to sufficient statistics? (This question was already considered, in a restricted context of regular estimation cases, in Theorem 8.7.9. We now raise the question in the general case.)

Theorem 8.8.32. *Suppose Assumption* 8.8.3 *holds. If* $t(X_1, \ldots, X_n)$ *is a sufficient statistic for* θ *and a unique MLE* $\hat{\theta}$ *of* θ *exists, then* $\hat{\theta}$ *is a function of* $t(X_1, \ldots, X_n)$. *If any MLE exists, an MLE* $\hat{\theta}$ *can be found which is a function of* $t(X_1, \ldots, X_n)$.

Proof: Since $t(X_1, \ldots, X_n)$ is a sufficient statistic for θ, by the Neyman Factorization Theorem (Theorem 8.8.4) it follows that for all $x_1, \ldots, x_n$, θ

and for some functions $h(\ )$ and $g(\ )$ we have

$$f(x_1, \ldots, x_n|\theta) = h(x_1, \ldots, x_n)g(t(x_1, \ldots, x_n)|\theta). \tag{8.8.33}$$

In seeking an MLE we hold $x_1, \ldots, x_n$ fixed and seek to find a $\hat{\theta}(x_1, \ldots, x_n)$ which maximizes (8.8.33). Thus, if there is a unique MLE, it will maximize $g(t(x_1, \ldots, x_n)|\theta)$ and hence be a function of $t(x_1, \ldots, x_n)$. In any case if any MLE exists, an MLE can be found which is a function only of $t(x_1, \ldots, x_n)$.

Note that Theorem 8.8.32 does *not* say that "any MLE must be a function of any sufficient statistic"; that remark need not be true if the MLE is not unique (for then the class of MLE's includes some which depend on other information about $X_1, \ldots, X_n$ than the information provided by $t(X_1, \ldots, X_n)$). This was first noted by D. S. Moore, "Maximum likelihood and sufficient statistics," *American Mathematical Monthly*, Vol. 78 (1971), pp. 50–52. The erroneous statement occurs in many books.

For a certain class of probability distributions a nontrivial sufficient (in fact, minimal sufficient) statistic is easy to find. This class, the "exponential class" or "Koopman-Darmois class" of distributions, will now be introduced and studied.

Definition 8.8.34. *Suppose Assumption* 8.8.3 *holds. Then the density of* $(X_1, \ldots, X_n)$ *is said to be a member of the* **exponential class** [or a member of the **Koopman-Darmois class**; or is said to have **Koopman-Darmois form**] if

$$f(x_1, \ldots, x_n|\theta) = e^{\sum_{i=1}^{r} \varphi_i(x_1, \ldots, x_n)c_i(\theta) + c_0(\theta) + \varphi_0(x_1, \ldots, x_n)} \tag{8.8.35}$$

for all $\theta = (\theta_1, \ldots, \theta_r) \in \Theta \subseteq \mathscr{R}^r$. [Here $\varphi_0(x_1, \ldots, x_n), \ldots, \varphi_r(x_1, \ldots, x_n)$ are some functions of $x_1, \ldots, x_n$; and $c_0(\theta), \ldots, c_r(\theta)$ are some functions of $\theta = (\theta_1, \ldots, \theta_r)$. If $n = 1$ we talk of a **univariate exponential family**, while if $r = 1$ we talk of a **one-parameter exponential family**.]

Example 8.8.36. Show that if $X_1, \ldots, X_n$ are independent r.v.'s, each $N(\mu, \sigma^2)$ where $\mu(-\infty < \mu < +\infty)$ and $\sigma^2(\sigma^2 > 0)$ are both unknown, then $(X_1, \ldots, X_n)$ is a member of the two-parameter exponential family.

Starting from (8.8.25) (a result in Example 8.8.24, in which we also studied this $(X_1, \ldots, X_n)$ and found a minimal sufficient statistic for (μ, σ^2)) we have

$$\begin{aligned} f(x_1, \ldots, x_n|\mu, \sigma^2) &= \left(\frac{1}{\sqrt{2\pi}\sigma}\right)^n e^{-\frac{1}{2\sigma^2}\left\{\sum_{i=1}^{n} x_i^2 - 2\mu \sum_{i=1}^{n} x_i + n\mu^2\right\}} \\ &= e^{\left(\sum_{i=1}^{n} x_i\right)\frac{\mu}{\sigma^2} + \left(\sum_{i=1}^{n} x_i^2\right)\left(\frac{-1}{2\sigma^2}\right) + \left\{-\frac{n\mu^2}{2\sigma^2} - n\ln(\sqrt{2\pi}\sigma)\right\}}, \end{aligned} \tag{8.8.37}$$

which is clearly of form (8.8.35) for all $(\mu, \sigma^2) \in \Theta = \{(x, y): -\infty < x < +\infty,$

$y > 0\}$ if we take

$$\begin{cases} \varphi_1(x_1, \ldots, x_n) = \sum_{i=1}^{n} x_i, \; c_1(\mu, \sigma^2) = \dfrac{\mu}{\sigma^2}, \\ \varphi_2(x_1, \ldots, x_n) = \sum_{i=1}^{n} x_i^2, \; c_2(\mu, \sigma^2) = \dfrac{-1}{2\sigma^2}, \\ c_0(\mu, \sigma^2) = \dfrac{-n\mu^2}{2\sigma^2} - n \ln(\sqrt{2\pi}\sigma) \\ \varphi_0(x_1, \ldots, x_n) = 1. \end{cases} \tag{8.8.38}$$

From Examples 8.8.36 and 8.8.24 it appears that $(\varphi_1(X_1, \ldots, X_n), \ldots, \varphi_r(X_1, \ldots, X_n))$ will be our minimal sufficient statistic for $\theta = (\theta_1, \ldots, \theta_r)$, and this is the case (Problem 8.52). The question then arises: If we have a nontrivial sufficient statistic, must our $(X_1, \ldots, X_n)$ be a member of the exponential class? As is shown by Example 8.8.22 and Problem 8.24H, the answer is "no." However, under certain conditions of "regularity" (here we go again!), the answer is "yes." [For example, see E. B. Andersen, "Sufficiency and exponential families for discrete sample spaces," *Journal of the American Statistical Association*, Vol. 65 (1970), pp. 1248–1255 and J. Pfanzagl, "Transformation groups and sufficient statistics," *Annals of Mathematical Statistics*, Vol. 43 (1972), pp. 553–568, and their references.]

Our last consideration in this section on sufficiency will be to prove a theorem which shows how to combine an unbiased estimator U with a sufficient statistic T to obtain an improved estimator. While previously we were sometimes able to obtain *b*est *r*egular *u*nbiased *e*stimators (Definition 8.6.23) or *b*est *l*inear *u*nbiased *e*stimators (Problem 8.8), this study will yield (in Section 8.9) a method of obtaining *b*est *u*nbiased *e*stimators (with no regularity or linearity assumptions). First, a basic theorem.

Theorem 8.8.39. (Rao-Blackwell Theorem) *Let X and Y be r.v.'s with $E(Y) = \mu$ and $\operatorname{Var}(Y) = \sigma_Y{}^2 < +\infty$. Let $\varphi(x) = E(Y|X = x)$. Then $E\varphi(X) = \mu$, and $\operatorname{Var} \varphi(X) \leqslant \sigma_Y{}^2$ with equality iff $P[Y = \varphi(X)] = 1$.*

Proof: [We shall only give the proof for the case where the distribution of (X, Y) is absolutely continuous. The student should review conditional densities (Equation (4.4.5)) and conditional expectation (Theorem 5.3.20) at this point.] Let $f(x, y)$ denote the joint density of (X, Y), $f_1(x)$ the density of X, $f_2(y)$ the density of Y, and $h(y|x)$ the conditional density of Y given $X = x$.

We will *first* show that $E\varphi(X) = \mu$. For since

$$E(Y|X = x) = \int_{-\infty}^{\infty} yh(y|x)\,dy = \frac{\int_{-\infty}^{\infty} yf(x, y)\,dy}{f_1(x)} \equiv \varphi(x),$$

we have $\int_{-\infty}^{\infty} yf(x, y)\, dy = \varphi(x)f_1(x)$. Thus,

$$\begin{aligned} E\varphi(X) &= \int_{-\infty}^{\infty} \varphi(x)f_1(x)\, dx = \int_{-\infty}^{\infty} \left[\int_{-\infty}^{\infty} yf(x, y)\, dy\right] dx \\ &= \int_{-\infty}^{\infty} y \int_{-\infty}^{\infty} f(x, y)\, dx\Big]\, dy = \int_{-\infty}^{\infty} yf_2(y)\, dy = \mu. \end{aligned} \tag{8.8.40}$$

We will *next* show that $\operatorname{Var}\varphi(X) \leqslant \sigma_Y{}^2$ with equality iff $P[Y = \varphi(X)] = 1$. For,

$$\begin{aligned} \sigma_Y{}^2 &= E(Y - \mu)^2 = E[(Y - \varphi(X)) + (\varphi(X) - \mu)]^2 \\ &= E(Y - \varphi(X))^2 + E(\varphi(X) - \mu)^2 + 2E(Y - \varphi(X))(\varphi(X) - \mu) \\ &= E(Y - \varphi(X))^2 + \operatorname{Var}\varphi(X) + 0. \end{aligned} \tag{8.8.41}$$

Hence,

$$\operatorname{Var}\varphi(X) = \sigma_Y^2 - E(Y - \varphi(X))^2 \leqslant \sigma_Y^2, \tag{8.8.42}$$

with equality iff $E(Y - \varphi(X))^2 = 0$; the last holds iff $P[Y = \varphi(X)] = 1$ (the simple proof being similar to that of Lemma 5.1.13).

Note that we must still show (for (8.8.41) to be demonstrated) that $E(Y - \varphi(X))(\varphi(X) - \mu) = 0$. However,

$$\begin{aligned} E(Y - \varphi(X))(\varphi(X) - \mu) &= \int_{-\infty}^{\infty}\int_{-\infty}^{\infty} (y - \varphi(x))(\varphi(x) - \mu)f(x, y)\, dy\, dx \\ &= \int_{-\infty}^{\infty} (\varphi(x) - \mu)\left[\int_{-\infty}^{\infty} (y - \varphi(x))f(x, y)\, dy\right] dx \\ &= \int_{-\infty}^{\infty} (\varphi(x) - \mu)\left[\int_{-\infty}^{\infty} (y - \varphi(x))h(y|x)\, dy\right] f_1(x)\, dx \\ &= 0 \end{aligned}$$

because $\int_{-\infty}^{\infty} (y - \varphi(x))h(y|x)\, dx = 0$ since $\varphi(x)$ *is* $\int_{-\infty}^{\infty} yh(y|x)\, dy$.

The application of the Rao-Blackwell Theorem to reduce the variance of an unbiased estimator by using a sufficient statistic will now be given. (The result on the *best* unbiased estimator will be given in Section 8.9; in this section we obtain an improved unbiased estimator. For this reason, Lemma 8.8.43 is sometimes called the **Rao-Blackwell Improvement Theorem**.)

Lemma 8.8.43. *Suppose Assumption* 8.8.3 *holds. Let T be a sufficient statistic for θ, and U an unbiased estimator of θ (which is not a function of T alone). Then*

$$\varphi_\theta(t) = E_\theta(U|T = t) \tag{8.8.44}$$

is in fact not a function of θ (hence we could use $\varphi(T)$ as an estimator of θ) and: $E_\theta\varphi(T) = \theta$, $\operatorname{Var}_\theta\varphi(T) < \operatorname{Var}_\theta U$.

Proof: First, $\varphi_\theta(t) = E_\theta(U|T = t) = \int_{-\infty}^{\infty} ug(u|t)\, du$. But $g(u|t) = f(u, t|\theta)/h(t|\theta)$ is independent of θ because T is a sufficient statistic (recall Definition 8.8.1). Hence, $\varphi_\theta(t)$ is in fact not a function of θ.

Next, applying the Rao-Blackwell Theorem (Theorem 8.8.39) with $X = T$ and $Y = U$, it follows that in fact $\varphi(T)$ is an unbiased estimator of θ ($E_\theta \varphi(T) = \theta$) with reduced variance compared to U ($\mathrm{Var}_\theta\, \varphi(T) < \mathrm{Var}_\theta\, U$).

Thus (Lemma 8.8.43), we can improve an unbiased estimator U_1 using a sufficient statistic T_1, obtaining the estimator $\varphi_1(T_1)$ (say). Can $\varphi_1(T_1)$, itself an unbiased estimator, be improved (perhaps using another sufficient statistic T_2)? The answer, with some examples, is given in Section 8.9.

8.9. COMPLETENESS

In Section 8.8 the Rao-Blackwell Improvement Theorem (Lemma 8.8.43) showed how to combine an unbiased estimator and a sufficient statistic (obtaining an unbiased estimator with smaller variance than the original unbiased estimator). We shall now see that (in certain circumstances) the resulting estimator is a *b*est *u*nbiased *e*stimator. The necessary condition is called completeness:

Definition 8.9.1. *Suppose Assumption* 8.8.3 *holds. Then the family of density or probability functions* $f(x|\theta)$, $\theta \in \Theta$, *is called* **complete** *if, for every function* $u(x)$, *the identity*

$$E_\theta u(X) \overset{\theta}{\equiv} 0 \tag{8.9.2}$$

implies $u(x) = 0$ *at all* x *(more precisely, at all* x *for which* $f(x|\theta) > 0$ *for some* θ*).*

Definition 8.9.3. *If a sufficient statistic* T *in some problem has a family of density or probability functions* $g(x|\theta)$ *which is complete, then* T *is called a* **complete sufficient statistic**.

Definition 8.9.4. *Suppose Assumption* 8.8.3 *holds. Then (for* $i = 1, \ldots, r$*)* t_i *is called a* **best unbiased estimator (BUE)** *or* **minimum variance unbiased estimator (MVUE)** *of* θ_i *if* $E_\theta t_i \overset{\theta}{\equiv} \theta_i$ *and, among all unbiased estimators of* θ_i, t_i *has the smallest variance for each possible value of* θ. *We will then call* t *a BUE of* θ.

We can now state and prove our general BUE result (which furnishes a method of obtaining BUE's in many cases).

Theorem 8.9.5. *Suppose Assumption* 8.8.3 *holds. Let* T *be a complete sufficient statistic for* θ. *If* $E_\theta \varphi(T) \equiv \theta$ *for some function* $\varphi(\cdot)$, *then* $\varphi(T)$: **(1)** *is unique; and* **(2)** *has smaller variance than any other unbiased estimator of* θ *(that is, is a MVUE).*

Proof: First, to show that $\varphi(\cdot)$ is unique, let $\psi(T)$ be any other function of T alone, such that $E_\theta \psi(T) \equiv \theta$. Then

$$E_\theta[\varphi(T) - \psi(T)] \equiv 0, \tag{8.9.6}$$

so (by the completeness of T) $\varphi(t) = \psi(t)$ for all t (more precisely, for all t for which $g(t|\theta) > 0$ for some θ).

Second, to show that $\varphi(T)$ has smaller variance than any other unbiased estimator of θ (whether based on T alone or not) we proceed as follows. Now $\varphi(T)$ is, by our first result, the only function of T alone which is an unbiased estimator of θ. Hence, if another unbiased estimator exists, say U, it is not a function of T alone. By the Rao-Blackwell Improvement Theorem (Lemma 8.8.43), the estimator defined by $\psi(t) = E_\theta(U|T = t)$ is unbiased and $\mathrm{Var}_\theta\, \psi(T) < \mathrm{Var}_\theta\, U$; and by our first result $\psi(\cdot) \equiv \varphi(\cdot)$. Hence, $\mathrm{Var}_\theta\, \varphi(T) < \mathrm{Var}_\theta\, U$.

Thus, we are now in a position to use the Rao-Blackwell Improvement Theorem to obtain unique MVU estimators.

Example 8.9.7. Suppose that $X_1, \ldots, X_n$ are independent r.v.'s, each having the Poisson distribution with unknown parameter $\mu \geqslant 0$. It can be shown that $\bar{X} = (X_1 + \cdots + X_n)/n$ is an unbiased estimator of μ (use Theorem 8.1.5 and Theorem 5.2.3, or a direct proof) and that $X_1 + \cdots + X_n$ is a sufficient statistic for μ (see Problem 8.51). Show that $X_1 + \cdots + X_n$ is in fact a complete statistic (and hence, by Theorem 8.9.5, $\bar{X}$ is the unique MVU estimator of μ).

Here $X_1 + \cdots + X_n$ has (by Theorem 6.1.3) a Poisson distribution with parameter $n\mu$. Hence, it suffices to show that the Poisson family of probability functions is complete. Let X have a Poisson distribution with arbitrary parameter $\mu > 0$. Then

$$E_\mu u(X) \equiv 0 \tag{8.9.8}$$

means

$$\sum_{k=0}^{\infty} u(k) e^{-\mu} \frac{\mu^k}{k!} \equiv 0. \tag{8.9.9}$$

The case $\mu = 0$ implies $u(0) = 0$. If $\mu > 0$, we can rewrite (8.9.9) as

$$\sum_{k=0}^{\infty} \frac{u(k)}{k!} \mu^k \equiv 0. \tag{8.9.10}$$

By the result on power series quoted following Equation (8.1.21) it follows that $u(k) = 0$ for $k = 0, 1, 2, \ldots$. Hence, the Poisson family of distributions is complete.

Example 8.9.11. Suppose that $X_1, \ldots, X_n$ are independent r.v.'s, each having the $N(\mu, \sigma^2)$ distribution with μ unknown $(-\infty < \mu < +\infty)$ and

$\sigma^2 > 0$ known. It is easy to show that $\bar{X} = (X_1 + \cdots + X_n)/n$ is an unbiased estimator of μ and (see Example 8.8.20) that $X_1 + \cdots + X_n$ is a sufficient statistic for μ. Show that $\bar{X}$ is the unique MVU estimator of μ.

By Theorem 8.9.5 the desired result follows if we can show that $X_1 + \cdots + X_n$ is in fact a complete statistic. Here $X_1 + \cdots + X_n$ has the $N(n\mu, n\sigma^2)$ distribution. Hence, it suffices to show that the normal family of probability distributions (with μ unknown but σ^2 fixed) is complete. Let X be $N(\mu, \sigma^2)$ with μ arbitrary and $\sigma^2 > 0$ known. Then

$$E_\mu u(X) \equiv 0 \tag{8.9.12}$$

means

$$\int_{-\infty}^{\infty} u(x) \frac{1}{\sqrt{2\pi}\sigma} e^{-\frac{1}{2}\left(\frac{x-\mu}{\sigma}\right)^2} dx \overset{\mu}{\equiv} 0 \tag{8.9.13}$$

or

$$\int_{-\infty}^{\infty} u(\sigma y + \mu) e^{-\frac{1}{2}y^2} dy \overset{\mu}{\equiv} 0. \tag{8.9.14}$$

That 8.9.14 implies $u(x) \equiv 0$ for all x is a mathematical theorem (see Problem 8.37H).

8.10. OTHER ESTIMATION METHODS

In this section we wish to briefly introduce several additional methods of estimation: the methods of moments, of minimum chi-square, and of modified minimum chi-square. We shall regard them as methods of generating possibly good estimators. (As with the method of maximum likelihood, under appropriate conditions certain good properties can be proven about these methods.)

Definition 8.10.1. *Let $X_1, \ldots, X_n$ be independent and identically distributed r.v.'s, each with d.f. $F(x|\theta)$ for some fixed $\theta \in \Theta \subseteq \mathscr{R}^r$. If the moments indicated exist, then the* **method of moments estimators** *of $\theta_1, \ldots, \theta_r$ are the solutions of the equations (see Theorem* 8.1.5)

$$\mu_k' = m_k' \quad (k = 1, \ldots, r). \tag{8.10.2}$$

Thus, the method of moments attempts to equate the first r population moments to the first r sample moments (each noncentral) and take the resulting solution in $\theta_1, \ldots, \theta_r$ (if one exists) as an estimator of $\theta_1, \ldots, \theta_r$.

Example 8.10.3. Suppose that $X_1, \ldots, X_n$ $(n \geq 2)$ are independent, each being a $N(\mu, \sigma^2)$ r.v. with $\mu(-\infty < \mu < +\infty)$ and $\sigma^2(\sigma^2 > 0)$ both unknown. Find the method of moments estimators of μ and σ^2.

Here the $r = 2$ equations to be solved are

$$\begin{cases} \mu_1' = m_1, \\ \mu_2' = m_2, \end{cases} \tag{8.10.4}$$

or

$$\begin{cases} \mu = EX_1 = \dfrac{\sum_{i=1}^{n} X_i}{n}, \\ \sigma^2 + \mu^2 = EX_1^2 = \dfrac{\sum_{i=1}^{n} X_i^2}{n}. \end{cases} \tag{8.10.5}$$

Solving system (8.10.5) yields

$$\begin{cases} \hat{\mu} = \bar{X}, \\ \hat{\sigma}^2 = \dfrac{\sum_{i=1}^{n} X_i^2}{n} - \hat{\mu}^2 = \dfrac{\sum_{i=1}^{n} (X_i - \bar{X})^2}{n}. \end{cases} \tag{8.10.6}$$

The estimators of (8.10.6) are also (see Problem 8.6) the maximum-likelihood estimators. This is often the case.

Definition 8.10.7. *Suppose we have available $s(\geqslant 1)$ sets of observations, and that set i consists of $n_i > 0$ observations ($i = 1, 2, \ldots, s$). Suppose each observation in set i is classified into one of k_i mutually exclusive and exhaustive categories ($1 \leqslant i \leqslant s$). Let*

$$\Pi_{ij} = P[\text{An observation in set } i \text{ falls into category } j] \tag{8.10.8}$$

[so, of course, $\Pi_{i1} + \cdots + \Pi_{ik} = 1$ for $1 \leqslant i \leqslant s$], *and assume $\Pi_{ij} > 0$ for all i, j. Let*

$$f_{ij} = \begin{pmatrix} \text{The proportion of the observations} \\ \text{in set } i \text{ which fall into category } j \end{pmatrix}. \tag{8.10.9}$$

Now, in general, the Π_{ij}'s will be functions of some unknown parameter $\theta \in \Theta \subseteq \mathscr{R}^r$. A statistic $t = (t_1, \ldots, t_r)$ is called a **minimum chi-square (MCS) estimator** *of $\theta = (\theta_1, \ldots, \theta_r)$ if it minimizes*

$$\begin{aligned} \chi^2(\theta) &= \sum_{i=1}^{s} \frac{1}{n_i} \sum_{j=1}^{k_i} \frac{(f_{ij} - \Pi_{ij})^2}{\Pi_{ij}} \\ &= \sum_{i=1}^{s} \frac{1}{n_i} \left(\sum_{j=1}^{k_i} \frac{f_{ij}^2}{\Pi_{ij}} - 1 \right). \end{aligned} \tag{8.10.10}$$

Note that the statistic (8.10.10), $\chi^2(\theta)$, which is to be minimized, is a weighted sum of deviations of f_{ij}'s from Π_{ij}'s. The MCS method, as we have given it, applies only in the cases of (1) a discrete distribution, or (2) a continuous distribution where the observations are categorized.

Definition 8.10.11. *Assume the set-up of Definition* 8.10.7. *A statistic* $t = (t_1, \ldots, t_r)$ *is called a* **modified MCS (MMCS) estimator** *of* $\theta = (\theta_1, \ldots, \theta_r)$ *if it minimizes*

$$(\chi'(\theta))^2 = \sum_{i=1}^{s} \frac{1}{n_i} \sum_{j=1}^{k_i} \frac{(f_{ij} - \Pi_{ij})^2}{f_{ij}} = \sum_{i=1}^{s} \frac{1}{n_i} \left(\sum_{j=1}^{k_i} \frac{\Pi_{ij}^2}{f_{ij}} - 1 \right). \tag{8.10.12}$$

[While the MCS method needs all $\Pi_{ij} > 0$, the MMCS method needs all $f_{ij} > 0$.]

Example 8.10.13. Suppose $X_1, \ldots, X_n$ are independent r.v.'s, each Poisson with parameter $\lambda > 0$ unknown. Find the MMCS estimator of λ.

Here we have $s = 1$ sample and $r = 1$ parameter. If we take our $k_1 = k$ categories as: category i meaning a value of $i - 1$ $(i = 1, \ldots, k - 1)$, and category k meaning a value $\geqslant k - 1$; then

$$\begin{cases} \Pi_{1j} = e^{-\lambda} \dfrac{\lambda^{j-1}}{(j-1)!} \quad (j = 1, \ldots, k-1) \\ \Pi_{1k} = \displaystyle\sum_{z=k-1}^{\infty} e^{-\lambda} \frac{\lambda^z}{z!}. \end{cases} \tag{8.10.14}$$

Denote $f_{11}, \ldots, f_{1k}$ simply by $f_1, \ldots, f_k$. We now seek to choose λ to minimize

$$(\chi'(\theta))^2 = \sum_{i=1}^{s} \frac{1}{n_i} \sum_{j=1}^{k_i} \frac{(f_{ij} - \Pi_{ij})^2}{f_{ij}} = \frac{1}{n} \left(\sum_{j=1}^{k} \frac{\Pi_{1j}^2}{f_j} - 1 \right). \tag{8.10.15}$$

To minimize (8.10.15) we need only minimize

$$\sum_{j=1}^{k} \frac{\Pi_{1j}^2}{f_j} = \sum_{j=1}^{k-1} \frac{1}{f_j} \left(e^{-\lambda} \frac{\lambda^{j-1}}{(j-1)!} \right)^2 + \frac{1}{f_k} \left(\sum_{z=k-1}^{\infty} e^{-\lambda} \frac{\lambda^z}{z!} \right)^2. \tag{8.10.16}$$

At this point derivatives and numerical work come into play. Usually results are harder to obtain than (but comparable to) MLE's.

PROBLEMS

8.1 Let $X_1, \ldots, X_n$ be independent and identically distributed r.v.'s. Assume that (for all c, $-\infty < c < +\infty$) $P[X_1 = c] < 1$. Prove that $\max(X_1, \ldots, X_n)$ is not an unbiased estimator of $\theta^* = EX_1$. [*Hint*:

$$\max(a, b) = \frac{a+b}{2} + \frac{|a-b|}{2}.\Big]$$

8.2 Let X_1, X_2 be independent and identically distributed r.v.'s. Prove that $\max(X_1, X_2)$ is an unbiased estimator of $\theta^* = EX_1$ iff (for some c, $-\infty < c < +\infty$) $P[X_1 = c] = 1$. [*Note*: Since part (f) of Problem 7.3, we have been talking of unbiased estimators of $g(\theta)$ for $\theta \in \Theta$, which is a slightly more general concept than that given in Definition 7.2.2. The same applies to asymptotic unbiasedness, consistency, and so on, and is usually assumed without so stating in works on statistics. Note that in Problems 8.1 and 8.2, $g(\theta) = \theta^*$ and $\Theta = \{F(\cdot): F(\cdot)$ is a d.f. whose expected value exists$\}$. Thus, EX_1 should more properly be written $E_\theta X_1$ (or, if we denote a typical member of Θ by F, $E_F X_1$). Since *many* distributions F have the *same* $E_F X_1$, it does *not* suffice to let EX_1 index Θ.]

8.3 **(a)** Find $Em_3 = E \dfrac{\sum_{i=1}^{n} (X_i - \bar{X})^3}{n}$.

(b) In part (a) you found $Em_3 = c(n)\mu_3$. How can you use this fact to find an unbiased estimator of μ_3?

8.4 Consider the newsboy of Example 8.1.13 who wishes to set K $(0 \leqslant K \leqslant U)$ to maximize the average profit $\bar{P}_K$.

(a) If $s < p < r$ and demand X is known to have a uniform distribution on $0, 1, 2, \ldots, M$, what is the best K? (First assume $M \leqslant U$; then do not assume $M \leqslant U$.) What is the value of K if $M = 200$, $U = 200$, $s = 5$, $p = 8$, and $t = 10$?

(b) Suppose $s = 5$, $p = 8$, $r = 10$, $U = 200$, and X is known to have a Poisson distribution with mean 100. Compute (via slide rule, desk calculator, or computer) the best K. Compare this with the value found in part (a) [both (a) and (b) specify $EX = 100$].

8.5 Suppose that $X_1, \ldots, X_n$ are independent r.v.'s, each having the Poisson distribution (see Definition 3.2.7) with unknown mean (see Theorem 5.2.3) $\lambda \in \Theta = \{x: x > 0\}$.

(a) Find "the" MLE of λ. Does an MLE of λ exist for all possible values of $X_1, \ldots, X_n$? If not, how could you alter Θ so that the MLE would always exist?

(b) The correct answer to the last question in part (a) is undesirable, in that it allows our MLE $\hat{\lambda}$ to take on a value *known* to be false. One attempt sometimes made to solve this problem is to define any $\hat{\theta} = \hat{\theta}_n(X_1, \ldots, X_n) \in \Theta$ such that (for some $\varepsilon > 0$)

$$L(\hat{\theta}) \geqslant (1 - \varepsilon) \sup\{L(\theta): \theta \in \Theta\}$$

to be an **ε-MLE of $\boldsymbol{\theta}$** [that is, it comes within ε of maximizing the likelihood]. Find all the ε-MLE's of λ. (Specify your answer implicitly, involving the solution of a certain equation. Specify it explicitly for the case $X_1 = X_2 = \cdots = X_n = 0$.)

8.6 Suppose that $X_1, \ldots, X_n$ are independent r.v.'s, each $N(\mu, \sigma^2)$ with both μ and σ^2 unknown, $(\mu, \sigma^2) \in \Theta = \{(x, y): -\infty < x < +\infty, y > 0\}$.

(a) Suppose $n \leqslant 2$. Find the MLE of (μ, σ^2). [*Hint*: Show that, regardless of σ^2, $L(\mu, \sigma^2)$ may be maximized by setting $\mu = \hat{\mu}$ for a certain $\hat{\mu}$; then maximize $L(\hat{\mu}, \sigma^2)$ with respect to σ^2.]

(b) Suppose $n = 1$. Does the MLE of (μ, σ^2) exist? (Compare with Examples 8.2.8 and 8.2.13.) Can the ε-MLE of Problem 8.5 be used here? Why or why not?

(c) Suppose that $n = 88$ and that $X_1, \ldots, X_{88}$ are taken to be the seasonal snowfalls observed at Rochester, New York, since the 1884–85 winter, which are given in Table 8.P–1. Find the numerical values of $\hat{\mu}$ and $\hat{\sigma}^2$ for this set of data.

8.7 Suppose that we observe r.v.'s $Y_1, \ldots, Y_n$ and we know that (for $i = 1, \ldots, n$)

$$Y_i = \alpha + \beta x_i + V_i,$$

where: $x_1, \ldots, x_n$ are known constants; α and β are unknown constants; and $V_1, \ldots, V_n$ are independent r.v.'s, each $N(0, \sigma^2)$ with σ^2 known. Find the MLE of (α, β).

8.8 Suppose that X and Y are independent r.v.'s, each with a normal distribution, and that $EX = \theta$, $EY = 3\theta$, $\mathrm{Var}(X) = \mathrm{Var}(Y) = 1$. Here $\theta \in \Theta = \{x: -\infty < x < +\infty\}$ is unknown.

(a) Find the MLE of θ.

(b) If an estimator of the form $aX + bY$ is to be used to estimate θ, what restrictions must the constants a and b satisfy in order that the estimator be unbiased?

(c) Of all unbiased estimators of θ which have the form $aX + bY$, which one has the smallest variance? (This estimator is called a *b*est (or *m*inimum *v*ariance) *l*inear *u*nbiased *e*stimator, usually abbreviated to BLUE or MVLUE.)

8.9 Let $\mathbf{p} \in \mathscr{R}^k$ and $c \in \mathscr{R}$ be fixed. Then, by definition, $\{\mathbf{x}: \mathbf{p}'\mathbf{x} = c, \mathbf{x} \in \mathscr{R}^k\}$ is a *hyperplane* in $\mathscr{R}^k$. Define

$$S_1 = \{\mathbf{x}: \mathbf{p}'\mathbf{x} > c\},$$
$$S_2 = \{\mathbf{x}: \mathbf{p}'\mathbf{x} < c\},$$

(these are the two sides of the hyperplane $\mathbf{p}'\mathbf{x} = c$; see the discussion following Figure 8.3-3).

(a) Prove that S_1 and S_2 are each convex sets in $\mathscr{R}^k$.

(b) Prove that S_1 and S_2 are disjoint (that is, $S_1 \cap S_2 = \emptyset$).

(c) Prove that the hyperplane $\mathbf{p}'\mathbf{x} = c$ divides $\mathscr{R}^k$ into two disjoint parts, in one of which $\mathbf{p}'\mathbf{x} > c$ and in the other of which $\mathbf{p}'\mathbf{x} < c$. Note that both parts are nonempty iff $\mathbf{p} \neq \mathbf{0}$.

Table 8.P–1. Seasonal Snowfall (in Inches) at Rochester, New York

Winter	*Amount*	*Winter*	*Amount*	*Winter*	*Amount*
1884–85	90.3	1914–15	60.8	1944–45	94.7
1885–86	64.7	1915–16	106.9	1945–46	49.5
1886–87	72.6	1916–17	92.9	1946–47	75.5
1887–88	50.7	1917–18	79.3	1947–48	63.4
1888–89	82.2	1918–19	36.0	1948–49	50.9
1889–90	65.8	1919–20	97.5	1949–50	81.7
1890–91	84.6	1920–21	61.3	1950–51	75.8
1891–92	77.1	1921–22	59.7	1951–52	75.8
1892–93	96.6	1922–23	102.5	1952–53	41.7
1893–94	102.4	1923–24	65.6	1953–54	77.5
1894–95	93.4	1924–25	60.8	1954–55	69.2
1895–96	123.8	1925–26	55.3	1955–56	121.4
1896–97	66.5	1926–27	80.6	1956–57	79.2
1897–98	91.0	1927–28	68.5	1957–58	130.8
1898–99	96.0	1928–29	55.1	1958–59	140.6
1899–1900	131.3	1929–30	59.5	1959–60	161.7
1900–01	141.5	1930–31	77.3	1960–61	89.4
1901–02	79.0	1931–32	75.7	1961–62	65.6
1902–03	68.3	1932–33	29.2	1962–63	76.4
1903–04	95.7	1933–34	77.1	1963–64	92.0
1904–05	68.3	1934–35	49.5	1964–65	71.1
1905–06	44.3	1935–36	90.0	1965–66	103.2
1906–07	63.7	1936–37	65.9	1966–67	74.0
1907–08	67.3	1937–38	54.7	1967–68	76.7
1908–09	66.5	1938–39	79.1	1968–69	79.8
1909–10	84.9	1939–40	54.5	1969–70	119.6
1910–11	107.4	1940–41	73.7	1970–71	142.7
1911–12	62.0	1941–42	66.3	1971–72	97.8
1912–13	59.2	1942–43	70.6		
1913–14	86.7	1943–44	46.1		

(d) What is the conclusion of the Separating Hyperplane Theorem (Theorem 8.3.5) for the disjoint convex sets S_1 and S_2? Are all the allowable $\mathbf{p}_1$'s in the conclusion of that theorem precisely $\mathbf{p}_1 = k\mathbf{p}$ (k any constant > 0)? (Prove or give a counterexample.)

(e) Illustrate the results of parts (a) through (d) by a *carefully* drawn graph of the case $k = 2$, $\mathbf{p}' = (-1, 1)$, $c = 1$. (Be sure to prove what the sets S_1 and S_2 are in this case.)

8.10 By definition (see Definition 8.3.1) a set $S \subseteq \mathscr{R}^k$ is *convex* if (for all $\mathbf{x}, \mathbf{y} \in S$)

$$\beta\mathbf{x} + (1 - \beta)\mathbf{y} \in S \quad (\text{all } \beta, 0 < \beta < 1).$$

Prove that if S is convex, $\mathbf{x}_1, \ldots, \mathbf{x}_n \in S$, and $q_1 + \cdots + q_n = 1$ ($q_1 \geq 0, \ldots,$

$q_n \geqslant 0$), then

$$q_1\mathbf{x}_1 + \cdots + q_n\mathbf{x}_n \in S.$$

[*Hint*: Do the trivial proofs for $n = 1, 2$ and then proceed by induction.]

8.11 Let S be any convex subset of $\mathscr{R}$. Prove that $f_1(x) = x^2$, $f_2(x) = |x|$, and $f_3(x) = e^x$ are each convex functions on S.

8.12 Let $\mathbf{p} \in \mathscr{R}^k$, $c \in \mathscr{R}$. Let $S = \{\mathbf{x}: \mathbf{p}'\mathbf{x} > c\}$. (In Problem 8.9 we proved S is convex in $\mathscr{R}^k$.) Prove that the function $f(\mathbf{x})$, defined on the $S \subseteq \mathscr{R}^k$ given above by $f(\mathbf{x}) = \mathbf{p}'\mathbf{x}$, is convex. For the case $k = 1$, graph S, $f(\mathbf{x})$, and the set

$$S_1 = \{(z_1, z_2)': \text{For some } x \in S, x = z_1 \text{ and } f(x) \leqslant z_2\}.$$

Is S_1 a convex subset of $\mathscr{R}^2$?

8.13 Suppose $f(\mathbf{x})$ is a convex function on $S \subseteq \mathscr{R}^k$. Let

$$S_1 = \{(z_1, \ldots, z_k, z_{k+1})': \quad \text{For some } \mathbf{x} \in S, \mathbf{x}' = (z_1, \ldots, z_k) \text{ and } f(\mathbf{x}) \leqslant z_{k+1}\}.$$

Prove that S_1 is a convex set in $\mathscr{R}^{k+1}$. [This fact gives $f(\mathbf{x})$ the name *convex function*; for a special case, see Problem 8.12.]

8.14 **(Geometric and arithmetic means)** Let $x_1, \ldots, x_n$ be any n real numbers. If $x_1 \geqslant 0, \ldots, x_n \geqslant 0$, then it is well-known that

$$(x_1 \cdots x_n)^{1/n} \leqslant \frac{x_1 + \cdots + x_n}{n} \qquad (*)$$

(the geometric mean is less than or equal to the arithmetic mean). Prove (*) using Jensen's Inequality (Theorem 8.3.7). [*Hint*: Let a r.v. Z be $\ln x_i$ w.p. $1/n$, and let $f(x) = e^x$.]

8.15 Assuming that $EZ \in S$, prove Jensen's Inequality (Theorem 8.3.7) for the case $k = 1$. [*Hint*: Since the point $(EZ, f(EZ))$ is on the boundary of the convex set S_1 of Problem 8.13, there is (by Theorem 8.3.4) a supporting hyperplane at $(EZ, f(EZ))$; call it $y = mx + b$. Since $(EZ, f(EZ))$ is on this hyperplane, the latter must be $y = f(EZ) + m(x - EZ)$. Compare this line with the curve $y = f(x)$, replace x by a random point Z, and take expected values to complete the proof.] A formal proof for any k is given in, for example, T. S. Ferguson, *Mathematical Statistics: A Decision Theoretic Approach*, Academic Press Inc., New York, 1967, pp. 76–77.

8.16 In the proof of Lemma 8.3.12, we used Jensen's Inequality for the function $f(x) = e^x$ which is (Problem 8.11) convex on $\mathscr{R}^1$, and for a r.v. U for which we knew EU exists and $P[U \in \mathscr{R}] = 1$. We could thus conclude that $EU \in \mathscr{R}$ and $e^{EU} \leqslant Ee^U$ (or $EU \leqslant \ln Ee^U$). In fact, we concluded that $EU < \ln Ee^U$ since U is *not* {constant w.p. 1}. Prove that $e^{EU} = Ee^U$ iff U is {constant w.p. 1}.

8.17 In the proof of Theorem 8.3.14, we introduced a quantity h and, just below (8.3.20), used the fact that $0 < h < 1$. Why did we need $0 < h < 1$? Prove that $0 < h < 1$.

8.18 Prove that any CAN estimator is consistent.

8.19 Prove that, if an estimator t_n of θ is such that $n^\delta(t_n - \theta)$ is asymptotically $N(0, J^2(\theta))$ for some $\delta > 0$, then t_n is consistent.

8.20 Let t_n be an estimator of θ such that $n^\delta(t_n - \theta)$ is asymptotically $N(0, J^2(\theta))$ for some $\delta > 0$. Show that (in a reasonable sense which you should define) such an estimator is: **(1)** more efficient than any CAN estimator if $\delta > \frac{1}{2}$; **(2)** a CAN estimator if $\delta = \frac{1}{2}$; **(3)** less efficient than any CAN estimator if $0 < \delta < \frac{1}{2}$.

8.21 Let X_n be a r.v. with d.f. $F_n(\cdot)$, and let X be a r.v. with d.f. $F(\cdot)$. Give examples where $F_n \to F$ and:

(a) $\lim_{n\to\infty} \text{Var}(X_n) = +\infty$ while $\text{Var}(X) = 1$;

(b) $\lim_{n\to\infty} \text{Var}(X_n) = \sigma^2 > 1$ while $\text{Var}(X) = 1$.

8.22 Suppose we observe $X_1, \ldots, X_n$, independent r.v.'s each $N(\mu, \sigma^2)$ with μ known and σ^2 unknown. The MLE $\hat{\theta}_n$ of $\theta = \sigma^2$ was found in Example 8.2.13. Show that $\sqrt{n}(\hat{\theta}_n - \theta)$ is asymptotically normal. [*Hint*: Use the CLT, Theorem 6.4.6.]

8.23 Suppose that $X_1, \ldots, X_n$ are independent r.v.'s, each with the uniform distribution on $[\theta - \frac{1}{2}, \theta + \frac{1}{2}]$ where θ is unknown. Any $t_n = t_n(X_1, \ldots, X_n)$ such that

$$\max(X_1, \ldots, X_n) - \tfrac{1}{2} \leqslant t_n \leqslant \min(X_1, \ldots, X_n) + \tfrac{1}{2}$$

was found (in Example 8.2.17) to be an MLE.

(a) For the particular choice of t_n as $t_n^* = \{\max(X_1, \ldots, X_n) + \min(X_1, \ldots, X_n)\}/2$, find the limiting distribution of $n(t_n^* - \theta)$.

(b) Compare t_n^* to any CAN estimator t_n'. Which is preferable (for large n) and why? Give an example of such a CAN estimator t_n'. [*Hint*: Try $\bar{X}$.]

(c) For a general MLE t_n, find the limiting distribution of $n(t_n - \theta)$.

8.24 Suppose that the MLE $\hat{\theta}_n$ of θ based on $X_1, \ldots, X_n$ is a CAN estimator, so that $\sqrt{n}(\hat{\theta}_n - \theta)$ is asymptotically $N(0, J^2(\theta))$. Let $\varepsilon > 0$ be fixed. Exhibit an alternate estimator t_n such that $\sqrt{n}(t_n - \theta)$ is asymptotically $N(0, J_2^2(\theta))$ with $J_2^2(\theta) = J^2(\theta)$ for $\theta \neq 0$ and $J_2^2(\theta) = \varepsilon$ for $\theta = 0$. (This shows that there is no nontrivial lower bound to the variance of a CAN estimator. Does it also show, or can it be extended to show, that there is no "best" CAN estimator—that is, that any CAN estimator can be improved by Hodges' Superefficienciation?) [*Hint*: Use Example 8.4.17.]

8.25 **(a)** Work through Example 8.4.17 in detail with the choice $\alpha = 0$. What does the result you obtain mean? Is it true that then $e_\theta(\hat{\theta}_n, t_n) = 1$ for $\theta \neq 0$ and $= +\infty$ for $\theta = 0$? [*Hint*: Is t_n then a CAN estimator?]

(b) Modify Example 8.4.17 so as to obtain an alternate estimator t_n^* which is superefficient at $\theta = 0$ and at $\theta = 1$.

(c) Example 8.4.17 implicitly assumed that $\theta \in \Theta$, where Θ is the set in which the unknown θ lies. Can a similar example be given if $\theta \notin \Theta$? Explain.

8.26 Suppose r.v.'s $X_1, \ldots, X_n$ are independent and have d.f. $F(x)$ and density $f(x)$. Denote the ranked $X_1, \ldots, X_n$ by $Y_1 \leqslant \cdots \leqslant Y_n$. Prove that for any k $(1 \leqslant k \leqslant n)$

$$f_{Y_k}(y) = \frac{n!}{(k-1)!(n-k)!}[F(y)]^{k-1}[1 - F(y)]^{n-k} f(y).$$

[*Hint*:

$$\begin{aligned}F_{Y_k}(y) &= P[Y_k \leqslant y] = P[\text{At least } k \text{ of } X_1, \ldots, X_n \text{ are } \leqslant y]\\ &= \sum_{i=k}^{n} P[\text{Exactly } i \text{ of } X_1, \ldots, X_n \text{ are } \leqslant y]\\ &= \sum_{i=k}^{n} \binom{n}{i} [F(y)]^i [1 - F(y)]^{n-i}\\ &= \frac{n!}{(k-1)!(n-k)!} \int_0^{F(y)} t^{k-1}(1-t)^{n-k}\, dt,\end{aligned}$$

where we make use of integration by parts.]

8.27 At (8.5.7) and (8.5.8) we implicitly used rules of the calculus of "$\sim$". That is, we say "$a \sim b$" iff $\lim_{n\to\infty} a/b = 1$. Then we can (and please do) prove such results as:
(a) If $a \sim b$ and $c \sim d$, then $ac \sim bd$;
(b) If $a \sim b$ and $c \sim d$, then $a + b \sim c + d$.
Which other such results were used in the proof referred to at (8.5.7) and (8.5.8)?

8.28 Suppose that $a_n \sim c_n$ and that $\lim_{n\to\infty} c_n = c$. Show that then $\lim_{n\to\infty} a_n = c$. Was this fact used in the first proof (Proof 1) of Theorem 8.5.3? [*Hint*: Inspect (8.5.9) and (8.5.10) carefully.]

8.29 Prove that if $\lim_{n\to\infty} \ln a_n = c$, then $\lim_{n\to\infty} a_n = e^c$.

8.30 Prove that for a_n and b_n as defined at (8.5.29) it follows that $\lim_{n\to\infty} a_n = 1$ and $\lim_{n\to\infty} b_n = 0$. Prove that, if Y_n is a r.v. such that $F(\mathrm{t}) = \lim_{n\to\infty} P[Y_n \leqslant t]$ exists and is a continuous function of t, then

$$\lim_{n\to\infty} P[Y_n \leqslant a_n t + b_n] = \lim_{n\to\infty} P[Y_n \leqslant t].$$

Is this result true if $F(t)$ is not a continuous function of t? (Prove or give a counterexample.)

8.31 At (8.5.31) we noted that, since $N_n^*(y)$ is a standardized binomial r.v. based on n trials and probability of success $p = 1 - F(y)$, the CLT implies

$$\lim_{n\to\infty} P[N_n^*(y) \leqslant t] = \frac{1}{\sqrt{2\pi}} \int_{-\infty}^{t} e^{-z^2/2}\, dz.$$

At (8.5.32) we needed (and used) this result with $y = y_p = \xi_p + \sqrt{pq/n}\,(t/f(\xi_p))$; that is, the probability of success is now a function of n, say $p_n = 1 - F(y_p) = 1 - p - \sqrt{(pq/n)}\, t + o(1/\sqrt{n})$ [where (8.5.27) was used to eliminate $F(\cdot)$]. Prove that

$$\lim_{n\to\infty} P[N_n^*(y_p) \leqslant t] = \frac{1}{\sqrt{2\pi}} \int_{-\infty}^{t} e^{-z^2/2}\, dz,$$

where $N_n^*(y_p)$ is a standardized binomial r.v. based on n trials and probability of success p_n. [*Hint*: Use characteristic function methods.]

8.32 Let $X_1, \ldots, X_n$ be independent $N(\mu, \sigma^2)$ r.v.'s. Show that, for the sample median M_n of Definition 8.1.28, $E_\mu M_n \equiv \mu$ (that is, M_n is an unbiased estimator of μ). [*Hint*: Consider the cases "n even" and "n odd" separately.]

8.33 Let $X_1, \ldots, X_n$ be independent $N(\mu, \sigma^2)$ r.v.'s. Show that the sample mean $\bar{X}$ is a CAN estimator of μ by using the CLT. Next, show that $\bar{X}$ is a CAN estimator of μ by using footnote 21 of Problem 8.14H (which can be used to determine that $\bar{X}$ is $N(\mu, \sigma^2/n)$).

8.34 **(a)** Verify that Assumption 8.6.2 holds in Example 8.6.20 [estimation of σ^2 based on $X_1, \ldots, X_n$ which are independent and each $N(\mu, \sigma^2)$ with μ known].

(b) Verify that $EY^2 = 1$ and $EY^4 = 3$ if Y is a $N(0, 1)$ r.v.

(c) Note that, even if we did not bother to verify that $\hat{\sigma}^2$ is a fully regular estimator of σ^2 in Example 8.6.20 [that is, to verify that Assumption 8.6.5 holds], we would still have a useful result: $\hat{\sigma}^2$ is as good as any regular estimator of σ^2. Verify that Assumption 8.6.5 holds.

(d) For the alternate unbiased estimator of σ^2 given by (see the proof of Theorem 8.1.5)

$$\frac{n}{n-1} m_2 = \frac{\sum_{i=1}^{n} (X_i - \bar{X})^2}{n-1},$$

show that

$$\mathrm{Var}_{\sigma^2}\left(\frac{n}{n-1} m_2\right) = \frac{2\sigma^4}{n-1}.$$

Why might we use $nm_2/(n-1)$ to estimate σ^2, even though its variance exceeds that of $\hat{\sigma}^2$? [*Hint*: How will an error in the supposedly known value of μ affect each of these estimators?]

8.35 Investigate whether the estimator $\varphi_n(X_1, \ldots, X_n) = 0$ satisfies Assumption 8.6.2.

8.36 In the proof of Theorem 8.6.10 we used the relation

$$\frac{dw(\theta)}{d\theta} = \frac{d \ln w(\theta)}{d\theta} w(\theta) \tag{$*$}$$

for any positive function $w(\theta)$. Now in the application made, $w(\theta) = f(x|\theta)$ is a nonnegative function (not a positive function), that is, sometimes $f(x|\theta) = 0$. How can we justify our use of relation $(*)$ in this case?

8.37 In obtaining Equation (8.6.13) we used the fact that

$$\frac{\partial}{\partial\theta} |J|^{-1} = 0.$$

Prove this result. [*Hint*: J depends only on the transformation employed, and not on θ.]

8.38 Let $X_1, \ldots, X_n$ be $n(\geqslant 1)$ independent r.v.'s, each Poisson with parameter $\lambda > 0$ unknown. In Example 8.7.14 we showed that $\bar{X}$ is an efficient estimator of λ. Show that $\sqrt{n}(\bar{X} - \lambda)$ is asymptotically $N(0, \lambda)$; hence, $\bar{X}$ is a CAN estimator.

8.39 Let $X_1, \ldots, X_n$ be $n(\geqslant 1)$ independent r.v.'s, with $P[X_i = 1] = p$ and $P[X_i = 0] = 1 - p$ $(i = 1, \ldots, n)$ where $p(0 \leqslant p \leqslant 1)$ is unknown. Suppose that this is a regular estimation case, and find the lower bound (8.6.19) on the variance of any regular unbiased estimator φ_n of p.

Find the MLE of p, and show that $\sqrt{n}(\text{MLE} - p)$ is asymptotically $N(0, \sigma^2(p))$ (and hence is a CAN estimator). Compare $\sigma^2(p)$ with

$$1/E_p\left(\frac{\partial \ln f(X|p)}{\partial p}\right)^2.$$

8.40 In our proof of the Neyman Factorization Theorem (Theorem 8.8.4), Equation (8.8.9) was developed by excluding the cases $t(X) = c \neq c_0$ and $P_\theta[t(X) = c_0] = 0$. Complete the proof by showing how to handle these cases.

8.41 Formally prove that if $t(\ )$ is a sufficient statistic, then so is $k(t(\ ))$ for any one-to-one function $k(\ )$. [*Hint*: See the paragraph after the proof of Theorem 8.8.4.]

8.42 Use the Neyman Factorization Theorem to prove that (under Assumption 8.8.3) the identity statistic is a sufficient statistic. (See the paragraph preceding Figure 8.8–3.)

8.43 Prove Theorem 8.8.12. [*Hint*: If there are no ties among $X_1, \ldots, X_n$, then $n!$ possible orderings exist and the conditional distribution on any contour assigns probability $1/n!$ to each ordering.]

8.44 At the end of Example 8.8.14 we claimed (8.8.17) could hold for at most 4 values of θ. However, it holds for all θ in the special case $b = c = d = e = 0$. Show the latter is true iff (x_1, x_2) is a permutation of (y_1, y_2) (which was excluded at the start of the example).

8.45 Develop considerations, for the proof of Theorem 8.8.18, like those developed in Problem 8.40 for the proof of the Neyman Factorization Theorem.

8.46 In the case of Example 8.8.20 find a minimal sufficient statistic for θ.

8.47 In the case of Example 8.8.22 find a minimal sufficient statistic for θ.

8.48 In the case of Example 8.8.24, use the Neyman Factorization Theorem to show that $t(X_1, \ldots, X_n) = \left(\sum_{i=1}^{n} X_i, \sum_{i=1}^{n} X_i^2\right)$ is a sufficient statistic for (μ, σ^2).

8.49 In Example 8.8.31 the likelihood-ratio statistic was not well-defined since we did not take account of the possibility that $f(x|\theta_0) = 0$. Take account of this in the course of using a Neyman Factorization Theorem argument to show that the likelihood-ratio statistic is a sufficient statistic.

8.50 Suppose $X_1, \ldots, X_n$ are independent r.v.'s, each with the uniform distribution on $[\theta - \frac{1}{2}, \theta + \frac{1}{2}]$, where θ is unknown, $\theta \in \Theta = \{x: -\infty < x < +\infty\}$.

(a) Show that $t(X_1, \ldots, X_n) = (\min(X_1, \ldots, X_n), \max(X_1, \ldots, X_n))$ is a sufficient statistic for θ.

(b) In Example 8.2.17 we found that the MLE of θ was not unique, but that any $\hat{\theta}$ such that

$$\max(X_1, \ldots, X_n) - \tfrac{1}{2} \leqslant \hat{\theta}(X_1, \ldots, X_n) \leqslant \min(X_1, \ldots, X_n) + \tfrac{1}{2}$$

is an MLE. Show how to choose an MLE which is not a function of $t(X_1, \ldots, X_n)$. (For example, D. S. Moore gives one in his paper referenced following the proof of Theorem 8.8.32.)

8.51 If $X_1, \ldots, X_n$ are independent r.v.'s, each having the Poisson distribution with unknown parameter $\mu \geqslant 0$, find a minimal sufficient statistic for μ. Is the MLE of μ (see Problem 8.5) itself a minimal sufficient statistic?

8.52 Show that if $(X_1, \ldots, X_n)$ is a member of the exponential class (see Definition 8.8.34), then $(\varphi_1(X_1, \ldots, X_n), \ldots, \varphi_r(X_1, \ldots, X_n))$ is a minimal sufficient statistic for θ. (See the discussion following Example 8.8.36.)

8.53 Show that, if equality holds in the Cramér-Rao Inequality for unbiased estimators (Corollary 8.6.18), then $(X_1, \ldots, X_n)$ is of one-parameter Koopman-Darmois form. (This result is also true in the case of $\theta = (\theta_1, \ldots, \theta_r)$ mentioned at the end of Section 8.7.)

8.54 Prove that if we have $n > 2$ independent observations from the Cauchy distribution (8.8.15), then no nontrivial sufficient statistic exists. (Example 8.8.14 proved this for the case $n = 2$.)

8.55 If X has the Poisson distribution with unknown parameter $\mu \geqslant 0$, prove that no nontrivial sufficient statistic exists. Compare this with the results of Problem 8.51.

8.56 In Example 8.8.24, prove that (for $n \geqslant 2$) $\left(\sum_{i=1}^{n} X_i/n, \sum_{i=1}^{n} (X_i - \bar{X})^2/(n-1)\right)$ is a minimal sufficient statistic.

8.57 Let $X_1, \ldots, X_n$ be independent r.v.'s, each Bernoulli (see Definition 3.2.1) with unknown probability of success $p(0 \leqslant p \leqslant 1)$. Prove that: $\sum_{i=1}^{n} X_i$ is a sufficient statistic; $\sum_{i=1}^{n} X_i/n$ is a sufficient statistic; $\left(\sum_{i=1}^{n} X_i\right)^2$ is a sufficient statistic; and $\sum_{i=1}^{n} X_i$ is a minimal sufficient statistic.

8.58 Let $X_1, \ldots, X_n$ be independent r.v.'s, each with $P[X_j = z_i] = p_i(i = 1, \ldots, k; j = 1, \ldots, n)$ where $0 \leqslant p_1, \ldots, p_k \leqslant 1$, $p_1 + \cdots + p_k = 1$, and no two of $z_1, \ldots, z_k$ are equal. Find a minimal sufficient statistic for $(p_1, \ldots, p_k)$, the unknown vector of probabilities. [*Hint*: Compare with the multinomial distribution of Definition 3.5.1, which bears essentially the same relation to these $X_1, \ldots, X_n$ as the binomial distribution bears to Bernoulli r.v.'s.]

8.59 Show that the family of Bernoulli distributions with unknown probability of success $p(0 \leqslant p \leqslant 1)$ is complete.

8.60 Show that the family of uniform distributions on $(0, \theta)$ (where $\theta > 0$ is unknown) is complete.

8.61 If X has the Bernoulli distribution with unknown probability of success $p(0 \leqslant p \leqslant 1)$, find: a complete sufficient statistic T; the unique MVU estimator $\varphi(T)$ of p; and $\operatorname{Var}_\theta \varphi(T)$.

8.62 As in Problem 8.61, but with X having a uniform distribution on $(0, \theta)$ where $\theta > 0$ is unknown.

8.63 Suppose we observe $X = (X_1, \ldots, X_n, Y_1, \ldots, Y_m)$ where $X_1, \ldots, X_n, Y_1, \ldots, Y_m$ are independent r.v.'s, and X_i is $N(\mu, \sigma_1^2)$ $(1 \leqslant i \leqslant n)$ while Y_j is $N(\mu, \sigma_2^2)$ $(1 \leqslant j \leqslant m)$. Show that (assuming μ, σ_1^2, σ_2^2 are unknown with $-\infty < \mu < +\infty$, $\sigma_1^2 > 0$, $\sigma_2^2 > 0$) $T = (X_1 + \cdots + X_n, X_1^2 + \cdots + X_n^2, Y_1 + \cdots + Y_m, Y_1^2 + \cdots + Y_m^2)$ is a sufficient statistic. Show that T is not a complete sufficient statistic. [*Hint*: Consider $u(T) = \bar{X} - \bar{Y}$.]

8.64 Let $X_1, \ldots, X_n$ be independent r.v.'s, each with density

$$f(x|\theta) = \begin{cases} \dfrac{x}{\theta} e^{-.5x^2/\theta}, & x > 0 \\ 0, & x \leqslant 0, \end{cases}$$

where θ is unknown, $\theta \in \Theta = \{z: z > 0\}$. [This is called the **Rayleigh distribution**.] Find the method of moments estimator of θ.

8.65 As in Problem 8.64, but with $f(x|\theta)$ the uniform density on $(0, \theta)$.

8.66 Show that the two expressions given for $(\chi'(\theta))^2$ in Equation (8.10.12) are equal.

8.67 Let $X_1, \ldots, X_n$ be independent r.v.'s, each Bernoulli with parameter $p (0 \leqslant p \leqslant 1)$, that is, with $P[X_i = 1] = p$ and $P[X_i = 0] = 1 - p$ $(i = 1, \ldots, n)$. Find the MMCS estimator of p. Find the MCS estimator of p.

HONORS PROBLEMS

8.1H Suppose that $X_1, \ldots, X_n$ are independent and identically distributed r.v.'s (and that all moments referred to do exist). We defined (Definition 8.1.2)

$$m_k = \frac{\sum_{i=1}^{n} (X_i - \bar{X})^k}{n},$$

and of course $\mu_k = E(X_i - \mu_1')^k$.

(a) Find Em_4.

(b) Find A and B such that $E(Am_4 + Bm_2^2) = \mu_4$. [*Note*: In principle A and B may be functions of $X_1, \ldots, X_n$, say $A(X_1, \ldots, X_n)$ and $B(X_1, \ldots, X_n)$; however, they may not be functions of any population moments μ_k, since the latter are unknown. This problem claims we can, in fact, choose A and B as functions of n only, independent of $X_1, \ldots, X_n$.]

(c) Find $\operatorname{Var}(m_2)$ and $\operatorname{Var}\left(\sum_{i=1}^{n} (X_i - \mu)^2/n\right)$.

8.2H In the newsboy problem (Example 8.1.13), suppose we wish to set K $(0 \leqslant K < +\infty)$ to maximize the average profit $\bar{P}_K$. Suppose that X is known to have a uniform distribution on the integers $L, L + 1, \ldots, L + Q - 1$ (Q possible demands), and that $s < p < r$. Show that

$$\bar{P}_K = \begin{cases} (r - p)K, & \text{if } K < L \\ (r - p)K + \dfrac{r - s}{2Q}[K(K + 1) - L(L - 1)] \\ \qquad + K(s - r)\dfrac{K - L + 1}{Q}, & \text{if } L \leqslant K \leqslant L + Q - 1 \\ (r - p)K + \dfrac{r - s}{2Q}[(L + Q - 1)(L + Q) \\ \qquad - (L - 1)L] + K(s - r), & \text{if } K > L + Q - 1. \end{cases}$$

Hence, deduce that the K which maximizes $\bar{P}_K$ is the integer closest to $Q(r - p)/(r - s) + L - .5$.

8.3H In the proof that $m_k \xrightarrow{P} \mu_k$ (Theorem 8.1.5) we immediately concluded from the fact of continuity of a polynomial Π jointly in its arguments that stochastic convergence of its arguments implied stochastic convergence of the polynomial itself (to the same polynomial evaluated at the limits of its arguments). Prove that this is so as formalized below in parts (a), (b), and (c).

Let $X_1, X_2, \ldots$ be any r.v.'s (not necessarily independent or identically distributed). Let $t_1 = t_1(X_1, \ldots, X_n), \ldots, t_r = t_r(X_1, \ldots, X_n)$ be r statistics. Suppose $t_1 \xrightarrow{P} a_1, \ldots, t_r \xrightarrow{P} a_r$ (as $n \to \infty$) with $a_1, \ldots, a_r$ constants.

(a) $\Pi(t_1, \ldots, t_r) \xrightarrow{P} \Pi(a_1, \ldots, a_r)$ for any polynomial $\Pi(\cdot, \ldots, \cdot)$.

(b) $R(t_1, \ldots, t_r) \xrightarrow{P} R(a_1, \ldots, a_r)$ for any rational function R of r variables with $|R(a_1, \ldots, a_r)| < +\infty$.

(c) $g(t_1, \ldots, t_r) \xrightarrow{P} g(a_1, \ldots, a_r)$ if $g(\cdot, \ldots, \cdot)$ is a continuous function of r variables and $|g(a_1, \ldots, a_r)| < +\infty$.

Notes: **(1)** (c) $\Rightarrow$ (b) $\Rightarrow$ (a). **(2)** (b) is called **Slutsky's Theorem** (for example, see the book by Fisz listed in the Selected References, p. 238, or S. S. Wilks, *Mathematical Statistics*, John Wiley & Sons, Inc., New York, 1962, p. 104). **(3)** (c) may be generalized to the case where $a_1, \ldots, a_r$ are r.v.'s (see, for example, p. 103 of Wilks' book). Note that if we also have a random parameter to contend with, things are a bit more complex; thus, Wilks' 4.3.8 on p. 105 is slightly wrong (see Yukio Suzuki, "On uniform convergence in probability," *Proceedings of the Institute of Statistical Mathematics*, Vol. 12 (1964–1965), pp. 221–224; this is in Japanese with an English Summary).

For a broad generalization of (c), see E. Sverdrup, "The limit distribution of a continuous function of random variables," *Skandinavisk Aktuarietidskrift*, Vol. 35 (1952), pp. 1–10.

8.4H Prove Theorem 8.1.30, Part (2) (the result that the sample median M_n is a consistent estimator of the population median ν) by using the Glivenko-Cantelli Theorem (Theorem 8.1.11) in any form.

8.5H In Theorem 8.2.15, we noted that it makes sense to talk of "the MLE of $g(\theta)$" if $g(\theta)$ is a one-to-one function, and that $\widehat{g(\theta)} = g(\hat{\theta})$. This is somewhat analogous to the situation discussed in Problem 8.2, in that we often talk of "the MLE of $g(\theta)$" without ever having defined it. If $g(\theta)$ is *not* one-to-one, both P. W. Zehna (see "Invariance of maximum likelihood estimators," *Annals of Mathematical Statistics*, Vol. 37 (1966), p. 744) and R. H. Berk (see "Review # 1922," *Mathematical Reviews*, Vol. 33 (1967), pp. 342–343) propose to call $g(\hat{\theta})$ an MLE anyway. But E. J. Dudewicz ("Maximum likelihood estimators for non–1–1 functions," *Trabajos de Estadística y de Investigación Operativa*, Vol. 22 (1971), pp. 65–70) proposes a different solution.

(a) Why does Zehna call $g(\hat{\theta})$ an MLE of $g(\theta)$? (Include the **"induced likelihood function"** in your discussion.)

(b) Why does Berk (who says Zehna "misses the point") wish to call $g(\hat{\theta})$ an MLE of $g(\theta)$?

(c) Why does Dudewicz think Berk's reasoning questionable? Study the

IMLE of Dudewicz in a case not covered by his paper, and compare it with $g(\hat{\theta})$.

8.6H Prove that it is not true that equality holds in Jensen's Inequality (8.3.7) iff **Z** is {constant w.p. 1}. Compare your result with that of Problem 8.16, where (in a special case) we showed equality does hold iff **Z** is {constant w.p. 1}. How do these results differ? [*Hint*: If the requirement that $f(\mathbf{x})$ be convex is strengthened to require that $f(\mathbf{x})$ be **strictly convex** in the sense that $\mathbf{x}, \mathbf{y} \in S$ and $0 < \beta < 1$ implies

$$f(\beta\mathbf{x} + (1 - \beta)\mathbf{y}) < \beta f(\mathbf{x}) + (1 - \beta)f(\mathbf{y}),$$

then equality will hold in Jensen's Inequality iff **Z** is {constant w.p. 1}. Prove this result.]

8.7H Prove Lemma 8.3.13.

8.8H In our proof of Theorem 8.3.14 (consistency of MLE's), we proved only the first part of (8.3.18). Prove the second part. Also, in our proof of (8.3.18) we made no explicit use of part (6) of Assumption 8.3.11. Explicitly consider the question of existence of $E_{\theta_0}[\ln f(X_j, \theta_i, \rho_\theta) - \ln f(X_j|\theta_0)]$. [*Hint*: By part (5) of (8.3.11), $E_{\theta_0} \ln f(X_j|\theta_0)$ exists, so we need only consider the effects of existence or non-existence (and in what sense nonexistence) of $E_{\theta_0} \ln f(X_j, \theta_i, \rho_{\theta_i})$ upon the proof.]

8.9H In Definition 7.2.4 we defined *consistency* of an estimator in terms of a "convergence in probability" or "weak convergence" (see Definition 6.3.2). We can similarly define **strong consistency** by replacing the requirement of "weak convergence" in Definition 7.2.4 by a requirement of "strong convergence" (see Definition 6.3.1). In fact, in his original article "Note on the consistency of the maximum likelihood estimate," *Annals of Mathematical Statistics*, Vol. 20 (1949), pp. 595–601, A. Wald actually proved strong consistency of the MLE. His proof is the same as the proof of Theorem 8.3.14 up to a certain point. Find that point and give in detail the remainder of his proof of strong consistency of the MLE.

8.10H Let X_n be a r.v. with d.f. F_n $(n = 1, 2, \ldots)$ and suppose $F_n \to F$ where F is a d.f. of some r.v. X. At Theorem 8.4.14, in establishing the reasonableness of looking at moments of limiting distributions rather than at limits of moments, we proved that (if $EX_n = 0$ for $n = 1, 2, \ldots$ and $\text{Var}(X) = \sigma^2 < +\infty$) the limit of $\text{Var}(X_n)$ [if the limit exists; limit inferior otherwise] is $\geqslant \text{Var}(X)$. We now seek to remove the conditions $\mathscr{C}$: "$EX_n = 0$ for $n = 1, 2, \ldots$ and $\text{Var}(X) = \sigma^2 < +\infty$."

(a) Show that if $\text{Var}(X_n)$ does not exist when $n \geqslant N$ for some integer N, then the result of Theorem 8.4.14 follows without assuming condition $\mathscr{C}$.

(b) Show that if $\text{Var}(X_n)$ exists for infinitely many $n \geqslant 1$, then for the purpose of investigating whether or not the result of Theorem 8.4.14 follows we may assume $\text{Var}(X_n)$ exists for all $n \geqslant 1$.

(c) Show that if $\text{Var}(X_n)$ exists for all $n \geqslant 1$, and if $\lim_{n\to\infty} \inf\{\text{Var}(X_j): j \geqslant n\} = +\infty$, then the result of Theorem 8.4.14 follows without assuming condition $\mathscr{C}$.

(d) Suppose that $\text{Var}(X_n)$ exists for all $n \geqslant 1$, and that $\lim_{n\to\infty} \inf\{\text{Var}(X_j):$

$j \geqslant n\} < +\infty$. Determine whether or not the result of Theorem 8.4.14 follows without assuming condition $\mathscr{C}$.

8.11H Let X_n be a r.v. with d.f. F_n $(n = 1, 2, \ldots)$ and suppose $F_n \to F$ where F is a d.f. of some r.v. X. If $0 = E(X_1) = E(X_2) = \cdots$, does $E(X) = 0$?

8.12H Suppose that the MLE $\hat{\theta}_n$ of θ based on $X_1, \ldots, X_n$ is such that $\sqrt{n}(\hat{\theta}_n - \theta)$ is asymptotically distributed according to a d.f. $L(\cdot|\theta)$. Attempt superefficienciation (using Hodges' method as given in Example 8.4.17); the aim here is to provide an alternate estimator t_n such that $\sqrt{n}(t_n - \theta)$ has a limiting d.f. $L'(\cdot|\theta)$, say, which is more concentrated about 0 than is $L(\cdot|\theta)$.

8.13H **(a)** For the MLE $\hat{d}$ of d in Example 8.2.16, find $E(\hat{d})$. Is $\hat{d}$ an unbiased estimator of d? Can you find a function of n, say $a(n)$, such that $a(n)\hat{d}$ is an unbiased estimator of d?

(b) In Example 8.2.17 we found that any $\hat{c}$ such that $L(X_1, \ldots, X_n) \leqslant \hat{c} \leqslant U(X_1, \ldots, X_n)$ is an MLE of c; here

$$L(X_1, \ldots, X_n) = \max(X_1, \ldots, X_n) - d,$$
$$U(X_1, \ldots, X_n) = \min(X_1, \ldots, X_n) + d.$$

Consider those $\hat{c}$'s which have the form

$$\hat{c} = a(n)L + b(n)U. \tag{$*$}$$

What restrictions on $a(n)$ and $b(n)$ are necessary and sufficient for $\hat{c}$ to be an unbiased MLE? Consider all $\hat{c}$'s of the form $(*)$; what restrictions on $a(n)$ and $b(n)$ are necessary and sufficient for $\hat{c}$ to be unbiased? Among all such unbiased $\hat{c}$'s, which one has the smallest variance? Find that smallest variance.

(c) In example 8.2.18 find $E(\hat{c})$. Is $\hat{c}$ an unbiased estimator of c?

8.14H **(a)** Let U, V, W be independent r.v.'s whose respective distributions are $N(0, \theta)$, $N(0, 1)$, $N(0, 1)$ where $\theta > 0$ is unknown. Let $X = U + V$ and $Y = U + W$. Show[21] that the joint distribution of (X, Y) is (see Definition 4.5.2 and Theorem 5.3.17) bivariate normal with $EX = EY = 0$, $\text{Var}(X) = \text{Var}(Y) = 1 + \theta$, and $\text{Cov}(X, Y) = \theta$ (hence $\rho(X, Y) = \theta/(1 + \theta)$).

(b) Let $(X_1, Y_1), \ldots, (X_n, Y_n)$ be n independent pairs of r.v.'s where (X_i, Y_i) have the bivariate normal distribution found in part (a). Find the MLE of $\lambda = 1/(1 + 2\theta)$ based on $(X_1, Y_1), \ldots, (X_n, Y_n)$. Find the MLE of θ.

(c) For the MLE of θ, say $\hat{\theta}$, found in part (b) find the exact probability distribution. [*Hint*: $(a + b)^2 = a^2 + 2ab + b^2$.]

8.15H Suppose Definition 8.5.2 is modified to let

$$Z_p = \begin{cases} Y_{[np]+1}, & \text{if } np \text{ is not an integer} \\ aY_{np} + (1 - a)Y_{np+1}, & \text{if } np \text{ is an integer} \end{cases}$$

---

[21] This result is a special case of the result [for example, see Fisz (1963), p. 162] that: ***If*** $(X_1, \ldots, X_n)$ ***has a multivariate normal distribution, then so does*** $(Y_1, \ldots, Y_r)$ ***where*** $Y_1, \ldots, Y_r$ ***are*** $r \leqslant n$ ***linear functions of*** $X_1, \ldots, X_n$. Here you are to prove this result directly in a special case.

where a is fixed ($0 < a \leqslant 1$). Show that Theorem 8.5.3 is still valid for this definition of Z_p. Find a natural choice of a. [*Hint*: Why do we choose $a = .5$ in Definition 8.1.28 for the case $p = .5$?]

8.16H Prove the two limits immediately preceding Equation (8.5.18) are in fact 0 and 1 as was stated.

8.17H The **Lebesgue Dominated Convergence Theorem (LDCT)** states that: *If* $E \subseteq \mathscr{R}$,

$$\lim_{n\to\infty} f_n(x) = f(x) \quad (\text{all } x \in E),$$

and

$$|f_n(x)| \leqslant g(x) \quad (\text{all } x \in E;\ n = 1, 2, 3, \ldots)$$

where $\int_{\{x:x\in E\}} g(x)\,dx$ *exists, then*

$$\lim_{n\to\infty} \int_{\{x:x\in E\}} f_n(x)\,dx = \int_{\{x:x\in E\}} f(x)\,dx.$$

Use this theorem to justify the interchange of limit and integral at (8.5.21).

8.18H Prove the result of Problem 8.26 without using integration by parts; rather, directly differentiate

$$F_{Y_k}(y) = \sum_{i=k}^{n} \binom{n}{i} [F(y)]^i [1 - F(y)]^{n-i}.$$

8.19H Fix $p (0 < p < 1)$. Let $X_1, \ldots, X_n$ be independent r.v.'s with d.f. $F(x)$ and density $f(x)$. Suppose that

$$\lim_{x\uparrow F^{-1}(p)} f(x) = A > 0 \quad \text{and} \quad \lim_{x\downarrow F^{-1}(p)} f(x) = B > 0$$

with $A \neq B$. (The case $A = B$ was dealt with in Theorem 8.5.3.) Find the limiting distribution of

$$\sqrt{n}(Z_p - F^{-1}(p)).$$

[*Hint*: This is a nonstandard case, and a proof like Proof 2 of Theorem 8.5.3 is recommended. Two references on this method are D. S. Moore, "Limiting distributions for sample quantiles," *American Mathematical Monthly*, Vol. 76 (1969), pp. 927–929, and L. Weiss, "Asymptotic distributions of quantiles in some nonstandard cases," *Nonparametric Techniques in Statistical Inference*, Cambridge University Press, London, 1970, pp. 343–348.] Note that this study involves the case of a discontinuity of $f(x)$ at $F^{-1}(p) = \xi_p$.

8.20H Show that Example 8.6.20 dealt with a regular estimation case.

8.21H Show that Example 8.7.14 dealt with a regular estimation case.

8.22H Show that Problem 8.39 dealt with a regular estimation case.

8.23H In Dynkin's Theorem (Theorem 8.8.27) we did not completely define $t(X)$, since we did not consider what to do if $f(X|\theta_0) = 0$. Show how to handle this problem. [*Hint*: First modify (8.8.29) appropriately. You may wish to consult E. B. Dynkin, "Necessary and sufficient statistics for a family of probability distributions," *Selected Translations in Mathematical Statistics and Probability, Vol. 1*, American Mathematical Society, Providence, Rhode Island, 1961, pp. 17–40.]

8.24H Show that with $n = 1$ neither the Cauchy distribution studied in Example

8.8.14, nor the uniform distribution on $(0, \theta)$ studied in Example 8.8.22, are members of the univariate one-parameter exponential family. [*Hint*: First prove that if X_1 is a member of the univariate one-parameter exponential family, then $(X_1, \ldots, X_n)$ where $X_1, \ldots, X_n$ are independent r.v.'s with the same distribution as X_1 is a member of the one-parameter exponential family.]

8.25H The difficulties we must deal with in trying to prove that existence of a nontrivial sufficient statistic implies Koopman-Darmois form [see the discussion following Example 8.8.36] can be illustrated as follows. Let $X_1, \ldots, X_n$ be independent r.v.'s, each with the same density function $f(x|\theta)$ where $\theta \in \Theta \subseteq \mathscr{R}$ is unknown. Let $d_i(j)$ be the digit in the 10^i place when X_j is put in decimal notation ($i = 0, \pm 1, \pm 2, \ldots$) for $j = 1, \ldots, n$. Form the statistic

$$t(X_1, \ldots, X_n) = \cdots d_2(1) \cdots d_2(n) d_1(1) \cdots d_1(n) d_0(1) \cdots d_0(n) d_{-1}(1) \cdots d_{-1}(n) \cdots. \qquad (*)$$

(For example, if $n = 2$ and we find $X_1 = 1.67$ and $X_2 = 36.85$, then $t(X_1, X_2) = 0316.6875$.) Prove that $t(X_1, \ldots, X_n)$ is a sufficient statistic for θ, in fact, that it is a trivial sufficient statistic.

Thus, we can "reduce the data" to a single number *without* uniting any points on contours, so in a real sense $t(X_1, \ldots, X_n)$ still has n dimensions! In using a statistic like $(*)$, we can run into practical problems, since in practice the "computer" usually has a finite memory. Some aspects of this have been studied by T. M. Cover, "Hypothesis testing with finite statistics," *Annals of Mathematical Statistics*, Vol. 40 (1969), pp. 828–835.

8.26H In our proof of Theorem 8.8.39 (the Rao-Blackwell Theorem) we wrote $h(y|x)$ as $f(x, y)/f_1(x)$; however, if $f_1(x) = 0$, then $h(y|x) = 0$ by definition. Revise the proof to allow for $f_1(x) = 0$, and show that this causes no problems.

8.27H Show that if $(X_1, \ldots, X_n)$ is a member of the exponential class, then (under certain conditions, which you should specify) $(\varphi_1(X_1, \ldots, X_n), \ldots, \varphi_r(X_1, \ldots, X_n))$ is a complete sufficient statistic. [*Hint*: See Problem 8.52. Problem 8.63 gives a counterexample to the statement that $(\varphi_1(X_1, \ldots, X_n), \ldots, \varphi_r(X_1, \ldots, X_n))$ is complete without any conditions being needed.]

8.28H Let $X_1, \ldots, X_n$ be independent r.v.'s, each with the same density $f(x|\theta)$ where $\theta \in \Theta = \{f(\cdot): f(\cdot) \text{ is a density function}\}$ (and $f(x|\theta)$ is the density function specified by θ). In Theorem 8.8.12 we showed that the order statistic $(Y_1, \ldots, Y_n)$ is a sufficient statistic in another context. Show that here, also, $(Y_1, \ldots, Y_n)$ is a sufficient statistic. [This is a distinct case since here Θ is not a subset of $\mathscr{R}^r$.] Prove that $(Y_1, \ldots, Y_n)$ is also a complete sufficient statistic. [*Hint*: See E. L. Lehmann, *Testing Statistical Hypotheses*, John Wiley & Sons, Inc., New York, 1959, p. 133.]

8.29H In Definition 8.9.1 we defined a *complete* family. *A family is called* **boundedly complete** *if (for every bounded function* $u(x)$*)* (8.9.2) *implies* $u(x) = 0$ *at all* x. Show that a family may be boundedly complete without being complete.

8.30H Construct a collection of probability distributions, indexed by a parameter θ, which has the following properties: **(1)** The collection consists of infinitely many distributions, no two of which are identical; **(2)** For every θ, $P_\theta[X = x] =$

0 except perhaps when $x = 1, 2, 3, \ldots$ [here X is a r.v. with the distribution indexed by θ]; and **(3)** A nontrivial sufficient statistic exists (that is, $t(X) = X$ is not the only sufficient statistic).

8.31H Let X_1, X_2 be two independent observations each with the same distribution. Give an example of a distribution (besides the Cauchy covered in Example 8.8.14) where no nontrivial sufficient statistic exists.

8.32H As in Problem 8.31H, but the distribution required to be such that $E_\theta X_1$ and $\text{Var}_\theta(X_1)$ exist and do not depend on θ.

Problems 8.33H and 8.34H present comprehensive studies of estimation in particular contexts, tying together the results of the various sections of Chapter 8. (Some aspects of these problems have been treated individually in the text.)

8.33H Let $X_1, \ldots, X_n$ $(n \geq 1)$ be independent r.v.'s, each with density

$$f(x|\theta) = \begin{cases} \dfrac{1}{\theta}, & 0 \leq x \leq \theta \\ 0, & \text{otherwise,} \end{cases}$$

where θ is unknown but $\theta \in \Theta = \{z: z > 0\}$.

(a) Find the maximum-likelihood estimator (say $\hat{\theta}_n$) of θ. [This wording does not necessarily imply that $\hat{\theta}_n$ exists or is unique.]

(b) Determine whether or not $\hat{\theta}_n$ is: consistent; unbiased; asymptotically unbiased.

(c) Find the iterated MLE (IMLE) (see Problem 8.5H) of θ. Find the method of moments estimator of θ.

(d) Let $Y = \hat{\theta}_n$. Find the limiting distribution of $n(Y - \theta)$.

(e) Prove that $Y = \max(X_1, \ldots, X_n)$ is a sufficient statistic for θ, via the Neyman Factorization Theorem. Prove that Y is a minimal sufficient statistic for θ.

(f) Prove that the family of probability density functions

$$f(y|\theta) = \begin{cases} n\dfrac{y^{n-1}}{\theta^n}, & 0 \leq y \leq \theta \\ 0, & \text{otherwise} \end{cases}$$

for $\theta \in \Theta = \{z: z > 0\}$ is complete.

(g) Prove that

$$\frac{2}{n} S_n = \frac{2}{n}(X_1 + \cdots + X_n)$$

is an unbiased estimator of θ, and find $\text{Var}_\theta((2/n)\, S_n)$. Give an expression for an estimator which has minimum variance among all unbiased estimators of θ.

(h) Suppose you are told that, for any unbiased estimator $\varphi_n = \varphi_n(X_1, \ldots, X_n)$ of θ, we have

$$E_\theta(\varphi_n - \theta)^2 = \text{Var}_\theta(\varphi_n) \geq \frac{\theta^2}{n(n + 2)}.$$

Then prove that $[(n + 1)/n]Y$ is the "unique" MVU estimator of θ. [A reference is J. Kiefer, "On minimum variance estimators," *Annals of Mathematical Statistics*, Vol. 23 (1952), pp. 627–629.]

(i) If the Cramér-Rao lower bound on the variance of unbiased estimators $\varphi_n(X_1, \ldots, X_n)$ of θ held here, it would say that

$$\mathrm{Var}_\theta(\varphi_n) \geqslant \underline{\qquad\qquad}.$$

8.34H Let $X_1, \ldots, X_n (n \geqslant 1)$ be independent random variables, each with density

$$f(x|\theta) = \begin{cases} e^{-(x-\theta)}, & x \geqslant \theta \\ 0, & \text{otherwise,} \end{cases}$$

where $\theta \in \Theta = \mathscr{R}$.

(a) through (i). Precisely as in (a) through (i) of Problem 8.33H, but with: $Y = \min(X_1, \ldots, X_n)$ in parts (d), (e), (h); $S_n/n - 1$ in part (g); $\mathrm{Var}_\theta(\varphi_n) \geqslant 1/n^2$ in part (h); and part (f) replaced by the corresponding problem for the density of the new Y.

8.35H Let $X_1, \ldots, X_n$ be independent random variables with density $f(x|\theta)$ for $\theta \in \Theta = \{\theta_1, \theta_2\}$.

(a) Prove directly [that is, without using Dynkin's Theorem (Theorem 8.8.27)] that

$$t_n = \frac{\prod_{i=1}^{n} f(X_i|\theta_2)}{\prod_{i=1}^{n} f(X_i|\theta_1)}$$

is a sufficient statistic for θ.

(b) Prove that t_n is a minimal sufficient statistic.

8.36H Suppose that, from a population of N items (for example, automobiles) which contains an unknown number θ of defective items with $\theta \in \Theta = \{0, 1, 2, \ldots, N\}$, a random sample of size n $(1 \leqslant n \leqslant N)$ is drawn under SRS without replacement (see Example 3.2.6). Let

$$X_i = \begin{cases} 1, & \text{if the } i\text{th item drawn is defective} \\ 0, & \text{otherwise} \end{cases}$$

$(i = 1, 2, \ldots, n)$. Show that (based on observations $X_1, \ldots, X_n$) the statistic $t(X_1, \ldots, X_n) = X_1 + \cdots + X_n$ is a sufficient statistic. Is it minimal sufficient? [$X_1 + \cdots + X_n$ is the number of defectives in the sample, and its distribution is hypergeometric with n, θ, N]

8.37H Prove that (8.9.14) implies $u(x) \equiv 0$ for all x (that is, that the normal family with μ unknown but σ^2 fixed is complete).

Consider generalizing Example 8.9.11 to the case where $\sigma^2 > 0$ is also unknown. Is $\bar{X}$ then still the unique MVU estimator of μ?

9

TESTS OF HYPOTHESES

In this chapter we cover the subject of "tests of hypotheses." This subject is covered in detail in most books on statistics at every level. Often this detail is given at the expense of point estimation, interval estimation, ranking and selection, and decision theory (which are sometimes even omitted). Usually, also, the detail is technical, with insufficient consideration being given to the underlying philosophy and to decision theory (which is treated in detail in general in Chapter 12). Below we give a balanced view of the subject, giving the motivation as well as the technical detail, and covering the contribution of decision theory to an integrated view of the topic as well as the contribution of the "classical" way of looking at the subject.

9.1. SIMPLE AND COMPOSITE NULL AND ALTERNATIVE HYPOTHESES

A hypothesis-testing problem usually results from questions like: "Does smoking cause cancer?"; "Do seat belts reduce car accident injuries?"; "Do atomic power plants increase radiation levels?"; or "Does adding putative catalyst A to a certain chemical process increase the yield?" Thus, we have an underlying parameter θ (in our four examples this might be: a cancer rate; an injury rate; a radiation level; or a process yield) and we wish to determine whether it changes in specified ways (for example, "Does it increase?") when an element of a system is changed (for example: smoking is added; seat belts are added; atomic power plants are added; or a putative catalyst is added).

Formally, let us make the following assumption (which represents no restriction, but gives us a precise notation).

Assumption 9.1.1. *Suppose that X is a r.v. with d.f. $F(x|\theta)$ where θ is unknown, $\theta \in \Theta$.* [Here X and θ may be multivariate.]

Let us now talk in terms of the question "Do atomic power plants increase radiation levels?" Then it may be that a long history of precise study has shown that to date the average level of radiation per year per person is $\theta_0 = 1.2$ units on some standard scale in the area within a 25-mile radius of a proposed model atomic power plant site. In fact, study has shown that the level of radiation per year in this area is random and normally distributed with a mean level of $\theta_0 = 1.2$ and a variance of $\sigma^2 = .09$; thus, a person chosen at random has a level X_1 which is an $N(\theta_0, \sigma^2)$ r.v. If n persons' radiation exposure is measured in this area next year we will observe independent r.v.'s $X_1, \ldots, X_n$ which are each $N(\theta, \sigma^2)$, *where*: $\theta = \theta_0$ if an atomic power plant *does not* operate in the area; *and* θ is unknown, $\theta \in \Theta = \{x: -\infty < x < +\infty\}$, if an atomic power plant *does* operate in the area.

The hypothesis that the atomic power plant has no effect (or a null effect) is called the "null hypothesis," denoted H_0, and is the assertion that $\theta = \theta_0$. The "alternative hypothesis," denoted H_1, that it has an effect is the assertion that $\theta \neq \theta_0$. Note that each "hypothesis" is a subset of Θ. It is also useful to distinguish whether, for any specified hypothesis $H \subseteq \Theta$, the distribution of X is completely determined or not.

Definition 9.1.2. *Suppose Assumption* 9.1.1 *holds. Any $H \subseteq \Theta$ is called a* **hypothesis**. *Often one hypothesis H_0 is called the* **null hypothesis**, *and another hypothesis H_1 is called the* **alternative hypothesis**.

Definition 9.1.3. *A hypothesis $H \subseteq \Theta$ is called* **simple** *if knowing $\theta \in H$ completely specifies $F(x|\theta)$. Otherwise, H is called a* **composite** *hypothesis.*

Thus, in our atomic power plant example, H_0 is a simple null hypothesis, while H_1 is a composite alternative hypothesis. (Note that it is not the case that H_0 and H_1 are unique for a particular problem. Thus, the alternative hypothesis in the atomic power plant example could just as well be taken to be $H_1^* = \{\theta: \theta \in \Theta, \theta > \theta_0\}$. This would be an alternative of increase, while H_1 was an alternative of change ($+$ or $-$).)

It is important to note that it is not as simple to "test" a hypothesis as the above simplified discussion might imply. Thus, we talked of sampling n people during a year of atomic power plant operation and comparing their radiation levels to θ_0. Yet the n people we choose will usually only be a sample of the N people living within 25 miles of the atomic power plant. For the N people the level θ may vary not only randomly (being a $N(\theta, \sigma^2)$

r.v.) but also by class of person (for example, it might be: higher for dentists and X-ray technicians; different—higher or lower—for new residents, people who were away part of the year on business or vacation, and babies; lower as distance from the atomic power plant increases; higher downwind from the plant; lower for people who dwell in basement apartments; and so on). It could also vary in general with wind velocities and directions, and with precipitation (for example, precipitation might tend to take radioactive particles to ground level, while wind would tend to take them to certain sections of the area or out of the area). It could also vary with the correlation experienced between peak load periods (when the most radiation might be emitted) and wind velocity. (Other factors which we would like to control include the level of background radiation, which varies as Sun spots vary, and so on.)

These problems of experimental validity also occur in other problem areas. If the question is "Does smoking cause cancer?" and we just compare smokers and nonsmokers, the retort is "Smokers may tend to be people with a predisposition to cancer." If one "smokes up" nonsmokers, the retort is "Those who let you do it are different"; if one "desmokes" smokers, the same retort calls the experimental results' validity into question. (Cigarette companies are experts at such retorts!) (To effectively totally answer such questions, we really need to find the causal mechanism. Nonetheless, an experiment was performed on dogs[1] which seems to obviate all retorts except "Humans are not dogs." To this we can say "Some are . . . especially those who, for economic gain, lead others to smoke." However, dogs (as well as guinea pigs, and so on) are customarily used to test drugs, heart disease remedies and theories, and so on, and are felt to be sufficiently comparable to humans to validate the results. Only in a Hitlerite society could many of the important medical hypotheses be tested on human populations.)

Figuring out how to conduct an experiment so that factors such as outlined above do not nullify its validity is called **designing an experiment** (or **experimental design**) and is a specialty of statisticians. It is for this reason that a statistician is usually employed on good research projects before any observations are made: in conjunction with experts in the subject matter of the specific area where the question arose, he formulates an experimental design which will yield data valid for the study of the question.

9.2. TESTS; ERRORS OF TYPE I AND TYPE II; POWER

Suppose Assumption 9.1.1 holds, that is, X is a r.v. with d.f. $F(x|\theta)$ where θ is unknown, $\theta \in \Theta$. Suppose we have a null hypothesis H_0 and an alternative hypothesis H_1. Our inference as to whether H_0 or H_1 contains the true θ will be based on observing X (whose d.f. $F(x|\theta)$ varies as $\theta \in \Theta$ varies).

(In terms of the atomic power plant example of Section 9.1, we can think of $\Theta = \mathscr{R}$, $H_0 = \{\theta_0\}$, and $H_1 = \{\theta: \theta \in \Theta, \theta \neq \theta_0\}$. We then wish to decide whether the true $\theta = \theta_0$ or the true $\theta \neq \theta_0$.) So, for some outcomes $X = x$ we will decide $\theta \in H_0$, while for the other outcomes $X = x$ we will decide $\theta \in H_1$. A "test" is a specification of these subsets of the set of possible outcomes $\mathscr{X}$ for some specification of H_0, H_1, Θ, X:

Definition 9.2.1. *Suppose that X is a r.v. with d.f. $F(x|\theta)$ where θ is unknown, $\theta \in \Theta$, and that a null hypothesis H_0 and an alternative hypothesis H_1 have been specified. The problem of deciding (after observing X) either that $\theta \in H_0$ (called* **accepting the null hypothesis**) *or that $\theta \in H_1$ (called* **rejecting the null hypothesis**) *is called a* **hypothesis-testing problem**. *H_0 is called* **true** *if $\theta \in H_0$, and* **false** *if $\theta \notin H_0$.*

If the set of possible outcomes of X is $\mathscr{X}$, then a division of $\mathscr{X}$ into two disjoint and exhaustive subsets A (called the **acceptance region**, *or* **region of acceptance of** H_0) *and R (called the* **critical region**, *or* **region of rejection of** H_0) *such that if we find $X \in A$ we accept H_0 and if we find $X \in R$ we reject H_0, is called a* (**nonrandomized**) **test of** H_0 **versus** H_1. [See Figure 9.2–1.]

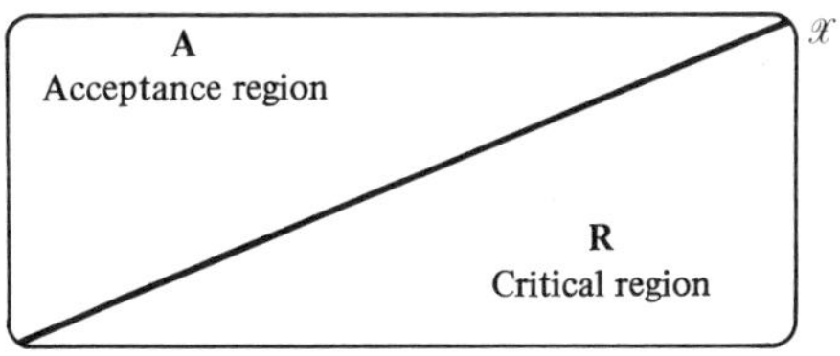

Figure 9.2–1. A nonrandomized test

Note that if we let $\phi(x)$ denote the probability that we reject H_0 after observing $X = x$, then for a nonrandomized test,

$$\phi(x) = \begin{cases} 1, & \text{if } x \in R \\ 0, & \text{if } x \in A. \end{cases}$$

However, for some points x it may be reasonable to flip a coin with some probability $\phi(x)$ of heads and decide to reject H_0 if heads turns up, otherwise accepting H_0. This is called a "randomized" test:

Definition 9.2.2. *A* (**randomized**) **test** *for a hypothesis-testing problem is a function (called a* **test function**) $\boldsymbol{\phi}(\mathbf{x})$, *defined for $x \in \mathscr{X}$, such that $0 \leqslant \phi(x) \leqslant 1$ for all $x \in \mathscr{X}$. If we observe X, we flip a coin with probability $\phi(X)$ of heads*

---

[1] See E. C. Hammond, O. Auerbach, D. Kirman, and L. Garfinkel, "Effects of cigarette smoking on dogs," *Archives of Environmental Health*, Vol. 21 (1970), pp. 740–768.

and reject (accept) H_0 *if heads (tails) occurs.* [Note that $\phi(\cdot)$ is also called a **critical function.**]

Such tests are usually needed only if X is a discrete r.v.; this need will be studied in later sections and can be ignored for now. (We will talk in terms of test functions, but these can be considered to take only values 0 or 1.)

Now when we conduct a test in a hypothesis-testing problem, two errors are possible: we can reject H_0 when it is true, and we can accept H_0 when it is false. These errors, and their probabilities of occurrence, are of central importance and hence have special names associated with them:

Definition 9.2.3. *In a hypothesis-testing problem, rejecting* H_0 *when it is true is called an* **error of Type I**. [*The probability of rejecting a true* H_0 *depends on the test* $\phi(\cdot)$ *employed and the true* $\theta \in H_0$, *and is called the* **level (of significance)** *of the test or the* **size of the critical region**.]

Accepting H_0 *when it is false is called an* **error of Type II**. [*The probability of rejecting a false* H_0 *depends on the test* $\phi(\cdot)$ *employed and the true* $\theta \in \Theta \cap \bar{H}_0$, *and is called the* **power** *of the test.*]

Example 9.2.4. Let us now illustrate these concepts using the atomic power plant example of Section 9.1. There we observe $X = (X_1, \ldots, X_n) \in \mathscr{X} = \mathscr{R}^n$ where $X_1, \ldots, X_n$ are independent r.v.'s, each $N(\theta, \sigma^2)$ with θ unknown, $\theta \in \Theta = \mathscr{R}$, and $\sigma^2 = .09$. (The case of unknown σ^2 will be considered later in this chapter, since it is more complicated.) Our null hypothesis is $H_0 = \{\theta_0\}$ where $\theta_0 = 1.2$, and our alternative hypothesis is $H_1 = \{\theta: \theta \in \mathscr{R}, \theta \neq \theta_0\}$. (As we noted in Section 9.1, H_0 is a simple hypothesis while H_1 is a composite hypothesis.) In our decision to accept or reject H_0 we will (if we restrict ourselves to nonrandomized tests; in Section 9.3 we will see that this restriction is possible here without loss) need to decide, for each possible value $x = (x_1, \ldots, x_n)$ of $X = (X_1, \ldots, X_n)$, whether $x \in A$ or $x \in R$. A reasonable way to do this would be to base our decision on a sufficient (hopefully, minimal sufficient) statistic for the problem, since it retains all information about θ which is possessed by $X = (X_1, \ldots, X_n)$. From Example 8.8.20 we know that $X_1 + \cdots + X_n$ is such a statistic. But by Example 8.9.11 we know that in this situation $\bar{X} = (X_1 + \cdots + X_n)/n$ is a BUE of θ, so it may make sense to compare $\bar{X}$ with θ_0 in order to make our acceptance or rejection decision.

Suppose we desire to have the level of our test be $\alpha = .05$. Two possible tests are: the test which rejects H_0 iff $|\bar{X} - \theta_0| > a$, that is, the test with critical function

$$\phi_1(x) = \begin{cases} 1, & \text{if } |\bar{x} - \theta_0| > a \\ 0, & \text{otherwise;} \end{cases} \tag{9.2.5}$$

and the test which rejects H_0 iff $\bar{X} > \theta_0 + b$, that is, the test with critical function

$$\phi_2(x) = \begin{cases} 1, & \text{if } \bar{x} > \theta_0 + b \\ 0, & \text{otherwise.} \end{cases} \tag{9.2.6}$$

In both cases a and b are to be set to give level $\alpha = .05$. What a and b are these, and how do the tests compare with regard to power?

For the test with critical function $\phi_1(\cdot)$ given in Equation (9.2.5), the level is

$$\begin{aligned} &P_{\theta_0}[|\bar{X} - \theta_0| > a] \\ &\quad = 1 - P_{\theta_0}[|\bar{X} - \theta_0| \leqslant a] = 1 - P_{\theta_0}[-a \leqslant \bar{X} - \theta_0 \leqslant a] \\ &\quad = 1 - P_{\theta_0}\left[-\frac{a}{\sigma/\sqrt{n}} \leqslant \frac{\bar{X} - \theta_0}{\sigma/\sqrt{n}} \leqslant \frac{a}{\sigma/\sqrt{n}}\right] \\ &\quad = 1 - P\left[-a\frac{\sqrt{n}}{\sigma} \leqslant Y \leqslant a\frac{\sqrt{n}}{\sigma}\right] \\ &\quad = 1 - \left[\Phi\left(a\frac{\sqrt{n}}{\sigma}\right) - \Phi\left(-a\frac{\sqrt{n}}{\sigma}\right)\right] \\ &\quad = 1 - \left[\Phi\left(a\frac{\sqrt{n}}{\sigma}\right) - \left\{1 - \Phi\left(a\frac{\sqrt{n}}{\sigma}\right)\right\}\right] \\ &\quad = 2\left[1 - \Phi\left(a\frac{\sqrt{n}}{\sigma}\right)\right], \end{aligned} \tag{9.2.7}$$

where: Y is a $N(0, 1)$ r.v. since (for example, see Problem 8.33 or Theorem 6.1.3) $\bar{X}$ is $N(\theta_0, \sigma^2/n)$; and $\Phi(z) = P[Y \leqslant z]$ (for example, review especially the last paragraph of Example 6.4.7). To obtain level $\alpha = .05$ for our test, we set

$$\begin{cases} .05 = 2\left[1 - \Phi\left(a\frac{\sqrt{n}}{\sigma}\right)\right], \\ .025 = 1 - \Phi\left(a\frac{\sqrt{n}}{\sigma}\right), \\ \Phi\left(a\frac{\sqrt{n}}{\sigma}\right) = .975, \\ a\frac{\sqrt{n}}{\sigma} = \Phi^{-1}(.975) = 1.96, \\ a = 1.96\frac{\sigma}{\sqrt{n}} = \frac{.588}{\sqrt{n}}, \end{cases} \tag{9.2.8}$$

where $\Phi^{-1}(.975) = 1.96$ (that is, $\Phi(1.96) = .975$) was determined using Table 11 and we used the fact that $\sigma^2 = .09$ (so $\sigma = .3$) in this example. So if, say, $n = 100$, then $a = .0588$; that is, we reject H_0 iff $|\bar{X} - \theta_0| > .0588$.

The power of the test with critical function $\phi_1(\cdot)$ is (for $\theta \in \Theta \cap \bar{H}_0$, that is, for $\theta \neq \theta_0$) the probability

$$
\begin{aligned}
P_\theta[|\bar{X}-\theta_0|>a] &= 1-P_\theta[|\bar{X}-\theta_0|\leqslant a] \\
&= 1-P_\theta[-a\leqslant \bar{X}-\theta_0\leqslant a] \\
&= 1-P_\theta[-a+\theta_0-\theta\leqslant \bar{X}-\theta\leqslant a+\theta_0-\theta] \\
&= 1-P_\theta\left[\frac{-a+\theta_0-\theta}{\sigma/\sqrt{n}}\leqslant\frac{\bar{X}-\theta}{\sigma/\sqrt{n}}\leqslant\frac{a+\theta_0-\theta}{\sigma/\sqrt{n}}\right] \\
&= 1-P_\theta\left[\frac{-a+\theta_0-\theta}{\sigma/\sqrt{n}}\leqslant Y\leqslant\frac{a+\theta_0-\theta}{\sigma/\sqrt{n}}\right] \\
&= 1-\left[\Phi\left(\frac{a+\theta_0-\theta}{\sigma/\sqrt{n}}\right)-\Phi\left(\frac{-a+\theta_0-\theta}{\sigma/\sqrt{n}}\right)\right] \\
&= 1-\left[\Phi\left(1.96+\frac{\theta_0-\theta}{\sigma}\sqrt{n}\right)-\Phi\left(-1.96+\frac{\theta_0-\theta}{\sigma}\sqrt{n}\right)\right],
\end{aligned}
\tag{9.2.9}
$$

where Y is a $N(0, 1)$ r.v. and $a = 1.96\sigma/\sqrt{n}$ was substituted. As a function of θ, say $f(\theta)$, the rejection probability (9.2.9) has the properties: $f(\theta_0) = \alpha = .05$; $f(\theta_0 + x) = f(\theta_0 - x)$ for all x; $f(\theta_2) > f(\theta_1)$ for all $\theta_2 > \theta_1 \geqslant \theta_0$; and $\lim_{\theta\to+\infty} f(\theta) = 1$. (See Problem 9.2.)

Now, for the test with critical function $\phi_2(\cdot)$ given in Equation (9.2.6) the probability of rejection of H_0 is

$$
\begin{aligned}
P_\theta[\bar{X} > \theta_0 + b] &= 1 - P_\theta[\bar{X} \leqslant \theta_0 + b] \\
&= 1 - P_\theta\left[\frac{\bar{X}-\theta}{\sigma/\sqrt{n}} \leqslant \frac{b+\theta_0-\theta}{\sigma/\sqrt{n}}\right] \\
&= 1 - P_\theta\left[Y \leqslant \frac{b+\theta_0-\theta}{\sigma/\sqrt{n}}\right] \\
&= 1 - \Phi\left(\frac{b+\theta_0-\theta}{\sigma/\sqrt{n}}\right),
\end{aligned}
\tag{9.2.10}
$$

where Y is a $N(0, 1)$ r.v. To set the level to $\alpha = .05$ requires that

$$
\left\{
\begin{aligned}
.05 &= P_{\theta_0}[\bar{X} > \theta_0 + b], \\
.05 &= 1 - \Phi\left(\frac{b+\theta_0-\theta_0}{\sigma/\sqrt{n}}\right), \\
.05 &= 1 - \Phi\left(b\frac{\sqrt{n}}{\sigma}\right), \\
\Phi\left(b\frac{\sqrt{n}}{\sigma}\right) &= .95, \\
b\frac{\sqrt{n}}{\sigma} &= 1.64, \\
b &= 1.64\frac{\sigma}{\sqrt{n}} = \frac{.492}{\sqrt{n}}.
\end{aligned}
\right.
\tag{9.2.11}
$$

So if, say, $n = 100$, then $b = .0492$; that is, we reject H_0 iff $\bar{X} - \theta_0 > .0492$. As a function of θ, say $g(\theta)$, the rejection probability (9.2.10) has the properties: $g(\theta_0) = .05$; $g(\theta_2) > g(\theta_1)$ for all $\theta_2 > \theta_1$; and $\lim_{\theta \to +\infty} g(\theta) = 1$. (See Problem 9.2.)

The rejection probabilities of the tests with critical functions $\phi_1(\cdot)$ and $\phi_2(\cdot)$ are plotted in Figure 9.2–2 for $n = 100$. From this graph we see that:

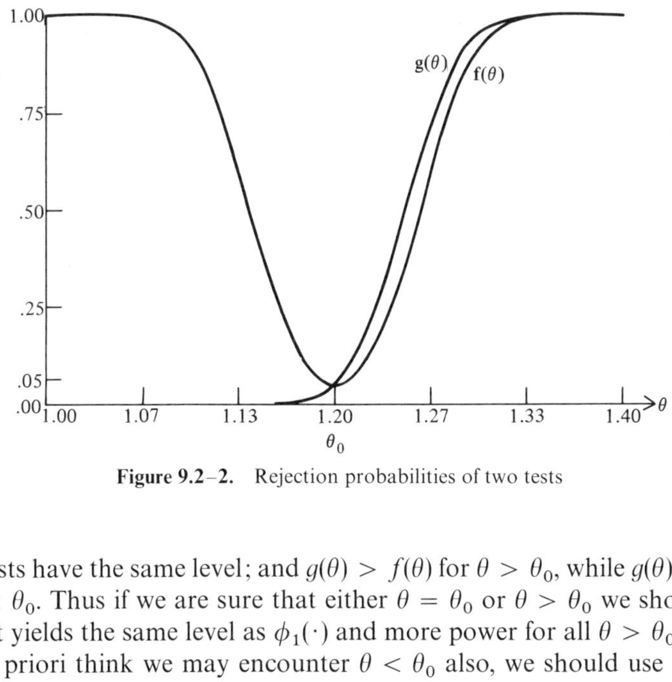

Figure 9.2–2. Rejection probabilities of two tests

both tests have the same level; and $g(\theta) > f(\theta)$ for $\theta > \theta_0$, while $g(\theta) < f(\theta)$ for $\theta < \theta_0$. Thus if we are sure that either $\theta = \theta_0$ or $\theta > \theta_0$ we should use $\phi_2(\cdot)$ (it yields the same level as $\phi_1(\cdot)$ and more power for all $\theta > \theta_0$); while if we a priori think we may encounter $\theta < \theta_0$ also, we should use $\phi_1(\cdot)$ in order to have a suitably high probability of rejection when θ is less than θ_0. The graphs in Figure 9.2–2 would be useful in the situation where we could only afford $n = 100$ and wanted to see what power this would yield for the two tests considered. Often n is set so that the power at a specified θ has at least a certain specified value; for example, if for the test using $\phi_1(\cdot)$ we wanted to have power $\geqslant .95$ when $\theta = 1.23$, we would set n as follows:

$$\begin{cases} .95 \leqslant P_{\theta=1.23}[|\bar{X} - \theta_0| > a], \\ .95 \leqslant 1 - \left[\Phi\left(1.96 + \dfrac{1.2 - 1.23}{.3}\sqrt{n}\right) - \Phi\left(-1.96 + \dfrac{1.2 - 1.23}{.3}\sqrt{n}\right)\right], \\ \Phi(1.96 - .1\sqrt{n}) - \Phi(-1.96 - .1\sqrt{n}) \leqslant .05. \end{cases} \tag{9.2.12}$$

Solving (9.2.12) by trial and error (that is, choose an n, calculate $\Phi(1.96 - .1\sqrt{n}) - \Phi(-1.96 - .1\sqrt{n})$, and so on) we find that the smallest n which will satisfy (9.2.12) is $n = 1296$.

Remark 9.2.13. Note that above we have at times needed to flip a coin with some probability of heads (for example, just before and in Definition 9.2.2 with a general probability of heads and later at (9.3.22) with probability .05 of heads). Where can such coins be obtained? A first answer to this question is that experiments have been done in the past (for example, rolling dice, drawing from a well-shuffled deck of cards or a well-stirred urn, and so on) and tables of the results prepared; since these were prepared with care and extensively tested, their use is usually preferable to performing your own experiment. The first such tables were prepared by L. H. C. Tippett (*Random Sampling Numbers*, Cambridge University Press, England, 1927), and the most extensive were prepared by The RAND Corporation (*A Million Random Digits with 100,000 Normal Deviates*, The Free Press, Glencoe, Illinois, 1955). These tables gave **random digits** chosen from 0, 1, 2, 3, 4, 5, 6, 7, 8, 9 rather than numbers on the interval 0 to 1. However if we wish to flip a coin with probability .95, for example, of heads, choose two digits from the table. If the digits are 01, 02, . . . , 95 say heads occurred (while if the digits are 96, 97, 98, 99, 00 say tails occurred). Tippett gave 41,600 digits, while The RAND Corporation gave a million. A similar table, with 50 random digits, is given in Table 9.2–1.

Table 9.2–1. 50 Random Digits

1	8	6	2	8	5	2	8	2	4	2
4	4	7	1	3	8	4	0	3	7	1
2	2	3	2	6	1	4	3	7	5	3
4	6	8	6	8	0	9	7	7	8	6
5	0	3	7	4	9	3	8	7	3	8

Since such tables are often insufficient and/or cumbersome for large sampling experiments, sample surveys, and simulations, procedures for generating numbers with similar properties by numerical methods have been developed and are usually available in a so-called "**random number generator**" in many digital computer languages and subroutine libraries. Most such routines are of adequate quality for the uses contemplated in our discussion. For simulations, however, we often need better generators (for example, see E. J. Dudewicz and J. S. Ramberg, *Random Variable Generation, Digital Computers and Simulation*, in preparation); a discussion of details is beyond the scope of this book. For a package of random number generators with documentation on quality and speed see *IRCCRAND—The Ohio State University Random Number Generator Package*, Statistics Laboratory Publication No. 1 (Technical Report No. 104), Department of Statistics, The Ohio State University, Columbus, Ohio, October 1974. For details of simulation methods (which rely heavily on probability and statistics for their proper use and interpretation) see G. S. Fishman, *Concepts and Methods* *in Discrete Event Digital Simulation*, John Wiley & Sons, Inc., New York, 1973.

9.3. THE NEYMAN-PEARSON LEMMA; ITS GENERALIZATIONS AND USES

To this point in this chapter we have considered both intuitively and precisely null and alternative, simple and composite, hypotheses (Section 9.1) and the hypothesis-testing problem including tests, level, and power with examples (Section 9.2). In the present section we will present a general method (the Neyman-Pearson Lemma) of obtaining "best" tests of simple hypotheses versus simple alternatives, see how the method sometimes provides "best" tests for composite hypotheses, and consider an application to legislative apportionment. (Succeeding sections consider a decision-theoretic view of the hypothesis-testing problem, giving a preview of the methods of Chapter 12 for this special case (Section 9.4), and consider specific hypothesis-testing problems which often occur in practice (Sections 9.5 ff).)

First we formulate a definition of a "best" (or "most powerful") test of a simple null hypothesis versus a simple alternative hypothesis, and then state[2] and prove the Neyman-Pearson Lemma.

Definition 9.3.1. *In a hypothesis-testing problem where H_0 and H_1 are both simple, a test with critical function $\phi(\cdot)$ is called* **most powerful** (*or* **most powerful of level α**) *if it has level α* [that is, P_{θ_0}[Rejecting H_0 using $\phi(\cdot)$] $= \alpha$] *and no other test of level α has a larger power at H_1* [that is,

$$P_{\theta_1}[\text{Rejecting } H_0 \text{ using } \phi(\cdot)] \geqslant P_{\theta_1}[\text{Rejecting } H_0 \text{ using } \psi(\cdot)]$$

for all critical functions $\psi(\cdot)$ such that P_{θ_0}[Rejecting H_0 using $\psi(\cdot)$] $= \alpha$].

Note that P_{θ_0}[Rejecting H_0 using $\phi(\cdot)$] $= E_{\theta_0}\phi(X)$. This relationship between rejection probability and expected value of the critical function will be used extensively to obtain a simple notation.

Theorem 9.3.2. **(Neyman-Pearson Lemma)** *Suppose that X is a r.v. with d.f. $F(x|\theta)$ where θ is unknown, $\theta \in \Theta$ and $x \in \mathscr{X}$. Assume $F(x|\theta)$ is either absolutely continuous* [in which case let $f(x|\theta)$ denote its density at x] *or discrete* [in which case let $f(x|\theta)$ denote $P_\theta[X = x]$]. *Suppose that $H_0 = \{\theta_0\}$ and $H_1 = \{\theta_1\}$* [often written as $H_0\colon \theta = \theta_0$ and $H_1\colon \theta = \theta_1$ in this case] *are simple null and alternative hypotheses, respectively.*

---

[2] Although we state it for the absolutely continuous and discrete cases, it holds in general. See, for example, E. L. Lehmann, *Testing Statistical Hypotheses*, John Wiley & Sons, Inc., New York, 1959, pp. 65–68.

For testing H_0 versus H_1 after observing X, if a test with critical function $\phi(\cdot)$ satisfies (for some k)

$$E_{\theta_0}\phi(X) = \alpha \tag{9.3.3}$$

and

$$\phi(x) = \begin{cases} 1, & \text{when } f(x|\theta_1) > kf(x|\theta_0) \\ 0, & \text{when } f(x|\theta_1) < kf(x|\theta_0), \end{cases} \tag{9.3.4}$$

then it is most powerful for testing H_0 versus H_1 at level α.

Proof: [We give the proof for the absolutely continuous case, the discrete case being analogous with summations replacing integrals.] We write, for example, $\int_{-\infty}^{\infty} g(x)\,dx$ as $\int_{x\in\mathscr{R}} g(x)\,dx$ as a shorthand (since the sets over which we are integrating below are notationally complex). For this use, define

$$\begin{cases} c_1 = \{x: f(x|\theta_1) > kf(x|\theta_0)\}, \\ c_2 = \{x: f(x|\theta_1) < kf(x|\theta_0)\}, \\ c_3 = \{x: f(x|\theta_1) = kf(x|\theta_0)\}. \end{cases} \tag{9.3.5}$$

Suppose $\phi(\cdot)$ satisfies (9.3.3) and (9.3.4). Let $\psi(\cdot)$ be the critical function of any other test with level α. Then the difference of the powers of the two tests at H_1 is

$$\begin{aligned} E_{\theta_1}\phi(X) - E_{\theta_1}\psi(X) &= E_{\theta_1}[\phi(X) - \psi(X)] \\ &= \int_{x\in\mathscr{X}} [\phi(x) - \psi(x)]f(x|\theta_1)\,dx \\ &= \int_{x\in c_1} [\phi(x) - \psi(x)]f(x|\theta_1)\,dx \\ &\quad + \int_{x\in c_2} [\phi(x) - \psi(x)]f(x|\theta_1)\,dx \\ &\quad + \int_{x\in c_3} [\phi(x) - \psi(x)]f(x|\theta_1)\,dx. \end{aligned} \tag{9.3.6}$$

Using the definitions of c_1, c_2, and c_3, we know that: $\phi(x) = 1$ for $x \in c_1$; and $\phi(x) = 0$ for $x \in c_2$. Substituting these in (9.3.6) and then using the facts $f(x|\theta_1) > kf(x|\theta_0)$ for $x \in c_1$, $f(x|\theta_1) < kf(x|\theta_0)$ for $x \in c_2$, and $f(x|\theta_1) = kf(x|\theta_0)$ for $x \in c_3$, we find

$$\begin{aligned} E_{\theta_1}\phi(X) - E_{\theta_1}\psi(X) &= \int_{x\in c_1} [1 - \psi(x)]f(x|\theta_1)\,dx \\ &\quad + \int_{x\in c_2} [-\psi(x)]f(x|\theta_1)\,dx \\ &\quad + \int_{x\in c_3} [\phi(x) - \psi(x)]f(x|\theta_1)\,dx \\ &\geqslant \int_{x\in c_1} [1 - \psi(x)]kf(x|\theta_0)\,dx \\ &\quad + \int_{x\in c_2} [-\psi(x)]kf(x|\theta_0)\,dx \\ &\quad + \int_{x\in c_3} [\phi(x) - \psi(x)]kf(x|\theta_0)\,dx \end{aligned} \tag{9.3.7}$$

$$\begin{aligned}
&= \int_{x\in c_1} [\phi(x) - \psi(x)]kf(x|\theta_0)\,dx \\
&\quad + \int_{x\in c_2} [\phi(x) - \psi(x)]kf(x|\theta_0)\,dx \\
&\quad + \int_{x\in c_3} [\phi(x) - \psi(x)]kf(x|\theta_0)\,dx \\
&= k\int_{x\in\mathscr{X}} [\phi(x) - \psi(x)]f(x|\theta_0)\,dx \\
&= kE_{\theta_0}[\phi(X) - \psi(X)] = k(\alpha - \alpha) = 0.
\end{aligned}$$

Hence $E_{\theta_1}\phi(X) \geqslant E_{\theta_1}\psi(X)$, as was to be shown.

Example 9.3.8. Suppose we have (or wish to find) a critical function which satisfies (9.3.4). If $f(x|\theta_1)$ and $f(x|\theta_0)$ are each >0 for all $x \in \mathscr{X}$, *then* (9.3.4) *can be rewritten as*

$$\phi(x) = \begin{cases} 1, & \text{when } \dfrac{f(x|\theta_1)}{f(x|\theta_0)} > k \\[2ex] 0, & \text{when } \dfrac{f(x|\theta_1)}{f(x|\theta_0)} < k. \end{cases} \tag{9.3.9}$$

Thus, the classification of a point x into the rejection region ($\phi(x) = 1$) or acceptance region ($\phi(x) = 0$) depends "only" on the likelihood ratio $f(X|\theta_1)/f(X|\theta_0)$ which is (see Example 8.8.31) a minimal sufficient statistic in this case. (We put "only" in quotes since, on the "boundary" $f(X|\theta_1)/f(X|\theta_0) = k$, $\phi(x)$ may be selected in many ways in general. In an absolutely continuous case, usually $P_\theta[f(X|\theta_1)/f(X|\theta_0) = k] = 0$ ($\theta = \theta_0, \theta_1$). In a discrete case this probability may be >0 and, to make $\phi(\cdot)$ have level α, we may need to use randomization ($0 < \phi(x) < 1$ for some x's).)

A motivation often given for putting in the critical region R those x such that $f(x|\theta_1)/f(x|\theta_0) > k$ is the following. In the discrete case, putting x in R contributes $f(x|\theta_0) = P_{\theta_0}[X = x]$ to level and $f(x|\theta_1) = P_{\theta_1}[X = x]$ to power. Now subject to a restriction on the level (level $= \alpha$) we want the most power, so we put in R those x which give us the most power-per-level, that is, those which have the highest values of the power-to-level ratio $f(x|\theta_1)/f(x|\theta_0)$. That is, for some k put in R all x such that $f(x|\theta_1)/f(x|\theta_0) > k$.

Note that the Neyman-Pearson Lemma was proven with essentially no "regularity" or other restrictions of any sort. (Many authors write (9.3.4) in terms of the ratio $f(x|\theta_1)/f(x|\theta_0)$ as at (9.3.9), and must therefore exclude cases where $f(x|\theta_0)$ may be zero. Some of these cases are important. One such case is discussed below in Example 9.3.17.)

Example 9.3.10. Suppose we observe $X = (X_1, \ldots, X_n)$ where $X_1, \ldots, X_n$ are independent r.v.'s, each $N(\mu, \sigma^2)$ where μ is unknown $(-\infty < \mu < +\infty)$

and $\sigma^2 > 0$ is known. Find a critical function for a most powerful level $\alpha(0 < \alpha < 1)$ test of $H_0\colon \mu = \mu_0$ versus $H_1 : \mu = \mu_1$ where $\mu_1 > \mu_0$.

By the Neyman-Pearson Lemma, we know that if a test satisfies (9.3.4) and (9.3.3) it is most powerful of level α for this hypothesis-testing problem. Now here $f(x|\mu_1)$ and $f(x|\mu_0)$ are each > 0 for all $x \in \mathscr{X} = \mathscr{R}^n$, so we wish to find those $x = (x_1, \ldots, x_n)$ for which $f(x|\mu_1)/f(x|\mu_0) > k$. Using (8.8.21) we find

$$\frac{f(x|\mu_1)}{f(x|\mu_0)} = \frac{\prod_{i=1}^{n}\left\{\frac{1}{\sqrt{2\pi}\sigma} e^{-\frac{1}{2}\left(\frac{x_i-\mu_1}{\sigma}\right)^2}\right\}}{\prod_{i=1}^{n}\left\{\frac{1}{\sqrt{2\pi}\sigma} e^{-\frac{1}{2}\left(\frac{x_i-\mu_0}{\sigma}\right)^2}\right\}}$$

$$= \frac{\left(\frac{1}{\sqrt{2\pi}\sigma}\right)^n e^{-\frac{1}{2\sigma^2}\sum_{i=1}^{n} x_i^2} e^{-\frac{1}{2\sigma^2}\left\{-2\mu_1 \sum_{i=1}^{n} x_i + n\mu_1^2\right\}}}{\left(\frac{1}{\sqrt{2\pi}\sigma}\right)^n e^{-\frac{1}{2\sigma^2}\sum_{i=1}^{n} x_i^2} e^{-\frac{1}{2\sigma^2}\left\{-2\mu_0 \sum_{i=1}^{n} x_i + n\mu_0^2\right\}}} \tag{9.3.11}$$

$$= e^{\frac{1}{\sigma^2}(\mu_1-\mu_0)\sum_{i=1}^{n} x_i - \frac{n}{2\sigma^2}(\mu_1^2-\mu_0^2)}$$

Since $f(x|\mu_1)/f(x|\mu_0) > 0$ for all $x \in \mathscr{X}$, we may take $k > 0$ (otherwise the rejection region is $\mathscr{X}$ and level = power = 1). Thus, this ratio is $>k$ iff

$$\left\{\begin{aligned}
&\frac{1}{\sigma^2}(\mu_1 - \mu_0)\sum_{i=1}^{n} x_i - \frac{n}{2\sigma^2}(\mu_1^2 - \mu_0^2) > \ln k \\
\Leftrightarrow\ & 2(\mu_1 - \mu_0)\sum_{i=1}^{n} x_i - n(\mu_1^2 - \mu_0^2) > 2\sigma^2 \ln k \\
\Leftrightarrow\ & 2(\mu_1 - \mu_0)\sum_{i=1}^{n} x_i > 2\sigma^2 \ln k + n(\mu_1^2 - \mu_0^2) \\
\Leftrightarrow\ & \sum_{i=1}^{n} x_i > \frac{2\sigma^2 \ln k + n(\mu_1^2 - \mu_0^2)}{2(\mu_1 - \mu_0)} \\
\Leftrightarrow\ & \bar{x} > \frac{2\sigma^2 \ln k + n(\mu_1^2 - \mu_0^2)}{2n(\mu_1 - \mu_0)},
\end{aligned}\right. \tag{9.3.12}$$

where we used the fact that $\mu_1 > \mu_0$, when we divided by $\mu_1 - \mu_0$, to establish that we did not need to reverse the inequality. Now if $k, \sigma^2, \mu_1, \mu_0, n$ are fixed, then

$$k' = \frac{2\sigma^2 \ln k + n(\mu_1^2 - \mu_0^2)}{2n(\mu_1 - \mu_0)} \tag{9.3.13}$$

is just another constant and $f(x|\mu_1)/f(x|\mu_0) > k$ iff $\bar{x} > k'$. So if $\phi(\cdot)$ sets $\phi(x) = 1$ iff $\bar{x} > k'$, then $\phi(\cdot)$ satisfies (9.3.4). What value should k' then have if we wish $\phi(\cdot)$ to satisfy (9.3.3)? We find

$$E_{\mu_0}\phi(X) = P_{\mu_0}[\bar{X} > k'] = P_{\mu_0}\left[\frac{\bar{X} - \mu_0}{\sigma/\sqrt{n}} > \frac{k' - \mu_0}{\sigma/\sqrt{n}}\right]$$
$$= 1 - \Phi\left(\frac{k' - \mu_0}{\sigma/\sqrt{n}}\right), \tag{9.3.14}$$

which will equal α iff

$$\begin{cases} \alpha = 1 - \Phi\left(\dfrac{k' - \mu_0}{\sigma/\sqrt{n}}\right), \\ \Phi\left(\dfrac{k' - \mu_0}{\sigma/\sqrt{n}}\right) = 1 - \alpha, \\ \dfrac{k' - \mu_0}{\sigma/\sqrt{n}} = \Phi^{-1}(1 - \alpha) \\ k' = \mu_0 + \dfrac{\sigma}{\sqrt{n}}\Phi^{-1}(1 - \alpha). \end{cases} \tag{9.3.15}$$

Thus, the test which rejects iff $\bar{X} > \mu_0 + (\sigma/\sqrt{n})\Phi^{-1}(1 - \alpha)$ is most powerful of level α in our case. Note that the test does not depend on μ_1, that is, the *same* test is most powerful for every $\mu_1 > \mu_0$ (although of course the power differs for different μ_1's). Such a test will be called "uniformly most powerful (UMP)" for testing $H_0: \mu = \mu_0$ versus $H_1: \mu_1 > \mu_0$, or UMP one-sided (since it is not UMP for testing $H_0: \mu = \mu_0$ versus $H_1: \mu = \mu^* < \mu_0$). This is the test we were led to in Example 9.2.4 (see $\phi_2(\cdot)$ in that example) by considerations of sufficiency.

Our development in Example 9.3.10 justifies the statement of the first paragraph of Example 9.2.4 that in this situation we could restrict ourselves without loss to nonrandomized tests: a nonrandomized test is most powerful.

We next define precisely the "uniformly most powerful" test concept (which arose in Example 9.3.10) and then consider an example where $f(x|\theta_0)$ may be zero.

Definition 9.3.16. *A test $\phi(\cdot)$ is called* **uniformly most powerful (UMP) (of level α)** *for testing a simple hypothesis H_0 against a composite hypothesis H, if this test is most powerful of level α for testing H_0 versus H_1 for each $H_1 \in H$.*

Example 9.3.17. (In this example we consider an important case in which $f(x|\theta_0)$ may be zero. We see below that there is not a unique most powerful test in general, and that the way in which we randomize when $f(x|\theta_1) = kf(x|\theta_0)$ can be very important.)

Suppose $X_1, \ldots, X_n$ are independent r.v.'s, each uniform on $(0, \theta)$ where θ is unknown, $\theta \in \Theta = \{z: z > 0\}$. Find a most powerful level $\alpha = .05$ test of $H_0: \theta = 1$ versus $H_1: \theta = 2$ and plot $E_\theta\phi_1(X)$ as a function of θ. Find the most powerful level $\alpha = .05$ test of $H_0: \theta = 1$ versus $H_1: \theta = 2$ based on observing the sufficient (see Example 8.8.22) statistic $Y = \max(X_1, \ldots, X_n)$, and plot $E_\theta\phi_2(X)$ as a function of θ. [We use $\phi_1(\cdot)$ and $\phi_2(\cdot)$ as notation for the critical functions of the two respective tests.]

First, suppose we observe $X = (X_1, \ldots, X_n)$ and want to find a most powerful level $\alpha = .05$ test of $H_0: \theta = 1$ versus $H_1: \theta = 2$. By the Neyman-Pearson Lemma, a $\phi_1(\cdot)$ satisfying (9.3.4) and (9.3.3) will be most powerful. Now (see (8.8.23))

$$f(x|\theta = 1) = \begin{cases} 1, & \text{if } 0 \leqslant \min(x_1, \ldots, x_n) \leqslant \max(x_1, \ldots, x_n) \leqslant 1 \\ 0, & \text{otherwise} \end{cases} \tag{9.3.18}$$

and

$$f(x|\theta = 2) = \begin{cases} \dfrac{1}{2^n}, & \text{if } 0 \leqslant \min(x_1, \ldots, x_n) \leqslant \max(x_1, \ldots, x_n) \leqslant 2 \\ 0, & \text{otherwise.} \end{cases} \tag{9.3.19}$$

For various conditions on $x = (x_1, \ldots, x_n)$ the values of $f(x|\theta = 1)$ and $f(x|\theta = 2)$ are given in Table 9.3–1.

Table 9.3–1. $f(x|\theta = 1)$ and $f(x|\theta = 2)$

Case	$f(x\|\theta = 1)$	$f(x\|\theta = 2)$
I. $\min(x_1, \ldots, x_n) < 0$	0	0
II. $\min(x_1, \ldots, x_n) \geqslant 0$ and $\max(x_1, \ldots, x_n) \leqslant 1$	1	$\frac{1}{2^n}$
III. $\min(x_1, \ldots, x_n) \geqslant 0$ and $1 < \max(x_1, \ldots, x_n) \leqslant 2$	0	$\frac{1}{2^n}$
IV. $\min(x_1, \ldots, x_n) \geqslant 0$ and $2 < \max(x_1, \ldots, x_n)$	0	0

Now, when is $f(x|\theta = 2) > kf(x|\theta = 1)$, when is $f(x|\theta = 2) < kf(x|\theta = 1)$, and when is $f(x|\theta = 2) = kf(x|\theta = 1)$? It is simple to verify (using Table 9.3–1) that the answer is as given in Table 9.3–2.

Thus, no matter what our choice of k, to satisfy (9.3.4): we must set $\phi_1(x) = 1$ when Case III occurs; we can set $\phi_1(x)$ as we like when Cases I or IV occur; and in Case II, we must set $\phi_1(x) = 1$ if we choose $k < 1/2^n$, we can choose $\phi_1(x)$ as we like if we choose $k = 1/2^n$, and we must set $\phi_1(x) = 0$ if we choose $k > 1/2^n$. Now

$$P_{\theta=1}[\text{Case I, or Case III, or Case IV}] = 0, \tag{9.3.20}$$

so even if we *always* rejected upon observing an x which fell into Cases I, III, IV our level would be zero. So, for level $\alpha = .05$ some rejection in Case II must be made possible. $k < 1/2^n$ does not allow any, and $k > 1/2^n$

Table 9.3–2. $f(x|\theta = 2)$ *versus* $kf(x|\theta = 1)$

Case	$f(x\|\theta = 2) > kf(x\|\theta = 1)$	$f(x\|\theta = 2) = kf(x\|\theta = 1)$	$f(x\|\theta = 2) < kf(x\|\theta = 1)$
I.	Never	Always	Never
II.	Iff $k < \frac{1}{2^n}$	Iff $k = \frac{1}{2^n}$	Iff $k > \frac{1}{2^n}$
III.	Always	Never	Never
IV.	Never	Always	Never

requires always rejecting (which yields level = 1). Hence, we need $k = 1/2^n$ if we are to be able to obtain any $0 < \alpha < 1$ (in particular $\alpha = .05$).

So suppose $k = 1/2^n$ is chosen. Then $\phi_1(x) = 1$ for x in Case III satisfies (9.3.4), and we can set $\phi_1(x)$ in Cases I, II, IV as we like. Since when Cases I or IV occur $H_0\colon \theta = 1$ is clearly false, set $\phi_1(x) = 1$ in those cases. Now let us set $\phi_1(x)$ for Case II so as to satisfy (9.3.3): $E_{\theta=1}\phi_1(X) = \alpha$ (in particular, we desire $\alpha = .05$). One way to accomplish this is to set $\phi_1(x) = \gamma$ for all x in Case II, γ then being chosen to yield level α:

$$\begin{aligned} E_{\theta=1}\phi_1(X) & \\ &= \gamma P_{\theta=1}[\text{Case II}] + P_{\theta=1}[\text{Case I, or Case III, or Case IV}] \\ &= \gamma \cdot 1 + 0 = \gamma. \end{aligned} \tag{9.3.21}$$

Thus, choose $\gamma = \alpha = .05$.

So a most powerful level $\alpha = .05$ test of $H_0\colon \theta = 1$ versus $H_1\colon \theta = 2$ is given by

$$\phi_1(x) = \begin{cases} .05, & \text{if } \min(x_1, \ldots, x_n) \geqslant 0 \text{ and } \max(x_1, \ldots, x_n) \leqslant 1 \\ 1, & \text{otherwise.} \end{cases} \tag{9.3.22}$$

The power function of this test is calculated as follows. For $0 < \theta \leqslant 1$,

$$\begin{aligned} E_\theta\phi_1(X) &= .05\, P_\theta[\min(X_1, \ldots, X_n) \geqslant 0 \text{ and } \max(X_1, \ldots, X_n) \leqslant 1] \\ &\quad + 1 \cdot P_\theta[\overline{\min(X_1, \ldots, X_n) \geqslant 0 \text{ and } \max(X_1, \ldots, X_n) \leqslant 1}] \\ &= (.05)(1) + (1)(0) = .05. \end{aligned} \tag{9.3.23}$$

For $1 < \theta$,

$$\begin{aligned} E_\theta\phi_1(X) &= .05\, P_\theta[\min(X_1, \ldots, X_n) \geqslant 0 \text{ and } \max(X_1, \ldots, X_n) \leqslant 1] \\ &\quad + 1 \cdot P_\theta[\overline{\min(X_1, \ldots, X_n) \geqslant 0 \text{ and } \max(X_1, \ldots, X_n) \leqslant 1}] \\ &= .05\left(\frac{1}{\theta}\right)^n + 1 \cdot \left(1 - \left(\frac{1}{\theta}\right)^n\right) = 1 - \frac{.95}{\theta^n}. \end{aligned} \tag{9.3.24}$$

The resulting curve is plotted as a solid line in Figure 9.3–1. It has

$$E_{\theta=1}\phi_1(X) = .05 \quad \text{and} \quad E_{\theta=2}\phi(X) = 1 - \frac{.95}{2^n};$$

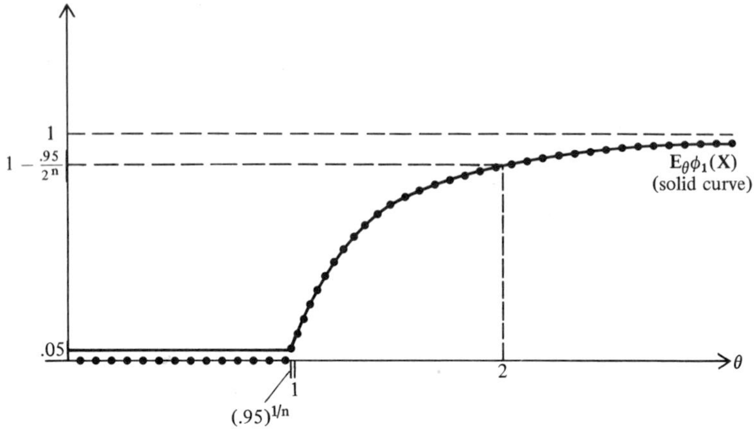

Figure 9.3–1. $\mathbf{E}_\theta\phi_1(\mathbf{X})$ and $\mathbf{E}_\theta\phi_2(\mathbf{X})$ [Graph has $n = 3$]

the Neyman-Pearson Lemma assures us that any other test $\psi(\cdot)$ with $E_{\theta=1}\psi(X) = .05$ has power $E_{\theta=2}\psi(X) \leqslant E_{\theta=2}\phi(X)$.

Now, for the *second* part, suppose we observe $Y = \max(X_1, \ldots, X_n)$ and want to find a most powerful level $\alpha = .05$ test of H_0: $\theta = 1$ versus H_1: $\theta = 2$. Again we will use the Neyman-Pearson Lemma. Here

$$\begin{aligned} P_\theta[Y \leqslant y] &= P_\theta[\max(X_1, \ldots, X_n) \leqslant y] \\ &= P_\theta[X_1 \leqslant y, \ldots, X_n \leqslant y] = (P_\theta[X_1 \leqslant y])^n \\ &= \begin{cases} 0, & \text{if } y < 0 \\ 1^n/\theta^n, & \text{if } 0 \leqslant y \leqslant \theta \\ 1, & \text{if } y > \theta. \end{cases} \end{aligned} \tag{9.3.25}$$

Hence, the density involved is

$$g(y|\theta) = \begin{cases} n\dfrac{y^{n-1}}{\theta^n}, & \text{if } 0 \leqslant y \leqslant \theta \\ 0, & \text{otherwise,} \end{cases} \tag{9.3.26}$$

and

$$g(y|\theta = 1) = \begin{cases} ny^{n-1}, & \text{if } 0 \leqslant y \leqslant 1 \\ 0, & \text{otherwise,} \end{cases} \tag{9.3.27}$$

while

$$g(y|\theta = 2) = \begin{cases} n\dfrac{y^{n-1}}{2^n}, & \text{if } 0 \leqslant y \leqslant 2 \\ 0, & \text{otherwise.} \end{cases} \tag{9.3.28}$$

Hence, a table analogous to Table 9.3–1 may be obtained. See Table 9.3–3.

Table 9.3–3. $g(y|\theta = 1)$ and $g(y|\theta = 2)$

Case	$g(y\|\theta = 1)$	$g(y\|\theta = 2)$
I. $y < 0$	0	0
II. $0 \leqslant y \leqslant 1$	ny^{n-1}	$n\dfrac{y^{n-1}}{2^n}$
III. $1 < y \leqslant 2$	0	$n\dfrac{y^{n-1}}{2^n}$
IV. $y > 2$	0	0

Here, analyzing as before, we find that to satisfy (9.3.4) of the Neyman-Pearson Lemma: we can choose $\phi_2(y)$ as we like if $y < 0$ or $y \geqslant 2$ (no matter what value k has); we must set $\phi_2(y) = 1$ for $1 < y \leqslant 2$ (no matter what value k has); and we set $\phi_2(y) = 1$ when $0 < y < 1$ iff

$$n\frac{y^{n-1}}{2^n} > kny^{n-1} \Leftrightarrow k < \frac{1}{2^n}. \tag{9.3.29}$$

As before, only $k = 1/2^n$ allows us to have $\alpha = .05$. Now in Case II we may feel intuitively that large values of Y are more likely to indicate that $H_1: \theta = 2$ is true than are small values of Y; then instead of choosing $\phi_2(y) = .05$ for y in Case II, we will wish to set

$$\phi_2(y) = \begin{cases} 1, & \text{if } y > c \\ 0, & \text{if } y \leqslant c, \end{cases} \tag{9.3.30}$$

where c is chosen to obtain level $\alpha = .05$:

$$\begin{aligned} .05 &= E_{\theta=1}\phi_2(Y) = P_{\theta=1}[Y > c] \\ &= 1 - P_{\theta=1}[Y \leqslant c] = 1 - \frac{c^n}{1^n} \end{aligned} \tag{9.3.31}$$

so $c = (.95)^{1/n}$. The function $E_\theta\phi_2(Y)$ is now

$$\begin{aligned} E_\theta\phi_2(Y) &= P_\theta[Y > (.95)^{1/n}] \\ &= \begin{cases} 0, & \text{if } \theta \leqslant (.95)^{1/n} \\ 1 - \dfrac{.95}{\theta^n}, & \text{if } \theta > (.95)^{1/n}, \end{cases} \end{aligned} \tag{9.3.32}$$

which is graphed in Figure 9.3–1 as a series of dots.

The tests $\phi_1(X)$ and $\phi_2(Y)$ obtained above are different and (see Figure 9.3–1) have different rejection probability functions. $\phi_1(X)$ is most powerful among tests based on $X = (X_1, \ldots, X_n)$; $\phi_2(Y)$ is most powerful among tests based on $Y = \max(X_1, \ldots, X_n)$. Since Y is a sufficient statistic, we should question why the two tests should be different: wouldn't we expect to obtain the *same* test as best even if we first reduced the data to a sufficient statistic? The answer is yes, we would, and *in the sense of the Neyman-Pearson Lemma* they *are* equivalent since they have the same level (.05) and the same power at $H_1{:}\theta = 2\,(1 - .95/2^n)$. In fact, the same test could have been chosen based on either $X = (X_1, \ldots, X_n)$ or $Y = \max(X_1, \ldots, X_n)$; we chose different ones because when looking at X our intuition led us to randomize over all $(X_1, \ldots, X_n)$ with $\max(X_1, \ldots, X_n) \leqslant 1$, while when looking at Y our intuition led us to set $\phi_2(\cdot)$ to 0 at some such $(X_1, \ldots, X_n)$ and to 1 at others. Thus, looking at the sufficient statistics for a problem may lead our intuition to choose more appropriately on the "boundary" $f(x|\theta_1) = kf(x|\theta_0)$.

The considerations of the above paragraph are not trivial. In practice, we often wish to know if θ is "small" or "large" (these are our intuitive H_0 and H_1), and simple H_0 and H_1 are set up as a way of obtaining a "best" test with hopefully reasonable properties, such as $E_0\phi(X)\downarrow$as $\theta\uparrow$. Thus, in the above example we would usually prefer $\phi_2(Y)$ to $\phi_1(X)$, since for the latter $E_\theta\phi_1(X)$ remains constant over $0 < \theta \leqslant 1$ while $E_\theta\phi_2(Y)$ goes quickly to 0 as $\theta\downarrow$.

With regard to UMP tests (see Definition 9.3.16) if we observe a r.v. X with d.f. $F(x|\theta)$ which has a *monotone likelihood ratio* (see Definition 9.3.33), then certain UMP tests for composite hypotheses are easy to obtain. We now prove this (after giving Definition 9.3.33). Then (as promised at the beginning of this section) we consider an application to legislative apportionment.

Definition 9.3.33. *Suppose that X is a r.v. with d.f. $F(x|\theta)$ where θ is unknown, $\theta \in \Theta \subseteq \mathscr{R}$* [so θ is a one-dimensional parameter]. *Assume $F(x|\theta)$ is either absolutely continuous* [in which case let $f(x|\theta)$ denote its density at x] *or discrete* [in which case let $f(x|\theta)$ denote $P_\theta[X = x]$]. *Then the d.f. $F(x|\theta)$ of X is said to have a* **monotone likelihood ratio in T(x) (MLR)** *if, for some statistic $T(X)$,*

$$\frac{f(x|\theta_2)}{f(x|\theta_1)}\uparrow \textit{as } T(x)\uparrow \tag{9.3.34}$$

for all $\theta_2 > \theta_1 (\theta_2, \theta_1 \in \Theta)$ and for all $x \in \mathscr{X}$.

Theorem 9.3.35. *Suppose that X is a r.v. with d.f. $F(x|\theta)$ where θ is unknown, $\theta \in \Theta$, $x \in \mathscr{X}$. Assume $F(x|\theta)$ has a monotone likelihood ratio in $T(x)$. Suppose that $H_0{:}\,\theta = \theta_0$ and $H_1{:}\,\theta > \theta_0$ are simple null and composite alternative hypotheses, respectively.*

For testing H_0 versus H_1 after observing X, if a test with critical function

$\phi(\cdot)$ satisfies (for some C)

$$E_{\theta_0}\phi(X) = \alpha \tag{9.3.36}$$

and

$$\phi(x) = \begin{cases} 1, & \text{when } T(x) > C \\ 0, & \text{when } T(x) < C, \end{cases} \tag{9.3.37}$$

then it is UMP of level α for testing H_0 versus H_1.

Proof: Consider the problem of testing $H_0: \theta = \theta_0$ versus $H_1^*: \theta = \theta_1$ for a fixed $\theta_1 > \theta_0$. Then both H_0 and H_1^* are simple, so by the Neyman-Pearson Lemma (Theorem 9.3.2) a test which satisfies (for some k) (9.3.25) and

$$\phi(x) = \begin{cases} 1, & \text{if } f(x|\theta_1) > kf(x|\theta_0) \\ 0, & \text{if } f(x|\theta_1) < kf(x|\theta_0) \end{cases} \tag{9.3.38}$$

is most powerful for testing H_0 versus H_1^* at level α. Since we assumed $F(x|\theta)$ had a MLR in $T(x)$, (9.3.38) can be rewritten as

$$\phi(x) = \begin{cases} 1, & \text{if } \dfrac{f(x|\theta_1)}{f(x|\theta_0)} > k \\ 0, & \text{if } \dfrac{f(x|\theta_1)}{f(x|\theta_0)} < k \end{cases} \tag{9.3.39}$$

or, since $f(x|\theta_2)/f(x|\theta_1)\uparrow$ as $T(x)\uparrow$ for all $\theta_2 > \theta_1$, as

$$\phi(x) = \begin{cases} 1, & \text{if } T(x) > C \\ 0, & \text{if } T(x) < C \end{cases} \tag{9.3.40}$$

for some C. Thus, a test satisfying (9.3.36) and (9.3.37) is most powerful for testing H_0 versus H_1^* at level α. As in Example 9.3.10, we find (see Problem 9.15) that the test is the same regardless of which θ_1 was chosen (as long as $\theta_1 > \theta_0$). Hence, a test satisfying (9.3.36) and (9.3.37) is most powerful for testing $H_0: \theta = \theta_0$ versus $H_1^*: \theta = \theta_1$ at level α for every $\theta_1 > \theta_0$, that is, is UMP of level α for testing H_0 versus H_1.

Example 9.3.41. (In this example we obtain a UMP test via MLR considerations.) Suppose $X_1, \ldots, X_n$ are independent r.v.'s, each with the density of an exponential distribution with mean θ (see Theorem 5.2.5), that is,

$$g(x|\theta) = \begin{cases} \dfrac{1}{\theta} e^{-x/\theta}, & \text{if } x \geqslant 0 \\ 0, & \text{if } x < 0, \end{cases}$$

where θ is unknown, $\theta \in \Theta = \{y: y > 0\}$. Based on $X = (X_1, \ldots, X_n)$ with $n = 31$, find a UMP level $\alpha = .05$ test of $H_0: \theta = 10$ versus $H_1: \theta > 10$.

[In Problem 9.4H we see that this test is also a UMP level $\alpha = .05$ test of $H_0: \theta \leqslant 10$ versus $H_1: \theta > 10$.]

First, does the d.f. $F(x|\theta)$ of X have a MLR (and, if so, in what)? Here the density of X is

$$f(x|\theta) = g(x_1|\theta) \cdots g(x_n|\theta) = \begin{cases} \dfrac{1}{\theta^n} e^{-(x_1 + \cdots + x_n)/\theta}, & \text{if } x_1, \ldots, x_n \geqslant 0 \\ 0, & \text{otherwise,} \end{cases} \tag{9.3.42}$$

so (for $x \in \mathscr{X}$ and $\theta_1, \theta_2 \in \Theta$)

$$\frac{f(x|\theta_2)}{f(x|\theta_1)} = \left(\frac{\theta_1}{\theta_2}\right)^n e^{-(x_1 + \cdots + x_n)\left(\frac{1}{\theta_2} - \frac{1}{\theta_1}\right)} = \left(\frac{\theta_1}{\theta_2}\right)^n e^{(x_1 + \cdots + x_n)(\theta_2 - \theta_1)/(\theta_1 \theta_2)}, \tag{9.3.43}$$

which (for all $\theta_2 > \theta_1$; $\theta_2, \theta_1 \in \Theta$) is an increasing function of $T(x) = x_1 + \cdots + x_n$. Thus (by Definition 9.3.33) $F(x|\theta)$ has a MLR in $x_1 + \cdots + x_n$.

Using Theorem 9.3.35, it follows that any test with $E_{\theta_0}\phi(X) = .05$ and (for some C)

$$\phi(x) = \begin{cases} 1, & \text{when } x_1 + \cdots + x_n > C \\ 0, & \text{when } x_1 + \cdots + x_n < C \end{cases} \tag{9.3.44}$$

is a UMP level .05 test of $H_0: \theta = 10$ versus $H_1: \theta > 10$. Now if $\phi(\cdot)$ satisfies (9.3.44), then

$$E_{\theta_0}\phi(X) = P_{\theta_0}[X_1 + \cdots + X_n > C]. \tag{9.3.45}$$

Here (by Example 5.4.24, using $n = 31$ and $\theta_0 = 10$) $Y = X_1 + \cdots + X_n$ has the gamma distribution $G(x|n - 1, \theta_0, 0) = G(x|30, 10, 0)$. To set C so that (9.3.45) holds is to set C so that $P[Y > C] = .05$, or $P[Y \leqslant C] = .95$. Using Table 9, we find this requires

$$\frac{C}{\theta_0} = 40.691 \tag{9.3.46}$$

or

$$C = 406.91. \tag{9.3.47}$$

Hence, we have found the desired UMP test of $H_0: \theta = 10$ versus $H_1: \theta > 10$, and it rejects H_0 iff $X_1 + \cdots + X_{31} > 406.91$, that is, iff

$$\bar{X} > \frac{406.91}{31} = 13.13. \tag{9.3.48}$$

We now proceed to the last business of this section, an application[3] of (a slight generalization of) the Neyman-Pearson Lemma to legislative apportionment. The generalization we need (others are given by, for example,

---

[3] This application is due to H. E. Reinhardt ("A maximization problem suggested by *Baker vs. Carr*," *American Mathematical Monthly*, Vol. 73 (1966), pp. 1069–1073).

Rao (1965), pp. 375–377 and are useful in maximization or minimization subject to constraints) is the following:

Theorem 9.3.49. *Let $f_1, \ldots, f_n$, $g_1, \ldots, g_n$, A, B be ≥ 0 with $A < B$. If $w_1, \ldots, w_n$ are such that $A \leq w_i \leq B$ $(i = 1, \ldots, n)$ and (for some k) satisfy*

$$\sum_{i=1}^{n} g_i w_i = c, \tag{9.3.50}$$

then by choosing

$$w_i = \begin{cases} B, & \text{if } f_i > kg_i \\ A, & \text{if } f_i < kg_i \end{cases} \tag{9.3.51}$$

we maximize $\sum_{i=1}^{n} f_i w_i$ subject to the restraint $\sum_{i=1}^{n} g_i w_i = c$.

The proof of Theorem 9.3.49 is essentially that of Theorem 9.3.2 with $\phi(x)$ now restricted to the interval $[A, B]$ rather than the interval $[0, 1]$ and is left as an exercise (Problem 9.6H) while we now proceed to our example.

Example 9.3.52. (Legislative apportionment) Suppose that a state is divided into n voting units called "districts" and that p_i is the proportion of the state's population in district i $(i = 1, \ldots, n)$. If we let $w_i^* = \{$The number of legislators-per-person in district $i\}$ $(i = 1, \ldots, n)$, then the set of numbers $\{w_1^*, \ldots, w_n^*\}$ represents any apportionment. [If we apportioned exactly in proportion to district population we would have $w_1^* = \cdots = w_n^*$ and Lp_i of the L legislators would be from district i $(1 \leq i \leq n)$.]

Following the Supreme Court's 1962 *Baker vs. Carr* decision that apportionment of state legislatures must be essentially proportional to population, it was suggested that "essentially proportional to population" be taken to mean

$$\frac{\max(w_1^*, \ldots, w_n^*)}{\min(w_1^*, \ldots, w_n^*)} \leq 1 + \varepsilon \tag{9.3.53}$$

for a specified $\varepsilon > 0$ [since w_i is the number of legislators-per-person in district i, criterion (9.3.53) means the best-represented district is not to be more than 100ε percent better represented than the worst-represented district—where "best" and "worst" apply respectively to w_i^* being "large" or "small," and not to the quality of the legislators], and $\varepsilon = .5$ was spoken for strongly.

Suppose r_i is the fraction of the legislators from district i whom one party (party R, say) can expect to control, while $d_i = 1 - r_i$ is the fraction the other party (party D, say) can expect to control $(i = 1, \ldots, n)$. [We assume only two parties.] *We then ask*: "If criterion (9.3.53) is to be satisfied, how effective can partisan choice of $\{w_1^*, \ldots, w_n^*\}$ be?"

A measure of the success of partisan tinkering by a party (let us say party R to be specific) is the fraction of the legislature expected to be controlled by party R,

$$\frac{\sum_{i=1}^{n} w_i^* r_i p_i}{\sum_{i=1}^{n} w_i^* p_i}, \tag{9.3.54}$$

and our question is one of the maxima and minima of (9.3.54) over choices of $\{w_1^*, \ldots, w_n^*\}$ subject to restraint (9.3.53).

In Theorem 9.3.49 by taking $f_i = r_i p_i$, $g_i = p_i$, and $w_i = w_i^*/\min(w_1^*, \ldots, w_n^*)$ $(i = 1, \ldots, n)$, we find $A = 1$ and $B = 1.5$ are our limits on w_i $(i = 1, \ldots, n)$ and the result is: If $w_1^*, \ldots, w_n^*$ are such that $1 \leqslant w_i^*/\min(w_1^*, \ldots, w_n^*) \leqslant 1.5$ $(i = 1, \ldots, n)$ and (for some k) satisfy

$$\frac{\sum_{i=1}^{n} w_i^* p_i}{\min(w_1^*, \ldots, w_n^*)} = c, \tag{9.3.55}$$

then by choosing

$$\frac{w_i^*}{\min(w_1^*, \ldots, w_n^*)} = \begin{cases} 1.5, & \text{if } r_i > k \\ 1, & \text{if } r_i < k \end{cases} \tag{9.3.56}$$

we maximize $\sum_{i=1}^{n} w_i^* r_i p_i/\min(w_1^*, \ldots, w_n^*)$ subject to the restraint

$$\sum_{i=1}^{n} w_i^* p_i/\min(w_1^*, \ldots, w_n^*) = c.$$

Since $\{w_1^*, \ldots, w_n^*\}$ which maximize (9.3.54) [subject to (9.3.53) with $\varepsilon = .5$] will satisfy (9.3.55) for *some* c, it follows that choosing $\{w_1^*, \ldots, w_n^*\}$ to satisfy (9.3.56) for *some* k will maximize (9.3.54). (The minimization follows similarly by taking $f_i = p_i$ and $g_i = r_i p_i$, since to maximize $1/x$ is to minimize $x (x > 0)$.)

While Example 9.3.52 is of interest in states where "districts" (or "counties") are fixed and are assigned a variable number of representatives, in other states "districts" may also be manipulated (Gerrymandering). In such cases a party may be interested in maximizing the proportion of districts which are considered "safe" for it, and other considerations than the Neyman-Pearson Lemma come into play.

9.4. A DECISION-THEORETIC VIEW OF THE HYPOTHESIS-TESTING PROBLEM

In this section we cover the contribution of decision theory to an integrated view of the hypothesis-testing problem, thus also giving a preview of the methods of Chapter 12 for what amounts to an important special case (the hypothesis-testing problem). The discussion in this section is largely informal.

In a hypothesis-testing problem (see Definition 9.2.1) where $H_0 = \{\theta_0\}$ and $H_1 = \{\theta_1\}$ are both simple we observe a r.v. X with d.f. $F(x|\theta)$ where θ is unknown, $\theta \in \Theta$. Then any test $\phi(\cdot)$ has two error probabilities associated with it: $\alpha(\phi) = E_{\theta_0}\phi(X)$ (probability of error of Type I, that is, probability of rejecting a true H_0) and $1 - \beta(\phi) = 1 - E_{\theta_1}\phi(X)$ (probability of error of Type II, that is, probability of accepting a false H_0). Thus, the point $(\alpha(\phi), 1 - \beta(\phi))$ is a measure of the goodness, or badness, of the test $\phi(\cdot)$: the smaller the coordinates, the better the test. This point

$$r(\phi) = (\alpha(\phi), 1 - \beta(\phi)) \tag{9.4.1}$$

is called the **risk point** of the test $\phi(\cdot)$, and if we plot the risk points $(\alpha(\phi), 1 - \beta(\phi))$ for *every possible* test $\phi(\cdot)$ based on X for the specific hypothesis-testing problem we are considering, we will fill a convex region as in Figure 9.4–1. Recalling (see Definition 8.3.1) that a subset S of $\mathscr{R}^k$ is called

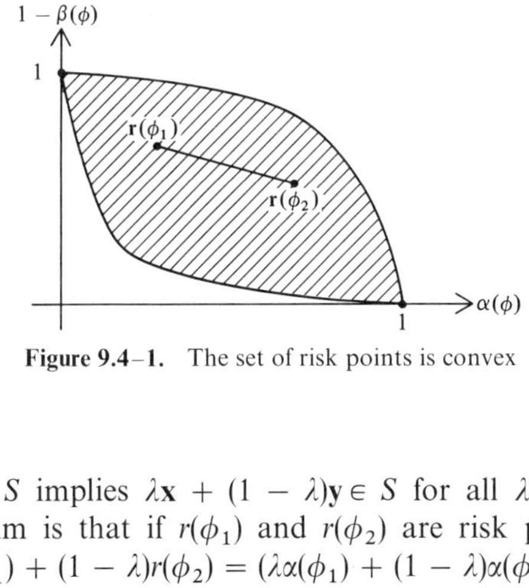

Figure 9.4–1. The set of risk points is convex

convex if $\mathbf{x}, \mathbf{y} \in S$ implies $\lambda\mathbf{x} + (1 - \lambda)\mathbf{y} \in S$ for all λ $(0 < \lambda < 1)$, we see that our claim is that if $r(\phi_1)$ and $r(\phi_2)$ are risk points of tests ϕ_1 and ϕ_2, then $\lambda r(\phi_1) + (1 - \lambda)r(\phi_2) = (\lambda\alpha(\phi_1) + (1 - \lambda)\alpha(\phi_2), \lambda(1 - \beta(\phi_1)) + (1 - \lambda)(1 - \beta(\phi_2)))$ is the risk point of some test ϕ_3 as long as $0 < \lambda < 1$. Formally:

Theorem 9.4.2. *For any hypothesis-testing problem where*[4] *H_0 and H_1 are both simple, the set $R = \{r(\phi): \phi(\cdot)$ is a test$\}$ of risk points of all possible tests is a convex subset of the unit square $I_2 = \{(x, y): 0 \leqslant x, y \leqslant 1\}$.*

[4] The considerations of this section can be generalized to composite hypotheses, but are no longer as simple (and so are best left as a special case of the full generality in Chapter 12).

Proof: First, for any test $\phi(\cdot)$, $\alpha(\phi)$ and $1 - \beta(\phi)$ are probabilities. Hence, $0 \leqslant \alpha(\phi) \leqslant 1$ and $0 \leqslant 1 - \beta(\phi) \leqslant 1$. Hence, $r(\phi) \in I_2$ for all $\phi(\cdot)$. Thus, $R \subseteq I_2$.

Next, suppose $r(\phi_1)$ and $r(\phi_2)$ are the risk points of tests $\phi_1(\cdot)$ and $\phi_2(\cdot)$. Then $\phi_3(\cdot) = \lambda\phi_1(\cdot) + (1 - \lambda)\phi_2(\cdot)$ is a test for $0 < \lambda < 1$ (check this claim; that is, verify that $0 \leqslant \phi_3(x) \leqslant 1$ for all $x \in \mathscr{X}$) and has (check this) risk point

$$\begin{aligned} r(\phi_3) &= (\lambda\alpha(\phi_1) + (1 - \lambda)\alpha(\phi_2), \lambda(1 - \beta(\phi_1)) + (1 - \lambda)(1 - \beta(\phi_2))) \\ &= \lambda r(\phi_1) + (1 - \lambda)r(\phi_2). \end{aligned} \tag{9.4.3}$$

Hence, R is a convex set.

Note that the points (0, 1) and (1, 0) are always in the convex set R:

Lemma 9.4.4. *For any hypothesis-testing problem where H_0 and H_1 are both simple, the set $R = \{r(\phi)\colon \phi(\cdot)$ a test$\}$ includes* (0, 1) *and* (1, 0).

Proof: If $\phi_1(\cdot) \equiv 1$, then $r(\phi_1) = (1, 0)$; while if $\phi_2(\cdot) \equiv 0$, then $r(\phi_2) = (0, 1)$.

Can we, by considering risk points of tests, restrict ourselves to a "few" tests (instead of the tests which lead to the whole convex set of risk points)? Now if the risk point of a test ϕ_1 is such that for some other test ϕ_2 we have $\alpha(\phi_1) > \alpha(\phi_2)$ and $1 - \beta(\phi_1) > 1 - \beta(\phi_2)$, then ϕ_2 has smaller Type I error *and* smaller Type II error. In such a case we would surely prefer ϕ_2. (See Figure 9.4–2.) Of course, even if $\alpha(\phi_1) \geqslant \alpha(\phi_2)$ and $1 - \beta(\phi_1) \geqslant 1 - \beta(\phi_2)$

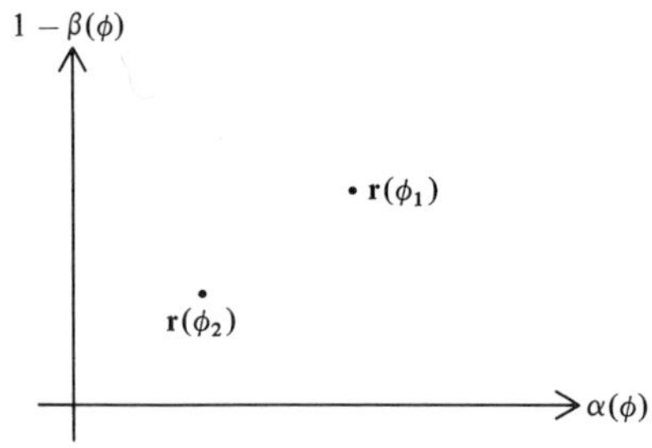

Figure 9.4–2. ϕ_2 preferred to ϕ_1

we have no use for ϕ_1: ϕ_2 is at least as good (and perhaps better) with respect to both types of error. In fact, if $\alpha(\phi_1) = \alpha(\phi_2)$ and $1 - \beta(\phi_1) = 1 - \beta(\phi_2)$ we can dispense with one of ϕ_1, ϕ_2 since they have the same risk point ($r(\phi_1) = r(\phi_2)$). These considerations are formalized as follows.

Definition 9.4.5. *For any hypothesis-testing problem where H_0 and H_1 are simple, test ϕ_2 is said to be* **better than** *test ϕ_1 (or test ϕ_1 is said to be* **worse**

than *test* ϕ_2) *if* $\alpha(\phi_2) \leqslant \alpha(\phi_1)$ *and* $1 - \beta(\phi_2) \leqslant 1 - \beta(\phi_1)$ *with at least one strict inequality* ($<$). *We say* ϕ_1 *is* **no better than** ϕ_2 *if* $\alpha(\phi_2) \leqslant \alpha(\phi_1)$ *and* $1 - \beta(\phi_2) \leqslant 1 - \beta(\phi_1)$.

Definition 9.4.6. *For any hypothesis-testing problem where* H_0 *and* H_1 *are both simple, let* $T = \{\phi(\cdot): \phi(\cdot) \text{ is a test}\}$. *Any class of tests* $C \subseteq T$ *is called* **complete** (*or a* **complete class of tests**) *if for each test* $\phi_1 \notin C$ *there is a test* $\phi_2 \in C$ *such that* ϕ_2 *is better than* ϕ_1. *A class* C *is called* **essentially complete** *if for each test* $\phi_1 \notin C$ *there is a test* $\phi_2 \in C$ *such that* ϕ_1 *is no better than* ϕ_2. *A class* C *is called* **minimal complete** *if* C *is complete, but if any* $\phi_2 \in C$ *is removed from* C *the resulting class is not complete. A class* C *is called* **minimal essentially complete** *if* C *is essentially complete, but if any* $\phi_2 \in C$ *is removed from* C *the resulting class is not essentially complete.*

Thus, by considering risk points of tests we can restrict ourselves to a "few" tests instead of the tests which lead to the whole convex set of risk points, since no "worse" tests (Definition 9.4.5) should be considered and, among equivalent tests, only one need be considered: those tests whose risk points fall on the lower boundary of the convex set of risk points form[5] a minimal complete class C_1. (See Figure 9.4–3.) But this class C_1 may still

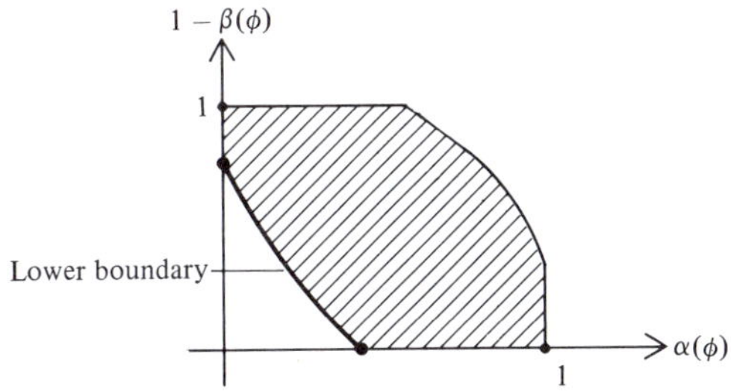

Figure 9.4–3. Risk points of minimal complete class

contain tests ϕ_1 and ϕ_2 such that $r(\phi_1) = r(\phi_2)$. If we then retain only ϕ_1 in each such case (that is, if a number of tests have the same risk point, eliminate all but one of these tests) we obtain a minimal essentially complete class C_2. Of course, C_2 contains tests which can give us any "desirable" risk point.

Now that we have seen how (by consideration of risk points of tests) we can restrict ourselves to a "few" tests, how do we do this in practice? (That is, how do we obtain a minimal essentially complete class of tests?) By the

[5] It is implicitly assumed that the convex set of risk points is closed (that is, contains its boundary). This is true but difficult to prove. For the mathematical details see E. L. Lehmann, *Testing Statistical Hypotheses*, John Wiley & Sons, Inc., New York, 1959, p. 67.

Neyman-Pearson Lemma (Theorem 9.3.2 and Problem 9.3H) for testing simple H_0 versus simple H_1 after observing X, a test $\phi(\cdot)$ is most powerful of level α iff it has level α and satisfies (9.3.4) for some k. Thus, we have the following.

Theorem 9.4.7. *For testing simple H_0 versus simple H_1 after observing X, the collection of tests $\{\phi_\alpha(\cdot), \alpha \leqslant \alpha_0\}$ is minimal essentially complete. Here $\phi_\alpha(\cdot)$ is a test of level α which satisfies* (9.3.4), *and α_0 is the smallest α for which* $1 - \beta(\phi_\alpha) = 0$ [that is, with regard to Figure 9.4–3, the smallest α at which the convex set of risk points touches the horizontal axis; usually $\alpha_0 = 1$].

Ultimately, the question arises: Once we have obtained the minimal essentially complete class in Theorem 9.4.7 (that is, a class of tests which yields the lower boundary shown in Figure 9.4–3), how should we select one member out of this class? This is a disputed question and several general principles have been proposed, each of which will lead to *an* answer; however, the choice may have to be made on pragmatic grounds (since there may be no quantity called *the* answer[6]) . . . it depends on which $(\alpha, 1 - \beta)$ on the lower boundary is suitable for our particular situation.

The several proposed methods of choosing one of the tests of the minimal essentially complete class clarify the problems involved, and we now discuss them. Each method essentially attempts to reduce the risk point $r(\phi) = (\alpha(\phi), 1 - \beta(\phi))$, which is two-dimensional (a point in the plane) to a one-dimensional number $f(r(\phi))$ (a point on a line). If the resulting number $f(r(\phi))$ is such that one can agree to prefer ϕ_2 to ϕ_1 whenever $f(r(\phi_2)) < f(r(\phi_1))$, we are done: we choose that ϕ which minimizes $f(r(\phi))$. [Unfortunately no function $f(\cdot)$ has yet been proposed for which statisticians can agree to prefer ϕ_2 to ϕ_1 whenever $f(r(\phi_2)) < f(r(\phi_1))$ *in every problem.* Probably none exists.] The proposed $f(\cdot)$ are as follows.

Definition 9.4.8. *When testing simple H_0 versus simple H_1 after observing X, we choose a test from the minimal essentially complete class $\mathbf{C}_\alpha = \{\phi_\alpha(\cdot), \alpha \leqslant \alpha_0\}$ of Theorem* 9.4.7. *A test which minimizes* (*for some p with* $0 \leqslant p \leqslant 1$)

$$f(r(\phi)) = p\alpha(\phi) + (1 - p)(1 - \beta(\phi)) \tag{9.4.9}$$

is called **Bayes (with respect to the a priori distribution** $(p, 1 - p)$), *or a* **Bayes**

[6] Of course, proponents of the various methods might like us to think they were "the" (not "an") answer. And if, as some say, "Philosophers come after the scientists have left and make principles," then the debate may grow bitter and the voices strident. A lucid introduction to the lurid debates now raging in some statistical circles was given by a great statistician, J. Wolfowitz ("Bayesian inference and axioms of consistent decision," *Econometrica*, Vol. 30 (1962), pp. 470–479), and is recommended reading for those wishing to pursue this avenue.

rule. *A test which minimizes*

$$f(r(\phi)) = \max\{\alpha(\phi), 1 - \beta(\phi)\} \tag{9.4.10}$$

is called **minimax**, *or a* **minimax (risk) rule**. *A test which minimizes*

$$f(r(\phi)) = \max\{\alpha(\phi) - \min\{\alpha(\phi): \phi \in C_\alpha\}, \\ (1 - \beta(\phi)) - \min\{1 - \beta(\phi): \phi \in C_\alpha\}\} \tag{9.4.11}$$

is called **minimax regret**,[7] *or a* **minimax regret rule**.

A geometrical interpretation of the minimax rule can be given with respect to the convex set of risk points, as follows. On the dotted curve in Figure 9.4–4, $f(x, y) = \max(x, y)$ is constant, at say value c. As $c\uparrow$, the quantity $f(x, y)\uparrow$. Thus, since we desire to minimize $f(r(\phi)) = \max\{\alpha(\phi), 1 - \beta(\phi)\}$, we choose a test corresponding to the first risk point (of those corresponding to tests in the minimal essentially complete class) touched (by the dotted curve) as $c\uparrow$.

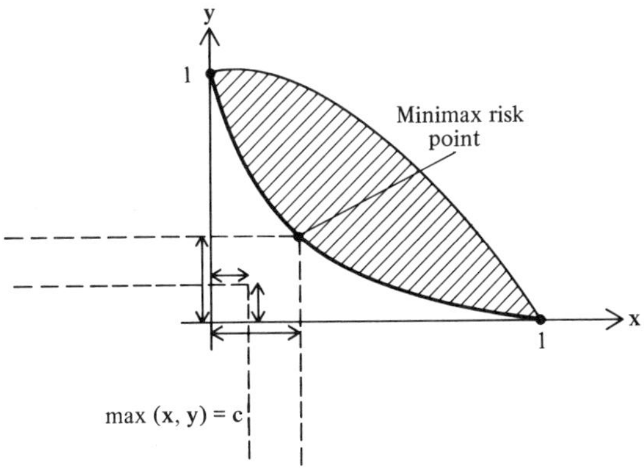

Figure 9.4–4. The minimax criterion

The geometrical interpretation of the Bayes rule is similar to that of the minimax rule. The quantity $f(x, y) = px + (1 - p)y$ is constant, say at value c, on the dotted line in Figure 9.4–5. Since we wish to minimize $f(r(\phi)) = p\alpha(\phi) + (1 - p)(1 - \beta(\phi))$, we choose a test corresponding to any risk point (of those corresponding to tests in the minimal essentially complete class) first touched (by the dotted line) as $c\uparrow$.

[7] If $(\alpha(\phi), 1 - \beta(\phi))$ is the *risk* in terms of errors of Types I and II which we incur by using ϕ, then the *regret* we incur is $(\alpha(\phi) - b_1, (1 - \beta(\phi)) - b_2)$, where b_1 is the smallest $\alpha(\phi)$ could be for any ϕ, and b_2 is the smallest $1 - \beta(\phi)$ could be for any ϕ. If H_0 is true, we must suffer at least b_1 as our Type I error; the increment of $\alpha(\phi) - b_1$ is due to using ϕ, while b_1 is common to all rules (similarly for Type II error).

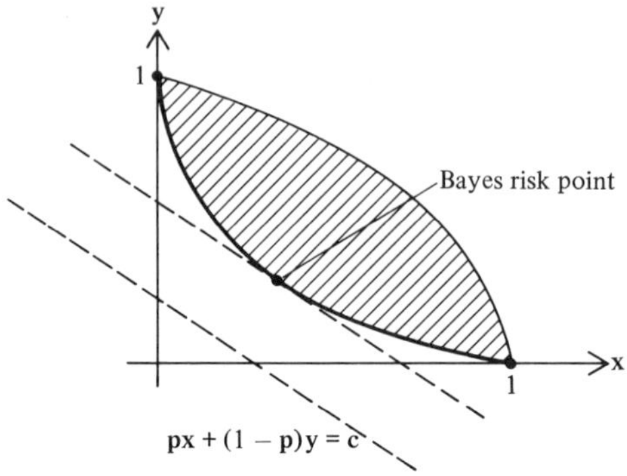

Figure 9.4–5. The Bayes criterion

From Figures 9.4–4 and 9.4–5, it is clear that the minimax rule is a Bayes rule with respect to $(p, 1 - p)$ for some p $(0 \leqslant p \leqslant 1)$. [Note that it is easy to show that it is not the case that "$p = 0.5$ always." However, it is the case that $\alpha(\phi) = 1 - \beta(\phi)$ for the minimax rule ϕ.] So, finally we ask: **How do we find a Bayes rule**? A rule ϕ is Bayes with respect to $(p, 1 - p)$ (where p is fixed with $0 < p < 1$; if $p = 0$ or $p = 1$ the Bayes rule is obvious) iff it minimizes (9.4.9). But (using the notation of Theorem 9.3.2 and its proof)

$$\begin{aligned} f(r(\phi)) &= p\alpha(\phi) + (1 - p)(1 - \beta(\phi)) \\ &= pE_{\theta_0}\phi(X) + (1 - p)E_{\theta_1}(1 - \phi(X)) \\ &= p\int_{x \in \mathscr{X}} \phi(x)f(x|\theta_0)\,dx + (1 - p)\int_{x \in \mathscr{X}} (1 - \phi(x))f(x|\theta_1)\,dx \\ &= \int_{x \in \mathscr{X}} [p\phi(x)f(x|\theta_0) + (1 - p)(1 - \phi(x))f(x|\theta_1)]\,dx. \end{aligned} \tag{9.4.12}$$

However, the quantity being integrated is $\geqslant 0$, so we minimize $f(r(\phi))$ by (if possible) choosing $\phi(x)$ for each x so as to minimize

$$p\phi(x)f(x|\theta_0) + (1 - p)(1 - \phi(x))f(x|\theta_1). \tag{9.4.13}$$

But for a specified x, either $pf(x|\theta_0) > (1 - p)f(x|\theta_1)$ (in which case $\phi(x) = 0$ minimizes (9.4.13)) or $pf(x|\theta_0) < (1 - p)f(x|\theta_1)$ (in which case $\phi(x) = 1$ minimizes (9.4.13)) or $pf(x|\theta_0) = (1 - p)f(x|\theta_1)$ (in which case $\phi(x)$ can be arbitrary with $0 \leqslant \phi(x) \leqslant 1$), so *the Bayes rule sets*

$$\phi(x) = \begin{cases} 1, & \text{if } (1 - p)f(x|\theta_1) > pf(x|\theta_0) \\ 0, & \text{if } (1 - p)f(x|\theta_1) < pf(x|\theta_0). \end{cases} \tag{9.4.14}$$

This is (since $p \neq 1$) essentially the Neyman-Pearson test (see Theorem 9.3.2) with $k = p/(1 - p)$.

Thus, in testing simple $\mathbf{H_0}$ versus simple $\mathbf{H_1}$ after observing X, any rule which satisfies (9.4.14) for all $\mathbf{x \in \mathscr{X}}$ is Bayes with respect to (p, 1 − p). This holds for any p with $0 \leqslant p \leqslant 1$. Recall that in Definition 9.4.8, **(p, 1 − p) was called an a priori (or prior) distribution, for reasons we will now make clear.**

Suppose we have a hypothesis-testing problem with simple $H_0 = \{\theta_0\}$ and simple $H_1 = \{\theta_1\}$, and observe X. Suppose that, before the experiment, we find out that θ is a r.v. such that $P[\theta = \theta_0] = p$ and $P[\theta = \theta_1] = 1 - p$ [this is the case if, say, a person (usually called "nature") chooses the d.f. $F(x|\theta)$ of X by picking at random from a bin full of d.f.'s $100p$ percent of which are $F(x|\theta_0)$ and $100(1 - p)$ percent of which are $F(x|\theta_1)$]. Then (using the notation of Theorem 9.3.2 in the discrete case)

$$P[X = x] = pf(x|\theta_0) + (1 - p)f(x|\theta_1). \tag{9.4.15}$$

So, before observing X, we knew $P[\theta = \theta_0] = p$ and $P[\theta = \theta_1] = 1 - p$. After observing X, we can find the *a posteriori* (or *posterior*) probability that $\theta = \theta_0$ or $\theta = \theta_1$ (using Bayes' rule (Theorem 2.3.6)) as

$$\begin{aligned} P[\theta = \theta_0|X = x] &= \frac{P[\theta = \theta_0, X = x]}{P[X = x]} \\ &= \frac{P[X = x|\theta = \theta_0]P[\theta = \theta_0]}{P[X = x|\theta = \theta_0]P[\theta = \theta_0] + P[X = x|\theta = \theta_1]P[\theta = \theta_1]} \\ &= \frac{pf(x|\theta_0)}{pf(x|\theta_0) + (1 - p)f(x|\theta_1)}, \end{aligned} \tag{9.4.16}$$

$$P[\theta = \theta_1|X = x] = \frac{(1 - p)f(x|\theta_1)}{pf(x|\theta_0) + (1 - p)f(x|\theta_1)}. \tag{9.4.17}$$

If we then decide θ_0 or θ_1 (that is, $\phi(x) = 0$ or $\phi(x) = 1$) according respectively as (9.4.16) or (9.4.17) is the larger, we are simply stating to be "true" that one of $\{\theta_0, \theta_1\}$ which has the larger posterior probability. This is (see (9.4.14)) *precisely what a Bayes rule with respect to* $(p, 1 - p)$ *does.*

9.5. TESTING NORMAL MEANS (ONE POPULATION)

One hypothesis-testing problem (see Definition 9.2.1) which occurs frequently in practice has $X = (X_1, \ldots, X_n)$ where $X_1, \ldots, X_n$ are independent r.v.'s, each $N(\mu, \sigma^2)$ where μ is unknown with $\mu \in \Theta = \{x: -\infty < x < +\infty\}$, $\sigma^2 > 0$ is known, $H = \{\mu_0\}$, and H_1 may have several forms. *The case* where $H_1 = \{\mu_1\}$ with $\mu_1 > \mu_0$ was studied in Example 9.3.10. Using the Neyman-Pearson Lemma, we found that the test which rejects iff

$$\bar{X} > \mu_0 + \frac{\sigma}{\sqrt{n}}\Phi^{-1}(1 - \alpha) \tag{9.5.1}$$

is most powerful of level α in this case. We also showed that *if* $H_1 = \{\mu_1 : \mu_1 > \mu_0\}$, the same test is UMP one-sided. *If* $H_1 = \{\mu_1\}$ with $\mu_1 < \mu_0$, or *if* $H_1 = \{\mu_1 : \mu_1 < \mu_0\}$, analogous results are easily shown (Problem 9.26).

Thus, for each of simple and one-sided alternative hypotheses, there is a "best" test available to us. In the case of *two-sided* alternatives (for example, $H_1 = \{\mu_1 : \mu_1 < \mu_0, \text{ or } \mu_1 > \mu_0\}$) we discussed two tests in a special case in Example 9.2.4. For this case there is no "best" (UMP) test (Problem 9.12H), but tests which reject iff

$$\bar{X} - \mu_0 < a \quad \text{or} \quad \bar{X} - \mu_0 > b \tag{9.5.2}$$

(for some a and b which give the proper level) form an essentially complete class of tests.

Now suppose our hypothesis-testing problem is as above $[X = (X_1, \ldots, X_n)$ where $X_1, \ldots, X_n$ are independent r.v.'s, each $N(\mu, \sigma^2)$ where μ is unknown with $\mu \in \Theta = \{x : -\infty < x < +\infty\}$, $H_0 = \{\mu_0\}]$ for one of the possible choices of H_1 studied, but that (instead of having $\sigma^2 > 0$ known) we have $\sigma^2 > 0$ *unknown*. Since even the null hypothesis $H_0 : \mu = \mu_0$ is no longer a simple hypothesis (because the distribution of X will be different for different values of σ^2), the Neyman-Pearson Lemma does not solve this problem by providing a "best" test. However (see Example 8.8.24 and Problem 8.56) for $n \geqslant 2$

$$\left(\bar{X}, \sum_{i=1}^{n} (X_i - \bar{X})^2/(n-1)\right) \tag{9.5.3}$$

is a minimal sufficient statistic for the problem and hence (see Problem 9.13H) a test need be based only on the sufficient statistic (9.5.3) in order to have any "good" properties. Now *what is a reasonable way to base a test on the sufficient statistic* (9.5.3)? First, let us reexamine the case where σ^2 is known. There, to be specific considering the case $H_1 = \{\mu_1\}$ with $\mu_1 > \mu_0$, the Neyman-Pearson test rejects (see Equation (9.5.1)) iff

$$\begin{aligned} & \bar{X} > \mu_0 + \frac{\sigma}{\sqrt{n}} \Phi^{-1}(1-\alpha) \\ \Leftrightarrow \quad & \bar{X} - \mu_0 > \frac{\sigma}{\sqrt{n}} \Phi^{-1}(1-\alpha) \\ \Leftrightarrow \quad & \frac{\bar{X} - \mu_0}{\sigma/\sqrt{n}} > \Phi^{-1}(1-\alpha). \end{aligned} \tag{9.5.4}$$

That is, the Neyman-Pearson test rejects iff $\bar{X} - \mu_0$ is "too large." This is intuitively reasonable since when $H_0 : \mu = \mu_0$ is true, $\bar{X} - \mu_0$ should be "close" to zero (it is then a $N(0, \sigma^2/n)$ r.v.), while when $H_1 : \mu = \mu_1$ with $\mu_1 > \mu_0$ is true, $\bar{X} - \mu_0$ should be "far" from zero (it is then a $N(\mu_1 - \mu_0, \sigma^2/n)$ r.v.); hence, it seems reasonable to reject iff

$$\bar{X} - \mu_0 > c \tag{9.5.5}$$

and set c for level α. In setting c in (9.5.5) for level α we desire

$$P_{\mu_0}[\bar{X} - \mu_0 > c] = \alpha, \tag{9.5.6}$$

and divide by $\sigma/\sqrt{n}$ so that we will be dealing with the $N(0, 1)$ distribution (for which tables, for example Table 11, are available):

$$P_{\mu_0}\left[\frac{\bar{X} - \mu_0}{\sigma/\sqrt{n}} > \frac{c}{\sigma/\sqrt{n}}\right] = \alpha. \tag{9.5.7}$$

Solving (9.5.7) for c, we find (essentially as in Example 9.3.10)

$$\begin{cases} 1 - \Phi\left(\dfrac{c}{\sigma/\sqrt{n}}\right) = \alpha, \\ \Phi\left(\dfrac{c}{\sigma/\sqrt{n}}\right) = 1 - \alpha, \\ \dfrac{c}{\sigma/\sqrt{n}} = \Phi^{-1}(1 - \alpha), \\ c = \dfrac{\sigma}{\sqrt{n}}\Phi^{-1}(1 - \alpha). \end{cases} \tag{9.5.8}$$

Now if it is reasonable (in fact via use of the Neyman-Pearson Lemma we know it is "best") to reject when

$$\frac{\bar{X} - \mu_0}{\sqrt{\sigma^2/n}} \tag{9.5.9}$$

is "too large" in the case of a known σ^2, perhaps in the case of an unknown σ^2 we should consider doing similarly *after replacing σ^2 by a good estimator of σ^2*. However (see Example 8.6.20) the second component of our sufficient statistic (9.5.3) is a good estimator of σ^2; **since it is used often it is usually denoted by**

$$\mathbf{s}^2 \equiv \frac{\sum_{\mathbf{i}=1}^{n} (\mathbf{X}_\mathbf{i} - \bar{\mathbf{X}})^2}{\mathbf{n} - 1}. \tag{9.5.10}$$

So, for the σ^2 unknown case our intuition leads us (by analogy to the σ^2 known case) to consider the test which rejects when

$$\frac{\bar{X} - \mu_0}{s/\sqrt{n}} \tag{9.5.11}$$

is "too large." In order to obtain level α for this test we need to set c so that

$$P_{\mu_0}\left[\frac{\bar{X} - \mu_0}{s/\sqrt{n}} > c\right] = \alpha. \tag{9.5.12}$$

We could do this *if* we knew the probability distribution of (9.5.11) and that distribution was tabled. In Theorem 9.5.13 and Lemma 9.5.15 we find that that distribution is one already studied in Chapter 4.

Theorem 9.5.13. *If Y_1 has the $N(0, 1)$ distribution* [see Definition 4.2.13], *if Y_2 has the $\chi^2_m(0)$ distribution* [see Definition 4.2.9], *and if Y_1 and Y_2 are independent r.v.'s, then*

$$Z = \frac{Y_1}{\sqrt{Y_2/m}} \tag{9.5.14}$$

has the Student's t-distribution with m degrees of freedom [see Definition 4.2.17].

Proof: This follows by using the methods studied in Section 4.6, and is left as a review exercise (Problem 9.27).

Lemma 9.5.15. *If $X_1, \ldots, X_n$ are independent r.v.'s, each $N(\mu_0, \sigma^2)$, then $\bar{X} = (X_1 + \cdots + X_n)/n$ and $s^2 = \sum_{i=1}^{n} (X_i - \bar{X})^2/(n-1)$ are independent r.v.'s. It follows that*

$$\frac{\bar{X} - \mu_0}{s/\sqrt{n}} \tag{9.5.16}$$

has the Student's t-distribution with $m = n - 1$ degrees of freedom.

Proof: *Before giving a precise proof, we note the following motivation.* Consider $\bar{X}$ and $X_i - \bar{X}$ for any fixed i ($1 \leqslant i \leqslant n$). Now $(\bar{X}, X_i - \bar{X})$ has a bivariate normal distribution [see the footnote to Problem 8.14H] since $(X_1, \ldots, X_n)$ has a multivariate normal distribution (see Problem 9.28). Also an easy computation yields (see Problem 9.29)

$$\begin{aligned}\rho(\bar{X}, X_i - \bar{X}) &= \frac{\operatorname{Cov}(\bar{X}, X_i - \bar{X})}{\sigma(\bar{X})\sigma(X_i - \bar{X})}\\ &= \frac{\operatorname{Cov}(\bar{X}, X_i) - \operatorname{Cov}(\bar{X}, \bar{X})}{\sigma(\bar{X})\sigma(X_i - \bar{X})}\\ &= \frac{\operatorname{Cov}((X_1 + \cdots + X_n)/n, X_i) - \operatorname{Var}(\bar{X})}{\sigma(\bar{X})\sigma(X_i - \bar{X})} = 0.\end{aligned} \tag{9.5.17}$$

Hence, by Theorem 5.3.17 $\bar{X}$ and $X_i - \bar{X}$ are independent. It follows easily (see Problem 9.30) that $\bar{X}$ and $(X_i - \bar{X})^2$ are independent. Since i was arbitrary, $\bar{X}$ is independent of each of $(X_1 - \bar{X})^2, \ldots, (X_n - \bar{X})^2$. (See Problem 9.14H.)

A precise proof proceeds as follows. Based on the r.v.'s $X_1, \ldots, X_n$ we define n new r.v.'s

$$\begin{cases} Y_1 = \dfrac{1}{\sqrt{n}}(X_1 + X_2 + X_3 + X_4 + \cdots + X_{n-2} + X_{n-1} + X_n), \\ Y_2 = \dfrac{1}{\sqrt{2}}(X_1 - X_2), \\ Y_3 = \dfrac{1}{\sqrt{3}\sqrt{2}}(X_1 + X_2 - 2X_3), \\ Y_4 = \dfrac{1}{\sqrt{4}\sqrt{3}}(X_1 + X_2 + X_3 - 3X_4), \\ \cdot \\ \cdot \\ \cdot \\ Y_{n-1} = \dfrac{1}{\sqrt{n-1}\sqrt{n-2}}(X_1 + X_2 + X_3 + X_4 + \cdots + X_{n-2} - (n-2)X_{n-1}), \\ Y_n = \dfrac{1}{\sqrt{n}\sqrt{n-1}}(X_1 + X_2 + X_3 + X_4 + \cdots + X_{n-2} + X_{n-1} - (n-1)X_n). \end{cases} \tag{9.5.18}$$

([9.5.18] is called the **Helmert orthogonal transformation**.) Since $(X_1, \ldots, X_n)$ has a multivariate normal distribution so does $(Y_1, \ldots, Y_n)$. Now a useful general form for Y_i $(2 \leqslant i \leqslant n)$ is

$$Y_i = \frac{1}{\sqrt{i}\sqrt{i-1}}\left(\sum_{j=1}^{i-1} X_j - (i-1)X_i\right). \tag{9.5.19}$$

Then we can find *which* multivariate normal distribution $(Y_1, \ldots, Y_n)$ has by calculating (see Problem 9.31)

$$\begin{cases} EY_1 = \sqrt{n}\mu_0, \quad EY_2 = \cdots = EY_n = 0, \\ \operatorname{Var}(Y_1) = \operatorname{Var}(Y_2) = \cdots = \operatorname{Var}(Y_n) = \sigma^2, \\ \operatorname{Cov}(Y_1, Y_i) = 0 \quad (2 \leqslant i \leqslant n), \\ \operatorname{Cov}(Y_i, Y_j) = 0 \quad (2 \leqslant i < j \leqslant n). \end{cases} \tag{9.5.20}$$

Hence, $Y_1, \ldots, Y_n$ are n independent r.v.'s so Y_1 is independent of $Y_2, \ldots, Y_n$ and hence (see Problem 9.32) any function of Y_1 is independent of any function of $Y_2, \ldots, Y_n$. Since (see Problem 9.33)

$$\begin{cases} Y_1 = \sqrt{n}\bar{X}, \\ Y_2^2 + \cdots + Y_n^2 = \displaystyle\sum_{i=1}^{n} (X_i - \bar{X})^2 = (n-1)s^2 \end{cases} \tag{9.5.21}$$

it follows that $\bar{X}$ is independent of s^2.

To complete the proof we wish to show that (9.5.16) has a certain Student's t-distribution. Following the hint of Theorem 9.5.13, we write

$$\frac{\bar{X} - \mu_0}{s/\sqrt{n}} = \frac{\frac{\sqrt{n}}{\sigma}(\bar{X} - \mu_0)}{s/\sigma}, \tag{9.5.22}$$

where $(\sqrt{n}/\sigma)(\bar{X} - \mu_0)$ is a $N(0, 1)$ r.v. which is independent (by the first part of the present theorem) of s/σ. Thus, if we can show that s/σ is $\sqrt{Y_2/(n-1)}$ where Y_2 is a $\chi^2_{n-1}(0)$ r.v., we are (by use of Theorem 9.5.13) done. Now using (9.5.21)

$$\begin{aligned} \frac{s}{\sigma} = \sqrt{\frac{s^2}{\sigma^2}} &= \sqrt{\frac{Y_2^2 + \cdots + Y_n^2}{(n-1)\sigma^2}} \\ &= \sqrt{\left\{\left(\frac{Y_2}{\sigma}\right)^2 + \cdots + \left(\frac{Y_n}{\sigma}\right)^2\right\}\Big/(n-1)}, \end{aligned} \tag{9.5.23}$$

where $Y_2/\sigma, \ldots, Y_n/\sigma$ are independent $N(0, 1)$ r.v.'s. Using Problem 9.34 the proof is complete. [Note that these results also justify the nomenclature "loss of 1 degree of freedom" noted following Equation (8.1.6).]

We are now in a position to provide a c which solves (9.5.12), since (via Lemma 9.5.15) we know that (9.5.11) has the Student's t-distribution with $n - 1$ degrees of freedom. First we need some notation. [For the $N(0, 1)$ distribution we also needed a notation for that number c such that if X is a $N(0, 1)$ r.v., then $P[X \leqslant c] = x$; namely $\Phi(c) = x$, or $c = \Phi^{-1}(x)$.]

Definition 9.5.24. *If a r.v. X has the Student's t-distribution with m degrees of freedom, we will say* **X has the t_m distribution**, *and will (for $0 < \alpha < 1$) denote the solution c of the equation*

$$P[X \leqslant c] = \alpha \tag{9.5.25}$$

by writing $t_m(c) = \alpha$, or $c = t_m^{-1}(\alpha)$. Thus, we have the following.

Theorem 9.5.26. *For the hypothesis-testing problem with $X = (X_1, \ldots, X_n)$ where $X_1, \ldots, X_n$ are independent r.v.'s, each $N(\mu, \sigma^2)$ where $\mu(-\infty < \mu < +\infty)$ and $\sigma^2(\sigma^2 > 0)$ are both unknown, $H_0 = \{\mu_0\}$, and $H_1 = \{\mu_1\}$ with $\mu_1 > \mu_0$, the test which rejects iff*

$$\frac{\bar{X} - \mu_0}{s/\sqrt{n}} > t_{n-1}^{-1}(1 - \alpha) \tag{9.5.27}$$

has level α. Similar tests are easily obtained [see Problem 9.35] *for other alternatives H_1.*

Although we were not able to use the Neyman-Pearson Lemma to show that this test in Theorem 9.5.26, called **the t-test**, has good properties [we reasoned to the test by using analogy to the $\sigma^2 > 0$ known case and sufficient statistics], in fact it has good properties. Since their development is mathematically intricate we will not discuss them here.

9.6. TESTING NORMAL VARIANCES (ONE POPULATION)

The hypothesis-testing problem of concern in this section is the one with $X = (X_1, \ldots, X_n)$ where $X_1, \ldots, X_n$ are independent r.v.'s, each $N(\mu, \sigma^2)$ where μ is known $(-\infty < \mu < +\infty)$, σ^2 is unknown with $\sigma^2 \in \Theta = \{x: x > 0\}$, $H_0 = \{\sigma_0^2\}$, and H_1 may have several forms. First we will study *the case* where $H_1 = \{\sigma_1^2\}$ with $\sigma_1^2 > \sigma_0^2$, and will reason intuitively to a test as follows. For this problem we know from Problem 8.34 that

$$\hat{\sigma}^2 = \frac{\sum_{i=1}^{n} (X_i - \mu)^2}{n} \tag{9.6.1}$$

is a good estimator of σ^2. In fact, $\hat{\sigma}^2$ is a minimal sufficient statistic for σ^2 (Problem 9.38) and is the unique MVU estimator of σ^2 (Problem 9.16H). Thus, it seems intuitively reasonable to reject H_0 iff

$$\hat{\sigma}^2 > k, \tag{9.6.2}$$

where k is a constant chosen (if such a choice is possible) to give level α. Now

$$\begin{aligned} P_{\sigma_0^2}[\hat{\sigma}^2 > k] &= P_{\sigma_0^2}\left[\frac{\sum_{i=1}^{n} (X_i - \mu)^2}{n} > k\right] \\ &= P_{\sigma_0^2}\left[\sum_{i=1}^{n} \left(\frac{X_i - \mu}{\sigma_0}\right)^2 > \frac{nk}{\sigma_0^2}\right] \end{aligned} \tag{9.6.3}$$

and since (by Problem 9.34) $\sum_{i=1}^{n} ((X_i - \mu)/\sigma_0)^2$ has the $\chi_n^2(0)$ distribution, Table 10 can be used to find k. Thus, we have the following:

Theorem 9.6.4. *For the hypothesis-testing problem with $X = (X_1, \ldots, X_n)$ where $X_1, \ldots, X_n$ are independent r.v.'s, each $N(\mu, \sigma^2)$ where μ is known $(-\infty < \mu < +\infty)$, σ^2 is unknown $(\sigma^2 > 0)$, $H_0 = \{\sigma_0^2\}$, and $H_1 = \{\sigma_1^2\}$*

with $\sigma_1^2 > \sigma_0^2$, the test which rejects iff

$$\frac{\sum_{i=1}^{n} (X_i - \mu)^2}{n} > k \tag{9.6.5}$$

(*where k satisfies* $P[Y > nk/\sigma_0^2] = \alpha$ *for* Y *a* $\chi_n^2(0)$ *r.v.*) *has level* α. *Similar tests are easily obtained* [as in Problem 9.35] *for other alternatives* H_1.

In fact, the test of Theorem 9.6.4 is most powerful (see Problem 9.39). **Now if $\mu(-\infty < \mu < +\infty)$ is unknown, what do we do?** As in Section 9.5, we cannot use the Neyman-Pearson Lemma (since now not even H_0 is a simple hypothesis), but we do know a minimal sufficient statistic for the problem. Hence, we may consider replacing μ in (9.6.5) by a good estimator based on the minimal sufficient statistic, namely the estimator $\bar{X}$, and rejecting iff

$$\frac{\sum_{i=1}^{n} (X_i - \bar{X})^2}{n} > k'. \tag{9.6.6}$$

Since, in the proof of Lemma 9.5.15, we established that $(n - 1)s^2/\sigma^2$ has a $\chi_{n-1}^2(0)$ distribution in this case, we may (by choosing k' to satisfy

$$P\left[Y > \frac{nk'}{\sigma_0^2}\right] = \alpha \tag{9.6.7}$$

for Y a $\chi_{n-1}^2(0)$ r.v.) obtain our level α test. [This again justifies the "loss of 1 degree of freedom" nomenclature noted following Equation (8.1.6).]

9.7. TESTING NORMAL VARIANCES (TWO POPULATIONS)

In Section 9.6 we considered testing whether the observations from one source (each such source is called a **population**) have a specified variance ($\sigma^2 = \sigma_0^2$). In the present section we consider the problem of testing whether observations from two sources (or populations) have the same variance. (For example, we might be concerned with whether the variability in height of people who smoked before their twenty-first birthday is equal to the variability in height of people who did not smoke until after their twenty-first birthday.)

Formally the hypothesis-testing problem of concern is one with $X = (X_1, \ldots, X_n, Y_1, \ldots, Y_m)$ where $X_1, \ldots, X_n, Y_1, \ldots, Y_m$ are $n + m$ independent r.v.'s, each of $X_1, \ldots, X_n$ is $N(\mu_1, \sigma_1^2)$, each of $Y_1, \ldots, Y_m$ is $N(\mu_2, \sigma_2^2)$, μ_1 and μ_2 are known $(-\infty < \mu_1, \mu_2 < +\infty)$, σ_1^2 and σ_2^2 are unknown $(\sigma_1^2, \sigma_2^2 \in \Theta = \{(x, y): x > 0, y > 0\})$, $H_0 = \{(\sigma_1^2, \sigma_2^2): \sigma_1^2 = \sigma_2^2 > 0\}$, and

H_1 may have several forms. Let us consider first *the case* where $H_1 = \{(\sigma_1^2, \sigma_2^2): \sigma_1^2 > \sigma_2^2\}$. We can reason to a test intuitively as follows. We know (see Section 9.6) that $\hat{\sigma}_i^2$ is a good estimator of σ_i^2 $(i = 1, 2)$ where

$$\hat{\sigma}_1^2 = \frac{\sum_{j=1}^{n} (X_j - \mu_1)^2}{n}, \quad \hat{\sigma}_2^2 = \frac{\sum_{j=1}^{m} (Y_j - \mu_2)^2}{m}. \tag{9.7.1}$$

So, we can consider rejecting H_0 if $\hat{\sigma}_1^2/\hat{\sigma}_2^2$ is much larger than 1; that is, reject H_0 iff

$$\frac{\hat{\sigma}_1^2}{\hat{\sigma}_2^2} > k, \tag{9.7.2}$$

where k is a constant chosen (if such a choice is possible) to give level α. Now (using the results of Problems 9.43 and 9.32)

$$\begin{aligned} P_{\sigma_1^2, \sigma_2^2}\left[\frac{\hat{\sigma}_1^2}{\hat{\sigma}_2^2} > k\right] &= P_{\sigma_1^2, \sigma_2^2}\left[\frac{\sum_{j=1}^{n} (X_j - \mu_1)^2/n}{\sum_{j=1}^{m} (Y_j - \mu_2)^2/m} > k\right] \\ &= P_{\sigma_1^2, \sigma_2^2}\left[\frac{\sum_{j=1}^{n} \left(\frac{X_j - \mu_1}{\sigma_1}\right)^2 \Big/ n}{\sum_{j=1}^{m} \left(\frac{Y_j - \mu_2}{\sigma_2}\right)^2 \Big/ m} > \frac{\sigma_2^2}{\sigma_1^2} k\right] \\ &= P\left[Y > \frac{\sigma_2^2}{\sigma_1^2} k\right], \end{aligned} \tag{9.7.3}$$

where Y has the $F(n, m)$ distribution, so Table 12 can be used to find a k which gives level α. Thus, we have the following:

Theorem 9.7.4. *For the hypothesis-testing problem with* $X = (X_1, \ldots, X_n, Y_1, \ldots, Y_m)$ *where* $X_1, \ldots, X_n, Y_1, \ldots, Y_m$ *are* $n + m$ *independent r.v.'s, each of* $X_1, \ldots, X_n$ *is* $N(\mu_1, \sigma_1^2)$, *each of* $Y_1, \ldots, Y_m$ *is* $N(\mu_2, \sigma_2^2)$, μ_1 *and* μ_2 *are known* $(-\infty < \mu_1, \mu_2 < +\infty)$, σ_1^2 *and* σ_2^2 *are unknown* $(\sigma_1^2, \sigma_2^2 > 0)$, $H_0 = \{\sigma_1^2 = \sigma_2^2\}$, *and* $H_1 = \{\sigma_1^2 > \sigma_2^2\}$, *the test which rejects iff*

$$\frac{\sum_{j=1}^{n} (X_j - \mu_1)^2/n}{\sum_{j=1}^{m} (Y_j - \mu_2)^2/m} > k \tag{9.7.5}$$

(*where* k *satisfies* $P[Y > \sigma_2^2 k/\sigma_1^2] = \alpha$ *for* Y *an* $F(n, m)$ *r.v.*) *has level* α. *Similar tests are easily obtained* [as in Problem 9.35] *for other alternatives* H_1.

Now **if μ_1 and μ_2 are unknown**, we may (see Section 9.6) consider estimating them by $\bar{X}$ and $\bar{Y}$ respectively, then rejecting iff

$$\frac{\sum_{j=1}^{n} (X_j - \bar{X})^2/(n-1)}{\sum_{j=1}^{m} (Y_j - \bar{Y})^2/(m-1)} > k'. \tag{9.7.6}$$

Since the left-hand side of Equation (9.7.6) has the $F(n-1, m-1)$ distribution when multiplied by σ_2^2/σ_1^2, we may (by choosing k' to satisfy

$$P\left[Y > \frac{\sigma_2^2 k'}{\sigma_1^2}\right] = \alpha \tag{9.7.7}$$

for Y an $F(n-1, m-1)$ r.v.) obtain our level α test. [Note that when μ_1 and μ_2 are unknown we must have $n \geqslant 2$ and $m \geqslant 2$ (otherwise we do not obtain a proper F-distribution).]

Note that in the case of (9.7.3) k [in the case of (9.7.7) k'] is chosen so the quantity in question equals α *when H_0 is true*, that is, when $\sigma_1^2 = \sigma_2^2$. Hence, our solution does not involve the (unknown) quantities σ_1^2 and σ_2^2.

9.8. TESTING NORMAL MEANS (TWO POPULATIONS WITH EQUAL VARIANCES)

In Section 9.5 we considered testing whether the observations from one source have a specified mean ($\mu = \mu_0$). In the present section we consider the problem of testing whether observations from two sources (or populations) have the same mean, when the sources have equal variability. (The case of unequal variabilities is considered in Section 9.9.) In terms of the example cited in the first paragraph of Section 9.7, we are concerned here with whether the "early smokers" have the same mean height as the "late and nonsmokers," and are assuming the variabilities of the heights to be equal.

Formally, the hypothesis-testing problem of concern is one with $X = (X_1, \ldots, X_n, \ Y_1, \ldots, Y_m)$ where $X_1, \ldots, X_n, \ Y_1, \ldots, Y_m$ are $n + m$ independent r.v.'s, each of $X_1, \ldots, X_n$ is $N(\mu_1, \sigma^2)$, each of $Y_1, \ldots, Y_m$ is $N(\mu_2, \sigma^2)$, μ_1 and μ_2 are unknown $(-\infty < \mu_1, \mu_2 < +\infty)$, σ^2 is known $(\sigma^2 > 0)$, $H_0 = \{\mu_1 = \mu_2\}$, and H_1 may have several forms. Let us consider first *the case* where $H_1 = \{\mu_1 > \mu_2\}$. We can reason to a test intuitively as follows. We know that $\bar{X}$ and $\bar{Y}$ are respectively good estimators of μ_1 and μ_2. So consider rejecting H_0 if $\bar{X}$ is much larger than $\bar{Y}$; that is, reject H_0 iff

$$\bar{X} - \bar{Y} > c, \tag{9.8.1}$$

where c is a constant to be chosen (if such a choice is possible) to give level α. Now (by work paralleling that following Equation (9.5.5)) we solve for c:

$$\begin{cases} P_{\mu_1=\mu_2}[\bar{X} - \bar{Y} > c] = \alpha, \\ P_{\mu_1=\mu_2}\left[\dfrac{\bar{X} - \bar{Y}}{\sqrt{\sigma^2\left(\frac{1}{n} + \frac{1}{m}\right)}} > \dfrac{c}{\sqrt{\sigma^2\left(\frac{1}{n} + \frac{1}{m}\right)}}\right] = \alpha, \\ 1 - \Phi\left(\dfrac{c}{\sigma\sqrt{\frac{1}{n} + \frac{1}{m}}}\right) = \alpha, \\ c = \sigma\sqrt{\dfrac{1}{n} + \dfrac{1}{m}}\,\Phi^{-1}(1 - \alpha). \end{cases} \tag{9.8.2}$$

So c can be found as desired. Similar tests are easily obtained for *other alternatives* H_1. Note that we can rewrite our test (using (9.8.1) and (9.8.2)) as one which rejects iff

$$\frac{\bar{X} - \bar{Y}}{\sqrt{\sigma^2\left(\frac{1}{n} + \frac{1}{m}\right)}} > \Phi^{-1}(1 - \alpha). \tag{9.8.3}$$

(This will help us in the case of unknown σ^2.)

Now **if σ^2 is unknown** we would consider estimating it and rejecting iff the left-hand side of Equation (9.8.3) [with σ^2 replaced by an appropriate estimator] exceeds some constant [which can hopefully be set to obtain level α]. Now we could estimate σ^2 by

$$s_1^2 = \frac{\sum_{j=1}^{n} (X_j - \bar{X})^2}{n - 1} \tag{9.8.4}$$

or by

$$s_2^2 = \frac{\sum_{j=1}^{m} (Y_j - \bar{Y})^2}{m - 1}, \tag{9.8.5}$$

and our estimators would have reasonable properties. However, s_1^2 and s_2^2 are (since $X_1, \ldots, X_n$ are independent of $Y_1, \ldots, Y_m$) independent r.v.'s, so we might ask: Can we somehow combine s_1^2 and s_2^2 to obtain an even better estimator of σ^2? If we combine them according to their respective d.f. of $n - 1$ and $m - 1$ into a weighted average, we obtain a third estimator

$$s_P^2 = \frac{\sum_{j=1}^{n} (X_j - \bar{X})^2 + \sum_{j=1}^{m} (Y_j - \bar{Y})^2}{n + m - 2}. \tag{9.8.6}$$

So, what about the test that rejects iff

$$\frac{\bar{X} - \bar{Y}}{\sqrt{\dfrac{\sum_{j=1}^{n} (X_j - \bar{X})^2 + \sum_{j=1}^{m} (Y_j - \bar{Y})^2}{n + m - 2}\left(\dfrac{1}{n} + \dfrac{1}{m}\right)}} > k'? \tag{9.8.7}$$

From the results of Section 9.5 it follows that when H_0 is true, the left-hand side of Equation (9.8.7) has Student's t-distribution with $n + m - 2$ d.f. [this is because the left-hand side is the ratio of the $N(0, 1)$ r.v.

$$\frac{\bar{X} - \bar{Y}}{\sigma\sqrt{\dfrac{1}{n} + \dfrac{1}{m}}} \tag{9.8.8}$$

and the square root of the χ^2_{n+m-2} r.v.

$$\sum_{j=1}^{n}\left(\frac{X_j - \bar{X}}{\sigma}\right)^2 + \sum_{j=1}^{m}\left(\frac{Y_j - \bar{Y}}{\sigma}\right)^2 \tag{9.8.9}$$

divided by its d.f., and these two r.v.'s are independent]. Hence, a proper choice for k' is

$$k' = t^{-1}_{n+m-2}(1 - \alpha). \tag{9.8.10}$$

Similar tests are easily obtained (see Problem 9.35) for *other alternatives* H_1.

9.9. TESTING NORMAL MEANS (TWO POPULATIONS WITH UNEQUAL VARIANCES; THE BEHRENS-FISHER PROBLEM)

In Section 9.8 we considered hypothesis-testing problems involving the means of two populations which have the same variance (which may be known or unknown). In this section we consider such problems when we cannot assume the variabilities of the two populations (for example, "early smokers" and "late smokers and nonsmokers") to be the same.

Formally, the hypothesis-testing problem of concern is one with $X = (X_1, \ldots, X_n, Y_1, \ldots, Y_m)$ where $X_1, \ldots, X_n, Y_1, \ldots, Y_m$ are $n + m$ independent r.v.'s, each of $X_1, \ldots, X_n$ is $N(\mu_1, \sigma_1^2)$, each of $Y_1, \ldots, Y_m$ is $N(\mu_2, \sigma_2^2)$, μ_1 and μ_2 are unknown $(-\infty < \mu_1, \mu_2 < +\infty)$, σ_1^2 and σ_2^2 are known (σ_1^2, $\sigma_2^2 > 0$ but $\sigma_1^2 \neq \sigma_2^2$), $H_0 = \{\mu_1 = \mu_2\}$, and H_1 may have several forms. Let us consider first *the case* where $H_1 = \{\mu_1 > \mu_2\}$. As in the development leading to (9.8.1) in Section 9.8, we consider rejecting H_0 iff

$$\bar{X} - \bar{Y} > c, \tag{9.9.1}$$

where c is a constant to be chosen (if such a choice is possible) to give level α. Now as at (9.8.2) we can find such a c:

$$\begin{cases} P_{\mu_1=\mu_2}[\bar{X} - \bar{Y} > c] = \alpha, \\ P_{\mu_1=\mu_2}\left[\dfrac{\bar{X} - \bar{Y}}{\sqrt{\dfrac{\sigma_1^2}{n} + \dfrac{\sigma_2^2}{m}}} > \dfrac{c}{\sqrt{\dfrac{\sigma_1^2}{n} + \dfrac{\sigma_2^2}{m}}}\right] = \alpha, \\ 1 - \Phi\left[\dfrac{c}{\sqrt{\dfrac{\sigma_1^2}{n} + \dfrac{\sigma_2^2}{m}}}\right] = \alpha, \\ c = \sqrt{\dfrac{\sigma_1^2}{n} + \dfrac{\sigma_2^2}{m}}\,\Phi^{-1}(1 - \alpha). \end{cases} \tag{9.9.2}$$

Similar tests are easily obtained for *other alternatives* H_1.

Now **if σ_1^2 and σ_2^2 are unknown** we would consider estimating them by s_1^2 (see (9.8.4)) and s_2^2 (see (9.8.5)) respectively, and rejecting iff

$$\frac{\bar{X} - \bar{Y}}{\sqrt{\dfrac{s_1^2}{n} + \dfrac{s_2^2}{m}}} > k'. \tag{9.9.3}$$

However (see Problem 9.47), the distribution of the left-hand side of (9.9.3) is not Student's t-distribution when H_0 is true; in fact, the distribution of the left-hand side depends on n, m, and σ_1^2/σ_2^2. Hence, it is not possible to choose a k' such that the test given by (9.9.3) has level α (since the probability of rejection varies as σ_1^2/σ_2^2 varies, instead of remaining constant at α). *The distribution of the quantity*

$$\frac{\bar{X} - \bar{Y}}{\sqrt{\dfrac{s_1^2}{n} + \dfrac{s_2^2}{m}}} \tag{9.9.4}$$

(when H_0 is true) is called the **Behrens-Fisher distribution** *with parameters n, m, σ_1^2/σ_2^2. The hypothesis-testing problem we are considering (with σ_1^2 and σ_2^2 unknown) is called the* **Behrens-Fisher problem**, and represents (in Sections 9.5–9.9) the first time we have encountered a situation where we could not obtain a level α test by "substituting a good estimator of an unknown parameter in the procedure used when that parameter is known."

One proposed solution to the Behrens-Fisher problem, which we will call the **Neyman-Bartlett solution**, *proceeds as follows.* Either $n < m$ or $n \geqslant m$, so let us suppose without any loss of generality that $n \geqslant m$. Define new r.v.'s $Z_1, \ldots, Z_m$ by

$$Z_i = X_i - Y_i \quad (i = 1, \ldots, m). \tag{9.9.5}$$

Then $Z_1, \ldots, Z_m$ are independent $N(\mu_1 - \mu_2, \sigma_1^2 + \sigma_2^2)$ r.v.'s, and we can solve the problem of testing $H_0: \mu_1 - \mu_2 = 0$ versus $H_1: \mu_1 - \mu_2 > 0$ by (see Theorem 9.5.26) rejecting H_0 iff

$$\frac{\bar{Z} - 0}{\sqrt{\dfrac{\sum_{i=1}^{m} (Z_i - \bar{Z})^2}{(m-1)m}}} > t_{m-1}^{-1}(1 - \alpha), \tag{9.9.6}$$

that is, iff (letting $\bar{X}(m) = (X_1 + \cdots + X_m)/m$ which is to be distinguished from $\bar{X} = (X_1 + \cdots + X_m + X_{m+1} + \cdots + X_n)/n$)

$$\frac{\bar{X}(m) - \bar{Y}}{\sqrt{\dfrac{\sum_{i=1}^{m} ((X_i - Y_i) - (\bar{X}(m) - \bar{Y}))^2}{(m-1)m}}} > t_{m-1}^{-1}(1 - \alpha). \tag{9.9.7}$$

As Scheffé has pointed out,[8] the Neyman-Bartlett solution essentially has two drawbacks. First, it throws away the $n - m$ observations $X_{m+1}, \ldots, X_n$ completely. Second, the outcome of the test depends on the order of the observations in the samples. *Scheffé has proposed another solution (which we will call the* **Scheffé solution** *as follows*: Reject H_0 iff

$$\frac{\bar{X} - \bar{Y}}{\sqrt{\dfrac{\sum_{i=1}^{m} \left(\left(X_i - \sqrt{\dfrac{m}{n}}\, Y_i\right) - \left(\bar{X}(m) - \sqrt{\dfrac{m}{n}}\, \bar{Y}\right)\right)^2}{(m-1)m}}} > t_{m-1}^{-1}(1 - \alpha). \tag{9.9.8}$$

When H_0 is true, the left-hand side of (9.9.8) has (Problem 9.20H) Student's t-distribution with $m - 1$ d.f., so the test is of level α. Note that this test uses $\bar{X}$ (not $\bar{X}(m)$) in the numerator and hence does not suffer from one of the maladies of the Neyman-Bartlett solution. However, the order of the observations still affects the outcome of the test. Scheffé has shown[9] that, in a certain framework, this dependence on order cannot be eliminated, and has recommended[10] that because of this dependence his solution should *not* be used.

A third solution is to return to the Behrens-Fisher statistic (9.9.4) *and attempt to use it as in* (9.9.3) *for some* k'. *What we will call the* **Hsu solution**

---

[8] See H. Scheffé, "On Solutions of the Behrens-Fisher problem, based on the t-distribution," *Annals of Mathematical Statistics*, Vol. 14 (1943), pp. 35–44. References to the Neyman-Bartlett solution are given on p. 35.

[9] See H. Scheffé, "A note on the Behrens-Fisher problem," *Annals of Mathematical Statistics*, Vol. 15 (1944), pp. 430–432.

[10] See p. 1503 of H. Scheffé, "Practical solutions of the Behrens-Fisher problem," *Journal of the American Statistical Association*, Vol. 65 (1970), pp. 1501–1508.

(see p. 1502 of Scheffé (1970) *op. cit.*) *takes* $k' = t_{f_0-1}^{-1}(1-\alpha)$ *where* $f_0 =$ *min*(m, n), *and has level* $\leqslant \alpha$. Other authors have suggested replacing $f_0 - 1$ with other (larger) quantities in order to obtain higher power (and also, of course, higher level hopefully $\approx \alpha$ and hopefully never $\gg \alpha$). For example, **Welch suggested**

$$\mathscr{F} = \frac{\left(\dfrac{s_1^2}{n} + \dfrac{s_2^2}{m}\right)^2}{\dfrac{(s_1^2/n)^2}{n-1} + \dfrac{(s_2^2/m)^2}{m-1}}, \tag{9.9.9}$$

and we can (Problem 9.49) show that

$$f_0 - 1 = \min(n-1, m-1) \leqslant \mathscr{F} \leqslant (n-1) + (m-1). \tag{9.9.10}$$

Such tests do not guarantee the stated level of test α, and for reasonably large $f_0 - 1$ (for example, 10 or more) do not reduce the $t_f^{-1}(1-\alpha)$ value appreciably (and hence do not greatly affect the probabilities of acceptance and rejection). Hence, we would usually choose $f = f_0 - 1$ rather than $f = \mathscr{F}$ or some other quantity.

PROBLEMS

9.1 For the question "Do seat belts reduce accident injuries?" noted in Section 9.1, specify several sources of variability which, if not controlled through careful experimental design, could vitiate any study's results.

9.2 In Example 9.2.4 prove that the two rejection probabilities considered (following (9.2.9) and following (9.2.11)) have the respective properties: $f(\theta_0) = \alpha = .05$, $f(\theta_0 + x) = f(\theta_0 - x)$ for all x, $f(\theta_2) > f(\theta_1)$ for all $\theta_2 > \theta_1 \geqslant \theta_0$, and $\lim_{\theta\to\infty} f(\theta) = 1$; and $g(\theta_0) = .05$, $g(\theta_2) > g(\theta_1)$ for all $\theta_2 > \theta_1$, and $\lim_{\theta\to\infty} g(\theta) = 1$.

9.3 In Example 9.2.4, if we desire the test using critical function $\phi_2(\cdot)$ to have level $\alpha = .05$ and power $\geqslant .95$ when $\theta = 1.23$, what is the smallest n which we can use?

9.4 For the hypothesis-testing problem discussed in Example 9.2.4, a third test sets $\phi_3(x) = .05$ for all $x \in \mathscr{X}$. (Thus, if this test is to be utilized, then we can just as well take no observations since this test ignores them, and simply flips a coin with probability .05 of heads.) Show that this test has level $\alpha = .05$.

Thus, **obtaining a test with level α is trivial in general**. Much more important is obtaining a test of level α which has "good" properties with regard to power; this is often overlooked in "cookbook" statistics courses and texts. Graph the power of the test utilizing $\phi_3(\cdot)$ as a function of θ, say $h(\theta)$. Compare it with $f(\theta)$ and $g(\theta)$ as shown in Figure 9.2–2.

9.5 For the hypothesis-testing problem discussed in Example 9.2.4, a fourth test sets

$$\phi_4(x) = \begin{cases} 1, & \text{if } |\bar{x} - \theta_0| \leqslant c \\ 0, & \text{otherwise.} \end{cases}$$

What value of c should be used to obtain level $\alpha = .05$ using this test? Graph the power of the test utilizing $\phi_4(\cdot)$ as a function of θ, say $i(\theta)$. Compare it with $f(\theta)$, $g(\theta)$, and $h(\theta)$ (see Figure 9.2–2 and Problem 9.4). Does this show that we can, with a little ingenuity, construct a test which is worse than the one based only on flipping a coin? Explain.

9.6 Suppose that $X_1, \ldots, X_n$ are independent r.v.'s, each $N(\mu, 1)$. If we desire to test $H_0: \mu = 0$ versus $H_1: \mu \neq 0$ at level $\alpha = .05$, using the test which rejects when $|\bar{X}| > c$ [where c is set to obtain level .05], then what is the smallest value of n which will guarantee power $\geqslant .95$ when $|\mu| = 1$?

9.7 For the hypothesis-testing problem considered in Theorem 9.3.2, suppose $f(x|\theta_1)$ and $f(x|\theta_0)$ are each > 0 for all $x \in \mathscr{X}$. Show that if t is a sufficient statistic for the problem, then in order to satisfy (9.3.4) a test need be based only on t.

9.8 Suppose we observe $X = (X_1, \ldots, X_n)$ where $X_1, \ldots, X_n$ are independent r.v.'s, each $N(\mu, \sigma^2)$ where μ is unknown $(-\infty < \mu < +\infty)$ and $\sigma^2 = 1$ is known. For the problem of testing $H_0: \mu = 0$ versus $H_1: \mu = 1$ at level $\alpha = .05$ using a most powerful test, find the smallest value of n which will guarantee power $\geqslant$ .90. Using this n, graph the power curve. (Calculate the power at least for $\mu = 0$, $\pm.5$, ±1, ±1.5, ±2 for your graph.)

9.9 Suppose we observe $X = (X_1, \ldots, X_n)$ where $X_1, \ldots, X_n$ are independent r.v.'s, each $N(\mu, \sigma^2)$ where μ is unknown $(-\infty < \mu < +\infty)$ and σ^2 is known $(\sigma^2 > 0)$. For the problem of testing $H_0: \mu = \mu_0$ versus $H_1: \mu = \mu_1 (>\mu_0)$ at level α $(0 < \alpha < 1)$ using a most powerful test, show that using (for $0 < \gamma < 1$)

$$n \geqslant \left[\frac{\Phi^{-1}(1-\gamma) - \Phi^{-1}(1-\alpha)}{\mu_0 - \mu_1} \sigma \right]^2$$

will yield level $\leqslant \alpha$ and power $\geqslant \gamma$.

9.10 In Example 9.3.10 the test which rejects H_0 when $\bar{X} > \mu_0 + (\sigma/\sqrt{n})\Phi^{-1}(1-\alpha)$ has level α $(0 < \alpha < 1)$. If this test has power γ at $H_1: \mu = \mu_1 (> \mu_0)$, then at what value of μ does this test have power $1 - \gamma$?

9.11 Show that in the statement of the Neyman-Pearson Lemma we may assume without loss that $k \geqslant 0$. [*Hint*: If $k < 0$, what is the value of $E_{\theta_0}\phi(X)$?] Can $k = 0$ also be excluded?

9.12 Suppose that, in Example 9.3.17, we choose to set $\phi(x)$ to 0 in Case I and Case IV. The resulting test is (say)

$$\phi_3(x) = \begin{cases} .05, & \text{if } \min(x_1, \ldots, x_n) \geqslant 0 \text{ and } \max(x_1, \ldots, x_n) \leqslant 1 \\ 1, & \text{if } \min(x_1, \ldots, x_n) \geqslant 0 \text{ and } 1 < \max(x_1, \ldots, x_n) \leqslant 2 \\ 0, & \text{otherwise.} \end{cases}$$

Is $\phi_3(\cdot)$ a most powerful level $\alpha = .05$ test of $H_0: \theta = 1$ versus $H_1: \theta = 2$? Find and graph $E_\theta\phi_3(X)$ on a facsimile of Figure 9.3–1.

9.13 Suppose that we observe X and wish to test H_0 versus H_1 (where H_0 and/or H_1 may be composite hypotheses). *The* **likelihood-ratio test** *is defined as: reject* H_0

iff (at the observed value of X)

$$\frac{\sup_{\theta \in H_1} L(\theta)}{\sup_{\theta \in H_0} L(\theta)} > k, \qquad (*)$$

where $L(\theta)$ is the likelihood function of (8.2.2) *and k is arbitrary*. Show that if H_0 and H_1 are simple hypotheses, then the likelihood-ratio test is most powerful. [*Hint*: Use the Neyman-Pearson Lemma.] How should $(*)$ be restated if we wish to allow cases where $L(\theta)$ may be 0 to be considered?

9.14 Suppose we are to observe X and have a hypothesis-testing problem of H_0 versus H_1, where H_0 and/or H_1 may be composite. Suppose we have a test of level α for each α $(0 \leqslant \alpha \leqslant 1)$, say $\phi_\alpha(X)$. Our intention is to use test $\phi_{\alpha^*}(X)$ if we decide to test the hypothesis at level $\alpha = \alpha^*$. *If we observe $X = x$, then the* **significance probability of x**, *say* $\boldsymbol{\alpha}_{\text{SP}}(\mathbf{x}, \boldsymbol{\phi}_\alpha)$, *is defined as the smallest α for which* $\phi_\alpha(x) = 1$ [the smallest α for which we would surely reject if we observed $X = x$]. Often an experimenter will report $\alpha_{\text{SP}}(x, \phi_\alpha)$ in his work. Then an experimenter with a possibly different desired level, say α', will *often* know: that he would also decide to reject if the reported $\alpha_{\text{SP}}(x, \phi_\alpha) \leqslant \alpha'$; and that he would not decide to reject if the reported $\alpha_{\text{SP}}(x, \phi_\alpha) > \alpha'$. (We say often, not always, because the assumption that the tests $\phi_\alpha(X)$ have the property

$$\phi_{\alpha_1}(x) = 1 \quad \Rightarrow \quad \phi_{\alpha_2}(x) = 1, \quad \text{for all } \alpha_2 > \alpha_1, \quad \text{for all } x \qquad (*)$$

is needed and is not always satisfied.)

In Example 9.3.10 a test of level α, say $\phi_\alpha(X)$, was given for each α $(0 < \alpha < 1)$. [$\alpha = 0$ can be obtained by never rejecting, and $\alpha = 1$ can be obtained by always rejecting.] Suppose there that $\sigma^2 = 100$, $\mu_0 = 0$, $\mu_1 = 1$, $n = 100$, and $\bar{X} = 2.68$ is found. What is $\alpha_{\text{SP}}(2.68, \phi_\alpha)$? Does it change if we change μ_1 to 3? What if $\sigma^2 = 10$? What if $\sigma^2 = 1$?

9.15 In the proof of Theorem 9.3.35 (UMP tests for MLR cases) we stated that the given test was the same regardless of which θ_1 was chosen (as long as $\theta_1 > \theta_0$). Prove this; that is, show that $\phi(x)$ does not depend on the particular $\theta_1 > \theta_0$ chosen.

9.16 Show that if the density of $X = (X_1, \ldots, X_n)$ is a member of the one-parameter exponential family (see Definition 8.8.34), then the d.f. $F(x|\theta)$ of X has a MLR (in which $T(x)$?) iff $c_1(\theta)$ is a nondecreasing function of θ (that is, iff $c_1(\theta_2) - c_1(\theta_1) \geqslant 0$ for all $\theta_2 > \theta_1$). Show that X in Example 9.3.41 is a member of the one-parameter exponential family, and conclude immediately that $F(x|\theta)$ has a MLR in $x_1 + \cdots + x_n$.

9.17 Prove that for any hypothesis-testing problem where H_0 and H_1 are both simple, if the set $R = \{r(\phi): \phi(\cdot) \text{ a test}\}$ includes the point $(\alpha, 1 - \beta)$, then it also includes the point $(1 - \alpha, \beta)$. [A special case was covered in Lemma 9.4.4.]

9.18 Prove that the class of all tests is a complete class of tests.

9.19 Prove that any complete class of tests is also an essentially complete class of tests.

9.20 How does the figure below indicate that, if we specify interest in a Bayes rule with respect to $(p, 1 - p)$, there may be several risk points which qualify?

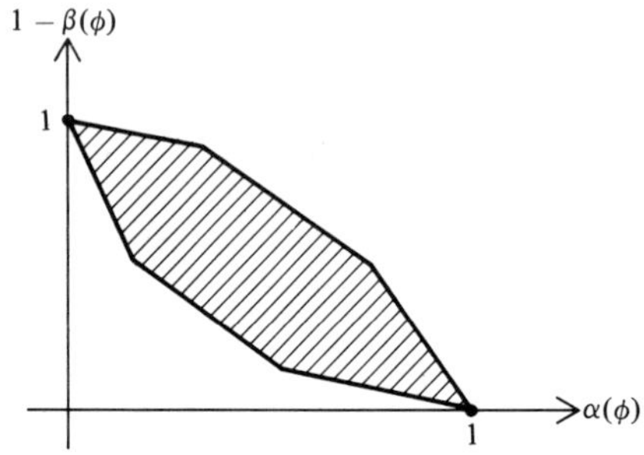

9.21 Show that the quantities (9.4.10) and (9.4.11) are equal. Deduce that the minimax risk and minimax regret rules are the same. [This is not so in general decision theory problems.]

9.22 Prove that the slope of the line $px + (1 - p)y = c$, used in our explanation of the Bayes criterion (see Figure 9.4–5), is negative when $0 < p < 1$.

9.23 If $p = 0$, which tests ϕ are Bayes (that is, what restrictions on $\phi(x)$ besides $0 \leqslant \phi(x) \leqslant 1$ must a test satisfy to be Bayes)? Similarly for $p = 1$. [*Hint*: See the discussion preceding (9.4.14).]

9.24 In Example 9.3.41 we considered observing $X = (X_1, \ldots, X_{31})$ where $X_1, \ldots, X_{31}$ are independent r.v.'s, each exponential with unknown mean $\theta > 0$. Then (for any $\theta_1 > 10$) a most powerful level $\alpha = .05$ test of $H_0: \theta = 10$ versus $H_1: \theta = \theta_1$ rejects H_0 iff $\bar{X} > 13.13$.

Fix p $(0 < p < 1)$. When does a Bayes rule with respect to $(p, 1 - p)$ reject H_0? Does the Bayes rule with respect to $(p, 1 - p)$ depend on θ_1? For what value(s) of p does the Bayes rule have level $\alpha = .05$? For what value of θ_1 does the level $\alpha = .05$ Bayes rule reject iff $\bar{X} > 13.13$?

9.25 Consider the case of Problem 9.24. When does the minimax rule reject H_0? What is the level of the minimax rule? For what value(s) of θ_1 is this level .05? For what value(s) of θ_1 does the minimax rule reject iff $\bar{X} > 13.13$? What is the level of the minimax rule which rejects iff $\bar{X} > 13.13$?

9.26 In Example 9.3.10 we found a most powerful test in a hypothesis-testing problem with $H_1 = \{\mu_1\}$ $(\mu_1 > \mu_0)$, and we showed that this test is UMP for $H_1 = \{\mu_1 : \mu_1 > \mu_0\}$. Find a UMP test of level α in the same hypothesis-testing problem but with $H_1 = \{\mu_1 : \mu_1 < \mu_0\}$.

9.27 Prove Theorem 9.5.13.

9.28 Verify that, in the terms of Definition 4.5.3, the r.v. $(X_1, \ldots, X_n)$ in Lemma 9.5.15 actually does have a multivariate normal distribution.

9.29 Verify (see (9.5.17)) that in the proof of Lemma 9.5.15 we have

$$\operatorname{Cov}\left(\frac{X_1 + \cdots + X_n}{n}, X_i\right) - \operatorname{Var}(\bar{X}) = 0.$$

9.30 Prove that if two r.v.'s X_1 and X_2 are independent, then $g_1(X_1)$ and $g_2(X_2)$ are also independent (for any fixed functions $g_1(\cdot)$ and $g_2(\cdot)$). [*Hint*: See Theorem 5.3.8 and Theorem 5.3.12.]

9.31 Verify that (9.5.20) is true.

9.32 Prove that if $X_1, \ldots, X_n$ are independent r.v.'s, then $g_1(X_1)$ and $g_2(X_2, \ldots, X_n)$ are also independent. (This generalizes Problem 9.30.)

9.33 Verify that (9.5.21) is true.

9.34 Let $X_1, \ldots, X_n$ be independent $N(0, 1)$ r.v.'s. Prove that $X_1^2 + \cdots + X_n^2$ has the $\chi_n^2(0)$ distribution. [*Hint*: Prove the result for $n = 1$ and then use part (4) of Theorem 6.1.3. A laborious application of the methods of Section 4.6 would also work.]

9.35 Fix p $(0 < p < 1)$. For the hypothesis-testing problem of Theorem 9.5.26 with $H_1: \mu = \mu_1$ $(>\mu_0)$ replaced by $H_1: \mu \neq \mu_0$, show that the test which rejects iff

$$\frac{\bar{X} - \mu_0}{s/\sqrt{n}} < t_{n-1}^{-1}(\alpha p) \quad \text{or} \quad \frac{\bar{X} - \mu_0}{s/\sqrt{n}} > -t_{n-1}^{-1}(\alpha(1 - p)) \qquad (*)$$

has level α. (A similar class of tests is noted in Problem 9.12H.)

9.36 The test stated in Theorem 9.5.26 [that is, reject iff (9.5.27) occurs] has level α no matter what our alternative H_1. The reason we changed to another rejection region when we changed H_1 [see Problem 9.35] was to obtain "good" power against "far" points in the new alternative. Actually a graph of the power of each test would be similar to Figure 9.2–2. Evaluating the power explicitly involves using tables of what is called the **noncentral t-distribution** (with $n - 1$ degrees of freedom); these tables are like Table 13 (for the t-distribution), but involve one such table for *every noncentrality value we need* (hence, for a large number of values if the tables are to be useful); hence, they are voluminous and are not given below. However, if $g(\mu)$ denotes the probability that (9.5.27) holds, and if $f(\mu)$ denotes the probability that $(*)$ of Problem 9.35 holds, prove that

$$\begin{cases} f(\mu_0) = g(\mu_0) = \alpha; \\ g(\mu) > f(\mu) \text{ for } \mu > \mu_0,\ g(\mu) < f(\mu) \text{ for } \mu < \mu_0; \\ f(\mu_0 + x) = f(\mu_0 - x) \text{ for all } x; \\ f(\mu_2) > f(\mu_1) \text{ for all } \mu_2 > \mu_1 \geqslant \mu_0; \\ g(\mu_2) > g(\mu_1) \text{ for all } \mu_2 > \mu_1; \text{ and} \\ \lim_{\mu \to \infty} f(\mu) = \lim_{\mu \to \infty} g(\mu) = 1. \end{cases}$$

This shows that our picture here is indeed similar to that of Figure 9.2–2.

9.37 If Y_1 has the $N(\delta, 1)$ distribution, if Y_2 has the $\chi_m^2(0)$ distribution, and if Y_1 and Y_2 are independent r.v.'s, then

$$Y = \frac{Y_1}{\sqrt{Y_2/m}}$$

is said to have the **noncentral t-distribution** *with m degrees of freedom and*

noncentrality δ. Show that, in the context of Theorem 9.5.26,

$$\frac{\bar{X} - \mu_0}{s/\sqrt{n}}$$

has a noncentral t-distribution when H_0 is false. (Find δ and m.)

9.38 Let $X_1, \ldots, X_n$ be independent r.v.'s, each $N(\mu, \sigma^2)$ where μ is known $(-\infty < \mu < +\infty)$ and σ^2 is unknown $(\sigma^2 > 0)$. Prove that $\hat{\sigma}^2$ given in Equation (9.6.1) is a minimal sufficient statistic.

9.39 Use the Neyman-Pearson Lemma to show that the test of Theorem 9.6.4 is most powerful. If we take H_1 to be $\{\sigma^2 : \sigma^2 > \sigma_0^2\}$, do we find that the test of Theorem 9.6.4 is UMP one-sided?

9.40 Show that if Y has the $\chi_n^2(0)$ distribution, then $EY = n$. [*Hint*: Write Y as $X_1^2 + \cdots + X_n^2$ where $X_1, \ldots, X_n$ are independent $N(0, 1)$ r.v.'s.]

9.41 Show that the power, when $\sigma^2 = \sigma_1^2$, of the level α test specified in Theorem 9.6.4 can be written as

$$g(\sigma_1^2) = P\left[Y > \frac{\sigma_0^2}{\sigma_1^2} L_{1-\alpha}^{-1}\right],$$

where Y is a $\chi_n^2(0)$ r.v. and $L_{1-\alpha}^{-1}$ is the number such that

$$P[Y < L_{1-\alpha}^{-1}] = 1 - \alpha.$$

Prove that $g(\sigma_1^2) > g(\sigma_2^2)$ for $\sigma_1^2 > \sigma_2^2$, that $\lim_{\sigma_1 \to 0} g(\sigma_1^2) = 0$, and that

$$\lim_{\sigma_1 \to \infty} g(\sigma_1^2) = 1.$$

9.42 Show that the power at each specified σ_1^2, of the level α test specified in Theorem 9.6.4, increases if n is increased.

9.43 Show that if Y_1 has the $\chi_n^2(0)$ distribution, Y_2 has the $\chi_m^2(0)$ distribution, and Y_1 and Y_2 are independent r.v.'s (with $n \geqslant 2$, $m \geqslant 2$), then

$$Y = \frac{Y_1/n}{Y_2/m}$$

has the F-distribution $F(n, m)$.

9.44 Consider the hypothesis-testing problem of Theorem 9.7.4 with H_0 and H_1 replaced respectively by $H_0 = \{\sigma_1^2 = c\sigma_2^2\}$ and $H_1 = \{\sigma_1^2 > c\sigma_2^2\}$ where $c > 0$ is a specified number. Obtain an explicit formula (comparable to (9.7.5)) for a level α test in this case, and show how to choose any constants which you use. If $n = 10$, $m = 20$, $c = 16$, and $\alpha = .05$, when will your test reject H_0?

9.45 Let $X_1, \ldots, X_n, Y_1, \ldots, Y_m$ be $n + m$ independent r.v.'s. Suppose that: $X_1, \ldots, X_n$ are each $N(\mu_1, \sigma^2)$; $Y_1, \ldots, Y_m$ are each $N(\mu_2, \sigma^2)$; μ_1 and μ_2 are unknown $(-\infty < \mu_1, \mu_2 < +\infty)$; and $\sigma^2 > 0$ is known. Find a complete minimal sufficient statistic for (μ_1, μ_2). Find the unique MVU estimator of μ_1; of μ_2. (This justifies the reasoning which led to test (9.8.1) in Section 9.8.)

9.46 Let ϕ_P denote the test specified by (9.8.7) and (9.8.10). Derive similar tests ϕ_1 and ϕ_2 based respectively on the left-hand side of Equation (9.8.3) with σ^2 replaced by s_1^2 and s_2^2. (Be sure to give the k_1' and k_2' for these tests.) Prove that ϕ_P is uniformly more powerful than either of ϕ_1 and ϕ_2. Prove that ϕ_1 is uniformly more powerful than ϕ_2 if $n > m$, that ϕ_1 has the same power as ϕ_2 if $n = m$, and that ϕ_2 is uniformly more powerful than ϕ_1 if $n < m$.

9.47 Show that (when H_0 is true) (9.9.4), called the **Behrens-Fisher statistic**, has the same distribution as (for some a and b) does

$$\frac{Y}{aC_1 + bC_2}$$

where Y is a $N(0, 1)$ r.v., C_1 is a χ^2_{n-1} r.v., C_2 is a χ^2_{m-1} r.v., and Y, C_1, C_2 are independent. For what values of a and b is the preceding true?

Prove that $aC_1 + bC_2$ (where C_1 is a χ^2_{n-1} r.v., C_2 is a χ^2_{m-1} r.v., and C_1 and C_2 are independent) has the chi-square distribution iff $a = b = 1$. Deduce that (9.9.4) does not have a Student's t-distribution unless $n = m = 2$ and $\sigma_1^2 = \sigma_2^2$.

9.48 Consider the two hypothesis-testing problems of Section 9.8 (that is, σ^2 known and σ^2 unknown) with H_0 and H_1 replaced by $H_0 = \{\mu_1 = \mu_2 + c\}$ and $H_1 = \{\mu_1 > \mu_2 + c\}$ where $c > 0$ is a specified number. Obtain an explicit formula for a level α test in each case, and show how to choose any constants which you use.

9.49 Prove that Equation (9.9.10) is true.

9.50 In this chapter we have seen that in cases of two-sided alternatives we usually obtain a class of tests rather than one "best" test. (For example, see Problem 9.12H.) *That one of these tests which devotes $\alpha/2$ of α as a probability of rejecting due to "large" values of the test statistic (and $\alpha/2$ as a probability of rejecting due to "small" values of the test statistic) when H_0 is true, is called a* **two-sided equal-tail test**. (For example, $\phi_\alpha^{(p)}(\cdot)$ in Problem 9.12H with $p = \frac{1}{2}$.)

Let $X = (X_1, \ldots, X_n)$ where $X_1, \ldots, X_n$ are independent $N(\mu, \sigma^2)$ r.v.'s and μ is known. For $n = 10$ and $\alpha = .10$ compute the power curve of the two-sided equal-tail test of $H_0\colon \sigma^2 = 1$ versus $H_1\colon \sigma^2 \neq 1$. For $n = 10$ and $\alpha = .10$ compute the power curve of the one-sided test of $H_0\colon \sigma^2 = 1$ versus $H_1\colon \sigma^2 > 1$. Plot the two power curves on the same graph.

9.51 Man A (a track star) has run the mile twice in competitions, with times of 3.8, 3.4. Man B (another track star) has run the mile thrice in competitions, with times of 3.7, 3.6, 4.1. Assume that man A and man B ran independently of each other, and that each one's times are independent observations of his running ability. Let $\mu_A(\mu_B)$ denote the average time it takes man $A(B)$ to run the mile in competitions. Similarly, let his variance be $\sigma_A^2(\sigma_B^2)$.

(a) Based on these results perform a two-sided, equal-tail (see Problem 9.50) test of $\mu_A = \mu_B$ at level $\alpha = .05$. Assume $\sigma_A^2 = \sigma_B^2 = .04$. State precisely the statistical assumptions that you are making.

(b) As in part (a), but with $\sigma_A^2 = .04$, $\sigma_B^2 = .09$.

(c) As in part (a), but with $\sigma_A^2 = \sigma_B^2$ unknown.

(d) For the problem as in part (a), but with σ_A^2 and σ_B^2 both unknown, what can be said?

9.52 In Problem 8.6 and Example 8.5.35 we considered the seasonal snowfall in Rochester, New York, for the last 88 years in the context of point estimation.

Now it turns out that the snowfalls for 1939–40 and prior years were measured at the Federal Building at Church and North Fitzhugh Streets in downtown Rochester, but that from 1940–41 onward the snowfalls have been measured by the National Weather Service at the Rochester-Monroe County airport. This poses an interesting question: "Should snowfalls from 1939–40 and previous years be considered in a statement about 'average Rochester snowfall'?"

If $X_1, \ldots, X_{88}$ are independent $N(\mu, \sigma^2)$ r.v.'s, the answer is (based on our estimation work in Chapter 8) clearly "Yes," for our estimator will be more concentrated about μ if it is based on 88 observations than if it is based on only 33 observations. Yet if $X_1, \ldots, X_{88}$ are independent r.v.'s, but $X_1, \ldots, X_{55}$ are $N(\mu_1, \sigma^2)$ while $X_{56}, \ldots, X_{88}$ are $N(\mu_2, \sigma^2)$ where $\mu_1 \neq \mu_2$, then the observations would be measuring different phenomena and should probably not be combined. Thus, test (at level $\alpha = .05$) $H_0: \mu_1 = \mu_2$ versus $H_1: \mu_1 \neq \mu_2$, assuming σ^2 is unknown. What is the significance probability of the data (see Problem 9.14)? [Use a two-sided equal-tail test (see Problem 9.50).]

Now a skeptical person (for example, a scientist) might question the assumption of equal variances. If $X_1, \ldots, X_{55}$ are $N(\mu_1, \sigma_1^2)$ while $X_{56}, \ldots, X_{88}$ are $N(\mu_2, \sigma_2^2)$, test (via a two-sided equal-tail test) $H_0: \sigma_1^2 = \sigma_2^2$ versus $H_1: \sigma_1^2 \neq \sigma_2^2$. Use level $\alpha = .05$. Find the significance probability of the data.

Now, what is your estimate of the average seasonal snowfall in Rochester, and why have you chosen a particular estimator?

[Reference: *The Times-Union*, Rochester, New York, March 10, 1972, p. 1B.]

HONORS PROBLEMS

9.1H In Definition 9.3.1 we stated that in a hypothesis-testing problem where H_0 and H_1 are both simple, a test with critical function $\phi(\cdot)$ is called *most powerful of level* α if $E_{\theta_0}\phi(X) = \alpha$ and, for all critical functions $\psi(\cdot)$ such that $E_{\theta_0}\psi(X) = \alpha$, $E_{\theta_1}\phi(X) \geqslant E_{\theta_1}\psi(X)$. Then in the Neyman-Pearson Lemma we proved that if $\phi(\cdot)$ has level α and satisfies (9.3.2), then $\phi(\cdot)$ is most powerful of level α.

We will now call a test with critical function $\phi(\cdot)$ **most powerful of level $\leqslant \alpha$** *if* $E_{\theta_0}\phi(X) \leqslant \alpha$ *and, for all critical functions* $\psi(\cdot)$ *such that* $E_{\theta_0}\psi(X) \leqslant \alpha$, $E_{\theta_1}\phi(X) \geqslant E_{\theta_1}\psi(X)$.

Prove that if $\phi(\cdot)$ satisfies (9.3.3) and (9.3.4), then it is most powerful of level $\leqslant \alpha$. (This is a slight extension of the Neyman-Pearson Lemma.) [*Hint*: A slight modification of the proof of Theorem 9.3.2 will suffice to prove the desired result.]

9.2H Prove that for testing simple H_0 versus simple H_1 after observing X (the context of the Neyman-Pearson Lemma) there exists a critical function $\phi(\cdot)$ and a constant k such that (9.3.3) and (9.3.4) hold (that is, you can obtain a level α test if you choose $\phi(\cdot)$ according to (9.3.4) as long as you choose k properly).

9.3H Prove that for testing simple H_0 versus simple H_1 after observing X (the context of the Neyman-Pearson Lemma) if $\phi(\cdot)$ is most powerful of level α, then it satisfies (9.3.4) for some k.

Also prove that if $\phi(\cdot)$ is most powerful of level $\leqslant \alpha$ (see Problem 9.1H), then it satisfies (9.3.3) (that is, has level α) unless there exists a test of level $< \alpha$ which has power 1.

9.4H In Definition 9.3.16 we defined a UMP test of level α for simple H_0 versus composite H_1, and in Theorem 9.3.35 we showed how to obtain UMP tests of $H_0: \theta = \theta_0$ versus $H_1: \theta > \theta_0$ in MLR cases. Thus, this work, as well as our definition of a most powerful test (Definition 9.3.1), is specifically for the case of simple H_0. *For composite H_0, we call a test* **most powerful of level α** *if Definition* 9.3.1 *is satisfied, with "level α" replaced by*

$$\sup_{\theta \in H_0} P_\theta[\text{Rejecting } H_0 \text{ using } \phi(\cdot)] = \alpha. \qquad (*)$$

Suppose that in Theorem 9.3.35 we replace: $H_0:\theta = \theta_0$ by $H_0^*:\theta \leqslant \theta_0$; and (9.3.36) by $(*)$, that is, by

$$\sup_{\theta \in H_0} E_\theta \phi(X) = \alpha.$$

Then show that the theorem remains true (and hence shows us how to obtain UMP level α tests of $H_0^*: \theta \leqslant \theta_0$ versus $H_1: \theta > \theta_0$ in MLR cases; in fact, the UMP level α test of $H_0: \theta = \theta_0$ versus $H_1: \theta > \theta_0$ *is* the UMP level α test of $H_0^*: \theta \leqslant \theta_0$ versus $H_1: \theta > \theta_0$).

9.5H Let A represent the collection of all one-parameter exponential families, and let B represent the collection of all MLR families. Then in Example 9.3.41 (see also Problem 9.16) we showed that $X = (X_1, \ldots, X_n)$ where $X_1, \ldots, X_n$ are independent r.v.'s with the same exponential distribution density with mean θ is in $A \cap B$. Give examples of X in $\bar{A} \cap B$, and in $A \cap \bar{B}$.

9.6H Prove Theorem 9.3.49.

9.7H In connection with Example 9.3.52, Reinhardt (in the article referenced in footnote 3) states that the choice of the r_i's is crucial since for party R in the Montana legislature he found minima and maxima of (9.3.54) of .50 and .68 for one "reasonable" set of r_i's, and .23 and .40 for another "reasonable" set. However, since the respective spreads of (9.3.54) indicated as possible are then .18 and .17, the approach does seem to show how much can be gained by the party in power apportioning (and it is substantial!).

Work out the minimum and maximum for some state other than Montana. How sensitive is the solution to $r_1, \ldots, r_n$? Since infinitely many values of k are possible in (9.3.56), how can we ever compute the answer? [*Hint*: If $r_{[1]} \leqslant \cdots r_{[n]}$ denote $r_1, \ldots, r_n$ in numerical order, then as we lower k no change is noted in (9.3.56) except when k "crosses" an r_i.]

9.8H Prove that the maximum of (9.3.54) referred to in Example 9.3.52 is the value k for which (9.3.56) yields the maximum.

9.9H Specify a hypothesis-testing problem where we find a convex set of risk points similar to the one displayed in Problem 9.20. [*Hint*: Consider a case where X takes on only finitely many values.]

9.10H **(Repeated tests of significance)** In some hypothesis-testing problems (especially in complex ones) there may be several tests we would consider using, and whose relationship to each other is not clear. (Even in simple problems, the Bayes rule with respect to $(p, 1 - p)$, the minimax rule, and the most powerful level α test may not bear a clear relation to one another until some effort is invested; see Problems 9.24 and 9.25.) Suppose we observe X and have available tests $\phi_1, \phi_2, \ldots, \phi_n$ each with level α; that is,

$$\alpha = E_{\theta_0}\phi_1(X) = E_{\theta_0}\phi_2(X) = \cdots = E_{\theta_0}\phi_n(X).$$

Then we sometimes say that we want to be "extra sure" H_0 is not false, and reject if *any one* of $\phi_1, \ldots, \phi_n$ calls for rejection (for example, see Cramér (1946), pp. 422–423). What relation does the level of this combined test, say ϕ with

$$\phi(x) = 1 - (1 - \phi_1(x)) \cdots (1 - \phi_n(x)),$$

bear to α? [*Hint*: Show that

$$\alpha \leqslant E_{\theta_0}\phi(X) \leqslant \min(n\alpha, 1),$$

and that $E_{\theta_0}\phi(X) = \alpha$ and $E_{\theta_0}\phi(X) = \min(n\alpha, 1)$ can occur. Hence, if we use even $n = 2$, the true (but unknown) level may be unacceptably large if we really desire level α.]

9.11H **(Data snooping)** Suppose we observe X and run a level α test $\phi(X)$ of some simple hypothesis H_0. Then even if H_0 is true, we have probability α of rejecting H_0. If we observe $X_1, \ldots, X_n$ (independent r.v.'s, each with the same distribution as X, for which H_0 is true) and test H_0 independently n times, the number Z of sets of data in which H_0 is rejected is a binomial r.v. with n trials and $p = \alpha$, and has expected value $EZ = np = n\alpha$. (For example, if $n = 1000$ and $\alpha = .01$, we expect to declare H_0 false in 10 of the 1000 data sets just due to the random nature of the data.) Hence, if we were to use the test to find which of $X_1, \ldots, X_n$ seem "odd" and then "test" the "odd" ones with the same test, all would be rejected *and the test would not be of level* α; its level would instead be 1!

Does the above apply to the situation in which a person, looking through an almanac or other source of data, is struck by some data given (thinks it "odd") and then tests it? Explain. (For example, we may be struck,[11] upon reading a list of how many games were played in each World Series, by the fact that more than half of the series played since World War II's end have lasted a full 7 games... 15/27 = .555 in 1945–1971. If we then test the hypothesis that nothing "unusual" is going on, we will reject it... since we formulated the hypothesis precisely because it looked as though something odd was happening. Thus, while *it may well be* that we have discovered something profound while reading the almanac, we have *no* statistical backing *unless we gather future data to test the hypothesis suggested by the past data.*)

9.12H In Definition 9.4.6 we considered complete classes for testing simple H_0 versus

---

[11] W. Simon ["'Back-to-the-wall' effect?," *Science*, Vol. 174 (1971), pp. 774–775] was so stuck. While he admits that "There are . . . difficulties associated with statistical interpretation of rare events defined after the fact . . . ," he goes on to gainsay the difficulties. Many others do not even realize that difficulties exist! [Even excellent books sometimes run afoul of this difficulty; for example, see Cramér (1946), pp. 457ff.]

simple H_1. *For any hypothesis-testing problem where H_0 is simple (but H_1 may be composite) call a test ϕ_2* **better than** *a test ϕ_1 if $\alpha_{\theta_0}(\phi_2) \leqslant \alpha_{\theta_0}(\phi_1)$ and $1 - \beta_\theta(\phi_2) \leqslant 1 - \beta_\theta(\phi_1)$ for all $\theta \in H_1$ with at least one strict inequality* ($<$). Extend the rest of Definition 9.4.5, and Definition 9.4.6, to simple H_0 versus composite H_1.

Suppose we observe $X = (X_1, \ldots, X_n)$ where $X_1, \ldots, X_n$ are independent r.v.'s, each $N(\mu, \sigma^2)$ where μ is unknown $(-\infty < \mu < +\infty)$ and $\sigma^2 > 0$ is known. Suppose it is desired to test $H_0: \mu = \mu_0$ at level α $(0 < \alpha < 1)$. In Example 9.2.4 we considered $\phi_1(\cdot)$ (reject iff $|\bar{X} - \mu_0| > a$) and $\phi_2(\cdot)$ (reject iff $\bar{X} - \mu_0 > b$), showing how to choose a and b to obtain level α in each case. In Example 9.3.10 we saw $\phi_2(\cdot)$ is UMP for $H_0: \mu = \mu_0$ versus $H_1: \mu > \mu_0$.

Prove that there is no UMP test for testing $H_0: \mu = \mu_0$ versus $H_1: \mu \neq \mu_0$. [*Hint*: Compare any such supposed UMP test with the Neyman-Pearson test of $H_0: \mu = \mu_0$ versus $H_1: \mu = \mu_1$ $(\mu_1 > \mu_0)$, which is unique.]

Consider the class of tests $C = \{\phi_\alpha^{(p)}(\cdot), 0 \leqslant p \leqslant 1, 0 \leqslant \alpha \leqslant 1\}$ where

$$\phi_\alpha^{(p)}(x) = \begin{cases} 1, & \text{if } \bar{x} - \mu_0 < a(p) \\ 1, & \text{if } \bar{x} - \mu_0 > b(p) \\ 0, & \text{otherwise;} \end{cases}$$

here $a(p)$ and $b(p)$ are chosen to satisfy

$$P_{\mu_0}[\bar{X} - \mu_0 < a(p)] = \alpha p,$$
$$P_{\mu_0}[\bar{X} - \mu_0 > b(p)] = \alpha(1 - p).$$

Find $a(p)$ and $b(p)$ explicitly. Prove that $\phi_\alpha^{(p)}(\cdot)$ has level α. Is C an essentially complete class of tests? (Figure 9.2–2 plotted the rejection probabilities for $\phi_\alpha^{(p)}(\cdot)$ when $\alpha = .05$ for $p = .00$ and for $p = .50$.) If C is essentially complete, then in this testing problem we must simply decide what proportion of level should go toward protecting against $\mu_1 < \mu_0$ (the rest going toward protecting against $\mu_1 > \mu_0$).

9.13H Generalizing Problem 9.7 (which treated the case of simple null and alternative hypotheses), show that for a general hypothesis-testing problem the class C of all tests based on a sufficient statistic t is (see Problem 9.12H) essentially complete. [*Hint*: See Problem 11 of E. L. Lehmann, *Testing Statistical Hypotheses*, John Wiley & Sons, Inc., New York, 1959, p. 57, and the references given there.]

9.14H The motivation given at the beginning of the proof of Lemma 9.5.15 would, in fact, yield a precise proof that $\bar{X}$ and s^2 are independent r.v.'s *if* the result "If (for $1 \leqslant i \leqslant n$) Y and W_i are independent, then Y and $W_1 + \cdots + W_n$ are independent" were true. Either prove this or give a counterexample.

9.15H Let $X_1, \ldots, X_n$ be independent $N(\mu, \sigma^2)$ r.v.'s. While proving Lemma 9.5.15 we showed that $\sum_{i=1}^{n} (X_i - \bar{X})^2/\sigma^2$ has the $\chi^2_{n-1}(0)$ distribution. (Verify this.)

P. Nemenyi ("A handy set of $n - 1$ contrasts . . . ," *Research Report No. 1*, Department of Statistics, Virginia State College, Petersburg, Virginia) has suggested that instead of the Helmert orthogonal transformation (9.5.18) we use the transformation

$$Y_j = (X_j - \bar{X}) + A(X_n - \bar{X}) \quad (1 \leqslant j \leqslant n - 1)$$

with A equal to a real root (does not exist?) of

$$(n-1)A^2 - 2A = 1.$$

Show that $Y_1, \ldots, Y_{n-1}$ are then independent $N(0, \sigma^2)$ r.v.'s, and that $Y_1^2 + \cdots + Y_{n-1}^2 = \sum_{i=1}^{n} (X_i - \bar{X})^2$. Deduce that $\sum_{i=1}^{n} (X_i - \bar{X})^2/\sigma^2$ has the $\chi^2_{n-1}(0)$ distribution.

If we also define $Y_n = (X_1 + \cdots + X_n)$, can we use Nemenyi's transformation to prove that $\bar{X}$ and s^2 are independent r.v.'s? If so, is Nemenyi's transformation a simpler way to prove Lemma 9.5.15 than is Helmert's transformation? [*Hint*: Review Problem 9.33.]

9.16H Let $X_1, \ldots, X_n$ be independent r.v.'s, each $N(\mu, \sigma^2)$ where μ is known $(-\infty < \mu < +\infty)$ and σ^2 is unknown $(\sigma^2 > 0)$. Prove that $\hat{\sigma}^2$ given in Equation (9.6.1) is the unique MVU estimator of σ^2. [*Hint*: Use Theorem 8.9.5.]

9.17H Despite the results of Problems 9.38 and 9.16H, $\hat{\sigma}^2$ is not perfect as an estimator of σ^2 in the μ known context, as the following considerations demonstrate. Suppose we wish to consider a class of estimators defined by

$$\hat{\sigma}^2(c) = c\hat{\sigma}^2 = c \sum_{i=1}^{n} \frac{(X_i - \mu)^2}{n}.$$

Show that

$$E_{\sigma^2}(\hat{\sigma}^2(c) - \sigma^2)^2 = \left(c^2 \frac{n+2}{n} - 2c + 1\right)\sigma^4,$$

which is minimized by choosing $c = n/(n+2)$, not by choosing $c = 1$. Hence, if the average squared error [which was considered in another estimation context at Definition 8.1.23 ff] is our criterion of goodness of an estimator, $\hat{\sigma}^2$ is beaten by $\hat{\sigma}^2(n/(n+2))$ for every value of σ^2 (that is, uniformly in σ^2).

9.18H Let $X_1, X_2, \ldots$ be a sequence of r.v.'s. Suppose that X_n has Student's t-distribution with n d.f. Show (first by characteristic function methods, and then directly) that the limiting distribution of X_n is $N(0, 1)$; that is, for every x (see Definition 6.4.1)

$$\lim_{n \to \infty} P[X_n \leqslant x] = \Phi(x).$$

9.19H Use Slutsky's Theorem (see Problem 8.3H) to prove the result of Problem 9.18H. [*Hint*: Use Slutsky's Theorem to show that

$$\sqrt{n}\frac{\bar{X} - \mu}{\sqrt{\dfrac{\sum_{i=1}^{n} (X_i - \bar{X})^2}{n-1}}} - \sqrt{n}\frac{\bar{X} - \mu}{\sigma} \xrightarrow{P} 0$$

and (see Definition 6.3.2) immediately deduce the desired result.]

9.20H Prove that when H_0 is true, the left-hand side of Equation (9.9.8) has Student's t-distribution with $m - 1$ d.f.

9.21H Compare the power of the Neyman-Bartlett and Scheffé solutions to the Behrens-Fisher problem. In particular, prove the Scheffé solution is more powerful.

10

INTERVAL ESTIMATION

In Chapter 8 we considered how to estimate an unknown parameter (for example, θ) by a point (for example, $\hat{\theta}$) in such a way that $\hat{\theta}$ is "close" to θ (for example, $E\hat{\theta} \equiv \theta$, $\text{Var}(\hat{\theta})$ small, and so on). In the present chapter we consider "interval estimation;" that is, how to estimate θ by an interval of values $\{x: \hat{\theta}_L < x < \hat{\theta}_U\}$ (abbreviated $(\hat{\theta}_L, \hat{\theta}_U)$) which has high probability of including θ but which also has, say, small average length $E(\hat{\theta}_U - \hat{\theta}_L)$. This subject is intimately related to tests of hypotheses (Chapter 9), and is grasped most easily in the context of general confidence sets (which may, but need not, be intervals) (Section 10.1). Some applications to various important cases are given (Sections 10.2ff), while others are treated in the problems, and still others can be constructed from the related tests by using the techniques of Section 10.1.

10.1. CONFIDENCE SETS AND TESTS OF HYPOTHESES

Some situations in which hypothesis-testing problems arise were described in Section 9.1. Generally, these were of the type "*Does* making a certain modification to some system or process change a particular parameter θ?" or "*Is* my parameter θ equal to θ_0?" In interval estimation we are concerned with the same situations, but instead ask "*How much* does making a certain modification to some system or process change a particular parameter θ?" or "*How large* is my parameter θ?" (Another approach to these latter questions is that of point estimation.)

Before introducing confidence sets in full generality, let us consider a special case; the results here will be helpful to our intuition. Suppose X is a univariate r.v. with d.f. $F(x|\theta)$ where θ is an unknown univariate parameter, $\theta \in \Theta$. Suppose we have available a test of the simple hypothesis $H_0: \theta = \theta_0$ *versus* the composite alternative $H_1: \theta \neq \theta_0$, and that test rejects iff $X < a$ or $X > b$, where a and b are specified numbers such that

$$P_{\theta_0}[X < a, \text{ or } X > b] = \alpha; \tag{10.1.1}$$

that is, such that the test has level α. Then usually a and b will be functions of θ_0, say $a = a(\theta_0)$ and $b = b(\theta_0)$. Consider plotting $a(\cdot)$ and $b(\cdot)$, as in Figure 10.1–1, for every θ. Then for a specified $\theta = \theta_0$, we have heights $a(\theta_0) < b(\theta_0)$, and (by (10.1.1))

$$P_{\theta_0}[a(\theta_0) \leqslant X \leqslant b(\theta_0)] = 1 - \alpha. \tag{10.1.2}$$

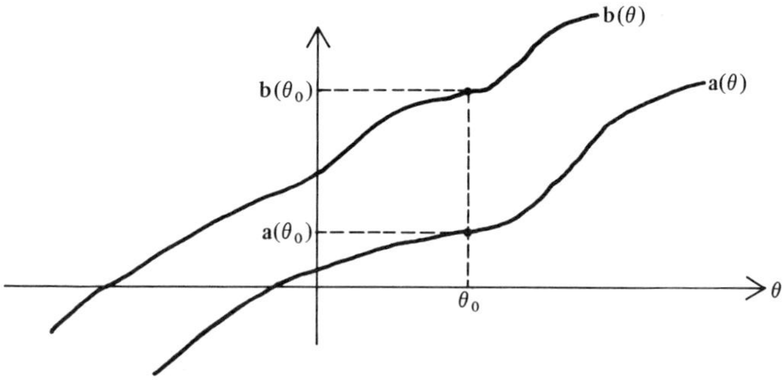

Figure 10.1–1. The functions **a**($\cdot$) and **b**($\cdot$)

Now (see Figure 10.1–2) plot the r.v. X on the vertical axis and draw a horizontal dotted line through it. Let $L(X)$ denote the value of θ at the intersection of the dotted line with the graph of $b(\theta)$, and $U(X)$ the value of θ at the intersection with the graph of $a(\theta)$. (We assume for now that $a(\theta)$ and $b(\theta)$ are such that there is, for each possible value of X, exactly one intersection of each type. More general cases will be considered below, in the context of confidence sets.) Then we claim that

$$P_\theta[L(X) \leqslant \theta \leqslant U(X)] \stackrel{\theta}{\equiv} 1 - \alpha; \tag{10.1.3}$$

that is, θ has probability $1 - \alpha$ (for example, .95 if $\alpha = .05$) of lying between $L(X)$ and $U(X)$. The proof is as follows: If θ^* is the true value of θ, then by (10.1.2), $P_{\theta^*}[a(\theta^*) \leqslant X \leqslant b(\theta^*)] = 1 - \alpha$. But $a(\theta^*) \leqslant X \leqslant b(\theta^*)$ iff $L(X) \leqslant \theta^* \leqslant U(X)$ (see Figure 10.1–3), which proves the result.

The interval $(L(X), U(X))$ is then called a "confidence interval" for θ with "confidence coefficient" $1 - \alpha$. The concept extends to the multivariate case:

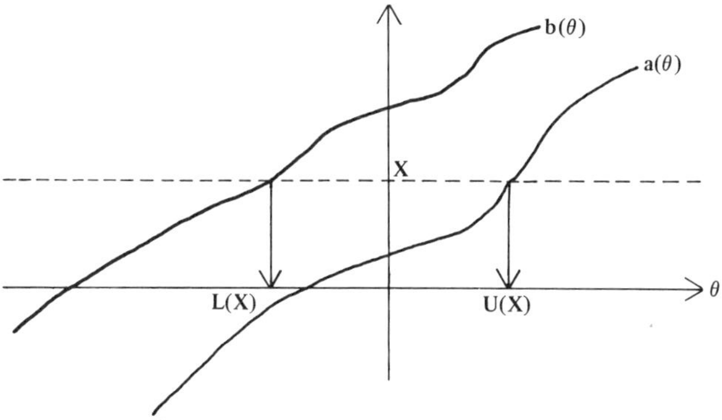

Figure 10.1–2. Construction of the interval (**L**(**X**), **U**(**X**))

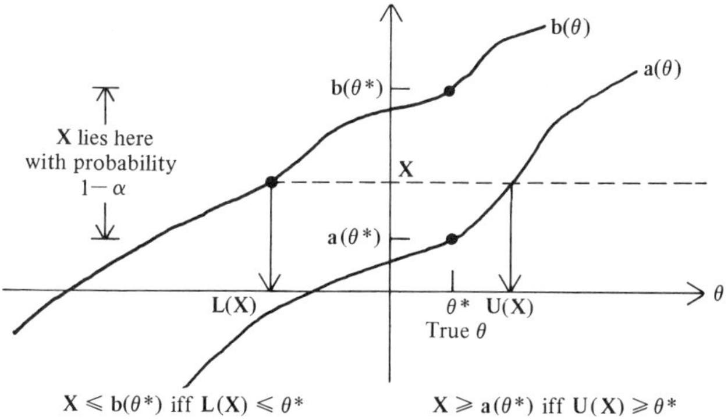

Figure 10.1–3. Proof of (10.1.3)

Definition 10.1.4. *Let X be a r.v. with d.f. $F(x|\theta)$ where θ is unknown, $\theta \in \Theta$.* [Here X and θ may be multivariate.] *If* (10.1.3) *holds, then the interval* $(L(X), U(X))$ *is called a* **confidence interval** *for θ with* **confidence coefficient** $1 - \alpha$.

If we have a multivariate, for example two-dimensional, parameter $\theta = (\theta_1, \theta_2)$, then "$\theta \in (L(X), U(X))$" means "$L_1(X) \leqslant \theta_1 \leqslant U_1(X)$ and $L_2(X) \leqslant \theta_2 \leqslant U_2(X)$," which is a rectangle (see Figure 10.1–4). But why shouldn't other shapes (for example, the dotted figure in Figure 10.1–4) be allowed; perhaps they will have an even higher confidence coefficient. Thus, we are led to consider general confidence sets, which are subsets of Θ in which we have (hopefully high) confidence θ lies.

We now generalize the development which led, from a univariate r.v. X and a test (see (10.1.1)) for a univariate parameter θ, to a confidence interval $(L(X), U(X))$ with property (10.1.3) (confidence coefficient $1 - \alpha$) as follows.

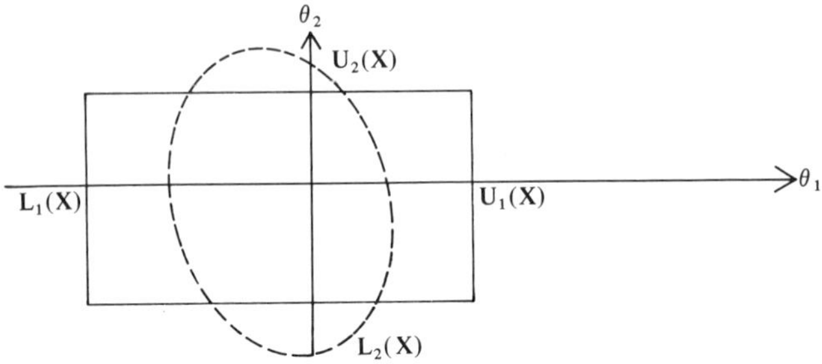

Figure 10.1–4. Confidence sets in two dimensions

Theorem 10.1.5. *Let X be a r.v. with d.f. $F(x|\theta)$ where θ is unknown, $\theta \in \Theta$.* [Here X and θ may be multivariate.] *Suppose that there is a set C_θ of values of X such that for every $\theta \in \Theta$*

$$P_\theta[X \in C_\theta] = 1 - \alpha. \tag{10.1.6}$$

[That is, C_{θ_0} is the complement of the rejection region of a level α test of $H_0: \theta = \theta_0$.] *Define*

$$S(X) = \{\theta^*: X \in C_{\theta^*}\}. \tag{10.1.7}$$

[That is, $S(X)$ is the collection of all θ^*'s for which observing X does not lead to rejection of $H_0: \theta = \theta^*$.] *Then*

$$P_\theta[\theta \in S(X)] \overset{\theta}{\equiv} 1 - \alpha. \tag{10.1.8}$$

Proof: We easily find, using (10.1.7) and (10.1.6), that

$$\begin{aligned} P_\theta[\theta \in S(X)] &\overset{\theta}{\equiv} P_\theta[\theta \in \{\theta^*: X \in C_{\theta^*}\}] \\ &\overset{\theta}{\equiv} P_\theta[X \in C_\theta] \overset{\theta}{\equiv} 1 - \alpha. \end{aligned} \tag{10.1.9}$$

Definition 10.1.10. *Let X be a r.v. with d.f. $F(x|\theta)$ where θ is unknown, $\theta \in \Theta$.* [Here X and θ may be multivariate.] *If* (10.1.8) *holds, then the set $S(X)$ is called a* **confidence set** *for θ with* **confidence coefficient** $1 - \alpha$.

This definition is to confidence sets as Definition 10.1.4 is to confidence intervals.

10.2. CONFIDENCE INTERVALS FOR ONE NORMAL MEAN

In this section we consider confidence intervals (and make some remarks about **confidence sets) for one normal mean, first in the case where σ^2 is known, and then in the case where σ^2 is unknown.**

Example 10.2.1. Let $X_1, \ldots, X_n$ be independent $N(\mu, \sigma^2)$ r.v.'s where μ is unknown with $\mu \in \Theta = \{x: -\infty < x < +\infty\}$ and $\sigma^2 > 0$ is known. Find a confidence interval for μ with confidence coefficient $1 - \alpha = .95$.

Here $\bar{X} = (X_1 + \cdots + X_n)/n$ is a minimal sufficient statistic, and hence we should let our confidence interval for μ depend on $X_1, \ldots, X_n$ solely through $\bar{X}$. In this case good[1] tests of the hypothesis $H_0: \mu = \mu_0$ were studied in Section 9.5 and (see (9.5.2)) a typical member of a reasonable class rejects $H_0: \mu = \mu_0$ iff

$$\bar{X} - \mu_0 < \frac{\sigma}{\sqrt{n}} \Phi^{-1}(\alpha p) \quad \text{or} \quad \bar{X} - \mu_0 > \frac{\sigma}{\sqrt{n}} \Phi^{-1}(1 - \alpha(1 - p)), \qquad (10.2.2)$$

where p $(0 \leqslant p \leqslant 1)$ is fixed but arbitrary. Since when H_0 is true (10.2.2) occurs with probability α, its complement occurs with probability $1 - \alpha$:

$$P_{\mu_0}\left[\mu_0 + \frac{\sigma}{\sqrt{n}} \Phi^{-1}(\alpha p) \leqslant \bar{X} \leqslant \mu_0 + \frac{\sigma}{\sqrt{n}} \Phi^{-1}(1 - \alpha(1 - p))\right] = 1 - \alpha. \qquad (10.2.3)$$

Thus (see (10.1.3)) here our functions $a(\mu)$ and $b(\mu)$ are

$$\begin{cases} a(\mu) = \mu + \dfrac{\sigma}{\sqrt{n}} \Phi^{-1}(\alpha p), \\ b(\mu) = \mu + \dfrac{\sigma}{\sqrt{n}} \Phi^{-1}(1 - \alpha(1 - p)). \end{cases} \qquad (10.2.4)$$

To construct our confidence interval $(L(\bar{X}), U(\bar{X}))$ we proceed as in Figure 10.1–2 (see Figure 10.2–1), finally obtaining the solution via solving $\bar{X} = b(\mu)$

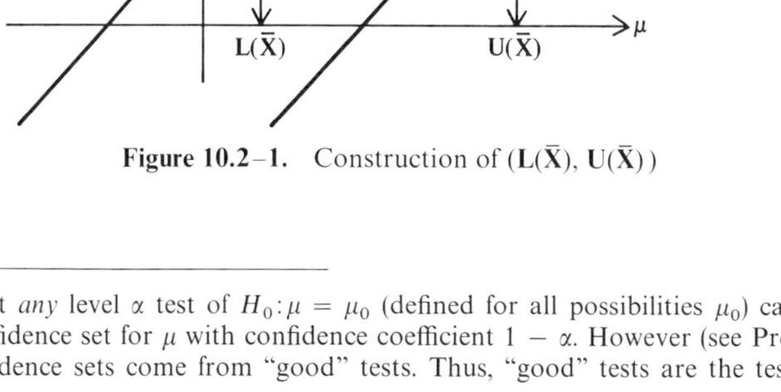

Figure 10.2–1. Construction of $(\mathbf{L}(\bar{\mathbf{X}}), \mathbf{U}(\bar{\mathbf{X}}))$

[1] Note that *any* level α test of $H_0: \mu = \mu_0$ (defined for all possibilities μ_0) can be used to obtain a confidence set for μ with confidence coefficient $1 - \alpha$. However (see Problem 10.1H) "good" confidence sets come from "good" tests. Thus, "good" tests are the tests of interest to us.

for $L(\bar{X})$ and $\bar{X} = a(\mu)$ for $U(\bar{X})$:

$$\begin{cases} L(\bar{X}) = \bar{X} - \dfrac{\sigma}{\sqrt{n}}\Phi^{-1}(1 - \alpha(1 - p)), \\ U(\bar{X}) = \bar{X} - \dfrac{\sigma}{\sqrt{n}}\Phi^{-1}(\alpha p). \end{cases} \tag{10.2.5}$$

Thus,

$$P_\mu\left[\bar{X} - \frac{\sigma}{\sqrt{n}}\Phi^{-1}(1 - \alpha(1 - p)) \leqslant \mu \leqslant \bar{X} - \frac{\sigma}{\sqrt{n}}\Phi^{-1}(\alpha p)\right] \overset{\mu}{\equiv} 1 - \alpha. \tag{10.2.6}$$

In the desired case of $1 - \alpha = .95$, if we choose $p = .5$ [one of many possible choices] and use Table 11, we find (see Problem 10.2) our interval is

$$\left(\bar{X} - 1.96\frac{\sigma}{\sqrt{n}}, \bar{X} + 1.96\frac{\sigma}{\sqrt{n}}\right). \tag{10.2.7}$$

In fact, $p = .5$ is a reasonable choice since of all p $(0 \leqslant p \leqslant 1)$ it yields (see Problem 10.3) the one with the smallest length, namely length equal to

$$\frac{\sigma}{\sqrt{n}}\left\{\Phi^{-1}\left(1 - \frac{\alpha}{2}\right) - \Phi^{-1}\left(\frac{\alpha}{2}\right)\right\} = 2\frac{\sigma}{\sqrt{n}}\Phi^{-1}\left(1 - \frac{\alpha}{2}\right). \tag{10.2.8}$$

Note that if we were concerned with a confidence interval which would put an *upper bound* on μ, we could choose $p = 1$ in (10.2.5) and obtain the interval

$$\left(-\infty, \bar{X} + \frac{\sigma}{\sqrt{n}}\Phi^{-1}(1 - \alpha)\right). \tag{10.2.9}$$

We can (see Problem 10.6) similarly obtain a *lower bound* on μ.

Also note that we obtained parallel lines $a(\mu)$ and $b(\mu)$ in Figure 10.2–1 due to our holding p fixed independent of μ. If we chose different p's for various μ values we would obtain other curves.

We next consider the case where σ^2 is unknown.

Example 10.2.10. Let $X_1, \ldots, X_n$ be independent $N(\mu, \sigma^2)$ r.v.'s where μ is unknown $(-\infty < \mu < +\infty)$ and σ^2 is unknown $(\sigma^2 > 0)$. Find a confidence interval for μ with confidence coefficient $1 - \alpha = .95$.

In this case good tests of the hypothesis $H_0\colon \mu = \mu_0$ were obtained in Theorem 9.5.26 and (see Problem 9.35) a typical test rejects $H_0\colon \mu = \mu_0$ iff

$$\bar{X} - \mu_0 < \frac{s}{\sqrt{n}}t_{n-1}^{-1}(\alpha p) \quad \text{or} \quad \bar{X} - \mu_0 > \frac{s}{\sqrt{n}}t_{n-1}^{-1}(1 - \alpha(1 - p)), \tag{10.2.11}$$

where p $(0 \leqslant p \leqslant 1)$ is fixed but arbitrary. By a development similar to that following (10.2.2) we find (see Problem 10.8)

$$\begin{cases} L = \bar{X} - \dfrac{s}{\sqrt{n}}\, t_{n-1}^{-1}(1 - \alpha(1 - p)), \\ U = \bar{X} - \dfrac{s}{\sqrt{n}}\, t_{n-1}^{-1}(\alpha p). \end{cases} \tag{10.2.12}$$

In the desired case of $1 - \alpha = .95$, if we choose $p = .5$ [one of many possible choices] and use Table 13, we find (see Problem 10.9) our interval is

$$\left(\bar{X} - 2.57\frac{s}{\sqrt{6}}, \bar{X} + 2.57\frac{s}{\sqrt{6}}\right) \quad \text{if } n = 6. \tag{10.2.13}$$

Here we needed to assume a value of n (we took $n = 6$ for purposes of illustration) since the quantity $t_{n-1}^{-1}(\cdot)$ is a function of n. [At (10.2.5) we had $\Phi^{-1}(\cdot)$, which was not a function of n, in the case of known $\sigma^2 > 0$.]

In fact, $p = .5$ is a reasonable choice here also since, as was the case in the instance of a known $\sigma^2 > 0$, it yields (see Problem 10.10) the shortest length

$$2\frac{s}{\sqrt{n}}\, t_{n-1}^{-1}\left(1 - \frac{\alpha}{2}\right). \tag{10.2.14}$$

Note that in the case of known σ^2 the intervals had fixed length, so by an appropriate choice of n we could specify the length of the interval on μ. (For example, if we have $p = .5$, we simply take n to be the smallest integer greater than or equal to the solution x of the equation

$$2\frac{\sigma}{\sqrt{x}}\,\Phi^{-1}\left(1 - \frac{\alpha}{2}\right) = L_0$$

and obtain an interval of length L_0 [or slightly less].) In this case the length is a random variable (for example, see (10.2.14) for the length in the case $p = .5$) since s is a r.v., and hence cannot be specified in advance of the experiment unless we have some idea of how large s will be (which we might if we had an idea of the size of σ^2).

In each of Examples 10.2.1 and 10.2.10, and in Section 10.1, **we have obtained confidence intervals from tests of hypotheses. However, another method of obtaining confidence intervals** (which is not always feasible, but when it is often yields answers quickly and neatly) **is available**, and will now be explained. (Since the result is closely related to Theorem 10.1.5 we could call it a corollary. However, it makes no use of hypothesis-testing ideas.)

Lemma 10.2.15. *Let X be a r.v. with d.f. $F(x|\theta)$ where θ is unknown, $\theta \in \Theta$.* [Here X and θ may be multivariate.] *Suppose there is a function $g(X, \theta)$*

whose distribution is known, and does not depend on θ. Then it is easy to find sets S of possible values of $g(X, \theta)$ such that

$$P[g(X,\theta) \in S] \equiv 1 - \alpha. \tag{10.2.16}$$

However, each such set S yields us a confidence set for θ since

$$P_\theta[\theta \in \{\theta: g(X,\theta) \in S\}] \overset{\theta}{\equiv} 1 - \alpha. \tag{10.2.17}$$

We will now demonstrate the use of this new method of finding confidence intervals.

Example 10.2.18. As in Example 10.2.1 let $X_1, \ldots, X_n$ be independent $N(\mu, \sigma^2)$ r.v.'s with μ unknown and σ^2 known. Then

$$g(X_1, \ldots, X_n, \mu) = \frac{\bar{X} - \mu}{\sigma/\sqrt{n}} \tag{10.2.19}$$

has the $N(0, 1)$ distribution, which does not depend on μ. One possible choice of S, $S = \{x: -1.96 \leqslant x \leqslant 1.96\}$, yields

$$\begin{aligned} P[g(X_1,\ldots,X_n,\mu) \in S] &= .95 \\ &= P\left[-1.96 \leqslant \frac{\bar{X} - \mu}{\sigma/\sqrt{n}} \leqslant 1.96\right] \\ &= P\left[-1.96\frac{\sigma}{\sqrt{n}} \leqslant \bar{X} - \mu \leqslant 1.96\frac{\sigma}{\sqrt{n}}\right] \\ &= P\left[\bar{X} - 1.96\frac{\sigma}{\sqrt{n}} \leqslant \mu \leqslant \bar{X} + 1.96\frac{\sigma}{\sqrt{n}}\right]. \end{aligned} \tag{10.2.20}$$

This is the same interval as at (10.2.5).

10.3. CONFIDENCE INTERVALS FOR ONE NORMAL VARIANCE

In this section we consider confidence intervals for one normal variance in the case where μ is known. The case where μ is unknown is handled by a simple modification.

Example 10.3.1. Let $X_1, \ldots, X_n$ be independent $N(\mu, \sigma^2)$ r.v.'s where μ is known ($-\infty < \mu < +\infty$) and $\sigma^2 > 0$ is unknown. Find a confidence interval for σ^2 with confidence coefficient $1 - \alpha = .95$ when $n = 10$.

By Problem 9.34 we know that

$$g(X_1, \ldots, X_n, \sigma) = \sum_{i=1}^{n} \left(\frac{X_i - \mu}{\sigma}\right)^2 \tag{10.3.2}$$

has the $\chi_n^2(0)$ distribution. We can thus use the method of Lemma 10.2.15 to find the desired confidence interval. From Table 10,

$$\begin{aligned} P\left[3.25 \leqslant \frac{\sum_{i=1}^{n}(X_i-\mu)^2}{\sigma^2} \leqslant 20.48\right] &= .95 \\ = P\left[\frac{\sum_{i=1}^{n}(X_i-\mu)^2}{20.48} \leqslant \sigma^2 \leqslant \frac{\sum_{i=1}^{n}(X_i-\mu)^2}{3.25}\right]. \end{aligned} \tag{10.3.3}$$

In the case of *unknown* μ we know (from the proof of Lemma 9.5.15) that

$$g^*(X_1, \ldots, X_n, \sigma) = \sum_{i=1}^{n}\left(\frac{X_i - \bar{X}}{\sigma}\right)^2 \tag{10.3.4}$$

has the $\chi_{n-1}^2(0)$ distribution, so the desired result is (see Problem 10.13) easily obtained.

In the case of unknown μ, (10.3.4) requires $n \geqslant 2$ in order to yield a confidence interval. However, even in the case of $n = 1$ an interval for σ which has probability at least $1 - \alpha$ of including σ can be obtained, as we will now see.

Example 10.3.5. Let X be a $N(\mu, \sigma^2)$ r.v. with μ $(-\infty < \mu < +\infty)$ and $\sigma^2 > 0$ both unknown. Let $c > 0$ be chosen to satisfy

$$\Phi\left(\frac{1}{c}\right) - \Phi\left(-\frac{1}{c}\right) = \alpha. \tag{10.3.6}$$

Then

$$\begin{aligned} P[\sigma \in (0, c|X|)] &= P\left[\frac{\sigma}{c} \leqslant |X|\right] = 1 - P\left[\frac{\sigma}{c} > |X|\right] \\ &= 1 - P\left[-\frac{\sigma}{c} < X < \frac{\sigma}{c}\right] \\ &= 1 - P\left[-\frac{1}{c} - \frac{\mu}{\sigma} < \frac{X-\mu}{\sigma} < \frac{1}{c} - \frac{\mu}{\sigma}\right] \\ &= 1 - \left\{\Phi\left(\frac{1}{c} - \frac{\mu}{\sigma}\right) - \Phi\left(-\frac{1}{c} - \frac{\mu}{\sigma}\right)\right\}. \end{aligned} \tag{10.3.7}$$

Since $\Phi(a - b) - \Phi(-a - b)$ is largest when $b = 0$ (see Problem 10.16), using (10.3.6) we find

$$P[\sigma \in (0, c|X|)] \geqslant 1 - \left\{\Phi\left(\frac{1}{c}\right) - \Phi\left(-\frac{1}{c}\right)\right\} = 1 - \alpha. \tag{10.3.8}$$

Hence, the interval $(0, c|X|)$ has probability at least $1 - \alpha$ of covering σ.

[For further considerations along these lines see Judah Rosenblatt, "Confidence interval for standard deviation from a single observation," *Technometrics*, Vol. 8 (1966), pp. 367–368.]

10.4. CONFIDENCE INTERVALS FOR A SUCCESS PROBABILITY

Let $X_1, \ldots, X_n$ be independent r.v.'s, each Bernoulli with unknown probability of success $p (0 \leqslant p \leqslant 1)$. In this section we study confidence intervals for p.

Here (by an easy modification of Problem 8.57) $\bar{X} = (X_1 + \cdots + X_n)/n$ is a minimal sufficient statistic, so we would expect to be able to base a good test on $\bar{X}$ alone. For testing $H_0: p = p_0$ it is clear intuitively that a good test will reject iff

$$\bar{X} < c_1, \quad \text{or} \quad \bar{X} > c_2, \tag{10.4.1}$$

where c_1 and c_2 are chosen (if possible) to give level α. (See Problem 10.17.)

When we try to use Theorem 10.1.5 (see also Figure 10.1–2) to construct our confidence interval, we encounter a problem studied in Problem 10.17 and alluded to in Problem 10.3H: since $\bar{X}$ is a discrete r.v. (its possible values are 0, $1/n$, $2/n, \ldots, 1$), the level α test of $H_0: p = p_0$ with rejection region (10.4.1) is usually randomized. Hence, for "most" values of p $(0 \leqslant p \leqslant 1)$ there do not exist $a(p)$ and $b(p)$ (see (10.1.2)) such that

$$P_p[a(p) \leqslant \bar{X} \leqslant b(p)] = 1 - \alpha. \tag{10.4.2}$$

Suppose we desire an equal-tail (see Problem 10.15) interval. Then we would like to solve

$$P_p[\bar{X} < a(p)] = \frac{\alpha}{2} \tag{10.4.3}$$

and

$$P_p[\bar{X} > b(p)] = \frac{\alpha}{2} \tag{10.4.4}$$

for $a(p)$ and $b(p)$ (respectively) for each p $(0 \leqslant p \leqslant 1)$. To see that this is not possible consider, say, (10.4.3). For a fixed p, $P_p[\bar{X} < c]$ jumps each time c passes one of the points 0, $1/n$, $2/n, \ldots, 1$ and remains constant between (see Figure 10.4–1). Hence, for most $\alpha/2$ (for example, $\alpha/2 = \alpha_1^*$ in Figure 10.4–1) there is no solution $a(p)$ to (10.4.3), though for different $\alpha/2$ (for example, $\alpha/2 = \alpha_2^*$ in Figure 10.4–1) there are many solutions.

However, for each a $(0 < a \leqslant 1)$ there is a unique p for which (10.4.3) is solved, and for each b $(0 \leqslant b < 1)$ there is a unique p for which (10.4.4) is solved. The points (p, b) are plotted in Figure 10.4–2, and should be compared with the curve $b(\theta)$ in Figure 10.1–1 to see graphically the nature of our present difficulties.

Now suppose we graph the points (p, b) for $b = 0, 1/n, \ldots, (n-1)/n$ (the dots in Figure 10.4–2) as well as the points (p, a) for $a = 1/n, 2/n, \ldots, 1$. (See Figure 10.4–3.) Then for any observed value of $\bar{X}$ except 0 or 1 (that is, for $\bar{X} = 1/n, \ldots, (n-1)/n$) an interval is determined. This interval does not $\{$contain the true p with probability $\geqslant 1 - \alpha\}$; the proof is as follows. Suppose that p_0 is the true value of p. Then [talking in terms of Figure 10.4–3 for the case $n = 7$, without loss of generality since the same argument works for any n] the interval generated by the procedure specified fails to cover p_0 iff $\bar{X} \geqslant 6/7$ or $\bar{X} \leqslant 1/7$. Since at p_* $P[\bar{X} \geqslant 6/7] = \alpha/2$, at p_0 $P[\bar{X} \geqslant 6/7] > \alpha/2$ (since $p_0 > p_*$). Similarly, since at p^* $P[\bar{X} \leqslant 1/7] = \alpha/2$, at p_0 $P[\bar{X} \leqslant 1/7] >$

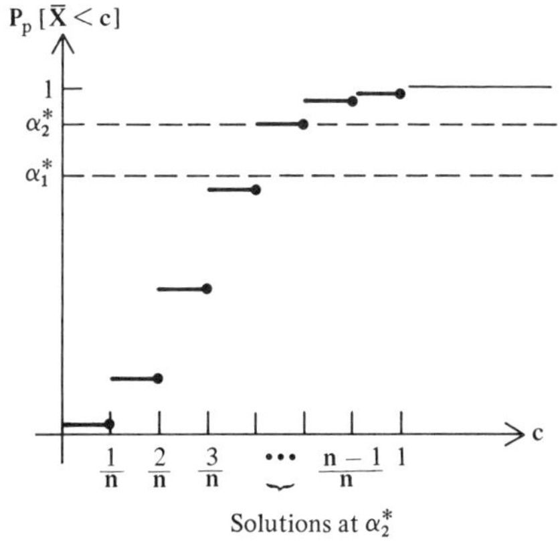

Figure 10.4–1. Solutions **c** of $\mathbf{P_p}[\bar{X} < \mathbf{c}] = \alpha/2$ [Graph for **n** = 7, **p** = .40, $\boldsymbol{\alpha}$ = .05]

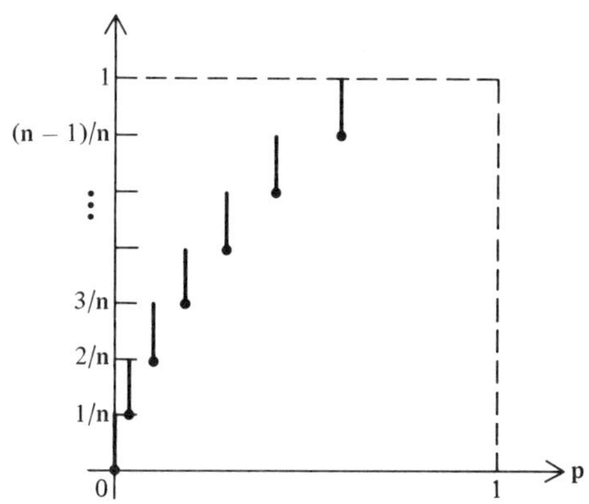

Figure 10.4–2. The points (**p**, **b**) [Graph for **n** = 7, $\boldsymbol{\alpha}$ = .05]

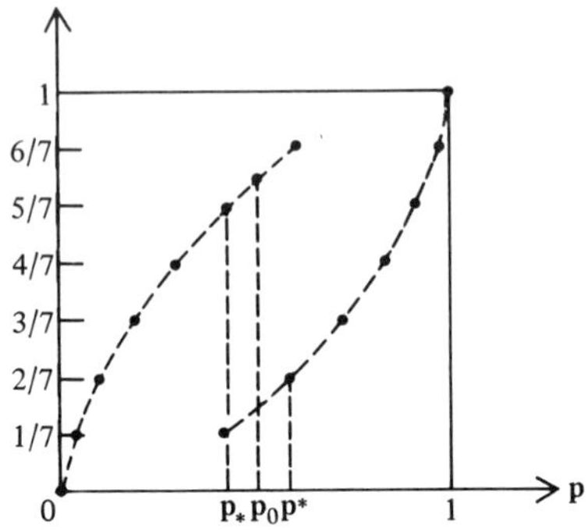

Figure 10.4–3. Points (**p**, **a**) and (**p**, **b**) [Graph for **n** = 7, **α** = .05]

$\alpha/2$ (since $p_0 < p^*$). (See Problem 10.19.) Hence,

$$\begin{aligned} P_{p_0}[2/7 \leqslant \bar{X} \leqslant 5/7] &= 1 - P_{p_0}[\bar{X} \geqslant 6/7] - P_{p_0}[\bar{X} \leqslant 1/7] \\ &< 1 - P_{p_*}[\bar{X} \geqslant 6/7] - P_{p^*}[\bar{X} \leqslant 1/7] \qquad (10.4.5) \\ &\leqslant 1 - \alpha. \end{aligned}$$

However, if we expand the intervals produced by our procedure by moving all b points up by $1/n$ (and all a points down by $1/n$), resulting in Figure 10.4–4, the proposed procedure will yield an interval with probability $\geqslant 1 - \alpha$ of containing the true value p_0. For *now* the interval generated by the procedure specified fails to cover p_0 iff $\bar{X} \geqslant 1.0$ or $\bar{X} \leqslant .0$. Since for a value of $p^* > p_0$ $P_{p^*}[\bar{X} \geqslant 1.0] = \alpha/2$, $P_{p_0}[\bar{X} \geqslant 1.0] < \alpha/2$. Similarly, for a value of $p_* < p_0$ $P_{p_*}[\bar{X} \leqslant .0] = \alpha/2$, so $P_{p_0}[\bar{X} \leqslant .0] < \alpha/2$. Hence (as at (10.4.5) but with new points p_* and p^* as shown in Figure 10.4–4) our confidence coefficient is $\geqslant 1 - \alpha$. (In fact, sometimes the probability of coverage is substantially greater than $1 - \alpha$.) It is easy to show that the probability of coverage is

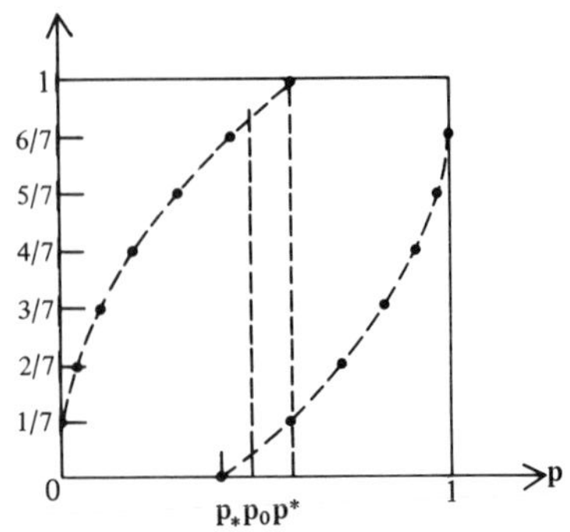

Figure 10.4–4. Points (**p**, **a**) [lowered] and (**p**, **b**) [raised] [Graph for **n** = 7, **α** = .05]

still $\geqslant 1 - \alpha$ if we complete the proposed procedure by adding the points (0, 0) and (1, 1) to the upper and lower "curves," respectively. (See Problem 10.20.)

Thus, the case of discrete r.v.'s can be quite complex, as our above analysis of the success probability case shows. In fact, even good texts often either skip these cases entirely (for example, Hogg and Craig (1965)), or use an approximate analysis via treating $\bar{X}$ as approximately normally distributed (for example, Lindgren (1968), pp. 289–290), or do not make it clear that the confidence is not $= 1 - \alpha$ but rather $\geqslant 1 - \alpha$ with $\gg 1 - \alpha$ being a not uncommon occurrence (for example, Mood and Graybill (1963)).

Excellent charts for $1 - \alpha = .95$ and .99 for $n = 8, 10, 12, 16, 20, 24, 30, 40, 60, 100, 200, 400, 1000$ have been given on pp. 228–229 of *Biometrika Tables for Statisticians, Volume I* (Third edition reprinted with additions) edited by E. S. Pearson and H. O. Hartley, Cambridge University Press, London, 1970. There it was possible (due to the result shown in Problem 10.18) to use only the "bottom half" of graphs like that in Figure 10.4–4 and still have the complete situation covered. One of these charts (the one for $1 - \alpha = .95$) is reproduced in Figure 10.4–5, on page 336, by the kind permission of the Biometrika Trustees. This chart is used as follows: If we observe $c = 26$ successes in $n = 100$ trials (so $c/n = .26$), an (at least) 95 percent confidence interval for p is $.18 \leqslant p \leqslant .36$ (this is read along the left and bottom axes); while if we observe $c = 88$ successes in $n = 100$ trials (so $c/n = .88$), an (at least) 95 percent confidence interval for p is $.80 \leqslant p \leqslant .93$ (this is read along the right and top axes).

PROBLEMS

10.1 Verify that in (10.2.4) we have $a(\mu) \leqslant b(\mu)$ for all α, p.

10.2 Verify that when $1 - \alpha = .95$ and $p = .5$

$$\begin{cases} \bar{X} - \dfrac{\sigma}{\sqrt{n}}\Phi^{-1}(1 - \alpha(1 - p)) = \bar{X} - 1.96\dfrac{\sigma}{\sqrt{n}}, \\ \bar{X} - \dfrac{\sigma}{\sqrt{n}}\Phi^{-1}(\alpha p) = \bar{X} + 1.96\dfrac{\sigma}{\sqrt{n}}. \end{cases}$$

10.3 Prove that of all intervals (10.2.5), the one with $p = .5$ has the smallest length. [Related considerations were contained in Problem 6.13.]

10.4 Prove the equality in Equation (10.2.8).

10.5 Verify that if $p = 1$ in (10.2.5), the interval becomes (10.2.9).

10.6 Obtain a lower bound on μ analogous to upper bound (10.2.9).

10.7 Suppose that in Example 10.2.1 we choose $a(\mu)$ and $b(\mu)$ as at (10.2.4), but with

$$p = p(\mu) = \begin{cases} 1, & \text{if } \mu \geqslant 0 \\ 0, & \text{if } \mu < 0. \end{cases}$$

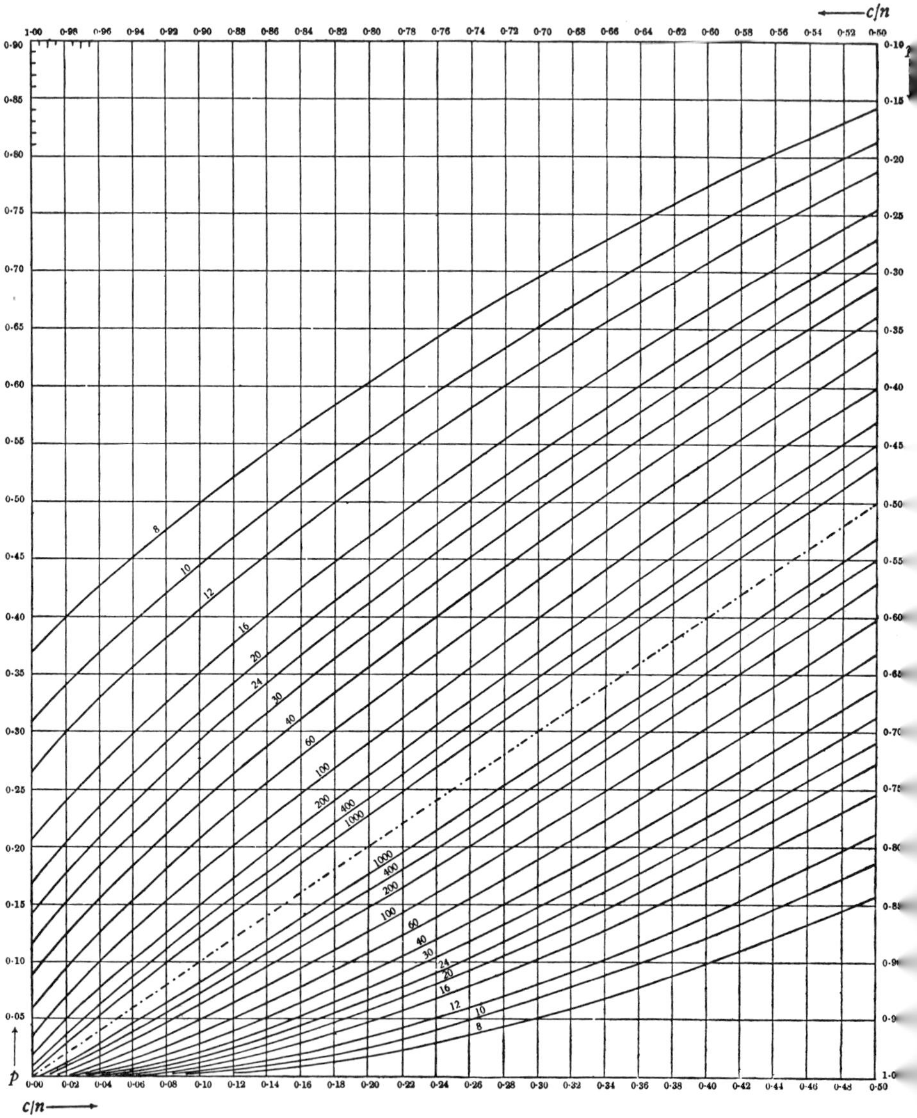

Figure 10.4–5. Chart providing confidence limits for a success probability **p** (**c** represents the observed number of successes in **n** trials. The numbers printed along the curves indicate the sample size **n**. From the *Biometrika Tables for Statisticians*.

Construct the graph which now replaces Figure 10.2–1. For the case $1 - \alpha = .95$, what confidence interval do we now obtain (that is, the new $(L(\bar{X}), U(\bar{X}))$ which replaces (10.2.7))? Be explicit. Why might we (or why might we not) want to use such an interval? (Speak in terms of real-world problems.)

10.8 Derive (10.2.12) in full detail.

10.9 Verify that when $1 - \alpha = .95$, $p = .5$, and $n = 6$,

$$\begin{cases} \bar{X} - \dfrac{s}{\sqrt{n}} t_{n-1}^{-1}(1 - \alpha(1 - p)) = \bar{X} - 2.57 \dfrac{s}{\sqrt{6}}, \\ \bar{X} - \dfrac{s}{\sqrt{n}} t_{n-1}^{-1}(\alpha p) = \bar{X} - 2.57 \dfrac{s}{\sqrt{6}}. \end{cases}$$

10.10 Prove that of all intervals (10.2.12), the one with $p = .5$ has the smallest length. Show that this length is given by (10.2.14).

10.11 Example 10.2.18 used the method of Lemma 10.2.15 of finding confidence intervals, for the problem previously studied in Example 10.2.1. Do the same for the problem of Example 10.2.10.

10.12 In the example of seasonal snowfall in Rochester, New York (see Problem 9.52), what confidence interval on μ do we obtain if we assume that:

(a) $X_1, \ldots, X_{88}$ are independent $N(\mu, \sigma^2)$ r.v.'s with μ and σ^2 unknown;

(b) $X_1, \ldots, X_{55}$ are independent $N(\mu_1, \sigma^2)$ r.v.'s, $X_{56}, \ldots, X_{88}$ are independent $N(\mu, \sigma^2)$ r.v.'s, and μ_1, μ, σ^2 are unknown;

(c) $X_1, \ldots, X_{55}$ are independent $N(\mu_1, \sigma_1^2)$ r.v.'s, $X_{56}, \ldots, X_{88}$ are independent $N(\mu, \sigma_2^2)$ r.v.'s, and $\mu_1, \mu, \sigma_1^2, \sigma_2^2$ are unknown.

[*Hint*: Your interval in (b) should be "better" than your interval in (c) since in (b) more is known about σ^2.] Based on Problem 9.52, which interval do you recommend should be used to guide the Highway Department in its purchases of road salt for the next season?

10.13 In Example 10.3.1 we found an interval (10.3.3) for σ^2 when μ was known. Find the comparable interval for the case of unknown μ. Also find upper and lower bounds (see (10.2.9)ff) on σ^2 for the cases of both known and unknown μ.

10.14 Derive confidence intervals for σ^2 in the contexts of **(1)** μ known and **(2)** μ unknown, as was done in Section 10.3, but using the tests for corresponding problems obtained in Section 9.6 (and not using Lemma 10.2.15).

10.15 Man A (a track star) has run the mile twice in competitions, with times of 3.8, 3.4. Man B (another track star) has run the mile thrice in competitions, with times of 3.7, 3.6, 4.1. Assume that man A and man B ran independently of each other, and that each one's times are independent observations of his running ability. Let $\mu_A(\mu_B)$ denote the average time it takes man $A(B)$ to run the mile in competitions. Similarly, let his variance be $\sigma_A^2(\sigma_B^2)$.

Assuming all observations are normally distributed, give a two-sided equal-tail confidence interval for $\mu_B - \mu_A$ (with confidence coefficient .95) if:

(a) $\sigma_A^2 = \sigma_B^2 = .04$.

(b) $\sigma_A^2 = .04, \sigma_B^2 = .09$.

(c) $\sigma_A^2 = \sigma_B^2$ but the common value is unknown.

(d) What can be said if σ_A^2, σ_B^2 are both unknown?

Note by "equal-tail" (see Problem 9.50 for a similar concept in hypothesis testing) we mean the interval comes from "cutting off" two tails of a distribution with equal probability in each tail. For example, in (10.3.3) we put probability .025 in each tail of a $\chi_n^2(0)$ distribution.

[*Hint*: Corresponding tests were studied in Sections 9.8 and 9.9. Also see Problem 9.51.]

10.16 Prove that for any fixed $a > 0$, $\Phi(a - b) - \Phi(-a - b)$ is largest when $b = 0$.

10.17 Let $X_1, \ldots, X_n$ be independent r.v.'s, each Bernoulli with unknown probability of success p $(0 \leqslant p \leqslant 1)$. Find a most powerful level $\alpha = .05$ test of $H_0: p = p_0$ versus $H_1: p = p_1$ where $1 \geqslant p_1 > p_0 \geqslant 0$. Is this test UMP for testing $H_0: p = p_0$ versus $H_1: p > p_0$? Note that the test is randomized (due to the r.v.'s being discrete) for all but a few of the possible values of α (which ones?).

10.18 In Figure 10.4–2 a relation between the points (p, a) and (p, b) can be given as follows. Prove that if (for some l with $1 \leqslant l \leqslant n - 1$) p_1 is such that $P_{p_1}[\bar{X} < l/n] = \alpha/2$ and p_2 is such that $P_{p_2}[\bar{X} > (n - l)/n] = \alpha/2$, then $p_2 = 1 - p_1$.

10.19 If $X_1, \ldots, X_n$ are independent Bernoulli r.v.'s with parameter p $(0 \leqslant p \leqslant 1)$, prove that $P_p[\bar{X} \geqslant c]$ increases as p increases, while $P_p[\bar{X} \leqslant c]$ decreases as p increases.

10.20 Prove that the confidence interval procedure for a success probability proposed in Section 10.4 still has probability of coverage $\geqslant 1 - \alpha$ when the points (0, 0) and (1, 1) are added as noted following Figure 10.4–4.

HONORS PROBLEMS

10.1H In Theorem 10.1.5 we saw how to construct a confidence set $S(X)$ for a parameter θ based on a r.v. X by utilizing any level α test of the hypothesis $H_0: \theta = \theta_0$ (the test being defined for all possibilities θ_0), the confidence coefficient being $1 - \alpha$. We now wish to justify the notion that "good" tests yield "good" confidence sets.

One good property we could look for in a confidence set with confidence coefficient $1 - \alpha$ would be that

$$P_\theta[\theta^* \in S(X)] \qquad (*)$$

be small for $\theta^* \neq \theta$ (small probability of including false values). **Prove that**: *If a test ϕ of level α of $H_0: \theta = \theta_0$ is UMP* [see Definition 9.3.16] *against the alternative $H(\theta_0)$, then (among all confidence sets for θ with confidence coefficient $1 - \alpha$) its related confidence set $S_\phi(X)$ minimizes $(*)$ for all $\theta \in H(\theta^*)$.*

10.2H Since the test corresponding to the upper-bound interval (10.2.9) in Example 10.2.1 has a "good" property (it is UMP one-sided), by Problem 10.1H interval (10.2.9) has a good property. State that property precisely and explicitly.

10.3H Suppose that the tests used to obtain (10.1.6) in Theorem 10.1.5 are randomized. Can we then obtain different confidence sets $S(x)$ from independent experiments which yield the same value $X = x$? [In such cases, which arise in the case of discrete r.v.'s, for example, we would often change α slightly in order to eliminate the need for randomization.]

10.4H In Example 10.2.10 $X_1, \ldots, X_n$ were independent $N(\mu, \sigma^2)$ r.v.'s with both $\mu\ (-\infty < \mu < +\infty)$ and $\sigma^2\ (\sigma^2 > 0)$ unknown, and we desired a confidence interval for μ with confidence coefficient $1 - \alpha = .95$. Such an interval was obtained [in fact, a whole class of such intervals was obtained] (see (10.2.12)) but had the undesirable property that its length was a r.v. [except in the cases $p = 0$ and $p = 1$, where the length was infinite]. We now show that this is unavoidable.

(a) For the case of known σ^2, use a sufficiency argument and Problem 10.3 to show that of all fixed-length confidence intervals with confidence coefficient $1 - \alpha$, (10.2.5) with $p = .5$ has the smallest length, namely,

$$d(\sigma) = 2\Phi^{-1}\left(1 - \frac{\alpha}{2}\right)\frac{\sigma}{\sqrt{n}}.$$

Using Lebesgue measure as the "length" of a set, extend this result to confidence sets.

(b) For the case of unknown σ^2, show there is no fixed-length confidence interval with confidence coefficient $1 - \alpha$. [*Hint*: Suppose there is such an interval, with length l (say). For each known σ we must have $l \geqslant d(\sigma)$.] Extend this result to confidence sets.

[Note that when we say random length of the confidence interval with confidence coefficient $1 - \alpha$ is "unavoidable," we mean unavoidable (as proven in (b)) if a sample size n is fixed before any experimentation. If, instead, we take a preliminary sample of n_0 observations (essentially for purposes of estimating σ^2) followed by a second sample of a random number of observations (the larger σ^2 appears to be based on the first n_0 observations, the more observations we take in this second stage) a fixed-length interval with confidence coefficient $1 - \alpha$ *can* be obtained. See C. Stein, "A two-sample test for a linear hypothesis whose power is independent of the variance," *Annals of Mathematical Statistics*, Vol. 16 (1945), pp. 243–258.]

10.5H At the end of Problem 10.12 we asked ". . . which interval . . . should be used to guide the Highway Department in its purchases of road salt for the next season?" However, the intervals under consideration were all intervals on μ, the average seasonal snowfall. For the purpose of a purchase for the next season it is more appropriate to base our decision on an interval on X, the next seasonal snowfall. Such an interval with the property

$$P[g(X_1, \ldots, X_n) \leqslant X \leqslant h(X_1, \ldots, X_n)] \equiv 1 - \alpha$$

is called a **prediction interval** for X based on $X_1, \ldots, X_n$. Find such intervals in the context of Problem 10.12. [*Hint*: See G. J. Hahn, "Additional factors for calculating prediction intervals for samples from a normal distribution," *Journal of the American Statistical Association*, Vol. 65 (1970), pp. 1668–1676.]

10.6H Let $X_1, \ldots, X_n$ be independent Poisson r.v.'s with unknown parameter $\mu \geqslant 0$. Based on the equal-tail test which rejects $H_0: \mu = \mu_0$ iff $\bar{X} < c_1$ or $\bar{X} > c_2$, find an equal-tail confidence interval for μ with confidence coefficient $\geqslant 1 - \alpha$. [*Hint*: See Problem 8.51. This study will parallel that of Section 10.4 (for a success probability) to a large degree. A reference is pp. 80–83, 227 of *Biometrika Tables for Statisticians, Volume I* (Third edition reprinted with additions) edited by E. S. Pearson and H. O. Hartley, Cambridge University Press, London, 1970.]

11

RANKING AND SELECTION PROCEDURES

In this chapter we cover the subject of "ranking and selection procedures." This subject has been developed so recently (its start is usually traced to a 1954 paper by R. E. Bechhofer) that it is as yet not covered in any but a few books on statistics (and then often only very briefly) even though its practical applicability is such that it has immediate and important applications to almost every field. (In many cases problems which are not actually problems of hypothesis testing were forced into that framework due to a lack of ranking and selection procedures, whose beginnings commenced in 1954. Since a proper treatment can now be given to such problems, it seems especially important to cover this material in courses at every level.) The "ranking and selection" problem formulation is very suggestive of decision theory (to be covered in Chapter 12), whose formulation by Abraham Wald[1] in the 1940s stimulated statistics to move more fully into its present role as "the science of decision making."

11.1. THE GENERAL RANKING PROBLEM

A ranking and selection problem usually results from questions like: "Which brand of cigarettes is least likely to cause cancer?"; "Which type of seat belt reduces car accident injuries the most?"; "Which type of atomic

[1] For historical notes and references see pp. 28–31 of A. Wald, *Statistical Decision Functions*, John Wiley & Sons, Inc., New York, 1950.

power plant increases radiation levels the least?"; or "Which type of catalyst increases a certain chemical process yield the most?" (Compare these with the corresponding questions, given in the first paragraph of Section 9.1, which led to hypothesis-testing problems.) Other questions leading to ranking and selection problems include: "Which television shows captured the highest proportions of the prime audience last night, and in what order?"; and "How do the candidates for President of the United States rank at present in popularity?"

Thus, we have several alternatives (in our first four examples these might be: brands of cigarettes; types of seat belts; types of atomic power plants; or types of catalysts), each of which has an underlying parameter θ (in our first four examples this might be: a cancer rate; an injury rate; a radiation level; or a process yield) and we wish to determine which alternative has the "best" (for example, largest or smallest in many cases) θ. This is a problem of **selection**. In our last two examples, on the other hand, we wish to order the alternatives (television shows or political candidates) based on their θ-values. This is a problem of **ranking**. [Thus, "selection" refers to choosing the best (for example, in baseball this is the goal since there is only a first prize) while "ranking" refers to one of the many other possible goals (for example, in the Olympic Games the goal may be to select the best three in order since there are three medals: gold, silver, and bronze).]

The general formulation which we use below is as follows. We have $k\ (\geqslant 2)$ sources of observations (each such source is called a **population**) denoted by $\pi_1, \pi_2, \ldots, \pi_k$. From source i we may obtain independent and identically distributed observations $X_{i1}, X_{i2}, X_{i3}, X_{i4}, \ldots$ whose distribution involves an unknown parameter θ_i $(1 \leqslant i \leqslant k)$; except for the value of θ_i the distribution is assumed not to differ from population to population. Our **goal** is to select that population[2] which has the largest parameter $\max(\theta_1, \theta_2, \ldots, \theta_k)$. We are to perform the selection in such a way that the probability $P(\mathrm{CS})$ of selecting the correct population is at least P^* (where P^* is a specified number between $1/k$ and 1; see Problem 11.1) whenever the largest and next largest of $\theta_1, \theta_2, \ldots, \theta_k$ are "sufficiently far" apart (it being usually impossible to satisfy $P(\mathrm{CS}) \geqslant P^*$ for all $\theta_1, \theta_2, \ldots, \theta_k$, since the θ's could then be arbitrarily close together); the demand that any proposed procedure satisfy this criterion is called the **probability requirement**. [Without the "probability requirement" the situation would be as trivial as a level α test with no power requirement (see Problem 9.4), where a simple flip of a coin suffices.] More general formulations can be considered (and will sometimes be noted below), but this one (with minor variations such as a goal of selecting that population which has the smallest parameter

[2] If several populations have the parameter $\max(\theta_1, \theta_2, \ldots, \theta_k)$, selection of any one of these is regarded as satisfying the goal.

$\min(\theta_1, \theta_2, \ldots, \theta_k)$) accounts for more than 75 percent of the work in this area to date.

Let us now illustrate this general formulation in terms of a specific question which arises[3] in poultry science: "Which stock of chickens lays the most eggs?" (New chicken stocks are under constant development, and the question of selecting the best is considered at agricultural experiment stations year after year. Similar applications arise in other fields.) So, suppose it is desired to select that one of $k = 10$ stocks of chickens which has the highest mean egg production. As would be expected, we take chickens of each type (n of each type, say), observe the egg laying of the $kn = 10n$ chickens for a period of time [this happens to be 500 days; at that age a pullet is eliminated] in a controlled environment, and select a stock (population) based on the observations obtained. If (as is often assumed) the observations from π_i ($X_{i1}, X_{i2}, \ldots, X_{in}$) are each $N(\mu_i, \sigma^2)$ r.v.'s with $\sigma = 72$, we select that stock (population) with the highest average (over its n pullets) of eggs-per-500-days in the trial. If we desire $P(\text{CS}) \geqslant P^* = .90$, say, whenever the best-laying stock out-lays the next-best-laying stock by 10 or more eggs per chicken in 500 days on the average [that is, whenever

$$\mu_{[10]} - \mu_{[9]} \geqslant \delta^* = 10 \tag{11.1.1}$$

where $\mu_{[1]} \leqslant \cdots \leqslant \mu_{[9]} \leqslant \mu_{[10]}$ denote $\mu_1, \ldots, \mu_{10}$ in numerical order], it can be shown [and will be in Section 11.2] that $n = 462$ pullets of each type will be required.

11.2. SELECTION OF THE BEST (LARGEST MEAN) NORMAL POPULATION

One selection problem (see Section 11.1) which occurs frequently in practice is that in which the $k \geqslant 2$ **populations** $\pi_1, \ldots, \pi_k$ (sources of observations) are such that if $X_{i1}, X_{i2}, \ldots$ are n independent **observations** from π_i, then each X_{ij} is $N(\mu_i, \sigma^2)$ where μ_i is unknown and σ^2 is known ($1 \leqslant i \leqslant k$). Let

$$\mu_{[1]} \leqslant \cdots \leqslant \mu_{[k]} \tag{11.2.1}$$

denote the ordered values of $\mu_1, \ldots, \mu_k$. Our **goal** is to select a population with mean $\mu_{[k]}$. If we formulate a procedure for selecting one of the k populations and declaring the selected population to be "best," we would like a reasonable probability of having it be, in fact, best.

[3] See W. A. Becker, "Comparing entries in random sample tests," *Poultry Science*, Vol. 40 (1961), pp. 1507–1514. Also see E. J. Dudewicz, "Selection procedures: an introduction for practitioners," *Estadística, Journal of the Inter-American Statistical Institute*, Vol. 27 (1969), pp. 377–382.

One intuitively reasonable procedure is to proceed as follows. Take independent samples of n independent observations from each of the k populations, use the observations from π_i to estimate μ_i, say by $\hat{\mu}_i$ ($1 \leqslant i \leqslant k$), and select the[4] population π_l for which

$$\hat{\mu}_l = \max(\hat{\mu}_1, \ldots, \hat{\mu}_k). \tag{11.2.2}$$

Since (see Example 8.8.20)

$$\bar{X}_i = \frac{X_{i1} + \cdots + X_{in}}{n} \tag{11.2.3}$$

is a sufficient statistic for μ_i, it seems reasonable (see Problem 11.1H) to let

$$\hat{\mu}_i = \bar{X}_i \quad (1 \leqslant i \leqslant k). \tag{11.2.4}$$

Thus, our **procedure** (called the **means procedure**) is to take k independent sets of n independent observations each (n observations per population), compute the sample means, and select (as being best) that population which yielded the largest sample mean.

We now wish to choose n (the sample size per population) in a rational way. One such way is to choose the smallest n which satisfies a **probability requirement** such as

$$P(\text{CS}) \geqslant P^* \quad \text{whenever} \quad \mu_{[k]} - \mu_{[k-1]} \geqslant \delta^* \tag{11.2.5}$$

where "CS" (for correct selection) denotes the final selection of a population with population mean $\mu_{[k]}$, and P^* ($1/k < P^* < 1$; see Problem 11.1) and δ^* ($\delta^* > 0$) are specified by the experimenter. With probability requirement (11.2.5) we will be at least $100P^*$ percent sure (95 percent sure when $P^* = .95$) of selecting the best population whenever the best mean $\mu_{[k]}$ is at least δ^* better than the second best mean $\mu_{[k-1]}$.

In order to solve for the smallest n which satisfies (11.2.5), as we do in Theorem 11.2.11, it will be convenient to make the following:

Definition 11.2.6. *$\bar{\mathbf{X}}_{(i)}$ is the sample mean obtained from the population which has population mean $\mu_{[i]}$ ($1 \leqslant i \leqslant k$), while $\bar{\mathbf{X}}_{[1]} \leqslant \cdots \leqslant \bar{\mathbf{X}}_{[k]}$ denote the sample means $\bar{X}_1, \ldots, \bar{X}_k$ in numerical order.*

Theorem 11.2.7. *The P(CS) of the means procedure equals*

$$\int_{-\infty}^{\infty} \Phi\left(y + \frac{\mu_{[k]} - \mu_{[1]}}{\sigma/\sqrt{n}}\right) \cdots \Phi\left(y + \frac{\mu_{[k]} - \mu_{[k-1]}}{\sigma/\sqrt{n}}\right) \phi(y)\, dy. \tag{11.2.8}$$

[4] In general, this could fail to be unique, although below we will see that in this section it is unique with probability 1 (see Problem 11.2).

Proof: The $P(\text{CS})$ is the probability that the sample mean ($\bar{X}_{(k)}$) from the population with the largest mean $\mu_{[k]}$ is the largest sample mean ($\bar{X}_{[k]}$):

$$\begin{aligned}
P(\text{CS}) &= P[\bar{X}_{(k)} = \bar{X}_{[k]}] \\
&= P[\bar{X}_{(k)} = \max(\bar{X}_{(1)}, \ldots, \bar{X}_{(k)})] \\
&= P[\bar{X}_{(1)} < \bar{X}_{(k)}, \ldots, \bar{X}_{(k-1)} < \bar{X}_{(k)}] \\
&= P[\bar{X}_{(i)} < \bar{X}_{(k)}, \quad i = 1, \ldots, k-1] \\
&= P\left[\frac{\bar{X}_{(i)} - \mu_{[i]}}{\sigma/\sqrt{n}} < \frac{\bar{X}_{(k)} - \mu_{[k]}}{\sigma/\sqrt{n}} + \frac{\mu_{[k]} - \mu_{[i]}}{\sigma/\sqrt{n}}, \quad i = 1, \ldots, k-1\right] \qquad (11.2.9) \\
&= P\left[Y_i < Y_k + \frac{\mu_{[k]} - \mu_{[i]}}{\sigma/\sqrt{n}}, \quad i = 1, \ldots, k-1\right],
\end{aligned}$$

where $Y_1, \ldots, Y_k$ are independent $N(0, 1)$ r.v.'s. [Here we used Problem 8.33 or Theorem 6.1.3 to show $\bar{X}_{(i)}$ is $N(\mu_{[i]}, \sigma^2/n)$. The $Y_1, \ldots, Y_k$ are independent since the $\bar{X}_{(1)}, \ldots, \bar{X}_{(k)}$ are.] Thus (since Y_k is a $N(0, 1)$ r.v. and $Y_1, \ldots, Y_{k-1}, Y_k$ are independent r.v.'s and $\phi(y) = (\sqrt{2\pi})^{-1} \exp(-.5y^2)$)

$$\begin{aligned}
P(\text{CS}) &= \int_{-\infty}^{\infty} P\left[Y_i < y + \frac{\mu_{[k]} - \mu_{[i]}}{\sigma/\sqrt{n}}, \quad i = 1, \ldots, k-1\right] \phi(y)\, dy \\
&= \int_{-\infty}^{\infty} P\left[Y_1 < y + \frac{\mu_{[k]} - \mu_{[1]}}{\sigma/\sqrt{n}}\right] \cdots P\left[Y_{k-1} < y + \frac{\mu_{[k]} - \mu_{[k-1]}}{\sigma/\sqrt{n}}\right] \phi(y)\, dy \qquad (11.2.10) \\
&= \int_{-\infty}^{\infty} \Phi\left(y + \frac{\mu_{[k]} - \mu_{[1]}}{\sigma/\sqrt{n}}\right) \cdots \Phi\left(y + \frac{\mu_{[k]} - \mu_{[k-1]}}{\sigma/\sqrt{n}}\right) \phi(y)\, dy.
\end{aligned}$$

Theorem 11.2.11. *The $P(\text{CS})$ of the means procedure is (among all sets of possible population means $\mu_1, \ldots, \mu_k$ which satisfy $\mu_{[k]} - \mu_{[k-1]} \geqslant \delta^*$) minimized when*

$$\mu_{[1]} = \cdots = \mu_{[k-1]} = \mu_{[k]} - \delta^* \qquad (11.2.12)$$

[this is called the **least favorable configuration (LFC)** of the population means]. *In the LFC, the $P(\text{CS})$, say $P_{\text{LFC}}(\text{CS})$, is given by*

$$P_{\text{LFC}}(\text{CS}) = \int_{-\infty}^{\infty} \Phi^{k-1}\left(x + \frac{\delta^*\sqrt{n}}{\sigma}\right) \phi(x)\, dx. \qquad (11.2.13)$$

This $P_{\text{LFC}}(\text{CS})$ is an increasing function of n and $\to 1$ as $n \to \infty$. Hence, there *is a unique smallest sample size n per population for which probability requirement* (11.2.5) *is satisfied:*

$$n = \left\{ \text{The smallest integer} \geqslant \frac{h_k^2(P^*)\sigma^2}{(\delta^*)^2} \right\}, \tag{11.2.14}$$

where $h_k(P^*)$ is the solution h of the equation

$$\int_{-\infty}^{\infty} \Phi^{k-1}(x+h)\phi(x)\,dx = P^*. \tag{11.2.15}$$

Proof: Since $\Phi(z)$ is an increasing function of z, by expression (11.2.8) for the $P(\text{CS})$ it follows that the $P(\text{CS})\downarrow$ as $\mu_{[k]} - \mu_{[i]}\downarrow$, for each of $i = 1, \ldots, k-1$. Subject to the restriction $\mu_{[k]} - \mu_{[k-1]} \geqslant \delta^*$, we make $\mu_{[k]} - \mu_{[i]}$ smallest when we set

$$\mu_{[1]} = \cdots = \mu_{[k-1]} = \mu_{[k]} - \delta^*.$$

Hence, the $P(\text{CS})$ is minimized (subject to $\mu_{[k]} - \mu_{[k-1]} \geqslant \delta^*$) when (11.2.12) holds. But when (11.2.12) holds, the $P(\text{CS})$ given by (11.2.8) reduces simply to

$$f(n) = \int_{-\infty}^{\infty} \Phi^{k-1}\left(x + \frac{\delta^*\sqrt{n}}{\sigma}\right)\phi(x)\,dx. \tag{11.2.16}$$

This is clearly an increasing function of n with $f(0) = 1/k$ (see Problem 11.3) and $f(n) \to 1$ as $n \to \infty$. In order to make the $P(\text{CS}) \geqslant P^*$ whenever $\mu_{[k]} - \mu_{[k-1]} \geqslant \delta^*$ we can proceed as follows. Find the smallest $P(\text{CS})$ can be when $\mu_{[k]} - \mu_{[k-1]} \geqslant \delta^*$ (this is given by (11.2.16)) and make n large enough that this minimum is $\geqslant P^*$. Thus, we set

$$\int_{-\infty}^{\infty} \Phi^{k-1}(x+h)\phi(x)\,dx = P^*, \tag{11.2.17}$$

solve for h, and then (since $\delta^*\sqrt{n}/\sigma$ is h) solve for n. However, n must be an integer (while $h^2\sigma^2/(\delta^*)^2$ need not be) so we may need to round up the quantity $h^2\sigma^2/(\delta^*)^2$ [for example, 5.46 becomes 6].

Remark 11.2.18. Tables of the solution $\mathbf{h_k(P^*)}$ of Equation (11.2.15) (for various k and P^* values) were given by R. E. Bechhofer in his pioneering paper "A single-sample multiple decision procedure for ranking means of normal populations with known variances," *Annals of Mathematical Statistics*, Vol. 25 (1954), pp. 16–39. [In fact, that paper also discussed, and gave tables for, much more general goals than simply selecting the best population. An indication of such other goals was given in Section 11.1 above.] These tables cover $k = 2(1)10$ and are given below in Table 11.2–1. (For $k > 10$ see Problem 11.5 and Remark 11.2.19.)

Remark 11.2.19. An approximation to the solution $\mathbf{h_k(P^*)}$ of Equation (11.2.15) is of interest because: (1) sometimes the k and P^* of interest to us

Table 11.2–1. Solutions h of $\int_{-\infty}^{\infty} \Phi^{k-1}(x + h)\phi(x)\, dx = P^*$

P^* \ k	2	3	4	5	6	7	8	9	10
.9995	4.6535	4.9163	5.0639	5.1661	5.2439	5.3066	5.3590	5.4039	5.4432
.9990	4.3703	4.6450	4.7987	4.9049	4.9856	5.0505	5.1047	5.1511	5.1917
.9950	3.6428	3.9517	4.1224	4.2394	4.3280	4.3989	4.4579	4.5083	4.5523
.99	3.2900	3.6173	3.7970	3.9196	4.0121	4.0861	4.1475	4.1999	4.2456
.98	2.9045	3.2533	3.4432	3.5722	3.6692	3.7466	3.8107	3.8653	3.9128
.97	2.6598	3.0232	3.2198	3.3529	3.4528	3.5324	3.5982	3.6543	3.7030
.96	2.4759	2.8504	3.0522	3.1885	3.2906	3.3719	3.4390	3.4961	3.5457
.95	2.3262	2.7101	2.9162	3.0552	3.1591	3.2417	3.3099	3.3679	3.4182
.94	2.1988	2.5909	2.8007	2.9419	3.0474	3.1311	3.2002	3.2590	3.3099
.93	2.0871	2.4865	2.6996	2.8428	2.9496	3.0344	3.1043	3.1637	3.2152
.92	1.9871	2.3931	2.6092	2.7542	2.8623	2.9479	3.0186	3.0785	3.1305
.91	1.8961	2.3082	2.5271	2.6737	2.7829	2.8694	2.9407	3.0012	3.0536
.90	1.8124	2.2302	2.4516	2.5997	2.7100	2.7972	2.8691	2.9301	2.9829
.88	1.6617	2.0899	2.3159	2.4668	2.5789	2.6676	2.7406	2.8024	2.8560
.86	1.5278	1.9655	2.1956	2.3489	2.4627	2.5527	2.6266	2.6893	2.7434
.84	1.4064	1.8527	2.0867	2.2423	2.3576	2.4486	2.5235	2.5868	2.6416
.82	1.2945	1.7490	1.9865	2.1441	2.2609	2.3530	2.4286	2.4926	2.5479
.80	1.1902	1.6524	1.8932	2.0528	2.1709	2.2639	2.3403	2.4049	2.4608
.75	.9539	1.4338	1.6822	1.8463	1.9674	2.0626	2.1407	2.2067	2.2637
.70	.7416	1.2380	1.4933	1.6614	1.7852	1.8824	1.9621	2.0293	2.0873
.65	.5449	1.0568	1.3186	1.4905	1.6168	1.7159	1.7970	1.8653	1.9242
.60	.3583	.8852	1.1532	1.3287	1.4575	1.5583	1.6407	1.7102	1.7700
.55	.1777	.7194	.9936	1.1726	1.3037	1.4062	1.4899	1.5604	1.6210
.50	—	.5565	.8368	1.0193	1.1526	1.2568	1.3418	1.4133	1.4748
.45	—	.3939	.6803	.8662	1.0019	1.1078	1.1941	1.2666	1.3289
.40	—	.2289	.5215	.7111	.8491	.9567	1.0443	1.1178	1.1810
.35	—	.0585	.3578	.5510	.6915	.8008	.8897	.9643	1.0284
.30	—	—	.1855	.3827	.5257	.6369	.7272	.8030	.8679
.25	—	—	—	.2014	.3472	.4604	.5523	.6292	.6951
.20	—	—	—	—	.1489	.2643	.3579	.4361	.5032
.15	—	—	—	—	—	.0364	.1319	.2117	.2800

* From the *Annals of Mathematical Statistics*, Vol. 25 (1959). By permission of The Institute of Mathematical Statistics.

will not be ones for which $h_k(P^*)$ is tabled[5]; (2) the tables may not be readily available when needed; and (3) in some circumstances a quick method of approximation will be desired. A number of approximations have been given in the literature. One which has been found[6] to be among the better ones is

$$h_k(P^*) \approx 2\sqrt{-\ln(1 - P^*)} \qquad (11.2.20)$$

or (since $\delta^* \sqrt{n}/\sigma$ is h)

$$n \approx n_1 = \frac{-4\sigma^2 \ln(1 - P^*)}{(\delta^*)^2}. \qquad (11.2.21)$$

This approximation n_1 has[7] the property that as $P^* \to 1$, $n_1/n \to 1$. Note that this approximation can be applied with the aid of only a slide rule, and allows us to readily assess approximately the effects (upon the sample size n required per population) of changes in δ^*, σ^2, and P^*.

We close this section by giving a numerical example of the use of the selection procedure developed above.

Example 11.2.22. An automobile safety restraint system (SRS) is a system designed to reduce injuries to passengers during crashes. Suppose that when such a system is built and tested, sensors in the test car (for example, in a mannequin) measure an index of cranial injury (for example, due to force applied to the skull and duration of the force), and that this measurement is subject to chance (random) variation which makes it a $N(\mu, \sigma^2)$ r.v.; here μ is unknown and varies from one SRS to another, while $\sigma^2 = 5$ is known and does not vary from one SRS to another (for example, it may depend primarily on the measuring process). If we have $k = 5$ SRS's and wish to select the best (here the one with the smallest index of cranial injury, $\mu_{[1]}$) with probability at least $P^* = .90$ whenever $\mu_{[2]} - \mu_{[1]} \geqslant \delta^* = 1.0$, how many observations [here test crashes under some standard conditions of speed, attitude, and so on] per population do we need? If our budget restricts us to $n = 25$ observations per population (since this is a very costly experiment)

---

[5] In addition to the tables noted in Remark 11.2.18 and Problem 11.5, others have been given by S. S. Gupta for $k = 2(1)51$ and $P^* = .75, .90, .95, .975, .99$ (see p. 810 of "Probability integrals of multivariate normal and multivariate t," *Annals of Mathematical Statistics*, Vol. 34 (1963), pp. 792–828; note that on p. 810 his n is our $k - 1$, his α is our $1 - P^*$, and his tabled value is $h_k(P^*)/\sqrt{2}$). Also, D. Teichroew tabled P^* as a function of h for $k = 2(1)10$ and $h/\sqrt{2} = .00(.01)6.09$ ("Probabilities associated with order statistics in samples from two normal populations with equal variance," *ENASR No. ES-3*, Chemical Corps Engineering Agency, Engineering Statistics Unit, Army Chemical Center, Maryland, December 7, 1955; note that his tables with his $r = 1$ are the ones appropriate here).

[6] See E. J. Dudewicz and N. A. Zaino, Jr., "Sample size for selection," *Statistical Decision Theory and Related Topics*, eds. S. S. Gupta and J. Yackel, Academic Press Inc., New York, 1971, pp. 363–376.

[7] See E. J. Dudewicz, "An approximation to the sample size in selection problems," *Annals of Mathematical Statistics*, Vol. 40 (1969), pp. 492–497.

is the experiment worth conducting at all? [Or should we perhaps arbitrarily select a few of the $k = 5$ alternatives and test them more extensively, perhaps retaining the deleted alternatives for testing in the future as funds become available?]

First, it is simple to show (by replacing each observation X by $-X$) that this problem is equivalent to selecting the population with the largest mean (largest of $-\mu_1, \ldots, -\mu_k$) (see Problem 11.7). Thus, from Table 11.2–1 we find our $h_k(P^*)$ is $h_5(.90) = 2.60$; hence, n must be the smallest integer greater than or equal to

$$\frac{h^2\sigma^2}{(\delta^*)^2} = \frac{(2.60)^2(5)}{(1.0)^2} = 33.80, \tag{11.2.23}$$

namely $n = 34$.

If we are restricted to $n = 25$, we will have an $h_5(P^*)$ equal to

$$\frac{\delta^*\sqrt{n}}{\sigma} = \frac{(1.0)(5)}{\sqrt{5}} = \sqrt{5} = 2.24. \tag{11.2.24}$$

From Table 11.2–1 we see that with $k = 5$ and $h = 2.24$ we have a P^* of about .84. Hence, the experiment would seem to be worth conducting (since $P^* = .84$ is much better than random choice, which yields $P^* = .20$ in the case $k = 5$).

We might also ask "What is the smallest δ^* for which $n = 25$ will guarantee probability $P^* = .90$?" This is the δ^* for which

$$\frac{\delta^*\sqrt{n}}{\sigma} = 2.60, \tag{11.2.25}$$

namely $\delta^* = (2.60)(\sqrt{5})/5 = 1.16$, and does not differ greatly from 1.0.

11.3. SELECTION OF THE BEST (SMALLEST VARIANCE) NORMAL POPULATION

In Section 11.2 we studied one selection problem which occurs frequently in practice, selection of the population with the largest mean. **In the present section we study the selection of the population with the least variability (as measured by variance).** For example, if we have available several measuring instruments such as scales, the most precise one is that with the least variance. As a second example, if several sources of supply are available for parts with the same mean, the source with the least variability will be the most desirable one (since its parts will meet the specifications the largest percent of the time). As a third example, if light bulbs produced by several processes have approximately the same average life, the process which produces bulbs with the least variability (variance) of life may be the most desirable in terms of customer satisfaction with the product.

We assume here that the $k \geqslant 2$ **populations** $\pi_1, \ldots, \pi_k$ (sources of observations) are such that if $X_{i1}, X_{i2}, \ldots$ are n independent **observations** from π_i, then each X_{ij} is $N(\mu_i, \sigma_i^2)$ where μ_i is unknown and σ_i^2 is unknown $(1 \leqslant i \leqslant k)$. Let

$$\sigma_{[1]}^2 \leqslant \cdots \leqslant \sigma_{[k]}^2 \tag{11.3.1}$$

denote the ordered values of $\sigma_1^2, \ldots, \sigma_k^2$. Our **goal** is to select a population with variance $\sigma_{[1]}^2$. If we formulate a procedure for selecting one of the k populations and declaring the selected population to be "best," we would (as in Section 11.2 in a different context) like a reasonable probability of having it be, in fact, best.

One intuitively reasonable procedure is to proceed as follows. Take independent samples of n independent observations from each of the k populations, use the observations from π_i to estimate σ_i^2, say by $\hat{\sigma}_i^2$ $(1 \leqslant i \leqslant k)$, and select the[8] population π_l for which

$$\hat{\sigma}_l^2 = \min(\hat{\sigma}_1^2, \ldots, \hat{\sigma}_k^2). \tag{11.3.2}$$

Since (see Problems 8.34 and 8.56), for $n \geqslant 2$,

$$s_i^2 = \frac{\sum_{j=1}^{n} (X_{ij} - \bar{X}_i)^2}{n-1} \quad \text{with} \quad \bar{X}_i = \frac{X_{i1} + \cdots + X_{in}}{n} \tag{11.3.3}$$

is a "good" unbiased estimator of σ_i^2 based on a minimal sufficient statistic for (μ_i, σ_i^2), it seems reasonable (see Problem 11.4H) to let

$$\hat{\sigma}_i^2 = s_i^2 \quad (1 \leqslant i \leqslant k). \tag{11.3.4}$$

Thus, our **procedure** is to take k independent sets of n independent observations each ($n \geqslant 2$ observations per population), compute s_i^2 $(1 \leqslant i \leqslant k)$, and select (as being best) that population which yielded the smallest of $s_1^2, \ldots, s_k^2$.

We now wish to choose n (the sample size per population) in a rational way. One such way is to choose the smallest n which satisfies a **probability requirement** such as

$$P(\text{CS}) \geqslant P^* \quad \text{whenever} \quad \sigma_{[2]}^2 \geqslant \theta^* \sigma_{[1]}^2, \tag{11.3.5}$$

where **CS** (for **c**orrect **s**election) denotes the final selection of a population with population variance $\sigma_{[1]}^2$, and P^* $(1/k < P^* < 1$; see Problem 11.10) and θ^* $(\theta^* > 1)$ are specified by the experimenter. With probability requirement (11.3.5) we will be at least $100P^*$ percent sure (90 percent sure when

[8] In general, this could fail to be unique, although below we will see that in this section it is unique with probability 1 (see Problem 11.9).

$P^* = .90$) of selecting the best population whenever the best variance $\sigma^2_{[1]}$ is exceeded by the second best variance $\sigma^2_{[2]}$ by at least $100(\theta^* - 1)$ percent.

In order to solve for the smallest n which satisfies (11.3.5), as we do in Theorem 11.3.11, it will be convenient to make the following:

Definition 11.3.6. $\mathbf{s}^2_{(i)}$ *is that one of* $s^2_1, \ldots, s^2_k$ *obtained from the population with population variance* $\sigma^2_{[i]}$ $(1 \leqslant i \leqslant k)$, *while* $\mathbf{s}^2_{[1]} \leqslant \cdots \leqslant \mathbf{s}^2_{[k]}$ *denote* $s^2_1, \ldots, s^2_k$ *in numerical order.*

Theorem 11.3.7. *The* $P(\mathrm{CS})$ *of the selection procedure equals*

$$\int_0^\infty \left[1 - G_{n-1}\left(y\frac{\sigma^2_{[1]}}{\sigma^2_{[2]}}\right)\right] \cdots \left[1 - G_{n-1}\left(y\frac{\sigma^2_{[1]}}{\sigma^2_{[k]}}\right)\right] g_{n-1}(y)\,dy, \qquad (11.3.8)$$

where $G_{n-1}(\cdot)$ *and* $g_{n-1}(\cdot)$ *are the d.f. and p.d.f. of a* $\chi^2_{n-1}(0)$ *r.v., respectively.*

Proof: The $P(\mathrm{CS})$ is the probability that the one of $s^2_1, \ldots, s^2_k$ (namely, $s^2_{(1)}$) from the population with the smallest variance $\sigma^2_{[1]}$ is the smallest of $s^2_1, \ldots, s^2_k$ (namely $s^2_{[1]}$):

$$\begin{aligned} P(\mathrm{CS}) &= P[s^2_{(1)} = s^2_{[1]}] \\ &= P[s^2_{(1)} = \min(s^2_{(1)}, \ldots, s^2_{(k)})] \\ &= P[s^2_{(1)} < s^2_{(2)}, \ldots, s^2_{(1)} < s^2_{(k)}] \\ &= P[s^2_{(i)} > s^2_{(1)}, \quad i = 2, \ldots, k] \\ &= P\left[\frac{s^2_{(i)}}{\sigma^2_{[i]}} \geqslant \frac{s^2_{(1)}}{\sigma^2_{[1]}}\frac{\sigma^2_{[1]}}{\sigma^2_{[i]}}, \quad i = 2, \ldots, k\right] \\ &= P\left[Y_i > Y_1\frac{\sigma^2_{[1]}}{\sigma^2_{[i]}}, \quad i = 2, \ldots, k\right], \end{aligned} \qquad (11.3.9)$$

where $Y_1, \ldots, Y_k$ are independent $\chi^2_{n-1}(0)$ r.v.'s. [Here we used the proof of Lemma 9.5.15 to show $s^2_{(i)}/\sigma^2_{[i]}$ is $\chi^2_{n-1}(0)$. The $Y_1, \ldots, Y_k$ are independent since the $s^2_{(1)}, \ldots, s^2_{(k)}$ are.] Thus,

$$\begin{aligned} P(\mathrm{CS}) &= \int_{-\infty}^\infty P\left[Y_i > y\frac{\sigma^2_{[1]}}{\sigma^2_{[i]}}, \quad i = 2, \ldots, k\right] g_{n-1}(y)\,dy \\ &= \int_0^\infty P\left[Y_2 > y\frac{\sigma^2_{[1]}}{\sigma^2_{[2]}}\right] \cdots P\left[Y_k > y\frac{\sigma^2_{[1]}}{\sigma^2_{[k]}}\right] g_{n-1}(y)\,dy \\ &= \int_0^\infty \left[1 - G_{n-1}\left(y\frac{\sigma^2_{[1]}}{\sigma^2_{[2]}}\right)\right] \cdots \\ &\qquad \left[1 - G_{n-1}\left(y\frac{\sigma^2_{[1]}}{\sigma^2_{[k]}}\right)\right] g_{n-1}(y)\,dy. \end{aligned} \qquad (11.3.10)$$

Theorem 11.3.11. *The* $P(\mathrm{CS})$ *of the selection procedure is (among all sets of possible population variances* $\sigma_1^2, \ldots, \sigma_k^2$ *which satisfy* $\sigma_{[2]}^2 \geqslant \theta^* \sigma_{[1]}^2$*) minimized when*

$$\theta^* \sigma_{[1]}^2 = \sigma_{[2]}^2 = \cdots = \sigma_{[k]}^2 \tag{11.3.12}$$

[this is called the **least favorable configuration (LFC)** of the population variances]. *In the* LFC, *the* $P(\mathrm{CS})$, *say* $P_{\mathrm{LFC}}(\mathrm{CS})$, *is given by*

$$P_{\mathrm{LFC}}(\mathrm{CS}) = \int_0^\infty \left[1 - G_{n-1}\left(\frac{y}{\theta^*}\right)\right]^{k-1} g_{n-1}(y)\, dy. \tag{11.3.13}$$

Proof: Since $G_{n-1}(z)$ is an increasing function of z, by expression (11.3.8) for the $P(\mathrm{CS})$ it follows that the $P(\mathrm{CS})\downarrow$ as $\sigma_{[1]}^2/\sigma_{[i]}^2\uparrow$, for each of $i = 2, \ldots, k$. Subject to the restriction $\sigma_{[2]}^2 \geqslant \theta^* \sigma_{[1]}^2$, we make $\sigma_{[1]}^2/\sigma_{[i]}^2$ largest when we set

$$\theta^* \sigma_{[1]}^2 = \sigma_{[2]}^2 = \cdots = \sigma_{[k]}^2.$$

Hence, the $P(\mathrm{CS})$ is minimized (subject to $\sigma_{[2]}^2 \geqslant \theta^* \sigma_{[1]}^2$) when (11.3.12) holds. But when (11.3.12) holds, the $P(\mathrm{CS})$ given by (11.3.8) reduces simply to (11.3.13).

Remark 11.3.14. Limited tables of the $P_{\mathrm{LFC}}(\mathrm{CS})$ of Equation (11.3.13) were given by R. E. Bechhofer and M. Sobel ["A single-sample multiple decision procedure for ranking variances of normal populations," *Annals of Mathematical Statistics*, Vol. 25 (1954), pp. 273–289]. Those tables cover the cases $n - 1 = 1(1)20$ and $\theta^* = 1.2(.2)2.2$ for $k = 2, 3, 4$. (An approximation is suggested for other cases.)

11.4. EXTENSIONS AND GENERALIZATIONS

In Section 11.1 we discussed what the concept of a ranking problem is, and in Sections 11.2 and 11.3 we discussed two important such problems (selection of the largest mean normal population and selection of the smallest variance normal population, respectively). Our discussion, while probably by far the most extensive yet available in any textbook, is necessarily only an introduction to what is by now a vast area[9] in which important new results

[9] Extensive references are available (for work through mid-1967) in R. E. Bechhofer, J. Kiefer, and M. Sobel, *Sequential Identification and Ranking Procedures, With Special Reference to Koopman-Darmois Populations*, University of Chicago Press, Chicago, Illinois, 1968. Another source is E. J. Dudewicz, "A categorized bibliography on multiple-decision (ranking and selection) procedures," unpublished bibliography, Department of Statistics, The University of Rochester, Rochester, New York, 1968. A third good source is the crisp and clear expository article by D. R. Barr and M. H. Rizvi, "An introduction to ranking and selection procedures," *Journal of the American Statistical Association*, Vol. 61 (1966), pp. 640–646.

are appearing each year. **In this section we attempt to note very briefly (and with appropriate references) some ways in which the results of Sections 11.2 and 11.3 have been and are being extended and generalized.** Some references to published accounts of applications in various fields are given throughout the chapter, and the student should be able to find additional applications in most fields of application.

The essential problem formulation given in Section 11.1 is that we have **populations** (sources of observations) $\pi_1, \ldots, \pi_k$ $(k \geqslant 2)$. From source i we obtain observations $X_{i1}, X_{i2}, \ldots$ which are independent and identically distributed r.v.'s, but whose d.f. contains an unknown parameter θ_i $(1 \leqslant i \leqslant k)$. We have a **goal** (for example, select the population associated with $\theta_{[k]} = \max(\theta_1, \ldots, \theta_k)$) and **probability requirement** (for example, $P(\text{CS}) \geqslant P^*$ whenever $\theta_{[k]} - \theta_{[k-1]} \geqslant \delta^*$, where $1/k < P^* < 1$ and $\delta^* > 0$), and desire a **procedure** which satisfies the probability requirement (for example, select the population yielding $\bar{X}_{[k]} = \max(\bar{X}_1, \ldots, \bar{X}_k)$, the largest of the sample means based on n observations). Then:

1. Section 11.2 discussed the case where for $i = 1, \ldots, k$ the observations from π_i are $N(\mu_i, \sigma^2)$ with σ^2 known (for example, $\sigma = 72$ eggs in 500 days) and $\theta_i = \mu_i$. In practice, often σ^2 is not known exactly. In that case a procedure is available[10] which proceeds in two stages; first it takes a few (n_0, say) observations from each of $\pi_1, \ldots, \pi_k$ and uses them to estimate σ^2, then it uses this estimate of σ^2 in order to determine how many additional observations to take from each population.
2. In Sections 11.2 and 11.3 it was assumed that the number of observations available from each of the k populations $\pi_1, \ldots, \pi_k$ would be the same; that is, if n_i denotes the number of observations to be taken from π_i $(1 \leqslant i \leqslant k)$ it was assumed we would take $n_1 = n_2 = \cdots = n_k = n$ (say). Indeed, in light of Problem 11.1H it seems reasonable to take the same sample size n from each population. However, sometimes, either because past observations are available for some of the populations, or because of the loss of some observations (for example, in the example of Section 11.1, some of the pullets may die before termination at 500 days), we will in fact have unequal sample sizes from the populations. This difficult problem has received little attention even in the basic cases of Sections 11.2 and 11.3 (see Problems 11.6H and 11.7H). Our intuitive feeling, that "acting as if we actually have the same sample size $n = \min(n_1, \ldots, n_k)$ from each population should underestimate the true $P(\text{CS})$," may not be true in all cases. Even if this feeling is proven correct

---

[10] See R. E. Bechhofer, C. W. Dunnett, and M. Sobel, "A two-sample multiple decision procedure for ranking means of normal populations with a common unknown variance," *Biometrika*, Vol. 41 (1954), pp. 170–176.

in some cases, however, we would still like to make use of the additional $n_1 - n, \ldots, n_k - n$ observations from $\pi_1, \ldots, \pi_k$ to make a stronger statement about the P(CS) than the statement which uses $n = \min(n_1, \ldots, n_k)$ only (as well as basing the decision on all available observations). Recently these problems have been receiving increasing attention.[11]

3. A variant of the problem of Section 11.2 is that in which we assume that the variance of observations from π_i is σ_i^2 $(1 \leqslant i \leqslant k)$ with the σ_i^2's *known* but *unequal*. We then ask: "If I have a total sample size n at my disposal, how should I allocate it among the k populations so as to do "best" in my P(CS)?" This has been studied recently (see Problems 11.14 and 11.15) but is far from solved.

4. Another variant of the problem of Section 11.2 is that in which we assume that the variance of observations from π_i is σ_i^2 $(1 \leqslant i \leqslant k)$ with the σ_i^2's *unknown* as well as *unequal*. It is known[12] that no procedure which requires only a single stage[13] of observations can satisfy the probability requirement in this case. A two-stage procedure has recently been given [see the paper referenced in part (c) of Problem 11.14], and also holds for subset selection formulations of the problem (see the next paragraph).

5. The approach to the selection problem taken in Sections 11.2 and 11.3 is called the "**indifference zone**" **approach** (because we specify we must have $P(\text{CS}) \geqslant P^*$ whenever $\mu_{[k]} - \mu_{[k-1]} \geqslant \delta^*$, but are apparently "indifferent" when $\mu_{[k]} - \mu_{[k-1]} < \delta^*$). This approach is relatively reasonable for small to moderate k (say $k \leqslant 15$). For moderate to large k (say $k \geqslant 100$) it requires more observations than we can afford, many of these observations being on inferior populations. (For example, if we have $k = 10{,}000$ possible cancer drugs, we may wish simply to select the few best for further extensive testing.) For such problems Gupta[14] pioneered an approach called **subset selection**. Here we desire to select a subset S of $\Pi = \{\pi_1, \ldots, \pi_k\}$ such that the probability that the best population (the one with mean $\mu_{[k]}$ in Section 11.2; the one with variance $\sigma_{[1]}^2$ in Section 11.3) is in S is $\geqslant P^*$. The subset size and composition are random, and we desire a rule which makes the number of populations

---

[11] See M. Sitek, "Application of the selection procedure R to unequal observation numbers," *Zastosowania Matematyki*, Vol. 12 (1972), pp. 355–371; and E. J. Dudewicz, "A note on selection procedures with unequal observation numbers," *Zastosowania Matematyki*, Vol. 14 (1974), pp. 31–35.

[12] See E. J. Dudewicz, "Nonexistence of a single-sample selection procedure whose P(CS) is independent of the variances," *South African Statistical Journal*, Vol. 5 (1971), pp. 37–39. Also see Section 4 of the paper referenced in part (c) of Problem 11.14 for a slight extension of this result.

[13] See problem "1" on page 353 for the notion of stages of observation.

[14] S. S. Gupta, *On a Decision Rule for a Problem in Ranking Means*, Ph.D. thesis, Department of Statistics, University of North Carolina, Chapel Hill, North Carolina, 1956.

in S ($\#(S)$) "small" (for example, $E(\#(S)) \leqslant 1 + \varepsilon$ for some $\varepsilon > 0$ under some restrictions on $\mu_1, \ldots, \mu_k$). Recently[15] Gupta has given a good introduction to this work and an extensive summary has been given by Gupta and Panchapakesan.

6. Other problems involve such considerations as the following: In what sense(s) might our selection procedures be optimal[16]? If our observations are obtained sequentially (one at a time, rather than all at once) it may become clear soon which one is best, or it may become clear that several are not best; how can we stop early or stop taking observations on inferior populations (respectively) and still guarantee a reasonable probability requirement?[17] If we do not know the distribution of our observations how can we proceed?[18]

The literature of this area is large, and the above only note a few of the many results recently obtained which make substantial contributions to the solution of meaningful and important problems in this area.

PROBLEMS

11.1 In the "probability requirement" $P(\text{CS}) \geqslant P^*$ discussed in Section 11.1, we required that $1/k < P^* < 1$. Show that such a requirement with $0 \leqslant P^* \leqslant 1/k$ is trivial (that is, can be achieved without taking any observations).

11.2 In footnote 4 we claimed that, if we choose $\hat{\mu}_i$ as in Equation (11.2.4), then $\max(\hat{\mu}_1, \ldots, \hat{\mu}_k)$ is unique with probability 1. Prove this result.

11.3 Prove that $f(0) = 1/k$, where $f(n)$ is specified by Equation (11.2.16). [Of course, if $n = 0$, there is no sampling. If we then choose a population as "best" at random, our $P(\text{CS})$ will be $1/k$. The derivation of Equation (11.2.16) assumed $n \geqslant 1$, but we thus see it has a simple interpretation even if $n = 0$.]

11.4 Show that, in the special case of just $k = 2$ populations, the $P(\text{CS})$ of the means procedure discussed in Section 11.2 is given by

$$\Phi\left(\frac{\mu_{[2]} - \mu_{[1]}}{\sqrt{2}\sigma/\sqrt{n}}\right).$$

11.5 R. C. Milton ("Tables of the equally correlated multivariate normal probability

---

[15] S. S. Gupta, "On some multiple-decision (selection and ranking) rules," *Technometrics*, Vol. 7 (1965), pp. 225–245. S. S. Gupta and S. Panchapakesan, "On multiple decision (subset selection) procedures," *Journal of Mathematical and Physical Sciences*, Vol. 6 (1972), pp. 1–71.

[16] See Problem 11.1H and E. L. Lehmann, "On a theorem of Bahadur and Goodman," *Annals of Mathematical Statistics*, Vol. 37 (1966), pp. 1–6.

[17] See the book by R. E. Bechhofer, J. Kiefer, and M. Sobel referenced in Footnote 9 of this section.

[18] See E. J. Dudewicz, "A nonparametric selection procedure's efficiency: largest location parameter case," *Journal of the American Statistical Association*, Vol. 66 (1971), pp. 152–161.

integral," *Technical Report No. 27*, Department of Statistics, University of Minnesota, Minneapolis, Minnesota, July 1963) in his Table II tabulated

$$\int_{-\infty}^{\infty} \Phi^{k-1}\left(\frac{x\sqrt{\rho}+H}{\sqrt{1-\rho}}\right)\phi(x)\,dx$$

for $H = .00(.05)5.15$, $k = 3(1)10(5)25$, $\rho = .00(.05)1.00$. How could we use these tables to find the h which solves Equation (11.2.15)? (We might wish to do so for $k > 10$, since Bechhofer's tables, noted in Remark 11.2.18, only extend to $k = 10$ for the problem of selecting the best.)

11.6 Man A and man B (track stars) have run the mile in competitions a number of times each. Assume that man A and man B ran independently of each other, and that each one's times are independent observations of his running ability. Let $\mu_A(\mu_B)$ denote the average time it takes man A (B) to run the mile in competitions. Similarly, let his variance be $\sigma_A^2(\sigma_B^2)$. We have available n_1 observations of A and n_2 observations of B, and know that $\sigma_A^2 = .04$, $\sigma_B^2 = .09$, and the observations are normally distributed. We wish to select the man (of A, B) who on the average has the lowest time for running the mile in competitions.

(a) If we select A when $\bar{X}_A < \bar{X}_B$ (and select B otherwise, where $\bar{X}_A$ is the average of the n_1 observations on A and $\bar{X}_B$ is the average of the n_2 observations on B), give an exact expression for the probability of a correct selection, $P(\mathrm{CS})$, as a function of $\delta = \mu_{[2]} - \mu_{[1]} \geqslant 0$. [Of course, $\mu_{[2]} = \max(\mu_A, \mu_B)$, and $\mu_{[1]} = \min(\mu_A, \mu_B)$.] Table the $P(\mathrm{CS})$ for $\delta = 0, .1, .2, .3, .4, .5$ if $n_1 = 2$, $n_2 = 3$ (still with $\sigma_A^2 = .04$, $\sigma_B^2 = .09$).

(b) If we have $n_1 = 2$ and $n_2 = 3$ (still with $\sigma_A^2 = .04$, $\sigma_B^2 = .09$) what is the largest P^* such that $P(\mathrm{CS}) \geqslant P^*$ whenever $\delta \geqslant .3$? If our observations are 3.8, 3.4 and 3.7, 3.6, 4.1 on A and B (respectively), do we select A or B?

[*Note*: Compare this problem with Problems 9.51 and 10.15, where hypothesis-testing and interval estimation were discussed for this same situation.]

11.7 Show in detail that (as was stated in Example 11.2.22) the problem of selecting the population with the smallest mean $\mu_{[1]}$ easily reduces to a problem of selecting a population with a largest mean, and that no new complexities are thus introduced.

11.8 In Example 11.2.22, suppose we are convinced (by previous experience in the field) that 2 of the $k = 5$ SRS's will be greatly inferior to the best. Use (11.2.8) to show we may then set n as for a problem with $k' = 5 - 2 = 3$ populations and still have a guarantee of approximately P^*. What n will now be required in Example 11.2.22?

11.9 In footnote 8 we claimed that if we choose $\hat{\sigma}_i^2$ as in Equation (11.3.4), then $\min(\hat{\sigma}_1^2, \ldots, \hat{\sigma}_k^2)$ is unique with probability 1. Prove this result. (See Problem 11.2.)

11.10 In the "probability requirement" $P(\mathrm{CS}) \geqslant P^*$ discussed in Section 11.3, we required that $1/k < P^* < 1$. Show that (as in Problem 11.1) such a requirement with $0 \leqslant P^* \leqslant 1/k$ is trivial.

11.11 Show that the $P_{\mathrm{LFC}}(\mathrm{CS})$ of Equation (11.3.13) is an increasing function of n

and $\to 1$ as $n \to \infty$. [Hence, there is a unique smallest sample size n per population for which probability requirement (11.3.5) is satisfied.]

11.12 In Section 11.1 we discussed an example from poultry science and stated that (for specific values of certain parameters) $n = 462$ pullets of each type would be required. Use the results of Section 11.2 to show that $n = 462$ is the required sample size per type.

11.13 An experimenter has been presented with $k = 10$ varieties of rice developed by researchers. They seem promising, and he wishes to run an experiment in order to select that variety which has the highest mean yield per acre. He plans to plant n acres with each variety (so nk total acres will be planted), the acres for any particular variety being selected at random from the nk acres available. (This is called a *Completely Randomized Design* of the experiment. The purpose of the randomization is to ensure that if the acres are not uniform in their effects on rice production, each variety will have an equal chance at both good and bad acreage; if one variety were to obtain all the "good" acres, the results of the experiment could be very misleading.) All nk observations of yield, in tons/acre, will be assumed independent.

The experimenter assumes that the yield of an acre planted with variety i is a $N(\mu_i, \sigma_i^2)$ r.v. $(1 \leqslant i \leqslant k)$ where $\mu_1, \ldots, \mu_k$ are unknown and that $\sigma_1^2 = \cdots = \sigma_k^2 = \sigma^2 = 1$. Intuitively, he decides to base his decision on sample means of yields and will select that variety which yields the largest sample mean yield. If he requires that he have probability at least $P^* = .90$ of selecting the variety which has the highest mean yield whenever $\mu_{[k]} - \mu_{[k-1]} \geqslant \delta^* = .1$ (here $\mu_{[1]} \leqslant \cdots \leqslant \mu_{[k]}$ denote the ranked values of $\mu_1, \ldots, \mu_k$), what is the smallest sample size n which he can use?

11.14 Suppose an experimenter has k populations $\pi_1, \ldots, \pi_k$ and wishes to run an experiment in order to select that one which has the largest mean. Observations from π_i are $N(\mu_i, \sigma_i^2)$ $(1 \leqslant i \leqslant k)$ where $\mu_1, \ldots, \mu_k$ are unknown and $\sigma_1^2, \ldots, \sigma_k^2$ are known (but not necessarily equal). He requires that he have probability $\geqslant P^*$ $(1/k < P^* < 1)$ of selecting the population which has the largest mean whenever $\mu_{[k]} - \mu_{[k-1]} \geqslant \delta^*$ $(\delta^* > 0)$. (See problem "3" of Section 11.4 and Section 11.2.)

(a) Show that if we set the number n_i of observations from π_i $(1 \leqslant i \leqslant k)$ so that

$$\frac{\sigma_1^2}{n_1} = \cdots = \frac{\sigma_k^2}{n_k} \qquad (*)$$

then the tables of Bechhofer (see Remark 11.2.18) of solutions of Equation (11.2.15) may be used to set $n_1, \ldots, n_k$ so as to guarantee the probability requirement. If $k = 3$, $\sigma_1^2 = \sigma_2^2 = 1$, $\sigma_3^2 = 9$, $P^* = .90$, and $\delta^* = .1$, what numbers of observations should we take on the populations if we adopt this method?

(b) Bechhofer[19] notes that his suggestion of a solution, for the case of $\sigma_1^2, \ldots, \sigma_k^2$ known but unequal, via $(*)$ is not optimal. For the case $k = 2$, find

---

[19] See p. 24 of his paper referenced in Remark 11.2.18.

the optimal allocation. [*Hint*: First find an exact expression for the $P(\text{CS})$ if n_1 observations are taken from π_1 and n_2 observations are taken from π_2. Show that subject to the restraint $n_1 + n_2 = n$ (n a positive integer) the $P(\text{CS})$ is uniformly maximized over all configurations where $\mu_1 \neq \mu_2$ by choosing n_1 and n_2 so that

$$\frac{\sigma_1}{n_1} = \frac{\sigma_2}{n_2}.\Big] \qquad (**)$$

(c) In part (b), if the pair (n_1, n_2) which $(**)$ and $n_1 + n_2 = n$ specify is not a pair of integers, then we must know how to move to the optimal pair of *integers* (n_1^*, n_2^*). (Sample sizes must be integers.) Let $[x]$ denote the largest integer $\leqslant x$. Show[20] that the optimal pair (n_1^*, n_2^*) is either $([n_1] + 1, [n_2])$ or $([n_1], [n_2] + 1)$ [here (n_1, n_2) is the solution of $(**)$ and $n_1 + n_2 = n$], and that $(n_1^*, n_2^*) = ([n_1], [n_2] + 1)$ iff

$$\frac{\sigma_1^2}{\sigma_2^2} \leqslant \frac{[n_1]([n_1] + 1)}{(n - [n_1])(n - ([n_1] + 1))}.$$

(d) Ignoring the fact that sample sizes must be integers, compare allocation of a sample size n via $(*)$ with allocation via $(**)$ in the case $k = 2$, as follows. Let n_U be the sample size needed for an allocation via $(*)$ to achieve a $P_{\text{LFC}}(\text{CS}) = P^*$ and let n_O be the sample size needed for an allocation via $(**)$ to achieve a $P_{\text{LFC}}(\text{CS}) = P^*$. Show that

$$\frac{n_U}{n_O} = 2\frac{1 + a}{(1 + \sqrt{a})^2},$$

where $a = \sigma_2^2/\sigma_1^2$. Compute the ratio for $a = 1(1)10$. [This gives us an idea of how much we lose by using allocation via $(*)$, instead of the optimal allocation $(**)$, in the case $k = 2$. The optimal allocation in the case $k > 2$ is not known. The comparison for the case $k = 2$ may give us a hint as to how much we lose by not knowing it.]

(e) Show by example that allocation via $(*)$ is not optimal for $k > 2$.

11.15 Consider the problem described in the introduction to Problem 11.14. There both proposed methods ($(*)$ and $(**)$) of assigning sample sizes to populations require (in general) different numbers of observations per population. However, there are cases (for example, when a statistician is consulted after an experiment has been run, or in special applications where the main cost of sampling from π_i is due to taking any observations at all from π_i with additional observations costing little $(1 \leqslant i \leqslant k)$) where we have $n_1 = \cdots = n_k = n$ (say), where n_i denotes the number of observations from π_i $(1 \leqslant i \leqslant k)$.

(a) The problem of finding the minimum of the $P(\text{CS})$ for $\mu_{[k]} - \mu_{[k-1]} \geqslant \delta^*$ in this case appears at least as complex as the problem of optimal allocation for $k > 2$ discussed in Problem 11.14. Thus, in practice we might be tempted

[20] This, and other parts of this problem, were discussed by E. J. Dudewicz and S. R. Dalal, "Allocation of observations in ranking and selection with unequal variances," *Sankhyā*, in press.

to evaluate the $P(\text{CS})$, acting as if all of $\pi_1, \ldots, \pi_k$ had variance equal to $\max(\sigma_1^2, \ldots, \sigma_k^2)$, and to assume that this furnishes a lower bound on the $P(\text{CS})$. Show that

$$P(\text{CS}) = \int_{-\infty}^{\infty} \left\{ \prod_{i=1}^{k-1} \Phi\left(\frac{\sigma_{(k)}}{\sigma_{(i)}} x + \frac{\mu_{[k]} - \mu_{[i]}}{\sigma_{(i)}} \sqrt{n} \right) \right\} \phi(x)\, dx.$$

(Here $\sigma_{(i)}^2$ is the variance of the population with mean $\mu_{[i]}$, $1 \leqslant i \leqslant k$.) Show that $dP(\text{CS})/d(\sigma_{(k)}) > 0$ in a neighborhood of $\mu_{[1]} = \cdots = \mu_{[k]}$. Deduce that if we have $\sigma_1^2 = \cdots = \sigma_{k-1}^2 = 1$ and $\sigma_k^2 = .5$ and [hoping to lower bound the $P(\text{CS})$] act as if $\sigma_k^2 = 1$, we could overestimate the $P(\text{CS})$.

(b) If we have $\sigma_1^2 = \cdots = \sigma_{k-1}^2 = .5$ and $\sigma_k^2 = 1$, can we upper bound the $P(\text{CS})$ by acting as if $\sigma_k^2 = .5$?

HONORS PROBLEMS

11.1H In Section 11.2, at (11.2.4) we invoked sufficiency in order to justify basing our decision on the sample means $\bar{X}_1, \ldots, \bar{X}_k$. It was shown by W. J. Hall ["The most-economical character of some Bechhofer and Sobel decision rules," *Annals of Mathematical Statistics*, Vol. 30 (1959), pp. 964–969] that **the resulting procedure is most-economical**. *No other procedure which takes a specified number* (m, *say*) *of observations per population can have both* **(1)** $m < n$ *where* n *is the means procedure sample size studied in Section* 11.2, *and* **(2)** *satisfy probability requirement* (11.2.5).

Assume the results stated above. Let $\mathscr{P}_M(n_1, \ldots, n_k)$ denote a procedure based on sample means of n_1 observations from $\pi_1, \ldots, n_k$ observations from π_k; thus, the most-economical procedure is $\mathscr{P}_M(n, \ldots, n)$ and it takes nk total observations in order to satisfy probability requirement (11.2.5). Let $\mathscr{P}(n_1, \ldots, n_k)$ denote any procedure based on n_1 observations from $\pi_1, \ldots, n_k$ observations from π_k. **(1)** Show that $\mathscr{P}_M(n, \ldots, n)$ is "better" (in some reasonable sense) than $\mathscr{P}_M(n_1, \ldots, n_k)$ when $n_1 + \cdots + n_k = nk$ (that is, the "best" allocation of nk observations for the means procedure is the allocation of the same number of observations to each population). **(2)** Show that $\mathscr{P}_M(n, \ldots, n)$ is "better" than any $\mathscr{P}(n_1, \ldots, n_k)$ with $n_1 + \cdots + n_k \leqslant nk$.

11.2H The procedure of Section 11.2 chooses the common sample size per population n in such a way that (under certain conditions of independence, normality, and so on, stated in Section 11.2) probability requirement (11.2.5) is satisfied: $P(\text{CS}) \geqslant P^*$ whenever $\mu_{[k]} - \mu_{[k-1]} \geqslant \delta^*$. Thus, with sample size n per population we are assured probability at least P^* of actually selecting the best population, as long as the best is at least δ^* better than the second best. However, even if the best is not δ^* better than the second best, it may be much more than δ^* better than the third best. Does the procedure of Section 11.2 offer any guarantee, in such situations, of obtaining one of the "few best"? Strengthening probability requirement (11.2.5), show that the procedure considered actually guarantees

$$P[\text{Selecting a population } \pi_l \text{ with } \mu_{[k]} - \mu_l < \delta^*] \geqslant P^*. \qquad (*)$$

[Related considerations in a more complex situation were studied by S. Sievers, "A Solution to an open problem of Bech[h]ofer-Kiefer-Sobel," Abstract, *Annals of Mathematical Statistics*, Vol. 42 (1971), p. 1795.]

11.3H Show that the $P(\mathrm{CS})$ of the means procedure of Section 11.2 (which was expressed as a single integral in Theorem 11.2.7) is equal to

$$P\left[Y_i < \frac{\mu_{[k]} - \mu_{[i]}}{\sqrt{2}\sigma/\sqrt{n}}, \quad i = 1, \ldots, k-1\right], \qquad (*)$$

where Y_i is a $N(0, 1)$ r.v. $(1 \leqslant i \leqslant k-1)$ and $\mathrm{Cov}(Y_i, Y_j) = .5$ $(i \neq j)$. Does $(Y_1, \ldots, Y_{k-1})$ have the multivariate normal distribution discussed in Section 4.5? (If so, what are $\boldsymbol{\mu}$ and $\boldsymbol{\Sigma}$ here?) Using $(*)$, find the minimum (among all sets of possible means $\mu_1, \ldots, \mu_k$ which satisfy $\mu_{[k]} - \mu_{[k-1]} \geqslant \delta^*$) of the $P(\mathrm{CS})$. (Compare your result with that of Theorem 11.2.11.)

11.4H In Section 11.3, at (11.3.4) we invoked sufficiency in order to justify basing our decision on $s_1^2, \ldots, s_k^2$. Develop considerations like those of Problem 11.1H to justify basing our decision on $s_1^2, \ldots, s_k^2$.

11.5H Show that the $P(\mathrm{CS})$ of the procedure of Section 11.3 (which was expressed as a single integral in Theorem 11.3.7) is equal to

$$P[Y_i \geqslant \sigma_{[1]}^2/\sigma_{[i+1]}^2, \quad i = 1, \ldots, k-1] \qquad (*)$$

where Y_i is a r.v. with the F-distribution $F(n-1, n-1)$ and the joint density of $Y_1, \ldots, Y_{k-1}$ is that studied in Example 4.6.9 with n replaced by $n - 1$. [*Hint*: See Problem 9.43.]

Compare your results with those of Problem 11.3H. Can you define a "multivariate F-distribution" by analogy?

11.6H On p. 276 of their paper (see Remark 11.3.14), Bechhofer and Sobel mention the case where one has n_1 observations from $\pi_1, \ldots, n_k$ observations from π_k. They also give an expression for the $P(\mathrm{CS})$ in that case [their Equation (11) with $t = 1$ and $n_{(i)}$ the number of observations from the population with population variance $\sigma_{[i]}^2$], say

$$f_{k, \theta^*}(n_{(1)}, \ldots, n_{(k)}). \qquad (*)$$

Can $(*)$ actually be evaluated? [*Hint*: If you are told $k = 3, \theta^* = 2.0, n_1 = 2, n_2 = 4, n_3 = 6$, do you know $n_{(1)}, n_{(2)}, n_{(3)}$?] Does $(*)$ increase as $n_{(i)}$ increases $(1 \leqslant i \leqslant k)$? [If so, we would be able to lower bound $(*)$ by $f_{k, \theta^*}(m, \ldots, m)$ where $m = \min(n_{(1)}, \ldots, n_{(k)}) = \min(n_1, \ldots, n_k)$; this lower bound can be calculated.]

11.7H In the context of Section 11.2, if we actually have $n_1, \ldots, n_k$ observations from $\pi_1, \ldots, \pi_k$ (respectively), and base our decision on the sample means $\bar{X}_1, \ldots, \bar{X}_k$ which utilize all observations, is our $P(\mathrm{CS})$ always greater than or equal to the $P(\text{CS})$ calculated assuming $n = \min(n_1, \ldots, n_k)$ observations from each of the populations?

11.8H Suppose we have experimented as specified in Section 11.2, but now wish to estimate $\mu_{[k]}$ (the largest population mean) rather than select the population associated with $\mu_{[k]}$. Is $\bar{X}_{[k]}$ a good estimator of $\mu_{[k]}$? [*Hint*: Find sup $E\bar{X}_{[k]}$ and inf $E\bar{X}_{[k]}$, where we are maximizing and minimizing over the set of all possible $(\mu_1, \ldots, \mu_k)$.][21]

11.9H Consider the situation of Problem 11.13, but now the experimenter avows that: while it is well-known that $\sigma_1^2 = \cdots = \sigma_k^2 = \sigma^2$, the common value σ^2 is unknown. The experiment can, however, be performed in two stages since at the test site climate is such that two crops can be grown in 1 year. Suppose that the two-stage procedure referenced in problem "1" of Section 11.4 is used, and that in the first stage $n_0 = 2$ observations are taken on each variety. What number n_1 of observations on each variety should be taken in the second stage? (n_1 will depend on the results of the first stage. Plot a graph of n_1 as a function of the results of the first stage.) [*Notes*: For solution of this problem you will probably have to use approximations to the multivariate t-distribution. For example, see C. W. Dunnett and M. Sobel, "A bivariate generalization of Student's t-distribution with tables for certain special cases," *Biometrika*, Vol. 41 (1954), pp. 153–169; C. W. Dunnett and M. Sobel, "Approximations to the probability integral and certain percentage points of a multivariate analogue of Student's t-distribution," *Biometrika*, Vol. 42 (1955), pp. 258–260; and J. V. Armitage and P. R. Krishnaiah, "Tables for multivariate t-distribution," *Sankhyā*, Series B, Vol. 28 (1966), pp. 31–56. For the goal of a complete ranking of $k = 3$, 4, or 5 populations, tables and an interesting application to ranking of competing consumer or producer goods were given by H. Freeman, A. Kuzmack, and R. Maurice in a study supported by Consumers Union, Mount Vernon, New York ("Multivariate t and the ranking problem," *Biometrika*, Vol. 54 (1967), pp. 305–308).]

11.10H The problems of: (1) selecting that one of k television programs (available to a television audience in a given area) which can claim the largest proportion of the total audience; (2) selecting that one of the 36 "bettable" numbers on an unbalanced roulette wheel which has the largest probability associated with it; (3) selecting that one of the 6 faces of a loaded die which has the largest probability of landing face up; and (4) selecting that one of k Presidential candidates who has the highest proportion of voters in favor of his election in a given state; are **selection problems involving a multinomial distribution**. (See Example 3.5.3.)

---

[21] This problem can arise when, for example, we wish to decide whether or not to continue research on the set of populations $\{\pi_1, \ldots, \pi_k\}$. See E. J. Dudewicz, "Research program evaluation (quality control of research) via estimation of ordered parameters," *Annual Technical Conference Transactions of the American Society for Quality Control*, Vol. 27 (1973), pp. 361–366. Also see S. Blumenthal and A. Cohen, "Estimation of the larger of two normal means," *Journal of the American Statistical Association*, Vol. 63 (1968), pp. 861–876, for additional applications and results on the case $k = 2$.

This problem was discussed by R. E. Bechhofer, S. Elmaghraby, and N. Morse,[22] who considered one's observation to be $\mathbf{X} = (X_1, \ldots, X_k)$ with the multinomial distribution of Definition 3.5.1. [Thus, we sample n people, each of whom has probability p_i of belonging to category i and let X_i be the number observed to belong to category i out of the n $(1 \leqslant i \leqslant k)$. Then $0 \leqslant p_1 \leqslant 1, \ldots, 0 \leqslant p_k \leqslant 1$, and $p_1 + \cdots + p_k = 1$.] They let $p_{[1]} \leqslant \cdots \leqslant p_{[k]}$ denote $p_1, \ldots, p_k$ in numerical order, let $X_{[k]}$ denote $\max(X_1, \ldots, X_k)$, and propose to select that component (television show, roulette number, die face, or presidential candidate, for example) which[23] yielded $X_{[k]}$ as being associated with $p_{[k]}$. If we desire $P(\text{CS}) \geqslant P^*$ whenever $p_{[k]} \geqslant \theta^* p_{[k-1]}$ (where $1/k < P^* < 1$ and $\theta^* > 1$ are set by the experimenter), the necessary number of people to be sampled, n, has been tabulated and an approximation given.

(a) In many cases of elections $k = 2$ and pollsters use[24] $n = 2000$. With $\theta^* = 1.1$ roughly what P^* does this guarantee? If we have $k = 3$ "serious" candidates (that is, candidates whose $p > .01$), what P^* does $n = 2000$ give us?

(b) How good are television ratings? [*Hint*: With $k = 6, n = 2000, \theta^* = 1.1$, roughly what P^* do we obtain? Now that is for simply selecting the best. How should the complete ranking case compare?]

11.11H Consider the problem discussed in Problems 9.51 and 10.15, from the ranking and selection point of view. [A special case was studied in Problem 11.6.]

---

[22] "A single-sample multiple-decision procedure for selecting the multinomial event which has the highest probability," *Annals of Mathematical Statistics*, Vol. 30 (1959), pp. 102–119.

[23] This may not be unique. In case of ties, one of the tied categories is selected at random.

[24] Pollsters use (or are presumed to use) the sophisticated statistical sampling methods described in, for example, D. Raj, *Sampling Theory*, McGraw-Hill Book Company, New York, 1968, which have been developed over the last 40 years. This should increase their P^* with a fixed n, but how much it does has apparently not yet been studied.

12

DECISION THEORY

One of the definitions of "statistics" (as it exists at present) calls it "the science of decision making." The author largely agrees with this definition. The present chapter has as its goal the elucidation and explication of the aims, axioms, terminology, and results (and how these can be seen as unifying concepts, bringing together, say, point estimation, statistical inference principles, hypothesis testing, interval estimation, ranking and selection procedures, and so on, under one roof) of what is called "statistical decision theory" (or sometimes simply "decision theory").

In Section 12.1 we cover the general formulation and principles of decision theory. The following three sections cover several special cases, studied already in detail in previous chapters, from the decision-theoretic viewpoint (hypothesis-testing in Section 12.2, interval estimation in Section 12.3, and ranking and selection in Section 12.4). Section 12.5 presents some general results of decision theory, while Section 12.6 introduces an overview of the area of statistics (and of the terminology of some of its subfields) through a discussion of a specific example.

12.1. THE GENERAL PROBLEM OF STATISTICS

In the "decision theory" view of the "general problem of statistics," two main components of the problem are: observations (data) with a usually not fully known distribution (for example, the distribution of the data may contain an unknown parameter θ which lies in some known set Θ); and possible numerical losses/gains due to choosing one of a number of possible decisions. The expected loss (gains are negative losses) is a primary consideration in evaluating decisions. [For example, in Section 9.4 we saw that for the problem of testing a simple hypothesis H_0 ($\theta = \theta_0$) against a simple alternative H_1 ($\theta = \theta_1$), where θ was unknown, we could decide either "H_0 is true" or "H_1 is true." Letting our loss be the "number of incorrect decisions" (which is either 0 or 1), the expected loss involved the error probabilities of Types I and II.]

In the general formulation of decision theory, we observe a r.v. X (which may be multivariate) with d.f. $F(x|\theta)$ where $\theta \in \Theta$ is unknown, and if we choose decision d from the set of all possible decisions $\mathscr{D}$, then we suffer a loss $l(\theta, d)$. A "decision rule" is a method of choosing d from $\mathscr{D}$ after observing $x \in \mathscr{X}$, that is, a function $s(x) = d$. Our average loss (called risk) $E_\theta l(\theta, s(X))$ is a function of both θ and the decision rule $s(\cdot)$, called the risk function $r_\theta(s)$, and is *the* criterion by which rules are compared. The problem is then to study general properties of "good" rules, and to find them in specific cases. **We will now define precisely the major quantities just introduced.**

Definition 12.1.1. *A general* **statistical decision problem** *is a triplet* $(\boldsymbol{\Theta}, \mathscr{D}, \boldsymbol{l})$ *and a r.v.* $\mathbf{X}$. *The r.v.* X *(called the* **data***) has a d.f.* $F(x|\theta)$ *where* θ *is unknown but it is known that* $\theta \in \Theta$. *[A problem with no such r.v.* X *is called a* **no-data decision problem**.] *The set of possible values of the r.v.* X *will be denoted by* $\mathscr{X}$.

$\boldsymbol{\theta}$ *is called the* **state of nature**, *while the nonempty set* $\boldsymbol{\Theta}$ *is called the* **parameter space**. *The nonempty set* $\mathscr{D}$ *is called the* **decision space** *or* **action space**.

Finally, l *is called the* **loss function** *and to each* $\theta \in \Theta$ *and* $d \in \mathscr{D}$ *it assigns a real number* $\boldsymbol{l(\theta, d)}$.

Definition 12.1.2. *For a statistical decision problem* $(\Theta, \mathscr{D}, l)$, X, *a* **(nonrandomized) decision rule** *is a function* $\mathbf{s}(\cdot)$ *which to each* $x \in \mathscr{X}$ *assigns a member* d *of* $\mathscr{D}$: $s(x) = d$.

A **randomized decision rule** *is a function* $\mathbf{s}(\cdot)$ *which for each* $x \in \mathscr{X}$ *specifies a probability distribution according to which a member* d *of* $\mathscr{D}$ *is to be chosen. This probability distribution will be denoted by* $\mathbf{s(x)}$.

Definition 12.1.3. *The* **risk function** $\mathbf{r}_\theta\mathbf{(s)}$ *of a decision rule* $s(X)$ *for a statistical decision problem* $(\Theta, \mathscr{D}, l)$, X *(the* **expected loss** *or* **average loss** *when* θ *is the true state of nature and a decision is chosen by rule* $s(\cdot)$*) is*

$$r_\theta(s) = E_\theta l(\theta, s(X)). \tag{12.1.4}$$

In Section 9.4 our decision rules were tests, which were also either randomized or nonrandomized (although this was not stressed). A fundamental precept of decision theory is that all of our losses (or gains) should be incorporated into the loss function $l(\cdot, \cdot)$ for the problem, so that then it is reasonable to consider *only* the risk function (average loss (12.1.4)) in seeking a "good" decision rule. As in Section 9.4 (see the paragraph preceding Definition 9.4.5 through Definition 9.4.6) **we can use the risk function to restrict ourselves to a "few" decision rules (instead of all decision rules) as follows**. Suppose $s_1(\cdot)$ and $s_2(\cdot)$ are decision rules for some statistical decision problem $(\Theta, \mathscr{D}, l)$, X. If

$$r_\theta(s_1) > r_\theta(s_2), \quad \text{all } \theta \in \Theta, \tag{12.1.5}$$

then $s_2(\cdot)$ has a smaller average loss no matter what the state of nature θ, and we will surely prefer $s_2(\cdot)$. (See Figure 9.4–2 for an illustration for the case where Θ contains only two points, in the hypothesis-testing case.) Even if

$$r_\theta(s_1) \geqslant r_\theta(s_2), \quad \text{all } \theta \in \Theta, \tag{12.1.6}$$

we have no use for $s_1(\cdot)$: $s_2(\cdot)$ is at least as good for all states of nature θ. Utilizing this analysis as motivation, we now make the following definitions.

Definition 12.1.7. *For any statistical decision problem* $(\Theta, \mathscr{D}, l)$, X, *decision rule* $s_2(\cdot)$ *is said to be* **better than** *decision rule* $s_1(\cdot)$ *(or decision rule* $s_1(\cdot)$ *is said to be* **worse than** *decision rule* $s_2(\cdot)$*) if*

$$\begin{cases} r_\theta(s_2) \leqslant r_\theta(s_1), & \text{all } \theta \in \Theta \\ r_{\theta^*}(s_2) < r_{\theta^*}(s_1), & \text{at least one } \theta^* \in \Theta. \end{cases} \tag{12.1.8}$$

We say $s_1(\cdot)$ *is* **no better than** $s_2(\cdot)$ *if*

$$r_\theta(s_2) \leqslant r_\theta(s_1), \quad \text{all } \theta \in \Theta. \tag{12.1.9}$$

Definition 12.1.10. *For any statistical decision problem* $(\Theta, \mathscr{D}, l)$, X, *let* $D = \{s(\cdot): s(\cdot) \text{ is a decision rule}\}$. *Any class of decision rules* $C \subseteq D$ *is called* **complete** *(or a* **complete class (of decision rules)***) if for each* $s_1 \notin C$, *there is an* $s_2 \in C$ *such that* s_2 *is better than* s_1. *A class C is called* **essentially complete** *if for each* $s_1 \notin C$, *there is an* $s_2 \in C$ *such that* s_1 *is no better than* s_2. *A class C is called* **minimal complete** *if C is complete but if any* $s_2 \in C$ *is removed from C, the resulting class is not complete. A class C is called* **minimal essentially complete** *if C is essentially complete but if any* $s_2 \in C$ *is removed from C, the resulting class is not essentially complete.*

The above notions (see Definition 12.1.7) of possible comparisons of two decision rules s_1 and s_2 allow us to *sometimes* state each of the following: "$\mathbf{s_2}$ **is better than** $\mathbf{s_1}$" (**written** $\mathbf{s_2 \succ s_1}$), "s_1 is better than s_2" (written $s_1 \succ s_2$), and "$\mathbf{s_2}$ **and** $\mathbf{s_1}$ **are equivalent**" (**written** $\mathbf{s_2 \sim s_1}$) (that is, "s_2 is at least as good as $s_1(\mathbf{s_2 \gtrsim s_1})$, and s_1 is at least as good as $s_2(s_1 \gtrsim s_2)$"). *However*, it is also possible that we may have "s_1 and s_2 are incomparable," that is, $r_\theta(s_1) < r_\theta(s_2)$ for at least one $\theta \in \Theta$, *and* $r_\theta(s_1) > r_\theta(s_2)$ for at least one $\theta \in \Theta$. For example, in Figure 12.1–1 we sketch risk functions leading to each of the above conclusions. (Here $\Theta = \{x: 0 \leqslant x \leqslant 1\}$.) Since the possibility of our

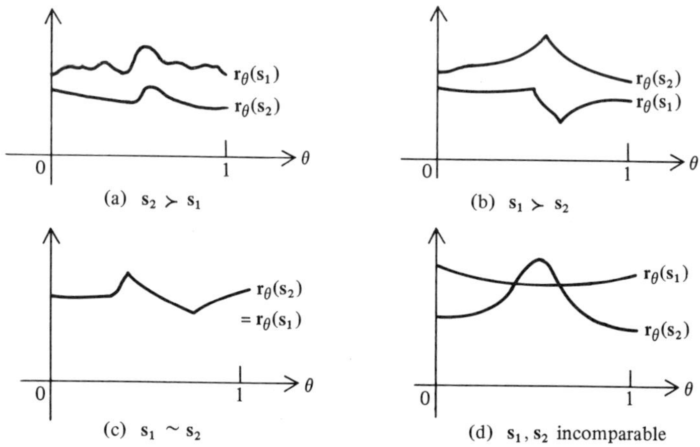

Figure 12.1–1. Comparisons of decision rules via risk functions

not being able to prefer s_1 or s_2 (simultaneously for all θ) exists (they may be incomparable), our ordering of decision rules is called a partial ordering (see Problem 12.1).

In Section 9.4 we were able to find a minimal essentially complete class of decision rules very easily, by use of the Neyman-Pearson Lemma (see Theorem 9.4.7), and then proceeded to the question of how to select one member out of that minimal essentially complete class. In the general case things are not so simple. Before studying this question further, we will first give **an example which will illustrate the concepts introduced so far and will motivate those still to come.**

Example 12.1.11. Suppose we have two possible states of nature (for example, a "bull" market and a "bear" market), that is, $\Theta = \{\theta_1, \theta_2\}$ with $\theta_1 \neq \theta_2$, and six possible decisions (for example, six possible investment strategies), that is $\mathscr{D} = \{d_1, d_2, d_3, d_4, d_5, d_6\}$, and that our loss function $l(\theta, d)$ is as given in the following table (for example, if we take decision d_5 (investment strategy 5) when $\theta = \theta_2$ (in a "bear" market), we suffer a loss of $l(\theta_2, d_5) = 8$).

	$l(\theta, d)$					
θ \ d	d_1	d_2	d_3	d_4	d_5	d_6
θ_1	17	19	14	10	9	9
θ_2	14	4	4	6	8	16

Suppose we observe no r.v. X, that is, we have a no-data decision problem. Find the set D of all possible decision rules and graph their risk functions. Find a minimal essentially complete class of decision rules.

In general (see Definition 12.1.2), a decision rule $s(x)$ specifies, for each $x \in \mathscr{X}$, a probability distribution according to which a member d of $\mathscr{D}$ is to be chosen. Since no X is observed in this example, a decision rule s for this problem simply specifies a probability distribution on $\mathscr{D}$, so the set D of all decision rules s is the set of all $s_{\mathbf{p}}$ where $\mathbf{p} = (p_1, p_2, p_3, p_4, p_5, p_6)$ with $p_1 \geqslant 0$, $p_2 \geqslant 0$, $p_3 \geqslant 0$, $p_4 \geqslant 0$, $p_5 \geqslant 0$, $p_6 \geqslant 0$, and $p_1 + p_2 + p_3 + p_4 + p_5 + p_6 = 1$. (Rule $s_{\mathbf{p}}$ takes decision d_i with probability p_i $(i = 1, 2, 3, 4, 5, 6)$.) Simply stated,

$$D = \{s_{\mathbf{p}} \colon \mathbf{p} = (p_1, \ldots, p_6) \text{ is a probability distribution}\}. \qquad (12.1.12)$$

Now the risk function $r_\theta(s_{\mathbf{p}})$ of a general decision rule $s_{\mathbf{p}}$ is a pair of numbers $(r_{\theta_1}(s_{\mathbf{p}}), r_{\theta_2}(s_{\mathbf{p}}))$ since there are only two possible θ's in Θ, and the risk function can thus be represented as a point in the plane (and is therefore often called a **risk point**). But

$$\begin{aligned} r_{\theta_1}(s_{\mathbf{p}}) &= E_{\theta_1} l(\theta_1, s_{\mathbf{p}}) \\ &= \sum_{i=1}^{6} l(\theta_1, d_i) P_{\theta_1}[s_{\mathbf{p}} \text{ selects } d_i] = \sum_{i=1}^{6} l(\theta_1, d_i) p_i \qquad (12.1.13) \\ &= 17p_1 + 19p_2 + 14p_3 + 10p_4 + 9p_5 + 9p_6. \end{aligned}$$

Similarly,

$$r_{\theta_2}(s_{\mathbf{p}}) = 14p_1 + 4p_2 + 4p_3 + 6p_4 + 8p_5 + 16p_6, \qquad (12.1.14)$$

so

$$\begin{aligned} (r_{\theta_1}(s_{\mathbf{p}}), r_{\theta_2}(s_{\mathbf{p}})) = (&17p_1 + 19p_2 + 14p_3 + 10p_4 + 9p_5 + 9p_6, \\ &14p_1 + 4p_2 + 4p_3 + 6p_4 + 8p_5 + 16p_6). \end{aligned} \qquad (12.1.15)$$

The nonrandomized decision rules are six in number: the one with $p_i = 1$, say s_i, always takes decision d_i $(1 \leqslant i \leqslant 6)$. Their risk points are easy to find, and are plotted in Figure 12.1.–2. (Note that while each of the six points is labeled s_i, it is actually $(r_{\theta_1}(s_i), r_{\theta_2}(s_i))$ which is plotted. The dotted lines are solely to aid in finding coordinates of the points plotted and have no other significance.) In order to plot the risk points of the randomized decision rules we need first to review some old material and introduce some new material. We will return to this problem in Example 12.1.18.

The material needed to finish Example 12.1.11 is connected with that dealing with convex sets which was discussed in Section 8.3 (from the beginning through Theorem 8.3.5). Recall that a point in the plane (x_1, x_2) can be represented as a **vector**

$$\mathbf{x} = \begin{pmatrix} x_1 \\ x_2 \end{pmatrix},$$

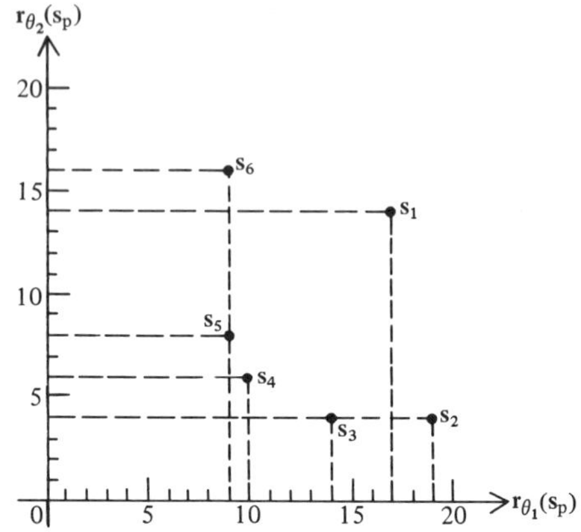

Figure 12.1–2. Nonrandomized risk points for Example 12.1.11

and that a point in $\mathscr{R}^k$ can be represented as a vector

$$\mathbf{x} = \begin{pmatrix} x_1 \\ x_2 \\ \cdot \\ \cdot \\ \cdot \\ x_k \end{pmatrix}$$

A subset S of $\mathscr{R}^k$ is called **convex** (Definition 8.3.1) if $\mathbf{x}, \mathbf{y} \in S$ implies $\lambda\mathbf{x} + (1 - \lambda)\mathbf{y} \in S$ for all λ with $0 < \lambda < 1$; that is, S is convex if all of the line segment joining two points of S, is itself in S. If we have a finite number of vectors $\mathbf{x}, \mathbf{y}, \mathbf{z}, \ldots, \mathbf{u}$, we will be interested in the smallest convex set which contains all of them, called **the convex set generated by** $\mathbf{x}, \mathbf{y}, \mathbf{z}, \ldots, \mathbf{u}$. A representation for this set (which is also called the **convex hull of** $\mathbf{x}, \mathbf{y}, \mathbf{z}, \ldots, \mathbf{u}$) is given in our next theorem.

Theorem 12.1.16. *The convex hull of the points* $\mathbf{x}, \mathbf{y}, \mathbf{z}, \ldots, \mathbf{u}$ *is the set of all points*

$$\alpha\mathbf{x} + \beta\mathbf{y} + \gamma\mathbf{z} + \cdots + \omega\mathbf{u}, \tag{12.1.17}$$

with $\alpha \geqslant 0, \beta \geqslant 0, \ldots, \omega \geqslant 0$, *and* $\alpha + \beta + \gamma + \cdots + \omega = 1$. [Each point of the form (12.1.17) is called a **convex combination of the points** $\mathbf{x}, \mathbf{y}, \mathbf{z}, \ldots, \mathbf{u}$.]

Proof: First, the set of all points of form (12.1.17) is a convex set. To show this, we must show that if $\mathbf{v}$ and $\mathbf{w}$ are two such points, then so is $\lambda\mathbf{v} + (1 - \lambda)\mathbf{w}$ as long as $0 < \lambda < 1$. Since $\mathbf{v}$ and $\mathbf{w}$ are such points, we can (if there are l

vectors **x**, **y**, **z**, ..., **u**) write

$$\begin{cases} \mathbf{v} = a_1\mathbf{x} + a_2\mathbf{y} + a_3\mathbf{z} + \cdots + a_l\mathbf{u}, \\ \mathbf{w} = b_1\mathbf{x} + b_2\mathbf{y} + b_3\mathbf{z} + \cdots + b_l\mathbf{u}, \end{cases}$$

where $a_i \geqslant 0\,(i = 1, 2, \ldots, l)$, $a_1 + a_2 + \cdots + a_l = 1$, $b_i \geqslant 0\,(i = 1, 2, \ldots, l)$, and $b_1 + b_2 + \cdots + b_l = 1$. Thus,

$$\lambda\mathbf{v} + (1-\lambda)\mathbf{w} = (\lambda a_1 + (1-\lambda)b_1)\mathbf{x} + (\lambda a_2 + (1-\lambda)b_2)\mathbf{y} + \cdots.$$

Since $\lambda a_i + (1-\lambda)b_i \geqslant 0$ (see Problem 12.4) and

$$\begin{aligned}(\lambda a_1 + (1-\lambda)b_1) + (\lambda a_2 + (1-\lambda)b_2) + \cdots + (\lambda a_l + (1-\lambda)b_l)\\ = \lambda(a_1 + a_2 + \cdots + a_l) + (1-\lambda)(b_1 + b_2 + \cdots + b_l)\\ = \lambda\cdot 1 + (1-\lambda)\cdot 1 = \lambda + 1 - \lambda = 1,\end{aligned}$$

$\lambda\mathbf{v} + (1-\lambda)\mathbf{w}$ is of form (12.1.17) for $0 < \lambda < 1$, so convexity is proven.

Second, we must show that no smaller convex set can contain the points **x**, **y**, **z**, ..., **u**. This follows if we show that a convex set H containing **x**, **y**, **z**, ..., **u** must contain

$$\alpha\mathbf{x} + \beta\mathbf{y} + \gamma\mathbf{z} + \cdots + \omega\mathbf{u}$$

whenever $\alpha \geqslant 0, \beta \geqslant 0, \gamma \geqslant 0, \ldots, \omega \geqslant 0$, and $\alpha + \beta + \gamma + \cdots + \omega = 1$, which we will now do. First since H is convex and contains **x** and **y**, H contains

$$\mathbf{q} = \frac{\alpha}{\alpha+\beta}\mathbf{x} + \frac{\beta}{\alpha+\beta}\mathbf{y}.$$

Next, since H is convex and contains **q** and **z**, H must contain

$$\frac{\alpha+\beta}{\alpha+\beta+\gamma}\mathbf{q} + \frac{\gamma}{\alpha+\beta+\gamma}\mathbf{z} = \frac{\alpha}{\alpha+\beta+\gamma}\mathbf{x} + \frac{\beta}{\alpha+\beta+\gamma}\mathbf{y} + \frac{\gamma}{\alpha+\beta+\gamma}\mathbf{z}.$$

Finally, we find by induction (see Problem 12.5) that H must contain

$$\frac{\alpha}{\alpha+\beta+\gamma+\cdots+\omega}\mathbf{x} + \cdots + \frac{\omega}{\alpha+\beta+\gamma+\cdots+\omega}\mathbf{u} = \alpha\mathbf{x} + \beta\mathbf{y} + \cdots + \omega\mathbf{u}.$$

Example 12.1.18. **(Continuation of Example 12.1.11.)** Continuing Example 12.1.11, we now wish to plot the risk points of the randomized decision rules. (Those of the nonrandomized rules were plotted in Figure 12.1–2.) For any such rule $s_\mathbf{p}$, the risk point is (see (12.1.15)) the vector

$$\begin{pmatrix} r_{\theta_1}(s_\mathbf{p}) \\ r_{\theta_2}(s_\mathbf{p}) \end{pmatrix} = \begin{pmatrix} 17p_1 + 19p_2 + 14p_3 + 10p_4 + 9p_5 + 9p_6 \\ 14p_1 + 4p_2 + 4p_3 + 6p_4 + 8p_5 + 16p_6 \end{pmatrix} \tag{12.1.19}$$

$$= p_1\begin{pmatrix}17\\14\end{pmatrix} + p_2\begin{pmatrix}19\\4\end{pmatrix} + p_3\begin{pmatrix}14\\4\end{pmatrix} + p_4\begin{pmatrix}10\\6\end{pmatrix} + p_5\begin{pmatrix}9\\8\end{pmatrix} + p_6\begin{pmatrix}9\\16\end{pmatrix}.$$

However, this is just a point in the convex hull of the risk points of the nonrandomized decision rules, and as **p** varies over all probability distributions (that is, as we consider all decision rules) we generate (by Theorem 12.1.16) the convex hull of the six nonrandomized decision rules' risk points $\binom{17}{14}, \binom{19}{4}, \binom{14}{4}, \binom{10}{6}, \binom{9}{8}$, and $\binom{9}{16}$. Thus, the set of all risk points is convex and closed, and is plotted in Figure 12.1–3. It is easy to see that the

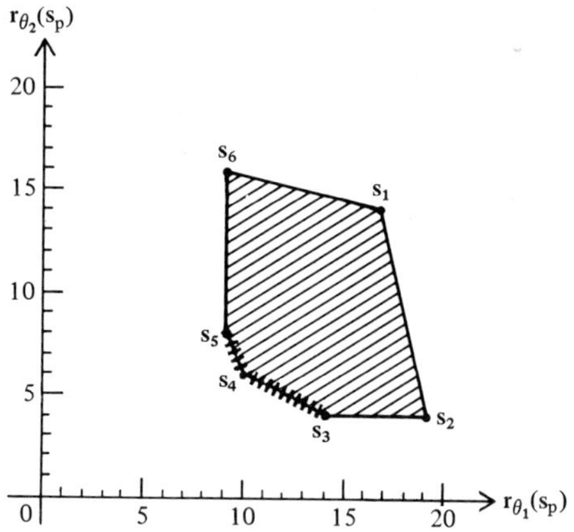

Figure 12.1–3. Closed convex set of risk points for Example 12.1.11

rules leading to the hatched (⧺⧺⧺) part of the boundary of the set of risk points (in Figure 12.1–3) form a minimal essentially complete class: any other rule s either yields a risk point on this hatched part of the boundary (in which case it is no better) or yields a risk point elsewhere in the set, in which case some decision rule t with a risk point on the hatched part of the boundary has $r_{\theta_1}(t) \leqslant r_{\theta_1}(s)$ and $r_{\theta_2}(t) \leqslant r_{\theta_2}(s)$ with at least one strict inequality ($<$). (To see this just note that if $(r_{\theta_1}(t), r_{\theta_2}(t))$ is the risk point of a decision rule t, then any rule s with a risk point $(r_{\theta_1}(s), r_{\theta_2}(s))$ in the shaded region in Figure 12.1–4 is inferior to t.) The rules forming this minimal essentially complete class are those with $(p_1, p_2, p_3, p_4, p_5, p_6)$ with

$$\begin{cases} (1) & p_1 = p_2 = p_3 = p_6 = 0, \quad p_4 \geqslant 0, \quad p_5 = 1 - p_4 \geqslant 0; \\ (2) & p_1 = p_2 = p_5 = p_6 = 0, \quad p_3 \geqslant 0, \quad p_4 = 1 - p_3 \geqslant 0; \end{cases} \tag{12.1.20}$$

that is, those which (1) randomize between d_4 and d_5; and (2) those which randomize between d_3 and d_4.

Returning to the problem stated immediately preceding Example 12.1.11, how do we find a minimal complete (or better, essentially complete) class of decision rules in the general decision problem (Θ, $\mathscr{D}, l$), X? (Here in the

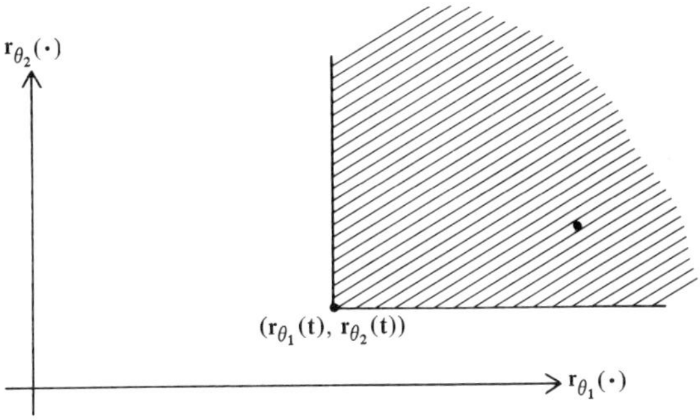

Figure 12.1–4. Rules worse than *t*

general case, we have no simple solution like use of the Neyman-Pearson Lemma in Section 9.4.) A helpful notion here is that of "admissibility." Recall (see (12.1.5), (12.1.6), Figure 12.1–1, and Figure (12.1–4)) that many decision rules are "beaten" by other decision rules. Could the set of those rules which are not so "beaten" (called "admissible" rules) be a minimal complete class? We will now investigate this question, first defining "admissibility" precisely and then proving some complete class results.

Definition 12.1.21. *For any statistical decision problem* $(\Theta, \mathscr{D}, l)$, X, *decision rule* $s(\cdot)$ *is said to be* **admissible** *if there is no decision rule* $t(\cdot)$ *which is better than* $s(\cdot)$. *A rule is said to be* **inadmissible** *if it is not admissible.*

Lemma 12.1.22. *If* $s_1(\cdot)$ *and* $s_2(\cdot)$ *are two admissible decision rules for the same statistical decision problem* $(\Theta, \mathscr{D}, l)$, X, *then either* $s_1 \sim s_2$ *or* s_1 *and* s_2 *are incomparable.*

Proof: See Problem 12.7.

Theorem 12.1.23. *For a statistical decision problem* $(\Theta, \mathscr{D}, l)$, X, *if a minimal complete class* C *exists, then* $C = A$ *where* A *is the set of all admissible rules.*

Proof: It is easy to show that A is a subset of every complete class (Problem 12.8); hence, $A \subseteq C$. It remains to show that $C \subseteq A$.

For suppose it is *not* the case that $C \subseteq A$. Then there is a rule s_0 such that $s_0 \in C$ and $s_0 \notin A$.

Since s_0 is inadmissible, there is a rule $s \succ s_0$. If $s \in C$, let $s_1 = s$. If $s \notin C$, there is (since C is complete) an $s_1 \in C$ such that $s_1 \succ s$ (hence, $s_1 \succ s_0$). Thus, there is a rule $s_1 \in C$ such that $s_1 \succ s_0$.

Now let C_1 be C with s_0 removed. We claim C_1 is then complete (which contradicts the assumption that C is minimal complete, and thereby shows that the assertion upon which the contradiction was based ("it is *not* the case that $C \subseteq A$") is false; that is, it shows that $C \subseteq A$).

For suppose $t \notin C_1$. If $t = s_0$, then $s_1 \in C_1$ and $s_1 \succ s_0$. If $t \neq s_0$, there must be a rule $t' \in C$ such that $t' \succ t$. If $t' = s_0$, then $s_1 \succ t$ and $s_1 \in C_1$. If $t' \neq s_0$, then $t' \in C_1$ and $t' \succ t$. Thus, there is always a rule in C_1 which is better than such a rule t, so C_1 is complete, the contradiction which completes the proof.

Thus, **for problems where a minimal complete class C exists (these comprise most of the problems of interest to us) the problem of finding that class C is equivalent to finding the set A of all admissible rules. The problem of finding the admissible rules** (it suffices to find only one rule for each risk point, for if several rules are admissible and have the same risk point, then all but one of the rules can be deleted from A, forming A', and A' will be an essentially complete class; in fact, A' will be minimal essentially complete (see Problem 12.10)) **is intimately connected with the notions of Bayes (and to some extent of minimax) rules**, which we will discuss after giving an example.

Example 12.1.24. In Example 12.1.18 the admissible rules are those corresponding to the hatched boundary of the set of risk points in Figure 12.1–3. It was shown there that there is exactly one rule corresponding to each such point (although other risk points may correspond to several rules (see Problem 12.11)), and that their class was minimal essentially complete.

We will now give the general definition of a Bayes decision rule.

Definition 12.1.25. *Let $(\Theta, \mathscr{D}, l)$, X be a statistical decision problem. Let $B(\theta)$ be a distribution function on Θ. The quantity*

$$R_{B(\theta)}(s) = Er_\theta(s) \tag{12.1.26}$$

is called the **Bayes risk of s(·) with respect to the a priori distribution B(θ).** [It is calculated as if θ were a r.v. with d.f. $B(\theta)$.] *A decision rule $t(\cdot)$ is called* **Bayes (with respect to the a priori distribution B(θ))** *if $t(\cdot)$ has the smallest*[1] *Bayes risk*:

$$R_{B(\theta)}(t) = \inf_{s \in D} R_{B(\theta)}(s). \tag{12.1.27}$$

Recall (see the discussion following Theorem 9.4.7) that once we have a class C (perhaps minimal essentially complete) of decision rules from which we desire to choose *one* rule, the risk function $r_\theta(s)$ is generally of little help since rules in C are incomparable (see Figure 12.1–1). Thus, **what is really needed is a way to assign a number to each decision rule, with the desirability**

[1] It may be helpful to review the footnote (on "sup" and "inf") to (7.1.2) at this point.

of a decision rule increasing as this number decreases. The Bayes risk is one proposal for such a number, and a Bayes rule is best when this method of assigning numbers is used.

We can interpret the Bayes risk (12.1.26) of $s(\cdot)$ as the average risk when in fact θ is a r.v. with d.f. $B(\theta)$. If this approach is taken, then $B(\theta)$ is usually said to involve **subjective (or personal) probability**: it incorporates our "feelings" about what θ may actually occur. The suggestion that we ought to attempt to formulate our "feelings" about θ into such a $B(\theta)$ and use the resulting Bayes rule has been responsible for insane debates and ill-will in statistics (one reason being the insistence of some proponents of this method that it is the only method any sensible person ought to use), and we shall leave that subject to those who have nothing better to do with their time. We should **note that there are several cases in which the proper approach may be to use a Bayes rule for a specified $B(\theta)$. First**, θ may actually be a r.v. and if the same situation has been faced many times before, we may have a good knowledge of $B(\theta)$. [For example, the severity of fires at alarms on Saturday nights in July is a random phenomenon θ, and past experience gives us a good idea of the distribution $B(\theta)$ over severities. It is reasonable to use a Bayes rule with respect to $B(\theta)$ in allocating fire-fighting apparatus and personnel to such alarms.] **Second**, there may be partial knowledge about θ which can be well-incorporated into such a $B(\theta)$. [For example, a business executive may have insight into a real-world system which can be expressed in terms of θ being a r.v. with a $B(\theta)$ in a certain class of possible distributions. It is then reasonable for him to give special consideration to the set of Bayes rules for such $B(\theta)$'s.] **Third**, it has been shown[2] that, if our preferences for combinations of θ's and decisions d satisfy certain axioms, then the decision we will make will be that which we would take if θ were known to be a r.v. with a certain $B(\theta)$. [The weakness of this third case is that the axioms involved are very strong. For example, if we prefer a prospect (future possible occurrence) P_1 to P_2, and prefer P_2 to P_3, then we are required to prefer P_1 to P_3. But if we are, say, choosing to bet on a certain player in a sports event, we could prefer P_1 to P_2, P_2 to P_3, yet P_3 to P_1 (since situations are well-known where P_1 usually beats P_2, P_2 usually beats P_3, yet P_3 usually beats P_1). Also, in subjects of scientific inference we may validly state we have no preference between whether a hypothesis is true or not, but that we simply wish to determine the truth of the matter. The third case then falls apart at the outset.]

Of course, in some situations it may be the case that no Bayes rule exists (for example, see Problem 12.13). For this reason we define "ε-Bayes" rules (rules which come within ε of the inf of $R_{B(\theta)}(s)$):

---

[2] See T. S. Ferguson, *Mathematical Statistics: A Decision Theoretic Approach*, Academic Press Inc., New York, 1967, pp. 17–19, for an introduction to this topic.

Definition 12.1.28. *Let $(\Theta, \mathscr{D}, l)$, X be a statistical decision problem. Let $B(\theta)$ be a d.f. on Θ, and let ε be > 0. A decision rule $t(\cdot)$ is called* **ε-Bayes (with respect to $B(\theta)$)** *if*

$$R_{B(\theta)}(t) \leqslant \inf_{s \in D} R_{B(\theta)}(s) + \varepsilon. \tag{12.1.29}$$

[However, even ε-Bayes rules may fail to exist.]

We now give the general definition of a minimax decision rule (and, for reasons similar to those leading to the introduction of the ε-Bayes rules, of the ε-minimax decision rules).

Definition 12.1.30. *Let $(\Theta, \mathscr{D}, l)$, X be a statistical decision problem. Let ε be > 0. A decision rule $t(\cdot)$ is called* **minimax** *if*

$$\sup_{\theta \in \Theta} r_\theta(t) = \inf_{s \in D} \sup_{\theta \in \Theta} r_\theta(s), \tag{12.1.31}$$

and **ε-minimax** *if*

$$\sup_{\theta \in \Theta} r_\theta(t) \leqslant \inf_{s \in D} \sup_{\theta \in \Theta} r_\theta(s) + \varepsilon. \tag{12.1.32}$$

Before continuing on to study the relationships between Bayes rules, admissible rules, and (minimal essentially) complete classes, **we will now give an example (which provides some insight into Bayes rules)**.

Example 12.1.33. Continuing Examples 12.1.11, 12.1.18, and 12.1.24, find a Bayes rule with respect to the a priori distribution $B(\theta)$ which assigns probability w to θ_1 and $1 - w$ to θ_2 (where $0 \leqslant w \leqslant 1$).

This no-data decision problem has two possible states of nature (θ_1, θ_2) and six possible decisions $(d_1, d_2, d_3, d_4, d_5, d_6)$, and a general decision rule $s_{\mathbf{p}}$ (see (12.1.12)) takes decision d_i with probability p_i $(1 \leqslant i \leqslant 6)$ where $\mathbf{p} = (p_1, p_2, p_3, p_4, p_5, p_6)$ is a probability distribution. Since (see (12.1.15))

$$\begin{pmatrix} r_{\theta_1}(s_{\mathbf{p}}) \\ r_{\theta_2}(s_{\mathbf{p}}) \end{pmatrix} = \begin{pmatrix} 17p_1 + 19p_2 + 14p_3 + 10p_4 + 9p_5 + 9p_6 \\ 14p_1 + 4p_2 + 4p_3 + 6p_4 + 8p_5 + 16p_6 \end{pmatrix}, \tag{12.1.34}$$

$$\begin{aligned} R_{B(\theta)}(s_{\mathbf{p}}) &= Er_\theta(s_{\mathbf{p}}) = wr_{\theta_1}(s_{\mathbf{p}}) + (1 - w)r_{\theta_2}(s_{\mathbf{p}}) \\ &= w(17p_1 + 19p_2 + 14p_3 + 10p_4 + 9p_5 + 9p_6) \\ &\quad + (1 - w)(14p_1 + 4p_2 + 4p_3 + 6p_4 + 8p_5 + 16p_6) \qquad (12.1.35) \\ &= p_1(14 + 3w) + p_2(4 + 15w) + p_3(4 + 10w) \\ &\quad + p_4(6 + 4w) + p_5(8 + w) + p_6(16 - 7w). \end{aligned}$$

However, this is minimized by choosing $p_i = 1$ (where p_i is that one of $p_1, p_2, p_3, p_4, p_5, p_6$ with the smallest coefficient) and the other p_j's $= 0$. If we graph the coefficients as functions of w (see Figure 12.1–5) we find that a Bayes rule is to choose $p_3 = 1$ when $0 < w \leqslant \frac{1}{3}$, $p_4 = 1$ when $\frac{1}{3} \leqslant w \leqslant \frac{2}{3}$, and $p_5 = 1$ when $\frac{2}{3} \leqslant w < 1$ (since, for example, $4 + 10w$ is the smallest multi-

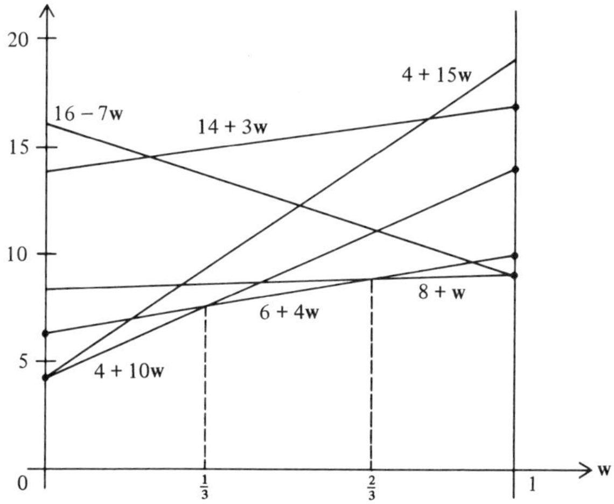

Figure 12.1–5. Components of Bayes risk in Example 12.1.33

plier of a p_i when $0 \leqslant w < \frac{1}{3}$, and it multiplies p_3). At the "boundary" w's ($w = \frac{1}{3}, \frac{2}{3}$) we can also randomize since more than one multiplier has the same minimum value (for example, at $w = \frac{1}{3}$ both $4 + 10w$ and $6 + 4w$, which multiply p_3 and p_4 respectively, have the same minimum value $7\frac{1}{3}$; hence, any rule with $p_3 + p_4 = 1$ is then Bayes). The points $\frac{1}{3}, \frac{2}{3}$ were determined graphically; for example, the lower point is the intersection of the lines $4 + 10w$ and $6 + 4w$:

$$\begin{cases} 4 + 10w = 6 + 4w, \\ 6w = 2, \\ w = \frac{1}{3}. \end{cases} \tag{12.1.36}$$

So by varying $0 < w < 1$ (that is, the a priori distribution $B(\theta)$) we find that the set of all **positive-Bayes rules** (that is, rules each of which is Bayes with respect to some $B(\theta)$ which sets neither w nor $1 - w$ to 0) consists of s_3, s_4, s_5, rules which randomize between d_3 and d_4, and rules which randomize between d_4 and d_5. These rules' risk points generate precisely the hatched portion of the boundary of the convex set of risk points (see Figure 12.1–3). The set of *all* Bayes rules consists of those rules plus the rules which are Bayes for $w = 0$ (s_2, s_3, and rules which randomize between d_2 and d_3) and for $w = 1$ (s_5, s_6, and rules which randomize between d_5 and d_6).

When Θ consists of only two points (here θ_1 and θ_2) the Bayes rule may be found geometrically as follows. We want a rule whose risk point minimizes (see (12.1.35))

$$w r_{\theta_1}(s_{\mathbf{p}}) + (1 - w) r_{\theta_2}(s_{\mathbf{p}}). \tag{12.1.37}$$

But this quantity is constant at value k on the line

$$wr_{\theta_1}(s_{\mathbf{p}}) + (1 - w)r_{\theta_2}(s_{\mathbf{p}}) = k. \tag{12.1.38}$$

If $w = 1$, this is the line

$$r_{\theta_1}(s_{\mathbf{p}}) = k \tag{12.1.39}$$

($x = b$ is the x-y plane); otherwise ($0 \leqslant w < 1$) it is the line (in standard $y = mx + b$ form)

$$r_{\theta_2}(s_{\mathbf{p}}) = \frac{-w}{1 - w} r_{\theta_1}(s_{\mathbf{p}}) + \frac{k}{1 - w}, \tag{12.1.40}$$

with intercept $k/(1 - w)$ and negative (unless $w = 0$) slope $-w/(1 - w)$. All rules whose risk points lie on this line will have Bayes risk k, but if k is "too small" there are no such rules. However, if we increase k until the line "just touches" the convex set of risk points, we will find the Bayes rules' risk point (and hence the Bayes rules) for the given a priori distribution. See Figure 12.1–6 for a typical case. (Also see Figure 9.4–5 for a similar construction in the case of hypothesis testing.) Note that what we called "positive-Bayes" rules are precisely the admissible rules in this example.

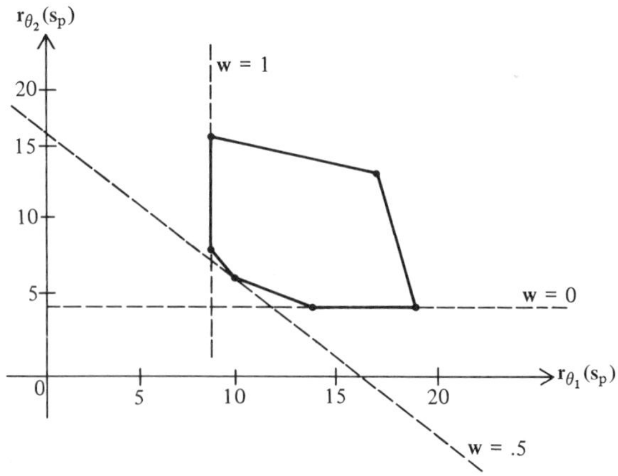

Figure 12.1–6. Graphical interpretation of Bayes rules

We continue now to the relationships between Bayes rules, admissible rules, and (minimal essentially) complete classes. Most of our work will now deal with the case where Θ is finite (although the results are true in considerably more generality, their proofs and statements in that generality are beyond the scope of this book). **We will first prove a lemma which will be used in proving that all admissible rules are Bayes rules.**

Lemma 12.1.41. *Let $(\Theta, \mathscr{D}, l)$, X be a statistical decision problem with Θ finite (that is, $\Theta = \{\theta_1, \ldots, \theta_k\}$). Then the set S of all risk points* [see (12.1.13)] *of decision rules s,*

$$S = \{(y_1, \ldots, y_k): y_1 = r_{\theta_1}(s), \ldots, y_k = r_{\theta_k}(s) \text{ for some decision rule } s\}, \quad (12.1.42)$$

is a convex subset of $\mathscr{R}^k$.

Proof: Let $u = (u_1, \ldots, u_k)$ and $v = (v_1, \ldots, v_k)$ be in S. Then by the definition of S, there are decision rules $s(\cdot)$ and $t(\cdot)$ such that

$$u_j = r_{\theta_j}(s), \quad v_j = r_{\theta_j}(t) \quad (1 \leqslant j \leqslant k). \quad (12.1.43)$$

Let λ be fixed $(0 < \lambda < 1)$. Let $\alpha(\cdot)$ be a decision rule which with probability λ takes the same decision as rule $s(\cdot)$ (and with probability $1 - \lambda$ takes the same decision as rule $t(\cdot)$). Then (for $j = 1, \ldots, k$)

$$r_{\theta_j}(\alpha) = \lambda r_{\theta_j}(s) + (1 - \lambda) r_{\theta_j}(t), \quad (12.1.44)$$

so $\lambda u + (1 - \lambda)v$ is in S. Hence, S is convex.

A partial converse of Lemma 12.1.41 states that any convex subset S of $\mathscr{R}^k$ is the set of risk points of at least one decision problem $(\Theta, \mathscr{D}, l)$, X. (See Problem 12.18.) Hence, counterexamples constructed using such sets are valid in general (for example, see Problem 12.6).

Theorem 12.1.45. *Let $(\Theta, \mathscr{D}, l)$, X be a statistical decision problem with Θ finite* (that is, $\Theta = \{\theta_1, \ldots, \theta_k\}$). *Then every admissible rule $t(\cdot)$ is also a Bayes rule with respect to some a priori distribution $B(\theta)$.*

Proof: Let $S \subseteq \mathscr{R}^k$ be the set of all risk points of decision rules $s(\cdot)$. Let

$$\mathbf{x}' = (r_{\theta_1}(t), \ldots, r_{\theta_k}(t)) \quad (12.1.46)$$

$$Q(\mathbf{x}) = \{\mathbf{y}: \mathbf{y} \in \mathscr{R}^k, y_1 \leqslant x_1, \ldots, y_k \leqslant x_k\}. \quad (12.1.47)$$

Then $Q(\mathbf{x}) \cap S = \{\mathbf{x}\}$ since $t(\cdot)$ is by assumption an admissible rule (for if any point $\mathbf{z}$ other than $\mathbf{x}$ were in the intersection, there would be a rule with that as its risk point, hence, there would be a better rule than $t(\cdot)$, which would render $t(\cdot)$ inadmissible, a contradiction). Hence, $S_1 = S$ and $S_2 = \{\mathbf{y}: \mathbf{y} \in Q(\mathbf{x}), \mathbf{y} \neq \mathbf{x}\} = Q(\mathbf{x}) - \{\mathbf{x}\}$ are disjoint convex sets, so by the Separating Hyperplane Theorem (Theorem 8.3.6) there is a vector $\mathbf{p} \neq \mathbf{0}$ such that

$$\mathbf{p}'\mathbf{y} \leqslant \mathbf{p}'\mathbf{z} \quad (\text{all } \mathbf{y} \in Q(\mathbf{x}) - \{\mathbf{x}\}, \mathbf{z} \in S). \quad (12.1.48)$$

Now $p_1 \geqslant 0, \ldots, p_k \geqslant 0$. For if at least one of $p_1, \ldots, p_k$ were negative, say $p_j < 0$, then by taking $\mathbf{y}$ such that y_j is a large (negative) number we could make

$$\mathbf{p}'\mathbf{y} = \sum_{i=1}^{k} p_i y_i > \sum_{i=1}^{k} p_i z_i = \mathbf{p}'\mathbf{z},$$

a contradiction. Now let

$$q_i = \frac{p_i}{p_1 + \cdots + p_k}. \tag{12.1.49}$$

Then $q_1 \geqslant 0, \ldots, q_k \geqslant 0, q_1 + \cdots + q_k = 1$ (so $\mathbf{q}$ is a probability distribution on Θ) and

$$\mathbf{q}'\mathbf{y} \leqslant \mathbf{q}'\mathbf{z} \quad (\text{all } \mathbf{y} \in Q(\mathbf{x}) - \{\mathbf{x}\}, \mathbf{z} \in S); \tag{12.1.50}$$

that is (letting $\mathbf{y} \in Q(\mathbf{x}) - \{\mathbf{x}\} \to \mathbf{x}$)

$$\sum_{j=1}^{k} q_j r_{\theta_j}(t) \leqslant \sum_{j=1}^{k} q_j r_{\theta_j}(s) \tag{12.1.51}$$

for any rule $s(\cdot)$ (since any such rule $s(\cdot)$ has a risk point $\mathbf{z} \in S$). Hence, $t(\cdot)$ is a Bayes rule with respect to the a priori distribution $\mathbf{q}$ on Θ.

We will now prove a partial converse to Theorem 12.1.45, after first making precise the notion of a "positive-Bayes" rule (see (12.1.36)ff in Example 12.1.33).

Definition 12.1.52. *Let* $(\Theta, \mathscr{D}, l)$, X *be a statistical decision problem with* Θ *finite (that is,* $\Theta = \{\theta_1, \ldots, \theta_k\}$*). A decision rule* $s(\cdot)$ *is called* **positive-Bayes** (**+ -Bayes**) *if it is Bayes with respect to some a priori distribution* $B(\theta)$ *which assigns positive probability to each* θ_j $(j = 1, \ldots, k)$.

Theorem 12.1.53. *Let* $(\Theta, \mathscr{D}, l)$, X *be a statistical decision problem with* Θ *finite (that is,* $\Theta = \{\theta_1, \ldots, \theta_k\}$*). Then every +-Bayes rule* $t(\cdot)$ *is admissible.*

Proof: Suppose that $t(\cdot)$ is +-Bayes for $B(\theta)$ which assigns probability p_j to θ_j $(j = 1, \ldots, k)$. If $t(\cdot)$ were inadmissible, then for some $s(\cdot)$,

$$\begin{cases} r_{\theta_j}(s) \leqslant r_{\theta_j}(t) & (j = 1, \ldots, k) \\ r_{\theta_i}(s) < r_{\theta_i}(t) & (\text{at least one } i). \end{cases} \tag{12.1.54}$$

Since $p_1 > 0, \ldots, p_k > 0$, this would imply

$$\sum_{j=1}^{k} p_j r_{\theta_j}(s) < \sum_{j=1}^{k} p_j r_{\theta_j}(t), \tag{12.1.55}$$

so that $t(\cdot)$ could not be Bayes with respect to $B(\theta)$, a contradiction.

The complete class theorem will now follow easily from Theorem 12.1.45, Problem 12.3H, and Problem 12.10.

Theorem 12.1.56. **(Complete Class Theorem)** *Let* $(\Theta, \mathscr{D}, l)$, X *be a statistical decision problem with* Θ *finite (that is,* $\Theta = \{\theta_1, \ldots, \theta_k\}$*). If the set* S *of risk points is closed*[3] *and bounded*[3] *then:* **(1)** *the set of all Bayes rules is a complete class;* **(2)** *the set of admissible Bayes rules is a minimal complete class; and* **(3)** *the set of admissible Bayes rules with duplication of risk points eliminated is a minimal essentially complete class.*

Other important theorems are available which consider Bayes and admissible rules, but the Complete Class Theorem (Theorem 12.1.56) is the major such theorem and should provoke great interest in specification of the class of Bayes rules. (Further detail on computation of Bayes rules will be given in Section 12.5.) For completeness we state two additional results, without proof. (The proof of Theorem 12.1.57 is simple; see Problem 12.19.)

Theorem 12.1.57. *Let* $(\Theta, \mathscr{D}, l)$, X *be a statistical decision problem. If every Bayes rule with respect to a priori distribution* $B(\theta)$ *has the same risk point, then each such rule is admissible.*

Theorem 12.1.58. *Let* $(\Theta, \mathscr{D}, l)$, X *be a statistical decision problem with* Θ *finite (that is,* $\Theta = \{\theta_1, \ldots, \theta_k\}$*). Assume the set* S *of risk points is bounded*[4] *and closed.*[4] *Then all possible* $+$*-Bayes rules exist (that is, for every a priori distribution* $B(\theta)$ *which assigns positive probability* p_j *to* θ_j, $j = 1, \ldots, k$, *a Bayes rule exists).*

12.2. HYPOTHESIS TESTING

In this section we begin our consideration of the relationships of decision theory with "conventional" (non decision-theoretic) statistical theory. The decision-theoretic view of the hypothesis-testing problem (especially the problem where H_0 and H_1 are both simple hypotheses, the case of composite hypotheses being noted in Problems 9.12H and 9.13H) has already been covered in Section 9.4 (it is helpful to review that section at this point), and the tie-in with the general results of Section 12.1 is clear: **a hypothesis-testing problem** where $H_0 = \{\theta_0\}$ and $H_1 = \{\theta_1\}$ are both simple and we observe a r.v. X with d.f. $F(x|\theta)$ where θ is unknown $(\theta \quad \Theta)$, **can be regarded as a decision problem** $(\Theta, \mathscr{D}, l)$, X where $\Theta = \{\theta_0, \theta_1\}$, X is unaltered, $\mathscr{D} = \{d_0, d_1\}$ where d_0 is the decision to accept H_0 and d_1 is the decision to reject H_0, and where $l(\theta_0, d_0) = l(\theta_1, d_1) = 0$ and $l(\theta_0, d_1) = l(\theta_1, d_0) = 1$. A general randomized decision rule $s(X)$ specifies (see Definition 12.1.2) for each

[3] This hypothesis can be weakened; see Problem 12.3H.
[4] This hypothesis can be weakened as in Theorem 12.1.56.

$x \in \mathscr{X}$ a probability distribution $s(x)$ according to which a member d of $\mathscr{D}$ is to be chosen. Thus, $s(X)$ is precisely a general randomized test, and the risk point of $s(X)$ (as at (9.4.1)) is simply a point in $\mathscr{R}^2$ giving the probabilities of Type I and Type II errors of $s(X)$. It then follows easily that the considerations of the rest of Section 9.4 were those of Section 12.1 for the (simple versus simple) hypothesis-testing case.

In Section 12.1 we used a no-data decision problem to illustrate and motivate the concepts of that section (see Examples 12.1.11, 12.1.18, 12.1.24, and 12.1.33). Now, **in order to illustrate the gains possible due to having data, we will consider a hypothesis-testing problem with data**.

Example 12.2.1. Consider the statistical decision problem $(\Theta, \mathscr{D}, l)$, X where $\Theta = \{\theta_0, \theta_1\}$ with $\theta_0 = 0$ and $\theta_1 = 2$, $\mathscr{D} = \{d_0, d_1\}$ with $d_0 \neq d_1$, $l(\theta_0, d_0) = l(\theta_1, d_1) = 0$, $l(\theta_0, d_1) = l(\theta_1, d_0) = 1$, and X has a normal distribution with unknown mean $\theta \in \Theta$ and known variance $\sigma^2 = 4$. (Thus, we have a hypothesis-testing problem.) Find a minimal essentially complete class of decision rules (tests) and graph their risk points. How would the graph change if we had no data X?

By Theorem 9.4.7, $\{\phi_\alpha(\cdot): \alpha \leqslant \alpha_0\}$ is a minimal essentially complete class of tests (decision rules) where α_0 is the smallest α for which $1 - \beta(\phi_\alpha) = 0$, where $\phi_\alpha(\cdot)$ is the (level α) Neyman-Pearson test which satisfies

$$E_{\theta_0}\phi(X) = \alpha \tag{12.2.2}$$

and

$$\phi(x) = \begin{cases} 1, & \text{when } f(x|\theta_1) > kf(x|\theta_0) \\ 0, & \text{when } f(x|\theta_1) < kf(x|\theta_0). \end{cases} \tag{12.2.3}$$

In our case (X a $N(\theta, 4)$ r.v.) $f(x|\theta_1)$ and $f(x|\theta_0)$ are each > 0 for all x, so let us consider the case $k > 0$ [$k \leqslant 0$ yields $\phi(x) \equiv 1$, which gives a (Type I error, Type II error) or risk point of $(1, 0)$]. Now (using the facts $\theta_0 = 0$, $\theta_1 = 2$, and $\sigma^2 = 4$)

$$\begin{aligned} & f(x|\theta_1) > kf(x|\theta_0) \\ \Leftrightarrow\quad & \frac{1}{\sqrt{2\pi}\sigma} e^{-\frac{1}{2}\left(\frac{x-\theta_1}{\sigma}\right)^2} > k\frac{1}{\sqrt{2\pi}\sigma} e^{-\frac{1}{2}\left(\frac{x-\theta_0}{\sigma}\right)^2} \\ \Leftrightarrow\quad & -\frac{1}{2}\left(\frac{x-\theta_1}{\sigma}\right)^2 > -\frac{1}{2}\left(\frac{x-\theta_0}{\sigma}\right)^2 + \ln k \\ \Leftrightarrow\quad & -(x-\theta_1)^2 > -(x-\theta_0)^2 + 2\sigma^2 \ln k \\ \Leftrightarrow\quad & 2x(\theta_1 - \theta_0) > (\theta_1^2 - \theta_0^2) + 2\sigma^2 \ln k \\ \Leftrightarrow\quad & 4x > 4 + 8 \ln k \\ \Leftrightarrow\quad & x > 1 + 2 \ln k \end{aligned} \tag{12.2.4}$$

and the test with critical function (12.2.3) then has

$$E_{\theta_0}\phi(X) = P_{\theta_0}[X > 1 + 2\ln k] = 1 - \Phi\left(\frac{1 + 2\ln k - 0}{2}\right) = 1 - \Phi\left(\frac{1 + 2\ln k}{2}\right), \tag{12.2.5}$$

which will equal α iff

$$1 + 2\ln k = 2\Phi^{-1}(1 - \alpha). \tag{12.2.6}$$

It follows that $\alpha_0 = 1$ (since no test $\phi(\cdot)$ with $k > 0$ has $1 - \beta(\phi) = 0$). Hence, the test $\phi_\alpha(\cdot)$ which rejects iff $X > 2\Phi^{-1}(1 - \alpha)$ has level α and $\{\phi_\alpha(\cdot): \alpha \leqslant 1\}$ is a minimal essentially complete class of tests. To plot the graph we also need

$$1 - \beta(\phi_\alpha) = P_{\theta_1}[X < 2\Phi^{-1}(1 - \alpha)] = \Phi\left(\frac{2\Phi^{-1}(1 - \alpha) - 2}{2}\right) = \Phi(\Phi^{-1}(1 - \alpha) - 1). \tag{12.2.7}$$

Hence, the plot consists of $(\alpha, \Phi(\Phi^{-1}(1 - \alpha) - 1))$ for $0 \leqslant \alpha \leqslant 1$ as in Figure 12.2–1. Without data, as in Example 12.1.18 we find the set of risk points is as in Figure 12.2–2, and the rise in $1 - \beta(\phi_\alpha)$ for a given level α is clear.

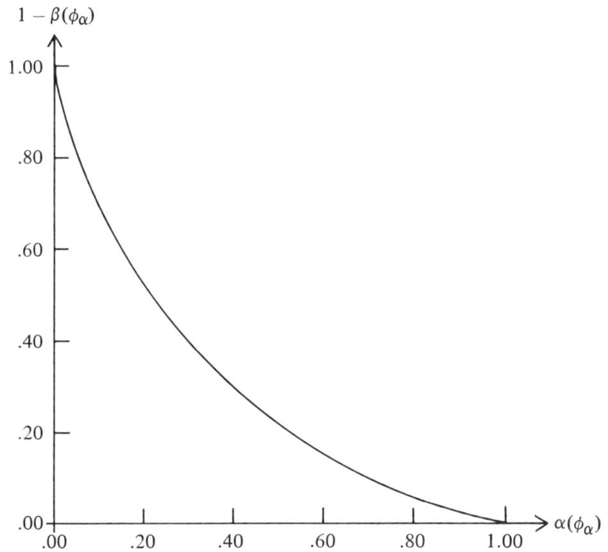

Figure 12.2–1. Risk points of the minimal essentially complete class with data in Example 12.2.1

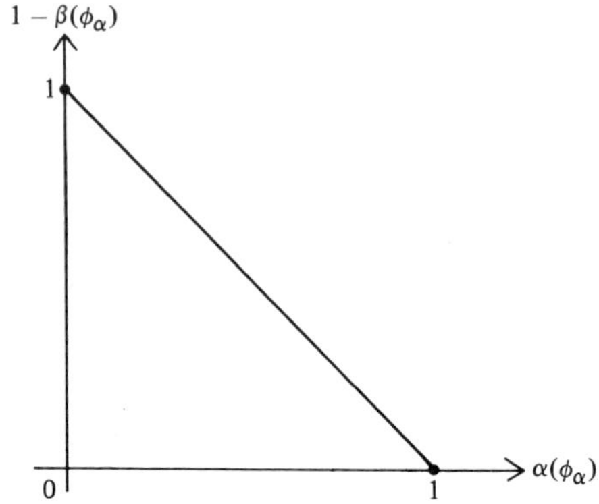

Figure 12.2–2. No-data risk points for Example 12.2.1

In hypothesis-testing problems, we can take a *more general loss function* than $l(\theta_0, d_0) = l(\theta_1, d_1) = 0$, $l(\theta_0, d_1) = l(\theta_1, d_0) = 1$. (For example, we might take $l(\theta_0, d_0) = l(\theta_1, d_1) = 0$ and $l(\theta_0, d_1) = 1$ but $l(\theta_1, d_0) = cl(\theta_0, d_1)$ with a $c > 1$. This would count Type II error as "worse" than Type I error; for example, if $c = 2$, it is "twice as bad.") This is rarely done, since it yields nothing new (it simply shifts the convex set of risk points by multiplying each y-coordinate by a constant c, while the minimal essentially complete class of tests remains a minimal essentially complete class of tests). We could also attempt to assess *monetary gains and losses* in constructing $l(\theta, d)$. However, in problems of scientific inference (for example, in the testing of whether a new drug is better than an old drug for the same condition) this is often impossible. In such cases the decision-theoretic approach yields insight into the testing problem.

Further consideration of hypothesis-testing problems could be given at this point (in particular, it would be possible to consider *composite hypotheses* in detail). However, in our view this is more a part of hypothesis testing (already covered in some detail in Chapter 9) than of decision theory, and once the student learns more advanced topics in hypothesis-testing he should have no trouble putting them into the decision-theoretic framework (in order to gain its insight) if he has understood this section. [See, for example, pp. 137–145 of Weiss (1961).]

12.3. INTERVAL ESTIMATION

In this section we continue our consideration of the relationships of decision theory with "conventional" (non decision-theoretic) statistical theory, showing how the problem of interval estimation fits into the decision-theoretic framework.

Recall (see Definition 10.1.10) that if X is a r.v. with d.f. $F(x|\theta)$ where θ is unknown and $\theta \in \Theta$ (here X and θ may be multivariate), then the set $S(X)$ is called a *confidence set* for θ with *confidence coefficient* $1 - \alpha$ if

$$P_\theta[\theta \in S(X)] \overset{\theta}{\equiv} 1 - \alpha. \tag{12.3.1}$$

The corresponding decision problem $(\Theta, \mathscr{D}, l)$, X has Θ and X unaltered, while $\mathscr{D}$ consists of all possible subsets of Θ. A general randomized decision rule $s(X)$ specifies (see Definition 12.1.2) for each $x \in \mathscr{X}$ a probability distribution $s(x)$ according to which a member $d_{s(X)}$ of $\mathscr{D}$ (that is, a subset of Θ) is to be chosen. If we take our loss function to be

$$l(\theta, d) = \begin{cases} 1, & \text{if } \theta \notin d_{s(X)} \\ 0, & \text{if } \theta \in d_{s(X)}, \end{cases} \tag{12.3.2}$$

then our risk function for a general randomized decision rule $s(X)$ is

$$r_\theta(s) = E_\theta l(\theta, s(X)) = P_\theta[\theta \notin d_{s(X)}]. \tag{12.3.3}$$

Hence, a confidence set with confidence coefficient $1 - \alpha$ corresponds to a decision rule $s(X)$ with risk function $r_\theta(x) \overset{\theta}{\equiv} \alpha$.

Once the above correspondence has been made, the problem of interval estimation emerges as just a special problem of decision theory, which may hence be treated by the general methods of decision theory. Generalizations of loss function (12.3.2) are possible, but we shall not go into those here. As in the case of hypothesis testing (see Section 12.2), the main value of the decision-theoretic approach may be the insight it yields into the interval estimation problem and the unity it gives to statistics (since statistical problems can be viewed as decision problems, whether they are problems of hypothesis testing, interval estimation, or other areas of statistics).

12.4. RANKING AND SELECTION

In this section we conclude our formal consideration of the relationships of decision theory with "conventional" statistical theory, showing how the problems of ranking and selection fit into the decision-theoretic framework. (An informal discussion of point estimation and decision theory will be given in Section 12.5, essentially through an example.)

In order to simplify the discussion we will consider only the problem of selection of the best (*largest mean*) *normal population* of Section 11.2. (The results carry over without any difficulty to general problems of ranking and selection.) Recall that we have $k \geqslant 2$ populations $\pi_1, \ldots, \pi_k$ such that if $X_{i1}, X_{i2}, \ldots$ are n independent observations from π_i, then each X_{ij} is $N(\mu_i, \sigma^2)$ where μ_i is unknown and σ^2 is known $(1 \leqslant i \leqslant k)$. The goal is to select that population with the largest mean $\mu_{[k]} = \max(\mu_1, \ldots, \mu_k)$, in such a way that

$$P(\text{CS}) \geqslant P^* \quad \text{whenever} \quad \mu_{[k]} - \mu_{[k-1]} \geqslant \delta^* \tag{12.4.1}$$

(see (11.2.5)). The corresponding decision problem $(\Theta, \mathscr{D}, l)$, X has $\Theta = \mathscr{R}^k$ (the set of all possible $(\mu_1, \ldots, \mu_k)$), $X = (X_{11}, \ldots, X_{1n}; \ldots; X_{k1}, \ldots, X_{kn})$, and $\mathscr{D} = \{d_1, \ldots, d_k\}$, where d_i denotes the decision that π_i is best $(1 \leqslant i \leqslant k)$. If we take our loss function to be

$$l((\mu_1, \ldots, \mu_k), d_i) = \begin{cases} 1, & \text{if } \mu_i \neq \max(\mu_1, \ldots, \mu_k) \\ 0, & \text{if } \mu_i = \max(\mu_1, \ldots, \mu_k), \end{cases} \tag{12.4.2}$$

then our risk function for a general randomized decision rule $s(X)$ is

$$r_\theta(s) = E_\theta l(\theta, s(X)) = 1 - P_\theta(\text{CS using } s(X)), \tag{12.4.3}$$

the probability of an incorrect decision when $\theta = (\mu_1, \ldots, \mu_k)$ is true and we use decision rule $s(X)$. Hence, a selection procedure with $P(\text{CS}) \geqslant P^*$ whenever $\mu_{[k]} - \mu_{[k-1]} \geqslant \delta^*$ corresponds to a decision rule $s(X)$ whose risk function satisfies

$$r_\theta(s) \leqslant 1 - P^* \quad \text{for all } \theta \text{ such that} \quad \mu_{[k]} - \mu_{[k-1]} \geqslant \delta^*. \tag{12.4.4}$$

With the above correspondence made, the problem of selection fits very neatly into the decision-theoretic framework (of which it was, as we noted at the beginning of Chapter 11, very suggestive). Such problems are sometimes called *multiple-decision* problems, and a number of papers have appeared which consider Bayes and minimax solutions for such problems. (The solutions presented in Chapter 11 generally are these rules for many loss functions.) For example, see M. L. Eaton, "Some optimum properties of ranking procedures," *Annals of Mathematical Statistics*, Vol. 38 (1967), pp. 124–137, and the references given there. Important Bayesian decision procedures for selection problems were discussed by C. W. Dunnett, "On selecting the largest of k normal population means," *Journal of the Royal Statistical Society, Series B*, Vol. 22 (1960), pp. 1–30 (also see the discussion on pp. 31–40).

12.5. GENERAL DECISION THEORY

In this section we continue the discussion of Section 12.1, discussing some aspects of general decision theory (as contrasted with the considerations of Sections 12.2, 12.3, and 12.4, which were concerned with relations of general decision theory with "conventional" statistical theory).

Our first concern in this section will be to discuss in more detail the construction (or computation) of Bayes rules. (This was promised in Section 12.1 just before Theorem 12.1.57, and is important because, as was seen in

Section 12.1, Bayes rules are important; hence, we need to know how to find them.)

Recall (see Definition 12.1.25) that a Bayes rule for a statistical decision problem $(\Theta, \mathscr{D}, l)$, X and a given a priori distribution $B(\theta)$ on Θ is a decision rule $s(X)$ which has the smallest Bayes risk:

$$R_{B(\theta)}(s) \leqslant R_{B(\theta)}(t) \tag{12.5.1}$$

for all decision rules $t(X)$. Now the Bayes risk of a decision rule $s(X)$ is

$$\begin{aligned} R_{B(\theta)}(s) &= E_\theta r_\theta(s) \\ &= E_\theta E_X l(\theta, s(X)), \end{aligned} \tag{12.5.2}$$

where the subscript on the expectation sign denotes the quantity with respect to which the expectation is being performed. Now suppose Θ is finite ($\Theta = \{\theta_1, \ldots, \theta_k\}$), $\mathscr{D}$ is finite ($\mathscr{D} = \{d_1, \ldots, d_m\}$), and X is discrete. Let $b(\theta)$ denote the probability assigned to θ by $B(\theta)$, and let $f(x|\theta)$ denote $P_\theta[X = x]$. Then the Bayes risk (12.5.2) of a rule $s(X)$ can be written as

$$R_{B(\theta)}(s) = \sum_{\theta\in\Theta} \sum_{x\in\mathscr{X}} \sum_{d\in\mathscr{D}} l(\theta, d) s_d(x) f(x|\theta) b(\theta), \tag{12.5.3}$$

where $s_d(x)$ denotes the probability that after observing $X = x$, decision rule $s(X)$ takes decision d. (Recall that a decision rule is a function $s(\cdot)$ which for each $x \in \mathscr{X}$ specifies a probability distribution, denoted by $s(x)$, according to which a member d of $\mathscr{D}$ is to be chosen.) We can now interchange summations in (12.5.3) and write

$$\begin{aligned} R_{B(\theta)}(s) &= \sum_{x\in\mathscr{X}} \sum_{d\in\mathscr{D}} s_d(x) \left[\sum_{\theta\in\Theta} l(\theta, d) f(x|\theta) b(\theta) \right] \\ &= \sum_{x\in\mathscr{X}} \sum_{d\in\mathscr{D}} s_d(x) K(d; x), \end{aligned} \tag{12.5.4}$$

where we have defined

$$K(d; x) = \sum_{\theta\in\Theta} l(\theta, d) f(x|\theta) b(\theta). \tag{12.5.5}$$

Now since our aim is to choose $s(\cdot)$ so as to minimize (12.5.4), that is, to choose the probability distribution $s(x)$ for each x in such a way that (12.5.4) is minimized, it is clear we should proceed as follows: For each $x \in \mathscr{X}$, let

$$s_{d_i}(x) = 0, \quad \text{if } K(d_i; x) > \min(K(d_1; x), \ldots, K(d_m; x)). \tag{12.5.6}$$

Then all weight of the probability distribution $s(x)$ will be put on decisions d_j which minimize $K(d; x)$. (If there are several such decisions, the Bayes rule is not unique.) Thus, we have the following.

Theorem 12.5.7. *Let $(\Theta, \mathscr{D}, l)$, X be a statistical decision problem. Suppose Θ and $\mathscr{D}$ are finite and X is discrete. Then a rule $s(X)$ is a Bayes rule* [with respect to a priori distribution $B(\theta)$ which assigns probability $b(\theta)$ to θ] *iff, for all* $x \in \mathscr{X}$, $s_{d_i}(x) = 0$ *for all* d_i *for which*

$$K(d_i; x) > \min(K(d_1; x), \ldots, K(d_m; x)), \tag{12.5.8}$$

where

$$K(d; x) = \sum_{\theta \in \Theta} l(\theta, d) f(x|\theta) b(\theta). \tag{12.5.9}$$

The above theorem also holds in the general case where Θ and $\mathscr{D}$ need not be finite and X need not be discrete (sums often being replaced by integrals and probabilities often being replaced by density functions in the obvious way) as long as a Bayes rule with respect to $B(\theta)$ exists for the given problem. Note that in any problem, if all we want to know is which action to take (after observing $X = x$) we need not compute the whole rule, but rather only the $K(d; x)$ for $d \in \mathscr{D}$ for the given x. Note also that a Bayes rule need never randomize: we can decide (for each $x \in \mathscr{X}$) to set $s_d(x) = 1$ for some one of the (in general several) d_i's which minimize $K(d; x)$.

Our second concern in this section is to note briefly other important aspects of decision theory. One such aspect is a consideration of facts about minimax decision rules. Since these have already been covered, we will just note that Problems 12.24 and 12.25 are relevant here, and that Problem 12.25 gives a widely used method of finding such rules. **A second aspect is a consideration of loss versus regret** (see Definition 9.4.8 for a definition of regret, in the special case of hypothesis testing). Essentially, the question is whether it might be appropriate to replace the loss function $l(\theta, d)$ by a new loss function

$$\text{reg}(\theta, d) = l(\theta, d) - l(\theta) \tag{12.5.10}$$

(called **regret**) where $l(\theta)$ is the smallest $l(\theta, d)$ can be for any d (with θ fixed). Indeed, the loss functions used with "conventional" statistical theory (see Sections 12.2, 12.3, and 12.4) are usually already regret functions. Appealing as the idea may seem to some, replacement of loss functions by regret functions (in cases where the loss functions are not already regret functions at the outset) has not gained wide popularity. One reason is that in a practical problem we will suffer losses (not regrets!) after the decision is taken, and decision theory is attempting to take proper account of those losses. Another reason is that (see Problem 12.27) whether we use losses or regrets, we obtain the same Bayes rule for a given a priori distribution $B(\theta)$. **A third aspect involves the uses of sufficient statistics.** Here it can be shown in great generality that if $(\Theta, \mathscr{D}, l)$, X is a statistical decision problem and $t(X)$ is a sufficient statistic, then: the set of all decision rules for the new statistical decision problem $(\Theta, \mathscr{D}, l)$, $t(X)$ is an essentially complete class of decision rules for the original decision problem $(\Theta, \mathscr{D}, l)$, X. Hence, we can reduce (by sufficiency) the data as much as possible (that is, use a minimal sufficient statistic if one exists) without "losing any information about θ" as far as the decision problem is concerned. (For example, the $N(\theta, 4)$ r.v. X in Example

12.2.1 would often be the average of several other independent normal r.v.'s.) For the proof (which is not hard, but which we omit) see, for example, p. 120 of T. S. Ferguson, *Mathematical Statistics: A Decision Theoretic Approach*, Academic Press Inc., New York, 1967. (For a related example see B. C. Arnold, "Some examples of minimum variance unbiased estimates," *The American Statistician*, Vol. 26, No. 4 (1972), pp. 34–36. The problem of whether we may omit randomization has light shed on it by an example in Section 2 of that paper.) **A fourth (and final) aspect of our discussion will be the relation between "point estimation" and decision theory**. Problems of point estimation are essentially decision-theory problems with $\mathscr{D} = \Theta$ (that is, we wish to "estimate θ"). One example of this sort is given in Problem 12.26, and we will now give another (more theoretical) one.

Example 12.5.11. Let $(\Theta, \mathscr{D}, l)$, X be a decision problem with $\Theta = \mathscr{D} = \mathscr{R}$ (that is, a problem of point estimation of a real-valued parameter). Suppose that

$$\begin{cases} l(\theta, d) = g(|\theta - d|) \\ g(x) < g(y), \quad \text{whenever } x < y \end{cases} \tag{12.5.12}$$

[that is, the loss is an increasing function of the absolute deviation of d from θ]. Suppose that X has a density function $f(\cdot)$ such that

$$\begin{cases} f(x) > 0, & (-\infty < x < +\infty) \\ f(\theta + x) = f(\theta - x), & (-\infty < x < +\infty) \\ f(\theta + c_1) > f(\theta + c_2), & \text{whenever } 0 \leqslant c_1 < c_2 \end{cases} \tag{12.5.13}$$

[that is, X has a positive density, symmetric about θ, with a unique maximum at θ]. Show that, among all nonrandomized decision rules $s(X)$ which satisfy the invariance condition $s(X + c) = s(X) + c$ for all c, $s(X) = X$ has the uniformly smallest risk $r_\theta(s) = El(\theta, s(X))$.

Since $s(\cdot)$ must increase its estimate of θ by c if X increases by c (that is, since $s(X + c) = s(X) + c$), we have $s(0) = s(X + (-X)) = s(X) + (-X)$, so $s(X) = X + s(0)$. Thus, the most general such rule estimates θ as $X + s$ for some constant s, and for any such nonrandomized rule $s(X)$,

$$\begin{aligned} r_\theta(s) &= El(\theta, s(X)) \\ &= \int_{-\infty}^{\infty} g(|s(x) - \theta|)f(x)\,dx \\ &= \int_{-\infty}^{\infty} g(|x + s - \theta|)f(x)\,dx. \end{aligned} \tag{12.5.14}$$

If we can show that $s = 0$ minimizes (12.5.14), we will have shown that $s(X) = X$ has uniformly smallest risk for all θ among the *translation invariant* estimators; that is, we wish to show that

$$\int_{-\infty}^{\infty} g(|x + s - \theta|)f(x)\,dx \geqslant \int_{-\infty}^{\infty} g(|x - \theta|)f(x)\,dx. \tag{12.5.15}$$

Now (using the transformation $y = x - \theta$ in the third step)

$$\begin{aligned}
\int_{-\infty}^{\infty} g(|x + s - \theta|)f(x)\,dx - \int_{-\infty}^{\infty} g(|x - \theta|)f(x)\,dx \\
&= \int_{-\infty}^{\infty} \{g(|x + s - \theta|) - g(|x - \theta|)\}f(x)\,dx \\
&= \int_{-\infty}^{\infty} \{g(|y + s|) - g(|y|)\}f(y + \theta)\,dy \qquad (12.5.16) \\
&= \int_{-\infty}^{-s/2} \{g(|y + s|) - g(|y|)\}f(y + \theta)\,dy \\
&\quad + \int_{-s/2}^{\infty} \{g(|y + s|) - g(|y|)\}f(y + \theta)\,dy.
\end{aligned}$$

We now assume without loss that $s \geqslant 0$ (see Problem 12.28). Then (making the transformation $z = -y - s$ in the second integral in the last expression of (12.5.16)) we find

$$\begin{aligned}
\int_{-\infty}^{\infty} g(|x + s - \theta|)f(x)\,dx - \int_{-\infty}^{\infty} g(|x - \theta|)f(x)\,dx \\
&= \int_{-\infty}^{-s/2} \{g(|y + s|) - g(|y|)\}f(y + \theta)\,dy \\
&\quad + \int_{-\infty}^{-s/2} \{g(|-z|) - g(|-z - s|)\}f(-z - s + \theta)\,dz \qquad (12.5.17) \\
&= -\int_{-\infty}^{-s/2} \{g(|y|) - g(|y + s|)\}f(y + \theta)\,dy \\
&\quad + \int_{-\infty}^{s/2} \{g(|z|) - g(|z + s|)\}f(z + s + \theta)\,dz \\
&= \int_{-\infty}^{-s/2} [g(|y|) - g(|y + s|)][f(y + s + \theta) - f(y + \theta)]\,dy.
\end{aligned}$$

But $y + s + \theta \geqslant y + \theta$ (since $s \geqslant 0$); and $\theta \geqslant y + s + \theta \geqslant y + \theta$ if $y \leqslant -s$ (hence, for $y \leqslant -s$ we have $f(y + s + \theta) \geqslant f(y + \theta)$) while for $-s < y \leqslant -s/2$ we have $y + s + \theta \geqslant \theta \geqslant y + \theta$ with $y + s + \theta$ closer to θ than is $y + \theta$ (hence, for $-s < y \leqslant -s/2$ we have $f(y + s + \theta) \geqslant f(y + \theta)$). Thus, for all possible values of y in (12.5.17),

$$f(y + s + \theta) - f(y + \theta) \geqslant 0. \qquad (12.5.18)$$

Also, for all possible y in (12.5.17),

$$g(|y|) - g(|y + s|) \geqslant 0. \qquad (12.5.19)$$

Hence, (12.5.17) (and thus (12.5.16)) is $\geqslant 0$, so (12.5.15) holds, proving the desired result. *This result can be stated as follows*: "For fairly general loss functions [those which are increasing functions of the absolute error], the uniformly minimum risk translation invariant decision rule sets $s(X) = X$ if X has a symmetric unimodal positive density with median θ."

12.6. AN OVERVIEW

Our purpose in this section is, through a discussion of a specific example, to give an overview of the area of statistics (and of its classification into subfields).

Two of these subfields (nonparametric methods and regression analysis) are discussed in Chapters 13 and 14; our main purpose here is to provide further integration of statistics into a coherent whole in the readers' minds. A secondary purpose is to note the terminology commonly used in referring to the various subfields so that a person wishing to pursue specific aspects in greater detail will know what key words to find in surveying books and journal articles. (Of course, not all areas are mentioned—statistics is too large for that—but enough are that the discussion should be enlightening.)

The example to be discussed deals with various aspects of the successful operation of a chain of pizza parlors. If the opening of a new establishment in a particular city (say Columbus, Ohio, to be specific) is under consideration, we are then faced with a problem of making a decision to open or not to open such a facility, with concomitant monetary gains or losses depending on the success of the establishment if opened (a problem of **decision theory**).

In order to assess the situation sufficiently precisely (that is, to make our risk function small) we will need data X appropriate to our situation. One part of such data might consist of a **sample survey** of a number of people in the Columbus area, asking each questions designed to allow us to gauge their expected consumption of our pizzas if we open the parlor. Some problems here are those of obtaining a representative sample large enough to yield estimates of small variance on expected consumption, those of how to proceed if a person we decide to include in the sample does not respond, and those of how to verify whether the questions are perceived by the respondents as asking what we desire to ask (and whether the respondents are answering truthfully).

Other data obtained might include observations on the consumption of pizzas experienced at the existing pizza purveying establishments in the area. Such data are random variables, and one question (that of **distribution theory**) is what probability distribution we should assume such data follow. If we have a specific distribution in mind, we may wish to test the hypothesis that it is the true distribution (**goodness-of-fit tests**). Alternatively, we may wish to only make some weak assumptions on the distribution involved (such as its being a member of a unimodal location parameter family as in Example 12.5.11) and attempt to draw inferences hence not dependent on strong distributional assumptions (**nonparametric methods**). If the number of observations is large, we may be able to use **asymptotic theory** to assess the distribution, for example via the empirical d.f. A question which could then arise would be the **efficiency (exact and asymptotic)** of this asymptotic theory with respect to exact theory based on the distribution we think may underlie the data.

If the data obtained with respect to existing establishments span a long time period, we might suspect trends or other changes due to time (since over a long time consumption is probably not a sequence of independent and identically distributed r.v.'s). In order to find and adjust for any time trends, periodicities in the data (for example, more pizzas sold in winter than in spring), and so on, the techniques of **statistical analysis of stochastic processes (or time series)** are needed.

Once the data has been adequately assessed and a decision taken, if that decision is to open the new parlor in Columbus new questions arise. One of these is the **estimation** of the demand to be experienced so that the facility may be of adequate size (yet not of excess size). This may also involve predicting not only the initial demand, but also the growth in demand to be expected over the near future. This prediction involves such items as population growth and income levels in the area, and trends in transportation, competition, and so on, and would utilize the methods of **linear statistical inference and regression analysis**. Since these data are comprised of several interrelated components, the techniques would involve **multivariate analysis**.

In operating the establishment the satisfaction of the customers will be of great importance, and methods of **quality control** will be needed to assure that we act as soon as possible when the production process starts to increase its defective (poor-tasting, burned, and so on) pizzas. It may also be necessary to assess the effects of the establishment's heat and waste on the surrounding environment, and the effects of our "special secret ingredients" on our customers' health (**biometrics**).

In operating the chain of establishments it will be necessary to buy large quantities of apparatus such as ovens, and their expected lifetimes (assessed by **life-testing) and reliability** will have to be studied quantitatively. We may even wish to oversee the design and manufacture of the items using **engineering statistics**.

At some time we will undoubtedly face the question of whether to change the recipe for the pizza or the sequence in which it is produced (what goes on first, for example), which will involve us in the **design and analysis of experiments**, for we will then wish to perform a taste-testing experiment to assess the desirability of the various alternatives.

Other problems to be faced will involve **ranking and selection** of vendors of goods; **testing of hypotheses** about the superiority of goods which are more expensive (for example, a special cheese); analysis of the sequence of daily sales (involving **sequential methods**); design of our facility to keep waiting time reasonable (**queueing theory and congestion**) and to keep an adequate inventory of supplies without having them kept so long as to spoil (**inventory and storage theory**); and **simulation** of proposed changes in our methods of operation of the chain of pizza parlors in order to pretest our decisions in a less costly environment than the real world.

PROBLEMS

12.1 *Given a set S, a relation ρ between pairs of members of S is said to be:* **reflexive** *if $x\rho x$ for all $x \in S$;* **antisymmetric** *if $x\rho y$ and $y\rho x \Rightarrow x$ and y denote the same element of S; and* **transitive** *if $x\rho y$ and $y\rho z \Rightarrow x\rho z$. If a relation ρ on a set S is reflexive, antisymmetric, and transitive, it is called a* **partial order** *and S is said to be* **partially ordered** *by ρ.*

Show that (if we consider two decision rules to be the same if they have the same risk function) the "at least as good as" relation $\gtrsim$ (see the paragraph following Definition 12.1.10) partially orders the set D of all decision rules.

12.2 Sketch the convex hull of the three vectors $\binom{1}{0}$, $\binom{0}{1}$, $\binom{5}{6}$.

12.3 In Example 12.1.18, is a rule which randomizes between d_3 and d_5 a member of the minimal essentially complete class?

12.4 Prove that if $a \geqslant 0$ and $b \geqslant 0$, then $\lambda a + (1 - \lambda)b \geqslant 0$ when $0 \leqslant \lambda \leqslant 1$. (Is this true for $\lambda < 0$? For $\lambda > 1$?)

12.5 Show in detail the induction referred to in the proof of Theorem 12.1.16.

12.6 For each of the following figures, suppose that the shaded area is the set of risk points corresponding to all decision rules for some decision problem $(\Theta, \mathscr{D}, l)$, X. Find a subset R_C of this set of risk points such that: if C is a set of decision rules which (for every $r \in R_C$) contains exactly one decision rule with risk point r, then C is a minimal essentially complete class.

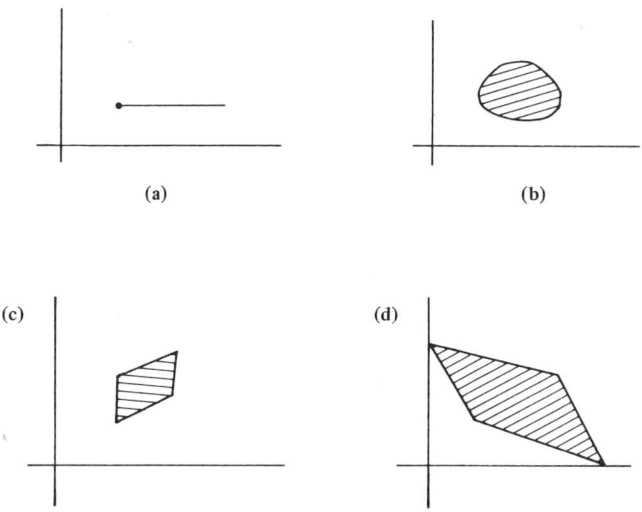

(a) (b) (c) (d)

12.7 Prove Lemma 12.1.22. [*Hint*: See the discussion preceding Figure 12.1–1.]

12.8 For a statistical decision problem $(\Theta, \mathscr{D}, l)$, X, let C be a complete class of decision rules and let A be the set of all admissible decision rules. Prove that $A \subseteq C$.

12.9 For a statistical decision problem $(\Theta, \mathscr{D}, l)$, X, let C be an essentially complete class of decision rules. Prove that if $s_1(\cdot)$ is an admissible decision rule and $s_1 \notin C$, then there is an $s_2(\cdot) \in C$ such that $s_1 \sim s_2$.

12.10 For a statistical decision problem $(\Theta, \mathscr{D}, l)$, X, suppose that the set A of all admissible rules is minimal complete. Let A' be A except that whenever several rules in A have the same risk point all but one of these rules are deleted. Show that A' is an essentially complete class. Show that A' is a minimal essentially complete class.

12.11 In Example 12.1.18, a risk point (that is, a point in Figure 12.1–3 which is the risk point of at least one decision rule) may correspond to more than one rule. Find all rules which have risk point $\binom{15}{10}$. [*Hint*: Since here a general decision rule is $s_{\mathbf{p}} = (p_1, \ldots, p_6)$ a probability distribution, we must find all $\mathbf{p}$ such that

$$\begin{pmatrix} r_{\theta_1}(s_{\mathbf{p}}) \\ r_{\theta_2}(s_{\mathbf{p}}) \end{pmatrix} = \begin{pmatrix} 15 \\ 10 \end{pmatrix},$$

that is, all probability distributions $\mathbf{p}$ such that

$$\begin{pmatrix} 17p_1 + 19p_2 + 14p_3 + 10p_4 + 9p_5 + 9p_6 \\ 14p_1 + 4p_2 + 4p_3 + 6p_4 + 8p_5 + 16p_6 \end{pmatrix} = \begin{pmatrix} 15 \\ 10 \end{pmatrix},$$

that is, all solutions $(p_1, \ldots, p_6)$ of the system of equations

$$\begin{cases} 17p_1 + 19p_2 + 14p_3 + 10p_4 + 9p_5 + 9p_6 = 15, \\ 14p_1 + 4p_2 + 4p_3 + 6p_4 + 8p_5 + 16p_6 = 10, \\ p_1 + p_2 + p_3 + p_4 + p_5 + p_6 = 1, \\ p_1 \geqslant 0, p_2 \geqslant 0, p_3 \geqslant 0, p_4 \geqslant 0, p_5 \geqslant 0, p_6 \geqslant 0. \end{cases}$$

To solve such a system, solve the first equation for p_1 in terms of p_2, p_3, p_4, p_5, p_6. Plug this into the second equation and solve for p_2 in terms of p_3, p_4, p_5, p_6. Use these to solve the third equation for p_3 in terms of p_4, p_5, p_6. Now express p_1 and p_2 in terms of p_4, p_5, p_6 alone (p_3 is already so expressed). Then the inequalities $p_1 \geqslant 0, p_2 \geqslant 0, p_3 \geqslant 0, p_4 \geqslant 0, p_5 \geqslant 0, p_6 \geqslant 0$ restrict the available set of p_4, p_5, p_6 to (say) $F \subseteq \mathscr{R}^3$. Having found F, a general solution consists of picking a point p_4, p_5, p_6 from F and finding the corresponding p_1, p_2, p_3.]

12.12 As in Problem 12.11, but for risk point $\binom{10}{6}$. What happens if you try the point $\binom{5}{5}$ (which is not a risk point)?

12.13 Consider the no-data decision problem $(\Theta, \mathscr{D}, l)$ with $\Theta = \mathscr{D} = \mathscr{R}$ and $l(\theta, d) = (\theta - d)^3$. Show that no matter what $B(\theta)$ is used, among the nonrandomized decision rules there is no Bayes rule. [*Hint*: In the plane plot graphs similar to those in Figure 12.1–1 for a few decision rules. Show that, no matter how small a rule s_1 may make the Bayes risk, it is easy to find another rule s_2 with a smaller risk. Since this risk can be made $< k$ for any specified number k $(-\infty < k < +\infty)$, conclude that among all decision rules there is no Bayes rule.]

12.14 In the context of Problem 12.13, show that no ε-Bayes rules exist.

12.15 In the context of Example 12.1.33 find the minimax rule and give a graphical interpretation (see Figure 9.4–4).

12.16 *A* **least-favorable a priori distribution B(θ)** *is one which makes the minimum Bayes risk as large as possible*. Deduce from Figure 12.1–5 that in Example 12.1.33 that distribution is $B(\theta)$ with $w = 1$. Show that the minimax rule is a rule which is Bayes for the least-favorable a priori distribution in this example.

12.17 On a graph similar to that in Figure 12.1–6, mark the risk points of all ε-Bayes decision rules, with $\varepsilon = 1$. Similarly for all ε-minimax decision rules, with $\varepsilon = 1$.

12.18 Let S be a nonempty convex subset of $\mathscr{R}^k$. Consider the no-data decision problem $(\Theta, \mathscr{D}, l)$ with $\Theta = \{\theta_1, \ldots, \theta_k\}$, $\mathscr{D} = S$, and $l(\theta_j, d) = d_j$ [the jth component of the decision

$$d = (d_1, \ldots, d_{j-1}, d_j, d_{j+1}, \ldots, d_k)].$$

Show that the set of all risk points of decision rules is S.

Deduce that any convex subset of $\mathscr{R}^k$ is the set of risk points of at least one decision problem $(\Theta, \mathscr{D}, l)$, X (irrespective of whether S is closed and bounded or not). Contrast this with the situation which prevailed in Section 9.4 (see Theorem 9.4.2 and footnote 5 of Chapter 9), where S was closed and bounded as well as convex.

12.19 Prove Theorem 12.1.57. [*Hint*: Assume the contrary and obtain a contradiction.]

12.20 Use the results of Problem 12.18 to show that Theorem 12.1.56 (the Complete Class Theorem) would sometimes be false if S were not required to be closed and bounded.

12.21 For the situation of Example 12.2.1, plot the set of risk points of a minimal essentially complete class for the case where the data X is replaced by $Y = (X_1, X_2, X_3, X_4)$ where X_1, X_2, X_3, X_4 are independent $N(\theta, \sigma^2)$ r.v.'s with $\sigma^2 = 4$. Compare this set with that of Figure 12.2–1; what gain arises from the three additional observations? [*Hint*: The calculations of Example 9.3.10 may be of use to you.]

12.22 Show that the usual loss functions for the hypothesis-testing problem's decision-theoretic statement,

$$\begin{cases} l(\theta_0, d_0) = l(\theta_1, d_1) = 0, \\ l(\theta_0, d_1) = 1, \quad l(\theta_1, d_0) = c > 1, \end{cases}$$

are already regret functions. [*Hint*: See footnote 7 in Chapter 9.]

12.23 Suppose that in a (simple versus simple) hypothesis-testing problem with loss function

$$\begin{cases} l(\theta_0, d_0) = l(\theta_1, d_1) = 0, \\ l(\theta_0, d_1) = 1, \quad l(\theta_1, d_0) = 1, \end{cases}$$

the minimax test has level α and Type II error $1 - \beta$. Show that if the loss function is altered by making

$$l(\theta_1, d_0) = c > 1,$$

then the minimax test for this new problem will have level $\alpha^* > \alpha$ and Type II error $1 - \beta^* < 1 - \beta$. Give a graphical interpretation involving a convex set of risk points.

12.24 Let $(\Theta, \mathscr{D}, l)$, X be a statistical decision problem. Prove that if every minimax rule has the same risk point, then each such rule is admissible. [*Hint*: This result is similar to that of Theorem 12.1.57, which dealt with Bayes rules.] *Note that such a minimax rule is often called* "**unique**."

12.25 Let $(\Theta, \mathscr{D}, l)$, X be a statistical decision problem. Prove that: *If a decision rule* $s(\cdot)$ *has constant risk (that is, if* $r_\theta(s)$ *is the same for all* θ *in* Θ*) and is a Bayes rule, then it is a minimax rule.* (Hence, if we can "guess" an a priori distribution $B(\theta)$ for which the corresponding Bayes rule has constant risk, we can find a minimax rule.)

12.26 In reliability (life-testing), it is sometimes assumed that the lifetime of an item is a r.v. which is $\geqslant x$ with probability

$$e^{-(x/\theta)^m}$$

$(x \geqslant 0)$ where $m > 0$ is known and $\theta > 0$ is unknown. Suppose n such items are tested, and yield independent r.v.'s $X_1, \ldots, X_n$ as their times of "death" (or "failure").

(a) Find an unbiased estimator of θ. Find the MLE of θ based on $X_1, \ldots, X_n$.

(b) Consider the statistical decision problem $(\Theta, \mathscr{D}, l)$, X with $\Theta = \{\theta : \theta > 0\}$, $\mathscr{D} = \Theta$, $X = (X_1, \ldots, X_n)$, and

$$l(\theta, d) = \left(\frac{d - \theta}{\theta}\right)^2.$$

Show that any decision rule which sets

$$d = a(X_1^m + \cdots + X_n^m)^{1/m}$$

(for some fixed $a > 0$) has constant risk. Is the decision rule which always sets d equal to the MLE admissible? Is the decision rule which always sets d equal to the unbiased estimator found in part (a) admissible? Find a minimax decision rule.

12.27 Show that: *Any rule is Bayes with respect to an a priori distribution* $B(\theta)$ *when we use the loss function, iff that same rule is Bayes when we use the regret function (12.5.10).* Show by example that minimax rules do not have the corresponding property. (Hence, Bayes rules are invariant under the change of a loss function to a regret function, but minimax rules are usually not.)

12.28 In Example 12.5.11, show that

$$\int_{-\infty}^{\infty} \{g(|z - s|) - g(|z|)\} f(z + \theta)\, dz = \int_{-\infty}^{\infty} \{g(|y + s|) - g(|y|)\} f(y + \theta)\, dy.$$

Explain how this justifies the statement following (12.5.16) that we can "assume without loss that $s \geqslant 0$."

12.29 Let $(\Theta, \mathscr{D}, l)$, X be a statistical decision problem with $\Theta = \mathscr{D} = \mathscr{R}$, $l(\theta, d) = (\theta - d)^2$, and X a $N(\theta, 1)$ r.v. Among all decision rules $s(X)$ such that $s(X + c) = s(X) + c$ for all c, there is one which minimizes $r_\theta(s)$ for all θ. Find it.

12.30 Let $(\Theta, \mathscr{D}, l)$, X be a statistical decision problem with $\Theta = \{-2, 3\}$, $\mathscr{D} = \{d_0, d_1\}$, $l(-2, d_0) = l(3, d_1) = 0$, $l(-2, d_1) = 6$, $l(3, d_0) = 4$, and $X = (X_1, X_2, X_3, X_4)$ where X_1, X_2, X_3, X_4 are independent $N(\theta, 4)$ r.v.'s. For the a priori distribution which assigns probability w to $\theta = -2$ and probability $1 - w$ to $\theta = 3$ $(0 \leqslant w \leqslant 1)$, find a Bayes rule $s_w(X)$. Plot the risk points $(r_{-2}(s_w), r_3(s_w))$ for $0 \leqslant w \leqslant 1$. [*Hint*: Use sufficiency to reduce $X = (X_1, X_2, X_3, X_4)$ to $\bar{X}$ in order to simplify your calculations.] On the same diagram plot the risk points of the Bayes rules in the no-data decision problem $(\Theta, \mathscr{D}, l)$.

12.31 Let $(\Theta, \mathscr{D}, l)$, X be a statistical decision problem with $\Theta = \mathscr{D} = \{x : 0 \leqslant x \leqslant 1\}$, $l(\theta, d) = (d - \theta)^2$, and $X = (X_1, \ldots, X_n)$ where $X_1, \ldots, X_n$ are independent r.v.'s with $P_\theta[X_i = 0] = 1 - \theta$ and $P_\theta[X_i = 1] = \theta$ $(1 \leqslant i \leqslant n)$. For the a priori distribution which puts a density function $b(\theta) = 2\theta$ for $0 \leqslant \theta \leqslant 1$ (and zero otherwise) on θ, find a Bayes decision rule $s(X)$. Compute $r_\theta(s)$. [*Hint*: Use sufficiency to reduce $X = (X_1, \ldots, X_n)$ to a binomial r.v.] Compare $s(\cdot)$ with the decision rule $t(X)$ which sets $t(X) = (X_1 + \cdots + X_n)/n$ (the observed proportion of successes in the sample). How is $r_\theta(s)$ related to $r_\theta(t)$ as $n \to \infty$? What is the Bayes rule for the no-data decision problem $(\Theta, \mathscr{D}, l)$?

12.32 For the decision problem $(\Theta, \mathscr{D}, l)$, X of Problem 12.30, find a minimax decision rule. [*Hint*: See Problem 12.25.]

12.33 An oil slick has spread from 0 (where it occurred) out to a radius $\theta > 0$. The slick is always spread uniformly over the area it covers. The slick is hard to spot from the air unless seagulls are stuck in it, and luckily (!) three are, at distances X_1, X_2, X_3 from the origin 0. We are thus motivated to consider the following estimation problem in decision-theoretic formulation: $\Theta = \mathscr{D} = \{x : 0 < x < \infty\}$, $l(\theta, d) = ((d/\theta) - 1)^2$, and $X = (X_1, X_2, X_3)$ where X_1, X_2, X_3 are independent r.v.'s, each with a uniform distribution on $(0, \theta)$. For the a priori distribution which puts a density function $b(\theta) = \theta^5 e^{-\theta}/5!$ for $\theta > 0$ (and zero otherwise) on θ, find a Bayes decision rule $s(X)$.

12.34 For the no-data decision problem $(\Theta, \mathscr{D}, l)$ with $\Theta = \{\theta_1, \theta_2, \theta_3, \theta_4\}$, $\mathscr{D} = \{d_1, d_2, d_3, d_4, d_5, d_6, d_7, d_8, d_9\}$, and $l(\theta, d)$ as specified in the following table

$l(\theta, d)$									
θ \ d	d_1	d_2	d_3	d_4	d_5	d_6	d_7	d_8	d_9
θ_1	80	25	40	89	23	53	74	5	71
θ_2	6	86	79	62	75	83	6	8	44
θ_3	63	86	75	75	87	73	3	12	42
θ_4	9	54	52	67	18	70	47	72	61

find (among all nonrandomized decision rules) the minimax rule.

12.35 Give example(s) to show that it may be the case that:

(a) a Bayes rule is not admissible;

(b) a minimax rule is not +-Bayes;

(c) an admissible rule is not +-Bayes.

[*Hint*: Take $\Theta = \{\theta_1, \theta_2\}$ and, with Problem 12.18 as justification, plot appropriate sets of risk points in $\mathscr{R}^2$.]

12.36 Each time a tornado alert is given, a certain community has a choice of three possible actions: d_1 (no mobilization), d_2 (partial mobilization), d_3 (full mobilization). Tornado severity is measured on a scale θ with three possible values: 0, .5, 1. The alert specifies a predicted severity X which is a r.v. for which $P_\theta[X = 0] = 1 - \theta$, $P_\theta[X = .5] = \theta/2 = P_\theta[X = 1]$. It is known that in this area tornadoes of severity 0 occur 50 percent of the time, with tornadoes of severity .5 and 1 occurring 25 percent of the time each. The loss incurred by the community by choosing decision d when θ is true is as given in the following table:

$l(\theta, d)$			
θ \ d	d_1	d_2	d_3
$\theta = 0$	0	-10	-20
$\theta = .5$	50	0	-10
$\theta = 1$	60	50	0

Formulate the problem of deciding upon an action as a statistical decision problem $(\Theta, \mathscr{D}, l)$, X with an a priori distribution on θ. (Specify each of Θ, $\mathscr{D}$, l, X, $B(\theta)$ carefully and precisely.) What is the Bayes decision rule's action if we observe $X = 0$?

12.37 Suppose for a certain statistical decision problem with $\Theta = \{\theta_1, \theta_2\}$ the set of all risk points is as drawn below. Indicate clearly all risk points corresponding to admissible decision rules.

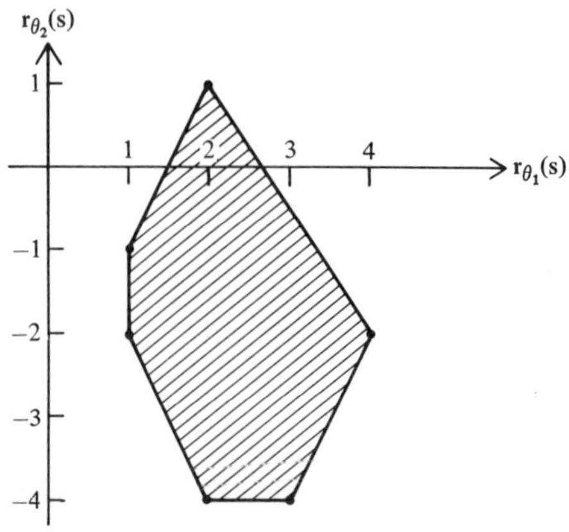

12.38 Suppose that, in a certain statistical decision problem with $\Theta = \{\theta : 0 < \theta\}$, three decision rules s_1, s_2, s_3 have risk functions as plotted below. Assuming no relation not indicated on the graph occurs outside the range of the graph, which decision rules (if any) can be definitely said to be: inadmissible? admissible? minimax? Bayes? better than s_1? better than s_2? better than s_3?

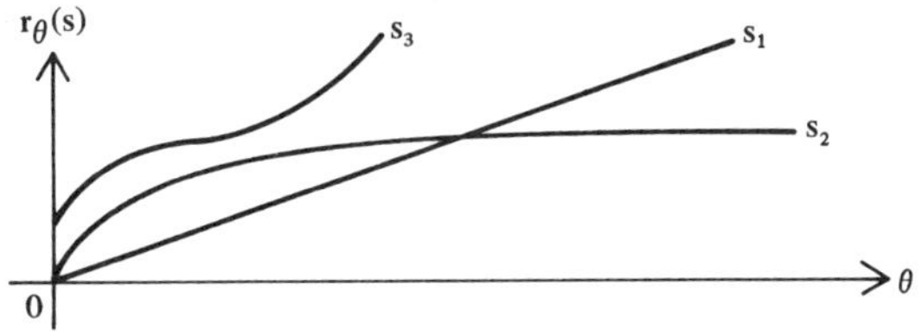

12.39 If we seek to minimize our maximum loss, we desire to use a minimax decision rule, that is, a rule $s(X)$ which achieves the[5]

$$\min_s \max_\theta r_\theta(s).$$

However, if one were a spy for a foreign power, one might seek to make the government worked upon follow a policy (decision rule) $t(X)$ which would maximize its minimum loss, that is a rule $t(X)$ which achieves the[5]

$$\max_t \min_\theta r_\theta(t).$$

Call such a rule an **NG strategy**. Suppose we have a decision problem $(\Theta, \mathscr{D}, l)$, X where $\Theta = \{\theta_1, \theta_2\}$ and the set of all risk points is as plotted below. Indicate clearly which risk point(s) correspond to NG strategies, and specify your method for finding those points.

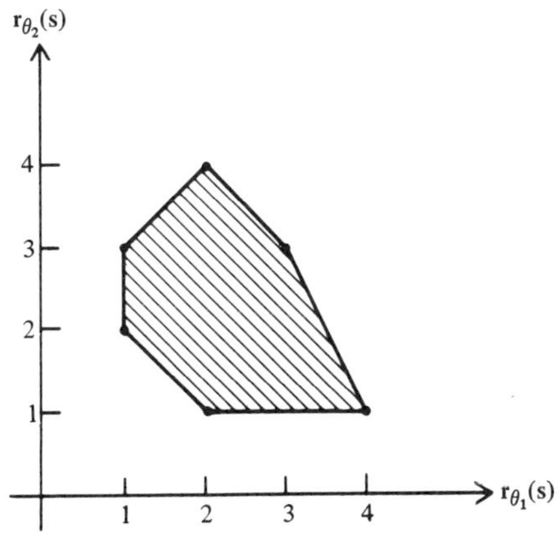

[5] More precisely we should write "inf" and "sup" for "min" and "max," respectively, as in Definition 12.1.30.

12.40 "**Two-finger Morra**" is a game which has been played in Italy since classical antiquity. The game is played by two people. Each person shows either one or two fingers and, at the same time, calls a guess as to the number of fingers his opponent will show. If exactly one player has a correct guess, he wins a number of dollars equal to the sum of the number of fingers shown (by himself and his opponent). In all other cases there is no gain or loss. If we let, say, (2, 1) denote the decision by a player to show two fingers while guessing one, fill in the following loss table.

$l(\theta, d)$					
d \ θ		Player I (Nature) $\theta_1 = (1, 1)$	$\theta_2 = (1, 2)$	$\theta_3 = (2, 1)$	$\theta_4 = (2, 2)$
Player II	$d_1 = (1, 1)$				
	$d_2 = (1, 2)$				
	$d_3 = (2, 1)$				
	$d_4 = (2, 2)$				

[*Hint*: Recall that this is a loss table for Player II, and that gains are negative losses.]

12.41 A certain organization (for example, a store, charitable agency, nonprofit organization, and so on) is trying to decide whether or not to make a mailing to a certain group of 1000 people. They assume that, if a person responds, the yield to them from his response is \$5. As a first approximation, assume that $10J$ percent will respond, where J is unknown ($J = 0, 1, 2, \ldots, 10$). If expenses connected with the mailing are \$1 per person if it is made (for example, cost of procuring names, printing, postage, and so on) complete the following loss table.

$l(J, d)$											
d \ J	0	1	2	3	4	5	6	7	8	9	10
d_1 (mail)											
d_2 (do not mail)											

12.42 Two companies (each with unlimited capital) are to competitively bid for a contract which, if it could be obtained at no cost, would have an economic value of $\$C$ ($C > 0$). The bidding process works as follows. Each company enters a sealed bid of $\$i$ for $i = 0, 1, 2, \ldots$ (of course, with i allowed to be different for each company). The higher bid wins the contract, but as compensation the high bidder must pay the amount of the bid to the low bidder (the other company, whose workers will now be unemployed). If the bids are equal, the contract is assigned by tossing a fair coin and no money changes hands. Complete the following loss table.

$l(\theta, d)$	
d \ θ	States of Nature (Company I's bid) 0 1 2 3 4 ...
Company II's bid 0	
1	
2	
3	
4	
. . .	

12.43 Suppose that you are an experimental scientist trying to estimate a new unknown physical constant θ. You are greeted by a Bayesian who tells you of the "personal probability approach." What criticism can you make of this approach as far as your case is concerned?

12.44 Suppose that in a certain decision problem there are five states of nature ($\Theta = \{\theta_1, \theta_2, \theta_3, \theta_4, \theta_5\}$) and that, relative to the a priori distribution $B(\theta)$ which assigns probability $b(i)$ to θ_i ($1 \leqslant i \leqslant 5$) rule $s(\cdot)$ is Bayes. Show that for the situation shown below $s(\cdot)$ is a minimax rule.

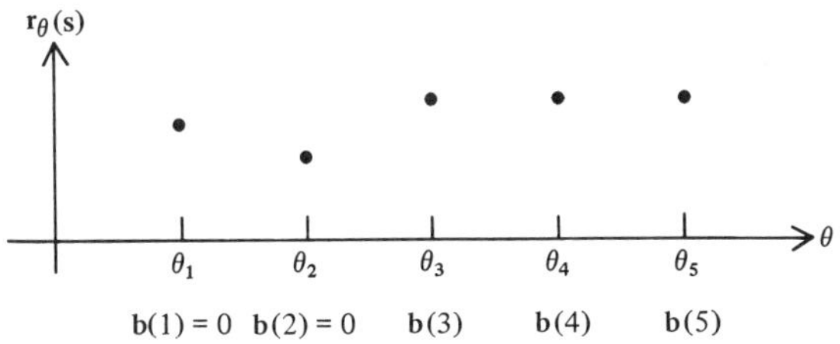

Generalizing this result show that: *for a decision problem with finite* Θ ($\Theta = \{\theta_1, \ldots, \theta_k\}$) *if* $s(\cdot)$ *is a Bayes rule with respect to the a priori distribution which assigns probability* $b(\theta)$ *to* θ ($\theta \in \Theta$), *and if*

$$r_{\theta_i}(s) = \max_{\theta} r_\theta(s)$$

for every θ_i *for which* $b(\theta_i) > 0$, *then* $s(\cdot)$ *is a minimax rule.* (This generalizes Problem 12.25.)

12.45 Let $X(t)$ denote the number of people who have arrived at a certain queue (waiting line) from time 0 to time t ($t > 0$). If arrivals occur in accord with what is called a Poisson process with parameter $\theta > 0$, then we know that

$$P[X(t) = k] = \frac{e^{-\theta t}(\theta t)^k}{k!} \quad (k = 0, 1, 2, \ldots).$$

For the decision problem of estimating θ on the basis of one observation of $X(t)$ for a fixed $t > 0$ (that is, we observe the number of arrivals from time 0 to fixed time t) with $l(\theta, d) = (d - \theta)^2$, what is the Bayes rule's estimate if the a priori distribution used has a density function $b(\theta) = e^{-\theta}$ for $\theta > 0$ (and zero otherwise)? Find $r_\theta(s)$ for this Bayes rule $s(\cdot)$. Discuss the problem of selecting t. (What is your basis for determining t? What is the ideal theoretical t? What difference would it make if you knew $0 < \theta \leqslant \theta_0$?)

12.46 Consider the decision problem $(\Theta, \mathscr{D}, l)$, X with $\Theta = \mathscr{D} = \{x: 0 < x \leqslant 1\}$, $l(\theta, d) = (d - \theta)^3$, and X a uniform r.v. on $(0, \theta)$. Find a Bayes rule with respect to a uniform a priori distribution on $(0, 1)$ (if one exists; otherwise find an ε-Bayes rule for $\varepsilon = .01$).

HONORS PROBLEMS

12.1H Our "randomized decision rule" for a statistical decision problem $(\Theta, \mathscr{D}, l)$, X (see Definition 12.1.2) is sometimes called a "**behavioral decision rule**" (for example, see T. S. Ferguson, *Mathematical Statistics: A Decision Theoretic Approach*, Academic Press Inc., New York, 1967, p. 25). Recall that such a rule $s(\cdot)$ specifies, for each possible value $x \in \mathscr{X}$ of X, a probability distribution $s(x)$ on $\mathscr{D}$ according to which a decision $d \in \mathscr{D}$ is selected.

Another method of specifying a decision rule is as follows, which we will call a **prerandomized decision rule**: s_p specifies a probability distribution on the collection of all nonrandomized decision rules. [Thus, the randomization can be performed before we observe X. The resulting nonrandomized rule specifies a unique decision to be taken after observing $X = x$.]

Usually it can be shown that the two methods of specifying a decision rule are equivalent: If s_p is a prerandomized decision rule, then there is a randomized decision rule $s(\cdot)$ such that

$$r_\theta(s) = r_\theta(s_p), \quad \text{all } \theta \in \Theta; \tag{$*$}$$

and if $s(\cdot)$ is a randomized decision rule, then there is a prerandomized decision rule s_p such that $(*)$ holds. Show this for the special case $\Theta = \{\theta_1, \theta_2\}$, $\mathscr{D} = \{d_1, d_2\}$, $\mathscr{X} = \{x_1, x_2\}$ (where $\theta_1 \neq \theta_2$, $d_1 \neq d_2$, $x_1 \neq x_2$).

12.2H Investigate the conditions under which "randomized decision rules" and "prerandomized decision rules" (see Problem 12.1H) are equivalent.

12.3H Let $(\Theta, \mathscr{D}, l)$, X be a statistical decision problem with Θ finite (that is, $\Theta = \{\theta_1, \ldots, \theta_k\}$). Show that if the set S of risk points is bounded from below[6]

[6] $A \subseteq \mathscr{R}^k$ is **bounded from below** *if there is a number* M *such that for every* $\mathbf{y} \in A$, $y_1 \geqslant M, \ldots, y_k \geqslant M$.

and closed from below,[7] then the set of all decision rules $s(\cdot)$ such that

$$(r_{\theta_1}(s), \ldots, r_{\theta_k}(s)) \in \lambda(S)$$

is a minimal complete class. Hence, use Theorem 12.1.23 to deduce that this set of rules consists of precisely the admissible rules. Thus, justify the statement, provoked by Figure 9.4–3, that "admissible rules yield the "lower boundary" of S."

12.4H A lot of $n = 1000$ items is received in a manufacturing facility. Out of the n items, an unknown number θ are defective and the other $n - \theta$ are good. A sample of size $c = 10$ is drawn at random without replacement, the number X of defectives in this sample is observed, and the lot is then either accepted (d_1) or rejected (d_2).

Defectives in the sample are returned for full credit and good units in the sample are used in production. Unsampled defectives in an accepted lot cause a loss of $c_1 = 3$ each, and sampling costs $c_2 = 1$ per unit sampled. Hence, the loss due to accepting a lot is

$$c_1(\theta - X) + c_2 c.$$

Each good item returned costs a penalty of $c_3 = 2$ per unit. If a lot is rejected, good items in the sample are kept, only $1000 - (c - X)$ items being returned. Hence, the loss due to rejecting a lot is

$$c_3(1000 - (c - X) - \theta) + c_2 c.$$

(a) If the number of defectives θ in the lot has an a priori distribution which assigns probability

$$g(\theta) = \binom{n}{\theta}(.05)^{\theta}(.95)^{n-\theta}$$

to θ ($\theta = 0, 1, \ldots, 1000$), find the Bayes decision rule.

(b) If the number of defectives θ in the lot has an a priori distribution which assigns probability $g(\theta) = (1001)^{-1}$ to θ ($\theta = 0, 1, \ldots, 1000$), find the Bayes decision rule.

(c) Suppose that a decision rule is to specify both c (the size of the sample) and the decision to be made if a sample of size c is used and X defectives are found in that sample. Find the Bayes rule under the assumption of part (a). Find the Bayes rule under the assumption of part (b).

12.5H In one conventional formulation of statistical decision theory (that used in Chapter 12, which we will now call **formulation $\mathscr{F}_1$**) we observe $X = x$ and then choose decision d with probability $s_d(x)$. The loss incurred is then some $W(\theta; d; x)$ [this generalizes Chapter 12 slightly; there it was assumed that $W(\theta; d; x) = l(\theta, d)$ was not a function of x]. If $f(x|\theta)$ denotes the $P_\theta[X = x]$

---

[7] *A convex set $A \subseteq \mathscr{R}^k$ is* **closed from below** *if $\lambda(A) \subseteq A$, where $\lambda(A)$ consists of all* **x** *such that* **x** *is a limit point of A and*

$$\mathbf{x} = Q(\mathbf{x}) \cap \mathrm{cl}(A) = \{\mathbf{y} : \mathbf{y} \in \mathscr{R}^k, y_1 \leqslant x_1, \ldots, y_k \leqslant x_k\} \cap \mathrm{cl}(A).$$

when X is discrete (and the corresponding density function when X is absolutely continuous) where θ is the true value of the unknown parameter, then the risk function of rule $s(\cdot)$ is

$$r_\theta(s) = \sum_x \sum_d W(\theta; d; x) f(x|\theta) s_d(x) \qquad (*)$$

(or a corresponding form with integrals).

In one nonconventional formulation of statistical decision theory (call it **formulation** $\mathscr{F}_2$) we observe $X = x$ and then choose decision d with probability $s_d(x)$, and we later observe that a further r.v. $Y = y$. The loss incurred is then some $W(y; d; x)$. If $f(x, y|\theta)$ denotes the $P_\theta[X = x, Y = y]$ when (X, Y) is a discrete r.v. (and the corresponding joint density function when (X, Y) is absolutely continuous) where θ is the true value of the unknown parameter, then the risk function of rule $s(\cdot)$ is

$$r_\theta(s) = \sum_x \sum_y \sum_d W(y; d; x) f(x, y|\theta) s_d(x) \qquad (**)$$

(or a corresponding form with integrals).

(a) A decision problem of type $\mathscr{F}_1$ is clearly also a decision problem of type $\mathscr{F}_2$ (simply take y to be θ). Prove that: *A decision problem of type $\mathscr{F}_2$ is also a decision problem of type $\mathscr{F}_1$* (that is, given a decision problem stated in the terms of $\mathscr{F}_2$, show how to choose a loss function in $\mathscr{F}_1$ such that $r_\theta(s)$ will not depend on the formulation used: $(*) = (**)$).

(b) Suppose we have a decision problem of type $\mathscr{F}_1$ with $\Theta = \mathscr{D} = \{\mu: 0 \leqslant \mu < +\infty\}$, X a $N(\theta, 1)$ r.v., and $W(\theta; d; x) = -(\theta - d)^3$. What is the best decision d? (There is a uniquely best decision in this problem.)

(c) Consider the following decision problem (which is stated in form $\mathscr{F}_2$). An automobile whose replacement cost is \$2000 can be insured against theft for 1 year for \$40. There are two possible decisions: to insure or not to insure. The unknown probability of the automobile being stolen during the year is denoted by θ. Before deciding on insurance, the number of automobiles that were stolen out of 5 automobiles, each with the same probability θ of being stolen, will be observed. If 1 of the 5 were stolen, which decision would be chosen by the Bayes decision rule with respect to a priori distribution $B(\theta)$ which sets $P[\theta \leqslant x] = 1 - (1 - x)^3$ for $0 \leqslant x \leqslant 1$? [*Hint*:

$$\int_0^1 x^a(1 - x)^b \, dx = \frac{a!b!}{(a + b + 1)!}$$

for all positive integers a and b.]

(d) Express the decision problem of part (c) in formulation $\mathscr{F}_1$ (in particular, finding the appropriate loss function $W(\theta; d; x)$).

13

NONPARAMETRIC STATISTICAL INFERENCE

"Nonparametric statistics" is the subfield of statistics which provides statistical inference procedures which rely on weaker assumptions about the underlying distribution(s) of the population(s) than do the ordinary (for example, normal distribution) procedures for the same problems. Since such nonparametric procedures assume less about the underlying distribution(s), errors in correct assessment of the nature of the underlying distribution(s) will usually have less effect on nonparametric procedures than on parametric procedures (since the latter usually rely more heavily on the correct assessment of the nature of the underlying distribution(s)). However, the more information we have (and use well) about the underlying distribution(s), the better our statistical inference procedures will be; thus, for a given situation, a nonparametric procedure will usually be the one with greater variance (in point estimation), with less power (in hypothesis testing), with wider intervals (in confidence interval estimation), with lower probability of correct selection (in ranking and selection), and with higher risk (in decision theory), when compared with a corresponding parametric procedure whose assumptions are not violated. Hence, it is usually advisable to use procedures which eliminate assumptions about the underlying distribution(s) only in so far as those assumptions are seriously doubtful in their validity. In this way we capture the greatest gain from both the parametric and nonparametric approaches to a problem.

In previous chapters we have studied general methods which turn out to be useful in providing nonparametric inference procedures (such as order statistics or quantiles in Section 8.5) and will soon study others (such as the probability-integral transformation in Problem 13.1). Using these methods we can then study various sorts of decision problems (estimation, hypothesis testing, confidence intervals, selection procedures, and so on) in more detail. This avoids the "bag of tricks" approach to nonparametric statistics, which leads to an unordered view of the subject.

13.1 NONPARAMETRIC CONFIDENCE INTERVALS

In Section 10.2 we found that if $X_1, \ldots, X_n$ are independent $N(\mu, \sigma^2)$ r.v.'s where μ is unknown with $\mu \in \Theta = \{x: -\infty < x < +\infty\}$, then a confidence interval for μ with confidence coefficient $1 - \alpha$ is

$$\left(\bar{X} - \frac{\sigma}{\sqrt{n}}\Phi^{-1}\left(1 - \frac{\alpha}{2}\right),\ \bar{X} + \frac{\sigma}{\sqrt{n}}\Phi^{-1}\left(1 - \frac{\alpha}{2}\right)\right) \tag{13.1.1}$$

if $\sigma^2 > 0$ is known, and

$$\left(\bar{X} - \frac{s}{\sqrt{n}}t_{n-1}^{-1}\left(1 - \frac{\alpha}{2}\right),\ \bar{X} + \frac{s}{\sqrt{n}}t_{n-1}^{-1}\left(1 - \frac{\alpha}{2}\right)\right) \tag{13.1.2}$$

if $\sigma^2 > 0$ is unknown. As we discussed in Example 8.5.34, while for the normal distribution the mean μ and the population median $\xi_{.5}$ coincide (that is, $\mu = \xi_{.5}$), in general this is not the case and (in many instances) we then want a confidence interval on $\xi_{.5}$ (the point which "splits the population in half"; that is, an observation is equally likely to fall above or below $\xi_{.5}$ if we are dealing with a continuous r.v.). **We will now develop a nonparametric interval for $\xi_p (0 < p < 1)$**, and then specialize to the case $p = .5$ in order to compare the nonparametric interval with the parametric ones (13.1.1) and (13.1.2).

As in Definition 8.5.2, **for n r.v.'s $X_1, \ldots, X_n$ let $Y_1, \ldots, Y_n$ denote the ordered X_i's (that is, the X_i's ordered so that $Y_1 \leqslant \cdots \leqslant Y_n$**; if several X_i's are equal, we can choose one at random for a place Y_i). Then **the sample quantile of order p ($0 < p < 1$) is**

$$\mathbf{Z_p = Y_{[np]+1}}, \tag{13.1.3}$$

where $[np]$ denotes the largest integer $< np$. We now wish to determine the probability that ξ_p lies between Y_i and Y_j (with $1 \leqslant i < j \leqslant n$); if we can control this probability without knowing the distribution of the r.v.'s $X_1, \ldots, X_n$, then we may wish (Y_i, Y_j) (for some i, j) to be our confidence interval on ξ_p.

Theorem 13.1.4. *Fix p ($0 < p < 1$). Let $X_1, \ldots, X_n$ be independent r.v.'s with a continuous d.f. $F(x)$. Then for all i and j ($1 \leqslant i < j \leqslant n$)*

$$P[Y_i \leqslant \xi_p \leqslant Y_j] = Q(i, j; p, n), \tag{13.1.5}$$

where

$$Q(i, j; p, n) = \int_p^1 \int_0^p \frac{n!}{(i-1)!(j-i-1)!(n-j)!} \cdot s^{i-1}(t-s)^{j-i-1}(1-t)^{n-j}\, ds\, dt. \tag{13.1.6}$$

Proof: We have (for $1 \leqslant i < j \leqslant n$)

$$\begin{aligned} P[Y_i \leqslant \xi_p \leqslant Y_j] &= P[F(Y_i) \leqslant F(\xi_p) \leqslant F(Y_j)] \\ &= P[F(Y_i) \leqslant p \leqslant F(Y_j)] = P[S < p < T], \end{aligned} \tag{13.1.7}$$

where we have let $S \equiv F(Y_i)$ and $T \equiv F(Y_j)$. However, it can be shown (see Problem 13.1H) that the joint density function of S and T is

$$f_{S,T}(s, t) = \begin{cases} \dfrac{n!}{(i-1)!(j-i-1)!(n-j)!} \cdot s^{i-1}(t-s)^{j-i-1}(1-t)^{n-j}, & \text{if } 0 \leqslant s < t \leqslant 1 \\ 0, & \text{otherwise.} \end{cases} \tag{13.1.8}$$

Hence,

$$P[S < p < T] = \int_p^1 \int_0^p f_{S,T}(s, t)\, ds\, dt = Q(i, j; p, n) \tag{13.1.9}$$

and the theorem follows.

From Theorem 13.1.4 we immediately obtain the following.

Corollary 13.1.10. *Fix p $(0 < p < 1)$. Let $X_1, \ldots, X_n$ be independent r.v.'s with an unknown, but continuous, d.f. Then (Y_i, Y_j) is a confidence interval for ξ_p with confidence coefficient $Q(i, j; p, n)$.*

In fact, Corollary 13.1.10 holds even if the common d.f. of $X_1, \ldots, X_n$ is not continuous; however, in this case the confidence coefficient is $\geqslant Q(i, j; p, n)$ (see Problem 13.4H).

Corollary 13.1.11. *Fix p $(0 < p < 1)$. Let $X_1, \ldots, X_n$ be independent r.v.'s with an unknown d.f. Then (Y_i, Y_j) is a confidence interval for ξ_p with confidence coefficient $\geqslant Q(i, j; p, n)$.*

Rather than computing $Q(i, j; p, n)$ from (13.1.6), it is simpler to use the following expression.

Theorem 13.1.12. *For $1 \leqslant i < j \leqslant n$ and $0 < p < 1$,*

$$Q(i, j; p, n) = \sum_{k=i}^{j-1} \binom{n}{k} p^k (1-p)^{n-k}. \tag{13.1.13}$$

Proof: By Problem 13.2,

$$Q(i; j; p, n) = P[U_i \leqslant p \leqslant U_j], \tag{13.1.14}$$

where $U_1, \ldots, U_n$ are the order statistics in a sample of size n from a uniform distribution on (0, 1). However (see Problem 13.3),

$$P[U_i \leqslant p \leqslant U_j] = 1 - P[U_i > p] - P[U_j < p]. \tag{13.1.15}$$

Since for all l ($1 \leqslant l \leqslant n$)

$$P[U_l \leqslant p] = \sum_{k=l}^{n} \binom{n}{k} p^k (1-p)^{n-k} \tag{13.1.16}$$

(the event $\{U_l \leqslant p\}$ occurs iff at least l of the basic uniform r.v.'s on (0, 1) are $\leqslant p$), the theorem follows on using (13.1.16) in (13.1.15).

Example 13.1.17. If $p = .5$ and $n = 16$, then (using Table 5 to evaluate $Q(i, j; p, n)$ since Theorem 13.1.12 allows us to write $Q(i, j; p, n)$ as a sum of binomial probabilities) we find that

$$\begin{cases} P[Y_5 \leqslant \xi_{.5} \leqslant Y_{12}] = .9234, \\ P[Y_4 \leqslant \xi_{.5} \leqslant Y_{12}] = .9512, \\ P[Y_5 \leqslant \xi_{.5} \leqslant Y_{13}] = .9512; \end{cases} \tag{13.1.18}$$

hence, if we desired a 95 percent confidence interval via a nonparametric method, we would use either (Y_4, Y_{12}) or (Y_5, Y_{13}), each of which has approximately 95 percent confidence (actually 95.12 percent). If we chose (Y_4, Y_{12}), our interval would have length

$$Y_{12} - Y_4, \tag{13.1.19}$$

a r.v. If we instead used interval (13.1.1), we would have nonrandom length

$$.98\sigma, \tag{13.1.20}$$

while interval (13.1.2) would have random length

$$1.07s. \tag{13.1.21}$$

It would be of interest to compare the average lengths (13.1.19), (13.1.20), and (13.1.21) (Problem 13.5H); such a comparison shows that we obtain longer intervals on the average with the nonparametric interval; as a compensation, we are protected against nonnormality of $X_1, \ldots, X_n$. Thus, the nonparametric interval should be used if (and only if) normality is subject to serious question in our experiment (and we do not know what other distribution governs $X_1, \ldots, X_n$).

Example 13.1.22. In Problem 8.6 we gave $X_1, \ldots, X_{88}$, the $n = 88$ seasonal snowfalls observed at Rochester, New York, since the 1884–85 winter. Find a nonparametric 95 percent confidence interval on the median snowfall. (Note that a similar question was considered in Problem 10.12(a) under the dubious assumption of a normal distribution for amount of snowfall.)

Since ξ_p with $p = .5$ and $n = 88$ is of interest here, we wish to choose (Y_i, Y_j) to be our interval where

$$P[Y_i \leqslant \xi_{.5} \leqslant Y_j] = .95. \tag{13.1.23}$$

Since by Theorem 13.1.4 and Theorem 13.1.12

$$P[Y_i \leqslant \xi_{.5} \leqslant Y_j] = \sum_{k=i}^{j-1} \binom{n}{k} p^k(1-p)^{n-k}, \tag{13.1.24}$$

we may (in order to obtain (13.1.23)) try to set i and j so that

$$\begin{cases} P[\xi_{.5} > Y_j] = \sum_{k=j}^{n} \binom{n}{k} p^k(1-p)^{n-k} = .025 \\ P[\xi_{.5} < Y_i] = \sum_{k=0}^{i-1} \binom{n}{k} p^k(1-p)^{n-k} = .025; \end{cases} \tag{13.1.25}$$

that is, set i and j so that

$$\begin{cases} \sum_{k=j}^{88} \binom{88}{k} 2^{-88} = .025 \\ \sum_{k=0}^{i-1} \binom{88}{k} 2^{-88} = .025. \end{cases} \tag{13.1.26}$$

While Table 5 does not go beyond $n = 20$ (and in fact most tables a student will find do not go beyond $n = 50$), extensive tables are available [*Tables of the Cumulative Binomial Probability Distribution*, Harvard University Press, Cambridge, Massachusetts, 1955, where $n = 88$ occurs on p. 432] and yield

$$\begin{cases} \sum_{k=54}^{88} \binom{88}{k} 2^{-88} = .02111, \quad \sum_{k=53}^{88} \binom{88}{k} 2^{-88} = .03467 \\ \sum_{k=0}^{34} \binom{88}{k} 2^{-88} = .02111, \quad \sum_{k=0}^{35} \binom{88}{k} 2^{-88} = .03467. \end{cases} \tag{13.1.27}$$

Hence

$$P[Y_{35} \leqslant \xi_{.5} \leqslant Y_{54}] = .95778. \tag{13.1.28}$$

By reference to the data $X_1, \ldots, X_{88}$ in Problem 8.6 we find $Y_{35} = 69.2$ and $Y_{54} = 79.3$, whence with 95.8 percent confidence we state that the median snowfall in Rochester, New York, is between 69.2 inches and 79.3 inches.

Remark 13.1.29. Note that if the binomial distribution tables needed in Example 13.1.22 had not been available, we would have used a normal approximation. We leave it as an exercise for the student to do so in Example 13.1.22 and compare the results with the exact ones given in that example.

13.2. NONPARAMETRIC TESTING OF QUANTILES ξ_p

In this chapter we have found it convenient to alter our usual order of presentation (point estimation; hypothesis testing; confidence intervals; ranking and selection; decision theory) slightly. **Nonparametric point estimation** was already discussed through a discussion of the properties of a nonparametric estimator of $\xi_{.50}$ in Chapter 8 (Example 8.5.34); this is indicative and nonparametric point estimation will not be discussed further here. We have considered **nonparametric confidence intervals** already (Section 13.1) and will now consider **nonparametric hypothesis-testing**, which divides into problems of two types (for numerical and ordinal data, respectively, in Sections 13.2 and 13.3). (**Nonparametric ranking and selection** was covered briefly in Section 11.4, part 6; see the reference cited there for details. **Nonparametric decision theory** has received a great impetus in recent years through the work of T. S. Ferguson,[1] but that work is beyond the scope of the present book.)

Since (see Corollary 13.1.10) (Y_i, Y_j) is a confidence interval for ξ_p with confidence coefficient $Q(i,j;p,n)$, by the relation given in Chapter 10 between tests of hypotheses and confidence intervals (see Theorem 10.1.5), it follows (see Problem 13.5) that **to test the hypothesis H_0 that ξ_p equals a specified value ξ_p^0 at level $1 - Q(i, j; p, n)$ we simply accept H_0 if $Y_i \leqslant \xi_p^0 \leqslant Y_j$ and reject H_0 otherwise.**

Example 13.2.1. Since (see Example 13.1.17) in a sample of size $n = 16$ the interval (Y_4, Y_{12}) is a 95.12 percent confidence interval for $\xi_{.5}$, we can obtain a level $.0488 \approx .05$ test of the hypothesis that $\xi_{.5} = 24.62$ (or any other number of interest) as follows: Form the confidence interval (Y_4, Y_{12}) on $\xi_{.5}$ and

$$\begin{cases} \text{Accept the hypothesis } \xi_{.5} = 24.62, & \text{if } Y_4 \leqslant 24.62 \leqslant Y_{12} \\ \text{Reject the hypothesis } \xi_{.5} = 24.62, & \text{otherwise.} \end{cases} \tag{13.2.2}$$

13.3. NONPARAMETRIC TESTS FOR ORDINAL DATA

Many of the situations we have dealt with thus far have been ones in which the r.v.'s of interest are observed directly, and hence situations which lend themselves to parametric analysis (if we are able to specify the underlying distribution as a member of a certain family of distributions characterized by one or more parameters; for example, the normal distributions

---

[1] See T. S. Ferguson, "A Bayesian analysis of some nonparametric problems," *Annals of Statistics*, Vol. 1 (1973), pp. 209–230.

$N(\mu, \sigma^2)$ characterized by the mean μ and variance σ^2). However, **situations also arise where the basic r.v.'s are not able to be observed directly, but rather, we can only obtain information about their order among a certain set of r.v.'s. This is the situation of *ordinal* (as opposed to *numerical*) *data*,** and lends itself very naturally to a nonparametric analysis. (In a sense we could say that the natural parametric analysis is nonparametric!) **Since we can always replace numerical data by their ranks, the following treatment can be used for numerical data as well** (and this may be appropriate in some situations where little is known about the distribution of the data).

While there are many ways in which ordinal data can arise, we will cover one (the one which leads to the **sign test**); once the principles which govern this are understood, the multiplicity of tests available for different ways in which ordinal data can arise are easily accessible. Finally, we will cover tests which require that numerical data be ordinalized before it is analyzed (this results in a need for fewer distributional assumptions on the underlying distribution(s)), and will illustrate this class of tests with the **Mann-Whitney-Wilcoxon test**.

Consider the situation where two objects A and B are presented to a judge selected at random from a pool of possible judges, who states his preference according to certain well-defined criteria. (For example, in a marketing test of a new product, or an old product being marketed in a new way or in a new area, the brand of interest (say brand A) is often presented to a consumer along with a competitor (say brand B); the consumer then states which brand he prefers.) Thus, we can regard the situation as one where the basic data is *ordinal*: the judge simply orders the two alternatives (for example, says, "I prefer A to B" in the marketing example). If we define a r.v. X_1 to be 0 if A is preferred to B and 1 if B is preferred to A, then X_1 is a binomial r.v. (see Definition 3.2.3) with $n = 1$ and p equal to the probability that B is preferred to A. If we wish to test the composite null hypothesis that $p \leqslant .1$ against the composite alternative that $p > .1$ (the alternative says brand B is preferred by more than 10 percent of the population of interest), we could ask n judges independently, thus obtaining $X_1, \ldots, X_n$, and reject H_0 if $\bar{X}$ is "large." More precisely,

Theorem 13.3.1. *For the hypothesis-testing problem with $X = (X_1, \ldots, X_n)$ where $X_1, \ldots, X_n$ are independent r.v.'s, each binomial with one trial and success probability p where p $(0 \leqslant p \leqslant 1)$ is unknown, $H_0 = \{p: p \leqslant p_0\}$, and $H_1 = \{p: p > p_0\}$ with p_0 fixed $(0 < p_0 < 1)$, the test which rejects iff*

$$X_1 + \cdots + X_n \geqslant j \tag{13.3.2}$$

has level

$$\sum_{k=j}^{n} \binom{n}{k} (p_0)^k (1 - p_0)^{n-k}. \tag{13.3.3}$$

Proof: The probability that the given test rejects H_0 is

$$P_p[X_1 + \cdots + X_n \geqslant j] = \sum_{k=j}^{n} \binom{n}{k} p^k(1-p)^{n-k}. \tag{13.3.4}$$

For all $p \leqslant p_0$, (13.3.4) is largest when $p = p_0$ (see Problem 10.19), and the result follows by the definition of level of significance in the case of a composite null hypothesis (see Problem 9.4H): the largest of the levels attained for any simple $H_0' \in H_0$.

It is clear that we can specify the n needed for a specified level and power in Theorem 13.3.1; see Problem 13.6, for example. **The situation studied in Theorem 13.3.1 can also arise as follows.** Suppose n individuals each test two treatments (A, B) and return respective numerical values on them of V_i and W_i ($i = 1, \ldots, n$). If we let $D_i = W_i - V_i$ ($i = 1, \ldots, n$) and finally set

$$X_i = \begin{cases} 1, & \text{if } D_i \geqslant 0 \\ 0, & \text{if } D_i < 0 \end{cases} \tag{13.3.5}$$

($i = 1, \ldots, n$), then Theorem 13.3.1 can be used to test hypotheses about $p = P[X_i = 1] = P[D_i \geqslant 0] = P[W_i - V_i \geqslant 0] = P[W_i \geqslant V_i]$, namely about the probability that W_i exceeds V_i (if the "individuals" are agricultural plots and the "treatments" are two wheat varieties, we are dealing with hypotheses on the probability that variety B will have a higher yield than variety A). **Since in this type of example we simply see whether $\mathbf{W_i - V_i}$ is positive (+) or negative (−) in each individual, the test of Theorem 13.3.1 has come to be called the *sign test*.** If we have only ordinal data, this is the natural parametric test. If, in fact, we observe W_i, V_i and only look at whether $W_i \geqslant V_i$, we have discarded some information in order to dispense with the need to know the distributions of W_i and V_i. This is often appropriate. On the other hand, if we know the distributions to be, say, normal and follow such a procedure, then our efficiency (in terms of the ratio n_1/n_2 of the respective sample sizes n_1 and n_2 needed for a specified level and power of the t-test and of the sign test) will be[2] approximately 95 percent if $n = 4$, 5, or 6 and decreases as n increases (for example, it is approximately 75 percent for $n = 13$).

Now we wish to consider, as a final topic in this section, cases in which numerical data is ordinalized, our illustration being what is called the Mann-Whitney-Wilcoxon test. Suppose that we have available some number N of subjects, of whom s will be selected at random to receive a new treatment

[2] See J. E. Walsh, "On the power function of the sign test for slippage of means," *Annals of Mathematical Statistics*, Vol. 17 (1946), pp. 358–362, for details of the comparison.

while the other $t = N - s$ receive a standard treatment (for example, for a type of heart disease). (While much data may be available on patients on whom the "standard" treatment was used in the past, results often differ substantially from one population of patients to another and over time, and the "standard" treatment may not be the same today as it was a year ago. Hence, if we wish to make sure we are comparing results of the "new" and "standard" treatments, and not time- or population-dependent changes, it will usually be necessary to have some number $N - s$ of the patients receive the "standard" treatment while the other s receive the "new" treatment. This is called **controlled experimentation**, and the $N - s$ patients are called the **control group**.) If the responses of the s "new" treatment patients are $Y_1, \ldots, Y_s$ while the responses of the t "standard" treatment patients are $X_1, \ldots, X_t$, we will want to know whether the "new" treatment has changed the response (for example, the X_i's and Y_j's could be survival times of patients after treatment). While it may be a tenable assumption that $X_1, \ldots, X_t$, $Y_1, \ldots, Y_s$ are independent r.v.'s, they are probably not known to have a normal (or other specified) distribution; hence, a method of testing the null hypothesis that the X_i's and Y_j's have the same d.f. is desired which does not assume a specific (say, normal) distribution for $X_1, \ldots, X_t$, $Y_1, \ldots, Y_s$. Such a method can be given as follows. Consider all $N = s + t$ observations ordered in increasing numerical order. Replace the smallest by the value 1, the next to smallest by $2, \ldots,$ the largest by $s + t$. Let the value by which X_i or Y_j is replaced be denoted by $R(X_i)$ or $R(Y_j)$, respectively. Let

$$W_X = R(X_1) + \cdots + R(X_t), \quad W_Y = R(Y_1) + \cdots + R(Y_s). \tag{13.3.6}$$

[Note that $W_X + W_Y = 1 + 2 + \cdots + N = N(N + 1)/2$.] Then W_Y will tend to be "large" if the new treatment has increased survival times. Hence, we may decide to reject the null hypothesis iff

$$W_Y \geqslant c. \tag{13.3.7}$$

In order to set c so that we have a level α test we will need to know the distribution of W_Y when in fact (H_0) the standard and new treatments lead to the same survival time distribution. However, in that case all $s + t$ observations are equally likely to have any specified rank in the pooled observations; hence, since there are $\binom{N}{t}$ equally likely ways to assign ranks to the Y_j's, the $P_{H_0}[W_Y \geqslant c]$ equals the number of such ways which lead to a rank sum of at least c, divided by the total number of ways to assign ranks to the Y_j's:

$$P_{H_0}[W_Y \geqslant c] = \frac{\#(W_Y \geqslant c)}{\binom{N}{t}}. \tag{13.3.8}$$

Example 13.3.9. A new treatment for increasing egg production has been proposed by a poultry scientist. To test this treatment, it is applied to 7 hens while 7 others receive a similar-looking placebo. (In fact, the experiment is run double-blind.) After a period of time, the following egg yields are observed to have occurred:

i	1	2	3	4	5	6	7
Treatment Y_i	260	248	240	245	255	258	220
Control X_i	230	227	242	244	261	250	265

Does a Mann-Whitney-Wilcoxon test at level .05 accept or reject the hypothesis that the treatment has no beneficial effect?

Here the ranked data are in the order of

Observations	$Y_7 <$	$X_2 <$	$X_1 <$	$Y_3 <$	$X_3 <$	$X_4 <$	$Y_4 <$	$Y_2 <$	$X_6 <$	$Y_5 <$	$Y_6 <$	$Y_1 <$	$X_5 <$	X_7
Ranks	*1*	2	3	*4*	5	6	*7*	*8*	9	*10*	*11*	*12*	13	*14*

where we have italicized (boldface) the ranks of the Y_j's. Hence, $W_Y = 67$ (and $W_X = 38$, $W_X + W_Y = 105 = 1 + 2 + \cdots + 14 = 14(15)/2$). Here, we can show (Problem 13.7) that

$$P_{H_0}(W_Y \geqslant 65) = .064, \quad P_{H_0}(W_Y \geqslant 66) = .049. \tag{13.3.10}$$

Hence, at level .05 (which is not exactly, but only approximately, attainable) the Mann-Whitney-Wilcoxon test rejects the hypothesis that the treatment has no beneficial effect.

Extensive tables of the critical values and related probabilities of the Mann-Whitney-Wilcoxon test are available (for example, see the reference cited in Problem 13.9H, where a 22-page table is given on pp. 272–293). However, unlike the situation with parametric tests (but rather typical of the situation with nonparametric tests), **the power of the test is hard to get at**: under what alternative are we to calculate this power? Comparisons of power as well as of level can be done under specific parametric alternative hypotheses. There computations show that the efficiency of the Mann-Whitney-Wilcoxon test relative to the t-test, for normal alternatives is asymptotically (as $\min(s, t) \to \infty$) $3/\pi \approx .96$, and for all shift alternatives is $\geqslant .864$. Thus, in the normal case (where we should use the t-test) we are essentially discarding 4 percent ($= 100 - 96$) of our sample by using the Mann-Whitney-Wilcoxon test; and for some nonnormal alternatives we can be essentially discarding 14 percent ($= 100 - 86$) of our sample by using the Mann-Whitney-Wilcoxon test instead of the t-test. However, for other types of alternatives the Mann-Whitney-Wilcoxon test can be substantially better than the t-test. Thus, our recommendation is still as before: **If you know you have a normal (or some other specific parametric) distribution for your data, use a parametric procedure; in cases of severe doubt consider nonparametric procedures.** (The border line between these two cases is ill-defined, in fact a matter of dispute, and we will not attempt to define it here.)

PROBLEMS

13.1 (**The probability-integral transformation**) Let X be a r.v. with a continuous monotone-increasing d.f. $F(x)$. Show that the r.v. $Y = F(X)$ has a uniform distribution over the interval $(0, 1)$. [*Hint*: $P[Y \leqslant y] = P[F(X) \leqslant y] = P[X \leqslant F^{-1}(y)] = F(F^{-1}(y))$, where $F^{-1}(\cdot)$ is the inverse function of $F(\cdot)$ (that is, $F^{-1}(y)$ is the value x such that $F(x) = y$) which exists since $F(x)$ is continuous and monotone-increasing.] Note that this result holds even if $F(x)$ is merely continuous; see Problem 13.2H.

13.2 Let $X_1, \ldots, X_n$ be independent r.v.'s with a continuous d.f. $F(x)$. Show that $F(Y_1), \ldots, F(Y_n)$ have the same joint distribution as the order statistics $U_1, \ldots, U_n$ in a sample of size n from a uniform distribution on $(0, 1)$. Show that (for $1 \leqslant i < j \leqslant n$ and $0 < p < 1$)

$$P[U_i \leqslant p \leqslant U_j] = Q(i, j; p, n).$$

13.3 Verify that (13.1.15) holds.

13.4 Show that $Q(1, n; p, n) = 1 - p^n - (1 - p)^n$ and find its limit as $n \to \infty$. What does this say about the confidence coefficient of the interval (Y_1, Y_n) for ξ_p for large n?

13.5 Show in full detail how Theorem 10.1.5 is used (see Section 13.2) to derive a test of level $\alpha = 1 - Q(i, j; p, n)$, for the hypothesis that $\xi_p = \xi_p^0$, from the confidence interval (Y_i, Y_j) on ξ_p.

13.6 Two tennis players (A and B, say) are to play n matches. Assume the matches are independent and that $P(B \text{ wins match } i) = p$ for $i = 1, \ldots, n$ for some p with $0 < p < 1$. Player A avows that he is at least as good as player B; that is, A states that $p \leqslant .5$. Using Theorem 13.3.1 we can test this hypothesis at level

$$\alpha = \sum_{k=j}^{n} \binom{n}{k} \Big/ 2^n$$

by rejecting iff $X_1 + \cdots + X_n \geqslant j$ (that is, reject if B wins j or more matches). What j and n should we choose for level $\alpha \approx .05$ and power (probability of rejecting H_0) $\approx .90$ when in fact $p = .6$?

13.7 Show that (13.3.10) holds via (**1**) direct calculation and (**2**) use of standard tables.

13.8 Suppose that Glopco (the General Glop Company), which manufactures breakfast cereals, is going to issue a new superfortified cereal called WheatBombs;

this cereal is superficially indistinguishable from their popular cereal WheatBoobs, but will retail for twice the price. These cereals appeal to students' parents.

In order to try to prove WheatBombs' superiority to WheatBoobs (so WheatBoob users can be convinced to switch to the higher-priced WheatBombs), the following experiment is run. Thirty pairs of identical twins are chosen, and one twin eats WheatBoobs for a year while the other twin eats WheatBombs for a year. (The packages are coded so they don't know which cereal they are receiving.) At the end of the year, the grade-point average (GPA) for that year of the WheatBoob and WheatBomb eater are compared in each pair. Let B denote the number of the 30 pairs in which the WheatBomb eater had a higher GPA than the WheatBoob eater. Supposing independence of the 30 pairs, we wish to test whether the hypothesis H: "WheatBombs are better than WheatBoobs" ["better" means induce a higher GPA] is true. If we perform a test which decides H is true if $B \geqslant 22$, what will our significance level be? [The null hypothesis is that the cereals are equivalent with regard to GPA.] Does this test accept H if we observe $B = 20$? What is the significance probability of the result $B = 20$?

If, in fact, the null hypothesis is true, and if 100 independent such experiments are run which each have significance level .05, what is the expected value of the number of experiments which will reject the null hypothesis?

HONORS PROBLEMS

13.1H Prove that the joint density of $S \equiv F(Y_i)$ and $T \equiv F(Y_j)$ $(1 \leqslant i < j \leqslant n)$ in Theorem 13.1.4 is given by (13.1.8). [*Hint*: See Problem 8.26.]

13.2H Let $F(x)$ be a continuous d.f. Define a function $G(y)$ as

$$G(y) = \begin{cases} F^{-1}(y), & \text{if } 0 < y < 1 \text{ and } F^{-1}(y) \text{ exists} \\ F^*(y), & \text{if } 0 < y < 1 \text{ and } F^{-1}(y) \text{ does not exist,} \end{cases}$$

where $F^*(y)$ is the smallest x such that $F(x) = y$. (Note that $G(y) = F^{-1}(y)$ for all y if $F(\cdot)$ is monotone-increasing.) Use $G(\cdot)$, instead of $F^{-1}(\cdot)$, to show that (see Problem 13.1) $Y = F(X)$ has a uniform distribution on the interval (0, 1) even if $F(\cdot)$ is merely continuous (and not necessarily monotone-increasing).

13.3H Let $F(x)$ be an arbitrary d.f. Show how to define a nondecreasing function $G(\cdot)$ such that, if U is a uniform r.v. on (0, 1), then $G(U)$ has d.f. $F(\cdot)$. [*Hint*: Let $G(\cdot)$ be as in Problem 13.2H, but also allow for jump points of $F(\cdot)$.]

13.4H Let $U_1, \ldots, U_n$ be the order statistics in a sample of size n from a uniform distribution on (0, 1). Prove that (for $0 < p < 1$)

$$\{U_i \leqslant p \leqslant U_j\} \Rightarrow \{G(U_i) \leqslant G(p) \leqslant G(U_j)\},$$

where $G(\cdot)$ is the function defined in Problem 13.3H. Hence, conclude that

$$P[U_i \leqslant p \leqslant U_j] \leqslant P[G(U_i) \leqslant G(p) \leqslant G(U_j)].$$

Use this result and Problem 13.2 to prove Corollary 13.1.11.

13.5H Compare the averages of the lengths (13.1.19), (13.1.20), and (13.1.21) numerically. (First, "wash out" σ by considering $X_1/\sigma, \ldots, X_n/\sigma$ as the basic r.v.'s in the experiment; justify this step.) [*Hint*: See H. L. Harter, "Expected values of normal order statistics," *Biometrika*, Vol. 48 (1961), pp. 151–165, for tables from which we can calculate $E(Y_j - Y_i)$.]

13.6H Numerically compare the power curve of the test in Example 13.2.1 with that of the usual test which is used when the observations are normally distributed. How much do we lose by using the nonparametric procedure?

13.7H As in Problem 13.6H, but assume the observations have a uniform density on $\xi_{.5} \pm 1$.

13.8H Using the results of Problems 13.6H and 13.7H, are we better off (when our observations follow a uniform density on $\xi_{.5} \pm 1$) using normal theory or nonparametric theory?

13.9H Suppose that each of n judges (see Section 13.3) now inspects k candidates where $k > 2$. Then each judge specifies some permutation of $(1, 2, \ldots, k - 1, k)$ as his ranking of the k candidates. In this case we can consider generalizing the sign test to a test of the hypothesis that all candidates are equally likely to receive each rank (that is, are "equally qualified" to be selected) (for example, see Chapter 7, pp. 138–184, of M. Hollander and D. A. Wolfe, *Nonparametric Statistical Methods*, John Wiley & Sons, Inc., New York, 1973); however, in many cases (for example, sports) it will be more appropriate to attempt to select[3] the "best" candidate. How can we define the "best" candidate? (For example, let $\mathbf{R}_i = (R_{i1}, \ldots, R_{ik})$ denote the random vector of ranks which arises when judge i orders the candidates. Then we might define

$$p_l = P[R_{il} = k]$$

$(1 \leqslant l \leqslant k)$ and define the "best" candidate to be the one who has the largest of $p_1, \ldots, p_k$. Will the same one still be best if we delete one or more other candidates? Not necessarily, and please explain why and give an example.)

[3] Such problems have been touched on in Chapter 11. For more details, see Y. J. Lee and E. J. Dudewicz, "Nonparametric ranking and selection procedures," *Technical Report No. 105*, Department of Statistics, The Ohio State University, Columbus, Ohio, August 1974.

14

REGRESSION AND LINEAR STATISTICAL INFERENCE

In the preceding chapters we occasionally assumed that we had available independent random variables $Y_1, \ldots, Y_n$ whose common distribution was known (for example, each of $Y_1, \ldots, Y_n$ might be a $N(\mu, \sigma^2)$ r.v. with μ and σ^2 both unknown as in Example 10.2.10) and then studied methods for making inferences based on the variables $Y_1, \ldots, Y_n$ (for example, to find a 95 percent confidence interval for μ, in Example 10.2.10; or to predict a future value Y from the same distribution, as in Problem 10.5H). In the present chapter we study similar inference problems for the case where each Y_i is related to some variable X_i $(i = 1, \ldots, n)$ which can be observed or controlled (for example, it might be that $Y = f(X)$ for some known or conjectured function $f(\cdot)$); in that case, we use the information about $X_1, \ldots, X_n$ in our inference procedure.

Some considerations of the question of relationships between two r.v.'s X and Y were given in Section 5.3 (for example, independence, covariance, correlation, and conditional expectation) and it may be helpful to review them when you start this chapter. The regression problem we will study in detail is that involving two variables X and Y and a linear relationship $Y = \alpha + \beta X$. Our aim is to illustrate the main ideas with this study, without the obscuring complexities of the cases with more than two variables (for example, $Y = \alpha + \beta X + \gamma Z$) or with a nonlinear relationship (for example, $Y = \alpha + \beta X + \beta^2 Z^2$ or $Y = \alpha + \alpha\beta \sin(X)$) or both (for example, $Y = \alpha + \beta X + \gamma X^2 + \delta Z + \delta^2 XZ$). (Once the main ideas are understood, the details of formulas and computations for such multivariate and/or nonlinear cases can be referenced in a number of sources which, in general, do not equip the reader to understand the material.)

Our study of the relationship $Y = \alpha + \beta X$ encompasses the questions of: estimating the unknown constants α and β (Section 14.1, which utilizes material from Chapter 8); testing hypotheses about α and β (Section 14.2, which utilizes material from Chapter 9); giving confidence intervals for α and β and prediction intervals for a new observation Y which is governed by the same relationship as are $Y_1, \ldots, Y_n$ (Section 14.3, which utilizes material from Chapter 10); and selecting the best of several regressions and associated experimental design questions (Section 14.4, which utilizes material from Chapter 11).

14.1. ESTIMATION OF REGRESSION PARAMETERS; LINEAR MODELS

In general, suppose that (X, Y) is a random variable with a bivariate distribution function $F_{X,Y}(x, y)$. If we can observe X, then it makes sense (if we are interested in predicting the value of Y) to predict Y by some function $f(X)$, since X and Y are related variables. We will now find the "best" such function $f(X)$, after first (in Definition 14.1.1) giving it the special name "regression curve of Y on X."

Definition 14.1.1. *The function $f(X)$ for which*

$$E(Y - f(X))^2 \tag{14.1.2}$$

is minimized is called the **regression curve of Y on X**.

Theorem 14.1.3. *Let (X, Y) have a bivariate d.f. such that* Var(X) *and* Var(Y) *both exist. Then the regression curve of Y on X is given (for all x) by*

$$f(x) = E(Y|X = x). \tag{14.1.4}$$

Proof: Let $f(x)$ denote the conditional expectation $E(Y|X = x)$ (see Definition 5.3.19), and let $\varphi(x)$ be a general function. Then

$$\begin{aligned} E(Y - \varphi(X))^2 &= E((Y - f(X)) + (f(X) - \varphi(X)))^2 \\ &= E(Y - f(X))^2 + E(f(X) - \varphi(X))^2 \\ &\quad + 2E(Y - f(X))(f(X) - \varphi(X)). \end{aligned} \tag{14.1.5}$$

However (see Theorem 5.3.20 and the discussion following it) $E(\varphi(X)Y) = E(\varphi(X)E(Y|X)) = E(\varphi(X)f(X))$; hence,

$$E(Y - f(X))\varphi(X) = 0 \tag{14.1.6}$$

and (taking $\varphi(\cdot)$ to be $f(\cdot)$ in (14.1.6))

$$E(Y - f(X))f(X) = 0. \tag{14.1.7}$$

From (14.1.6) and (14.1.7) we find that the last term in (14.1.5) is zero; hence,

$$E(Y - \varphi(X))^2 = E(Y - f(X))^2 + E(f(X) - \varphi(X))^2, \tag{14.1.8}$$

which is clearly minimized (as a function of $\varphi(\cdot)$) by taking $\varphi(x) = f(x)$ for all x.

Sometimes (for example, because the regression curve $E(Y|X = x)$ of Y on X may be very complex or hard to find, or because we expect an approximately linear relationship to hold between Y and X) we will wish to find the "best" linear function of X for predicting Y; we will now (in analogy to Definition 14.1.1 and Theorem 14.1.3) develop this notion.

Definition 14.1.9. *The α and β for which $\alpha + \beta X$ causes*

$$E(Y - (\alpha + \beta X))^2 \tag{14.1.10}$$

to be minimized are called the **regression intercept** *and* **regression slope** *(of Y on X), respectively. The line $y = \alpha + \beta x$ is called the* **regression line of Y on X**.

Theorem 14.1.11. *The regression line of Y on X is given by $y = \alpha + \beta x$ with α and β such that*

$$\alpha = E(Y) - \rho(X, Y)\frac{\sigma(Y)}{\sigma(X)}E(X), \tag{14.1.12}$$

$$\beta = \rho(X, Y)\frac{\sigma(Y)}{\sigma(X)}. \tag{14.1.13}$$

With this α and β, we find that the regression line of Y on X is

$$y = E(Y) - \rho(X, Y)\frac{\sigma(Y)}{\sigma(X)}E(X) + \rho(X, Y)\frac{\sigma(Y)}{\sigma(X)}x \tag{14.1.14}$$

or

$$\frac{y - E(Y)}{\sigma(Y)} = \rho(X, Y)\frac{x - E(X)}{\sigma(X)}, \tag{14.1.15}$$

which has mean squared prediction error

$$E(Y - (\alpha + \beta X))^2 = (1 - \rho^2(X, Y))\sigma^2(Y). \tag{14.1.16}$$

Proof: For a general α and β,

$$\begin{aligned}
E(Y-(\alpha+\beta X))^2 &= E((Y-E(Y))-\beta(X-E(X)) \\
&\qquad -(\alpha-E(Y)+\beta E(X)))^2 \\
&= E(Y-E(Y))^2+\beta^2E(X-E(X))^2 \\
&\quad +(\alpha-E(Y)+\beta E(X))^2 \\
&\quad -2\beta E(Y-E(Y))(X-E(X)) \\
&\quad -2(\alpha-E(Y)+\beta E(X))E(Y-E(Y)) \\
&\quad +2\beta(\alpha-E(Y)+\beta E(X))E(X-E(X)) \\
&= \sigma^2(Y)+\beta^2\sigma^2(X)+(\alpha-E(Y)+\beta E(X))^2 \\
&\quad -2\beta E(Y-E(Y))(X-E(X))-0+0.
\end{aligned} \tag{14.1.17}$$

Now (looking at the last equation in (14.1.17) we see that α only enters through the term $(\alpha - E(Y) + \beta E(X))^2$, so (for any β) the best α is $E(Y) - \beta E(X)$. With that choice of α (call it α^*) we then find

$$\begin{aligned}
&E(Y-(\alpha^*+\beta X))^2 \\
&\quad = \sigma^2(Y)+\beta^2\sigma^2(X)-2\beta E(Y-E(Y))(X-E(X)) \\
&\quad = \sigma^2(Y)+\beta^2\sigma^2(X)-2\beta\sigma(X)\sigma(Y)\rho(X,Y) \\
&\quad = (\beta\sigma(X)-\rho(X,Y)\sigma(Y))^2+(1-\rho^2(X,Y))\sigma^2(Y),
\end{aligned} \tag{14.1.18}$$

which is minimized by setting β as in (14.1.13) whence (using this β in α^*) we find α should be set as in (14.1.12). The mean squared prediction error follows easily from (14.1.18) where (with the best β) the first term is zero.

Example 14.1.19. Suppose that (X, Y) has a bivariate d.f. with $\sigma^2(Y) = 10$, $\rho(X, Y) = .5$. Then if we ignore X and predict Y as simply $E(Y)$, we will have a mean squared error of prediction equal to the variance of Y, namely 10. If we use the regression line of Y on X to predict Y, the mean squared error of prediction will (see (14.1.16)) be cut to

$$(1-\rho^2(X,Y))\sigma^2(Y) = (1-.25)(10) = \frac{(3)(10)}{4} = 7.5, \tag{14.1.20}$$

which is a 25 percent reduction. For this reason **it is often said that "the square of the correlation coefficient is the proportion of the variance (of Y) accounted for by linear regression on X," and (14.1.16) is called the residual variance**(after linear regression on X).

The regression curve of Y on X (determined in Theorem 14.1.3) and the regression line of Y and X (determined in Theorem 14.1.11) are both theoretical, and complete knowledge about the bivariate d.f. of (X, Y) is needed to calculate them. While in some situations (such as repetitive sampling from the same bivariate distribution with the X immediately available but the Y unavailable for a long period and in need of prediction; for example, X could be a high school grade-point average while Y could be a corresponding college grade-point average at a certain school) we know this d.f. and can calculate these regressions, **often we have available only n independent pairs of observations $(X_1, Y_1), \ldots, (X_n, Y_n)$ and need to estimate the regressions. We will consider this problem now for the linear regression case.**

One method of estimating the regression line, called the **method of least squares**, is to estimate α and β by the $\hat{\alpha}$ and $\hat{\beta}$ (say) which minimize the sums of the squares of the errors in predicting Y_i by $\alpha + \beta X_i$ $(i = 1, \ldots, n)$, that is, which minimize

$$S(\alpha, \beta) = \sum_{i=1}^{n} (Y_i - (\alpha + \beta X_i))^2. \tag{14.1.21}$$

We will now derive the $\hat{\alpha}$ and $\hat{\beta}$ which the method of least squares suggests we use; further justification for these estimators will be given later in this section (the Gauss-Markov Theorem).

Theorem 14.1.22. *Let $(X_1, Y_1), \ldots, (X_n, Y_n)$ be n independent observations from a bivariate d.f. Then the method of least squares estimates the regression line of Y on X [where (X, Y) is an observation from this same bivariate d.f.] by $y = \hat{\alpha} + \hat{\beta}x$ where*

$$\hat{\alpha} = \bar{Y} - r_{XY}\frac{s_Y}{s_X}\bar{X}, \tag{14.1.23}$$

$$\hat{\beta} = r_{XY}\frac{s_Y}{s_X}, \tag{14.1.24}$$

$$r_{XY} = \frac{n\sum_{i=1}^{n} X_i Y_i - \sum_{i=1}^{n} X_i \sum_{i=1}^{n} Y_i}{\sqrt{n\sum_{i=1}^{n} X_i^2 - \left(\sum_{i=1}^{n} X_i\right)^2}\sqrt{n\sum_{i=1}^{n} Y_i^2 - \left(\sum_{i=1}^{n} Y_i\right)^2}}, \tag{14.1.25}$$

and

$$\begin{cases} s_X^2 = \dfrac{n\sum_{i=1}^{n} X_i^2 - \left(\sum_{i=1}^{n} X_i\right)^2}{n^2}, & s_Y^2 = \dfrac{n\sum_{i=1}^{n} Y_i^2 - \left(\sum_{i=1}^{n} Y_i\right)^2}{n^2}, \\ \bar{X} = \dfrac{X_1 + \cdots + X_n}{n}, & \bar{Y} = \dfrac{Y_1 + \cdots + Y_n}{n}. \end{cases} \tag{14.1.26}$$

Proof: By definition of the method of least squares, we are to choose α and

β as the $\hat{\alpha}$ and $\hat{\beta}$ which minimize (14.1.21):

$$S(\alpha, \beta) = \sum_{i=1}^{n} (Y_i - (\alpha + \beta X_i))^2. \tag{14.1.27}$$

If we take the partial derivatives of (14.1.27) with respect to α and β, set them to zero, and solve, we then (see Problem 14.1H) find $\hat{\alpha}$ and $\hat{\beta}$:

$$\begin{cases} \dfrac{\partial S(\alpha, \beta)}{\partial \alpha} = \displaystyle\sum_{i=1}^{n} (-2)(Y_i - (\alpha + \beta X_i)) = 0, \\ \dfrac{\partial S(\alpha, \beta)}{\partial \beta} = \displaystyle\sum_{i=1}^{n} (-2X_i)(Y_i - (\alpha + \beta X_i)) = 0, \end{cases} \tag{14.1.28}$$

that is,

$$\begin{cases} \displaystyle\sum_{i=1}^{n} Y_i - n\alpha - \beta \sum_{i=1}^{n} X_i = 0, \\ \displaystyle\sum_{i=1}^{n} X_i Y_i - \alpha \sum_{i=1}^{n} X_i - \beta \sum_{i=1}^{n} X_i^2 = 0, \end{cases} \tag{14.1.29}$$

and (via elementary algebraic manipulations; see Problem 14.6) we find (14.1.23) and (14.1.24).

So far in this section we have been considering our observations $(\mathbf{X}_1, \mathbf{Y}_1), \ldots, (\mathbf{X}_n, \mathbf{Y}_n)$ to have been drawn independently from a bivariate population, and this is appropriate in many situations. However, in many other situations (such as in physical science) $\mathbf{X}_1, \ldots, \mathbf{X}_n$ are not r.v.'s (for example, they may be amounts of a catalyst provided by an experimenter, amounts of fertilizer, and so on). In this case an additional theoretical justification of the least-squares estimators of α and β (note that those estimators in no way made use of the assumption that $X_1, \ldots, X_n$ were r.v.'s, and hence the same estimators apply to the nonrandom $X_1, \ldots, X_n$ case) can be given. The general result which provides this justification is called the **Gauss-Markov Theorem**. Stated somewhat informally (for a general statement see p. 14 of H. Scheffé, *The Analysis of Variance*, John Wiley & Sons, Inc., New York, 1959) this theorem says that the best (that is, minimum variance) unbiased estimators of α and of β, among all estimators which are linear functions of $Y_1, \ldots, Y_n$ when $x_1, \ldots, x_n$ are fixed values, are those obtained by the method of least squares. We will now prove this result for the case of α and β.

Theorem 14.1.30. *Let $Y_1, \ldots, Y_n$ be independent observations (taken at respective fixed levels $x_1, \ldots, x_n$ of some independent variable) such that $E(Y_i) = \alpha + \beta x_i$ and $\mathrm{Var}(Y_i) = \sigma^2$ $(i = 1, \ldots, n)$. Then, of all estimators of*

α and of β which are linear functions of $Y_1, \ldots, Y_n$ and which are unbiased, the $\hat{\alpha}$ and $\hat{\beta}$ of Theorem 14.1.22 *have the smallest respective variances.*

Proof: Consider a general linear estimator of α, say $t = u_1 y_1 + \cdots + u_n Y_n$. (Note that this is the most general linear combination of $Y_1, \ldots, Y_n$.) Now t will be an unbiased estimator of α if $E(t) \equiv \alpha$; that is,

$$\begin{aligned} \alpha \equiv E(t) &= E\left(\sum_{i=1}^{n} u_i Y_i\right) = \sum_{i=1}^{n} u_i E(Y_i) \\ &= \sum_{i=1}^{n} u_i(\alpha + \beta X_i) = \alpha \sum_{i=1}^{n} u_i + \beta \sum_{i=1}^{n} u_i x_i. \end{aligned} \tag{14.1.31}$$

Since (14.1.31) must hold for all α, we find we must have

$$\sum_{i=1}^{n} u_i = 1, \qquad \sum_{i=1}^{n} u_i x_i = 0. \tag{14.1.32}$$

The variance of t, which is to be minimized subject to (14.1.32), is

$$\text{Var}(t) = \text{Var}\left(\sum_{i=1}^{n} u_i Y_i\right) = \left(\sum_{i=1}^{n} u_i^2\right)\sigma^2; \tag{14.1.33}$$

hence, it suffices to minimize $u_1^2 + \cdots + u_n^2$ subject to $u_1 + \cdots + u_n = 1$ and $u_1 x_1 + \cdots + u_n x_n = 0$. Let (for $i = 1, \ldots, n$)

$$v_i = \frac{\sum_{i=1}^{n} x_i^2 - \left(\sum_{i=1}^{n} x_i\right) x_i}{n \sum_{i=1}^{n} x_i^2 - \left(\sum_{i=1}^{n} x_i\right)^2}. \tag{14.1.34}$$

Then we can express u_i as

$$u_i = v_i + \Delta_i \qquad (i = 1, \ldots, n) \tag{14.1.35}$$

for some Δ_i. From (14.1.32), it follows that we must have

$$\sum_{i=1}^{n} v_i + \sum_{i=1}^{n} \Delta_i = 1, \qquad \sum_{i=1}^{n} v_i x_i + \sum_{i=1}^{n} \Delta_i x_i = 0,$$

but (as can be easily verified; see Problem 14.9) $\sum_{i=1}^{n} v_i = 1$ and $\sum_{i=1}^{n} v_i x_i = 0$; hence, we must have

$$\sum_{i=1}^{n} \Delta_i = 0, \qquad \sum_{i=1}^{n} \Delta_i x_i = 0.$$

Now using (14.1.35),

$$\begin{aligned}\sum_{i=1}^{n} u_i^2 &= \sum_{i=1}^{n} (v_i + \Delta_i)^2 = \sum_{i=1}^{n} v_i^2 + \sum_{i=1}^{n} \Delta_i^2 + 2\sum_{i=1}^{n} v_i\Delta_i \\ &= \sum_{i=1}^{n} v_i^2 + \sum_{i=1}^{n} \Delta_i^2 + 0 \geqslant \sum_{i=1}^{n} v_i^2\end{aligned} \tag{14.1.36}$$

and the theorem follows.

The considerations of this section and chapter deal with the area of statistics known as *linear models*, which deals with statistical inference problems which arise when we assume that a r.v. Y can be expressed as

$$Y = \beta_0 x_0 + \beta_1 x_1 + \cdots + \beta_m x_m + \varepsilon, \tag{14.1.37}$$

where: $x_0, x_1, \ldots, x_m$, and m are known constants; $\beta_0, \beta_1, \ldots, \beta_m$ are unknown constants; and ε is a r.v. with $E(\varepsilon) = 0$ whose distribution does not vary as the parameters $\beta_0, \beta_1, \ldots, \beta_m$ vary. (If we assume $m = 2$, $x_0 = 1$, $\beta_0 = \alpha$, $x_1 = x$, and $\beta_1 = \beta$, then (14.1.37) becomes the model $Y = \alpha + \beta x + \varepsilon$ studied above.) The general results, of which the results in this chapter are indicative, can be found in such books as *An Introduction to Linear Statistical Models, Vol. I*, by F. A. Graybill, McGraw-Hill Book Company, Inc., New York, 1961.

14.2. TESTING HYPOTHESES ABOUT REGRESSION PARAMETERS

In Section 14.1 we considered the problem of estimating α and β given n independent r.v.'s $Y_1, \ldots, Y_n$ (taken at respective fixed levels $x_1, \ldots, x_n$ of some independent variable) such that $E(Y_i) = \alpha + \beta x_i$ and $\operatorname{Var}(Y_i) = \sigma^2$ $(i = 1, \ldots, n)$; for example (see Theorem 14.1.30) the estimators $\hat{\alpha}$ and $\hat{\beta}$ of (14.1.23) and (14.1.24) are unbiased and have the smallest variance of all unbiased linear estimators. **In order to test hypotheses about $\boldsymbol{\alpha}$ and/or $\boldsymbol{\beta}$, it will be necessary to make some assumptions about the distribution of $\mathbf{Y_i}$ $(\mathbf{i = 1, \ldots, n})$** (up to this point we only talked of the mean and variance of Y_i $(i = 1, \ldots, n)$). If we assume that each Y_i has a normal distribution, then (see Problem 14.15) it turns out that $\hat{\alpha}$ and $\hat{\beta}$ are the respective MLE's of α and of β, and that we can find an MLE for σ^2:

Theorem 14.2.1. *Let $Y_1, \ldots, Y_n$ be independent r.v.'s where Y_i is $N(\alpha + \beta x_i, \sigma^2)$ where x_i is known and α, β, σ^2 are unknown $(1 \leqslant i \leqslant n)$. Then $\hat{\alpha}$ of* (14.1.23) *is the MLE of* α, $\hat{\beta}$ *of* (14.1.24) *is the MLE of* β, *and the MLE of* σ^2 *is*

$$\hat{\sigma}^2 = \frac{\sum_{i=1}^{n} (Y_i - \hat{\alpha} - \hat{\beta}x_i)^2}{n}. \tag{14.2.2}$$

If we examine formulas (14.1.23) and (14.1.24) carefully (Problem 14.16) we find that $\hat{\alpha}$ and $\hat{\beta}$ can be expressed simply as

$$\begin{cases} \hat{\alpha} = \bar{Y} - \dfrac{n \sum_{i=1}^{n} x_i Y_i - \sum_{i=1}^{n} x_i \sum_{i=1}^{n} Y_i}{n \sum_{i=1}^{n} x_i^2 - \left(\sum_{i=1}^{n} x_i\right)^2} \bar{x}, \\[2ex] \hat{\beta} = \dfrac{n \sum_{i=1}^{n} x_i Y_i - \sum_{i=1}^{n} x_i \sum_{i=1}^{n} Y_i}{n \sum_{i=1}^{n} x_i^2 - \left(\sum_{i=1}^{n} x_i\right)^2}. \end{cases} \tag{14.2.3}$$

However, we note that each of $\hat{\alpha}$ and $\hat{\beta}$ is a linear combination of $Y_1, \ldots, Y_n$ and hence (see footnote 21 in Chapter 8) the r.v. $(\hat{\alpha}, \hat{\beta})$ has a bivariate normal distribution. We already know the means of this distribution (since we already showed that $E(\hat{\alpha}) = \alpha$ and $E(\hat{\beta}) = \beta$ in Theorem 14.1.30) and it is easy to show (Problem 14.17) that the variances are

$$\begin{cases} \operatorname{Var}(\hat{\alpha}) = \dfrac{n \sum_{i=1}^{n} x_i^2}{n \sum_{i=1}^{n} x_i^2 - \left(\sum_{i=1}^{n} x_i\right)^2} \sigma^2, \\[2ex] \operatorname{Var}(\hat{\beta}) = \dfrac{n}{n \sum_{i=1}^{n} x_i^2 - \left(\sum_{i=1}^{n} x_i\right)^2} \sigma^2. \end{cases} \tag{14.2.4}$$

Thus, we have proven the theorem:

Theorem 14.2.5. *Let* $Y_1, \ldots, Y_n$ *be independent r.v.'s where* Y_i *is* $N(\alpha + \beta x_i, \sigma^2)$ $(1 \leqslant i \leqslant n)$. *Then* $(\hat{\alpha}, \hat{\beta})$ *has a bivariate normal distribution with means* (α, β) *and variances given by* (14.2.4).

Now it can be shown (Problem 14.7H) that $n\hat{\sigma}^2/\sigma^2$ is a $\chi^2_{n-2}(0)$ r.v. and (Problem 14.8H) that it is independent of each of $\hat{\alpha}$ and $\hat{\beta}$. **Hence, as in Section 9.6 in a simpler case, we may easily test the hypothesis** $\mathbf{H}_0 = \{\boldsymbol{\sigma}_0^2\}$ **about** $\boldsymbol{\sigma}^2$ **at level** $\boldsymbol{\alpha}$: reject iff

$$\hat{\sigma}^2 > k'', \tag{14.2.6}$$

where k'' satisfies $P[Y > nk''/\sigma_0^2] = \alpha$ for Y a $\chi^2_{n-2}(0)$ r.v.

Similarly, we can obtain tests of hypotheses like $\mathbf{H}_0 = \{\boldsymbol{\alpha}_0\}$ since (under the null hypothesis)

$$\frac{\hat{\alpha} - \alpha_0}{\sqrt{\dfrac{n \sum_{i=1}^{n} x_i^2}{n \sum_{i=1}^{n} x_i^2 - \left(\sum_{i=1}^{n} x_i\right)^2} \dfrac{n\hat{\sigma}^2}{n-2}}} \tag{14.2.7}$$

has the t_{n-2} distribution (see Definition 9.5.24). The test is entirely analogous to that of Theorem 9.5.26: reject iff (14.2.7) exceeds $t_{n-2}^{-1}(1 - \alpha)$. **Similar tests are easily obtained for other alternatives, and for $\boldsymbol{\beta}$** (see Problem 14.19).

14.3. PREDICTION AND CONFIDENCE INTERVALS IN REGRESSION

To summarize some results from previous sections, if $Y_1, \ldots, Y_n$ are independent r.v.'s where Y_i is $N(\alpha + \beta x_i, \sigma^2)$ $(1 \leqslant i \leqslant n)$, then $(\hat{\alpha}, \hat{\beta})$ given at (14.2.3) has a bivariate normal distribution with mean vector (α, β) and variances as given by (14.2.4), and σ^2 may be estimated by $\hat{\sigma}^2$ given by (14.2.2) which is stochastically independent of $\hat{\alpha}$ and of $\hat{\beta}$ and $n\hat{\sigma}^2/\sigma^2$ is a $\chi^2_{n-2}(0)$ r.v. **In light of these facts, confidence intervals for $\boldsymbol{\sigma}^2$ are obtained easily essentially as in Section 10.3 in a simpler case, and confidence intervals for $\boldsymbol{\alpha}$ and for $\boldsymbol{\beta}$ follow as in Section 10.2**; for example, a confidence interval for α with confidence coefficient $1 - \alpha$ is given by (L, U) where

$$L = \hat{\alpha} - \sqrt{\frac{n \sum_{i=1}^{n} x_i^2}{n \sum_{i=1}^{n} x_i^2 - \left(\sum_{i=1}^{n} x_i\right)^2} \frac{n\hat{\sigma}^2}{n-2}}\, t_{n-2}^{-1}\left(1 - \frac{\alpha}{2}\right) \tag{14.3.1}$$

and

$$U = \hat{\alpha} + \sqrt{\frac{n \sum_{i=1}^{n} x_i^2}{n \sum_{i=1}^{n} x_i^2 - \left(\sum_{i=1}^{n} x_i\right)^2} \frac{n\hat{\sigma}^2}{n-2}}\, t_{n-2}^{-1}\left(1 - \frac{\alpha}{2}\right). \tag{14.3.2}$$

Often we will not be interested in inference (for example, tests of hypotheses or confidence intervals) about $\boldsymbol{\alpha}$ and $\boldsymbol{\beta}$ directly, but rather in using the observations $\mathbf{Y}_1, \ldots, \mathbf{Y}_n$ (taken at respective values $\mathbf{x}_1, \ldots, \mathbf{x}_n$ of the independent variable) to predict the value of Y we might obtain in the future if the independent variable is set to value $\mathbf{x}^*$. An unbiased estimator of $E(Y) = \alpha + \beta x^*$ is, of course, $\hat{\alpha} + \hat{\beta}x^*$. This $\hat{\alpha} + \hat{\beta}x^*$ is actually an unbiased estimator of two quantities: first, of $\alpha + \hat{\beta}x^*$; second, of $E(Y)$, the mean of a future observation taken at x^*. The difference between these two quantities becomes clear when we seek confidence intervals about the point estimate $\hat{\alpha} + \hat{\beta}x^*$, as we shall now see.

If we desire a confidence interval for $\boldsymbol{\alpha} + \boldsymbol{\beta}\mathbf{x}^*$, we need to know

$$\operatorname{Var}((\hat{\alpha} + \hat{\beta}x^*) - (\alpha + \beta x^*)) = \sigma^2\left(\frac{1}{n} + \frac{(x^* - \overline{x})^2}{ns_x^2}\right); \tag{14.3.3}$$

a 95 percent confidence interval for $\alpha + \beta x^*$ is then (Y_L^*, Y_U^*) where

$$Y_L^* = (\hat{\alpha} + \hat{\beta}x^*) - t_{n-2}^{-1}(.975)\sqrt{\frac{n\hat{\sigma}^2}{n-2}}\sqrt{\frac{1}{n} + \frac{(x^* - \overline{x})^2}{ns_x^2}} \tag{14.3.4}$$

and

$$Y_U^* = (\hat{\alpha} + \hat{\beta}x^*) + t_{n-2}^{-1}(.975)\sqrt{\frac{n\hat{\sigma}^2}{n-2}}\sqrt{\frac{1}{n} + \frac{(x^* - \overline{x})^2}{ns_x^2}}. \tag{14.3.5}$$

On the other hand, if we desire a confidence interval for Y at x*, we need to know

$$\begin{aligned}
&\operatorname{Var}(Y - (\hat{\alpha} + \hat{\beta}x^*)) \\
&\quad = \operatorname{Var}((Y - (\alpha + \beta x^*)) - ((\hat{\alpha} + \hat{\beta}x^*) - (\alpha + \beta x^*))) \\
&\quad = \operatorname{Var}(Y - (\alpha + \beta x^*)) + \operatorname{Var}((\hat{\alpha} + \hat{\beta}x^*) - (\alpha + \beta x^*)) \\
&\quad = \sigma^2 + \sigma^2\left(\frac{1}{n} + \frac{(x^* - \overline{x})^2}{ns_x^2}\right) \\
&\quad = \sigma^2\left(1 + \frac{1}{n} + \frac{(x^* - \overline{x})^2}{ns_x^2}\right);
\end{aligned} \tag{14.3.6}$$

a 95 percent confidence interval for Y at x^* is then (Y_L, Y_U) where

$$Y_L = (\hat{\alpha} + \hat{\beta}x^*) - t_{n-2}^{-1}(.975)\sqrt{\frac{n\hat{\sigma}^2}{n-2}}\sqrt{1 + \frac{1}{n} + \frac{(x^* - \overline{x})^2}{ns_x^2}}, \tag{14.3.7}$$

and

$$Y_U = (\hat{\alpha} + \hat{\beta}x^*) + t_{n-2}^{-1}(.975)\sqrt{\frac{n\hat{\sigma}^2}{n-2}}\sqrt{1 + \frac{1}{n} + \frac{(x^* - \overline{x})^2}{ns_x^2}}. \tag{14.3.8}$$

Note that while the length of the interval on $\alpha + \beta x^*$ (that is, $Y_U^* - Y_L^*$) tends to zero as $n \to \infty$ (the mean of Y at x^* is pinned down very precisely), the length of the interval on Y at x^* (that is, $Y_U - Y_L$) does not tend to zero as $n \to \infty$ (what does it tend to?) since Y will still vary and cannot be predicted perfectly.

A plot of (Y_L^*, Y_U^*) for all x^* is called a **confidence band (for the regression line)**, while a plot of (Y_L, Y_U) for all x^* is called a **prediction band (for future values)**.

Example 14.3.9. A survey of agricultural yield against the amount of a pest allowed (the latter can be controlled by crop treatment) has produced the data of Table 14.3–1.

Table 14.3–1. Pest Rate (x_i) and Crop Yield (Y_i) Data

i	1	2	3	4	5	6	7	8	9	10	11	12
x_i	8	6	11	22	14	17	18	24	19	23	26	40
Y_i	59	58	56	53	50	45	43	42	39	38	30	27

Find $\hat{\alpha}$ and $\hat{\beta}$. Plot the regression line with confidence bands and prediction bands (each at the 95 percent confidence level), and with the data points superimposed on the graph. If the pest rate is allowed to go to $x^* = 50$

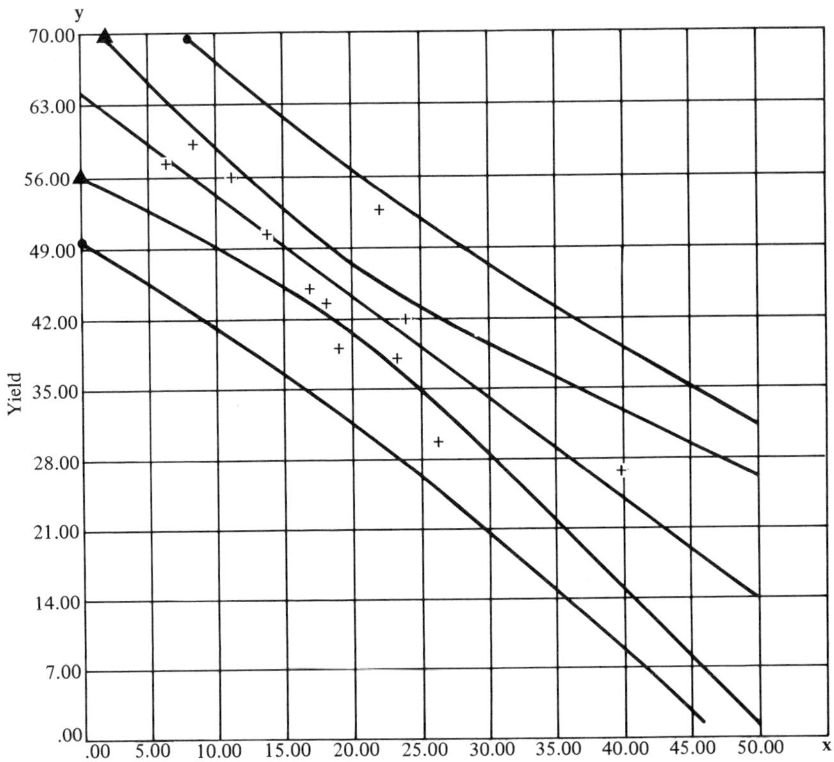

+ Data Point, ● Prediction Bands, ▲ Confidence Bands

Figure 14.3–1. Output for Example 14.3.9

next year, within what limits can we be 95 percent sure our crop yield will fall?

Calculations here yield $\hat{\alpha} = 64.247$ and $\hat{\beta} = -1.0130$, so the regression line is estimated as $y = 64.247 - 1.0130x$. The plot is given in Figure 14.3–1 and shows that if we let our pest reach level $x^* = 50$, we can be 95 percent sure our yield will be between 0 and 31.5 units. (The mean yield at $x^* = 50$ is between 1.4 and 26.3 units with 95 percent confidence.)

14.4. SELECTION OF THE BEST REGRESSION

Examples of problems where selection of the best of several regressions is the experimenter's main concern (rather than the testing of hypotheses or forming of confidence intervals) are not rare. In one class of such problems we assume that r.v.'s drawn from k populations $\pi_1, \ldots, \pi_k$ $(k \geqslant 2)$ are independent, and that those drawn from π_i are $N(\alpha_i + \beta_i x, \sigma^2)$ r.v.'s if the independent variable is set at level x. **If we know that $\boldsymbol{\alpha}_1 = \cdots = \boldsymbol{\alpha}_k$ (the regression lines have the same intercept), it may be our concern to select the regression with the largest slope $\max(\boldsymbol{\beta}_1, \ldots, \boldsymbol{\beta}_k)$**; for example, this would be so if $\pi_1, \ldots, \pi_k$ represented k treatments and observations represented yield. If we obtain observations Y_{ij} at level x_{ij} from π_i for $j = 1, \ldots, n_j$ $(1 \leqslant i \leqslant k)$, we will naturally (see (14.2.3)) estimate β_i by

$$\hat{\beta}_i = \frac{n_i \sum_{j=1}^{n_i} x_{ij} Y_{ij} - \sum_{j=1}^{n_i} x_{ij} \sum_{j=1}^{n_i} Y_{ij}}{n_i \sum_{j=1}^{n_i} x_{ij}^2 - \left(\sum_{j=1}^{n_i} x_{ij} \right)^2} \tag{14.4.1}$$

and select the population yielding $\max(\hat{\beta}_1, \ldots, \hat{\beta}_k)$ (the largest sample slope). Since $\hat{\beta}_i$ is normally distributed with mean β_i and variance (see (14.2.4))

$$\operatorname{Var}(\hat{\beta}_i) = \frac{n_i}{n_i \sum_{j=1}^{n_i} x_{ij}^2 - \left(\sum_{j=1}^{n_i} x_{ij} \right)^2} \sigma^2, \tag{14.4.2}$$

the problem is essentially one of ranking and selection of normal means as studied in Chapter 11. If we choose $n_1 = \cdots = n_k = n$ (say) and $x_{1j} = \cdots = x_{kj}$ $(j = 1, \ldots, n)$, all $\hat{\beta}_i$'s have the same variance and the results of Section 11.2 may be used directly if σ^2 is known, while if σ^2 is unknown we proceed as noted in item 1 of Section 11.4.

Similarly for parallel lines $(\boldsymbol{\beta}_1 = \cdots = \boldsymbol{\beta}_k)$ we could derive a procedure for selecting the one with the largest intercept $\max(\boldsymbol{\alpha}_1, \ldots, \boldsymbol{\alpha}_k)$.

If we have regression lines with different slopes *and* different intercepts, the definition of the "best" line may not be clear; for example, we may have to specify an x level of interest (since the line which on the average produces the largest value Y will not be the same for all x).

PROBLEMS

14.1 Show that the quantity $E(Y - c)^2$ is minimized when $c = E(Y)$. Thus, the "best" predictor of a r.v.'s value (in the sense of minimizing mean squared prediction error), in the absence of any auxiliary r.v. X, is the mean of the r.v. [*Hint*: Use Theorem 4.1.2.]

14.2 For the case of Example 5.3.21, find and graph both the regression curve of Y on X, and the regression line of Y on X. Calculate and compare the respective mean square errors of prediction (14.1.2) and (14.1.16).

14.3 Find the regression line of Y on X in the case $\rho(X, Y) = 0$. What is the mean square error of prediction in this case?

14.4 Find the regression line of Y on X in the case $\rho(X, Y) = 1.0$. Similarly, in the case $\rho(X, Y) = -1.0$. What is the mean square error of prediction in each case?

14.5 Suppose (as we did following Lemma 5.3.14) that $Y = X^2$. Find the regression curve of Y on X and the regression line of Y on X. Compare their mean square errors of prediction.

14.6 Prove that simultaneous solution of the system of equations (14.1.29) for α and β yields $\hat{\alpha}$ and $\hat{\beta}$ as at (14.1.23) and (14.1.24).

14.7 In a certain class $(X_1, Y_1), \ldots, (X_7, Y_7)$ were measured where X_i was the score on homework of student i and Y_i was student i's subsequent score on an examination. The data are

i	1	2	3	4	5	6	7
X_i	0	10	49	52	59	64	64
Y_i	12	20	60	43	37	58	63

Find the estimated regression line of Y on X (see Theorem 14.1.22). Plot the seven data points and the line $y = \hat{\alpha} + \hat{\beta}x$ on the same graph. If a similar student in a similar situation in the future obtained a homework score of 30, what would the best linear estimate of his subsequent examination score?

14.8 Suppose that the regression line of Y on X is $Y = \alpha + \beta X$, and that $X_1, \ldots, X_n$ are not r.v.'s, but are fixed values. If (as in Theorem 14.1.22) α and β are estimated from the data $(X_1, Y_1), \ldots, (X_n, Y_n)$, show that $E(\hat{\alpha}) = \alpha$ and $E(\hat{\beta}) = \beta$. Find $\operatorname{Var}(\hat{\alpha})$ and $\operatorname{Var}(\hat{\beta})$.

14.9 In the proof of Theorem 14.1.30, verify that $\sum_{i=1}^{n} v_i = 1$ and $\sum_{i=1}^{n} v_i x_i = 0$; that is, verify that

$$\sum_{i=1}^{n} \frac{\sum_{i=1}^{n} x_i^2 - \left(\sum_{i=1}^{n} x_i\right) x_i}{n \sum_{i=1}^{n} x_i^2 - \left(\sum_{i=1}^{n} x_i\right)^2} = 1$$

and

$$\sum_{i=1}^{n} \frac{\sum_{i=1}^{n} x_i^2 - \left(\sum_{i=1}^{n} x_i\right) x_i}{n \sum_{i=1}^{n} x_i^2 - \left(\sum_{i=1}^{n} x_i\right)^2} x_i = 0.$$

14.10 Let $x_1, \ldots, x_n$ be fixed numbers. Using calculus methods, find among all $u_1, \ldots, u_n$ such that $u_1 + \cdots + u_n = 1$ and $u_1 x_1 + \cdots + u_n x_n = 0$, those $(u_1, \ldots, u_n)$ which minimize $u_1^2 + \cdots + u_n^2$. Use this result to prove Theorem 14.1.30.

14.11 In completing the proof of Theorem 14.1.30, show that $\sum_{i=1}^{n} v_i \Delta_i = 0$, that $\sum_{i=1}^{n} v_i^2 + \sum_{i=1}^{n} \Delta_i^2 \geq \sum_{i=1}^{n} v_i^2$, and that (see (14.1.36)) hence the theorem follows. [*Hint*: Use (14.1.34) and the facts $\sum_{i=1}^{n} \Delta_i = 0$, $\sum_{i=1}^{n} \Delta_i x_i = 0$. Note that $\Delta_1 = \cdots = \Delta_n = 0$ satisfies the restrictions and minimizes (14.1.36). Note that v_i as in (14.1.34) yields $\hat{\alpha}$.]

14.12 Theorem 14.1.30 claimed results for $\hat{\alpha}$ and $\hat{\beta}$, but these were proven only for $\hat{\alpha}$. State and prove similar results for $\hat{\beta}$.

14.13 Suppose that the joint probability distribution of two discrete random variables (X, Y) assigns probability 1/67 to each of the 67 pairs of values (x, y) given in the table below (and probability zero to all other pairs of values).

(a) Find $E(X)$, $E(Y)$, $\operatorname{Var}(X)$, $\operatorname{Var}(Y)$, $\operatorname{Cov}(X, Y)$, $\operatorname{Corr}(X, Y)$.

(b) Plot (carefully and neatly) the 67 possible (x, y) pairs on coordinate axes.

(c) Denote $E(X)$ by μ_1, $E(Y)$ by μ_2, $\operatorname{Var}(X)$ by σ_1^2, $\operatorname{Var}(Y)$ by σ_2^2, and $\operatorname{Corr}(X, Y)$ by ρ. Plot the lines

$$x = \mu_1 + \rho \frac{\sigma_1}{\sigma_2}(y - \mu_2)$$

and

$$y = \mu_2 + \rho \frac{\sigma_2}{\sigma_1}(x - \mu_1)$$

on the same graph used in part (b).

(d) Which of the lines in part (c) is the "regression line of y on x"? Show that the other line is the "regression line of x on y."

Data From Two Examinations

1	(64, 41)	(94, 70)	(72, 22)	(73, 61)
2	(48, 15)	(47, 15)	(61, 20)	(100, 87)
3	(89, 39)	(83, 40)	(65, 28)	(95, 54)
4	(97, 64)	(93, 81)	(70, 67)	(97, 100)
5	(74, 72)	(87, 68)	(74, 24)	(70, 28)
6	(86, 38)	(68, 57)	(57, 31)	(84, 34)
7	(74, 50)	(80, 78)	(90, 68)	(94, 54)
8	(85, 63)	(90, 53)	(75, 45)	
9	(78, 62)	(90, 36)	(57, 50)	
10	(73, 30)	(76, 26)	(69, 45)	
11	(81, 39)	(100, 68)	(91, 61)	
12	(72, 57)	(70, 18)	(74, 29)	
13	(80, 32)	(80, 22)	(98, 81)	
14	(26, 0)	(81, 37)	(97, 78)	
15	(42, 23)	(66, 44)	(63, 18)	
16	(80, 35)	(75, 51)	(78, 75)	
17	(73, 25)	(67, 80)	(65, 40)	
18	(61, 31)	(70, 59)	(76, 38)	
19	(100, 71)	(81, 73)	(72, 70)	
20	(72, 38)	(71, 49)	(49, 52)	

14.14 Suppose that the joint probability distribution of two continuous random variables (X, Y) is bivariate normal with $E(X) = \mu_1$, $E(Y) = \mu_2$, $\text{Var}(X) = \sigma_1^2$, $\text{Var}(Y) = \sigma_2^2$, and $\text{Corr}(X, Y) = \rho$ where the values of μ_1, μ_2, σ_1^2, σ_2^2, ρ are the numbers calculated in Problem 14.13. Then the joint probability density function of (X, Y) is

$$f(x, y) = \frac{1}{2\pi\sigma_1\sigma_2\sqrt{1-\rho^2}} e^{-\frac{1}{2(1-\rho^2)}\left[\left(\frac{x-\mu_1}{\sigma_1}\right)^2 - 2\rho\left(\frac{x-\mu_1}{\sigma_1}\right)\left(\frac{y-\mu_2}{\sigma_2}\right) + \left(\frac{y-\mu_2}{\sigma_2}\right)^2\right]}.$$

For any $c(0 < c < 1/(2\pi\sigma_1\sigma_2\sqrt{1-\rho^2}))$, the locus of $f(x, y) = c$ consists of those points (x, y) on the ellipse.

$$\left(\frac{x-\mu_1}{\sigma_1}\right)^2 - 2\rho\left(\frac{x-\mu_1}{\sigma_1}\right)\left(\frac{y-\mu_2}{\sigma_2}\right) + \left(\frac{y-\mu_2}{\sigma_2}\right)^2 = -2(1-\rho^2)\ln(2\pi\sigma_1\sigma_2\sqrt{1-\rho^2}\, c). \qquad (*)$$

It can be shown that

$$\iint_A f(x, y)\, dx\, dy = 1 - 2\pi\sigma_1\sigma_2\sqrt{1-\rho^2}\, c, \qquad (**)$$

where A is the set of all points (x, y) such that

$$\left(\frac{x-\mu_1}{\sigma_1}\right)^2 - 2\rho\left(\frac{x-\mu_1}{\sigma_1}\right)\left(\frac{y-\mu_2}{\sigma_2}\right) + \left(\frac{y-\mu_2}{\sigma_2}\right)^2 \leqslant -2(1-\rho^2)\ln(2\pi\sigma_1\sigma_2\sqrt{1-\rho^2}\,c).$$

Below we table (**) as a function of $\sqrt{-2\ln(2\pi\sigma_1\sigma_2\sqrt{1-\rho^2}\,c)}$:

$\iint_A f(x, y)\,dx\,dy = 1 - 2\pi\sigma_1\sigma_2\sqrt{1-\rho^2}\,c$	$\sqrt{-2\ln(2\pi\sigma_1\sigma_2\sqrt{1-\rho^2}\,c)}$
.25	.7585
.50	1.177
.75	1.665
.90	2.146
.95	2.448
.99	3.035

(a) Plot (on the same graph as in Problem 14.13, which already contains 67 points and 2 lines) the two ellipses (*) associated with values of $\iint_A f(x, y)\,dx\,dy$ of .50 and .95.

(b) Of the 67 data points, how many are inside the ellipse associated with .50? What percent? How many are inside the ellipse associated with .95? What percent?

14.15 Prove Theorem 14.2.1.

14.16 Show that $\hat{\alpha}$ and $\hat{\beta}$ can be simplified from (14.1.23) and (14.1.24), respectively, to (14.2.3).

14.17 Derive the formulas for $\mathrm{Var}(\hat{\alpha})$ and $\mathrm{Var}(\hat{\beta})$ given in (14.2.4).

14.18 Prove that (14.2.6) yields a level α test of $H_0 = \{\sigma_0^2\}$. [*Hint*: See Theorem 9.6.4 ff.]

14.19 Develop a quantity similar to (14.2.7), but involving $\hat{\beta} - \beta_0$ in the numerator, which has the t_{n-2} distribution. Show how to use this quantity to test hypotheses about β.

14.20 For the data of Problem 14.7:

(a) Find $\hat{\alpha}$, $\hat{\beta}$, and $\hat{\sigma}^2$;

(b) Find two-sided equal-tail confidence intervals for each of α, β, σ^2 (each with confidence coefficient .95).

14.21 Show that, as $n \to \infty$, $Y_U^* - Y_L^* \xrightarrow{p} 0$ (see (14.3.4) and (14.3.5)).

14.22 Show that, as $n \to \infty$, $Y_U - Y_L \xrightarrow{p} c$ with $c \neq 0$. Find c. (See (14.3.7) and (14.3.8).)

14.23 Show that $\mathrm{Var}(\hat{\beta}_i)$ in (14.4.2) is the same for all $i = 1, \ldots, k$ if: $n_1 = \cdots = n_k = n$ (say) and $x_{1j} = \cdots = x_{kj}$ for $j = 1, \ldots, n$.

14.24 Derive in detail a procedure for selecting that one of k parallel regression lines which has the largest intercept $\max(\alpha_1, \ldots, \alpha_k)$.

14.25 If $n_1 = \cdots = n_k = n$ with n even and $x_{1j} = \cdots = x_{kj} = x_j$ (say) ($j = 1, \ldots, n$) in (14.4.2), what choice of $x_1, \ldots, x_n$ will produce the best P(CS) function? How does your answer change if you are only allowed to choose x's between some bounds, say $x_L \leqslant x \leqslant x_U$?

HONORS PROBLEMS

14.1H Show that the only critical point of (14.1.27) is $(\hat{\alpha}, \hat{\beta})$ as in Theorem 14.1.22. Quote and use appropriate theorems on minimization of bivariate functions to show that $(\hat{\alpha}, \hat{\beta})$ yields a unique minimum of $S(\alpha, \beta)$.

14.2H Find the method of least squares estimators of $\beta_0, \beta_1, \ldots, \beta_m$ for predicting Y by $\beta_0 + \beta_1 X + \beta_2 X^2 + \cdots + \beta_m X^m$. [*Hint*: Choose $\beta_0, \beta_1, \ldots, \beta_m$ to minimize

$$S(\beta_0, \beta_1, \ldots, \beta_m) = \sum_{i=1}^{n} (Y_i - (\beta_0 + \beta_1 X_i + \cdots + \beta_m X_i^m))^2.$$

Be sure you have a minimum $(\hat{\beta}_0, \hat{\beta}_1, \ldots, \hat{\beta}_m)$; see Problem 14.1H. The case $m = 1$ was covered in Theorem 14.1.22.]

14.3H Find the best quadratic ($m = 2$ in Problem 14.2H) for the data of Problem 14.7. Plot it on the graph used in Problem 14.7. How does the examination estimate for a homework score of 30 change?

14.4H Give a result and proof like those of Theorem 14.1.30 for the general model of Problem 14.2.H.

14.5H Derive Equation (**) of Problem 14.14.

14.6H Find the correlation coefficient in the bivariate normal distribution of $(\hat{\alpha}, \hat{\beta})$ (see Theorem 14.2.5).

14.7H **(a)** Prove (using a proof similar to that of Lemma 9.5.15) that $n\hat{\sigma}^2/\sigma^2$ (see Theorem 14.2.1) is a $\chi^2_{n-2}(0)$ r.v.

(b) Prove the result of (a) by using the theory of distributions of functions of r.v.'s.

(c) Prove the result of (a) by using Cochran's Theorem (for example, see Theorem 4 on p. 33 of *Statistical Design and Analysis of Experiments* by P.W.M. John, The Macmillan Company, New York, 1971).

14.8H **(a)** Prove directly that $\hat{\sigma}^2$ (see (14.2.2)) is stochastically independent of $\hat{\alpha}$ and of $\hat{\beta}$.

(b) Prove part (a) using Cochran's Theorem (see Problem 14.7H(c)).

14.9H [Refer to the discussion following (14.3.2).] We know that $\hat{\alpha} + \hat{\beta}x^*$ is an unbiased estimator of $\alpha + \beta x^*$, and that $\hat{\alpha}$ has the smallest variance of all unbiased linear estimators of α, while $\hat{\beta}$ has the smallest variance of all unbiased linear estimators of β.

(a) Does $\hat{\alpha}$ have the smallest variance of all unbiased estimators of α? (We seek to drop the linearity restriction in this example.) Similarly for $\hat{\beta}$ and β.

(b) Does $\hat{\alpha} + \hat{\beta}x^*$ have the smallest variance of all unbiased linear estimators of $\alpha + \beta x^*$? Of all unbiased estimators?

14.10H Derive the variance in (14.3.3).

14.11H Perform an analysis similar to that of Example 14.3.9 for the data

i	1	2	3	4	5
x_i	35	45	55	65	75
Y_i	114	124	143	158	166

Table 1. Auxiliary Function $\sqrt{x}$

x	$\sqrt{x}$	$\sqrt{10x}$	x	$\sqrt{x}$	$\sqrt{10x}$	x	$\sqrt{x}$	$\sqrt{10x}$
1	1.0000	3.1623	34	5.8310	18.439	67	8.1854	25.884
2	1.4142	4.4721	35	5.9161	18.708	68	8.2462	26.077
3	1.7321	5.4772	36	6.0000	18.974	69	8.3066	26.268
4	2.0000	6.3246	37	6.0828	19.235	70	8.3666	26.458
5	2.2361	7.0711	38	6.1644	19.494	71	8.4261	26.646
6	2.4495	7.7460	39	6.2450	19.748	72	8.4853	26.833
7	2.6458	8.3666	40	6.3246	20.000	73	8.5440	27.019
8	2.8284	8.9443	41	6.4031	20.248	74	8.6023	27.203
9	3.0000	9.4868	42	6.4807	20.494	75	8.6603	27.386
10	3.1623	10.000	43	6.5574	20.736	76	8.7178	27.568
11	3.3166	10.488	44	6.6332	20.976	77	8.7750	27.749
12	3.4641	10.954	45	6.7082	21.213	78	8.8318	27.928
13	3.6056	11.402	46	6.7823	21.448	79	8.8882	28.107
14	3.7417	11.832	47	6.8557	21.679	80	8.9443	28.284
15	3.8730	12.247	48	6.9282	21.909	81	9.0000	28.460
16	4.0000	12.649	49	7.0000	22.136	82	9.0554	28.636
17	4.1231	13.038	50	7.0711	22.361	83	9.1104	28.810
18	4.2426	13.416	51	7.1414	22.583	84	9.1652	28.983
19	4.3589	13.784	52	7.2111	22.804	85	9.2195	29.155
20	4.4721	14.142	53	7.2801	23.022	86	9.2736	29.326
21	4.5826	14.491	54	7.3485	23.238	87	9.3274	29.496
22	4.6904	14.832	55	7.4162	23.452	88	9.3808	29.665
23	4.7958	15.166	56	7.4833	23.664	89	9.4340	29.833
24	4.8990	15.492	57	7.5498	23.875	90	9.4868	30.000
25	5.0000	15.811	58	7.6158	24.083	91	9.5394	30.166
26	5.0990	16.125	59	7.6811	24.290	92	9.5917	30.332
27	5.1962	16.432	60	7.7460	24.495	93	9.6437	30.496
28	5.2915	16.733	61	7.8102	24.698	94	9.6954	30.659
29	5.3852	17.029	62	7.8740	24.900	95	9.7468	30.822
30	5.4772	17.321	63	7.9373	25.100	96	9.7980	30.984
31	5.5678	17.607	64	8.0000	25.298	97	9.8489	31.145
32	5.6569	17.889	65	8.0623	25.495	98	9.8995	31.305
33	5.7446	18.166	66	8.1240	25.690	99	9.9499	31.464

Table 2. Auxiliary Function $n!$

n	$n!$	n	$n!$
1	1	26	4.03291×10^{26}
2	2	27	1.08889×10^{28}
3	6	28	3.04888×10^{29}
4	24	29	8.84176×10^{30}
5	120	30	2.65253×10^{32}
6	720	31	8.22284×10^{33}
7	5040	32	2.63131×10^{35}
8	40320	33	8.68332×10^{36}
9	362880	34	2.95233×10^{38}
10	3.62880×10^{6}	35	1.03331×10^{40}
11	3.99168×10^{7}	36	3.71993×10^{41}
12	4.79002×10^{8}	37	1.37638×10^{43}
13	6.22702×10^{9}	38	5.23023×10^{44}
14	8.71783×10^{10}	39	2.03979×10^{46}
15	1.30767×10^{12}	40	8.15915×10^{47}
16	2.09228×10^{13}	41	3.34525×10^{49}
17	3.55687×10^{14}	42	1.40501×10^{51}
18	6.40237×10^{15}	43	6.04153×10^{52}
19	1.21645×10^{17}	44	2.65827×10^{54}
20	2.43290×10^{18}	45	1.19622×10^{56}
21	5.10909×10^{19}	46	5.50262×10^{57}
22	1.12400×10^{21}	47	2.58623×10^{59}
23	2.58520×10^{22}	48	1.24139×10^{61}
24	6.20448×10^{23}	49	6.08282×10^{62}
25	1.55112×10^{25}	50	3.04141×10^{64}

Table 3. Auxiliary Function $\binom{n}{r}$

n \ r	2	3	4	5	6	7	8	9	10	11	12	13
2	1											
3	3	1										
4	6	4	1									
5	10	10	5	1								
6	15	20	15	6	1							
7	21	35	35	21	7	1						
8	28	56	70	56	28	8	1					
9	36	84	126	126	84	36	9	1				
10	45	120	210	252	210	120	45	10	1			
11	55	165	330	462	462	330	165	55	11	1		
12	66	220	495	792	924	792	495	220	66	12	1	
13	78	286	715	1,287	1,716	1,716	1,287	715	286	78	13	1
14	91	364	1,001	2,002	3,003	3,432	3,003	2,002	1,001	364	91	14
15	105	455	1,365	3,003	5,005	6,435	6,435	5,005	3,003	1,365	455	105
16	120	560	1,820	4,368	8,008	11,440	12,870	11,440	8,008	4,368	1,820	560
17	136	680	2,380	6,188	12,376	19,448	24,310	24,310	19,448	12,376	6,188	2,380
18	153	816	3,060	8,568	18,564	31,824	43,758	48,620	43,758	31,824	18,564	8,568
19	171	969	3,876	11,628	27,132	50,388	75,582	92,378	92,378	75,582	50,388	27,132
20	190	1,140	4,845	15,504	38,760	77,520	125,970	167,960	184,756	167,960	125,970	77,520
21	210	1,330	5,985	20,349	54,264	116,280	203,490	293,930	352,716	352,716	293,930	203,490
22	231	1,540	7,315	26,334	74,613	170,544	319,770	497,420	646,646	705,432	646,646	497,420
23	253	1,771	8,855	33,649	100,947	245,157	490,314	817,190	1,144,066	1,352,078	1,352,078	1,144,066
24	276	2,024	10,626	42,504	134,596	346,104	735,471	1,307,504	1,961,256	2,496,144	2,704,156	2,496,144
25	300	2,300	12,650	53,130	177,100	480,700	1,081,575	2,042,975	3,268,760	4,457,400	5,200,300	5,200,300
26	325	2,600	14,950	65,780	230,230	657,800	1,562,275	3,124,550	5,311,735	7,726,160	9,657,700	10,400,600

Table 4. Auxiliary Function e^{-x}

x	e^{-x}	x	e^{-x}	x	e^{-x}	x	e^{-x}
.00	1.00000	.50	.60653	1.00	.36788	1.50	.22313
.01	.99005	.51	.60050	1.01	.36422	1.51	.22091
.02	.98020	.52	.59452	1.02	.36059	1.52	.21871
.03	.97045	.53	.58860	1.03	.35701	1.53	.21654
.04	.96079	.54	.58275	1.04	.35345	1.54	.21438
.05	.95123	.55	.57695	1.05	.34994	1.55	.21225
.06	.94176	.56	.57121	1.06	.34646	1.56	.21014
.07	.93239	.57	.56553	1.07	.34301	1.57	.20805
.08	.92312	.58	.55990	1.08	.33960	1.58	.20598
.09	.91393	.59	.55433	1.09	.33622	1.59	.20393
.10	.90484	.60	.54881	1.10	.33287	1.60	.20190
.11	.89583	.61	.54335	1.11	.32956	1.61	.19989
.12	.88692	.62	.53794	1.12	.32628	1.62	.19790
.13	.87810	.63	.53259	1.13	.32303	1.63	.19593
.14	.86936	.64	.52729	1.14	.31982	1.64	.19398
.15	.86071	.65	.52205	1.15	.31664	1.65	.19205
.16	.85214	.66	.51685	1.16	.31349	1.66	.19014
.17	.84366	.67	.51171	1.17	.31037	1.67	.18825
.18	.83527	.68	.50662	1.18	.30728	1.68	.18637
.19	.82696	.69	.50158	1.19	.30422	1.69	.18452
.20	.81873	.70	.49659	1.20	.30119	1.70	.18268
.21	.81058	.71	.49164	1.21	.29820	1.71	.18087
.22	.80252	.72	.48675	1.22	.29523	1.72	.17907
.23	.79453	.73	.48191	1.23	.29229	1.73	.17728
.24	.78663	.74	.47711	1.24	.28938	1.74	.17552
.25	.77880	.75	.47237	1.25	.28650	1.75	.17377
.26	.77105	.76	.46767	1.26	.28365	1.76	.17204
.27	.76338	.77	.46301	1.27	.28083	1.77	.17033
.28	.75578	.78	.45841	1.28	.27804	1.78	.16864
.29	.74826	.79	.45384	1.29	.27527	1.79	.16696
.30	.74082	.80	.44933	1.30	.27253	1.80	.16530
.31	.73345	.81	.44486	1.31	.26982	1.81	.16365
.32	.72615	.82	.44043	1.32	.26714	1.82	.16203
.33	.71892	.83	.43605	1.33	.26448	1.83	.16041
.34	.71177	.84	.43171	1.34	.26185	1.84	.15882
.35	.70469	.85	.42741	1.35	.25924	1.85	.15724
.36	.69768	.86	.42316	1.36	.25666	1.86	.15567
.37	.69073	.87	.41895	1.37	.25411	1.87	.15412
.38	.68386	.88	.41478	1.38	.25158	1.88	.15259
.39	.67706	.89	.41066	1.39	.24908	1.89	.15107
.40	.67032	.90	.40657	1.40	.24660	1.90	.14957
.41	.66365	.91	.40252	1.41	.24414	1.91	.14808
.42	.65705	.92	.39852	1.42	.24171	1.92	.14661
.43	.65051	.93	.39455	1.43	.23931	1.93	.14515
.44	.64404	.94	.39063	1.44	.23693	1.94	.14370
.45	.63763	.95	.38674	1.45	.23457	1.95	.14227
.46	.63128	.96	.38289	1.46	.23224	1.96	.14086
.47	.62500	.97	.37908	1.47	.22993	1.97	.13946
.48	.61878	.98	.37531	1.48	.22764	1.98	.13807
.49	.61263	.99	.37158	1.49	.22537	1.99	.13670

Table 4. Auxiliary Function e^{-x} (Continued)

x	e^{-x}	x	e^{-x}	x	e^{-x}	x	e^{-x}
2.00	.13534	2.50	.08208	3.00	.04979	5.50	.00409
2.01	.13399	2.51	.08127	3.05	.04736	5.60	.00370
2.02	.13266	2.52	.08046	3.10	.04505	5.70	.00335
2.03	.13134	2.53	.07966	3.15	.04285	5.80	.00303
2.04	.13003	2.54	.07887	3.20	.04076	5.90	.00274
2.05	.12873	2.55	.07808	3.25	.03877	6.00	.00248
2.06	.12745	2.56	.07730	3.30	.03688	6.10	.00224
2.07	.12619	2.57	.07654	3.35	.03508	6.20	.00203
2.08	.12493	2.58	.07577	3.40	.03337	6.30	.00184
2.09	.12369	2.59	.07502	3.45	.03175	6.40	.00166
2.10	.12246	2.60	.07427	3.50	.03020	6.50	.00150
2.11	.12124	2.61	.07353	3.55	.02872	6.75	.00117
2.12	.12003	2.62	.07280	3.60	.02732	7.00	.00091
2.13	.11884	2.63	.07208	3.65	.02599	7.25	.00071
2.14	.11765	2.64	.07136	3.70	.02472	7.50	.00055
2.15	.11648	2.65	.07065	3.75	.02352	7.75	.00043
2.16	.11533	2.66	.06995	3.80	.02237	8.00	.00034
2.17	.11418	2.67	.06925	3.85	.02128	8.25	.00026
2.18	.11304	2.68	.06856	3.90	.02024	8.50	.00020
2.19	.11192	2.69	.06788	3.95	.01925	8.75	.00016
2.20	.11080	2.70	.06721	4.00	.01832	9.00	.00012
2.21	.10970	2.71	.06654	4.05	.01742	9.25	.00010
2.22	.10861	2.72	.06587	4.10	.01657	9.50	.00007
2.23	.10753	2.73	.06522	4.15	.01576	9.75	.00006
2.24	.10646	2.74	.06457	4.20	.01500	10.00	.00005
2.25	.10540	2.75	.06393	4.25	.01426		
2.26	.10435	2.76	.06329	4.30	.01357		
2.27	.10331	2.77	.06266	4.35	.01291		
2.28	.10228	2.78	.06204	4.40	.01228		
2.29	.10127	2.79	.06142	4.45	.01168		
2.30	.10026	2.80	.06081	4.50	.01111		
2.31	.09926	2.81	.06020	4.55	.01057		
2.32	.09827	2.82	.05961	4.60	.01005		
2.33	.09730	2.83	.05901	4.65	.00956		
2.34	.09633	2.84	.05843	4.70	.00910		
2.35	.09537	2.85	.05784	4.75	.00865		
2.36	.09442	2.86	.05727	4.80	.00823		
2.37	.09348	2.87	.05670	4.85	.00783		
2.38	.09255	2.88	.05613	4.90	.00745		
2.39	.09163	2.89	.05558	4.95	.00708		
2.40	.09072	2.90	.05502	5.00	.00674		
2.41	.08982	2.91	.05448	5.05	.00641		
2.42	.08892	2.92	.05393	5.10	.00610		
2.43	.08804	2.93	.05340	5.15	.00580		
2.44	.08716	2.94	.05287	5.20	.00552		
2.45	.08629	2.95	.05234	5.25	.00525		
2.46	.08544	2.96	.05182	5.30	.00499		
2.47	.08458	2.97	.05130	5.35	.00475		
2.48	.08374	2.98	.05079	5.40	.00452		
2.49	.08291	2.99	.05029	5.45	.00430		

Table 5. The Binomial Distribution

A random variable X is said to have the binomial distribution if (for some positive integer n, and some p with $0 \leqslant p \leqslant 1$; see Definition 3.2.3)

$$p_X(x) = \begin{cases} \binom{n}{x} p^x(1-p)^{n-x}, & x = 0, 1, \ldots, n \\ 0, & \text{otherwise.} \end{cases}$$

The following table gives values of $p_X(x)$ (for various values of n, p).

n	x	p = .05	.10	.15	.20	.25	.30	.35	.40	.45	.50
1	0	.9500	.9000	.8500	.8000	.7500	.7000	.6500	.6000	.5500	.5000
	1	.0500	.1000	.1500	.2000	.2500	.3000	.3500	.4000	.4500	.5000
2	0	.9025	.8100	.7225	.6400	.5625	.4900	.4225	.3600	.3025	.2500
	1	.0950	.1800	.2550	.3200	.3750	.4200	.4550	.4800	.4950	.5000
	2	.0025	.0100	.0225	.0400	.0625	.0900	.1225	.1600	.2025	.2500
3	0	.8574	.7290	.6141	.5120	.4219	.3430	.2746	.2160	.1664	.1250
	1	.1354	.2430	.3251	.3840	.4219	.4410	.4436	.4320	.4084	.3750
	2	.0071	.0270	.0574	.0960	.1406	.1890	.2389	.2880	.3341	.3750
	3	.0001	.0010	.0034	.0080	.0156	.0270	.0429	.0640	.0911	.1250
4	0	.8145	.6561	.5220	.4096	.3164	.2401	.1785	.1296	.0915	.0625
	1	.1715	.2916	.3685	.4096	.4219	.4116	.3845	.3456	.2995	.2500
	2	.0135	.0486	.0975	.1536	.2109	.2646	.3105	.3456	.3675	.3750
	3	.0005	.0036	.0115	.0256	.0469	.0756	.1115	.1536	.2005	.2500
	4	.0000	.0001	.0005	.0016	.0039	.0081	.0150	.0256	.0410	.0625
5	0	.7738	.5905	.4437	.3277	.2373	.1681	.1160	.0778	.0503	.0312
	1	.2036	.3280	.3915	.4096	.3955	.3602	.3124	.2592	.2059	.1562
	2	.0214	.0729	.1382	.2048	.2637	.3087	.3364	.3456	.3369	.3125
	3	.0011	.0081	.0244	.0512	.0879	.1323	.1811	.2304	.2757	.3125
	4	.0000	.0004	.0022	.0064	.0146	.0284	.0488	.0768	.1128	.1562
	5	.0000	.0000	.0001	.0003	.0010	.0024	.0053	.0102	.0185	.0312
6	0	.7351	.5314	.3771	.2621	.1780	.1176	.0754	.0467	.0277	.0156
	1	.2321	.3543	.3993	.3932	.3560	.3025	.2437	.1866	.1359	.0938
	2	.0305	.0984	.1762	.2458	.2966	.3241	.3280	.3110	.2780	.2344
	3	.0021	.0146	.0415	.0819	.1318	.1852	.2355	.2765	.3032	.3125
	4	.0001	.0012	.0055	.0154	.0330	.0595	.0951	.1382	.1861	.2344
	5	.0000	.0001	.0004	.0015	.0044	.0102	.0205	.0369	.0609	.0938
	6	.0000	.0000	.0000	.0001	.0002	.0007	.0018	.0041	.0083	.0156
7	0	.6983	.4783	.3206	.2097	.1335	.0824	.0490	.0280	.0152	.0078
	1	.2573	.3720	.3960	.3670	.3115	.2471	.1848	.1306	.0872	.0547
	2	.0406	.1240	.2097	.2753	.3115	.3177	.2985	.2613	.2140	.1641
	3	.0036	.0230	.0617	.1147	.1730	.2269	.2679	.2903	.2918	.2734
	4	.0002	.0026	.0109	.0287	.0577	.0972	.1442	.1935	.2388	.2734
	5	.0000	.0002	.0012	.0043	.0115	.0250	.0466	.0774	.1172	.1641
	6	.0000	.0000	.0001	.0004	.0013	.0036	.0084	.0172	.0320	.0547
	7	.0000	.0000	.0000	.0000	.0001	.0002	.0006	.0016	.0037	.0078
8	0	.6634	.4305	.2725	.1678	.1001	.0576	.0319	.0168	.0084	.0039
	1	.2793	.3826	.3847	.3355	.2670	.1977	.1373	.0896	.0548	.0312
	2	.0515	.1488	.2376	.2936	.3115	.2965	.2587	.2090	.1569	.1094
	3	.0054	.0331	.0839	.1468	.2076	.2541	.2786	.2787	.2568	.2188
	4	.0004	.0046	.0185	.0459	.0865	.1361	.1875	.2322	.2627	.2734
	5	.0000	.0004	.0026	.0092	.0231	.0467	.0808	.1239	.1719	.2188
	6	.0000	.0000	.0002	.0011	.0038	.0100	.0217	.0413	.0703	.1094
	7	.0000	.0000	.0000	.0001	.0004	.0012	.0033	.0079	.0164	.0312
	8	.0000	.0000	.0000	.0000	.0000	.0001	.0002	.0007	.0017	.0039

Table 5. The Binomial Distribution (Continued)

$$p_X(x) = \binom{n}{x} p^x(1 - p)^{n-x}$$

n	x	.05	.10	.15	.20	.25	.30	.35	.40	.45	.50
9	0	.6302	.3874	.2316	.1342	.0751	.0404	.0207	.0101	.0046	.0020
	1	.2985	.3874	.3679	.3020	.2253	.1556	.1004	.0605	.0339	.0176
	2	.0629	.1722	.2597	.3020	.3003	.2668	.2162	.1612	.1110	.0703
	3	.0077	.0446	.1069	.1762	.2336	.2668	.2716	.2508	.2119	.1641
	4	.0006	.0074	.0283	.0661	.1168	.1715	.2194	.2508	.2600	.2461
	5	.0000	.0008	.0050	.0165	.0389	.0735	.1181	.1672	.2128	.2461
	6	.0000	.0001	.0006	.0028	.0087	.0210	.0424	.0743	.1160	.1641
	7	.0000	.0000	.0000	.0003	.0012	.0039	.0098	.0212	.0407	.0703
	8	.0000	.0000	.0000	.0000	.0001	.0004	.0013	.0035	.0083	.0176
	9	.0000	.0000	.0000	.0000	.0000	.0000	.0001	.0003	.0008	.0020
10	0	.5987	.3487	.1969	.1074	.0563	.0282	.0135	.0060	.0025	.0010
	1	.3151	.3874	.3474	.2684	.1877	.1211	.0725	.0403	.0207	.0098
	2	.0746	.1937	.2759	.3020	.2816	.2335	.1757	.1209	.0763	.0439
	3	.0105	.0574	.1298	.2013	.2503	.2668	.2522	.2150	.1665	.1172
	4	.0010	.0112	.0401	.0881	.1460	.2001	.2377	.2508	.2384	.2051
	5	.0001	.0015	.0085	.0264	.0584	.1029	.1536	.2007	.2340	.2461
	6	.0000	.0001	.0012	.0055	.0162	.0368	.0689	.1115	.1596	.2051
	7	.0000	.0000	.0001	.0008	.0031	.0090	.0212	.0425	.0746	.1172
	8	.0000	.0000	.0000	.0001	.0004	.0014	.0043	.0106	.0229	.0439
	9	.0000	.0000	.0000	.0000	.0000	.0001	.0005	.0016	.0042	.0098
	10	.0000	.0000	.0000	.0000	.0000	.0000	.0000	.0001	.0003	.0010
11	0	.5688	.3138	.1673	.0859	.0422	.0198	.0088	.0036	.0014	.0005
	1	.3293	.3835	.3248	.2362	.1549	.0932	.0518	.0266	.0125	.0054
	2	.0867	.2131	.2866	.2953	.2581	.1998	.1395	.0887	.0513	.0269
	3	.0137	.0710	.1517	.2215	.2581	.2568	.2254	.1774	.1259	.0806
	4	.0014	.0158	.0536	.1107	.1721	.2201	.2428	.2365	.2060	.1611
	5	.0001	.0025	.0132	.0388	.0803	.1321	.1830	.2207	.2360	.2256
	6	.0000	.0003	.0023	.0097	.0268	.0566	.0985	.1471	.1931	.2256
	7	.0000	.0000	.0003	.0017	.0064	.0173	.0379	.0701	.1128	.1611
	8	.0000	.0000	.0000	.0002	.0011	.0037	.0102	.0234	.0462	.0806
	9	.0000	.0000	.0000	.0000	.0001	.0005	.0018	.0052	.0126	.0269
	10	.0000	.0000	.0000	.0000	.0000	.0000	.0002	.0007	.0021	.0054
	11	.0000	.0000	.0000	.0000	.0000	.0000	.0000	.0000	.0002	.0005
12	0	.5404	.2824	.1422	.0687	.0317	.0138	.0057	.0022	.0008	.0002
	1	.3413	.3766	.3012	.2062	.1267	.0712	.0368	.0174	.0075	.0029
	2	.0988	.2301	.2924	.2835	.2323	.1678	.1088	.0639	.0339	.0161
	3	.0173	.0852	.1720	.2362	.2581	.2397	.1954	.1419	.0923	.0537
	4	.0021	.0213	.0683	.1329	.1936	.2311	.2367	.2128	.1700	.1208
	5	.0002	.0038	.0193	.0532	.1032	.1585	.2039	.2270	.2225	.1934
	6	.0000	.0005	.0040	.0155	.0401	.0792	.1281	.1766	.2124	.2256
	7	.0000	.0000	.0006	.0033	.0115	.0291	.0591	.1009	.1489	.1934
	8	.0000	.0000	.0001	.0005	.0024	.0078	.0199	.0420	.0762	.1208
	9	.0000	.0000	.0000	.0001	.0004	.0015	.0048	.0125	.0277	.0537
	10	.0000	.0000	.0000	.0000	.0000	.0002	.0008	.0025	.0068	.0161
	11	.0000	.0000	.0000	.0000	.0000	.0000	.0001	.0003	.0010	.0029
	12	.0000	.0000	.0000	.0000	.0000	.0000	.0000	.0000	.0001	.0002
13	0	.5133	.2542	.1209	.0550	.0238	.0097	.0037	.0013	.0004	.0001
	1	.3512	.3672	.2774	.1787	.1029	.0540	.0259	.0113	.0045	.0016
	2	.1109	.2448	.2937	.2680	.2059	.1388	.0836	.0453	.0220	.0095
	3	.0214	.0997	.1900	.2457	.2517	.2181	.1651	.1107	.0660	.0349
	4	.0028	.0277	.0838	.1535	.2097	.2337	.2222	.1845	.1350	.0873
	5	.0003	.0055	.0266	.0691	.1258	.1803	.2154	.2214	.1989	.1571
	6	.0000	.0008	.0063	.0230	.0559	.1030	.1546	.1968	.2169	.2095
	7	.0000	.0001	.0011	.0058	.0186	.0442	.0833	.1312	.1775	.2095
	8	.0000	.0000	.0001	.0011	.0047	.0142	.0336	.0656	.1089	.1571
	9	.0000	.0000	.0000	.0001	.0009	.0034	.0101	.0243	.0495	.0873
	10	.0000	.0000	.0000	.0000	.0001	.0006	.0022	.0065	.0162	.0349
	11	.0000	.0000	.0000	.0000	.0000	.0001	.0003	.0012	.0036	.0095
	12	.0000	.0000	.0000	.0000	.0000	.0000	.0000	.0001	.0005	.0016
	13	.0000	.0000	.0000	.0000	.0000	.0000	.0000	.0000	.0000	.0001

Table 5. The Binomial Distribution (Continued)

$$p_X(x) = \binom{n}{x} p^x(1 - p)^{n-x}$$

		p									
n	*x*	.05	.10	.15	.20	.25	.30	.35	.40	.45	.50
14	0	.4877	.2288	.1028	.0440	.0178	.0068	.0024	.0008	.0002	.0001
	1	.3593	.3559	.2539	.1539	.0832	.0407	.0181	.0073	.0027	.0009
	2	.1229	.2570	.2912	.2501	.1802	.1134	.0634	.0317	.0141	.0056
	3	.0259	.1142	.2056	.2501	.2402	.1943	.1366	.0845	.0462	.0222
	4	.0037	.0349	.0998	.1720	.2202	.2290	.2022	.1549	.1040	.0611
	5	.0004	.0078	.0352	.0860	.1468	.1963	.2178	.2066	.1701	.1222
	6	.0000	.0013	.0093	.0322	.0734	.1262	.1759	.2066	.2088	.1833
	7	.0000	.0002	.0019	.0092	.0280	.0618	.1082	.1574	.1952	.2095
	8	.0000	.0000	.0003	.0020	.0082	.0232	.0510	.0918	.1398	.1833
	9	.0000	.0000	.0000	.0003	.0018	.0066	.0183	.0408	.0762	.1222
	10	.0000	.0000	.0000	.0000	.0003	.0014	.0049	.0136	.0312	.0611
	11	.0000	.0000	.0000	.0000	.0000	.0002	.0010	.0033	.0093	.0222
	12	.0000	.0000	.0000	.0000	.0000	.0000	.0001	.0005	.0019	.0056
	13	.0000	.0000	.0000	.0000	.0000	.0000	.0000	.0001	.0002	.0009
	14	.0000	.0000	.0000	.0000	.0000	.0000	.0000	.0000	.0000	.0001
15	0	.4633	.2059	.0874	.0352	.0134	.0047	.0016	.0005	.0001	.0000
	1	.3658	.3432	.2312	.1319	.0668	.0305	.0126	.0047	.0016	.0005
	2	.1348	.2669	.2856	.2309	.1559	.0916	.0476	.0219	.0090	.0032
	3	.0307	.1285	.2184	.2501	.2252	.1700	.1110	.0634	.0318	.0139
	4	.0049	.0428	.1156	.1876	.2252	.2186	.1792	.1268	.0780	.0417
	5	.0006	.0105	.0449	.1032	.1651	.2061	.2123	.1859	.1404	.0916
	6	.0000	.0019	.0132	.0430	.0917	.1472	.1906	.2066	.1914	.1527
	7	.0000	.0003	.0030	.0138	.0393	.0811	.1319	.1771	.2013	.1964
	8	.0000	.0000	.0005	.0035	.0131	.0348	.0710	.1181	.1647	.1964
	9	.0000	.0000	.0001	.0007	.0034	.0116	.0298	.0612	.1048	.1527
	10	.0000	.0000	.0000	.0001	.0007	.0030	.0096	.0245	.0515	.0916
	11	.0000	.0000	.0000	.0000	.0001	.0006	.0024	.0074	.0191	.0417
	12	.0000	.0000	.0000	.0000	.0000	.0001	.0004	.0016	.0052	.0139
	13	.0000	.0000	.0000	.0000	.0000	.0000	.0001	.0003	.0010	.0032
	14	.0000	.0000	.0000	.0000	.0000	.0000	.0000	.0000	.0001	.0005
	15	.0000	.0000	.0000	.0000	.0000	.0000	.0000	.0000	.0000	.0000
16	0	.4401	.1853	.0743	.0281	.0100	.0033	.0010	.0003	.0001	.0000
	1	.3706	.3294	.2097	.1126	.0535	.0228	.0087	.0030	.0009	.0002
	2	.1463	.2745	.2775	.2111	.1336	.0732	.0353	.0150	.0056	.0018
	3	.0359	.1423	.2285	.2463	.2079	.1465	.0888	.0468	.0215	.0085
	4	.0061	.0514	.1311	.2001	.2252	.2040	.1553	.1014	.0572	.0278
	5	.0008	.0137	.0555	.1201	.1802	.2099	.2008	.1623	.1123	.0667
	6	.0001	.0028	.0180	.0550	.1101	.1649	.1982	.1983	.1684	.1222
	7	.0000	.0004	.0045	.0197	.0524	.1010	.1524	.1889	.1969	.1746
	8	.0000	.0001	.0009	.0055	.0197	.0487	.0923	.1417	.1812	.1964
	9	.0000	.0000	.0001	.0012	.0058	.0185	.0442	.0840	.1318	.1746
	10	.0000	.0000	.0000	.0002	.0014	.0056	.0167	.0392	.0755	.1222
	11	.0000	.0000	.0000	.0000	.0002	.0013	.0049	.0142	.0337	.0667
	12	.0000	.0000	.0000	.0000	.0000	.0002	.0011	.0040	.0115	.0278
	13	.0000	.0000	.0000	.0000	.0000	.0000	.0002	.0008	.0029	.0085
	14	.0000	.0000	.0000	.0000	.0000	.0000	.0000	.0001	.0005	.0018
	15	.0000	.0000	.0000	.0000	.0000	.0000	.0000	.0000	.0001	.0002
	16	.0000	.0000	.0000	.0000	.0000	.0000	.0000	.0000	.0000	.0000
17	0	.4181	.1668	.0631	.0225	.0075	.0023	.0007	.0002	.0000	.0000
	1	.3741	.3150	.1893	.0957	.0426	.0169	.0060	.0019	.0005	.0001
	2	.1575	.2800	.2673	.1914	.1136	.0581	.0260	.0102	.0035	.0010
	3	.0415	.1556	.2359	.2393	.1893	.1245	.0701	.0341	.0144	.0052
	4	.0076	.0605	.1457	.2093	.2209	.1868	.1320	.0796	.0411	.0182
	5	.0010	.0175	.0668	.1361	.1914	.2081	.1849	.1379	.0875	.0472
	6	.0001	.0039	.0236	.0680	.1276	.1784	.1991	.1839	.1432	.0944
	7	.0000	.0007	.0065	.0267	.0668	.1201	.1685	.1927	.1841	.1484
	8	.0000	.0001	.0014	.0084	.0279	.0644	.1134	.1606	.1883	.1855
	9	.0000	.0000	.0003	.0021	.0093	.0276	.0611	.1070	.1540	.1855
	10	.0000	.0000	.0000	.0004	.0025	.0095	.0263	.0571	.1008	.1484
	11	.0000	.0000	.0000	.0001	.0005	.0026	.0090	.0242	.0525	.0944
	12	.0000	.0000	.0000	.0000	.0001	.0006	.0024	.0081	.0215	.0472
	13	.0000	.0000	.0000	.0000	.0000	.0001	.0005	.0021	.0068	.0182
	14	.0000	.0000	.0000	.0000	.0000	.0000	.0001	.0004	.0016	.0052

Table 5. The Binomial Distribution (Continued)

$$p_X(x) = \binom{n}{x} p^x (1 - p)^{n-x}$$

n	x	.05	.10	.15	.20	.25	.30	.35	.40	.45	.50
17	15	.0000	.0000	.0000	.0000	.0000	.0000	.0000	.0001	.0003	.0010
	16	.0000	.0000	.0000	.0000	.0000	.0000	.0000	.0000	.0000	.0001
	17	.0000	.0000	.0000	.0000	.0000	.0000	.0000	.0000	.0000	.0000
18	0	.3972	.1501	.0536	.0180	.0056	.0016	.0004	.0001	.0000	.0000
	1	.3763	.3002	.1704	.0811	.0338	.0126	.0042	.0012	.0003	.0001
	2	.1683	.2835	.2556	.1723	.0958	.0458	.0190	.0069	.0022	.0006
	3	.0473	.1680	.2406	.2297	.1704	.1046	.0547	.0246	.0095	.0031
	4	.0093	.0700	.1592	.2153	.2130	.1681	.1104	.0614	.0291	.0117
	5	.0014	.0218	.0787	.1507	.1988	.2017	.1664	.1146	.0666	.0327
	6	.0002	.0052	.0301	.0816	.1436	.1873	.1941	.1655	.1181	.0708
	7	.0000	.0010	.0091	.0350	.0820	.1376	.1792	.1892	.1657	.1214
	8	.0000	.0002	.0022	.0120	.0376	.0811	.1327	.1734	.1864	.1669
	9	.0000	.0000	.0004	.0033	.0139	.0386	.0794	.1284	.1694	.1855
	10	.0000	.0000	.0001	.0008	.0042	.0149	.0385	.0771	.1248	.1669
	11	.0000	.0000	.0000	.0001	.0010	.0046	.0151	.0374	.0742	.1214
	12	.0000	.0000	.0000	.0000	.0002	.0012	.0047	.0145	.0354	.0708
	13	.0000	.0000	.0000	.0000	.0000	.0002	.0012	.0045	.0134	.0327
	14	.0000	.0000	.0000	.0000	.0000	.0000	.0002	.0011	.0039	.0117
	15	.0000	.0000	.0000	.0000	.0000	.0000	.0000	.0002	.0009	.0031
	16	.0000	.0000	.0000	.0000	.0000	.0000	.0000	.0000	.0001	.0006
	17	.0000	.0000	.0000	.0000	.0000	.0000	.0000	.0000	.0000	.0001
	18	.0000	.0000	.0000	.0000	.0000	.0000	.0000	.0000	.0000	.0000
19	0	.3774	.1351	.0456	.0144	.0042	.0011	.0003	.0001	.0000	.0000
	1	.3774	.2852	.1529	.0685	.0268	.0093	.0029	.0008	.0002	.0000
	2	.1787	.2852	.2428	.1540	.0803	.0358	.0138	.0046	.0013	.0003
	3	.0533	.1796	.2428	.2182	.1517	.0869	.0422	.0175	.0062	.0018
	4	.0112	.0798	.1714	.2182	.2023	.1491	.0909	.0467	.0203	.0074
	5	.0018	.0266	.0907	.1636	.2023	.1916	.1468	.0933	.0497	.0222
	6	.0002	.0069	.0374	.0955	.1574	.1916	.1844	.1451	.0949	.0518
	7	.0000	.0014	.0122	.0443	.0974	.1525	.1844	.1797	.1443	.0961
	8	.0000	.0002	.0032	.0166	.0487	.0981	.1489	.1797	.1771	.1442
	9	.0000	.0000	.0007	.0051	.0198	.0514	.0980	.1464	.1771	.1762
	10	.0000	.0000	.0001	.0013	.0066	.0220	.0528	.0976	.1449	.1762
	11	.0000	.0000	.0000	.0003	.0018	.0077	.0233	.0532	.0970	.1442
	12	.0000	.0000	.0000	.0000	.0004	.0022	.0083	.0237	.0529	.0961
	13	.0000	.0000	.0000	.0000	.0001	.0005	.0024	.0085	.0233	.0518
	14	.0000	.0000	.0000	.0000	.0000	.0001	.0006	.0024	.0082	.0222
	15	.0000	.0000	.0000	.0000	.0000	.0000	.0001	.0005	.0022	.0074
	16	.0000	.0000	.0000	.0000	.0000	.0000	.0000	.0001	.0005	.0018
	17	.0000	.0000	.0000	.0000	.0000	.0000	.0000	.0000	.0001	.0003
	18	.0000	.0000	.0000	.0000	.0000	.0000	.0000	.0000	.0000	.0000
	19	.0000	.0000	.0000	.0000	.0000	.0000	.0000	.0000	.0000	.0000
20	0	.3585	.1216	.0388	.0115	.0032	.0008	.0002	.0000	.0000	.0000
	1	.3774	.2702	.1368	.0576	.0211	.0068	.0020	.0005	.0001	.0000
	2	.1887	.2852	.2293	.1369	.0669	.0278	.0100	.0031	.0008	.0002
	3	.0596	.1901	.2428	.2054	.1339	.0716	.0323	.0123	.0040	.0011
	4	.0133	.0898	.1821	.2182	.1897	.1304	.0738	.0350	.0139	.0046
	5	.0022	.0319	.1028	.1746	.2023	.1789	.1272	.0746	.0365	.0148
	6	.0003	.0089	.0454	.1091	.1686	.1916	.1712	.1244	.0746	.0370
	7	.0000	.0020	.0160	.0545	.1124	.1643	.1844	.1659	.1221	.0739
	8	.0000	.0004	.0046	.0222	.0609	.1144	.1614	.1797	.1623	.1201
	9	.0000	.0001	.0011	.0074	.0271	.0654	.1158	.1597	.1771	.1602
	10	.0000	.0000	.0002	.0020	.0099	.0308	.0686	.1171	.1593	.1762
	11	.0000	.0000	.0000	.0005	.0030	.0120	.0336	.0710	.1185	.1602
	12	.0000	.0000	.0000	.0001	.0008	.0039	.0136	.0355	.0727	.1201
	13	.0000	.0000	.0000	.0000	.0002	.0010	.0045	.0146	.0366	.0739
	14	.0000	.0000	.0000	.0000	.0000	.0002	.0012	.0049	.0150	.0370
	15	.0000	.0000	.0000	.0000	.0000	.0000	.0003	.0013	.0049	.0148
	16	.0000	.0000	.0000	.0000	.0000	.0000	.0000	.0003	.0013	.0046
	17	.0000	.0000	.0000	.0000	.0000	.0000	.0000	.0000	.0002	.0011
	18	.0000	.0000	.0000	.0000	.0000	.0000	.0000	.0000	.0000	.0002
	19	.0000	.0000	.0000	.0000	.0000	.0000	.0000	.0000	.0000	.0000
	20	.0000	.0000	.0000	.0000	.0000	.0000	.0000	.0000	.0000	.0000

Table 6. The Hypergeometric Distribution

A random variable X is said to have the hypergeometric distribution if (for some integers n, a, N with $1 \leqslant n \leqslant N$ and $0 \leqslant a \leqslant n$; see Definition 3.2.5)

$$p_X(x) = \begin{cases} \dfrac{\dbinom{a}{x}\dbinom{N-a}{n-x}}{\dbinom{N}{n}}, & x = \max(0, n-(N-a)), \ldots, \min(n, a) \\ 0, & \text{otherwise.} \end{cases}$$

The following table gives values of $p_X(x)$ (for various values of n, a, N).

N	n	a	x	$p_X(x)$
2	1	1	0	.500000
2	1	1	1	.500000
3	1	1	0	.666667
3	1	1	1	.333333
3	2	1	0	.333333
3	2	1	1	.666667
3	2	2	1	.666667
3	2	2	2	.333333
4	1	1	0	.750000
4	1	1	1	.250000
4	2	1	0	.500000
4	2	1	1	.500000
4	2	2	0	.166667
4	2	2	1	.666667
4	2	2	2	.166667
4	3	1	0	.250000
4	3	1	1	.750000
4	3	2	1	.500000
4	3	2	2	.500000
4	3	3	2	.750000
4	3	3	3	.250000
5	1	1	0	.800000
5	1	1	1	.200000
5	2	1	0	.600000
5	2	1	1	.400000
5	2	2	0	.300000
5	2	2	1	.600000
5	2	2	2	.100000
5	3	1	0	.400000
5	3	1	1	.600000
5	3	2	0	.100000
5	3	2	1	.600000
5	3	2	2	.300000
5	3	3	1	.300000
5	3	3	2	.600000
5	3	3	3	.100000
5	4	1	0	.200000
5	4	1	1	.800000
5	4	2	1	.400000
5	4	2	2	.600000
5	4	3	2	.600000
5	4	3	3	.400000
5	4	4	3	.800000
5	4	4	4	.200000
6	1	1	0	.833333
6	1	1	1	.166667
6	2	1	0	.666667
6	2	1	1	.333333
6	2	2	0	.400000
6	2	2	1	.533333
6	2	2	2	.066667
6	3	1	0	.500000
6	3	1	1	.500000
6	3	2	0	.200000
6	3	2	1	.600000
6	3	2	2	.200000
6	3	3	0	.050000
6	3	3	1	.450000
6	3	3	2	.450000
6	3	3	3	.050000
6	4	1	0	.333333
6	4	1	1	.666667
6	4	2	0	.066667
6	4	2	1	.533333
6	4	2	2	.400000
6	4	3	1	.200000
6	4	3	2	.600000
6	4	3	3	.200000
6	4	4	2	.400000
6	4	4	3	.533333
6	4	4	4	.066667
6	5	1	0	.166667
6	5	1	1	.833333
6	5	2	1	.333333
6	5	2	2	.666667
6	5	3	2	.500000
6	5	3	3	.500000
6	5	4	3	.666667
6	5	4	4	.333333
6	5	5	4	.833333
6	5	5	5	.166667
7	1	1	0	.857143
7	1	1	1	.142857
7	2	1	0	.714286
7	2	1	1	.285714
7	2	2	0	.476190
7	2	2	1	.476190
7	2	2	2	.047619
7	3	1	0	.571429
7	3	1	1	.428571
7	3	2	0	.285714
7	3	2	1	.571429
7	3	2	2	.142857
7	3	3	0	.114286
7	3	3	1	.514286
7	3	3	2	.342857
7	3	3	3	.028571
7	4	1	0	.428571
7	4	1	1	.571429
7	4	2	0	.142857
7	4	2	1	.571429
7	4	2	2	.285714
7	4	3	0	.028571
7	4	3	1	.342857
7	4	3	2	.514286
7	4	3	3	.114286
7	4	4	1	.114286
7	4	4	2	.514286
7	4	4	3	.342857
7	4	4	4	.028571
7	5	1	0	.285714
7	5	1	1	.714286
7	5	2	0	.047619
7	5	2	1	.476190
7	5	2	2	.476190
7	5	3	1	.142857
7	5	3	2	.571429
7	5	3	3	.285714
7	5	4	2	.285714
7	5	4	3	.571429
7	5	4	4	.142857
7	5	5	3	.476190
7	5	5	4	.476190
7	5	5	5	.047619
7	6	1	0	.142857
7	6	1	1	.857143
7	6	2	1	.285714
7	6	2	2	.714286
7	6	3	2	.428571
7	6	3	3	.571429
7	6	4	3	.571429
7	6	4	4	.428571
7	6	5	4	.714286
7	6	5	5	.285714
7	6	6	5	.857143
7	6	6	6	.142857
8	1	1	0	.875000
8	1	1	1	.125000
8	2	1	0	.750000
8	2	1	1	.250000
8	2	2	0	.535714
8	2	2	1	.428571
8	2	2	2	.035714
8	3	1	0	.625000
8	3	1	1	.375000
8	3	2	0	.357143
8	3	2	1	.535714
8	3	2	2	.107143
8	3	3	0	.178571
8	3	3	1	.535714
8	3	3	2	.267857
8	3	3	3	.017857
8	4	1	0	.500000
8	4	1	1	.500000
8	4	2	0	.214286
8	4	2	1	.571429
8	4	2	2	.214286
8	4	3	0	.071429
8	4	3	1	.428571
8	4	3	2	.428571

Table 6. The Hypergeometric Distribution (Continued)

$$p_X(x) = \frac{\binom{a}{x}\binom{N-a}{n-x}}{\binom{N}{n}}$$

N	n	a	x	$p_X(x)$	N	n	a	x	$p_X(x)$	N	n	a	x	$p_X(x)$	N	n	a	x	$p_X(x)$
8	4	3	3	.071429	8	7	1	0	.125000	9	4	4	1	.317460	9	6	6	3	.238095
8	4	4	0	.014286	8	7	1	1	.875000	9	4	4	2	.476190	9	6	6	4	.535714
8	4	4	1	.228571	8	7	2	1	.250000	9	4	4	3	.158730	9	6	6	5	.214286
8	4	4	2	.514286	8	7	2	2	.750000	9	4	4	4	.007936	9	6	6	6	.011905
8	4	4	3	.228571	8	7	3	2	.375000	9	5	1	0	.444444	9	7	1	0	.222222
8	4	4	4	.014286	8	7	3	3	.625000	9	5	1	1	.555556	9	7	1	1	.777778
8	5	1	0	.375000	8	7	4	3	.500000	9	5	2	0	.166667	9	7	2	0	.027778
8	5	1	1	.625000	8	7	4	4	.500000	9	5	2	1	.555556	9	7	2	1	.388889
8	5	2	0	.107143	8	7	5	4	.625000	9	5	2	2	.277778	9	7	2	2	.583333
8	5	2	1	.535714	8	7	5	5	.375000	9	5	3	0	.047619	9	7	3	1	.083333
8	5	2	2	.357143	8	7	6	5	.750000	9	5	3	1	.357143	9	7	3	2	.500000
8	5	3	0	.017857	8	7	6	6	.250000	9	5	3	2	.476190	9	7	3	3	.416667
8	5	3	1	.267857	8	7	7	6	.875000	9	5	3	3	.119048	9	7	4	2	.166667
8	5	3	2	.535714	8	7	7	7	.125000	9	5	4	0	.007936	9	7	4	3	.555556
8	5	3	3	.178571	9	1	1	0	.888889	9	5	4	1	.158730	9	7	4	4	.277778
8	5	4	1	.071429	9	1	1	1	.111111	9	5	4	2	.476190	9	7	5	3	.277778
8	5	4	2	.428571	9	2	1	0	.777778	9	5	4	3	.317460	9	7	5	4	.555556
8	5	4	3	.428571	9	2	1	1	.222222	9	5	4	4	.039683	9	7	5	5	.166667
8	5	4	4	.071429	9	2	2	0	.583333	9	5	5	1	.039683	9	7	6	4	.416667
8	5	5	2	.178571	9	2	2	1	.388889	9	5	5	2	.317460	9	7	6	5	.500000
8	5	5	3	.535714	9	2	2	2	.027778	9	5	5	3	.476190	9	7	6	6	.083333
8	5	5	4	.267857	9	3	1	0	.666667	9	5	5	4	.158730	9	7	7	5	.583333
8	5	5	5	.017857	9	3	1	1	.333333	9	5	5	5	.007936	9	7	7	6	.388889
8	6	1	0	.250000	9	3	2	0	.416667	9	6	1	0	.333333	9	7	7	7	.027778
8	6	1	1	.750000	9	3	2	1	.500000	9	6	1	1	.666667	9	8	1	0	.111111
8	6	2	0	.015714	9	3	2	2	.083333	9	6	2	0	.083333	9	8	1	1	.888889
8	6	2	1	.428571	9	3	3	0	.238095	9	6	2	1	.500000	9	8	2	1	.222222
8	6	2	2	.535714	9	3	3	1	.535714	9	6	2	2	.416667	9	8	2	2	.777778
8	6	3	1	.107143	9	3	3	2	.214286	9	6	3	0	.011905	9	8	3	2	.333333
8	6	3	2	.535714	9	3	3	3	.011905	9	6	3	1	.214286	9	8	3	3	.666667
8	6	3	3	.357143	9	4	1	0	.555556	9	6	3	2	.535714	9	8	4	3	.444444
8	6	4	2	.214286	9	4	1	1	.444444	9	6	3	3	.238095	9	8	4	4	.555556
8	6	4	3	.571429	9	4	2	0	.277778	9	6	4	1	.047619	9	8	5	4	.555556
8	6	4	4	.214286	9	4	2	1	.555556	9	6	4	2	.357143	9	8	5	5	.444444
8	6	5	3	.357143	9	4	2	2	.166667	9	6	4	3	.476190	9	8	6	5	.666667
8	6	5	4	.535714	9	4	3	0	.119048	9	6	4	4	.119048	9	8	6	6	.333333
8	6	5	5	.107143	9	4	3	1	.476190	9	6	5	2	.119048	9	8	7	6	.777778
8	6	6	4	.535714	9	4	3	2	.357143	9	6	5	3	.476190	9	8	7	7	.222222
8	6	6	5	.428571	9	4	3	3	.047619	9	6	5	4	.357143	9	8	8	7	.888889
8	6	6	6	.035714	9	4	4	0	.039683	9	6	5	5	.047619	9	8	8	8	.111111

Table 6. The Hypergeometric Distribution (Continued)

$$p_X(x) = \frac{\binom{a}{x}\binom{N-a}{n-x}}{\binom{N}{n}}$$

N	n	a	x	$p_X(x)$
10	1	1	0	.900000
10	1	1	1	.100000
10	2	1	0	.800000
10	2	1	1	.200000
10	2	2	0	.622222
10	2	2	1	.355556
10	2	2	2	.022222
10	3	1	0	.700000
10	3	1	1	.300000
10	3	2	0	.466667
10	3	2	1	.466667
10	3	2	2	.066667
10	3	3	0	.291667
10	3	3	1	.525000
10	3	3	2	.175000
10	3	3	3	.008333
10	4	1	0	.600000
10	4	1	1	.400000
10	4	2	0	.333333
10	4	2	1	.533333
10	4	2	2	.133333
10	4	3	0	.166667
10	4	3	1	.500000
10	4	3	2	.300000
10	4	3	3	.033333
10	4	4	0	.071429
10	4	4	1	.380952
10	4	4	2	.428571
10	4	4	3	.114286
10	4	4	4	.004762
10	5	1	0	.500000
10	5	1	1	.500000
10	5	2	0	.222222
10	5	2	1	.555556
10	5	2	2	.222222
10	5	3	0	.083333
10	5	3	1	.416667
10	5	3	2	.416667
10	5	3	3	.083333
10	5	4	0	.023810

N	n	a	x	$p_X(x)$
10	5	4	1	.238095
10	5	4	2	.476190
10	5	4	3	.238095
10	5	4	4	.023810
10	5	5	0	.003968
10	5	5	1	.099206
10	5	5	2	.396825
10	5	5	3	.396825
10	5	5	4	.099206
10	5	5	5	.003968
10	6	1	0	.400000
10	6	1	1	.600000
10	6	2	0	.133333
10	6	2	1	.533333
10	6	2	2	.333333
10	6	3	0	.033333
10	6	3	1	.300000
10	6	3	2	.500000
10	6	3	3	.166667
10	6	4	0	.004762
10	6	4	1	.114286
10	6	4	2	.428571
10	6	4	3	.380952
10	6	4	4	.071429
10	6	5	1	.023810
10	6	5	2	.238095
10	6	5	3	.476190
10	6	5	4	.238095
10	6	5	5	.023810
10	6	6	2	.071429
10	6	6	3	.380952
10	6	6	4	.428571
10	6	6	5	.114286
10	6	6	6	.004762
10	7	1	0	.300000
10	7	1	1	.700000
10	7	2	0	.066667
10	7	2	1	.466667
10	7	2	2	.466667
10	7	3	0	.008333

N	n	a	x	$p_X(x)$
10	7	3	1	.175000
10	7	3	2	.525000
10	7	3	3	.291667
10	7	4	1	.033333
10	7	4	2	.300000
10	7	4	3	.500000
10	7	4	4	.166667
10	7	5	2	.083333
10	7	5	3	.416667
10	7	5	4	.416667
10	7	5	5	.083333
10	7	6	3	.166667
10	7	6	4	.500000
10	7	6	5	.300000
10	7	6	6	.033333
10	7	7	4	.291667
10	7	7	5	.525000
10	7	7	6	.175000
10	7	7	7	.008333
10	8	1	0	.200000
10	8	1	1	.800000
10	8	2	0	.022222
10	8	2	1	.355556
10	8	2	2	.622222
10	8	3	1	.066667
10	8	3	2	.466667
10	8	3	3	.466667
10	8	4	2	.133333
10	8	4	3	.533333
10	8	4	4	.333333
10	8	5	3	.222222
10	8	5	4	.555556
10	8	5	5	.222222
10	8	6	4	.333333
10	8	6	5	.533333
10	8	6	6	.133333
10	8	7	5	.466667
10	8	7	6	.466667
10	8	7	7	.066667
10	8	8	6	.622222

N	n	a	x	$p_X(x)$
10	8	8	7	.355556
10	8	8	8	.022222
10	9	1	0	.100000
10	9	1	1	.900000
10	9	2	1	.200000
10	9	2	2	.800000
10	9	3	2	.300000
10	9	3	3	.700000
10	9	4	3	.400000
10	9	4	4	.600000
10	9	5	4	.500000
10	9	5	5	.500000
10	9	6	5	.600000
10	9	6	6	.400000
10	9	7	6	.700000
10	9	7	7	.300000
10	9	8	7	.800000
10	9	8	8	.200000
10	9	9	8	.900000
10	9	9	9	.100000

Table 7. The Poisson Distribution

A random variable X is said to have the Poisson distribution if (for some $\mu > 0$; see Definition 3.2.7)

$$p_X(x) = \begin{cases} e^{-\mu}\dfrac{\mu^x}{x!}, & x = 0, 1, 2, \ldots \\ 0, & \text{otherwise.} \end{cases}$$

The following table gives values of $p_X(x)$ (for various values of μ).

	μ									
x	.1	.2	.3	4	.5	.6	.7	.8	.9	1.0
0	.9048	.8187	.7408	.6703	.6065	.5488	.4966	.4493	.4066	.3679
1	.0905	.1637	.2222	.2681	.3033	.3293	.3476	.3595	.3659	.3679
2	.0045	.0164	.0333	.0536	.0758	.0988	.1217	.1438	.1647	.1839
3	.0002	.0011	.0033	.0072	.0126	.0198	.0284	.0383	.0494	.0613
4	.0000	.0001	.0002	.0007	.0016	.0030	.0050	.0077	.0111	.0153
5	.0000	.0000	.0000	.0001	.0002	.0004	.0007	.0012	.0020	.0031
6	.0000	.0000	.0000	.0000	.0000	.0000	.0001	.0002	.0003	.0005
7	.0000	.0000	.0000	.0000	.0000	.0000	.0000	.0000	.0000	.0001

	μ									
x	1.1	1.2	1.3	1.4	1.5	1.6	1.7	1.8	1.9	2.0
0	.3329	.3012	.2725	.2466	.2231	.2019	.1827	.1653	.1496	.1353
1	.3662	.3614	.3543	.3452	.3347	.3230	.3106	.2975	.2842	.2707
2	.2014	.2169	.2303	.2417	.2510	.2584	.2640	.2678	.2700	.2707
3	.0738	.0867	.0998	.1128	.1255	.1378	.1496	.1607	.1710	.1804
4	.0203	.0260	.0324	.0395	.0471	.0551	.0636	.0723	.0812	.0902
5	.0045	.0062	.0084	.0111	.0141	.0176	.0216	.0260	.0309	.0361
6	.0008	.0012	.0018	.0026	.0035	.0047	.0061	.0078	.0098	.0120
7	.0001	.0002	.0003	.0005	.0008	.0011	.0015	.0020	.0027	.0034
8	.0000	.0000	.0001	.0001	.0001	.0002	.0003	.0005	.0006	.0009
9	.0000	.0000	.0000	.0000	.0000	.0000	.0001	.0001	.0001	.0002

	μ									
x	2.1	2.2	2.3	2.4	2.5	2.6	2.7	2.8	2.9	3.0
0	.1225	.1108	.1003	.0907	.0821	.0743	.0672	.0608	.0550	.0498
1	.2572	.2438	.2306	.2177	.2052	.1931	.1815	.1703	.1596	.1494
2	.2700	.2681	.2652	.2613	.2565	.2510	.2450	.2384	.2314	.2240
3	.1890	.1966	.2033	.2090	.2138	.2176	.2205	.2225	.2237	.2240
4	.0992	.1082	.1169	.1254	.1336	.1414	.1488	.1557	.1622	.1680
5	.0417	.0476	.0538	.0602	.0668	.0735	.0804	.0872	.0940	.1008
6	.0146	.0174	.0206	.0241	.0278	.0319	.0362	.0407	.0455	.0504
7	.0044	.0055	.0068	.0083	.0099	.0118	.0139	.0163	.0188	.0216
8	.0011	.0015	.0019	.0025	.0031	.0038	.0047	.0057	.0068	.0081
9	.0003	.0004	.0005	.0007	.0009	.0011	.0014	.0018	.0022	.0027
10	.0001	.0001	.0001	.0002	.0002	.0003	.0004	.0005	.0006	.0008
11	.0000	.0000	.0000	.0000	.0000	.0001	.0001	.0001	.0002	.0002
12	.0000	.0000	.0000	.0000	.0000	.0000	.0000	.0000	.0000	.0001

	μ									
x	3.1	3.2	3.3	3.4	3.5	3.6	3.7	3.8	3.9	4.0
0	.0450	.0408	.0369	.0334	.0302	.0273	.0247	.0224	.0202	.0183
1	.1397	.1304	.1217	.1135	.1057	.0984	.0915	.0850	.0789	.0733
2	.2165	.2087	.2008	.1929	.1850	.1771	.1692	.1615	.1539	.1465
3	.2237	.2226	.2209	.2186	.2158	.2125	.2087	.2046	.2001	.1954
4	.1734	.1781	.1823	.1858	.1888	.1912	.1931	.1944	.1951	.1954
5	.1075	.1140	.1203	.1264	.1322	.1377	.1429	.1477	.1522	.1563
6	.0555	.0608	.0662	.0716	.0771	.0826	.0881	.0936	.0989	.1042
7	.0246	.0278	.0312	.0348	.0385	.0425	.0466	.0508	.0551	.0595
8	.0095	.0111	.0129	.0148	.0169	.0191	.0215	.0241	.0269	.0298
9	.0033	.0040	.0047	.0056	.0066	.0076	.0089	.0102	.0116	.0132
10	.0010	.0013	.0016	.0019	.0023	.0028	.0033	.0039	.0045	.0053
11	.0003	.0004	.0005	.0006	.0007	.0009	.0011	.0013	.0016	.0019
12	.0001	.0001	.0001	.0002	.0002	.0003	.0003	.0004	.0005	.0006
13	.0000	.0000	.0000	.0000	.0001	.0001	.0001	.0001	.0002	.0002
14	.0000	.0000	.0000	.0000	.0000	.0000	.0000	.0000	.0000	.0001

Table 7. The Poisson Distribution (Continued)

$$p_X(x) = e^{-\mu}\frac{\mu^x}{x!}$$

	μ									
x	4.1	4.2	4.3	4.4	4.5	4.6	4.7	4.8	4.9	5.0
0	.0166	.0150	.0136	.0123	.0111	.0101	.0091	.0082	.0074	.0067
1	.0679	.0630	.0583	.0540	.0500	.0462	.0427	.0395	.0365	.0337
2	.1393	.1323	.1254	.1188	.1125	.1063	.1005	.0948	.0894	.0842
3	.1904	.1852	.1798	.1743	.1687	.1631	.1574	.1517	.1460	.1404
4	.1951	.1944	.1933	.1917	.1898	.1875	.1849	.1820	.1789	.1755
5	.1600	.1633	.1662	.1687	.1708	.1725	.1738	.1747	.1753	.1755
6	.1093	.1143	.1191	.1237	.1281	.1323	.1362	.1398	.1432	.1462
7	.0640	.0686	.0732	.0778	.0824	.0869	.0914	.0959	.1002	.1044
8	.0328	.0360	.0393	.0428	.0463	.0500	.0537	.0575	.0614	.0653
9	.0150	.0168	.0188	.0209	.0232	.0255	.0280	.0307	.0334	.0363
10	.0061	.0071	.0081	.0092	.0104	.0118	.0132	.0147	.0164	.0181
11	.0023	.0027	.0032	.0037	.0043	.0049	.0056	.0064	.0073	.0082
12	.0008	.0009	.0011	.0014	.0016	.0019	.0022	.0026	.0030	.0034
13	.0002	.0003	.0004	.0005	.0006	.0007	.0008	.0009	.0011	.0013
14	.0001	.0001	.0001	.0001	.0002	.0002	.0003	.0003	.0004	.0005
15	.0000	.0000	.0000	.0000	.0001	.0001	.0001	.0001	.0001	.0002

	μ									
x	5.1	5.2	5.3	5.4	5.5	5.6	5.7	5.8	5.9	6.0
0	.0061	.0055	.0050	.0045	.0041	.0037	.0033	.0030	.0027	.0025
1	.0311	.0287	.0265	.0244	.0225	.0207	.0191	.0176	.0162	.0149
2	.0793	.0746	.0701	.0659	.0618	.0580	.0544	.0509	.0477	.0446
3	.1348	.1293	.1239	.1185	.1133	.1082	.1033	.0985	.0938	.0892
4	.1719	.1681	.1641	.1600	.1558	.1515	.1472	.1428	.1383	.1339
5	.1753	.1748	.1740	.1728	.1714	.1697	.1678	.1656	.1632	.1606
6	.1490	.1515	.1537	.1555	.1571	.1584	.1594	.1601	.1605	.1606
7	.1086	.1125	.1163	.1200	.1234	.1267	.1298	.1326	.1353	.1377
8	.0692	.0731	.0771	.0810	.0849	.0887	.0925	.0962	.0998	.1033
9	.0392	.0423	.0454	.0486	.0519	.0552	.0586	.0620	.0654	.0688
10	.0200	.0220	.0241	.0262	.0285	.0309	.0334	.0359	.0386	.0413
11	.0093	.0104	.0116	.0129	.0143	.0157	.0173	.0190	.0207	.0225
12	.0039	.0045	.0051	.0058	.0065	.0073	.0082	.0092	.0102	.0113
13	.0015	.0018	.0021	.0024	.0028	.0032	.0036	.0041	.0046	.0052
14	.0006	.0007	.0008	.0009	.0011	.0013	.0015	.0017	.0019	.0022
15	.0002	.0002	.0003	.0003	.0004	.0005	.0006	.0007	.0008	.0009
16	.0001	.0001	.0001	.0001	.0001	.0002	.0002	.0002	.0003	.0003
17	.0000	.0000	.0000	.0000	.0000	.0001	.0001	.0001	.0001	.0001

	μ									
x	6.1	6.2	6.3	6.4	6.5	6.6	6.7	6 8	6.9	7.0
0	.0022	.0020	.0018	.0017	.0015	.0014	.0012	.0011	.0010	.0009
1	.0137	.0126	.0116	.0106	.0098	.0090	.0082	.0076	.0070	.0064
2	.0417	.0390	.0364	.0340	.0318	.0296	.0276	.0258	.0240	.0223
3	.0848	.0806	.0765	.0726	.0688	.0652	.0617	.0584	.0552	.0521
4	.1294	.1249	.1205	.1162	.1118	.1076	.1034	.0992	.0952	.0912
5	.1579	.1549	.1519	.1487	.1454	.1420	.1385	.1349	.1314	.1277
6	.1605	.1601	.1595	.1586	.1575	.1562	.1546	.1529	.1511	.1490
7	.1399	.1418	.1435	.1450	.1462	.1472	.1480	.1486	.1489	.1490
8	.1066	.1099	.1130	.1160	.1188	.1215	.1240	.1263	.1284	.1304
9	.0723	.0757	.0791	.0825	.0858	.0891	.0923	.0954	.0985	.1014
10	.0441	.0469	.0498	.0528	.0558	.0588	.0618	.0649	.0679	.0710
11	.0245	.0265	.0285	.0307	.0330	.0353	.0377	.0401	.0426	.0452
12	.0124	.0137	.0150	.0164	.0179	.0194	.0210	.0227	.0245	.0264
13	.0058	.0065	.0073	.0081	.0089	.0098	.0108	.0119	.0130	.0142
14	.0025	.0029	.0033	.0037	.0041	.0046	.0052	.0058	.0064	.0071
15	.0010	.0012	.0014	.0016	.0018	.0020	.0023	.0026	.0029	.0033
16	.0004	.0005	.0005	.0006	.0007	.0008	.0010	.0011	.0013	.0014
17	.0001	.0002	.0002	.0002	.0003	.0003	.0004	.0004	.0005	.0006
18	.0000	.0001	.0001	.0001	.0001	.0001	.0001	.0002	.0002	.0002
19	.0000	.0000	.0000	.0000	.0000	.0000	.0000	.0001	.0001	.0001

Table 7. The Poisson Distribution (Continued)

$$p_X(x) = e^{-\mu} \frac{\mu^x}{x!}$$

x	μ = 7.1	7.2	7.3	7.4	7.5	7.6	7.7	7.8	7.9	8.0
0	.0008	.0007	.0007	.0006	.0006	.0005	.0005	.0004	.0004	.0003
1	.0059	.0054	.0049	.0045	.0041	.0038	.0035	.0032	.0029	.0027
2	.0208	.0194	.0180	.0167	.0156	.0145	.0134	.0125	.0116	.0107
3	.0492	.0464	.0438	.0413	.0389	.0366	.0345	.0324	.0305	.0286
4	.0874	.0836	.0799	.0764	.0729	.0696	.0663	.0632	.0602	.0573
5	.1241	.1204	.1167	.1130	.1094	.1057	.1021	.0986	.0951	.0916
6	.1468	.1445	.1420	.1394	.1367	.1339	.1311	.1282	.1252	.1221
7	.1489	.1486	.1481	.1474	.1465	.1454	.1442	.1428	.1413	.1396
8	.1321	.1337	.1351	.1363	.1373	.1382	.1388	.1392	.1395	.1396
9	.1042	.1070	.1096	.1121	.1144	.1167	.1187	.1207	.1224	.1241
10	.0740	.0770	.0800	.0829	.0858	.0887	.0914	.0941	.0967	.0993
11	.0478	.0504	.0531	.0558	.0585	.0613	.0640	.0667	.0695	.0722
12	.0283	.0303	.0323	.0344	.0366	.0388	.0411	.0434	.0457	.0481
13	.0154	.0168	.0181	.0196	.0211	.0227	.0243	.0260	.0278	.0296
14	.0078	.0086	.0095	.0104	.0113	.0123	.0134	.0145	.0157	.0169
15	.0037	.0041	.0046	.0051	.0057	.0062	.0069	.0075	.0083	.0090
16	.0016	.0019	.0021	.0024	.0026	.0030	.0033	.0037	.0041	.0045
17	.0007	.0008	.0009	.0010	.0012	.0013	.0015	.0017	.0019	.0021
18	.0003	.0003	.0004	.0004	.0005	.0006	.0006	.0007	.0008	.0009
19	.0001	.0001	.0001	.0002	.0002	.0002	.0003	.0003	.0003	.0004
20	.0000	.0000	.0001	.0001	.0001	.0001	.0001	.0001	.0001	.0002
21	.0000	.0000	.0000	.0000	.0000	.0000	.0000	.0000	.0001	.0001

x	μ = 8.1	8.2	8.3	8.4	8.5	8.6	8.7	8.8	8.9	9.0
0	.0003	.0003	.0002	.0002	.0002	.0002	.0002	.0002	.0001	.0001
1	.0025	.0023	.0021	.0019	.0017	.0016	.0014	.0013	.0012	.0011
2	.0100	.0092	.0086	.0079	.0074	.0068	.0063	.0058	.0054	.0050
3	.0269	.0252	.0237	.0222	.0208	.0195	.0183	.0171	.0160	.0150
4	.0544	.0517	.0491	.0466	.0443	.0420	.0398	.0377	.0357	.0337
5	.0882	.0849	.0816	.0784	.0752	.0722	.0692	.0663	.0635	.0607
6	.1191	.1160	.1128	.1097	.1066	.1034	.1003	.0972	.0941	.0911
7	.1378	.1358	.1338	.1317	.1294	.1271	.1247	.1222	.1197	.1171
8	.1395	.1392	.1388	.1382	.1375	.1366	.1356	.1344	.1332	.1318
9	.1256	.1269	.1280	.1290	.1299	.1306	.1311	.1315	.1317	.1318
10	.1017	.1040	.1063	.1084	.1104	.1123	.1140	.1157	.1172	.1186
11	.0749	.0776	.0802	.0828	.0853	.0878	.0902	.0925	.0948	.0970
12	.0505	.0530	.0555	.0579	.0604	.0629	.0654	.0679	.0703	.0728
13	.0315	.0334	.0354	.0374	.0395	.0416	.0438	.0459	.0481	.0504
14	.0182	.0196	.0210	.0225	.0240	.0256	.0272	.0289	.0306	.0324
15	.0098	.0107	.0116	.0126	.0136	.0147	.0158	.0169	.0182	.0194
16	.0050	.0055	.0060	.0066	.0072	.0079	.0086	.0093	.0101	.0109
17	.0024	.0026	.0029	.0033	.0036	.0040	.0044	.0048	.0053	.0058
18	.0011	.0012	.0014	.0015	.0017	.0019	.0021	.0024	.0026	.0029
19	.0005	.0005	.0006	.0007	.0008	.0009	.0010	.0011	.0012	.0014
20	.0002	.0002	.0002	.0003	.0003	.0004	.0004	.0005	.0005	.0006
21	.0001	.0001	.0001	.0001	.0001	.0002	.0002	.0002	.0002	.0003
22	.0000	.0000	.0000	.0000	.0001	.0001	.0001	.0001	.0001	.0001

x	μ = 9.1	9.2	9.3	9.4	9.5	9.6	9.7	9.8	9.9	10
0	.0001	.0001	.0001	.0001	.0001	.0001	.0001	.0001	.0001	.0000
1	.0010	.0009	.0009	.0008	.0007	.0007	.0006	.0005	.0005	.0005
2	.0046	.0043	.0040	.0037	.0034	.0031	.0029	.0027	.0025	.0023
3	.0140	.0131	.0123	.0115	.0107	.0100	.0093	.0087	.0081	.0076
4	.0319	.0302	.0285	.0269	.0254	.0240	.0226	.0213	.0201	.0189
5	.0581	.0555	.0530	.0506	.0483	.0460	.0439	.0418	.0398	.0378
6	.0881	.0851	.0822	.0793	.0764	.0736	.0709	.0682	.0656	.0631
7	.1145	.1118	.1091	.1064	.1037	.1010	.0982	.0955	.0928	.0901
8	.1302	.1286	.1269	.1251	.1232	.1212	.1191	.1170	.1148	.1126
9	.1317	.1315	.1311	.1306	.1300	.1293	.1284	.1274	.1263	.1251

Table 7. The Poisson Distribution (Continued)

$$p_X(x) = e^{-\mu} \frac{\mu^x}{x!}$$

x	μ = 9.1	9.2	9.3	9.4	9.5	9.6	9.7	9.8	9.9	10
10	.1198	.1210	.1219	.1228	.1235	.1241	.1245	.1249	.1250	.1251
11	.0991	.1012	.1031	.1049	.1067	.1083	.1098	.1112	.1125	.1137
12	.0752	.0776	.0799	.0822	.0844	.0866	.0888	.0908	.0928	.0948
13	.0526	.0549	.0572	.0594	.0617	.0640	.0662	.0685	.0707	.0729
14	.0342	.0361	.0380	.0399	.0419	.0439	.0459	.0479	.0500	.0521
15	.0208	.0221	.0235	.0250	.0265	.0281	.0297	.0313	.0330	.0347
16	.0118	.0127	.0137	.0147	.0157	.0168	.0180	.0192	.0204	.0217
17	.0063	.0069	.0075	.0081	.0088	.0095	.0103	.0111	.0119	.0128
18	.0032	.0035	.0039	.0042	.0046	.0051	.0055	.0060	.0065	.0071
19	.0015	.0017	.0019	.0021	.0023	.0026	.0028	.0031	.0034	.0037
20	.0007	.0008	.0009	.0010	.0011	.0012	.0014	.0015	.0017	.0019
21	.0003	.0003	.0004	.0004	.0005	.0006	.0006	.0007	.0008	.0009
22	.0001	.0001	.0002	.0002	.0002	.0002	.0003	.0003	.0004	.0004
23	.0000	.0001	.0001	.0001	.0001	.0001	.0001	.0001	.0002	.0002
24	.0000	.0000	.0000	.0000	.0000	.0000	.0000	.0001	.0001	.0001

x	μ = 11	12	13	14	15	16	17	18	19	20
0	.0000	.0000	.0000	.0000	.0000	.0000	.0000	.0000	.0000	.0000
1	.0002	.0001	.0000	.0000	.0000	.0000	.0000	.0000	.0000	.0000
2	.0010	.0004	.0002	.0001	.0000	.0000	.0000	.0000	.0000	.0000
3	.0037	.0018	.0008	.0004	.0002	.0001	.0000	.0000	.0000	.0000
4	.0102	.0053	.0027	.0013	.0006	.0003	.0001	.0001	.0000	.0000
5	.0224	.0127	.0070	.0037	.0019	.0010	.0005	.0002	.0001	.0001
6	.0411	.0255	.0152	.0087	.0048	.0026	.0014	.0007	.0004	.0002
7	.0646	.0437	.0281	.0174	.0104	.0060	.0034	.0018	.0010	.0005
8	.0888	.0655	.0457	.0304	.0194	.0120	.0072	.0042	.0024	.0013
9	.1085	.0874	.0661	.0473	.0324	.0213	.0135	.0083	.0050	.0029
10	.1194	.1048	.0859	.0663	.0486	.0341	.0230	.0150	.0095	.0058
11	.1194	.1144	.1015	.0844	.0663	.0496	.0355	.0245	.0164	.0106
12	.1094	.1144	.1099	.0984	.0829	.0661	.0504	.0368	.0259	.0176
13	.0926	.1056	.1099	.1060	.0956	.0814	.0658	.0509	.0378	.0271
14	.0728	.0905	.1021	.1060	.1024	.0930	.0800	.0655	.0514	.0387
15	.0534	.0724	.0885	.0989	.1024	.0992	.0906	.0786	.0650	.0516
16	.0367	.0543	.0719	.0866	.0960	.0992	.0963	.0884	.0772	.0646
17	.0237	.0383	.0550	.0713	.0847	.0934	.0963	.0936	.0863	.0760
18	.0145	.0256	.0397	.0554	.0706	.0830	.0909	.0936	.0911	.0844
19	.0084	.0161	.0272	.0409	.0557	.0699	.0814	.0887	.0911	.0888
20	.0046	.0097	.0177	.0286	.0418	.0559	.0692	.0798	.0866	.0888
21	.0024	.0055	.0109	.0191	.0299	.0426	.0560	.0684	.0783	.0846
22	.0012	.0030	.0065	.0121	.0204	.0310	.0433	.0560	.0676	.0769
23	.0006	.0016	.0037	.0074	.0133	.0216	.0320	.0438	.0559	.0669
24	.0003	.0008	.0020	.0043	.0083	.0144	.0226	.0328	.0442	.0557
25	.0001	.0004	.0010	.0024	.0050	.0092	.0154	.0237	.0336	.0446
26	.0000	.0002	.0005	.0013	.0029	.0057	.0101	.0164	.0246	.0343
27	.0000	.0001	.0002	.0007	.0016	.0034	.0063	.0109	.0173	.0254
28	.0000	.0000	.0001	.0003	.0009	.0019	.0038	.0070	.0117	.0181
29	.0000	.0000	.0001	.0002	.0004	.0011	.0023	.0044	.0077	.0125
30	.0000	.0000	.0000	.0001	.0002	.0006	.0013	.0026	.0049	.0083
31	.0000	.0000	.0000	.0000	.0001	.0003	.0007	.0015	.0030	.0054
32	.0000	.0000	.0000	.0000	.0001	.0001	.0004	.0009	.0018	.0034
33	.0000	.0000	.0000	.0000	.0000	.0001	.0002	.0005	.0010	.0020
34	.0000	.0000	.0000	.0000	.0000	.0000	.0001	.0002	.0006	.0012
35	.0000	.0000	.0000	.0000	.0000	.0000	.0000	.0001	.0003	.0007
36	.0000	.0000	.0000	.0000	.0000	.0000	.0000	.0001	.0002	.0004
37	.0000	.0000	.0000	.0000	.0000	.0000	.0000	.0000	.0001	.0002
38	.0000	.0000	.0000	.0000	.0000	.0000	.0000	.0000	.0000	.0001
39	.0000	.0000	.0000	.0000	.0000	.0000	.0000	.0000	.0000	.0001

TABLE 8. THE BETA DISTRIBUTION

A random variable X is said to have the beta distribution $B(x|\alpha, \lambda)$ if (for some $\alpha, \lambda > -1$; see Definition 4.2.1)

$$f_X(x) = \begin{cases} \dfrac{1}{\beta(\alpha + 1, \lambda + 1)} x^{\alpha}(1 - x)^{\lambda}, & 0 \leqslant x \leqslant 1 \\ 0, & \text{otherwise.} \end{cases}$$

The following graphs[a] show beta distribution densities with various selections of parameters α, λ. (The symbol a, b means $\alpha = a, \lambda = b$.)

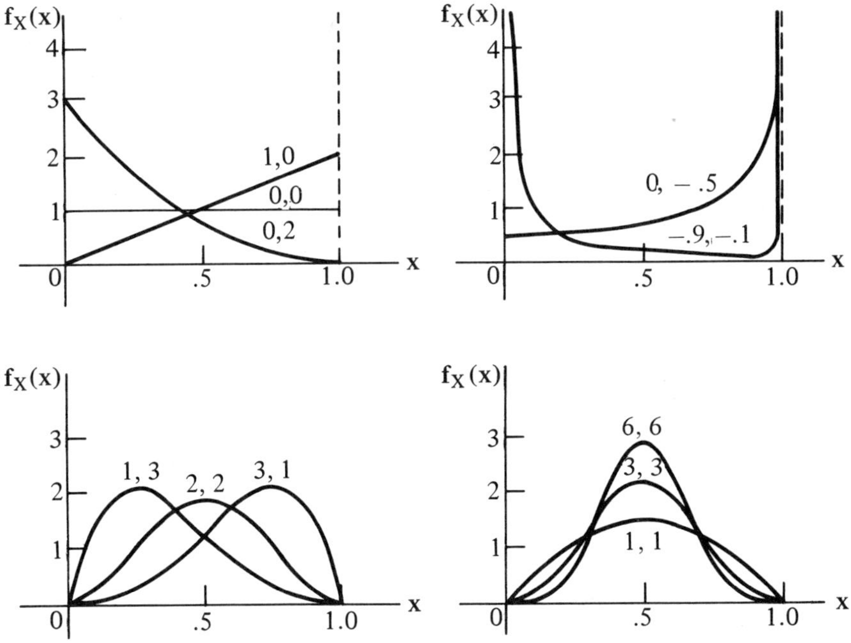

[a] Figures from *Introduction to Probability and Random Variables.* By G. P. Wadsworth and J. G. Bryan. Copyright © 1960 by McGraw-Hill Book Co., Inc. Used with permission.

Table 8. The Beta Distribution (Continued)

The following table gives values of y such that (for various values of α, λ)

$$P[X \leqslant y] = \int_0^y f_X(x)\,dx = .05$$

α \ λ	−·5	·0	·5	1·0	1·5	2·0	2·5	3·0	3·5
−·5	·0061558	·0025000	·0015429	·0011119	·0^{3}86820	·0^{3}71179	·0^{3}60300	·0^{3}52300	·0^{3}46170
·0	·097500	·050000	·033617	·025321	·020308	·016952	·014548	·012741	·011334
·5	·22852	·13572	·097308	·076010	·062413	·052962	·046007	·040671	·036447
1·0	·34163	·22361	·16825	·13535	·11338	·097611	·085727	·076440	·068979
1·5	·43074	·30171	·23553	·19403	·16528	·14409	·12778	·11482	·10427
2·0	·50053	·36840	·29599	·24860	·21477	·18926	·16927	·15316	·13989
2·5	·55593	·42489	·34929	·29811	·26063	·23182	·20890	·19019	·17461
3·0	·60071	·47287	·39607	·34259	·30260	·27134	·24613	·22532	·20783
3·5	·63751	·51390	·43716	·38245	·34080	·30777	·28082	·25835	·23930
4·0	·66824	·54928	·47338	·41820	·37553	·34126	·31301	·28924	·26894
4·5	·69425	·58003	·50546	·45033	·40712	·37203	·34283	·31807	·29677
5·0	·71654	·60696	·53402	·47930	·43590	·40031	·37044	·34494	·32286
5·5	·73583	·63073	·55958	·50551	·46219	·42635	·39604	·37000	·34732
6·0	·75268	·65184	·58256	·52932	·48626	·45036	·41980	·39338	·37025
6·5	·76754	·67070	·60333	·55102	·50836	·47255	·44187	·41521	·39176
7·0	·78072	·68766	·62217	·57086	·52872	·49310	·46242	·43563	·41196
7·5	·79249	·70297	·63933	·58907	·54750	·51217	·48159	·45474	·43094
8·0	·80307	·71687	·65503	·60584	·56490	·52991	·49949	·47267	·44880
8·5	·81263	·72954	·66944	·62131	·58103	·54645	·51624	·48951	·46564
9·0	·82131	·74113	·68271	·63564	·59605	·56189	·53194	·50535	·48152
9·5	·82923	·75178	·69496	·64894	·61004	·57635	·54669	·52027	·49652
10·0	·83647	·76160	·70632	·66132	·62312	·58990	·56056	·53434	·51071
10·5	·84313	·77067	·71687	·67287	·63536	·60263	·57363	·54764	·52415
11·0	·84927	·77908	·72669	·68366	·64684	·61461	·58596	·56022	·53689
11·5	·85494	·78690	·73586	·69377	·65764	·62590	·59761	·57213	·54898
12·0	·86021	·79418	·74444	·70327	·66780	·63656	·60864	·58343	·56048
12·5	·86511	·80099	·75249	·71219	·67738	·64663	·61909	·59416	·57141
13·0	·86967	·80736	·76004	·72060	·68643	·65617	·62900	·60436	·58183
13·5	·87394	·81334	·76715	·72854	·69499	·66522	·63842	·61407	·59177
14·0	·87794	·81896	·77386	·73604	·70311	·67381	·64738	·62332	·60125
19·0	·90734	·86089	·82447	·79327	·76559	·74053	·71758	·69636	·67663
29·0	·93748	·90497	·87881	·85591	·83517	·81606	·79824	·78150	·76569
59·0	·96837	·95130	·93720	·92458	·91290	·90192	·89148	·88150	·87191
∞	1·00000	1·00000	1·00000	1·00000	1·00000	1·00000	1·00000	1·00000	1·00000

Table 8. The Beta Distribution (Continued)

y such that $P[X \leqslant y] = \int_0^y f_X(x)\,dx = .05$

α \ λ	4·0	5·0	6·5	9·0	11·0	14·0	19·0	29·0	59·0
– ·5	$\cdot 0^{3}41325$	$\cdot 0^{3}34154$	$\cdot 0^{3}27098$	$\cdot 0^{3}20156$	$\cdot 0^{3}16727$	$\cdot 0^{3}13326$	$\cdot 0^{4}99535$	$\cdot 0^{4}66082$	$\cdot 0^{4}32904$
·0	·010206	·0085124	·0068158	·0051162	·0042653	·0034137	·0025614	·0017083	$\cdot 0^{3}85452$
·5	·033020	·027794	·022465	·017026	·014264	·011472	·0086511	·0057991	·0029157
1·0	·062850	·053376	·043541	·033319	·028053	·022679	·017191	·011585	·0058568
1·5	·095510	·081790	·067312	·051995	·043994	·035747	·027240	·018458	·0093841
2·0	·12876	·11111	·092207	·071870	·061103	·049898	·038224	·026043	·013317
2·5	·16142	·14029	·11733	·092238	·078783	·064651	·049781	·034103	·017540
3·0	·19290	·16875	·14216	·11267	·096658	·079695	·061676	·042481	·021976
3·5	·22292	·19618	·16638	·13288	·11449	·094827	·073748	·051068	·026572
4·0	·25137	·22244	·18984	·15272	·13211	·10991	·085885	·059786	·031288
4·5	·27823	·24746	·21244	·17207	·14943	·12484	·098008	·068575	·036094
5·0	·30354	·27125	·23413	·19086	·16636	·13955	·11006	·077394	·040967
5·5	·32737	·29383	·25492	·20908	·18288	·15401	·12199	·086209	·045889
6·0	·34981	·31524	·27481	·22669	·19895	·16818	·13377	·094994	·050847
6·5	·37095	·33554	·29382	·24370	·21457	·18203	·14539	·10373	·055827
7·0	·39086	·35480	·31199	·26011	·22972	·19556	·15682	·11240	·060821
7·5	·40965	·37307	·32936	·27594	·24441	·20877	·16805	·12099	·065820
8·0	·42738	·39041	·34596	·29120	·25865	·22164	·17908	·12950	·070818
8·5	·44414	·40689	·36183	·30591	·27244	·23418	·18989	·13791	·075809
9·0	·45999	·42256	·37701	·32009	·28580	·24639	·20050	·14622	·080789
9·5	·47501	·43746	·39154	·33375	·29874	·25828	·21088	·15442	·085753
10·0	·48925	·45165	·40544	·34693	·31126	·26985	·22106	·16252	·090698
10·5	·50276	·46518	·41877	·35964	·32340	·28112	·23102	·17051	·095622
11·0	·51560	·47808	·43154	·37190	·33515	·29208	·24077	·17838	·10052
11·5	·52782	·49040	·44379	·38373	·34653	·30275	·25032	·18615	·10539
12·0	·53945	·50217	·45554	·39516	·35756	·31314	·25966	·19379	·11024
12·5	·55054	·51343	·46683	·40619	·36826	·32325	·26880	·20133	·11505
13·0	·56112	·52420	·47768	·41685	·37862	·33309	·27775	·20875	·11983
13·5	·57122	·53452	·48812	·42715	·38867	·34267	·28650	·21606	·12458
14·0	·58088	·54442	·49816	·43711	·39842	·35200	·29507	·22326	·12930
19·0	·65819	·62459	·58083	·52099	·48175	·43321	·37136	·28936	·17453
29·0	·75070	·72282	·68535	·63185	·59522	·54807	·48477	·39458	·25416
59·0	·86266	·84504	·82047	·78342	·75661	·72016	·66738	·58326	·42519
∞	1·00000	1·00000	1·00000	1·00000	1·00000	1·00000	1·00000	1·00000	1·00000

Adapted from Table 16 on pp. 156–157 of *Biometrika Tables for Statisticians*, Vol. I (Third edition reprinted with additions), edited by E. S. Pearson and H. O. Hartley, Cambridge University Press, London, 1970. Used by permission of the Biometrika Trustees.

TABLE 9. THE GAMMA DISTRIBUTION

A random variable X is said to have the gamma distribution $G(x|\alpha, \beta, A)$ if (for some $\alpha > -1, \beta > 0, -\infty < A < +\infty$; see Definition 4.2.6)

$$f_X(x) = \begin{cases} \dfrac{1}{\beta^{\alpha+1}\Gamma(\alpha+1)}(x - A)^{\alpha} e^{-(x-A)/\beta}, & A \leqslant x \leqslant \infty \\ 0, & \text{otherwise.} \end{cases}$$

The following graphs[a] show gamma distribution densities with $A = 0, \beta = 1$ and various selections of α.

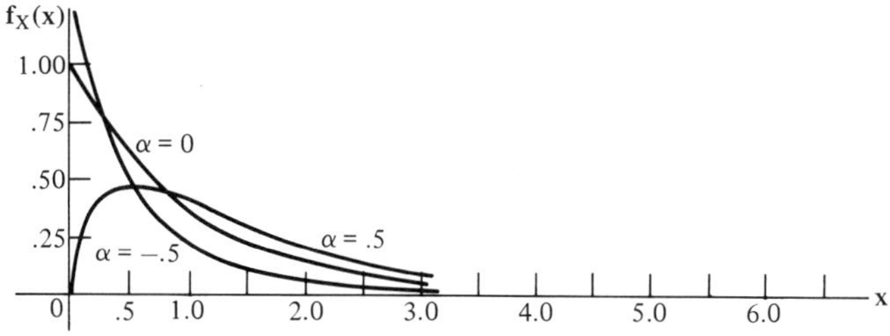

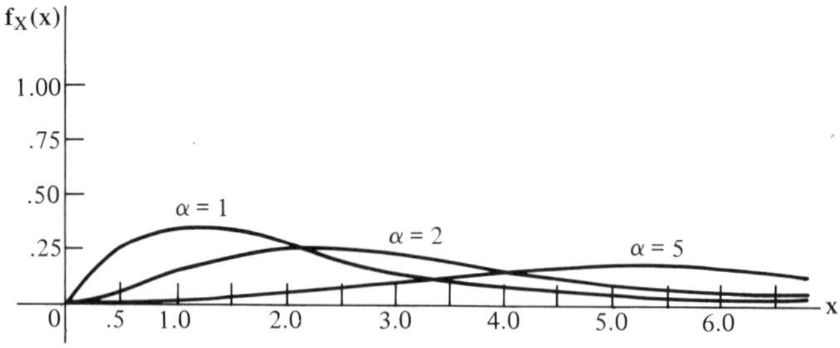

[a] Figures from *Introduction to Probability and Random Variables*. By G. P. Wadsworth and J. G. Bryan. Copyright © 1960 by McGraw-Hill Book Co., Inc. Used by permission.

TABLE 9. THE GAMMA DISTRIBUTION (continued)

The following graphs[a] show, respectively, gamma distribution densities with: $A = 0$, $\alpha = 1$ and various selections of β; and $A = 0$ and various selections of α, β with $\alpha\beta = 2$.

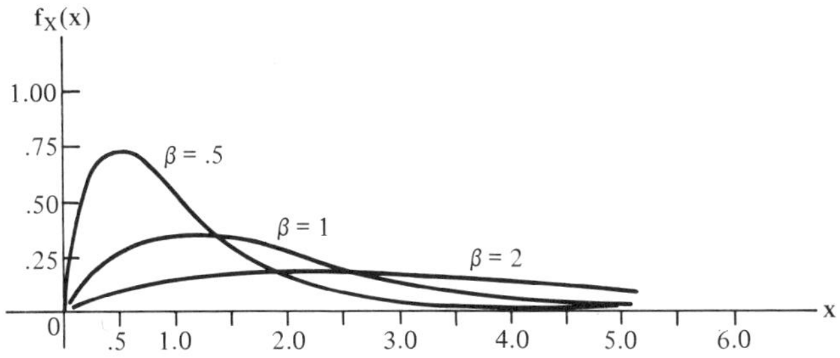

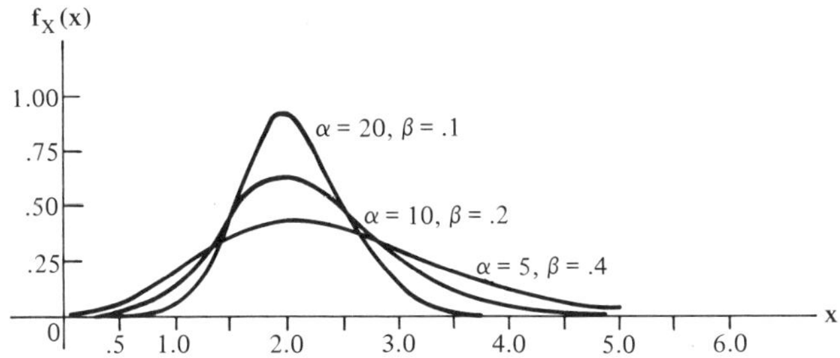

[a] Figures from *Introduction to Probability and Random Variables*. By G. P. Wadsworth and J. G. Bryan. Copyright © 1960 by McGraw-Hill Book Co., Inc. Used by permission.

Table 9. The Gamma Distribution (Continued)

The following table[a] gives values of y/β such that (for various values of α, γ)

$$P[X \leqslant y] = \int_A^y f_X(x)\,dx = \gamma$$

α \ γ	.01	.05	.10	.25	.50	.75	.90	.95	.99
−.5	.0000785	.00197	.00790	.0508	.227	.662	1.353	1.921	3.317
0	.01005	.0513	.105	.288	.693	1.386	2.303	2.996	4.605
.5	.0574	.176	.292	.606	1.183	2.054	3.126	3.907	5.672
1.0	.149	.355	.532	.961	1.678	2.693	3.890	4.744	6.638
1.5	.277	.573	.805	1.337	2.176	3.313	4.618	5.535	7.543
2.0	.436	.818	1.102	1.727	2.674	3.920	5.322	6.296	8.406
2.5	.620	1.084	1.417	2.127	3.173	4.519	6.008	7.034	9.238
3.0	.823	1.366	1.745	2.535	3.672	5.109	6.681	7.754	10.045
3.5	1.044	1.663	2.084	2.949	4.171	5.694	7.342	8.460	10.833
4.0	1.279	1.970	2.433	3.369	4.671	6.274	7.994	9.154	11.605
4.5	1.527	2.287	2.789	3.792	5.170	6.850	8.638	9.838	12.362
5.0	1.785	2.613	3.152	4.219	5.670	7.423	9.275	10.513	13.108
5.5	2.053	2.946	3.521	4.650	6.170	7.992	9.906	11.181	13.844
6.0	2.330	3.285	3.895	5.083	6.670	8.558	10.532	11.842	14.571
6.5	2.615	3.630	4.273	5.518	7.169	9.123	11.154	12.498	15.289
7.0	2.906	3.981	4.656	5.956	7.669	9.684	11.771	13.148	16.000
7.5	3.204	4.336	5.043	6.396	8.169	10.244	12.384	13.794	16.704
8.0	3.507	4.695	5.432	6.838	8.669	10.802	12.995	14.435	17.403
8.5	3.816	5.058	5.825	7.281	9.169	13.359	13.602	15.072	18.095
9.0	4.130	5.425	6.221	7.726	9.669	11.914	14.206	15.705	18.783
9.5	4.449	5.796	6.620	8.172	10.169	12.467	14.808	16.335	19.466
10.0	4.771	6.169	7.021	8.620	10.668	13.020	15.407	16.962	20.145
11.0	5.428	6.924	7.829	9.519	11.668	14.121	16.598	18.208	21.490
12.0	6.099	7.690	8.646	10.422	12.668	15.217	17.782	19.443	22.821
13.0	6.782	8.464	9.470	11.329	13.668	16.310	18.958	20.669	24.139
14.0	7.477	9.246	10.300	12.239	14.668	17.400	20.128	21.886	25.446
15.0	8.180	10.035	11.135	13.152	15.668	18.487	21.293	23.098	26.744
20.0	11.825	14.072	15.382	17.755	20.668	23.883	27.045	29.062	33.104
25.0	15.623	18.218	19.717	22.404	25.667	29.234	32.711	34.916	39.308
30.0	19.532	22.444	24.113	27.085	30.667	34.552	38.315	40.691	45.401

[a] Adapted from "Tables of percentage points of the chi-square distribution," by C. M. Thompson, *Biometrika*, Vol. 32 (1941), pp. 188–189. Used by permission of the Biometrika Trustees.

TABLE 10. THE CHI-SQUARE DISTRIBUTION

A random variable X is said to have the chi-square distribution with n degrees of freedom $\chi_n^2(0)$ if (for some integer $n > 0$; see Definition 4.2.9)

$$f_X(x) = \begin{cases} \dfrac{1}{2^{n/2}\Gamma(n/2)} x^{(n/2)-1} e^{-x/2}, & 0 \leqslant x < \infty \\ 0, & \text{otherwise.} \end{cases}$$

The following graph shows chi-square distribution densities with various selections of n.

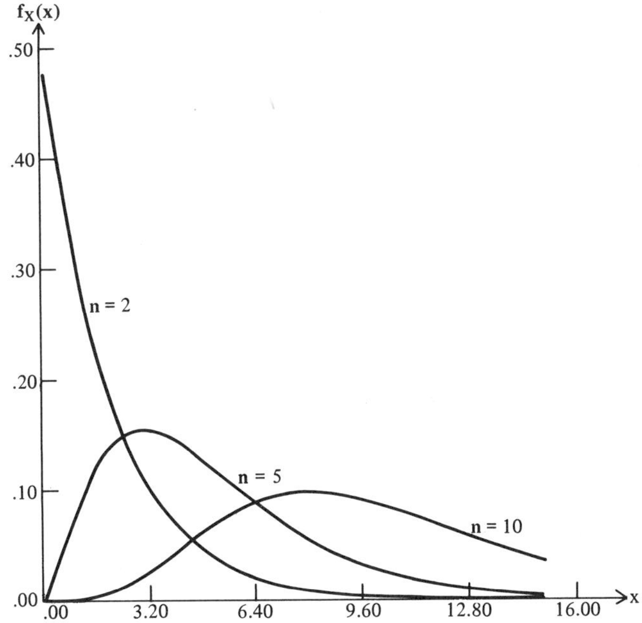

Table 10. The Chi-Square Distribution (Continued)

The following table gives values of y such that (for various values of n, γ)

$$P[X \leqslant y] = \int_0^y f_X(x)\,dx = \gamma$$

n \ γ	·005	·01	·025	·05	·10	·25	·50
1	392704.10^{-10}	157088.10^{-9}	982069.10^{-9}	393214.10^{-8}	·0157908	·1015308	·454936
2	·0100251	·0201007	·0506356	·102587	·210721	·575364	1·38629
3	·0717218	·114832	·215795	·351846	·584374	1·212534	2·36597
4	·206989	·297109	·484419	·710723	1·063623	1·92256	3·35669
5	·411742	·554298	·831212	1·145476	1·61031	2·67460	4·35146
6	·675727	·872090	1·23734	1·63538	2·20413	3·45460	5·34812
7	·989256	1·239043	1·68987	2·16735	2·83311	4·25485	6·34581
8	1·34441	1·64650	2·17973	2·73264	3·48954	5·07064	7·34412
9	1·73493	2·08790	2·70039	3·32511	4·16816	5·89883	8·34283
10	2·15586	2·55821	3·24697	3·94030	4·86518	6·73720	9·34182
11	2·60322	3·05348	3·81575	4·57481	5·57778	7·58414	10·3410
12	3·07382	3·57057	4·40379	5·22603	6·30380	8·43842	11·3403
13	3·56503	4·10692	5·00875	5·89186	7·04150	9·29907	12·3398
14	4·07467	4·66043	5·62873	6·57063	7·78953	10·1653	13·3393
15	4·60092	5·22935	6·26214	7·26094	8·54676	11·0365	14·3389
16	5·14221	5·81221	6·90766	7·96165	9·31224	11·9122	15·3385
17	5·69722	6·40776	7·56419	8·67176	10·0852	12·7919	16·3382
18	6·26480	7·01491	8·23075	9·39046	10·8649	13·6753	17·3379
19	6·84397	7·63273	8·90652	10·1170	11·6509	14·5620	18·3377
20	7·43384	8·26040	9·59078	10·8508	12·4426	15·4518	19·3374
21	8·03365	8·89720	10·28293	11·5913	13·2396	16·3444	20·3372
22	8·64272	9·54249	10·9823	12·3380	14·0415	17·2396	21·3370
23	9·26043	10·19567	11·6886	13·0905	14·8480	18·1373	22·3369
24	9·88623	10·8564	12·4012	13·8484	15·6587	19·0373	23·3367
25	10·5197	11·5240	13·1197	14·6114	16·4734	19·9393	24·3366
26	11·1602	12·1981	13·8439	15·3792	17·2919	20·8434	25·3365
27	11·8076	12·8785	14·5734	16·1514	18·1139	21·7494	26·3363
28	12·4613	13·5647	15·3079	16·9279	18·9392	22·6572	27·3362
29	13·1211	14·2565	16·0471	17·7084	19·7677	23·5666	28·3361
30	13·7867	14·9535	16·7908	18·4927	20·5992	24·4776	29·3360
40	20·7065	22·1643	24·4330	26·5093	29·0505	33·6603	39·3353
50	27·9907	29·7067	32·3574	34·7643	37·6886	42·9421	49·3349
60	35·5345	37·4849	40·4817	43·1880	46·4589	52·2938	59·3347
70	43·2752	45·4417	48·7576	51·7393	55·3289	61·6983	69·3345
80	51·1719	53·5401	57·1532	60·3915	64·2778	71·1445	79·3343
90	59·1963	61·7541	65·6466	69·1260	73·2911	80·6247	89·3342
100	67·3276	70·0649	74·2219	77·9295	82·3581	90·1332	99·3341

Table 10. The Chi-Square Distribution (Continued)

y such that $P[X \leqslant y] = \int_0^y f_X(x)\,dx = \gamma$

n \ γ	·75	·90	·95	·975	·99	·995	·999
1	1·32330	2·70554	3·84146	5·02389	6·63490	7·87944	10·828
2	2·77259	4·60517	5·99146	7·37776	9·21034	10·5966	13·816
3	4·10834	6·25139	7·81473	9·34840	11·3449	12·8382	16·266
4	5·38527	7·77944	9·48773	11·1433	13·2767	14·8603	18·467
5	6·62568	9·23636	11·0705	12·8325	15·0863	16·7496	20·515
6	7·84080	10·6446	12·5916	14·4494	16·8119	18·5476	22·458
7	9·03715	12·0170	14·0671	16·0128	18·4753	20·2777	24·322
8	10·2189	13·3616	15·5073	17·5345	20·0902	21·9550	26·125
9	11·3888	14·6837	16·9190	19·0228	21·6660	23·5894	27·877
10	12·5489	15·9872	18·3070	20·4832	23·2093	25·1882	29·588
11	13·7007	17·2750	19·6751	21·9200	24·7250	26·7568	31·264
12	14·8454	18·5493	21·0261	23·3367	26·2170	28·2995	32·909
13	15·9839	19·8119	22·3620	24·7356	27·6882	29·8195	34·528
14	17·1169	21·0641	23·6848	26·1189	29·1412	31·3194	36·123
15	18·2451	22·3071	24·9958	27·4884	30·5779	32·8013	37·697
16	19·3689	23·5418	26·2962	28·8454	31·9999	34·2672	39·252
17	20·4887	24·7690	27·5871	30·1910	33·4087	35·7185	40·790
18	21·6049	25·9894	28·8693	31·5264	34·8053	37·1565	42·312
19	22·7178	27·2036	30·1435	32·8523	36·1909	38·5823	43·820
20	23·8277	28·4120	31·4104	34·1696	37·5662	39·9968	45·315
21	24·9348	29·6151	32·6706	35·4789	38·9322	41·4011	46·797
22	26·0393	30·8133	33·9244	36·7807	40·2894	42·7957	48·268
23	27·1413	32·0069	35·1725	38·0756	41·6384	44·1813	49·728
24	28·2412	33·1962	36·4150	39·3641	42·9798	45·5585	51·179
25	29·3389	34·3816	37·6525	40·6465	44·3141	46·9279	52·618
26	30·4346	35·5632	38·8851	41·9232	45·6417	48·2899	54·052
27	31·5284	36·7412	40·1133	43·1945	46·9629	49·6449	55·476
28	32·6205	37·9159	41·3371	44·4608	48·2782	50·9934	56·892
29	33·7109	39·0875	42·5570	45·7223	49·5879	52·3356	58·301
30	34·7997	40·2560	43·7730	46·9792	50·8922	53·6720	59·703
40	45·6160	51·8051	55·7585	59·3417	63·6907	66·7660	73·402
50	56·3336	63·1671	67·5048	71·4202	76·1539	79·4900	86·661
60	66·9815	74·3970	79·0819	83·2977	88·3794	91·9517	99·607
70	77·5767	85·5270	90·5312	95·0232	100·425	104·215	112·317
80	88·1303	96·5782	101·879	106·629	112·329	116·321	124·839
90	98·6499	107·565	113·145	118·136	124·116	128·299	137·208
100	109·141	118·498	124·342	129·561	135·807	140·169	149·449

Adapted from Table 8 on pp. 136–137 of *Biometrika Tables for Statisticians, Vol. I* (Third edition reprinted with additions), edited by E. S. Pearson and H. O. Hartley, Cambridge University Press, London, 1970. Used by permission of the Biometrika Trustees.

TABLE 11. THE NORMAL DISTRIBUTION

A random variable X is said to have the normal distribution $N(\mu, \sigma^2)$ if (for some $-\infty < \mu < +\infty, \sigma^2 > 0$; see Definition 4.2.13)

$$f_X(x) = \frac{1}{\sqrt{2\pi}\sigma} e^{-\frac{1}{2}\left(\frac{x-\mu}{\sigma}\right)^2}, \quad -\infty < x < +\infty$$

The following graph shows normal distribution densities with $\mu = 0$ and various selections of σ.

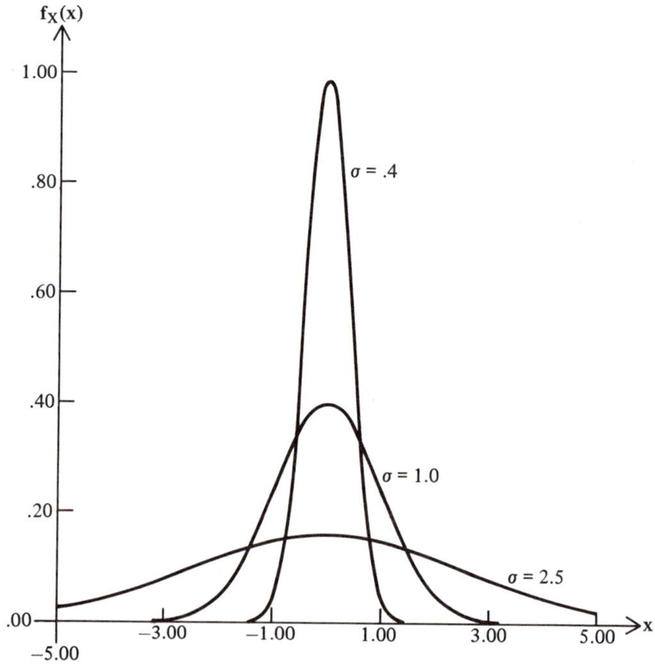

Table 11. The Normal Distribution (Continued)

The following table gives, for various values of y (when $\mu = 0$ and $\sigma^2 = 1$), the probability

$$P[X \leqslant y] = \int_{-\infty}^{y} f_X(x)\,dx$$

y	0	1	2	3	4	5	6	7	8	9
−3.0	.0013	.0010	.0007	.0005	.0003	.0002	.0002	.0001	.0001	.0000
−2.9	.0019	.0018	.0017	.0017	.0016	.0016	.0015	.0015	.0014	.0014
−2.8	.0026	.0025	.0024	.0023	.0023	.0022	.0021	.0021	.0020	.0019
−2.7	.0035	.0034	.0033	.0032	.0031	.0030	.0029	.0028	.0027	.0026
−2.6	.0047	.0045	.0044	.0043	.0041	.0040	.0039	.0038	.0037	.0036
−2.5	.0062	.0060	.0059	.0057	.0055	.0054	.0052	.0051	.0049	.0048
−2.4	.0082	.0080	.0078	.0075	.0073	.0071	.0069	.0068	.0066	.0064
−2.3	.0107	.0104	.0102	.0099	.0096	.0094	.0091	.0089	.0087	.0084
−2.2	.0139	.0136	.0132	.0129	.0126	.0122	.0119	.0116	.0113	.0110
−2.1	.0179	.0174	.0170	.0166	.0162	.0158	.0154	.0150	.0146	.0143
−2.0	.0228	.0222	.0217	.0212	.0207	.0202	.0197	.0192	.0188	.0183
−1.9	.0287	.0281	.0274	.0268	.0262	.0256	.0250	.0244	.0238	.0233
−1.8	.0359	.0352	.0344	.0336	.0329	.0322	.0314	.0307	.0300	.0294
−1.7	.0446	.0436	.0427	.0418	.0409	.0401	.0392	.0384	.0375	.0367
−1.6	.0548	.0537	.0526	.0516	.0505	.0495	.0485	.0475	.0465	.0455
−1.5	.0668	.0655	.0643	.0630	.0618	.0606	.0594	.0582	.0570	.0559
−1.4	.0808	.0793	.0778	.0764	.0749	.0735	.0722	.0708	.0694	.0681
−1.3	.0968	.0951	.0934	.0918	.0901	.0885	.0869	.0853	.0838	.0823
−1.2	.1151	.1131	.1112	.1093	.1075	.1056	.1038	.1020	.1003	.0985
−1.1	.1357	.1335	.1314	.1292	.1271	.1251	.1230	.1210	.1190	.1170
−1.0	.1587	.1562	.1539	.1515	.1492	.1469	.1446	.1423	.1401	.1379
− .9	.1841	.1814	.1788	.1762	.1736	.1711	.1685	.1660	.1635	.1611
− .8	.2119	.2090	.2061	.2033	.2005	.1977	.1949	.1922	.1894	.1867
− .7	.2420	.2389	.2358	.2327	.2297	.2266	.2236	.2206	.2177	.2148
− .6	.2743	.2709	.2676	.2643	.2611	.2578	.2546	.2514	.2483	.2451
− .5	.3085	.3050	.3015	.2981	.2946	.2912	.2877	.2843	.2810	.2776
− .4	.3446	.3409	.3372	.3336	.3300	.3264	.3228	.3192	.3156	.3121
− .3	.3821	.3783	.3745	.3707	.3669	.3632	.3594	.3557	.3520	.3483
− .2	.4207	.4168	.4129	.4090	.4052	.4013	.3974	.3936	.3897	.3859
− .1	.4602	.4562	.4522	.4483	.4443	.4404	.4364	.4325	.4286	.4247
− 0	.5000	.4960	.4920	.4880	.4840	.4801	.4761	.4721	.4681	.4641

Table 11. The Normal Distribution (Continued)

$P[X \leqslant y] = \int_{-\infty}^{y} f_X(x)\, dx$ for various y

y	0	1	2	3	4	5	6	7	8	9
.0	.5000	.5040	.5080	.5120	.5160	.5199	.5239	.5279	.5319	.5359
.1	.5398	.5438	.5478	.5517	.5557	.5596	.5636	.5675	.5714	.5753
.2	.5793	.5832	.5871	.5910	.5948	.5987	.6026	.6064	.6103	.6141
.3	.6179	.6217	.6255	.6293	.6331	.6368	.6406	.6443	.6480	.6517
.4	.6554	.6591	.6628	.6664	.6700	.6736	.6772	.6808	.6844	.6879
.5	.6915	.6950	.6985	.7019	.7054	.7088	.7123	.7157	.7190	.7224
.6	.7257	.7291	.7324	.7357	.7389	.7422	.7454	.7486	.7517	.7549
.7	.7580	.7611	.7642	.7673	.7703	.7734	.7764	.7794	.7823	.7852
.8	.7881	.7910	.7939	.7967	.7995	.8023	.8051	.8078	.8106	.8133
.9	.8159	.8186	.8212	.8238	.8264	.8289	.8315	.8340	.8365	.8389
1.0	.8413	.8438	.8461	.8485	.8508	.8531	.8554	.8577	.8599	.8621
1.1	.8643	.8665	.8686	.8708	.8729	.8749	.8770	.8790	.8810	.8830
1.2	.8849	.8869	.8888	.8907	.8925	.8944	.8962	.8980	.8997	.9015
1.3	.9032	.9049	.9066	.9082	.9099	.9115	.9131	.9147	.9162	.9177
1.4	.9192	.9207	.9222	.9236	.9251	.9265	.9278	.9292	.9306	.9319
1.5	.9332	.9345	.9357	.9370	.9382	.9394	.9406	.9418	.9430	.9441
1.6	.9452	.9463	.9474	.9484	.9495	.9505	.9515	.9525	.9535	.9545
1.7	.9554	.9564	.9573	.9582	.9591	.9599	.9608	.9616	.9625	.9633
1.8	.9641	.9648	.9656	.9664	.9671	.9678	.9686	.9693	.9700	.9706
1.9	.9713	.9719	.9726	.9732	.9738	.9744	.9750	.9756	.9762	.9767
2.0	.9772	.9778	.9783	.9788	.9793	.9798	.9803	.9808	.9812	.9817
2.1	.9821	.9826	.9830	.9834	.9838	.9842	.9846	.9850	.9854	.9857
2.2	.9861	.9864	.9868	.9871	.9874	.9878	.9881	.9884	.9887	.9890
2.3	.9893	.9896	.9898	.9901	.9904	.9906	.9909	.9911	.9913	.9916
2.4	.9918	.9920	.9922	.9925	.9927	.9929	.9931	.9932	.9934	.9936
2.5	.9938	.9940	.9941	.9943	.9945	.9946	.9948	.9949	.9951	.9952
2.6	.9953	.9955	.9956	.9957	.9959	.9960	.9961	.9962	.9963	.9964
2.7	.9965	.9966	.9967	.9968	.9969	.9970	.9971	.9972	.9973	.9974
2.8	.9974	.9975	.9976	.9977	.9977	.9978	.9979	.9979	.9980	.9981
2.9	.9981	.9982	.9982	.9983	.9984	.9984	.9985	.9985	.9986	.9986
3.	.9987	.9990	.9993	.9995	.9997	.9998	.9998	.9999	.9999	1.0000

TABLE 12. THE *F*-DISTRIBUTION

A random variable X is said to have the F-distribution $F(n_1, n_2)$ if (for some integers $n_1, n_2 \geqslant 1$; see Definition 4.2.14)

$$f_X(x) = \begin{cases} \dfrac{\Gamma\left(\dfrac{n_1+n_2}{2}\right)\left(\dfrac{n_1}{n_2}\right)^{.5n_1}}{\Gamma\left(\dfrac{n_1}{2}\right)\Gamma\left(\dfrac{n_2}{2}\right)} \dfrac{x^{.5(n_1-2)}}{\left(1+\dfrac{n_1}{n_2}x\right)^{.5(n_1+n_2)}}, & 0 < x < \infty \\ 0, & \text{otherwise.} \end{cases}$$

The following graphs show F-distribution densities with various selections of n_1 and n_2.

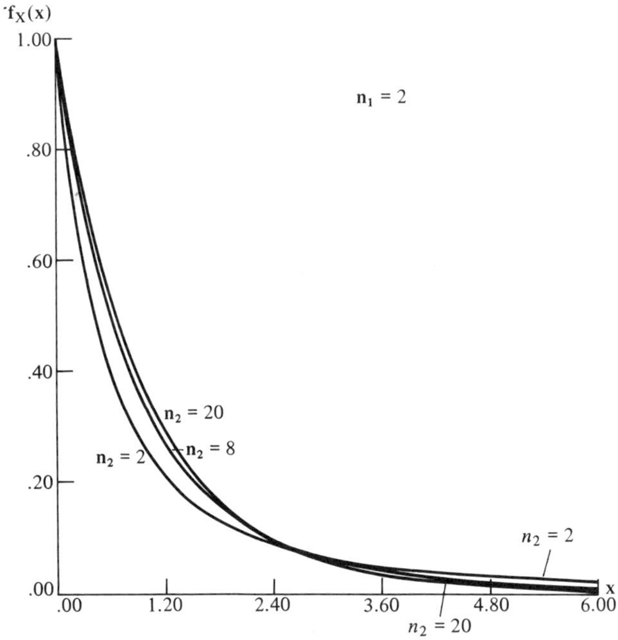

TABLE 12. THE *F*-DISTRIBUTION (continued)

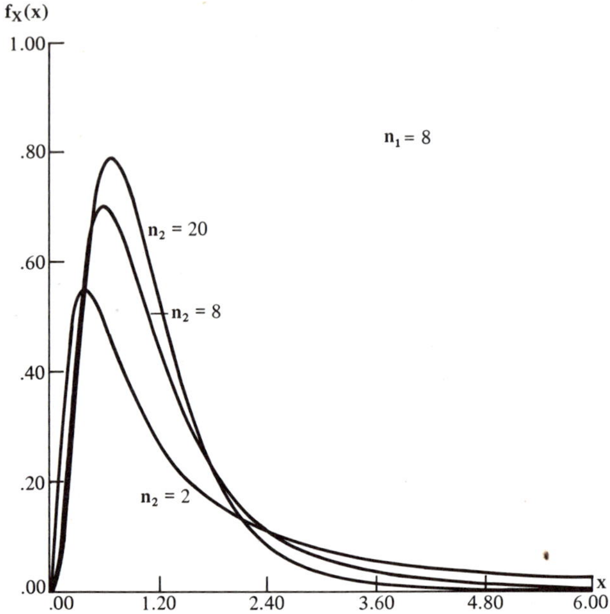

TABLE 12. THE *F*-DISTRIBUTION (continued)

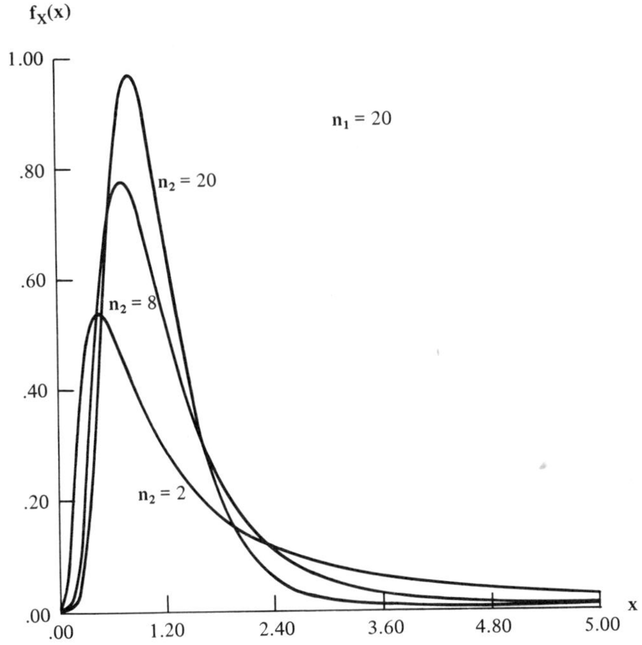

Table 12. The *F*-Distribution (Continued)

The following table gives values of y such that (for various values of n_1, n_2)

$$P[X \leqslant y] = \int_0^y f_X(x)\, dx = .95$$

n_2 \ n_1	1	2	3	4	5	6	7	8	9	10	12	15	20	24	30	40	60	120	∞
1	161·4	199·5	215·7	224·6	230·2	234·0	236·8	238·9	240·5	241·9	243·9	245·9	248·0	249·1	250·1	251·1	252·2	253·3	254·3
2	18·51	19·00	19·16	19·25	19·30	19·33	19·35	19·37	19·38	19·40	19·41	19·43	19·45	19·45	19·46	19·47	19·48	19·49	19·50
3	10·13	9·55	9·28	9·12	9·01	8·94	8·89	8·85	8·81	8·79	8·74	8·70	8·66	8·64	8·62	8·59	8·57	8·55	8·53
4	7·71	6·94	6·59	6·39	6·26	6·16	6·09	6·04	6·00	5·96	5·91	5·86	5·80	5·77	5·75	5·72	5·69	5·66	5·63
5	6·61	5·79	5·41	5·19	5·05	4·95	4·88	4·82	4·77	4·74	4·68	4·62	4·56	4·53	4·50	4·46	4·43	4·40	4·36
6	5·99	5·14	4·76	4·53	4·39	4·28	4·21	4·15	4·10	4·06	4·00	3·94	3·87	3·84	3·81	3·77	3·74	3·70	3·67
7	5·59	4·74	4·35	4·12	3·97	3·87	3·79	3·73	3·68	3·64	3·57	3·51	3·44	3·41	3·38	3·34	3·30	3·27	3·23
8	5·32	4·46	4·07	3·84	3·69	3·58	3·50	3·44	3·39	3·35	3·28	3·22	3·15	3·12	3·08	3·04	3·01	2·97	2·93
9	5·12	4·26	3·86	3·63	3·48	3·37	3·29	3·23	3·18	3·14	3·07	3·01	2·94	2·90	2·86	2·83	2·79	2·75	2·71
10	4·96	4·10	3·71	3·48	3·33	3·22	3·14	3·07	3·02	2·98	2·91	2·85	2·77	2·74	2·70	2·66	2·62	2·58	2·54
11	4·84	3·98	3·59	3·36	3·20	3·09	3·01	2·95	2·90	2·85	2·79	2·72	2·65	2·61	2·57	2·53	2·49	2·45	2·40
12	4·75	3·89	3·49	3·26	3·11	3·00	2·91	2·85	2·80	2·75	2·69	2·62	2·54	2·51	2·47	2·43	2·38	2·34	2·30
13	4·67	3·81	3·41	3·18	3·03	2·92	2·83	2·77	2·71	2·67	2·60	2·53	2·46	2·42	2·38	2·34	2·30	2·25	2·21
14	4·60	3·74	3·34	3·11	2·96	2·85	2·76	2·70	2·65	2·60	2·53	2·46	2·39	2·35	2·31	2·27	2·22	2·18	2·13
15	4·54	3·68	3·29	3·06	2·90	2·79	2·71	2·64	2·59	2·54	2·48	2·40	2·33	2·29	2·25	2·20	2·16	2·11	2·07
16	4·49	3·63	3·24	3·01	2·85	2·74	2·66	2·59	2·54	2·49	2·42	2·35	2·28	2·24	2·19	2·15	2·11	2·06	2·01
17	4·45	3·59	3·20	2·96	2·81	2·70	2·61	2·55	2·49	2·45	2·38	2·31	2·23	2·19	2·15	2·10	2·06	2·01	1·96
18	4·41	3·55	3·16	2·93	2·77	2·66	2·58	2·51	2·46	2·41	2·34	2·27	2·19	2·15	2·11	2·06	2·02	1·97	1·92
19	4·38	3·52	3·13	2·90	2·74	2·63	2·54	2·48	2·42	2·38	2·31	2·23	2·16	2·11	2·07	2·03	1·98	1·93	1·88
20	4·35	3·49	3·10	2·87	2·71	2·60	2·51	2·45	2·39	2·35	2·28	2·20	2·12	2·08	2·04	1·99	1·95	1·90	1·84
21	4·32	3·47	3·07	2·84	2·68	2·57	2·49	2·42	2·37	2·32	2·25	2·18	2·10	2·05	2·01	1·96	1·92	1·87	1·81
22	4·30	3·44	3·05	2·82	2·66	2·55	2·46	2·40	2·34	2·30	2·23	2·15	2·07	2·03	1·98	1·94	1·89	1·84	1·78
23	4·28	3·42	3·03	2·80	2·64	2·53	2·44	2·37	2·32	2·27	2·20	2·13	2·05	2·01	1·96	1·91	1·86	1·81	1·76
24	4·26	3·40	3·01	2·78	2·62	2·51	2·42	2·36	2·30	2·25	2·18	2·11	2·03	1·98	1·94	1·89	1·84	1·79	1·73
25	4·24	3·39	2·99	2·76	2·60	2·49	2·40	2·34	2·28	2·24	2·16	2·09	2·01	1·96	1·92	1·87	1·82	1·77	1·71
26	4·23	3·37	2·98	2·74	2·59	2·47	2·39	2·32	2·27	2·22	2·15	2·07	1·99	1·95	1·90	1·85	1·80	1·75	1·69
27	4·21	3·35	2·96	2·73	2·57	2·46	2·37	2·31	2·25	2·20	2·13	2·06	1·97	1·93	1·88	1·84	1·79	1·73	1·67
28	4·20	3·34	2·95	2·71	2·56	2·45	2·36	2·29	2·24	2·19	2·12	2·04	1·96	1·91	1·87	1·82	1·77	1·71	1·65
29	4·18	3·33	2·93	2·70	2·55	2·43	2·35	2·28	2·22	2·18	2·10	2·03	1·94	1·90	1·85	1·81	1·75	1·70	1·64
30	4·17	3·32	2·92	2·69	2·53	2·42	2·33	2·27	2·21	2·16	2·09	2·01	1·93	1·89	1·84	1·79	1·74	1·68	1·62
40	4·08	3·23	2·84	2·61	2·45	2·34	2·25	2·18	2·12	2·08	2·00	1·92	1·84	1·79	1·74	1·69	1·64	1·58	1·51
60	4·00	3·15	2·76	2·53	2·37	2·25	2·17	2·10	2·04	1·99	1·92	1·84	1·75	1·70	1·65	1·59	1·53	1·47	1·39
120	3·92	3·07	2·68	2·45	2·29	2·17	2·09	2·02	1·96	1·91	1·83	1·75	1·66	1·61	1·55	1·50	1·43	1·35	1·25
∞	3·84	3·00	2·60	2·37	2·21	2·10	2·01	1·94	1·88	1·83	1·75	1·67	1·57	1·52	1·46	1·39	1·32	1·22	1·00

Adapted from Table 18 on p. 171 of *Biometrika Tables for Statisticians*, Vol. I (Third edition

TABLE 13. THE *t*-DISTRIBUTION

A random variable X is said to have the t-distribution with n degrees of freedom if (for some integer $n \geqslant 1$; see Definition 4.2.17)

$$f_X(x) = \frac{\Gamma\left(\frac{n+1}{2}\right)}{\sqrt{n\pi}\,\Gamma\left(\frac{n}{2}\right)} \frac{1}{\left(1 + \frac{x^2}{n}\right)^{(n+1)/2}}, \quad -\infty < x < +\infty.$$

The following graph shows t-distribution densities with several selections of n, and the $N(0, 1)$ density (labeled $n = \infty$).

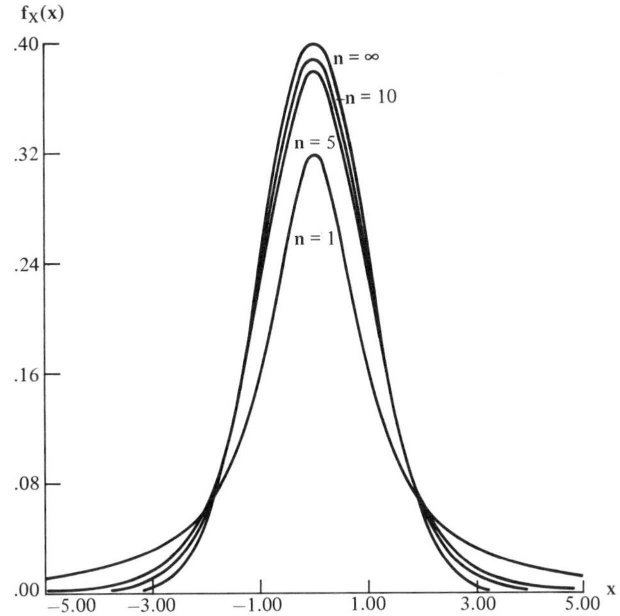

Table 13. The t-Distribution (Continued)

The following table gives values of y such that (for various values of n, γ)

$$P[X \leqslant y] = \int_{-\infty}^{y} f_X(x)\,dx = \gamma$$

n \ γ	·60	·75	·90	·95	·975	·99	·995	·9975	·999	·9995
1	·325	1·000	3·078	6·314	12·706	31·821	63·657	127·32	318·31	636·62
2	·289	·816	1·886	2·920	4·303	6·965	9·925	14·089	22·327	31·598
3	·277	·765	1·638	2·353	3·182	4·541	5·841	7·453	10·214	12·924
4	·271	·741	1·533	2·132	2·776	3·747	4·604	5·598	7·173	8·610
5	·267	·727	1·476	2·015	2·571	3·365	4·032	4·773	5·893	6·869
6	·265	·718	1·440	1·943	2·447	3·143	3·707	4·317	5·208	5·959
7	·263	·711	1·415	1·895	2·365	2·998	3·499	4·029	4·785	5·408
8	·262	·706	1·397	1·860	2·306	2·896	3·355	3·833	4·501	5·041
9	·261	·703	1·383	1·833	2·262	2·821	3·250	3·690	4·297	4·781
10	·260	·700	1·372	1·812	2·228	2·764	3·169	3·581	4·144	4·587
11	·260	·697	1·363	1·796	2·201	2·718	3·106	3·497	4·025	4·437
12	·259	·695	1·356	1·782	2·179	2·681	3·055	3·428	3·930	4·318
13	·259	·694	1·350	1·771	2·160	2·650	3·012	3·372	3·852	4·221
14	·258	·692	1·345	1·761	2·145	2·624	2·977	3·326	3·787	4·140
15	·258	·691	1·341	1·753	2·131	2·602	2·947	3·286	3·733	4·073
16	·258	·690	1·337	1·746	2·120	2·583	2·921	3·252	3·686	4·015
17	·257	·689	1·333	1·740	2·110	2·567	2·898	3·222	3·646	3·965
18	·257	·688	1·330	1·734	2·101	2·552	2·878	3·197	3·610	3·922
19	·257	·688	1·328	1·729	2·093	2·539	2·861	3·174	3·579	3·883
20	·257	·687	1·325	1·725	2·086	2·528	2·845	3·153	3·552	3·850
21	·257	·686	1·323	1·721	2·080	2·518	2·831	3·135	3·527	3·819
22	·256	·686	1·321	1·717	2·074	2·508	2·819	3·119	3·505	3·792
23	·256	·685	1·319	1·714	2·069	2·500	2·807	3·104	3·485	3·767
24	·256	·685	1·318	1·711	2·064	2·492	2·797	3·091	3·467	3·745
25	·256	·684	1·316	1·708	2·060	2·485	2·787	3·078	3·450	3·725
26	·256	·684	1·315	1·706	2·056	2·479	2·779	3·067	3·435	3·707
27	·256	·684	1·314	1·703	2·052	2·473	2·771	3·057	3·421	3·690
28	·256	·683	1·313	1·701	2·048	2·467	2·763	3·047	3·408	3·674
29	·256	·683	1·311	1·699	2·045	2·462	2·756	3·038	3·396	3·659
30	·256	·683	1·310	1·697	2·042	2·457	2·750	3·030	3·385	3·646
40	·255	·681	1·303	1·684	2·021	2·423	2·704	2·971	3·307	3·551
60	·254	·679	1·296	1·671	2·000	2·390	2·660	2·915	3·232	3·460
120	·254	·677	1·289	1·658	1·980	2·358	2·617	2·860	3·160	3·373
∞	·253	·674	1·282	1·645	1·960	2·326	2·576	2·807	3·090	3·291

Adapted from Table 12 on p. 146 of *Biometrika Tables for Statisticians, Vol. I* (Third edition reprinted with additions) edited by E. S. Pearson and H. O. Hartley, Cambridge University Press, London, 1970. Used by permission of the Biometrika Trustees.

SELECTED REFERENCES

The following books comprise those the author would suggest as supplementary reading for the student. Those which are introductory are starred (*); the others are either more advanced or cover specialized topics (or both). In the text these books are referred to by citing author(s) and year of publication; for example, Feller (1968), or Mood and Graybill (1963). For references to other books or to journal articles a complete citation is given in the text itself; for example, see p. 1 where a book by S. Feferman is mentioned.

CRAMÉR, H., *Mathematical Methods of Statistics,* Princeton University Press, Princeton, New Jersey, 1946.

* FELLER, W., *An Introduction to Probability Theory and Its Applications, Volume I* (Third Edition), John Wiley & Sons, Inc., New York, 1968.

FELLER, W., *An Introduction to Probability Theory and Its Applications, Volume II,* John Wiley & Sons, Inc., New York, 1966.

FISZ, M., *Probability Theory and Mathematical Statistics* (Third Edition), John Wiley & Sons, Inc., New York, 1963.

* FRASER, D. A. S., *Statistics: An Introduction,* John Wiley & Sons, Inc., New York, 1958.

* FREEMAN, H., *Introduction to Statistical Inference,* Addison-Wesley Publishing Co., Inc., Reading, Massachusetts, 1963.

HOGG, R. V. and CRAIG, A. T., *Introduction to Mathematical Statistics* (Second Edition), The Macmillan Co., New York, 1965.

* LARSON, H. J., *Introduction to Probability Theory and Statistical Inference,* John Wiley & Sons, Inc., New York, 1969.

* LINDGREN, B. W., *Statistical Theory* (Second Edition), The Macmillan Co., New York, 1968.

LOÈVE, M., *Probability Theory* (Third Edition), D. Van Nostrand Co., Inc., Princeton, New Jersey, 1963.

* MOOD, A. M. and GRAYBILL, F. A., *Introduction to the Theory of Statistics* (Second Edition), McGraw-Hill Book Co., Inc., New York, 1963.

* MOSTELLER, F., ROURKE, R. E. K., and THOMAS, G. B., JR., *Probability with Statistical Applications,* Addison-Wesley Publishing Co., Inc., Reading, Massachusetts, 1961.

MUNROE, M. E., *Theory of Probability,* McGraw-Hill Book Co., Inc., New York, 1951.

* PARZEN, E., *Modern Probability Theory and Its Applications,* John Wiley & Sons, Inc., New York, 1960.

RAO, C. R., *Linear Statistical Inference and Its Applications,* John Wiley & Sons, Inc., New York, 1965.

TUCKER, H. G., *A Graduate Course in Probability,* Academic Press Inc., New York, 1967.

VAN DER WAERDEN, B. L., *Mathematical Statistics,* Springer-Verlag, New York, 1969.

WADSWORTH, G. P. and BRYAN, J. G., *Introduction to Probability and Random Variables,* McGraw-Hill Book Co., Inc., New York, 1960.

* WALLIS, W. A. and ROBERTS, H. V., *The Nature of Statistics* (Paperback), The Free Press, New York, 1965.

WEISS, L., *Statistical Decision Theory,* McGraw-Hill Book Co., Inc., New York, 1961.

* WETHERILL, G. B., *Elementary Statistical Methods,* Methuen & Co., Ltd., London, 1967.

ANSWERS TO SELECTED PROBLEMS

In order to aid students working problems not assigned in their classes (as well as persons studying the material on their own) to check the correctness of their work, below we give fairly detailed answers (but not solutions) to those Problems (but not Honors Problems) whose problem numbers are divisible by five.

CHAPTER 1

1.5 (Poretsky's Law) First, to show that if $X = \emptyset$ then $T = (X \cap T^c) \cup (X^c \cap T)$, note that $X \cap T^c = \emptyset \cap T^c = \emptyset$ and $X^c \cap T = \emptyset^c \cap T = \Omega \cap T = T$, hence $(X \cap T^c) \cup (X^c \cap T) = \emptyset \cup T = T$. Next, to show that if $T = (X \cap T^c) \cup (X^c \cap T)$ then $X = \emptyset$, assume that $T = (X \cap T^c) \cup (X^c \cap T)$ and show that (for each $\omega \in \Omega$) $\omega \notin X$ because $\omega \in X \Rightarrow \omega \in T$, which yields a contradiction ($\omega \in T$ and $\omega \notin (X \cap T^c) \cup (X^c \cap T)$).

CHAPTER 2

2.5 If the vehicle from the U.S.A. has probability p_1 of succeeding ($0 \leqslant p_1 \leqslant 1$), the vehicle from the U.S.S.R. has probability p_2 of succeeding ($0 \leqslant p_2 \leqslant 1$), and the success or failure of one vehicle does not affect the other's chances, then the desired probability is $p_1(1-p_2)+p_2(1-p_1)$.

2.10 **(a)** Use mathematical induction on r.

(b) Use mathematical induction on k.

2.15 $P(B|A \cup \bar{B}) = 0.25$.

2.20 5/6.

2.25 **(a)** $\left(\frac{k}{N}\right)^n$.

(b) $\frac{1}{N+1} \frac{1^n + 2^n + \ldots + N^n}{N^n}$.

(c) $\dfrac{1}{N+1}\dfrac{1^{n+1}+2^{n+1}+\ldots+N^{n+1}}{N^{n+1}}.$

(d) $\dfrac{1}{N}\dfrac{1^{n+1}+2^{n+1}+\ldots+N^{n+1}}{1^{n}+2^{n}+\ldots+N^{n}}.$

(e) $\dfrac{n+1}{n+2}.$

2.30 $R(p) = 2p^2 - p^4.$

CHAPTER 3

3.5 If X is the number of questions directed to the first student, then

$$p_X(x) = \begin{cases} \binom{10}{x}\left(\frac{1}{4}\right)^x\left(\frac{3}{4}\right)^{10-x}, & x = 0,1,\ldots,10 \\ 0, & \text{otherwise} \end{cases}$$

and

$$F_X(x) = \begin{cases} 1, & x > 10 \\ \sum_{t=0}^{[x]} p_X(t), & 0 \leq x \leq 10 \\ 0, & x > 0, \end{cases}$$

where $[x]$ denotes the largest integer $\leq x$ (so, for example, $[5.3] = 5$ and $[6] = 6$).

3.10 This involves plotting $P[X = 1]$ as a function of p on coordinate axes, using the following values:

p	0	.1	.2	.3	.4	
$P[X=1]$	0	.0028	.0334	.1260	.2897	

p	.5	.6	.7	.8	.9	1
$P[X=1]$	.5	.7103	.8740	.9666	.9972	1

3.15 First find $P[X + Y = z]$ (thus showing that $X + Y$ is Poisson with parameter $\lambda_1 + \lambda_2$) and then calculate $P[X = x \mid X + Y = z]$ (using the definition of conditional probability) and recognize the result as a binomial distribution with z trials and probability $\lambda_1/(\lambda_1 + \lambda_2)$ of success on each trial.

3.20 $F_X\left(\dfrac{z-b}{a}\right).$

3.25 (a) 2^n.
(b) $n + 1$.

3.30 For the geometric distribution the function to be plotted is $q(x) = p$ for $x = 1,2,3, \ldots$ ($q(x) = 0$ otherwise), while for the uniform distribution the function to be plotted is $q(x) = \dfrac{1}{n-i+1}$ if x is the ith smallest of the possible values $x_1, \ldots, x_n$ ($q(x) = 0$ otherwise).

CHAPTER 4

4.5 **(a)** $1 - (.01)^{1/5}$ thousand gallons.
(b) 1000 gallons.

4.10 The d.f. of Y is

$$F_Y(y) = \begin{cases} F_X(y), & y > 0 \\ F_X(0), & y = 0 \\ 0, & y < 0. \end{cases}$$

4.15 The density of X is

$$f_X(x) = \begin{cases} \dfrac{1}{\pi} \dfrac{1}{\sqrt{1-x^2}}, & |x| < 1 \\ 0, & \text{otherwise.} \end{cases}$$

4.20 We find

$$P[Z_n \leqslant x] = \begin{cases} 0, & x \leqslant 0 \\ x^n, & 0 < x < 1 \\ 1, & x \geqslant 1, \end{cases}$$

and $n \geqslant 44$ is needed in order that $P[Z_n > .9] \geqslant .99$.

4.25 We find

$$f_V(v) = \begin{cases} \dfrac{\Gamma\left(\dfrac{m+n}{2}\right)}{\Gamma\left(\dfrac{m}{2}\right)\Gamma\left(\dfrac{n}{2}\right)} \dfrac{v^{\frac{m}{2}-1}}{(v+1)^{(m+n)/2}}, & v > 0 \\ 0, & v \leqslant 0 \end{cases}$$

and

$$f_U(u) = \begin{cases} \dfrac{\Gamma\left(\dfrac{m+n}{2}\right)}{\Gamma\left(\dfrac{m}{2}\right)\Gamma\left(\dfrac{n}{2}\right)} u^{\frac{m}{2}-1}(1-u)^{\frac{n}{2}-1}, & 0 < u < 1 \\ 0, & \text{otherwise.} \end{cases}$$

CHAPTER 5

5.5 The mode may be found by examination of the derivative of $f_X(x)$. When $n=1$, the mode is zero and $EX=1$.

5.10 (a) $\phi_X(t) = Ee^{itX} \quad (-\infty < t < +\infty)$.

(b) $\phi_{X_n}(t) = pe^{it} + q$.

(c) $\phi_{S_n}(t) = (pe^{it} + q)^n$.

(d) From $\phi_{S_n}(t)$ we conclude via the Uniqueness Theorem (Theorem 5.4.14) that S_n is binomial, and then $P[S_n = j]$ follows.

5.15 Use direct calculation.

5.20 $E(Y) = n/4$ and $\text{Var}(Y) = (5n-2)/16$.

CHAPTER 6

6.5 Using the CLT, $n \geqslant (128.8)^2$.

6.10 (a) Via Chebyshev's Inequality,

$$P\left[\left|\frac{W_1 + \ldots + W_{100}}{100} + \frac{13}{36}\right| > \frac{1}{2}\sqrt{\frac{1235}{1296}}\right] < \frac{1}{25}.$$

(b) (1) Show by Chebyshev's Inequality that

$$P\left[\left|\frac{W_1 + \ldots + W_n}{n} + \frac{13}{36}\right| > \delta\right] \to 0 \text{ as } n \to \infty.$$

(2) Show that

$$\phi_{\frac{W_1 + \ldots + W_n}{n}}(t) \to e^{-\frac{13}{36}it} \text{ as } n \to \infty.$$

(3) The conditions of the WLLN are satisfied.

(4) The conditions of Khinchin's Theorem are satisfied.

CHAPTER 7

7.5 (a) $(X_1 + \ldots + X_n)/n$, which is easily shown to have expected value θ.

(b) $(X_1 + \ldots + X_n)/n$, which is easily shown (e.g., use Khinchin's Theorem) to converge in probability to θ.

CHAPTER 8

8.5 (a) "The" MLE is $\hat{\lambda} = (X_1 + \ldots + X_n)/n$ if $X_1 + \ldots + X_n \neq 0$. If $X_1 + \ldots + X_n = 0$, the supremum of $L(\lambda)$ is the value at $\lambda = 0$, but $0 \notin \Theta$. If we redefine Θ as $\Theta = \{x{:}x \geqslant 0\}$ the MLE will always exist.

(b) λ^* is an ε-MLE iff

$$e^{-n\lambda^*}(\lambda^*)^{X_1+\ldots+X_n} \geqslant (1-\varepsilon)e^{-n\hat{\lambda}}(\hat{\lambda})^{X_1+\ldots+X_n}.$$

If $X_1 + \ldots + X_n = 0$, this means

$$\lambda^* \leqslant \frac{-\ln(1-\varepsilon)}{n}.$$

8.10 This is solved as noted in the hint.

8.15 This can be solved as noted in the hint.

8.20 Let φ_n be any CAN estimator, so that $\sqrt{n}\,(\varphi_n - \theta)$ is asymptotically $N(0,K^2(\theta))$. Then (see Lemma 8.4.12 and the discussion following it) for large n we can act as if t_n is $N(\theta,J^2(\theta)/n^{2\delta})$ and φ_n is $N(\theta,K^2(\theta)/n)$. Hence for large n t_n will have a distribution more concentrated about θ (than is the distribution of φ_n) iff $n^{2\delta} > n$, that is iff $\delta > \frac{1}{2}$. (t_n will be less concentrated if $\delta < \frac{1}{2}$, and t_n will be a CAN estimator if $\delta = \frac{1}{2}$.)

8.25 (a) $\sqrt{n}(t_n - \theta)$ is not asymptotically normal when $\theta = 0$. The result stated on $e_\theta(\hat{\theta}_n, t_n)$ is formally true (although strictly speaking $e_\theta(\hat{\theta}_n, t_n)$ is not defined since $\hat{\theta}_n$ is not a CAN estimator).

(b) Let

$$t_n^* = \begin{cases} \alpha\hat{\theta}_n, & \text{if } |\hat{\theta}_n| < n^{-1/4} \\ 1 + \alpha|\hat{\theta}_n - 1|, & \text{if } |\hat{\theta}_n - 1| < n^{-1/4} \text{ and } |\hat{\theta}_n| > n^{-1/4} \\ \hat{\theta}_n, & \text{otherwise.} \end{cases}$$

Then calculations similar to those in Example 8.4.17 show that $\sqrt{n}(t_n^* - \theta)$ is asymptotically $N(0,J_*^2(\theta))$ where

$$J_*^2(\theta) = \begin{cases} \alpha^2 J^2(\theta), & \text{if } \theta = 0 \text{ or } \theta = 1 \\ J^2(\theta), & \text{otherwise.} \end{cases}$$

(c) If θ lies outside Θ, we have no assurance of the properties of the estimator $\hat{\theta}$ which is a CAN estimator of θ *for* θ*'s in* Θ.

8.30 The limits of a_n and of b_n follow upon manipulating to place each $o(1/\sqrt{n})$ over the quantity $1/\sqrt{n}$ and using standard properties of limits. (Note that $\lim_{n\to\infty}\{o(1/\sqrt{n})/(1/\sqrt{n})\} = \lim_{n\to\infty} \sqrt{n}\, o(1/\sqrt{n}) = 0$. Use a continuity argument to show that

$$\lim_{n\to\infty} P[Y_n \leqslant a_n t + b_n] = \lim_{n\to\infty} P[Y_n \leqslant t].$$

If $F(t)$ is not continuous, the result can fail. For a counterexample, let Y_1 be a r.v. such that $P[Y_1 = 0] = 1$, and let $Y_1 = Y_2 = Y_3 = \ldots$, let $t=0$, let $1 = a_1 = a_2 = \ldots$, and let $b_n = -1/n$.

8.35 The specified φ_n does not satisfy Assumption 8.6.2 because the transformation is not one-to-one.

8.40 Note that $P_\theta[X = x_0 | t(X) = c] = 0$ if $c \neq c_0$, hence $P_\theta[X \in \mathscr{U} | t(X) = c] = 0$ if $c \neq c_0$ and is then independent of θ. The case $P[t(X) = c_0] = 0$ is not of interest since the conditional probability of interest is then undefined (hence nothing need be shown for it).

8.45 These considerations are essentially the same as those in Problem 8.40.

8.50 **(a)** Use the Neyman Factorization Theorem.

(b) For example, take

$$\hat{\theta}(X_1, \ldots, X_n) = \max(X_1, \ldots, X_n) - \frac{1}{2}$$
$$+ (\cos^2 X_1)\,\{\min(X_1, \ldots, X_n) - \max(X_1, \ldots, X_n) + 1\}.$$

8.55 X itself is sufficient, but is also trivial since X is the only data available. To complete the proof, one shows that any real reduction (e.g. $Y = g(X)$) cannot be sufficient because (8.8.2) would fail. More simply, one can use Lemma 8.8.13 to show that X is minimal sufficient.

8.60 Here one needs to show that

$$\int_0^\theta u(x)\frac{1}{\theta}dx \overset{\theta}{\equiv} 0 \qquad (*)$$

implies $u(x) = 0$ for $x > 0$. The equation $(*)$ is equivalent (since $\theta > 0$) to $\int_0^a u(x)\,dx = 0$ for all $a > 0$. Since $u(\cdot)$ is the derivative of its integral (and its integral is zero) the result follows.

8.65 $\hat{\theta} = 2\bar{X}$.

CHAPTER 9

9.5 $c = .07\sigma/\sqrt{n}$. This test is worse than "flipping a coin," since this test rejects precisely when we have the strongest evidence for accepting, and thus uses the data to our disadvantage.

9.10 At $\mu_2 = 2\mu_0 - \mu_1 + 2\Phi^{-1}(1-\alpha)\dfrac{\sigma}{\sqrt{n}}$.

9.15 Setting the level to α will specify the C at (9.3.40) at the same value regardless of the θ_1 chosen.

9.20 Several risk points may qualify since (see the geometrical interpretation of Bayes rules given following Definition 9.4.8) the line may hit the convex set along a whole side, hence at more than one risk point.

9.25 The minimax rule rejects when $X_1 + \ldots + X_n > C$ where C is such that

$$P[Y > nC/\theta_0] = P[Y < nC/\theta_1]$$

where Y has the gamma distribution $G(x|30,1,0)$.

9.30 This follows easily from the theorems mentioned in the hint.

9.35 This follows easily upon writing down the level of the test as

$$t_{n-1}(t_{n-1}^{-1}(\alpha p)) + t_{n-1}(-t_{n-1}^{-1}(\alpha(1-p))) = \alpha p + \alpha(1-p) = \alpha.$$

9.40 Use the hint, noting that if Z is a $N(0,1)$ r.v. then $EZ^2 = \text{Var}(Z) = 1$.

9.45 $(\bar{X}, \bar{Y})$ is a complete minimal sufficient statistic for (μ_1, μ_2). $\bar{X}$ is the unique MVU estimator of μ_1, and $\bar{Y}$ is the unique MVU estimator of μ_2.

9.50 The two-sided equal-tail test rejects if either $\sum_1^{10} (X_i - \mu)^2 < 3.94$ or $\sum_1^{10} (X_i - \mu)^2 > 18.3$. The one-sided test rejects if $\sum_1^{10} (X_i - \mu)^2 > 16.0$.

The respective power curves are as follows:

σ^2	0	.25	.50	.75	1.0
P[One-sided rejects]	0	.00	.00	.02	.10
P[Two-sided rejects]	1	.89	.35	.13	.10

σ^2	1.5	2.0	2.5	3.0
P[One-sided rejects]	.40	.63	.78	.88
P[Two-sided rejects]	.30	.51	.70	.82

CHAPTER 10

10.5 This follows upon noting that $\Phi^{-1}(1) = +\infty$ (so that $L(\bar{X}) = -\infty$ while $U(\bar{X}) = \bar{X} - (\sigma/\sqrt{n})\Phi^{-1}(\alpha) = \bar{X} + (\sigma/\sqrt{n})\Phi^{-1}(1-\alpha)$).

10.10 The length is

$$f(p) = -\frac{s}{\sqrt{n}}\left\{t_{n-1}^{-1}(\alpha p) - t_{n-1}^{-1}(1-\alpha(1-p))\right\}$$

which (via differentiation) is easily seen to have a minimum (for each fixed s) at $p = 1/2$, in which case its value is $f(1/2)$ as given at (10.2.14).

10.15 (a) $-.16 \leqslant \mu_B - \mu_A \leqslant .56$.
(b) $-.24 \leqslant \mu_B - \mu_A \leqslant .64$.
(c) $-.59 \leqslant \mu_B - \mu_A \leqslant .99$.
(d) Using the Hsu solution given in Section 9.9 just before equation (9.9.9), $-2.3 \leqslant \mu_B - \mu_A \leqslant 2.7$.

10.20 This is clear if one writes down precisely how the confidence interval procedure changes when the points (0, 0) and (1, 1) are added in the manner noted following Figure 10.4–4.

CHAPTER 11

11.5 Consider the case $\rho = 0.5$. Then

$$h = H/\sqrt{.5} = \sqrt{2}\, H.$$

11.10 Consider choosing a population at random without any observations being taken. Then one finds a P^* of $1/k$ trivially.

11.15 (a) Writing down an expression for the $P(CS)$ in terms of the decision procedure specified and simplifying, we come to the $P(CS)$ formula given in the statement of Problem 11.15. Noting that this is a continuous function with $dP(CS)/d(\sigma_{(k)}) > 0$ at $\mu_{[1]} = \ldots = \mu_{[k]}$, it follows that $dP(CS)/d(\sigma_{(k)})$ is positive in a neighborhood of $\mu_{[1]} = \ldots = \mu_{[k]}$, and the desired deduction follows.
(b) No.

CHAPTER 12

12.5 Use mathematical induction on the number of vectors.

12.10 If $s_1 \notin A'$, then there is a rule $t \epsilon A'$ which is at least as good as s_1, hence A' is essentially complete. A' is minimal essentially complete because $A' - \{t\}$ (where t is any rule in A') is not essentially complete (since t is admissible) while A' is essentially complete.

12.15 Any rule with risk point of the form $(9, 9p + 8q)$, for some $p, q > 0$ with $p + q = 1$, is minimax. Graphically, this is where a certain orthant intersects the set of risk points.

12.20 Let S be the set of risk points of Figure 12.1–3 but with all boundary points removed. Then (by Problem 12.18) S is the set of risk points of some statistical decision problem. However, the set of Bayes rules is empty, and the empty set of rules is not a complete class, let alone a minimal complete class or a minimal essentially complete class.

12.25 We know that, for any other rule t,

$$\sup_\theta r_\theta(t) \geqslant \sup_\theta r_\theta(s) = r_\theta(s)$$

(otherwise s could not be a Bayes rule: since s has constant risk, t would have a smaller Bayes risk); hence s is minimax.

12.30 A Bayes rule $s_w(X)$ chooses decision d_1 iff

$$\bar{X} \geqslant \frac{1}{2} - \frac{1}{5} \ln \left(\frac{2(1-w)}{3w}\right).$$

12.35 Use the hint and the geometrical interpretation of Bayes, + -Bayes, and minimax rules.

12.40

$l(\theta, d)$					
d \\ θ		Player I (Nature) θ_1	θ_2	θ_3	θ_4
Player II	d_1	0	−2	3	0
	d_2	2	0	0	−3
	d_3	−3	0	0	4
	d_4	0	3	−4	0

12.45 A Bayes rule's estimate is $(X(t) + 1)/(t+1)$. For this rule s,

$$r_\theta(s) = \frac{1}{(t+1)^2}\left(\theta^2 + (t-2)\theta + 1\right).$$

The basis for determining t is the risk function $r_\theta(s)$, which depends on the choice of t. The ideal theoretical t would be $+\infty$, which is unachievable. If we could bound θ by θ_0, we could choose t to make

$$\sup_{\theta \leqslant \theta_0} r_\theta(s) \leqslant K$$

where K is the largest risk we are willing to accept.

CHAPTER 13

13.5 Specify in detail the quantities in Theorem 10.1.5 and you have derived the desired test. (The quantities are $X = (X_1, \ldots, X_n)$, $\theta = F_p$, $\Theta = \{x{:}x\epsilon\mathscr{R}\}$, $C_\theta = \{X{:}Y_i \leqslant \xi_p \leqslant Y_j\}$, and $1 - \alpha = Q(i, j; p, n)$.)

CHAPTER 14

14.5 The regression curve is $f(x) = x^2$. The regression line is $y = \alpha + \beta x$ with

$$\begin{cases} \alpha = E(X^2) - \dfrac{EX^3 - EX\,EX^2}{EX^2 - (EX)^2}\, EX \\[2ex] \beta = \dfrac{EX^3 - EX\,EX^2}{EX^2 - (EX)^2}. \end{cases}$$

The mean square errors of prediction are, respectively, zero for the regression curve and

$$\mathrm{Var}(X^2) - \frac{(EX^3 - EX\,EX^2)^2}{\mathrm{Var}(X)}$$

for the regression line.

14.10 Using the method of Lagrange multipliers, we must solve the system of equations $\{2u_i + \lambda_1 + \lambda_2 x_i = 0\ (i = 1, \ldots, n),\ u_1 + \ldots + u_n - 1 = 0,\ u_1x_1 + \ldots + u_nx_n = 0\}$, which yields (after some manipulations) a solution u_i equal to the v_i of (14.1.34).

14.15 This follows by writing out the likelihood function and (considering it as a function of the three variables α, β, σ^2) maximizing it.

14.20 We find $\hat{\alpha} = 13.3$, $\hat{\beta} = .675$, and $\hat{\sigma}^2 = 74.34$ (hence $y = 13.3 + .675x$). The best linear estimate would be $\hat{y} = 33$ at $x = 30$.

14.25 The best $P(CS)$ function will be obtained by choosing half of the x_i's to equal x_U, the other half to equal x_L.

GLOSSARY OF SYMBOLS

This glossary is intended to aid the reader in finding the main place(s) in Chapters 1–14 of the text where a symbol is introduced. Most symbols which recur frequently are included in this glossary. The Glossary lists functions and operators first (for example, $A \cup B$ is listed in order to include the operator $\cup$; the "A" and "B" are incidental to the listing), followed by an alphabetical listing[1] (in which the Greek alphabet is alphabetized with its Latin equivalents). For each listing a brief descriptive phrase is given as well as a page number. (If the item occurs in a problem, the problem number is given in parentheses following the page number.) For example, "S(A,B)" can be briefly described as the "symmetric difference" and occurs on page 10 in Problem 1.6.

OPERATORS AND FUNCTIONS

$A \cup B$, $A + B$	Union of sets	2
$\bigcup_{A \in \mathcal{M}} A$	Extended union	6
$A \cap B$, AB, $A \cdot B$	Intersection of sets	2
$\bigcap_{A \in \mathcal{M}} A$	Extended intersection	6
A^c, $\bar{A}$, A'	Complement of a set	3
$A - B$	Difference of sets	3
$A = B$	Identity of sets	4
$A \neq B$	Inequality of sets	4
$A \subseteq B$, $B \supseteq A$, $A \subset B$, $B \supset A$	Subset relation	4
$A \underset{\neq}{\subset} B$, $B \underset{\neq}{\supset} A$	Proper subset relation	4
$\mathbf{E}$	Expectation operator	110
$\mathbf{n}(\cdot)$	Number of points operator	16
$(n)_0$, $(n)_r$	Product of successive integers	17
$n!$	n-factorial	17

[1] This listing is largely alphabetical (but not strictly alphabetical, since similar terms have been grouped together for the reader's ease of reference).

$\binom{n}{r}$, ${}^n C_r$	Binomial coefficients	17
$\#(\cdot)$	Number of points operator	20
Var	Variance operator	112
$\approx$	Approximate equality operator	150
$\sim$	Asymptotic equality operator	21
$\xrightarrow{2}$	Convergence in mean square operator	145
$\xrightarrow{p}$	Convergence in probability operator	145
$\succ, \succsim, \sim$	Relations comparing decision rules	365

a, A, α

$(\mathbf{a}_{j_1}, \mathbf{b}_{j_2}, \ldots, \mathbf{x}_{j_r})$	r-tuplets	16
A	Acceptance region	271
α, $\alpha(\phi)$	Level of a test	277, 291
$\alpha_{SP}(\mathbf{x}, \phi_\alpha)$	Significance probability of x	313 (9.14)
α	Regression intercept	418
$\hat{\alpha}$	Least squares estimator	420

b, B, β

$\mathbf{b}_n(\theta)$	Bias	159, 172
$(\mathbf{BC})_1$, $(\mathbf{BC})_i$, $(\mathbf{BC})_\infty$	Ball and cell models	18
$\mathbf{B}(\mathbf{x}\mid\alpha,\lambda)$	Beta distribution	80
$\mathbf{B}(\theta)$	Distribution function on Θ	372
$\beta(\mathbf{m}, \mathbf{n})$	Beta function	78
$1 - \beta(\phi)$	Probability of error of Type II	291
β	Regression slope	418
$\hat{\beta}$	Least squares estimator	420
$\beta_0, \beta_1, \ldots, \beta_m$	Unknown constants in a linear model	423

c, C

$\mathbf{Cov}(\mathbf{X}_1, \mathbf{X}_2)$	Covariance of X_1 and X_2	120

d, D, δ, Δ

$\mathbf{d}_y\mathbf{H}(\mathbf{y}\mid\mathbf{x})$	Probability density function or probability function	161
D*	Collection of subintervals of I	158
D	Set of all decision rules	365
$\mathscr{D}$	Decision space	364
$\boldsymbol{\Delta}$	Cell coordinate n-tuples	60
$\boldsymbol{\Delta}_{k,n}$	Restricted cell coordinate n-tuples	60

e, E, ϵ

$\mathbf{e}_n$	Average of a squared difference	163
$\mathbf{e}_\theta(\mathbf{t}_1, \mathbf{t}_2)$	Asymptotic relative efficiency	209
$\mathbf{e}_\theta(\boldsymbol{\varphi}_n)$	Efficiency of an unbiased regular estimator	230
ED50	Effective Dose for 50%	171
EX, E(X)	Expectation of X	110, 197
$\mathbf{E}(\mathbf{Y}\mid\mathbf{X}=\mathbf{x})$	Conditional mean of Y, given X	125
ε**-Bayes**	ε-Bayes	374
ε**-minimax**	ε-minimax	374
ε**-MLE**	ε-MLE	252 (8.5, 8.6)

f, F

$\mathbf{f}(\mathbf{x}\mid\boldsymbol{\theta})$	Density or probability function	208
$\mathbf{f}(\mathbf{x}, \boldsymbol{\theta}, \boldsymbol{\rho})$	Supremum of $f(x\mid\theta')$ over $\lvert\theta'-\theta\rvert \leq \rho$	198
$\mathbf{f}^*(\mathbf{x}, \boldsymbol{\theta}, \boldsymbol{\rho})$	$f(x, \theta, \rho)$ if $f(x, \theta, \rho) > 1$, 1 otherwise	198
$\mathbf{f}_X(\mathbf{y})$	Probability density function	77
$\mathbf{f}_X(\mathbf{x}_1, \ldots, \mathbf{x}_n)$	Probability density function	90
$\mathbf{f}_{X_2\mid X_1=x_1'}(\mathbf{x}_2\mid\mathbf{x}_1')$	Conditional density of X_2 given $X_1 = x_1'$	91
$\mathbf{F}(\mathbf{n}_1, \mathbf{n}_2)$	F-distribution	82
$\mathbf{F}(\mathbf{n}_1, \mathbf{n}_2, \boldsymbol{\lambda})$	Noncentral F-distribution	83
$\mathbf{F}_X(\mathbf{x})$	Distribution function	46
$\mathbf{F}_X(\mathbf{x}_1, \ldots, \mathbf{x}_n)$	Joint distribution function	59
$\mathbf{F}_{X\mid G}(\mathbf{x}\mid\mathbf{G})$	Conditional distribution function	60

$\mathbf{F}^{d}(\mathbf{x})$	Discrete d.f.	78
$\mathbf{F}^{ac}(\mathbf{x})$	Absolutely continuous d.f.	78
$\mathbf{F}^{sc}(\mathbf{x})$	Singularly continuous d.f.	78
$\mathbf{F} * \mathbf{F}, \mathbf{F}^{2*}$	Convolution of F with F	142
$\mathbf{F}^{n*}$	Convolution of F with F n − 1 times	142
$\mathbf{F}_{X_1} * \mathbf{F}_{X_2}$	Convolution of F_{X_1} with F_{X_2}	142
$\mathbf{F}_n \to \mathbf{L}$	Convergence of d.f.'s	147
$\mathbf{F}_n(\mathbf{x} \mid \mathbf{Y}_1, \ldots, \mathbf{Y}_n)$	Empiric distribution function	176
$\mathscr{F}$	Class of distribution functions, or Welch's degrees of freedom	158, 311

g, G, γ, Γ

$\hat{\mathbf{g}}(\boldsymbol{\theta})$	MLE of a function of θ	188
$\mathbf{G}(\mathbf{x} \mid \boldsymbol{\alpha}, \boldsymbol{\beta}, \mathbf{A})$	Gamma distribution	81
$\mathbf{g}_{n-1}(\cdot)$	P.d.f. of a $\chi^2_{n-1}(0)$ r.v.	351
$\mathbf{G}_{n-1}$	D.f. of a $\chi^2_{n-1}(0)$ r.v.	351
$\boldsymbol{\Gamma}(\mathbf{n})$	Gamma function	78

h, H

$\mathbf{h}_k\ (\mathbf{P}^*)$	Solution h of $\int_{-\infty}^{\infty} \Phi^{k-1}(x+h)\phi(x)dx = P^*$	346
$\mathbf{H}$	Hypothesis	269
$\mathbf{H}_0$	Null hypothesis	269
$\mathbf{H}_1$	Alternative hypothesis	269
$\mathbf{H}(\mathbf{y} \mid \mathbf{x})$	Distribution of responses	161

i, I

inf	Infimum	158
$\mathbf{I}$	$\{x : 0 \leq x \leq 1\}$	158
$\mathbf{I}(\boldsymbol{\theta})$	$E_\theta \left(\frac{\partial \ln f(x \mid \theta)}{\partial \theta} \right)^2$	208
Im z	Imaginary part of z	129

$\mathbf{I}_S(\mathbf{x})$	Indicator function of the set S	129
$\mathbf{I}(\mathbf{X}_1, \ldots, \mathbf{X}_n)$	Identity statistic $(X_1, \ldots, X_n)$	237

j, J

$\mathbf{J}\left(\frac{\mathbf{y}_1, \ldots, \mathbf{y}_n}{\mathbf{x}_1, \ldots, \mathbf{x}_n}\right)$	Jacobian	98
$\mathbf{J}^2(\boldsymbol{\theta})$	Variance of an asymptotic distribution	207

k, K, κ

$\mathbf{K}(\mathbf{d};\mathbf{x})$	Function to minimize for a Bayes rule	386

l, L, λ, Λ

$l(\mathscr{U}\|\mathbf{c})$	$P_\theta[(X_1, \ldots, X_n) \in \mathscr{U} \| t(X_1, \ldots, X_n) = c]$	234
$l(\boldsymbol{\theta}, \mathbf{d})$	Loss function	364
LD50	Lethal Dose for 50%	160, 171
L(f, s)	Length of an interval	158
L(x, y)	Lower bound on a bivariate d.f.	107 (4.26)
L(X)	Lower confidence limit	324
$\mathbf{L}(\boldsymbol{\theta})$	Likelihood function	184
$\boldsymbol{\Lambda}$	Indexing set	31, 36 (2.18)

m, M, μ

m	Median	112, 180
$\mathbf{m}_2$	Second central sample moment	174
$\mathbf{m}'_k$	k^{th} (noncentral) sample moment	173
$\mathbf{m}_k$	k^{th} central sample moment	173
$\mathbf{m}'_1$	Sample mean	173
$\mathbf{M}_n$	$E\|X\|^n$; or, Sample median	115, 181
$\mathbf{M}(\cdot)$	Function	157
$\boldsymbol{\mu}_2$	Variance	112

μ_3	Skewness	112
μ'_n	n^{th} (noncentral) moment	112, 173
μ_n	n^{th} central moment	112, 173
$\mu'_{1,1}$	Product moment	120
$\mu_{1,1}$	Covariance of X_1 and X_2	120
μ'_{n_1,n_2}	Joint moment of (X_1, X_2) of order $n_1 + n_2$	119
μ_{n_1,n_2}	Joint central moment of (X_1, X_2) of order $n_1 + n_2$	120
$\mu_{[1]}, \ldots, \mu_{[k]}$	Ranked means	343

n, N, ν

n_1	Approximation to sample size in selection	348
$N(0, 1)$	Standard normal distribution	82
$N(\mu, \sigma^2)$	Normal distribution	82, 92
$N(\mu_1, \mu_2; \sigma_1^2, \sigma_2^2, \rho)$	Bivariate normal distribution	92
$N(\mu, \Sigma)$	Multivariate normal distribution	92
$N_n(y)$	Number of $X_1, \ldots, X_n$ which are $> y$	219
$N_n^*(y)$	Standardized $N_n(y)$	220

O, 0, ω, Ω

$o(h)$	$o(h)/h \to 0$ as $h \to$ a limit	53
o_i	Multinomial outcome	65
$\emptyset$, 0	Empty set	4
Ω	Universal set	1
Ω_A	Assignment vector space	20
$(\Omega, \mathscr{B}, P)$	Probability space	40(2.4H)
Ω_o	Occupancy vector space	20

p, P, φ. ϕ, Φ, Ψ, π, Π

$\mathscr{P}_M(n_1, \ldots, n_k)$	Means procedure	359(11.1H)
$p_{[1]}, \ldots, p_{[k]}$	Ranked multinomial probabilities	362(11.10H)
$p(x)$, $p_X(x)$	Probability function	49, 50, 62

$\mathbf{p}_m(\mathbf{r}, \mathbf{n})$	Probability exactly m cells are empty	137
$\mathbf{P}(\cdot)$	Probability function	13
$\mathbf{P}(\cdot \mid \cdot)$	Conditional probability	23
$\mathbf{P}_n(\mathbf{T})$	Poisson process probability	53
$\underline{\mathbf{P}}(\mathbf{K}, \mathbf{X})$	Profit in newsboy problem	177
$\overline{\mathbf{P}}_K$	Average profit in newsboy problem	178
$\mathbf{P}(\mathbf{CS})$	Probability of correct selection	342
$\varphi_n(\mathbf{X}_1, \ldots, \mathbf{X}_n)$	Estimator	207
$\varphi(\mathbf{x}, \mathbf{r})$	Supremum of $f(x\|\theta')$ over $\|\theta'\| > r$	198
$\varphi^*(\mathbf{x}, \mathbf{r})$	$\varphi(x, r)$ if $\varphi(x, r) > 1$, 1 otherwise	198
$\boldsymbol{\phi}_X(\mathbf{t})$	Characteristic function	129
$\boldsymbol{\phi}(\mathbf{x})$	Test function, or N(0, 1) density	271, 344, 345
$\boldsymbol{\Phi}(\mathbf{x})$	Normal d.f.	104 (4.6), 151
$\boldsymbol{\Psi}_X(\mathbf{t})$	Moment-generating function	127
$\boldsymbol{\Psi}_{X_1, \ldots, X_n}(\mathbf{t}_1, \ldots, \mathbf{t}_n)$	Joint m.g.f.	128
$\boldsymbol{\pi}_1, \ldots, \boldsymbol{\pi}_k$	Populations	343
$\boldsymbol{\Pi}(\cdot, \ldots, \cdot)$	Polynomial	175
$\boldsymbol{\Pi}$	Set $\{\pi_1, \ldots, \pi_k\}$	354
+ -Bayes	Positive-Bayes	378

q, Q

$\mathbf{q}$	$1 - p$	55, 186
$\mathbf{q}(\mathbf{x})$	Failure rate	71 (3.30), 88
$\mathbf{Q}(\mathbf{i}, \mathbf{j}; \mathbf{p}, \mathbf{n})$	Nonparametric quantile coverage	405
$\mathbf{Q}(\mathbf{x})$	Lower orthant at x	377

r, R, $\boldsymbol{\rho}$

$\mathscr{R}$	The real line	45
$\mathscr{R}^n$	Euclidean n-space	46
$\mathbf{r}(\cdot)$	Risk point of a test	291
$\mathbf{reg}(\boldsymbol{\theta}, \mathbf{d})$	Regret function	386
$\mathbf{r}_\theta(\mathbf{s})$	Risk function	364
$\mathbf{r}_{XY}$	Sample correlation	420

$\mathbf{R}$	Critical region, or set of risk points	271, 291
$\mathbf{R(A)}$	Relative frequency	13
$\mathbf{R(p)}$	System reliability	38 (2.30, 2.31)
$\mathbf{R(A\|B)}$	Conditional relative frequency	23
$\mathbf{R}_{B(\theta)}\mathbf{(t)}$	Bayes risk of t	372
$\mathbf{Re\ z}$	Real part of z	129
$\boldsymbol{\rho}(\mathbf{X}_1, \mathbf{X}_2)$	Correlation coefficient	122

s, S, $\boldsymbol{\sigma}$, $\boldsymbol{\Sigma}$

$\mathbf{s}(\cdot)$	Statistical decision rule	364
$\mathbf{s}^2$	Unbiased estimator of σ^2	299
$\mathbf{s}_N^*$	Minimax decision procedure	158
sup	Supremum	158
$\mathbf{S}$	Subset of $\Pi = \{\pi_1, \ldots, \pi_k\}$	354
$\mathbf{S(A, B)}$	Symmetric difference	10 (1.6), 30
$\mathbf{S}(\boldsymbol{\alpha}, \boldsymbol{\beta})$	Sum of squared deviations	420
$\mathbf{S}(\boldsymbol{\theta}, \boldsymbol{\rho})$	Open sphere, center θ, radius ρ	201
$\mathbf{S}_n$	$X_1 + \ldots + X_n$	146
$\underline{\mathbf{S}}_N$	Class of procedures	158
$\overline{\mathbf{S}}_N$	Class of procedures	158
$\mathbf{S(X)}$	Confidence set	326
$\boldsymbol{\sigma}(\mathbf{X})$	Standard deviation of X	113
$\boldsymbol{\sigma}^2(\mathbf{X})$	Variance of X	112
$\boldsymbol{\sigma}^2_{[1]}, \ldots, \boldsymbol{\sigma}^2_{[k]}$	Ranked variances	350
$\boldsymbol{\Sigma}$	Symmetric positive definite matrix	92

t, T, $\boldsymbol{\tau}$, $\boldsymbol{\theta}$, $\boldsymbol{\Theta}$, $\boldsymbol{\Theta}$

$\mathbf{t}_m$	Student's t-distribution	302
$\mathbf{t}_m^{-1}(\boldsymbol{\alpha})$	$100\alpha\%$ point of the t_m distribution	302
$\mathbf{t}_n(\mathbf{x}_1, \ldots, \mathbf{x}_n)$	Estimate	159
$\mathbf{t}_n(\mathbf{X}_1, \ldots, \mathbf{X}_n)$	Statistic or estimator	159
$\mathbf{T(X)}$	Statistic	286
$\boldsymbol{\theta}$	State of nature	364
$\|\boldsymbol{\theta}\|$	$\sqrt{\theta_1^2 + \ldots + \theta_k^2}$	198
$\hat{\boldsymbol{\theta}}, \hat{\boldsymbol{\theta}}_n(\mathbf{X}_1, \ldots, \mathbf{X}_n)$	Maximum likelihood estimator	184
$\boldsymbol{\Theta}$	Set of parameter values	159, 364

$\boldsymbol{\Theta}$	Set of parameter values	198
$(\boldsymbol{\Theta}, \mathscr{D}, l)$, **X**	Statistical decision problem	364

u, U

U(x, y)	Upper bound on a bivariate d.f.	107 (4.26)
U(X)	Upper confidence limit	324

v, V

Var(X)	Variance of X	112, 197

w, W

$\mathbf{w}_i$	Legislators-per-person in district i	289
$\mathbf{W}_X$	Sum of ranks	411
$\mathbf{W}(\boldsymbol{\theta}; \mathbf{d}; \mathbf{x})$	Loss function	401

x, X, χ, ξ

$\mathbf{x}'$	Transpose of vector **x**	193
$\mathbf{x}'\mathbf{y}$	Inner product $x_1y_1 + \ldots + x_ky_k$	193
$\mathbf{x}^+$	Positive part of x	105 (4.10)
$\mathbf{x}^-$	Negative part of x	105 (4.10)
X	Random variable	46, 59
$\mathbf{X}^*$	Standardized X	114
$\overline{\mathbf{X}}$	$(X_1 + \ldots + X_n)/n$	144
$\mathscr{X}$	Set of possible values of X	271
$\chi_n^2(\mathbf{0})$	Chi-Square distribution	82
$\chi_n^2(\boldsymbol{\lambda})$	Noncentral chi-square distribution	82
$\chi^2(\boldsymbol{\theta})$	Criterion for MCS estimation	249
$(\chi'(\boldsymbol{\theta}))^2$	Criterion for MMCS estimation	250
ξ, ξ_p	Quantile of order p	215

y, Y

$Y_1, \ldots, Y_n$	Order statistics	181
$Y_{[np]+1}$	Sample quantile of order p	215
Y_L^*	Lower confidence limit	426
Y_U^*	Upper confidence limit	426
Y_L	Lower prediction limit	426
Y_U	Upper prediction limit	426

z, Z, ζ

Z_p	Sample quantile of order p	215
$\zeta(N)$	Estimator based on number of trials	179

AUTHOR INDEX

A

Andersen, E. B., 244
Arkin, H., 58
Armitage, J. V., 361
Arnold, B. C., 387
Auerbach, O., 271

B

Bahadur, 213, 355
Baker, 288, 289
Balintfy, J. L., 139
Barr, D. R., 352
Bartlett, 309, 310, 322
Bayes, 23, 25, 26, 30, 294, 295, 296, 297, 314, 320, 372, 373, 374, 375, 376, 377, 378, 379, 384, 385, 386, 392, 393, 394, 395, 396, 397, 399, 400, 401, 402, 408
Bechhofer, R. E., 103, 341, 346, 352, 353, 355, 356, 357, 359, 360, 362
Becker, W. A., 343
Behara, M., 214
Behrens, 308, 309, 310, 317, 322
Berk, R. H., 261
Bernoulli, 51, 56, 116, 259, 260, 332, 338
Biometrika Trustees, 335
Blackwell, 244, 245, 246, 247, 265
Bloom, D. M., 11
Blumenthal, S., 361
Blyth, C. R., 34, 39
Boltzmann, 20
Boole, 15, 26, 140, 204
Borel, 201
Bose, 20
Bray, 210
Brogan, D. R., 138
Brown, A., 108
Bryan, 79
Bryson, M. C., 107
Burdick, D. S., 139

C

Cantelli, 177, 183, 261
Cantor, 83
Carleman, 115, 139
Carr, 288, 289
Cauchy, 142, 238, 259, 264, 266
Chebyshev, 143, 144, 146, 152, 153, 154, 166, 180
Christopher, M., 155
Chu, K., 139
Chung, K.-L., 39
Cochran, 433
Cohen, A., 361
Colton, R. R., 58
Cover, T. M. 265
Craig, 335
Cramér, H., 197, 198, 208, 222, 225, 227, 228, 231, 233, 259, 267, 320

D

Dalal, S. R., 358
Darmois, 243, 259, 265
Deming, W. E., 19
DeMorgan, 6, 7, 10
Dirac, 20

This index lists each author or other person mentioned at least once in Chapters 1-14 of the text.

Doob, J. L., 198
Dudewicz, E. J., 261, 262, 276, 343, 348, 352, 354, 355, 358, 361, 415
Dulmage, H. T., 171
Dunnett, C. W., 353, 361, 384
Dykstra, R. L., 108
Dynkin, 241, 242, 264, 267

E

Eaton, M. L., 384
Einstein, 20
Elmaghraby, S., 362
Esary, J. D., 140

F

Feferman, S., 1
Feller, W., 20, 41, 43, 44, 58, 72, 78, 109, 114, 121, 137, 155
Ferguson, T. S., 194, 254, 373, 387, 400, 408
Fermi, 20
Fisher, R. A., 209, 211, 213, 308, 309 310, 317, 322
Fishman, G. S., 276
Fisz, M., 210, 261, 263
Fréchet, 107
Freeman, H., 361

G

Garfinkel, L., 271
Gauss, 420, 421
Gerber, H., 72
Gibbs, 107
Glivenko, 177, 183, 261
Goodman, 355
Gottlieb, P., 11
Graybill, 41, 114, 335, 423
Gupta, S. S., 348, 354, 355
Guttman, L., 167, 168
Guttman, R., 167, 168

H

Hahn, G. J., 339
Hall, W. J., 67, 359
Halmos, P. R., 10
Hammond, E. C., 271
Hansel, C. E. M., 155
Harter, H. L., 415
Hartley, H. O., 335, 340
Hausdorff, 115
Hayes, R. H., 178
Heine, 201
Helly, 210
Helmert, 301, 322
Heuer, C. V., 44
Hitler, 270
Hodges, J. L., 211, 213, 255, 263
Hoeffding, W., 107
Hogg, 335
Hollander, M., 415
Hsu, 310
Hughes, T., 44
Huzurbazar, 197, 198

J

Jensen, 196, 197, 199, 254, 262
John, P. W. M., 433
Johnson, N. L., 75

K

Kazarinoff, N. D., 73
Keilson, J., 72, 135
Kendall, M. G., 113
Khinchin, 147, 148, 154, 175, 177, 182, 201, 202
Kiefer, J., 168, 267, 352, 355, 360
Kirman, D., 271
Koopman, 243, 259, 265
Koopmans, L. H., 108
Krickeberg, K., 214
Krishnaiah, P. R., 361
Kuzmack, A., 361

L

Lachenbruch, P. A., 138
Lamperti, J., 108
Landon, 28, 29
Lass, H., 11
Lebesgue, 110, 264, 339

LeCam, 213
Lee, Y. J., 415
Lehmann, E. L., 235, 239, 265, 277, 293, 321
Lévy, 147, 148, 149, 154, 155
Lindgren, 335
Loève, 144
Lukacs, E., 130

M

Mann, 409, 410, 412, 413
Markov, 144, 420, 421
Marshall, A. W., 107
Martin, W. T., 54
Maurice, R., 361
Maxwell, 20
Milton, R. C., 355
Mitchelmore, M. C., 39
Monro, S., 170, 171
Mood, 41, 114, 335
Moore, D. S., 243, 258, 264
Morse, N., 362
Moser, L., 73

N

Nakano, H., 9
Naylor, T. H., 139
Nelson, B. A., 51
Nemenyi, P., 321, 322
Newton, 168
Neyman, 234, 238, 239, 240, 241, 242, 258, 266, 277, 279, 280, 282, 284, 285, 286, 287, 288, 290, 294, 296, 297, 298, 299, 303, 304, 309, 310, 312, 313, 316, 318, 321, 322, 366, 371, 380

O

Olkin, I., 107

P

Panchapakesan, S., 355
Parzen, E., 53, 111
Pearson, E. S., 335, 340
Pearson, 277, 279, 280, 282, 284, 285, 286, 287, 288, 290, 294, 296, 297, 298, 299, 303, 304, 312, 313, 316, 318, 321, 366, 371, 380
Perlman, M. D., 198
Pfanzagl, J., 244
Pierce, D. A., 108
Poisson, 52, 54, 56, 58, 68, 69, 70, 72, 73, 75, 117, 131, 137, 138, 142, 153, 156, 232, 247, 251, 257, 258, 259
Polya, 149, 207
Poretsky, 10
Proschan, F., 140

R

Raj, D., 362
Ramberg, J. S., 276
RAND Corporation, 276
Rao, C. R., 149, 222, 225, 227, 228, 231, 244, 245, 246, 247, 259, 265, 267, 289
Raphson, 168
Rayleigh, 260
Raynor, B., 39
Reinhardt, H. E., 288, 319
Reissner, E., 54
Rizvi, M. H., 352
Robbins, H., 170
Roberts, 29
Roosevelt, 28, 29
Rosenberg, L., 106
Rosenblatt, J., 332
Rudin, W., 47, 108
Russell, B., 9, 10

S

Samuelson, P. A., 144
Sarkar, T. K., 107
Scheffé, H., 93, 239, 310, 311, 322, 421
Schwarz, 122, 225
Seagram, J. E., 23
Sievers, S., 360
Simon, W., 320

Simons, G., 75, 156
Simpson, 33, 34, 39
Sitek, M., 354
Slutsky, 261, 322
Sobel, M., 103, 352, 353, 355, 359, 360, 361
Stein, C., 339
Stieltjes, 110
Stirling, 21, 216
Stuart, A., 113
Student, 83, 142, 238, 300, 302, 308, 309, 310, 317, 322, 361
Suzuki, Y., 261
Sverdrup, E., 261

T

Takeuti, G., 9
Taylor, 130, 217, 218, 220
Teichroew, D., 348
Thomas, G. B., Jr., 187
Tippett, L. H. C., 276
Tucker, H. G., 40, 110

V

van der Waerden, B. L., 171
Varberg, D. E., 97
Venn, 2, 5

W

Wadsworth, 79
Wald, A., 197, 198, 262, 341
Walkup, D. W., 140
Wallis, 29
Walsh, J. E., 410
Wasan, M. T., 162, 168
Weibull, 88, 89, 96, 97
Weiss, L., 152, 178, 214, 264, 382
Welch, 311
Whitney, 409, 410, 412, 413
Wilansky, A., 73
Wilcoxon, 409, 410, 412, 413
Wilks, S. S., 261
Wolfe, D. A., 415
Wolfowitz, J., 170, 197, 198, 213, 214, 222, 294
Woodroofe, M., 232

Y

Yackel, J., 348

Z

Zaino, N. A., Jr., 348
Zaring, W. M., 9
Zehna, P. W., 261
Zuckerman, P., 10

SUBJECT INDEX

A

Accident injuries, 268, 311, 341
Age of death, 74
Agricultural plots, 410
  yield, 427
Alcohol tasting, 23
Algebra of sets, 11, **34**
Almanac, 320
Approximation, nonstochastic, 157, 168
  root θ such that $M(\theta) = \alpha$, 157
  minimax decision procedure, 158
Arithmetic mean, 196, **254**
Assignment vectors, 19, **20,** 42
  $\#(\Omega_A)$, 20
Associative laws, **6**
Asymptotic theory, 389
Atomic power plants, 268
  do they increase radiation levels?, 269
  hypothesis test about, 272
  type that increases radiation levels the least, 342
Automobiles, 267
  accident injuries, 268, 341
  safety restraint systems (SRS's), 348, 356
  insurance, 402
Axiomatic set theory, 9, 10

B

Baby pictures, 41
Bad luck, 115, 133
Baker vs. Carr, 288
Ball and cell models, **18,** 36, 42
  relation with sampling, 18
  number of empty cells, 137
  estimation of number of cells, 138
Bank, 178
Barber of Seville, **10**
Baseball, 68
Bayes' rule, **25,** 26
Behrens-Fisher problem, **309**
  Behrens-Fisher distribution, **309,** 317
  Neyman-Bartlett solution, 309
  Scheffé solution, 310
  Hsu solution, 310
  Welch's suggestion, 311
  Hsu solution recommended, **311**
  comparison of solutions, 322
Bernoulli distribution, **51**
  traffic example, 51
  "successes in one trial" interpretation, 51
  sums leading to binomials, 56
  mean and variance, 116
  minimal sufficient statistic, 259
  MMCS estimator of p, 260
  MCS estimator of p, 260
  confidence intervals for p, 332
  chart for 95% intervals, 336
Beta distribution $B(x \mid \alpha,\lambda)$, **80**
  relation with uniform distribution, 81
  tables (Table 8), **451**
Beta functions, **78**
  expression in terms of gamma functions, 79
Bibliographies, 9
Bidding, 398
Binomial coefficients, 17
  tables (Table 3), **437**

In this index, the significant page reference(s) for many of the items (e.g., a main term's definition) are set in boldface type. Subentries are listed in order of first appearance within the text.

Binomial distribution, **51**
- traffic example, 52
- "occurrences in n trials" interpretation, 52
- as a sum of Bernoullis, 56
- limits leading to Poisson, 56, 75
- Poisson approximation to, 56
- as a limit of hypergeometrics, 57
- derived from Poissons, 69
- mean and variance, 116
- characteristic function, 131
- d.f. of $X_1 + \ldots + X_n$, 142
- in empiric d.f. convergence proof, 177
- use in sample quantile proofs, 219
- confidence intervals for p, 332
- chart for 95% intervals, 336
- evaluation of normal approximation to the binomial distribution, 407
- testing $H_0: p \leq p_0$ versus $H_1: p > p_0$, 409
- tables (Table 5), **440**

Binomial theorem, **27**

Biomathematics and Computer Science in the Life Sciences, Ninth Annual Symposium, 50

Biometrics, 390

Birth, 10

Birthdays, 21, 43, 304

Blood, 58, 160

Boole's inequality, **15,** 26, 140

Borel field of sets, **34**

Bose-Einstein statistical mechanics, 20

Business decision making, 373

C

Calculators, 69

Cancer, 268, 341
- screening of drugs, 354

Cantor-type distributions, **83**

Cards, 28, 43, 68, 276

Cargo weight, estimate of, 19

Casino, 59, 83, 150

Catalyst, 268, 342, 421

Categorical data, 249

Cauchy distribution, **142**
- d.f. of $X_1 + \ldots + X_n$, 142
- relation with t-distribution, 142, 238
- no nontrivial sufficient statistics, 238

Cell division, rate of, 167

Cell in $\mathscr{R}^n$, **59**
- vertices of, 60

Cells and balls, 19
- cell walls, 20
- cells with "two positions," 72

Central limit theorem (CLT), classical, **148**
- proof for common situations, 149

Cereal, breakfast, 413

Characteristic function (ch.f.) of a r.v., **129**
- properties of ch.f.'s, 130
- ch.f. of $aX + b$, 130
- expansion of a ch.f. in a series, 130
- uniqueness theorem, 130, 141
- moments from a ch.f., 131, 132
- of a binomial r.v., 131
- of a Poisson r.v., 131
- of a constant r.v., 131
- of a normal r.v., 131
- of a uniform r.v., 132
- of an exponential r.v., 132
- of a gamma r.v., 132
- use to find a distribution of a sum, 133
- of $X_1 + \ldots + X_n$, 142
- ch.f. convergence equivalent to d.f. convergence ,147

Charge, of a particle, 110

Chebyshev's inequality, **143, 144**
- equality in Chebyshev's inequality, 143
- equality in, with $\sigma^2 > 0$, 144
- use at a newspaper stand, 144
- use in proving WLLN's, 146
- use in stochastic approximation proof, 166

Cheese, 390

Chemical Process, 268, 323
- catalyst which increases yield the most, 342

Chickens, selection of best stock, 343

Children, probability of i, 24

Chi-square distribution $\chi_n^2(0)$, **82**
- relation with gamma distribution, 82
- relation with noncentral chi-square distribution, 82
- d.f. of $X_1 + \ldots + X_n$, 142
- use in testing a normal variance, 304
- ratios leading to F-distributions, 316
- mean, 316
- tables (Table 10), **457**

Chocolate bars, 35

Cigarettes, 270
  cause cancer when smoked, 270
  brand least likely to cause cancer, 341
Cities, 38
City council, 35
Classical definition of probability, **13**
Cleansing, 50
Closed set, **193**
Closet, 29
Clumps, 50, 192
Cochran's theorem, **433**
Code, 38
Coin melting, 178
Coin tossing, 15, 71, 86, 153, 399
  convergence of S_n/n to p, 146
  choice of n to estimate p, 150
  various estimators of p, 169
  use in a randomized test, 271, 276, 342
Columbus, Ohio, 389
Combination, 35
Combinatorial analysis, 16
Combinatorics, 16
  how we got involved, 29
Commutative laws, **5**
  proof of a commutative law, 6
Commuter, 29
Complement (of a set), **3**
Complete additivity, 41
Complete family (of distributions), **246**
  boundedly complete family, **265**
Complete set of statements, **5**
Complex exponential functions, 129
Complex numbers, 129
  real part, 129
  imaginary part, 129
Computer, 36
  computer consoles, 37
  computing facility, 37
Concentration of a r.v., **116,** 206
  implication for variances, 116
  implication for normal r.v.'s, 206
  asymptotic concentration (for CAN estimators), 209
  asymptotic relative efficiency (for CAN estimators), 209
Conditional d.f., **60**
  of X_2 given $X_1 = x_1'$, **61**
Conditional density of X_2 given $X_1 = x_1'$, **91**
Conditional failure probability function, 71
Conditional frequency, 24
Conditional mean of Y, given X, **125**
  conditional expectation, 125
  expectation as a sum of weighted conditional expectations, 125
  extensions of expectation as a sum of ..., 125, 164, 165
Conditional probability of A given B, **23**
  as a probability function, 24
Conditional probability function, **62**
Confidential-data, 37
Confidence interval, 154
  confidence coefficient, 154
Congestion, 390
Consumers Union, 361
Contamination, 214
Continuous distributions
  relations among, 90
  multivariate, **90**
Controlled experimentation, 411
  control group, 411
Convergence of a sequence of d.f.'s $F_1(\cdot), F_2(\cdot), F_3(\cdot),\ldots$ to a d.f. $L(\cdot)$ ($F_n \to L$), **147**
  weak convergence (of d.f.'s), 147
  complete convergence (of d.f.'s), 147
  equivalent to ch.f. convergence, 147
  example of t-distributions, 322
Convergence in mean square ($Z_n \xrightarrow{2} Z$), **145**
  $Z_n \to Z$ in $\mathscr{L}_2$, 145
Convergence in probability ($Z_n \xrightarrow{p} Z$), **145**
  stochastic convergence, 145
  convergence in measure, 145
  weak convergence, 145
  use in WLLN's statement, 145
  Chebyshev's inequality method of proving, 146
  criterion for $Z_n \xrightarrow{p} c$ (c a constant), 147
  statistical relations to convergence w.p. 1, 156
  Slutsky's theorem on, **261**
Convergence with probability one (w.p. 1), **145**
  almost sure (a.s.) convergence, 145
  almost everywhere (a.e. or p.p.) convergence, 145
  strong convergence, 145
Convex set, **193,** 253, 368
  convex function, **196,** 254
  strictly convex function, **262**
  convex set generated by **x,y,z,...,u, 368**

Convex set, (continued)
  convex hull of **x,y,z,...,u, 368**
  convex combination, 368
  convex hull as set of convex combinations, 368
Convolution, of discrete distributions, 67
  for continuous distributions, 106
  convolution of d.f.'s, 141
"Cookbook" statistics courses, 311
Correlation coefficient (of two r.v.'s), **122**
  a measure of linear relationship, 122
  not a measure of quadratic relationship, 122
  is a number in $(-1,1)$, 123
  is ± 1 iff standardized X_1 equals $\pm$ standardized X_2, 123
  of a bivariate normal, 124
  properties, 135
  matrix of correlation coefficients, 139
  regression interpretation, **419**
Councillor, 35
Countable additivity, **13,** 40
Covariance (of two r.v.'s), **120**
  expression in terms of $E(X_1X_2)$ and $E(X_1)E(X_2)$, 120
  role in variance of a sum, 120
  properties, 135
Cramér-Rao inequality, 222, **225**
  lemma of aid in evaluation, 227
  for unbiased estimators, **228**
  when lower bound is achieved, 230, 259
Cranial injury index, 348
Cricket, 39
Critical points (of a function), **187**
  relation to maxima and minima, 187
Cumulative distribution function (c.d.f.), **46**
CUPL, 36
Cure, 45
Customer satisfaction, 349
Cycles, 68

D

Death, 71, 74
  of experimental animals, 160
Decision theory, 152, 158, 178, **363**
  risk point of a test, 291
  complete class, 293, **365**
  essentially complete class, 293, **365**
  minimal complete class, 293, **365**
  minimal essentially complete class, 293, **365**
  Bayes in testing, 294
  minimax in testing, 295
  suggested by ranking and selection, 341
  statistical decision problem $(\Theta,\mathscr{D},l),X$, **364**
  data, **364**
  no-data decision problem, **364**
  state of nature, **364**
  parameter space, **364**
  decision space (action space), **364**
  loss function $l(\theta,d)$, **364**
  non-randomized decision rule $s(\cdot)$, **364**
  randomized decision rule $s(\cdot)$, **364**
  risk function $r_\theta(s)$, **364**
  expected loss (average loss), 364
  risk function as fundamental, 364
  better than, worse than, no better than, equivalent, 365
  "risk point", **367**
  role of convex sets, 367
  finding complete classes, **370, 371**
  admissible rules, **371**
  inadmissible rules, 371
  Bayes decision rule, **372**
  Bayes risk, **372**
  a priori distribution $B(\theta)$, **372**
  what Bayes risk is, 373
  Subjective (personal) probability, 373
  when a Bayes rule for fixed $B(\theta)$ may be appropriate, 373
  ε-Bayes rule, **374**
  minimax rule, **374,** 394
  ε-minimax rule, **374**
  positive-Bayes rules (+-Bayes), **375, 378**
  Bayes when $\Theta = \{\theta_1, \theta_2\}$, **375**
  Bayes, admissible, complete relationships, 376
  risk set convex (Θ finite), 377
  convex set in $\mathscr{R}^k$ is risk set, 377
  admissible$\Rightarrow$Bayes (Θ finite), **377**
  +-Bayes$\Rightarrow$admissible (Θ finite), **378**
  complete class theorem, **379,** 393
  hypothesis testing in decision theory framework, **379**
  interval estimation in decision theory framework, **382**
  ranking and selection in decision theory

framework, **383**
computation of Bayes rules, **384**
minimax, regret, sufficiency, 386
point estimation in decision theory framework, **387**
invariance, 387
least-favorable a priori distribution, **393**
NG strategy, **397**
Bayes and minimax same, **399**
behavioral decision rule, **400**
prerandomized decision rule, **400**
two formulations ($\mathscr{F}_1, \mathscr{F}_2$) of decision theory contrasted, **401**
Defective items, 267
Degrees of freedom, 83
loss of 1, 175, 302
Democrats, 35, 39
DeMorgan's laws, **6**
for arbitrary unions and intersections, 7
special cases, 7
Denumerable, 49
Derivative
proof of existence, 54
of an infinite sum as sum of derivatives, 127
under an integral sign, 130
partial, 223
interchange with integral, 225
Design and analysis of experiments, 390
Desk calculator, 36, 69
Destination, 31
Deterministic phenomena, 7
Dice, 8, 10, 34, 36, 45, 134, 136, 153, 276
weighting, 69, 73
selecting a face, 361
Difference (of sets), **3**
Differential equation system, 53, 54, 72
Digital computer, 36, 276
Discrete distributions
relations among, 56
Discrete failure rate, 71, 73, 74
Disjoint sets, **4**
Distance, **193**
Distilleries, 23
Distribution function (d.f.), **46**
properties of a d.f., 46, 47
discontinuities of a d.f., 47
characterization of one-dimensional d.f., 47
(joint) d.f. (of an n-dimensional r.v.), **59**
properties of a (joint) d.f., 59
characterization of n-dimensional d.f., 60, 73
discontinuities (of an n-dimensional d.f.),73
classification, 77
(absolutely) continuous d.f., 77
continuous d.f. and the p.d.f., 77
d.f. respresentation (decomposition), 78
singularly continuous d.f., 78, 108
example of a singular d.f., 84
n-dimensional (absolutely) continuous, 90
bound for 2-dimensional d.f. in terms of marginals, 107
α^{th} quantile, 170
Distributions of functions of r.v.'s
discrete case, 66
distribution of $X + Y$ via convolution if X, Y independent and discrete, 67
$aX + b$ (general case; special cases), 69, 94
$2W$-1 with W uniform, 86
continuous case, 93
$g(X)$, with $g(\cdot)$ differentiable, X continuous, 95
X^{β} with X Weibull, 96
X_1+X_2, (X_1, X_2) continuous, 97
X_1-X_2, X_1X_2, and X_1/X_2 with (X_1, X_2) continuous, 97
$g_i(X_1,\ldots,X_n)$ for $1 \leqslant i \leqslant m \leqslant n$, 97, 98
$(X_2/X_1,\ldots, X_k/X_1)$with X_i chi-square $(1 \leqslant i \leqslant k)$, 100, 108
X^+ with X general, 105
$\sin(2\pi U)$ with U uniform on (0,1), 106
$\min(X_1,\ldots, X_n)$ with X_i exponential $(1 \leqslant i \leqslant n)$, 107
$\max(x_1,\ldots,x_n)$ with x_i exponential $(1 \leqslant i \leqslant n)$, 107
$\sqrt{|X|}$ with X general, 107
$X/(X+Y)$ and X/Y with X, Y chi-square, 107
$(X, g(X))$ with X general, 107
$X_1+\ldots+X_n$, 142
$X_1+\ldots+X_n$, for n large (CLT), 149
local coordinates, 223
Distribution theory, 389
Distributive laws, **6**
District, 289
Division of Vital Statistics, 74
Domain, 46
Double-blind experiment, 412

Drugs, selection of, 103
  variability of effects, 103
  dosage, 160
  LD50 problem, 160
  tests on animals, 270
  screening of, 354
Dynkin's theorem, **241,** 264

E

Efficiency (exact and asymptotic), 389
Eggs, 343, 353, 412
Egress pattern, 22
Election, 28, 39
Electronic equipment, 33, 39
Elementary event, 12, 40
Elevator, 22
Empiric (empirical) d.f., **176**
  consistency at any point, 177
  uniform consistency, 177
Empirical estimates of probability, 13
Empty set, **4**
Engineering statistics, 390
Equally likely outcomes
  equally probable points, 13, **16**
ESP, 155
Estimation, 50, 390
  regular case, **225**
  of ordered parameters, 361
  of regression parameters, 420
Estimation, interval, 323
  confidence sets, special case, **324**
  relation to tests, 324, **326**
  confidence interval, 324, **325**
  confidence coefficient, **325, 326**
  confidence set, **326**
  "good" sets from "good" tests, 327, **338**
  sets from another method, **329**
  problems in discrete cases, 332
  equal-tail intervals, 338
  prediction intervals, **339**
  for regression parameters, **425**
  confidence band in regression, **426**
  prediction band in regression, **426**
Estimator, **159,** 172
  estimate, 159, 172
  unbiased estimator, **159,** 172
  bias of an estimator, **159**
  consistent sequence of estimators, **160,** 173, 262
  best linear unbiased estimator (BLUE), 169, 252, 421
  asymptotically unbiased estimator, 173
  sample moments as estimators of population moments, 174
  polynomials in sample moments as estimators of polynomials in population moments, 175
  of a distribution function, 176
  infinitely many unbiased estimators, 179
  one (absurd) unbiased estimator, 179
  consistent in mean square, **180**
  consistent in quadratic mean, 180
  criteria for mean square consistency, 180
  mean square consistency$\Rightarrow$consistency, 180
  consistent asymptotically normal (CAN), **207,** 254
  regular, **225**
  best regular unbiased (BRUE), **229**
  efficiency (of an unbiased regular estimator), **230**
  efficient, 230
  asymptotic efficiency (of an unbiased regular estimator), 230
  efficiency, sufficiency, and likelihood, 231
  best unbiased estimators (BUE's), 244, **246**
  reduction of variance of an unbiased estimator via sufficiency, 245
  minimum variance unbiased (MVUE), 246
  method of moments estimators, **248**
  minimum chi-square (MCS) estimator, **249**
  modified MCS (MMCS) estimator, **250**
Euclidean n-space ($\mathscr{R}^n$), **46**
Even positive integers, 3
Event, 12, 40
  elementary event, 12, 40
e^x, as a series, 28, 127
  as a limit, 148
  tables of e^{-x} (Table 4), **438**
Examination scores, 136, 429, 431
Expectation (of a r.v.), **110**
  necessary and sufficient conditions for its existence, 111

$Eg(X)$ in terms of the d.f. of X, 111, 119
absolute convergence and its uses, 111
properties, 111
as a measure of location, 112
of a Bernoulli r.v., 116
of a binomial r.v., 116
of a Poisson r.v., 117
of a uniform r.v., 117
of an exponential r.v., 118
of a normal r.v., 118
of a bivariate normal, 124
of an infinite sum as sum of expectations, 127
$P[X \in S]$ as an expectation, 129
$F_X(x)$ as an expectation, 129
for complex-valued functions, 129
examples of nonexistence, 138, 140

Experiment, 7, 10, 13
possible outcomes, 16
experimental design, 270, 311
completely randomized design, 357

Exponential class of distributions, 243
unvariate exponential family, 243
one-parameter exponential family, 243
relation to sufficiency, 244

Exponential distribution, **81**
relation with gamma distribution, 81, 288
light bulb life example, 86
failure rate of, 88
relation with Weibull, 89, 96, 97
$X_1 + X_2$ and X_1/X_2 with X_1, X_2 exponential, 99, 138
mean and variance, 118
characteristic function, 132
sums give gammas, 133
MLE and its mean, 191
a study of exponential location, 267
UMP test via MLR, 287

Extended intersections (of sets), **6,** 10
Extended unions (of sets), **6,** 10

F

Factorial, 17
approximation to, 21
Tables (Table 2), **436**

Failed, 38
Failure rate, **88**
of an exponential distribution, 88
of a Weibull distribution, 88
increasing failure rate (IFR), 88
decreasing failure rate (DFR), 88
constant, 89
of a series system, 107
of a parallel system, 107
of a $N(0,1)$ distribution, 108
of a $N(\mu,\sigma^2)$ distribution, 109
IFR nature of $N(\mu, \sigma^2)$ distribution, 109

Family (boys-girls composition), 24
F-distribution $F(n_1,n_2)$, **82**
relation with noncentral F-distribution, 83
use in testing two normal variances, 305
relation to chi-square distributions, 316
tables (Table 12), **463**

Federal building, 318
Feller's alpha distribution, **139**
Female, failure rate, 74
Fermi-Dirac statistical mechanics, 20
Fertilizer, 421
Field of sets, 34
Financial losses, 115
Finite additivity (of probability functions), 14
Fire-fighting, 373
Fish, 214
Fluctuations, 41
Force of mortality, 88
FORTRAN, 36
Frequency probability, 13
Fuel oil, 103
tank capacity, 104

Function algebras, 108
Functional, 38
Functional equations, 54
Fundamental theorem of statistics, **176**

G

Gambler, 110, 134
Gamma distribution $G(x|\alpha,\beta,A)$,**81**
relation with exponential distribution, 81, 288
relation with chi-square distribution, 82
characteristic function, 132
as a sum of exponentials, 133
tables (Table 9), **454**

Gamma functions, 78
  for integer arguments, 79
  $\Gamma(n) = n\Gamma(n-1)$ for $n > 1$, 79
Gauss-Markov theorem, **421**
Geometric distribution, **55,** 70
  "first occurrence in independent trials", 55
  relation to negative binomial, 56
  lack of memory property, 71
  MLE of success probability, 186
Geometric mean, 196, **254**
Gerrymandering, 290
Glivenko-Cantelli theorem, **177**
Glopco (General Glop Company), 413
Golf, 40
Good items, 33
Good luck, 133
Goodness-of-fit tests, 389
Government, 33
Grade-point average, 414
  college predicted from high school, 420
Green light, 51
Guinea pigs, 270

H

Hazard rate, 88
Heart attack studies, 160, 411
  tests on animals, 270
  survival times, 411
Heine-Borel theorem, 201
Helly-Bray theorem, **210**
Helmert orthogonal transformation, **301**
Hitlerite society, 270
Homework, relation to examination performance, 429
House, 103, 153
Hypergeometric distribution, **52**
  "type A items" interpretation, 52
  limits leading to binomials, 57
  limits leading to Poissons, 58
  verification as a distribution, 72
  tables (Table 6), **444**
Hyperplane, **194**
  two sides of the hyperplane, 194, 252
  Supporting Hyperplane Theorem, 195
  Separating Hyperplane Theorem, 195
Hypersurface, **223**
  family of hypersurfaces, 223
Hypothesis, **269**
  null hypothesis H_0, **269**
  alternative hypothesis H_1, **269**
  simple, **269**
  composite, **269**
  medical, 270
Hypothesis-testing problem, **271,** 390
  accepting the null hypothesis, 271
  "true" H_0, 271
  rejecting the null hypothesis, 271
  "false" H_0, 271
  nonrandomized test of H_0 versus H_1, **271**
  acceptance region, **271**
  region of acceptance of H_0, 271
  critical region, **271**
  region of rejection of H_0, 271
  randomized test, **271**
  test function $\phi(X)$, **271**
  critical function, 272
  error of Type I, **272**
  level (of significance), **272**
  size of the critical region, 272
  error of Type II, **272**
  power, **272**
  how to reason to a good test, 272
  most powerful test, **277**
  most powerful of level α, 277
  rejection probability as an expectation, **277**
  most powerful via Neyman-Pearson lemma, 277
  uniformly most powerful (UMP) test (of level α), **281**
  restriction to nonrandomized tests, 272, 281
  UMP tests via a MLR, **286**
  decision-theoretic view, 290
  risk point of a test, **291**
  set of risk points R is convex, 291
  (0,1) and (1,0) in R, 292
  ϕ_2 better than ϕ_1, 292
  ϕ_1 worse than ϕ_2, 292
  ϕ_1 no better than ϕ_2, 293
  minimal complete class of tests, 293
  set R of risk points closed, 293
  minimal essentially complete class of tests, **294**
  selection of a test, 294
  Bayes rule, **294**
  a priori distribution, 294

minimax (risk) rule, **295**
minimax regret rule, **295**
geometrical interpretations of minimax and Bayes, 295
computation of a Bayes rule, **296**
interpretation of prior distribution, 297
level α test trivial to obtain, **311**
two-sided equal-tail test, **317**
most powerful of level $\leqslant \alpha$, **318**
most powerful level α tests for composite hypotheses, **319**
repeated tests of significance, **319**
data snooping, **320**
"better than" for composite alternatives, **321**
two-sample tests, 339
forced by a lack of ranking and selection procedures, 341

I

Ice cream bars, 35
Idempotency laws, **5**
Identification procedures, 352
Identifying characteristics, 41
Identity
identity (of sets), **4**
identity elements, **6**
Independent events, **30,** 35, 36, 37
independent set of events, **31**
Independent family of r.v.'s, **61**
relation to independence of events, 61
characterization in terms of d.f.'s, 61
characterization in terms of probability functions (discrete case), 61
not implied by distribution of $X+Y$ as convolution in X, Y (discrete case), 67
characterization in terms of density functions (continuous case), 91
independence of X_1+X_2 and X_1/X_2 with X_1, X_2 exponential, 99
characterization in terms of p.d.f. factorization, 105
not implied by distribution of $X+Y$ as convolution in X, Y (continuous case), 106
independence versus pairwise independence, 108
mean of a product of independent r.v.'s as product of means, 120
characterization in terms of moments, 121
Indexing set, 31, 36, 41, 61
Indicator function (of a set), **129**
Infimum (inf), **158**
Information theory, 41
Inner product, **193**
Internal Revenue Service, 23
Intersection, 51
Intersection (of sets), **2**
extended intersections, 6, 10
Inventory theory, 390
Italy, 398

J

Jacobians of transformations, 79, **98**
$J(\frac{x}{y})$ versus $J(\frac{y}{x})$, 99
example with chi-square r.v.'s, 101
Jensen's inequality, **197,** 254
Join (of sets), **2**
Joseph E. Seagram & Sons, 23
Judges, 409, 415
Jumps (of a d.f.), 47

K

Key, 37
Khinchin's theorem, **148**
Kindergarten teacher, 35
Kinetin, 167
Koopman-Darmois class of distributions, **243**
Koopman-Darmois form, 243
ranking and selection procedures for, 352

L

Lack of memory property, of geometric, 71
of exponential distribution, 88, 106
Laplace's law of succession, 37
Las Vegas, 153
Late commuter, 29
Law of large numbers, weak (WLLN), **145**
general method of proving WLLN's, 146
Khinchin's theorem form, 148

Lawsuits, 33
Lead-time, 33
Least squares method, **420**
Lebesgue dominated convergence theorem (LDCT), **264**
 justifying interchanges of limits and integrals, 264
Lebesgue-Stieltjes integral, 110
Legislative apportionment, 289
 legislators-per-person, 289
 Montana example, 319
Lévy's continuity theorem, **147**
 use in proving CLT, 149
Librarian of Congress, 9
Library, of Congress, 9
 of subroutines, 276
Life distribution, 86
Life-testing, 390
Light bulbs, 86
 with least life variability, 349
Likelihood function, **184**
 likelihood equation, **197,** 208
 relation with efficient estimators, 231
 relation with sufficiency, 231
 likelihood-ratio and its sufficiency, **242,** 258, 267
 induced likelihood function, 261
 likelihood-ratio in testing, 279
 monotone likelihood-ratio in $T(x)$ (MLR), **286**
 likelihood-ratio test, composite hypotheses, **312**
Linear statistical inference, 390
Lion, appetite, 185
Literary Digest, 28
Local coordinates, **223**
Lock, 35
Log-concave distribution, discrete, **71,** 72

M

Machine, 68
Mailing, 398
Male, failure rate, 74
Mannequin, 348
Manufacturing methods, 33
Map, 50
Marginal density function, **90**
Marginal distribution function, 59
Marginal probability function, **62**
Marketing tests, 409
Markov's inequality, **144**
Martin Marietta Corporation, 50
Match, 40
Matching problem, **28,** 41, 42
Matrix algebra, 93
Maximum-likelihood estimator (MLE), 70, **184**
 maximizing $L(\theta)$ versus maximizing ln $L(\theta)$, 185
 invariance of MLE's, 188, 261
 consistency of the MLE, 197, **200, 262**
 heuristic proof of consistency, **205**
 MLE as a CAN estimator, 208
 is MLE most concentrated CAN estimator?, 209, 211
 J. L. Hodges' superefficienciation, 211, 255, 263
 "deprecation of the evil," 213, 222
 "exclusion of the evil," 213, 222
 generalized MLE, 214
 maximum probability estimator, 214
 MLE relates asymptotic efficiency and asymptotic relative efficiency, 232
 relation to sufficient statistics, 242
 ε-MLE, 252
 IMLE, 262
Maxwell-Boltzmann statistical mechanics, 20
Mean of a r.v., 112
 equals median in case of symmetry, 112
 fallacies about, 113
 sample mean, **173**
 sample mean consistent, unbiased, 181
 sample mean versus sample median, 221
 selecting population with largest, **343**
Mean squared error
 Cramér-Rao bound on, **225**
Median of a r.v., 112, **180,** 404
 equals mean in case of symmetry, 112
 fallacies about, 113
 sample median M_n, **181**
 sample median consistent, biased, 181
 sample mean versus sample median, 221
 nonparametric confidence interval, **404**
Meet (of sets), **2**
Methyl mercury, 214
Microorganisms, 50
 microbiological assays, 50

microbial insecticide, 171
MLE of distribution parameter, 192
Microscope, 58
Mile runners, 317, 337
selection of the best, 356
MLE, 70
Mode of a r.v., 112
fallacies about, 113
Model $(BC)_i$, 18
Moment generating function (m.g.f.) of a r.v., **127**
moment retrieval from a m.g.f., 127
joint m.g.f. (of several r.v.'s), **128**
moments from joint m.g.f.'s, 128
Moments, 110
n^{th} (noncentral) moment μ'_n, **112**
n^{th} central moment μ_n, **112**
the moment problem, 114
when moments determine a d.f., 114
Carleman's theorem on moments, 115, 139
Hausdorff's theorem on moments, 115
multivariate, 119
joint moment of (X_1, X_2) of order n_1+n_2, μ'_{n_1,n_2}, **119**
product moment $\mu'_{1,1}$, **120**
joint central moment of (X_1, X_2) of order $n_1 + n_2$, μ_{n_1,n_2}, **120**
k^{th} (noncentral) sample moment m'_k, **173**
k^{th} central sample moment m_k, **173**
why m'_k, m_k are called *sample* moments, 173
properties of sample moments, 174
Monotone likelihood ratio in $T(x)$ (MLR), **286**
for one-parameter exponential family, 313, 319
Moon, 35
Mortality, 74
Movie magazine, 41
Multinomial coefficients, **18,** 35
Multinomial distribution, **64**
"k possible outcomes" interpretation, 65
television show example, 65
selection problems involving, 361
Multiple decision procedures, 103, 352, 384
Multiple integrals, changes of variables, 97
Multiplets theorem, **16,** 35
Multiplication rule, 24
Multivariate analysis, 390
Multivariate distributions, 59, 90
special discrete, 64
special multivariate, 91
Mutually exclusive sets (events), **4,** 31, 35, 37
Mutually favorable events, 39

N

Names, 41
National Center for Health Statistics, 74
National Office of Vital Statistics, 73
National Weather Service, 318
Nature, 297
Negative binomial distribution, **55**
"no. failures needed for r successes," 55
Negative information, **31,** 37, 39
Negative part of x, 105
Neighborhood, 71
Nemenyi's transformation, **321**
Network, 38
Neutral grain alcohol, 23
Newspaper, 42
newspaper stand, 144
newsboy problem, 177, 251, 260
Newton-Raphson iteration, 168
Neyman-Pearson lemma, **277**
motivation for, 279
power-per-level interpretation, 279
likelihood-ratio relationship, 279
use where $f(x|\theta_0)$ may be zero, 281
generalized, **289**
relation to Bayes tests, 296
extensions, 318
Noncentral chi-square distribution $\chi^2_n(\lambda)$, **82**
relation with chi-square distribution, 82
Noncentral F-distribution $F(n_1,n_2,\lambda)$, **83**
relation with F-distribution, 83
Noncentral t-distribution, **315**
relation with t-distribution, 316
Noncentrality
of a noncentral chi-square distribution, 82
of a noncentral F-distribution, 83
of a noncentral t-distribution, 316
Nonparametric methods, 389

Nonparametric methods, (continued)
- nonparametric statistics, **403**
- when to use, **403,** 413
- order statistics, quantiles, 403, **404**
- probability-integral transformation, 403, **413**
- "bag of tricks" approach, 404
- nonparametric confidence interval for ξ_p, **404, 405**
- computation of confidence coefficient, 405
- comparison with parametric confidence intervals, 406
- nonparametric point estimation, **408**
- nonparametric hypothesis-testing, **408**
- nonparametric test of H_0: $\xi_p = \xi_p^0$, 408
- nonparametric ranking and selection, **408**
- nonparametric decision theory, **408**
- ordinal versus numerical data, 409
- sign test, **410,** 409
- efficiency: sign test versus t-test, 410
- Mann-Whitney-Wilcoxon test, **410**
- tables for Mann-Whitney-Wilcoxon, 412
- Mann-Whitney-Wilcoxon efficiency, 412

Nonwhite, failure rate, 74

Normal distribution $N(\mu,\sigma^2)$, **82**
- standard normal distribution $N(0,1)$, 82
- σ as positive square root of σ^2, 82, 92
- (univariate) normal $N(\mu,\sigma^2)$, **92**
- (bivariate) normal $N(\mu_1,\mu_2;\sigma_1^2,\sigma_2^2,\rho)$, **92,** 134, 300, 431
- multivariate normal $N(\mu, \ddagger)$, 91, **92,** 355
- bivariate as a case of multivariate, 93
- X_1 and X_2 normal does not imply (X_1, X_2) normal, 105
- linear combinations of normals, 106, **263**
- failure rate of, 108, 109
- mean, median, mode (univariate case), 113
- mean and variance (univariate), 118
- characteristic function (univariate), 131
- d.f. of $X_1 + \ldots + X_n$ (univariate), 142
- role in CLT, 149
- approximate normality, 152, 153
- confidence intervals for μ, 154
- MLE of μ, 186
- MLE of σ^2, 187
- Cramér-Rao in estimating σ^2, 228
- BRUE of σ^2, 229
- sufficient statistic for μ, 239
- minimal sufficient statistic for (μ,σ^2), 240
- in two-parameter exponential family, 243
- completeness, 248, 259, 267, 316
- unique MVU estimator of μ, 248, 316
- method of moments estimators of (μ,σ^2), 248
- MLE of (μ,σ^2), 252
- hypothesis tests about μ, σ^2 known, 272, 279
- UMP tests about μ, 281, 298
- tests about μ, σ^2 unknown, 298, 302
- t-distribution of $(\bar{X}-\mu_0)/(s/\sqrt{n})$, 300
- $\bar{X}$ independent of s^2, 301
- tests about σ^2, 303
- two-population tests on σ_1^2, σ_2^2, 304
- two-population tests on μ_1, μ_2 $(\sigma_1^2 = \sigma_2^2)$, 306
- two-population tests on μ_1, μ_2 $(\sigma_1^2 \neq \sigma_2^2)$, 308
- confidence intervals for μ, 326, 330
- upper bound on μ, 328
- lower bound on μ, 328
- confidence intervals for σ^2, 330
- intervals for σ^2 with one observation, 331
- intervals of minimal length, 339
- intervals of fixed-length, 339
- prediction intervals, 339
- selection of the largest mean, 343
- selection of the smallest variance, 349
- expected values of normal order statistics, 415
- MLE's in regression, **423**
- ranking normal regressions, **428**
- ellipse of concentration, **431**
- tables (Table 11), **460**

Normal person, 58

Nuclear weapons, 153

Null set, **4**

Number systems, 1

O

Occupancy vectors, 19, **20**
- $\#(\Omega_O)$, 20

Odd positive integers, 3, 11
o(*h*), 53
Oil slick, 395
Olympic Games, 342
Order statistics, **176, 181,** 255
  order statistic estimates, 178
  order statistics not independent, 181
  sufficiency, 237, 265
Ordered sample (of size *r*), 16
Organoleptic abilities, 23
Origin, 193
Outcomes of an experiment, 7
  when dice are rolled, 8
  in roulette, 59
  functions of outcomes (r.v.'s), 45

P

Pairwise disjoint sets, **4**
Pairwise independent events, 36
Paradoxes, 1, 9, 10, 33, 39
Parallel system, 39
  failure rate with exponentials, 107
Paramecium caudatum, 167
Parameter, 53
Parapsychology, 155
Partial order, **391**
  reflexive relation, 391
  antisymmetric relation, 391
  transitive relation, 391
Parts, 38
Passengers, 22
Path, 39
Patient, 45
Permutations, **17,** 35
Persistence, of bad luck, 115
  of good luck 133
Pest, effect on yield, 427
Pharmacology, 157, 160
Philosophers, 294
Physics, 20
  physicist, 110
  physical constant, 399
  physical science, 421
Pigs, similarity to humans, 160
Pizza parlors, **389**
  giving overview of statistics, 388
Placebo, 412
Plays, 59, 61, 134, 153
PL/I, 36
Point, **1**
Point estimation, 157, **159,** 172
Point probability function, 47
Poisons, 170
Poisson distribution, **53**
  events occurring in time leading to, 53, 72
  "average" number of occurrences, 54
  as a limit of binomials, 56, 75, 153
  as a limit of hypergeometrics, 57
  tables, 58
  leading to binomials, 69
  mean and variance, 117
  characteristic function, 131
  d.f. of $X_1 + \ldots + X_n$, 142
  Cramér-Rao in estimating λ, 232
  completeness, 247
  unique MVU estimator of λ, 247
  MMCS estimator of λ, 250
  confidence intervals for λ, 340
  tables (Table 7), **447**
Polar coordinates, 79
Political reasons, 33
  political parties, 289
  political candidates, ranking of, 342
Polls, 39, 362
Polya's theorem, 149, **207**
Positive information, **31,** 39
Popcorn, 35
Population, 16
  as a source of observations, 304
Poretsky's law, **10**
Positive integers, 3, 11, 36
Positive part of *x*, 105
Post Office operations, 1
  P.O. box size, **19,** 43
Poultry science, 343, 412
Prediction intervals, **339**
Prediction, of an election, 28
  of one r.v. from another, 122
Presidential poll, 28, 39
  ranking Presidential candidates on popularity, 342
  selection on popularity of a candidate, 361
Principles, 294
Probability density function (p.d.f.), **77**
  n-dimensional case, **90**
  p.d.f. symmetric about μ, **112**
  fallacies about symmetry, 113

Probability (function), 12, **13,** 40
 classical definition of probability,13
 empirical estimates of probability, 13
 axioms of probability functions, 14
 probability *m* of *N* occur, 29
 probability space $(\Omega, \mathcal{B}, P)$, 40
Probability function (of a discrete r.v.), **47**
 typical case, 48
 characterization of probability functions, 49
 of discrete *n*-dimensional r.v., **62**
 examples with 2-dimensional case, 63, 65
 binomial probability function, 68
 Poisson probability function, 68
Probability measure, 13
Problème des Rencontres, **28,** 41
Product (of sets), **2**
Production methods, 33, 151
Profits (of a business), 110, 177, 251, 260
Proper subset, **4**
Prunes (and their pitting), 68
Pullets, 353
Pumping station, 31
Push-button, 35

Q

Quadratic polynomials, 123
Quality control, 23, 390
 standard process versus new process, 151
 of research, 361
Quantile of Order *p*, 214, 263
 sample quantile of order p, Z_p, **215, 404**
 properties of Z_p, 215, 264
 modern proof of Z_p's properties, 219
 sample mean versus sample median, 221
Questions, 68
Queueing theory, 390

R

Radiation levels, 268
 relation to atomic power plants, **269,** 342
 background, 270
 relation to Sun spots, 270
Random digit table, **276**
Random number generators, 276
 speed, 276
 quality, 276
 package, **276**
Random phenomena, 7, 10
 "at random," 18
 random guessing, 23
 random arrangement in a line, 35
 random events in time, 53
 random point in (0,1), 104, 139
 average value of, 110
 random digits, 276
 random subset, 354
Random variable (r.v.), **45**
 n-dimensional r.v., **46**
 discrete case, 45, **47**
 discrete *n*-dimensional case, **62**
 (absolutely) continuous r.v., **77**
 example of continuous r.v., 77
 continuous *n*-dimensional case, **90**
 unobservable, 134, 169
 associated, **140**
Range, 46
Ranking and selection procedures, 341, 390
 types of questions leading to, 341
 selection versus ranking, **342**
 population (source of observations), **342**
 goal, **342**
 P(CS), **342**
 probability requirement, **342**
 poultry science example, 343
 largest mean normal population, 343
 smallest variance normal population, 103, 349
 extensions and generalizations, **352**
 essential problem formulation, 353
 indifference zone approach, **354**
 subset selection approach, **354**
 procedures for unknown unequal variances, 354
 optimal selection procedures, 355, 384
 nonparametric selection, 355, 415
 unequal sample sizes behave counterintuitively **359**
 most economical procedures, **359**
 multinomial selection problems, 361
 complete ranking applications, 361
 in regression, **428**
Rao-Blackwell theorem, **244**

Rao-Blackwell improvement theorem, **245**
use in finding MVUE's, 247
Rats, control program, 71
Rayleigh distribution, **260**
method of moments estimator of θ, 260
Red blood cells, 58
Red light, 51
Regression analysis, 390, 416
two variables and linear, 416
regression curve of Y on X, **417**
regression line of Y on X, **418**
regression intercept, 418
regression slope, 418
mean squared prediction error, 418
residual variance, 419
method of least squares, 420
estimation of regression parameters, **420, 421**
Gauss-Markov theorem, **421**
linear models, 423
MLE's in normal regression, **423**
tests of hypotheses about regression parameters, **425**
confidence intervals on regression parameters, **425**
confidence interval for $\alpha+\beta x^*$, **426**
confidence interval for Y at x^*, **426**
confidence band (for a regression line), **426**
prediction band (for future values), **426**
selection of the best regression, **428**
Relative frequency, 23
Reliability, 390
Reliability of a system, $R(p)$, 32, **38**
Rencontre at the ith place, 28
Republicans, 35, 39
Research projects, good, **270**
research program evaluation, 361
Rice, selection of best variety, 357
Roads, 38
Roosevelt versus Landon, 28
Roulette, 59
selection of number to bet on, 361
Runs, 40
Russell's paradox, 9, **10**

S

Sample space, 12, 40
sample space chosen in several ways, 15
discrete sample space, 16
denumerably infinite sample space, 16
Sample survey, 389
Samples, 16
sampling distribution, 50
with multinomial outcomes, 65
Sampling inspection, **401**
Satellite, communications, 104
Scales, most precise, 349
Schwarz' inequality, **122**
use in a lemma, 225
Scientists, 33, 294
experimental, 399
Scrambled order, 41
Scrappaper, 177
scrap more profitable than sales, 178
Seat belts, 268, 311
selection of the best type, 341
Second-generation, 69
Segment, 38
Selection, largest normal mean, **343**
populations, **343**
observations, **343**
goal, **343**
means procedure, **344**
probability requirement, **344**
P(CS) expression, **344**
least favorable configuration (LFC), **345**
P(CS) in the LFC, **345**
sample size needed, **346**
tables needed to compute sample size, **346, 347**
approximations to sample size, **348**
numerical example, **348**
Selection, smallest variance normal, **349**
populations, **350**
observations, **350**
goal, **350**
procedure, **350**
probability requirement, **350**
P(CS) expression, **351**
least favorable configuration (LFC), **352**
P(CS) in the LFC, **352**
tables and approximations, **352**
Semilog graph paper, 74
Sequential, experimentation, 168, 390
search, 168
Series system, 38
failure rate with exponentials, 107

Series system (continued)
  confidence bound for reliability, 107
Set, **1**
  set theory, 9
  bounded from below, **400**
  closed from below, **401**
Sex distributions, 24, 44
Ship, 19
Shoes in a closet, 29
Shuffling machine, 43
Sigma-algebra (σ-algebra) of sets, 11, **34,** 40
Significance probability $\alpha_{SP}(x,\phi_\alpha)$, **313**
  in snowfall example, 318
Silver dollar, 68
Simple random sampling, 16
  with replacement, 16
  without replacement, 16
Simpson's paradox, 33, **34,** 39
Simulation, 139, 276, 390
Skewness of a d.f., **112**
  zero for symmetric p.d.f.'s, 112
Slot machine, 153
Slutsky's theorem, **261,** 322
Smoking, 268
  "smoke up" nonsmokers, 270
  "desmoke" smokers, 270
  smoking dogs, **271**
  relating to age and height, 304, 306
Snowfall, 222, **253,** 317
  downtown versus at airport, 318
  confidence intervals for, 337
  implications for Highway Department purchases of road salt, 337
  prediction intervals and road salt, 339
  nonparametric confidence interval on median snowfall, 406
Source, 31
Space vehicles, 35, 50
Sphere, **193**
  ball, solid sphere, spherical cell, 193
Square roots, table (Table 1), **435**
SRS, 16, 267
  leading to hypergeometric distribution, 52
Standard deviation of a r.v., **113**
Standardized r.v., **114**
  has mean zero, 114
  has variance one, 114
Stars and bars, 20
States, 39
Statistic, **159,** 172
  statistician, 177, **270**
  great statistician, **294**
  "the science of decision making," **341,** 363
Statistical mechanics, 20
Stieltjes integral, 110, 161
Stimuli, 170
Stirling's formula, 21, 216
Stochastic approximation, 157
  in medical research, 160
  Assumptions A-1, A-2, A-3, 161
  method of generating n^{th} dosage X_n, 162
  $X_n \xrightarrow{P} \theta$, 162, **163**
  biological science example, 167
  in bio-assay, 170
  versus chemical assay, 171
Stochastic processes, 390
Stock market, 366
  investment strategy choice, 366
Storage theory, 390
Stratify, 24
Student's t-distribution, **83**
  relation with Cauchy distribution, 142
  related to chi-square and normal, **300**
  "X has the t_m distribution", **302**
  the t-test, 303
  relation with noncentral t-distribution, 315
  multivariate, 361
  tables (Table 13), **467**
Subpopulation (of size r), 17
Subset, **4**
Successes, 51
Sufficient statistic (for θ), **234**
  in a regular estimation case, **231**
  summarizes all information, 230, 233
  and conditional distribution on hypersurface, 230, 234
  critierion for (in regular cases), 231
  relation to likelihood equation, 231
  Neyman factorization theorem criterion, **234**
  contour of a sufficient statistic, 236
  at least one always exists, 236
  identity statistic is sufficient, 237
  minimal sufficient statistic, **237**
  trivial sufficient statistics, 237
  no nontrivial for Cauchy case, 238

method for finding a minimal sufficient statistic, 238
Dynkin's theorem, 241
sufficiency of the likelihood-ratio, 242
MLE related to sufficient statistics, 242
relation to exponential class, 244, 265
use to reduce variance of an unbiased estimator, 245
complete sufficient statistic, **246**
use to find BUE's, 246
Sum (of sets), **2**
Superefficiency, points of, **213**
Supply, source with least variability, 349
Supremum (sup), **158**
Sure-thing principle, 34, **39**
Surface, 50, 192
Surgical instruments, 50
Symmetric difference (of sets), **10,** 30
Symmetric positive definite matrix, 92
System, 38

T

Tagging (and recapture), 71
Taste-testing, 390
Taylor's theorem with remainder, **130,** 217
t-distribution, **83**
relation with Cauchy distribution, 142
related to chi-square and normal, **300**
"X has the t_m distribution", **302**
the t-test, 303
relation with noncentral t-distribution, 315
multivariate, 361
tables (Table 13), **467**
Teams, 68
Telephone, books, 28
switchboard, 69
call length, 104
intercontinental calls, 104
Television shows, 65
ranking on popularity, 342, 361
evaluation of ratings, 362
Tennis, 413
Third-generation, 69
Threshold, 70
Time homogeneity, 53
Time series, 390
Times-Union, 23, 318
Tobacco bales, 19
Tornado, 396
Total probabilities, theorem of, **25,** 26
use in stochastic approximation proof, 166
use in Neyman factorization theorem proof, 235
Tournament, 40
Track, 317, 337. 356
Transformation, 223
one-to-one, **223**
Transitivity, 37
Transpose (of a vector), **193**
Treatment, 45, 410
Trials (of an experiment), 13
t-test, the, **303**
efficiency, 410, 412
Tuplets, 16
Two-finger Morra, **398**
Two-stage statistical procedures, 339

U

Uncorrelated r.v.'s, **121**
uncorrelated iff covariance zero, 121
uncorrelated$\not\Rightarrow$independent, 121
independent$\Rightarrow$uncorrelated, 121
uncorrelated iff correlation zero, 122
if bivariate normal, uncorrelated$\Rightarrow$independent, 124
if merely each normal, uncorrelated$\not\Rightarrow$ independent, 125
Unethical experiments, 160
Uniform convergence, 149
Uniform distribution (continuous) on (b,c), **81**
example of a "random number", 77
relation with uniform on (0,1), 81
relation with beta distribution, 81
example of $2W$-1 with W uniform, 86
X_1+X_2, X_1-X_2, $|X_1-X_2|$, max(X_1, X_2), min(X_1,X_2), X_1X_2, and X_2/X_2 with X_1, X_2 uniform, 105
mean and variance, 117
characteristic function, 132
as a limit of $a_1X_1+ \ldots + a_nX_n$, 138, 155
uniform on $[-d, d]$:MLE of d, 188
uniform on $[c-d, c+d]$:MLE of c, 189, 255

Uniform distribution, (continued)
- uniform on $[c-d, c+d]$:MLE of (c, d), 190
- uniform on$(0,\theta)$: sufficient statistic for θ, 240
- uniform on$(0,\theta)$: a study, 266
- uniform on$(0,\theta)$: tests about θ, 281

Uniform distribution (discrete), **54**
- finite discrete sample space example, 55

Unimodal distribution, discrete, **71,** 72
Union (of sets), **2**
- extended unions, 6, 10
- probability of a union, **26,** 27, 28, 36

Universal set, **1**
Universe, **1**
University of Texas, Graduate School of Biomedical Sciences at Houston, 50
Unordered sample (of size r), 17
Urns, 37, 276
U. S. A., 35, 44, 73, 178
U. S. Presidency, 39, 342
U. S. S. R., 35

V

Variance (of a r.v.), **112**
- as a measure of dispersion, 112
- zero variance means constant r.v., 113
- expression in terms of $E(X^2)$ and $E(X)$, 113
- variance of $aX+b$, 114
- of a Bernoulli r.v., 116
- of a binomial r.v., 116
- of a Poisson r.v., 117
- of a uniform r.v., 117
- of an exponential r.v., 118
- of a normal r.v., 118
- variance of a sum, 120, **135**
- variance of a sum equals sum of variances if uncorrelated, 121
- of a bivariate normal, 124
- bounds on variance ("deviance"), 144
- unbiased estimator of, 175
- variance of a limiting distribution versus limit of the variance, 210, 255, 262
- Cramér-Rao bound on for unbiased estimators, 228
- estimator s^2, **299**
- alternative estimators, 322
- selecting population with least, **349**
- unknown unequal variances, **354**

Vector algebra, 93
- vector, **193,** 367

Vendors, selection of, 390
Venn diagrams, **2,** 4, 5, 27
Virginia State College, 321
Vital statistics, 73
Voting, 289

W

Waiting times, 115
Weak law of large numbers (WLLN), **145**
- general method of proving WLLN's, 146
- Khinchin's theorem form, 148

Wear-out, 88
Weibull distribution, **88**
- failure rate of, 88
- relation with exponential, 89, 96, 97

Wheat varieties, 410
White, failure rate, 74
Wickets, 40
Winnings, 59, 61, 83, 110, 134, 153
Woman, 68
Work-hardening, 88
World Series, 68
- number of games played, 320

Y

Yellow light, 51
Yield, 268, 342, 357, 412, 427